Advances in Heat Transfer and Engineering

Chuang Wen · Yuying Yan
Editors

Advances in Heat Transfer and Thermal Engineering

Proceedings of 16th UK Heat Transfer
Conference (UKHTC2019)

🐎 Springer

Editors
Chuang Wen
Faculty of Engineering
University of Nottingham
Nottingham, UK

Yuying Yan
Faculty of Engineering
University of Nottingham
Nottingham, UK

ISBN 978-981-33-4767-0   ISBN 978-981-33-4765-6 (eBook)
https://doi.org/10.1007/978-981-33-4765-6

This Springer imprint is published by the registered company Springer Nature Singapore Pte Ltd.
The registered company address is: 152 Beach Road, #21-01/04 Gateway East, Singapore 189721, Singapore

Preface

Nowadays, we are facing ever-severe crisis of conventional energy resources and pollutions, increasing demand for new energy resources and applications, and significant challenges for the technologies of efficient cooling, heat and mass transfer enhancement, and effective thermal management, etc. These have become crucially important for almost all engineering area and industries such as mechanical, aerospace, civil and building, chemical and process, electric and electronic, pharmaceutical and medical, as well as power industries.

Over the more than 100 years development, heat transfer has now become a cross-disciplinary subject. The study on micro-nano scale heat transfer has played an important role in the research progress of material science, and the development of biomedical engineering, etc. The 16th UK Heat Transfer Conference (UKHTC2019) addressed these challenges. The conference gathered the UK active and leading researchers typically many young and new academic colleagues, as well as the researchers from heat transfer communities of Europe, North Americans, Australia, South Africa, Kuwait, Japan, and China, etc.

The conference was aimed at a closer collaboration and cooperation between the UK and international scholars in the field of heat transfer. The UK National Heat Transfer Committee organised this conference biennially to provide an innovative platform for scholars in thermal engineering to share and exchange new ideas and solutions. Six plenary and three keynote lectures and more than 190 papers were presented at UKHTC2019 throughout two days in four sets of seven parallel oral sessions. The proceedings in the title of *Advances in Heat Transfer and Thermal Engineering* contains selected and the authors agreed extended abstracts or papers that cover almost all the topics in heat transfer and thermal engineering.

We would like to thank all authors for their contributions to UKHTC2019 and thank the staff members of the University of Nottingham to provide active assistance during the preparation stage of this conference. We send our sincere gratitude to the dedicated reviewers for their time and contribution to improve the scientific quality of the manuscripts. We also would like to acknowledge the support received from the sponsors. We hope that the emerging solutions described in the conference proceedings will inspire our academic and industrial communities to create, innovate, and build a more energy-efficient world.

Nottingham, UK   Chuang Wen
                                                                          Yuying Yan

Contents

Micro-Nano Heat Transfer

Convection

Natural Convection from Heated Surface-Mounted Circular Cylinder

H. Malah, Y. S. Chumakov, and S. Ramzani Movafagh

1 Introduction

In recent years, there are more efforts on natural convection heat transfer from a horizontal cylinder, because of its practical applications. However, unconfined cylinder is well studied; the effect of introducing end-walls on the heat transfer rate of cylinder is considerably less investigated [1]. By development of computers and enhancement of advanced computational techniques, many studies of flow over a bluff body relevant to solid wall have been performed numerically [2], although experimental studies keep their place among researchers' efforts because of their advantages [3]. All numerical and experimental studies confirmed the expected arise on the heat transfer rate in the upstream region of the cylinder. However, the flow configuration, bluff body geometry and applied conditions on solid walls affect the arising flow [4, 5]. In this study, a numerical model of a heated horizontal circular cylinder mounted on vertical isothermal plate is employed to evaluate the natural convection heat transfer. To quantify the effect of vertical plate on the heat transfer from the cylinder surface, the aspect ratio of the cylinder (H/D) is selected equal to 0.6, in order to immerse in the arisen boundary layer on the vertical plate entirely. This geometrical configuration is evaluated on the vertical plate at fixed Grashof number equals 3×10^8 that represents laminar Grashof number. As a result, we describe the three-dimensional characteristics of natural convection heat transfer, which affect flow around the circular cylinder mounted on vertical heated plate. The results proved the significant effect of height of cylinder on the heat transfer rate from circular cylinder surface in the case of

H. Malah (✉) · Y. S. Chumakov
Institute of Applied Mathematics and Mechanics, Peter the Great St. Petersburg Polytechnic University, St. Petersburg 195251, Russian Federation
e-mail: hamid.malah@gmail.com

S. Ramzani Movafagh
Department of Environmental Engineering, Faculty of Engineering and Technology, Saint-Petersburg State Institute of Technology, St. Petersburg 190013, Russian Federation

© Springer Nature Singapore Pte Ltd. 2021
C. Wen and Y. Yan (eds.), *Advances in Heat Transfer and Thermal Engineering* ,
https://doi.org/10.1007/978-981-33-4765-6_1

3

laminar natural convection flow. This study improves fundamental understanding of the buoyancy-induced flows around three-dimensional obstacles in different industrial applications to address the anticipated needs to enhance the rate of heat transfer and safety simultaneously.

2   Computational Methodology

In this work, in order to develop a laminar boundary layer on the heated vertical plate, the cylinder was mounted on an isothermal rectangular plate, which its dimensions considered equal to $100D$ in vertical direction (Y) and $7D$ in lateral direction (X). The vertical plate temperature is set to 333.15 K. In order to achieve a developed laminar incoming flow around the cylinder, the vertical position of cylinder from leading edge of rectangular plate is equal $30D$, which provides Grashof number for laminar flow equals to 3×10^8. In addition, the computational domain was extended $9D$ from leading and trailing edge of plate and $10D$ normal to the plate (Z-direction) in order to ensure impermeability and slip in these regions.

In the analysed case of high aspect ratio cylinder, which performed in the similar conditions as present study, the computed thickness of incoming boundary layer on the vertical plate was equal to $0.9D$ [4]. The cylinder diameter (D) was equal to 0.02 m with fixed surface temperature at 353.15 K. The cylinder height (H) is fixed to $0.6D$, in order to immerse in the laminar boundary layer entirely.

The schematic configuration of problem, its dimensions and imposed boundary conditions are shown in Fig. 1a. In Fig. 1a, the solid walls were applied no-slip boundary condition. The boundary condition, which called "Opening" in Fig. 1a, refers to penetrable side of computational domain in constant pressure.

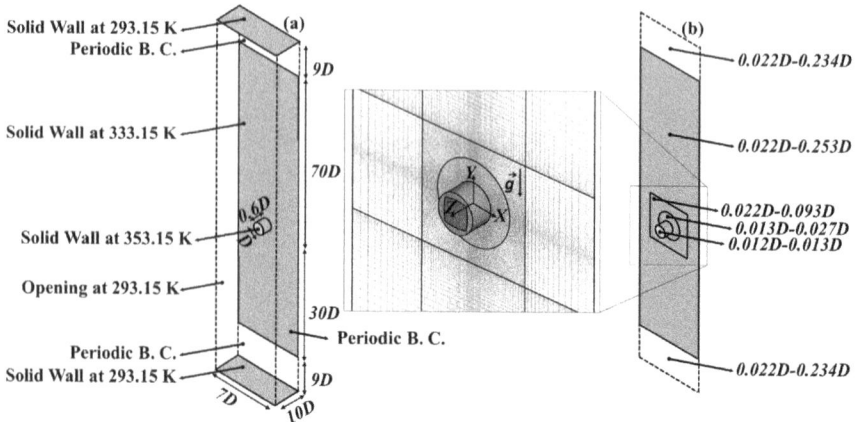

Fig. 1 Schematic of the case geometry and computational details: **a** problem configuration, **b** multi-blocked grid layout

In this work, the case geometry discretized by using body-fitted mesh. The multi-blocked grid in XY plane forms two-dimensional grid, and the range of its cells size in different region is shown in Fig. 1b. The two-dimensional grid, which consists of 41 thousand cells, was clustered to the vertical plate over Z-axis with a coefficient equals to one and generates the three-dimensional grid layout. The cells' size over Z spatial orientation is set to $0.02D$. The three-dimensional mesh grid consists of approximately 4.8 million hexahedron cells.

In this study, a time-based numerical simulation was performed by using a commercial code (ANSYS FLUENT 16.2). The numerical model is based on the momentum and the energy balance equations, which were coupled by considering the fractional step algorithm and solved by using the Boussinesq approximation. The governing equations were discretized using second-order accurate schemes for all the spatial derivatives. Lastly, the computations were run up to 300 s in physical time with a time step equal to 0.002 s.

3 Results

In this work, in order to survey on characteristics of natural convection heat transfer around surface-mounted circular cylinder, the localized Nusselt number related to angular coordinate (Nu_θ) is determined on the solid surface of circular cylinder. To aim this purpose, the spatial angular coordinates were considered as angles, on which zero angle refers to leading edge of circular cylinder on YZ plane. In order to investigate the effect of height of cylinder on the heat transfer rate, the local Nusselt numbers at different Z coordinates along height of cylinder within laminar boundary layer thickness were presented in Table 1. Table 1 illustrates the computational results

Table 1 Local Nusselt number (Nu_θ) comparison

Z/D	Source	0	30	60	90	120	150	180
0.1	Present	12.39	11.49	8.43	4.45	3.06	3.82	6.21
	[5]	12.56	11.60	8.39	4.31	5.16	10.63	7.58
0.2	Present	21.36	20.50	14.99	6.92	2.06	3.31	5.41
	[5]	21.38	20.46	14.77	6.42	4.59	5.91	3.90
0.3	Present	24.55	23.79	18.13	8.56	2.19	2.62	3.36
	[5]	24.45	23.69	17.89	8.02	3.64	4.83	3.18
0.4	Present	24.95	24.29	18.83	9.44	2.60	2.03	2.03
	[5]	24.36	23.66	18.16	8.71	2.82	3.89	3.32
0.5	Present	25.51	24.94	19.85	10.83	3.75	2.63	1.91
	[5]	23.10	22.47	17.62	9.23	2.67	2.95	3.26
0.6	Present	30.36	30.37	26.01	17.22	9.31	8.50	7.96
	[5]	21.50	20.94	16.78	9.73	3.02	2.61	3.31

at seven discrete angular coordinates on the cylinder surface. In addition, in order to investigate the effect of cylinder aspect ratio on transferred heated flow from cylinder, the computed local Nusselt numbers of the work [3], where the results of high aspect ratio cylinder were provided, are included in Table 1 for comparison.

Based on the presented values in Table 1, the local Nusselt numbers decrease from the leading edge ($\theta = 0°$) to the trailing edge ($\theta = 180°$) of the cylinder for each Z coordinate. In the downstream region of the cylinder ($\theta = 120°$), there is a rapid decline in value of local Nusselt numbers, and after that ($\theta = 150°$) the Nusselt number values experience a slight increase. Since the rate of natural convection heat transfer is proportional to the buoyancy of the fluid, the local Nusselt numbers increase along Z-axis from confined end-wall (rectangular vertical plate) to the unconfined end of circular cylinder.

Although there is a good agreement between the values of local Nusselt number in the present analysis and results of high aspect ratio study [5], Table 1 demonstrates two regions on the cylinder surface, where the rate of convection heat transfer is comparable between low and high aspect ratio cylinder. The first region is at the unconfined end of cylinder (around $Z/D = 0.6$), where incoming flow can bypass the cylinder in the case of low aspect ratio cylinder, so there is anticipated a dramatically increase in the local Nusselt numbers. The second zone is downstream region of cylinder (from $\theta = 120°$ to $\theta = 180°$), where the arisen heated flow from the high aspect ratio cylinder acts as a separate source of heat generation. Since the high aspect ratio cylinder crosses the formed boundary layer entirely, arisen heated flow interacts with the incoming boundary layer and leads to an obvious increase in the rate of convection heat transfer.

An overall view of the data in Table 1 demonstrates the fact that the leading edge of surface-mounted circular cylinder ($\theta = 0°$) is a specific line for this problem. Although the local Nusselt number is practically constant for a long cylinder [5], the local Nusselt number increases for a short cylinder on the leading edge ($\theta = 0°$) along cylinder height (Z-direction). The localized Nusselt number in the wake region of the cylinder is qualitatively similar for different variants, representing a slow, almost monotonic decrease in local Nusselt number with increasing spatial angular coordinates around the cylinder.

4   Conclusions

By comparing the numerical results of arisen convection heat transfer rate in present work (low aspect ratio) with the results of high aspect ratio cylinder [5], the significant effect of cylinder aspect ratio on the Nusselt number in the case of laminar natural convection flow is demonstrated.

Maximum values of local Nusselt number are observed at the confined end of cylinder near the vertical plate. These values are located in laminar incoming boundary layer, which arose on the heated vertical plate. Furthermore, heat transfer coefficient decreases from leading edge of cylinder in upstream region to trailing edge in the downstream of the cylinder.

References

1. C.E. Clifford, M.L. Kimber, Optimizing laminar natural convection for a heat generating cylinder in a channel. J. Heat Transfer **136**, 112502 (2014)
2. G. Delibra, K. Hanjalic, D. Borello, F. Rispoli, Vortex structures and heat transfer in a wall-bounded pin matrix: LES with a RANS wall-treatment. Int. J. Heat Fluid Flow **31**, 740–753 (2010)
3. J.K. Ostanek, K.A. Thole, Wake development in staggered short cylinder arrays within a channel. Exp. Fluids **53**, 673–697 (2012)
4. H. Malah, Y.S. Chumakov, A.M. Levchenya, A study of the vortex structures around circular cylinder mounted on vertical heated plate. AIP Conf. Proc. **2018**, 050018 (1959)
5. Y.S. Chumakov, A.M. Levchenya, H. Malah, The vortex structure formation around a circular cylinder placed on a vertical heated plate. St. Petersburg State Polytech. Univ. J. Phys. Math. **11**, 56–66 (2018)

Experimental Investigation of Transitional Flow Forced Convection Heat Transfer Through a Smooth Vertical Tube with a Square-Edged Inlet

Abubakar I. Bashir, Marilize Everts, and Josua P. Meyer

1 Introduction

Limited work has been done on forced convection heat transfer in the transitional flow regime, especially in horizontal tubes with higher heat fluxes where the uncertainties are low. Forced convection experiments in horizontal tubes are challenging to perform because of the difference in density between the fluid near the surface (hot) and near the center of the tube (cold) that cause buoyancy effects and lead to mixed convection. Mixed convection can change the heat transfer characteristics in the laminar and transitional flow regimes significantly. For forced convection, the theoretical fully developed laminar flow Nusselt number is 4.36 (for a constant heat flux boundary condition). For mixed convection, the Nusselt numbers can increase up to 180–520% higher than 4.36 [1–3] due to buoyancy effects. In vertical tubes, the buoyancy effects can be reduced as the flow is in the same direction as the buoyancy force and is mostly suppressed at higher Reynolds numbers. Therefore, forced convection conditions can be achieved in the laminar and transitional flow regimes of a smooth vertical tube, even at higher heat fluxes.

Ghajar and Tam [3] found that the boundaries and the heat transfer characteristics of the transitional flow regime were inlet dependent. Everts and Meyer [1] investigated the effect of buoyancy on the heat transfer in the transitional flow regime and found that the transition Reynolds numbers were significantly affected by the buoyancy effects. Furthermore, buoyancy effects increased with increase in heating, and therefore, heating also changes the transition boundaries. However, these analyses

A. I. Bashir · M. Everts · J. P. Meyer (✉)
Department of Mechanical and Aeronautical Engineering, University of Pretoria, Pretoria 0002, South Africa
e-mail: josua.meyer@up.ac.za

A. I. Bashir
Department of Mechanical Engineering, Bayero University, Kano, Nigeria

© Springer Nature Singapore Pte Ltd. 2021
C. Wen and Y. Yan (eds.), *Advances in Heat Transfer and Thermal Engineering* ,
https://doi.org/10.1007/978-981-33-4765-6_2

focused on mixed convection conditions in horizontal tubes. It is important to investigate the heat transfer characteristics for pure forced convection in the transitional flow regime in order to fundamentally understand the behavior of transition heat transfer without the influence of buoyancy. Therefore, the purpose of this study was to experimentally investigate the single-phase forced convection heat transfer characteristics of the transitional flow regime in a smooth vertical tube, with a square-edged inlet, heated at constant heat flux.

2   Experimental Setup

The schematic of the experimental facility is shown in Fig. 1 and water was used as working fluid. A magnetic gear pump was used to pump the water from a storage tank to the flow meters, flow-calming section and inlet section and then to the test section. After the test section, the heated water returned to the storage tank for cooling and recirculation. The flow-calming section was placed prior to the test section to ensure a uniform flow distribution through the inlet section and test section, because transition is inlet dependent. A square-edged inlet geometry was used for all the experiments.

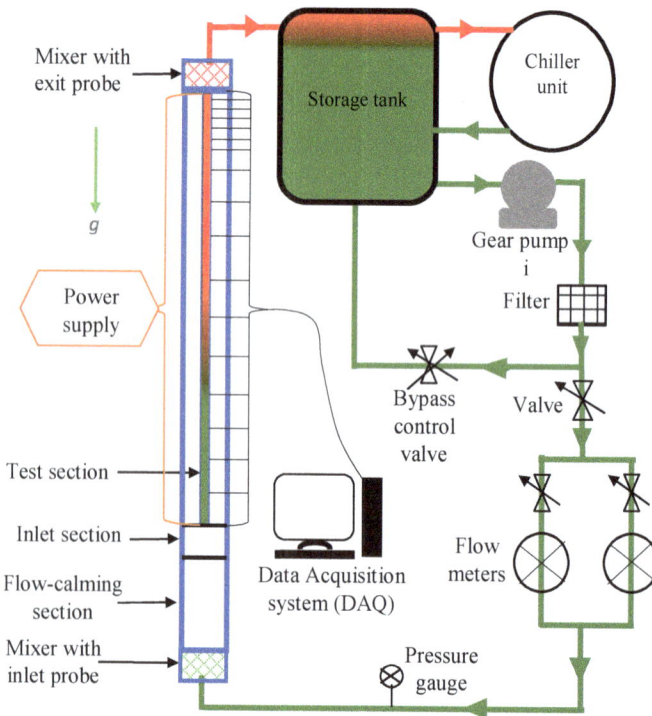

Fig. 1 Schematic of the experimental facility

The test section was a smooth hard drawn copper tube with an inner diameter of 5.1 mm and a heated length of 4.52 m (maximum length-to-diameter, x/D_i of 886). Twenty-one thermocouples were attached to the test section to measure the local wall temperatures. The inlet and exit bulk fluid temperatures were measured using two Pt100 probes placed inside the inlet and exit mixers, respectively. The test section was heated at heat fluxes of 1, 4, 6 and 8 kW/m^2 using a direct current (DC) power supply. At a Reynolds number of 2000, the flow was found to be fully developed from $x/D_i = 416$, because the heat transfer coefficients became relatively constant along the tube length. Therefore, the local results at $x/D_i = 592$ were used for the fully developed flow analyses. The test section was set at vertical upward flow direction to avoid the effect of buoyancy or free convection that might cause mixed convection. The experiments were performed for Reynolds numbers between 1000 and 6000 to cover the entire transitional flow regime as well as sufficient parts of the laminar and turbulent flow regimes.

The setup was validated against the literature by comparing laminar and turbulent flow heat transfer coefficients with well-known correlations. The laminar flow heat transfer results were compared using the flow regime map of Metais and Eckert [4] for constant heat flux in vertical tubes. All the heat transfer results fell within the forced convection region of the Metais and Eckert [4] map. Furthermore, at a Reynolds number of 1000, the laminar forced convection Nusselt number was 4.41, which is within 1.1% of 4.36. Thus, the forced convection condition was confirmed in the laminar flow regime up to the start of transition. In the turbulent flow regime, the maximum deviation of the heat transfer coefficients from Gnielinski [5] correlation was 3.9%.

3 Results

Figure 2a compares the local fully developed heat transfer results in terms of Nusselt number (Nu = hD_i/k) as a function of Reynolds number at $x/D_i = 592$. The Nusselt numbers for all the different heat fluxes in the laminar flow regime were approximately the same and approached the theoretical forced convection Nusselt number of 4.36 for a constant heat flux boundary condition. This indicated that there is negligible or no buoyancy effects and confirmed forced convection conditions for all the heat fluxes up to the start of transitional flow regime. However, as the Reynolds number increased and the flow approached the transitional flow regime, the laminar flow Nusselt numbers of all the heat fluxes increased slightly, which might be due to the effect of variable fluid property (viscosity). As expected, there was a negligible difference between the results of the different heat fluxes in the turbulent flow regime, therefore, the flow was also dominated by forced convection conditions. Because both the laminar and turbulent flow regimes were dominated by pure forced convection heat transfer, it confirmed that the entire transitional flow regime was also dominated by forced convection.

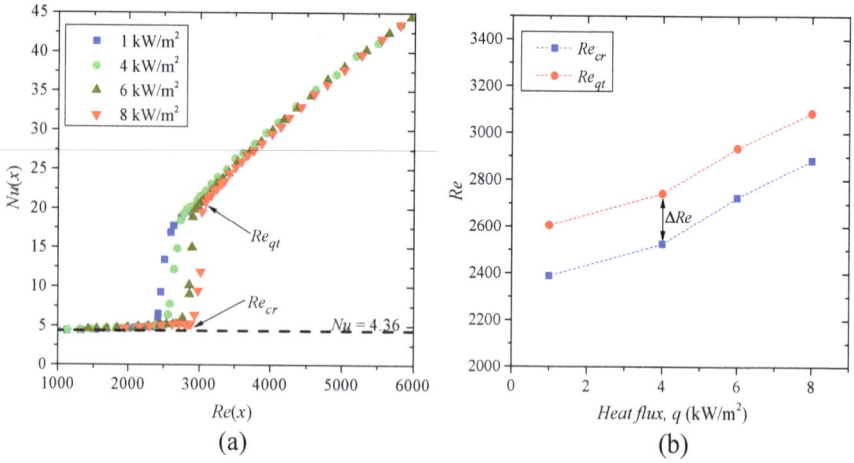

Fig. 2 Comparison of **a** fully developed local Nusselt numbers as a function of Reynolds numbers at $x/D = 592$ for the different heat fluxes and **b** Reynolds numbers at the start (Re_{cr}) and end (Re_{qt}) of transitional flow regime in **a** as a function heat flux

For all the heat fluxes in Fig. 2a, transition occurred at the same mass flow rate of approximately 0.00890 kg/s, while the critical Reynolds numbers at the start of transition increased with increase in heat flux. As the heat flux was increased for a constant mass flow rate, the increased fluid temperature led to a decreased viscosity which in turn caused the Reynolds numbers to increase. At a heat flux of 1 kW/m^2, transition occurred at a critical Reynolds number of 2388, while at 8 kW/m^2, the critical Reynolds number increased to 2883 for the same mass flow rate. Similarly, transition ended at approximately the mass flow rate but the Reynolds number at end of transition also increased with increased heat fluxes. Figure 2b compares the transition Reynolds numbers for the different heat fluxes in Fig. 2a. It followed that both the Reynolds numbers at the start (Re_{cr}) and end (Re_{qt}) of the transitional flow regime increased simultaneously with increasing heat flux. It also showed that the width of the transitional flow regime, defined by Everts and Meyer [1] as $\Delta Re = Re_{cr} - Re_{qt}$, for all the heat fluxes was approximately equal and ranged between 203 and 219. This is different from that of mixed convection condition as was found by Evert and Meyer [1] in horizontal tubes. For mixed convection condition, the width of the transitional flow regime was significantly affected by free convection effects and therefore decreased with increasing heat flux.

4 Conclusions

Single-phase forced convection heat transfer characteristics of the transitional flow was experimentally investigated using a smooth vertical tube with a square-edged inlet geometry. The test section was heated using constant heat fluxes of 1, 4, 6 and 8 kW/m^2 between Reynolds number of 1000–6000. It was found that pure forced convection condition was achieved at the fully developed part of the test section for vertically upward flow direction in all the flow regimes. Although the start and end of the transitional flow regime occurred at the same mass flow rate for all the heat fluxes, the corresponding Reynolds numbers increased with increase in heat flux, due to decreasing viscosity with increasing temperature. Because forced convection conditions existed, the width of the transitional flow was not affected by increasing heat fluxes and remained constant and approximately equal for all the heat fluxes.

Acknowledgements The authors acknowledge the funding from the European Union's Horizon 2020 research and innovation programme under the Marie Skłodowska-Curie grant agreement No. 778104 and the Tertiary Education Fund (TETFUND), Nigeria.

References

1. M. Everts, J.P. Meyer, Heat transfer of developing and fully developed flow in smooth horizontal tubes in the transitional flow regime. Int. J. Heat Mass Transf. **117**, 1331–1351 (2018)
2. J.P. Meyer, M. Everts, Single-phase mixed convection of developing and fully developed flow in smooth horizontal circular tubes in the laminar and transitional flow regimes. Int. J. Heat Mass Transf. **117**, 1251–1273 (2018)
3. A.J. Ghajar, L.M. Tam, Heat transfer measurements and correlations in the transition region for a circular tube with three different inlet configurations. Exp. Therm. Fluid Sci. **8**(1), 79–90 (1994)
4. B. Metais, E.R.G. Eckert, Forced, mixed and free convection regimes. ASME J. Heat Transf. **86**, 295–296 (1964)
5. V. Gnielinski, New equations for heat and mass-transfer in turbulent pipe and channel flow. Int. Chem. Eng. **16**(2), 359–368 (1976)

Effects of Rarefied Effect, Axial Conduction, and Viscous Dissipation on Convective Heat Transfer in 2D Parallel Plate Microchannel or Nanochannel with Walls at Uniform Temperature

Qiangqiang Sun, Kwing-So Choi, and Xuerui Mao

1 Introduction

Experiments have shown that flow and heat transfer characters in microchannel or nanochannel are quite different from their well-known macroscale counterparts [1] because the effects of rarefied effect, axial conduction, and viscous dissipation on heat transfer cannot be neglected. Significant efforts have been devoted to studying the separate effect of those three factors with non-slip boundary conditions. For example, Tso et al. [2–4] performed dimensional analyses and experiments to show the impact of Brinkman number on the microchannel flow. Their results stated that the Brinkman number has an important role in determining the flow transition point and the temperature distribution in spite of its relatively small values. A series of analytical solutions for heat transfer in one-dimensional microchannel when only axial conduction is considered have been extensively reported [5–12].

These may not exactly mimic convective heat transfer because slip and joint effects of axial conduction, viscous dissipation, and rarefied effects co-exist in the vicinity of the wall. In this paper, therefore, viscous dissipation and axial conduction terms are considered simultaneously in the process of deriving the analytical solution of the 2D energy equation with the first-order velocity slip model and the temperature jump boundary conditions. We give the analytical expression of dimensionless 2D temperature profile via much simpler separation of variables and substitution approaches and apply it to examine the impacts of viscous dissipation, axial conduction, and rarefied effect on heat transfer.

Q. Sun (✉) · K.-S. Choi · X. Mao
Faculty of Engineering, University of Nottingham, Nottingham NG7 2RD, UK
e-mail: Sun@nottingham.ac.uk

© Springer Nature Singapore Pte Ltd. 2021
C. Wen and Y. Yan (eds.), *Advances in Heat Transfer and Thermal Engineering* ,
https://doi.org/10.1007/978-981-33-4765-6_3

15

2 Methodology

2.1 Analytical Solution of the 2D Energy Equation

Assuming that fluid property including density, specific heat, thermal conductivity, and dynamic viscosity, are constants, the 2D energy equation for parallel plate microchannel or nanochannel including axial conduction and viscous dissipation, as well as boundary conditions can be established as

$$\rho c_p u \frac{\partial T}{\partial x} = k \frac{\partial^2 T}{\partial x^2} + k \frac{\partial^2 T}{\partial y^2} + \mu \left(\frac{\partial u}{\partial y} \right)^2 \tag{1}$$

$$T = T_i \ \ \text{at} \ x = 0, \quad \frac{\partial T}{\partial y} = 0 \ \text{at} \ y = 0, \quad T - T_w$$

$$= -\frac{2 - \sigma_T}{\sigma_T} \frac{2\gamma}{\gamma + 1} \frac{\lambda}{Pr} \frac{\partial T}{\partial y} \ \ \text{at} \ y = H \tag{2}$$

where ρ is the density, c_p is the specific heat at constant pressure, k is the thermal conductivity, μ is the dynamic viscosity, σ_T is the thermal accommodation coefficient, γ is the specific heat ratio, λ is the molecular free path, and Pr is the Prandtl number.

After defining dimensionless variables: $\theta = \frac{T - T_w}{T_i - T_w}$, $Br = \frac{\mu u_{ave}^2}{k(T_i - T_w)}$, $\xi = \frac{x}{Re \cdot Pr \cdot H}$, $\eta = \frac{y}{H}$, Eq. (1) becomes

$$\frac{1}{4} u^* \frac{\partial \theta}{\partial \xi} = \frac{1}{Pe^2} \frac{\partial^2 \theta}{\partial \xi^2} + \frac{\partial^2 \theta}{\partial \eta^2} + Br \left(\frac{\partial u^*}{\partial \eta} \right)^2 \tag{3}$$

where the non-dimensionless velocity is [13] $u^* = \frac{u}{u_{ave}} = \frac{3}{2} \frac{1 - \eta^2 + 8 \frac{2 - \sigma}{\sigma} Kn}{C_2}$ and $C_2 = 1 + 12 \frac{2 - \sigma}{\sigma} Kn$.

Boundary conditions are also non-dimensionalized as

$$\theta = 1 \ \ \text{at} \ \xi = 0, \quad \frac{\partial \theta}{\partial \eta} = 0 \ \text{at} \ \eta = 0, \quad \theta = -4C_1 \frac{\partial \theta}{\partial \eta} \ \ \text{at} \ \eta = 1 \tag{4}$$

θ can be decomposed as an asymptotic temperature θ_1 and a transient term θ_2. When $\xi \to +\infty$, there is $\frac{\partial \theta_1}{\partial \xi} = 0$, $\frac{\partial^2 \theta_1}{\partial \xi^2} = 0$. Then, Eqs. (3) and (4) can be rewritten as

$$\frac{\partial^2 \theta_1}{\partial \eta^2} = -Br \left(\frac{\partial u^*}{\partial \eta} \right)^2 \tag{5}$$

$$\frac{\partial \theta_1}{\partial \eta} = 0 \ \ \text{at} \ \eta = 0, \quad \theta_1 = -4C_1 \frac{\partial \theta_1}{\partial \eta} \ \ \text{at} \ \eta = 1 \tag{6}$$

$$\frac{1}{4}u^* \frac{\partial \theta_2}{\partial \xi} - \frac{1}{Pe^2}\frac{\partial^2 \theta_2}{\partial \xi^2} - \frac{\partial^2 \theta_2}{\partial \eta^2} = 0 \tag{7}$$

$$\frac{\partial \theta_2}{\partial \eta} = 0 \quad \text{at } \eta = 0, \quad \theta_2 = -4C_1 \frac{\partial \theta_2}{\partial \eta} \quad \text{at } \eta = 1 \tag{8}$$

where the coefficient C_1 is $\frac{2-\sigma_T}{\sigma_T}\frac{2\gamma}{\gamma+1}\frac{Kn}{Pr}$. The asymptotic temperature θ_1 can be solved based on Eqs. (5) and (6).

$$\theta_1 = \frac{Br}{C_2^2}\left(-\frac{3}{4} + \frac{3}{4} + 12C_1\right) \tag{9}$$

Equation (7) can be transformed to the standard confluent hypergeometric equation [14], and its solution is

$$\theta_2 = \sum_{n=1}^{\infty} A_n \left[\sum_{m=0}^{\infty} \frac{(a_n)^{(m)}\left(\beta_n\sqrt{3/(8C_2)}\eta^2\right)^m}{(b_n)^{(m)}m!} \right.$$

$$\left. \exp\left(-\frac{1}{2}\beta_n\sqrt{\frac{3}{8C_2}}\eta^2\right) \right] \exp(-\beta_n^2\xi) \tag{10}$$

where a_n is $a_n = \frac{-\beta_n^3 - k_1 Pe^2\beta_n - k_2 Pe^2\beta_n + Pe^2\sqrt{3/(8C_2)}}{4Pe^2\sqrt{3/(8C_2)}}$ and b_n is $\frac{1}{2}$.

Therefore, the dimensionless temperature θ is

$$\theta = \frac{Br}{C_2^2}\left(-\frac{3}{4} + \frac{3}{4} + 12C_1\right) + \sum_{n=1}^{\infty} A_n \left[\sum_{m=0}^{\infty} \frac{(a_n)^{(m)}\left(\beta_n\sqrt{3/(8C_2)}\eta^2\right)^m}{(b_n)^{(m)}m!} \right.$$

$$\left. \exp\left(-\frac{1}{2}\beta_n\sqrt{\frac{3}{8C_2}}\eta^2\right) \right] \exp(-\beta_n^2\xi) \tag{11}$$

Summation coefficients A_n are calculated via applying the Gram–Schmidt orthogonal approach in the literature [15].

2.2 Numerical Simulation of Forced Convective Heat Transfer in a Plane Parallel Channel

In order to validate the aforementioned analytical solution, a new extended heat transfer solver including viscous heat is developed based on the standard icoFOAM solver in OpenFOAM. The continuity, momentum, and energy equations [16] for the incompressible fluid in the new solver are

$$\nabla \cdot \boldsymbol{U} = 0 \tag{12}$$

$$\frac{\partial \boldsymbol{U}}{\partial t} + \nabla \cdot (\boldsymbol{U} \otimes \boldsymbol{U}) = -\nabla \frac{p}{\rho} + \nabla \cdot (\nu \nabla \boldsymbol{U}) \tag{13}$$

$$\frac{\partial (\rho e)}{\partial t} + \nabla \cdot (\rho \boldsymbol{U} e) = \rho r - \nabla \cdot \boldsymbol{q} - p \nabla \cdot \boldsymbol{U} + \boldsymbol{\tau} : \nabla \boldsymbol{U} \tag{14}$$

where e is internal energy and $\boldsymbol{\tau}$ is stress tensor.

Because the linear Navier velocity slip and first-order temperature jump, boundary conditions [17] are applied in the analytical solution of the 2D energy equation. Hence, these two slip boundary conditions shown as follows are also defined for the new solver in order to validate the correctness of the analytical solution.

$$\boldsymbol{U}_{ws} = -S_l \mu (\boldsymbol{I} - \boldsymbol{n} \otimes \boldsymbol{n}) \left(\frac{\partial \boldsymbol{U}}{\partial \delta_{ZC}} \right)_{\text{wall}} \tag{15}$$

$$T - T_w = -\frac{2 - F_t}{F_t} \frac{2\gamma}{\gamma + 1} \frac{\lambda}{Pr} \left(\frac{\partial T}{\partial \delta_{ZC}} \right)_{\text{wall}} \tag{16}$$

where $\boldsymbol{U}_{ws}$ is velocity of the fluid in the vicinity of the solid wall, S_l is the velocity slip coefficient, μ is dynamic viscosity, $\boldsymbol{I}$ is a unit diagonal matrix, and $\boldsymbol{n}$ is the normal vector of a boundary face of a mesh cell connecting with the solid wall. Subscripts Z and C represent the central point of the boundary face and the mesh cell, respectively.

3   Validation

In the test case, the length of the channel is 1.0 m, and the half-height is 0.05 m. The tangential momentum and thermal accommodation coefficient are set as $\sigma = 1$ and $\sigma_t = 1$, respectively. The thermal diffusivity is $\alpha = 0.02\,\text{m}^2/\text{s}$, kinematic viscosity is $\nu = 0.014\,\text{m}^2/\text{s}$, specific heat at constant pressure is $c_p = 0.07\,\text{m}^2/(\text{s}^2\,\text{K})$, specific heat ratio is $\gamma = 1.4$, and the density is $\rho = 1380\,\text{kg/m}^3$. The average velocity is 1 m/s. Dimensionless numbers are: Pe $= 10$, Re $= 14.29$, Pr $= 0.7$, Br $= -1$, Kn $= 0.10$. The inlet and outlet pressure are $p_{\text{in}} = 10538.18\,\text{Pa}$ and $p_{\text{out}} = 0\,\text{Pa}$ in the case simulated by the new solver.

The velocity and temperature profiles from the analytical solution and numerical simulation in the height direction of the channel are shown in Fig. 1. Results obtained from the two approaches are in good agreement. This indicates that the analytical solution has sufficient accuracy. Also, obvious velocity slip and temperature jump at the wall can be seen from this figure.

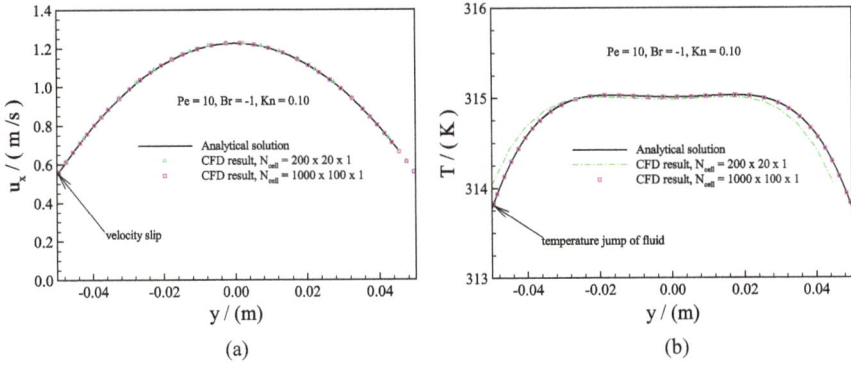

Fig. 1 Velocity profile (**a**) and temperature profile (**b**) at $x = 0.5$ m cross section

4   Results

Then, the non-dimensional analytical method is used to examine the impacts of Pe, Br, and Kn number on the temperature profile.

From Fig. 2a, one can see that fluid asymptotic temperature is higher than constant wall temperature irrespective of whether $T_{in} > T_{wall}$ or $T_{in} < T_{wall}$. The asymptotic dimensionless bulk temperature of the fluid θ_{bF} at various viscous and rarefied conditions is shown in Fig. 2b when considering axial conduction or not by setting Pe $= 10$ or Pe $= 10^6$ (the result is similar to one when Pe $= 10$, and it is not shown here to save pages). θ_{bF} is a linear function of Br number. It roughly drops with increasing of Kn and is independent on Pe number.

The critical length ξ_C, observed in Fig. 2a, deserves further studies. The variation of ξ_C at various Br, Pe, and Kn is illustrated in Fig. 3. Clearly, ξ_C reduces with larger Br and Pe and smaller Kn, corresponding to stronger viscous dissipation, weaker axial conduction, and weaker rarefied effect, respectively.

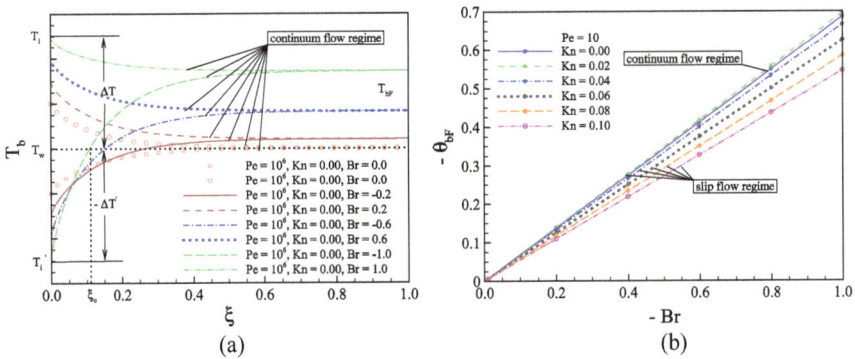

Fig. 2 Fluid bulk temperature (**a**) and asymptotic dimensionless bulk temperature (**b**) distribution

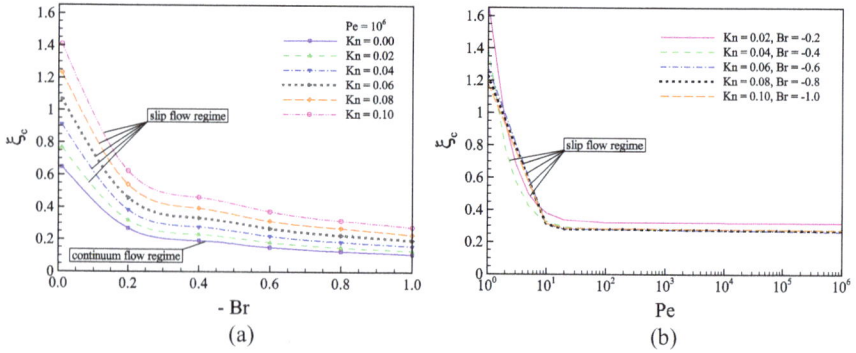

Fig. 3 Distribution of the critical length ξ_C. **a** ξ_C changes with Br and Kn when neglecting axial conduction. **b** ξ_C profile at different Pe

5 Conclusions

The analytical solution of the 2D energy equation with linear Navier velocity slip and first-order temperature jump boundary conditions for plane channel flow is performed, and summation coefficients in the analytical solution series are determined via using the Gram–Schmidt orthogonalization accompanied with Gauss–Legendre quadrature. The velocity and temperature profile obtained from the analytical solution approach for the case Pe $= 10$, Br $= -1$, Kn $= 0.10$ are in excellent agreement with numerical simulation results, which validates the accuracy of the analytical solution. The asymptotic fluid bulk temperature converges to a constant value higher than wall temperature if the viscosity dissipation is considered, and the asymptotic dimensionless bulk temperature of fluid increases linearly with Br number, drops with Kn number, and is roughly independent on Pe number. If a fluid is used to cool the wall with uniform temperature, there is a critical point where the fluid bulk temperature reaches the same value as wall temperature and the Nusselt number become ill-defined at this point. This point moves much closer to the channel entrance at increasing $|Br|$ or Pe. On the contrary, the point moves toward the end of the channel with increasing Kn.

Acknowledgements We would like to acknowledge financial support under the Engineering and Physical Sciences Research Council grant EP/M025039/2 and China Scholarship Council (CSC).

References

1. J.C. Harley, et al., Gas flow in micro-channels. J. Fluid Mech. **284**, 257–274 (1995)
2. C. Tso, S. Mahulikar, The use of the brinkman number for single phase forced convective heat transfer in microchannels. Int. J. Heat Mass Transf. **41**(12), 1759–1769 (1998)

3. C. Tso, S. Mahulikar, The role of the brinkman number in analysing flow transitions in microchannels. Int. J. Heat Mass Transf. **42**(10), 1813–1833 (1999)
4. C. Tso, S. Mahulikar, Experimental verification of the role of brinkman number in microchannels using local parameters. Int. J. Heat Mass Transf. **43**(10), 1837–1849 (2000)
5. M. Michelsen, J. Villadsen, The graetz problem with axial heat conduction. Int. J. Heat Mass Transf. **17**(11), 1391–1402 (1974)
6. J. Lahjomri, A. Oubarra, Analytical solution of the Graetz problem with axial conduction. J. Heat Transf. **121**(4), 1078–1083 (1999)
7. J. Lahjomri, A. Oubarra, A. Alemany, Heat transfer by laminar Hartmann flow in thermal entrance region with a step change in wall temperatures: the Graetz problem extended. Int. J. Heat Mass Transf. **45**(5), 1127–1148 (2002)
8. B. Weigand, D. Lauer, The extended graetz problem with piecewise constant wall temperature for pipe and channel flows. Int. J. Heat Mass Transf. **47**(24), 5303–5312 (2004)
9. I. Tiselj, G. Hetsroni, B. Mavko, A. Mosyak, E. Pogrebnyak, Z. Segal, Effect of axial conduction on the heat transfer in micro-channels. Int. J. Heat Mass Transf. **47**(12–13), 2551–2565 (2004)
10. G. Maranzana, I. Perry, D. Maillet, Mini-and micro-channels: influence of axial conduction in the walls. Int. J. Heat Mass Transf. **47**(17–18), 3993–4004 (2004)
11. W. Minkowycz, A. Haji-Sheikh, Heat transfer in parallel plates and circular porous passages with axial conduction. Int. J. Heat Mass Transf. **49**(13–14), 2381–2390 (2006)
12. A. Haji-Sheikh, J. Beck, D.E. Amos, Axial heat conduction effects in the entrance region of circular ducts. Heat Mass Transf. **45**(3), 331–341 (2009)
13. H.-E. Jeong, J.-T. Jeong, Extended Graetz problem including streamwise conduction and viscous dissipation in microchannel. Int. J. Heat Mass Transf. **49**(13–14), 2151–2157 (2006)
14. H.J. Weber, F.E. Harris, *Mathematical Methods for Physicists: A Comprehensive Guide* (Academic Press, 2013)
15. Q. Sun, K.-S. Choi, X. Mao, An analytical solution of convective heat transfer in microchannel or nanochannel. (Submitted to Journal)
16. J.H. Ferziger, M. Peric, A. Leonard, *Computational Methods for Fluid Dynamics* (1999)
17. C. Fernandes, L.L. Ferrás, F. Habla, O.S. Carneiro, J.M. Nóbrega, Implementation of partial slip boundary conditions in an open-source finite-volume-based computational library. J. Polym. Eng. **39**, 377–387 (2019)

Enhancement of Laminar Natural Convection Heat Transfer in Horizontal Annuli Using Two Fins

A. El Amraoui, A. Cheddadi, and M. T. Ouazzani

1 Introduction

The optimization of heat transfer in a finned annular cavity has been studied extensively in recent decades. The results of this research are used in various industrial applications such as the cooling of nuclear reactors, electronic equipment [1] and solar collectors.

Nag et al. [2] have studied the effect of a fin on heat transfer when it is placed on one of the active walls of a square cavity heated differentially. The results showed that, as the thickness of the fin decreases, the heat transfer through the cavity decreases until reaching a critical thickness for which the heat transfer rate is minimal. Beyond this value, the heat transfer rises as the fin thickness decreases further more. Elatar et al. [3] studied the effect of fin thickness with a different thermal conductivity on average Nusselt number in a square enclosure with active vertical walls. This study has shown that for fins with thermal conductivity ratios between 10 and 100, the thickness of the fins has a minor effect on heat transfer, while for fins with a thermal conductivity ratio of 1000 it has almost no effect. It has also been mentioned that the heat transfer rate increases with increasing fin height and thermal conductivity. Dindarloo et al. [4] studied the effect of the thickness and the angle of attachment of the fins to the hot wall in order to bring a maximum reduction of heat transfer in a square enclosure. The results obtained show that for the fin with a thermal conductivity ratio of 1, the minimum heat transfer rate is obtained at the greatest thickness whereas for a thermal conductivity ratio of 1000 the minimum heat transfer rate is obtained at the smallest thickness. For the fin with a thermal conductivity ratio of 10, the optimum thickness varies with the Rayleigh number. Taher et al. [5] carried out a study on the influence

A. El Amraoui · A. Cheddadi (✉) · M. T. Ouazzani
Modeling of Energy Systems, Materials and Mechanical Structures, and Industrial Processes, MOSEM2PI, Mohammadia School of Engineers, Mohammed V University, P. O. Box 765, Agdal, Rabat, Morocco
e-mail: cheddadi@emi.ac.ma

© Springer Nature Singapore Pte Ltd. 2021
C. Wen and Y. Yan (eds.), *Advances in Heat Transfer and Thermal Engineering* ,
https://doi.org/10.1007/978-981-33-4765-6_4

of the thickness of two fins on the heat transfer, and these fins have a height of $h = 0.2$ and been placed at a central position in a cylindrical annular cavity. It has been shown that increasing the thickness of the fins causes an increase in heat transfer rate and the heat exchange for $Ra = 10^3$ is greater than for $Ra = 10^4$.

The purpose of this work is to analyze the effect of varying the thickness of fins placed at a high position in a cylindrical annular cavity on fluid flow and heat transfer rate taking into account the effect of fin height for Rayleigh numbers varying from 10^3 to 10^4.

2 Methodology

The mathematical modeling of the problem is based on the principles of conservation of momentum, energy and mass. The system obtained is completed by introducing the boundary conditions on the hot and cold walls and on the boundaries of the fins assumed to be very conductive. The fluid occupying the annular space is air ($Pr = 0.7$) considered to be incompressible viscous, obeying the Boussinesq approximation. The problem considered is two-dimensional, steady and laminar. The non-dimensional governing equations written in vorticity-stream function formulation are:

$$\Delta \psi + \omega = 0 \tag{1}$$

$$\frac{\partial \omega}{\partial t} + U \frac{\partial \omega}{\partial r} + \frac{V}{r} \frac{\partial \omega}{\partial \varphi} = Pr \Delta \omega + Ra \, Pr \left(\sin \varphi \frac{\partial T}{\partial r} + \frac{\cos \varphi}{r} \frac{\partial T}{\partial \varphi} \right) \tag{2}$$

$$\frac{\partial T}{\partial t} + U \frac{\partial T}{\partial r} + \frac{V}{r} \frac{\partial T}{\partial \varphi} = \Delta T \tag{3}$$

The integration of the system (1)–(3) is based on the discretization of these equations by the centered finite difference method with an ADI scheme. This leads to solve tridiagonal matrix systems, using Thomas algorithm. The initialization of the calculations is carried out by the introduction of zero fields of the stream function and vorticity, and a pure conduction temperature field ($T = 1 - \ln r / \ln R) \ln r$. The fluid flows considered in this study are steady state and laminar. The radius ratio between the radius of the outer cylinder r_o and the inner cylinder r_i is $R = 2$. The fins held at the angular position $\varphi_m = 0.82\pi$ from downward have a dimensionless thickness w ranging from 0.015 to 0.203 and a dimensionless height h ranging from 0.015 to 0.953.

The heat transfer rate is characterized by the average Nusselt number defined by:

$$\overline{Nu} = -\frac{R}{\pi} \ln R \int_0^\pi \frac{\partial T}{\partial r} \bigg|_{r=R} d\varphi \tag{4}$$

3   Results

The flow structure at low heights, $h \leq 0.109$ for $Ra = 10^3$ and $h \leq 0.093$ for $Ra = 10^4$, when w varies from 0.015 to 0.203 indicates that the flow regime established is unicellular (UCR), characterized by a single main cell. At intermediate heights, the increase in the thickness of the fins corresponds to the appearance of a secondary cell in the upper part of the cavity with the establishment of the bicellular regime (BCR). This cell grows gradually for $Ra = 10^3$ and suddenly at $Ra = 10^4$. Figure 1 shows for $h = 0.218$ the UCR regime at $w = 0.015$ and the first appearance of the BCR regime at $w = 0.078$ for $Ra = 10^3$ and $Ra = 10^4$. At larger heights, $h \geq 0.328$ for $Ra = 10^3$ and $h \geq 0.703$ for $Ra = 10^4$, the secondary cell appears at the smallest thickness $w = 0.015$.

Figure 2 shows the iso-values of the heat transfer rate $\overline{Nu}$, in the w-h plane for $Ra = 10^3$ and $Ra = 10^4$. For $Ra = 10^3$ at a given height, the rate of heat transfer increases with the increase of the fin thickness. At low heights, $\overline{Nu}$ varies slightly as the thickness of the fins increases while it grows relatively quickly at high heights. It can thus be concluded that for $Ra = 10^3$ as the dimension of the fins h and w increases the heat transfer rate increases as well.

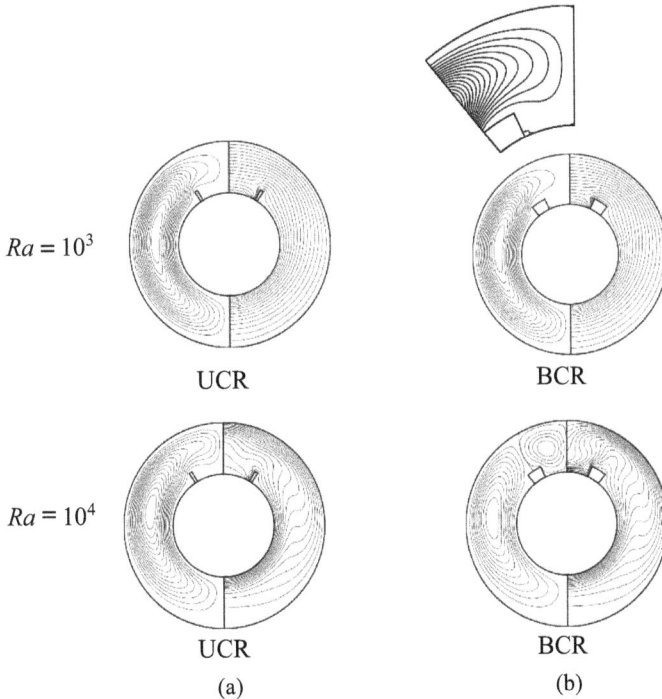

Fig. 1   Flow regimes for **a** $w = 0.015$ and **b** $w = 0.078$ at $h = 0.218$

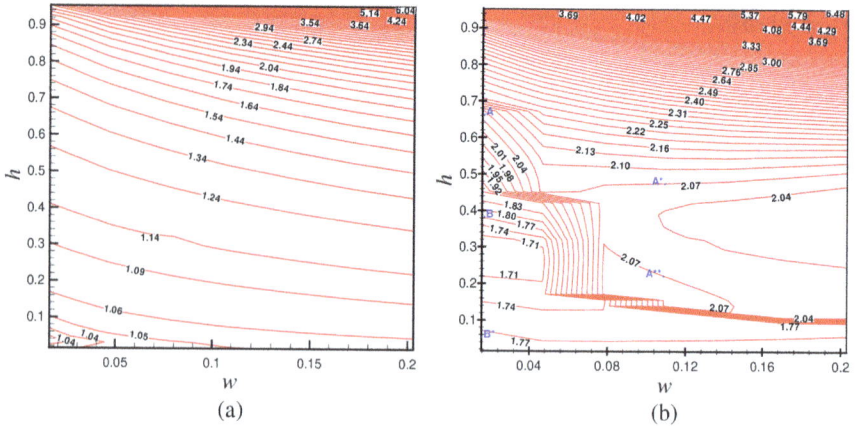

Fig. 2 Iso-values of $\overline{\mathrm{Nu}}$ in the w-h plane for **a** Ra $= 10^3$ and **b** Ra $= 10^4$

For Ra $= 10^4$ at low heights $h \leq 0.093$ when establishing the UCR regime, the Nusselt number remains almost constant when w increases. At the median heights, $0.109 \leq h \leq 0.453$, a small increase in w leads to a rise in the heat transfer. $\overline{\mathrm{Nu}}$ increases from 1.75 for $h = 0.109$ and 1.86 for $h = 0.453$ to the same value around 2.06, respectively, at $w = 0.171$ and $w = 0.046$. From these values of w, the Nusselt number remains almost constant (Fig. 2-b). However, combinations of heights and thicknesses where the heat transfer is almost the same are noted. For example, the points $A*$ ($w = 0.109$; $h = 0.453$), $A**$ ($w = 0.109$; $h = 0.218$) and $A(w = 0.015$; $h = 0.671$) have the same Nusselt number equal to 2.06 (± 0.005); similarly, the points B ($w = 0.015$; $h = 0.390$) and $B*$ ($w = 0.015$; $h = 0.062$) have the same Nusselt number equal to 1.77 (± 0.008).

4 Conclusion

The present work is based on a numerical simulation of natural convection in an annular cavity provided with two fins whose thickness and height, respectively, vary from 0.015 to 0.203 and 0.015 to 0.953. The effects of fin thickness and height as well as the Rayleigh number on heat transfer and fluid flow are analyzed. The results show that:

- For Ra $= 10^3$, several flow structures appear and the transition between them is gradual. The heat transfer increases with increasing thickness and height of the fins.
- For Ra $= 10^4$ at low h, the flow regime is unicellular and the increase in thickness has no significant effect on heat transfer. At median heights, combinations of heights and thicknesses offer the same heat transfer. For larger heights, the greater the thickness, the higher the heat transfer rate, and the flow regime is bicellular.

References

1. S. Acharya, S.K. Dash, Natural convection heat transfer from a hollow horizontal cylinder with external longitudinal fins: a numerical approach. Numer. Heat Transf., Part A: Appl. **74**, 1405–1423 (2018)
2. A. Nag, A. Sarkar, V.M.K. Sastri, Effect of thick horizontal partial partition attached to one of the active walls of a differentially heated square cavity. Numer. Heat Transf., Part A: Appl. **25**, 611–625 (1994)
3. A. Elatar, M.A. Teamah, M.A. Hassab, Numerical study of laminar natural convection inside square enclosure with single horizontal fin. Int. J. Therm. Sci. **99**, 41–51 (2016)
4. M.R. Dindarloo, S. Payan, Effect of fin thickness, grooves depth, and fin attachment angle to the hot wall on maximum heat transfer reduction in a square enclosure. Int. J. Therm. Sci. **136**, 473–490 (2019)
5. Y. Taher, A. Cheddadi, M.T. Ouazzani, Heat transfer in an annular space provided with fins: numerical simulation of the effect of the fins width. Phys. Chem. News **30**, 132–138 (2006)

A Computational Study of Ultrasound-Enhanced Convective Heat Transfer

Raheem Abbas Nabi, Yi Sui, and Xi Jiang

1 Introduction

Enhanced convective heat transfer is paramount to many engineering applications. If a method can be identified to increase the cooling effects of a convective fluid flow or a disrupted boundary layer, the systems can be designed with increased efficiency. This computational study was intended to understand the behaviour exhibited by convection in a closed cylindrical domain. The main purpose is to establish a validated methodology in order to identify a method to enhance heat transfer. Current work is being conducted to develop a solver that can resolve the acoustic streaming effects seen when a fluid is subject to ultrasonic oscillations. The disturbance of the fluid boundary layer encourages fluid mixing, and the resultant thermal mixing improves the heat transfer mechanism. Thermal resistance is subsequently reduced through alternation of boundary layer, and the resulting benefits for cooling in engineering applications are demonstrated [1].

Acoustic streaming consists of two primary components: the acoustically induced oscillatory flow seen in the form of the distinctive vortical shredding structures for a characteristic fluid subject to streaming, and stationary non-oscillatory flow induced by thermal buoyancy [1]. In order to apply such a solver to a computational setup, a base case needs to be established where no acoustic oscillations are applied and only the effects of buoyancy and convection are observed. Here, the Boussinesq approximation is used to calculate the convective density change. A previous set of results [2] is used as a baseline to measure the degree of validation for the methodology. The effects of heat transfer are identified through a submersed 0.25 mm diameter wire in a computational cylindrical domain at a water depth of 80 mm. A heat-flux boundary

R. A. Nabi · Y. Sui · X. Jiang (✉)
School of Engineering and Materials Science, Queen Mary University of London, London E1 4NS, UK
e-mail: xi.jiang@qmul.ac.uk

© Springer Nature Singapore Pte Ltd. 2021
C. Wen and Y. Yan (eds.), *Advances in Heat Transfer and Thermal Engineering* ,
https://doi.org/10.1007/978-981-33-4765-6_5

condition is applied to the wire surface to initiate heat transfer for convection, and a k-epsilon turbulent model was selected.

2 Methodology

The aim of the investigation is to understand the ultrasound-enhanced convective heat transfer behaviour of fluid flow. As part of the work carried out, an acoustic solver is also devised to resolve the effects of acoustic streaming. Both the continuity and momentum equations were solved for an incompressible flow, and the perturbation method was adopted to calculate the acoustic forcing term needed for the baseline convection solver that made use of the Boussinesq approximation.

To account for the ultrasonic effect, the continuity equation needs to be considered in expansion of equilibrium form where the initial equilibrium value corresponds to no acoustic excitation (i.e. zeroth order) [3]: $\rho = \rho^o + \rho^1 + \rho^2$ for density, $u = u^o + u^1 + u^2$ for velocity, and $p = p^o + p^1 + p^2$ for pressure, where each superscript denotes the order of equations for flow analysis. The first-order term in this case resolves the damped propagation of acoustic waves and provides analysis of the instantaneous acoustic flow. The acoustic forcing term is subsequently calculated and substituted into the second-order equations which then describe the acoustic streaming effects.

This study considers the first-order continuity and state equation:

$$\frac{\delta\rho^o}{\delta t} + \rho^o \nabla \cdot (u') = 0, \quad \rho^o \frac{\delta u^1}{\delta t} = -\nabla \cdot p^1 + \mu\nabla^2 u^1, \quad p^1 = c^2\rho^1$$

where ρ^o is the equilibrium density, ρ^1 is the first-order density, u^1 is the first-order velocity, and c is the speed of sound.

The time averaged acoustic force is calculated after solving the first-order system and is obtained by integrating the driving force term over one wave period. The analysis of acoustic streaming effects over large timescales is now possible by applying this time averaged driving force to the second-order system. This system will consider the solution to the first-order equations as known data and will describe the mean global flow for acoustic streamline. The second-order continuity equation becomes:
$\frac{\delta\rho^2}{\delta t} + \rho^o \nabla \cdot (u^2) = \langle -\nabla \cdot (\rho^1 u^1) \rangle$.

The second-order momentum equation becomes:

$$\rho^o \frac{\delta u^2}{\delta t} = -\nabla \cdot p^2 + \mu\nabla^2 u^2 + \left\langle -\rho^1\frac{\delta u^1}{\delta t} - \rho^o (u^1 \cdot \nabla)u^1 \right\rangle$$

In addition, we also have $p^2 = c^2\rho^2$. The term in cycled brackets represents the nonlinear driving source terms.

3 Results and Discussion

Figure 1 presents result for a simple 2D geometry using the solver developed that solves for the first-order terms. The plot shows the distinctive wave fronts associated with damped acoustic propagation.

This study was aimed at developing a second-order solver to resolve the acoustic streaming effects and to validate the solution against a set of existing results already solved for convection. Figure 2 shows such an example. The experimental setup is based on that used by Dehbani et al. [2] and shows a characteristic flow structure for convection. Contours are of temperature, and overlaid is a velocity vector plot. The overall setup consists of a 0.25 mm diameter wire submerged in a cylindrical domain full of water.

Distinctive vortical patterns are seen in the flow structure, and a solution was obtained for several heat fluxes applied to the wire boundary as displayed in Fig. 3.

A proportional increase in wire temperature is seen as heat flux is increased. The effect of this trend when the same geometry is subject to acoustic oscillations based on the solver developed that solves for both first and second order will determine the extent to which heat transfer has been enhanced and will provide for additional results on flow structure, characterization, and validation.

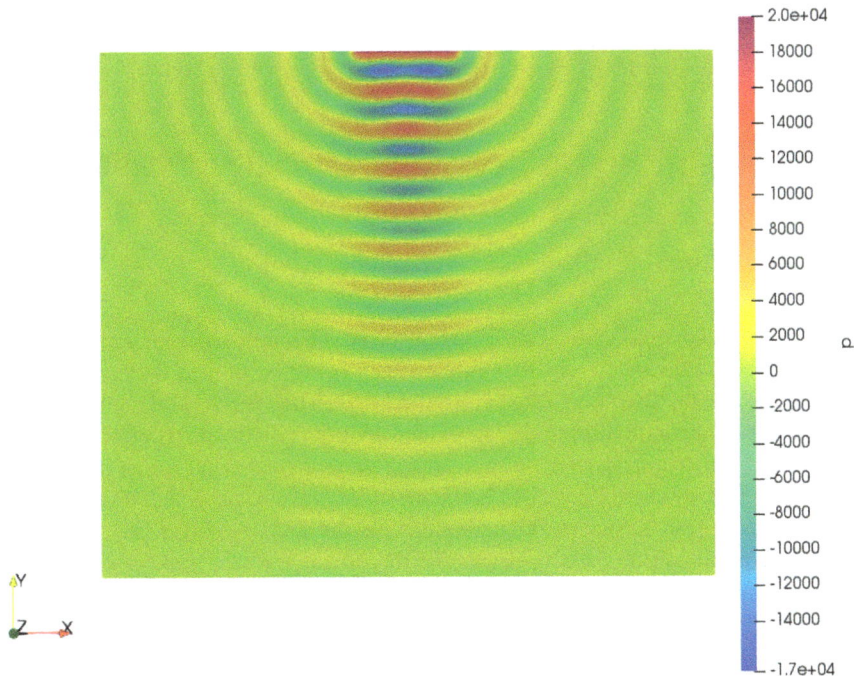

Fig. 1 First-order acoustic solution showing wave propagation for a simple 2D geometry

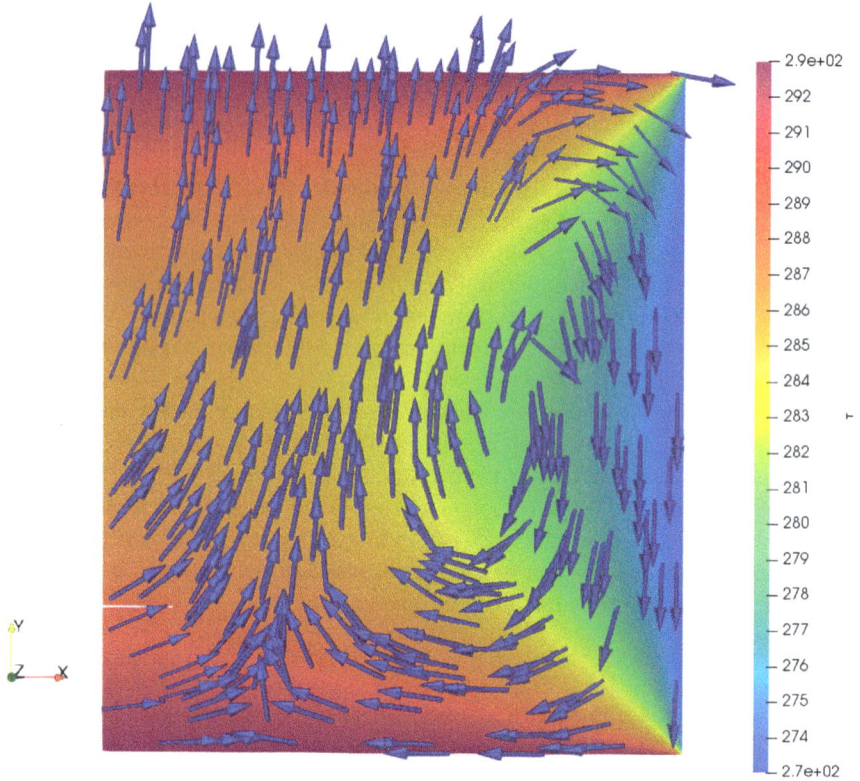

Fig. 2 Sample simulation results showing temperature contours overlaid with a velocity vector plot for a submersed platinum wire

4   Conclusions

In the present work, an investigation on convective heat transfer from a heated platinum wire was conducted whilst simultaneously being developed with an acoustic code to resolve the streaming effects in a fluid flow. The results lay the groundwork to initialize a calculation using the developed acoustic solver and its effects on heat transfer. A proportionate increase in average temperature is seen for an increase in heat flux, and distinctive convective vortical patterns are also observed. If the convective calculation can be successfully coupled with the acoustic solver, the changes in net temperature can be used to determine the effects on heat transfer coefficient. The foundations are therefore laid to further develop the research on acoustically enhanced heat transfer, and further investigations can be made.

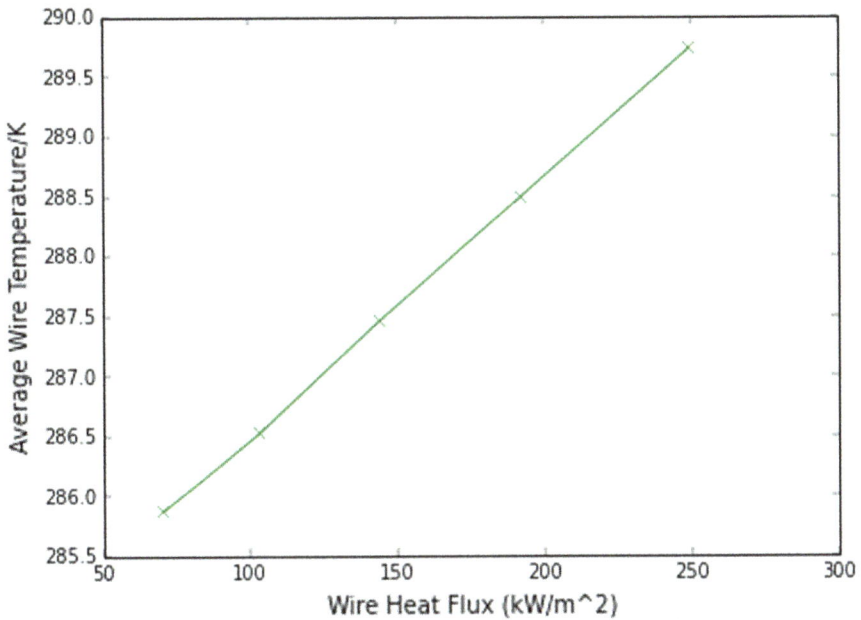

Fig. 3 Wire temperature variations as heat flux is increased

References

1. M. Legay, N. Gondrexon, S. Le Person, P. Boldo, A. Bontemps, Enhancement of heat transfer by ultrasound: review and recent advances. Int. J. Chem. Eng. **2011**, 670108 (2011)
2. M. Dehbani, M. Rahimi, M. Abolhasani, A. Maghsoodi, P.G. Afshar, A.R. Dpdmantipi, A.A. Alsairafi, CFD modelling of convection heat transfer using 1.7 MHz and 24 kHz ultrasonic waves: a comparitive study. Heat Mass Transf. **50**, 1319–1333 (2014)
3. P.B. Muller, R. Barnkob, M.J.H. Jensen, H. Bruus, A numerical study of microparticle acoustophoresis driven by acoustic radiation forces and streaming-induced drag forces. Lab Chip **12**, 4617–4627 (2012)

Boiling-Evaporation-Condensation

Wall Effects on the Thermocapillary Migration of Single Fluorinert Droplet in Silicon Oil Liquid

Yousuf Alhendal and Ali Turan

1 Introduction

Fluid particle (bubble or drop) will move to the hotter side when placed in another immiscible fluid under a temperature gradient. Surface tension generally decreases with increasing temperature and the non-uniform surface tension at the fluid interface leads to shear stresses that act on the outer fluid by viscous forces. This causes droplets in the fluid to move in the direction of the thermal gradient. The flow from a region of low surface tension to a region of higher surface tension is referred to as Marangoni flow.

A few microgravity experiments on the thermocapillary droplet flow in zero gravity have been performed on board the microgravity sounding rocket and Space laboratory, and noted that there are no numerical results with which to evaluate their data [1] mentioned the complex behaviour of thermocapillary droplet migration, confirming that further studies were still needed. They also confirmed that a longer experimental time in microgravity conditions is necessary for a droplet approaching its steady thermocapillary velocity. It is also difficult to obtain complete information about the behaviour of bubbles/droplets in space, and Computational Fluid Dynamics (CFD) studies have been undertaken by many researchers in order to compare and analyse their experimental results [2]. In recent years and with advances in numerical calculation, knowledge of thermocapillary flow has undergone a considerable change and new calculated results could be used to revise and adjust some previous results. On the other hand, numerical simulations have consequently become an

Y. Alhendal (✉)
College of Technological Studies (CTS), Public Authority for Applied Education and Training, Kuwait City, Kuwait
e-mail: Ya.alhendal@paaet.edu.kw

A. Turan
School of Mechanical, Aerospace and Civil Engineering, The University of Manchester, Manchester M13 9PL, UK

© Springer Nature Singapore Pte Ltd. 2021
C. Wen and Y. Yan (eds.), *Advances in Heat Transfer and Thermal Engineering* ,
https://doi.org/10.1007/978-981-33-4765-6_6

important tool in studies of two-phase flows in a microgravity environment and can help to clarify the basic fluid physics, as well as to assist in the design of the experiments or systems for the zero-gravity environment. Understanding thermocapillary droplet flow is very important for future research and for designing useful experiments; indeed, an understanding of these phenomena is highly desirable for the future design of space shuttles and equipment that might be employed in space condition. There is still much to be understood about two-phase flows in general and especially in zerogravity conditions. As a result, this paper will investigate the sensitivity of various parameters and/or scenarios that could not be investigated or fully covered previously, i.e. walls effect.

2 Methdology

The thermocapillary motion of a drop was first examined experimentally by Young, Block, and Goldstein [3], when Reynolds number (Re) and Marangoni number (Ma) are small, which means that both convective momentum and energy transport are negligible, who also found an analytical expression for its terminal velocity in the creeping flow limit:

$$V_{\text{YGB}} = \frac{2|\text{d}\sigma/\text{d}T|r_d \lambda \text{d}T/\text{d}x}{(2\mu + 3\mu')(2\lambda + \lambda')} \tag{1}$$

commonly called the YGB model, which is suitable for small Reynolds and Marangoni numbers:

$$\text{Re}_T = \frac{r_d V_T}{\nu} \tag{2}$$

$$\text{Ma}_T = r_d V_T / \alpha = \text{Re}_T \cdot \text{Pr} \tag{3}$$

where Prandtl number is the ratio of kinematic viscosity to thermal diffusivity:

$$\text{Pr} = \nu/\alpha \tag{4}$$

and ν is the kinematic viscosity in m^2/s:

$$\nu = \mu/\rho$$

The velocity V_T derived from the tangential stress balance at the free surface is used for scaling the migration velocity (m/s) in Eqs. (2) and (3):

$$V_T = \frac{(\text{d}\sigma/\text{d}T).(\text{d}T/\text{d}x).r_d}{\mu} \tag{5}$$

where μ and μ', λ, and λ' are the dynamic viscosity and thermal conductivity of continuous phase and droplet, respectively. ρ is the density and r_d is the radius of the droplet. The constant $d\sigma/dT$ or σ_T is the rate of change of interfacial tension and dT/dx is the temperature gradient imposed in the continuous phase fluid.

3 VOF Model and Computational Procedure

The governing continuum conservation equations for two-phase flow were solved using the ANSYS-FLUENT commercial software package [4], and the volume of fluid (VOF) method was used to track the liquid/gas interface. This method deals with completely separated phases with no diffusion. The geometric reconstruction scheme, based on the piece-wise linear interface calculation (PLIC) method of Young's [5] in Ansys-Fluent, was chosen for the current investigation. Geo-reconstruction is an added module to the already existing VOF scheme that allows for a more accurate definition of the free surface [6]. The movement of the gas/liquid interface is tracked based on the distribution of the volume fraction of the gas, i.e. α_G, in a computational cell, where the value of α_G is 0 for the liquid phase and 1 for the gas phase. Therefore, the gas/liquid interface exists in the cell where α_G lies between 0 and 1. A single momentum equation, which is solved throughout the domain and shared by all the phases, given by:

$$\frac{\partial}{\partial t}(\rho\vec{v}) + \nabla.(\rho\vec{v}\vec{v}) = -\nabla p + \nabla.[\mu(\nabla\vec{v} + \nabla\vec{v}^T)] + \vec{F} \qquad (6)$$

In Eq. 6, $\vec{F}$ represents volumetric forces at the interface, resulting from the surface tension force per unit volume. The continuum surface force (CSF) model proposed by Brackbill et al. [7] is used to compute the surface tension force for the cells containing the gas/liquid interface:

$$\vec{F} = \sigma \frac{\rho k \vec{n}}{\frac{1}{2}(\rho_L + \rho_G)} \qquad (7)$$

where σ is the coefficient of surface tension,

$$\sigma = \sigma_0 + \sigma_T(T_0 - T) \qquad (8)$$

and σ_0 is the surface tension at a reference temperature T_0, T is the liquid temperature.

4 Results

This section presents the results of an extensive numerical investigation of the thermocapillary flow of a Fluorient droplet, $d_b = 9$ mm, rising in stagnant silicon oil. The size of the computational wall bounded domain was chosen as 60×30 mm with zero permeability "no inflow or outflow" from the sides. For the simulations, silicone oil properties were taken as the ones given from Hadland et al. [8]. A numerical prescription for the viscosity, density, and surface tension variation against temperature is provided via user defined functions (UDFs). These UDFs are dynamically linked with the FLUENT solver. The initial rise velocity for the droplet is set to zero. The upper surface (top wall) of the model is hotter than the bottom surface (bottom wall); both top and bottom walls are set to no-slip solid walls. The final numerical results for a range of thermal Reynolds numbers and thermal Marangoni numbers are compared to the experimental measurements of SZ-4 Space Shuttle IML-2 Experiment and Space Shuttle LMS Experiment [1] as seen in Fig. 1 which are found to be in decent agreement.

The size and aspect ratios of the cylinders were varied by using three different columns with diameters of 15, 20, and 30 mm to test the effect of column width only on the time and speed of the droplet migration. Figure 2 shows that when the ratio of the droplet diameter to the column diameter is less than 0.45, the influence of the column diameter on the ascension velocity is negligible; however, as the AR increases, there is a significant reduction in the droplet's velocity. The results obtained here were considered as a factor having crucial effect on the droplet speed; consequently, the wall effect was removed from the calculations. Note that in all thermocapillary droplet flow calculations, the migration velocity is taken as the droplet migration in the radial direction, as seen in Figs. 3 and 4; moreover, the droplet remained spherical in shape and no deformation was noticed for any of the ARs.

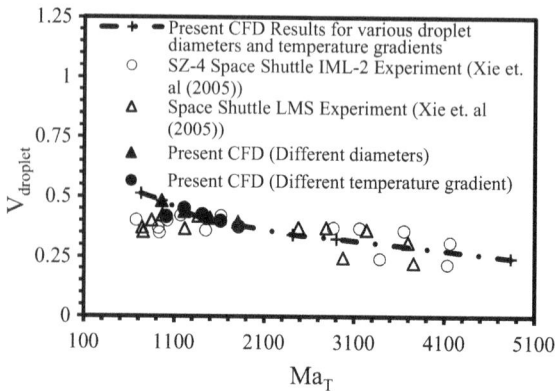

Fig. 1 Validation of present CFD with experimental data

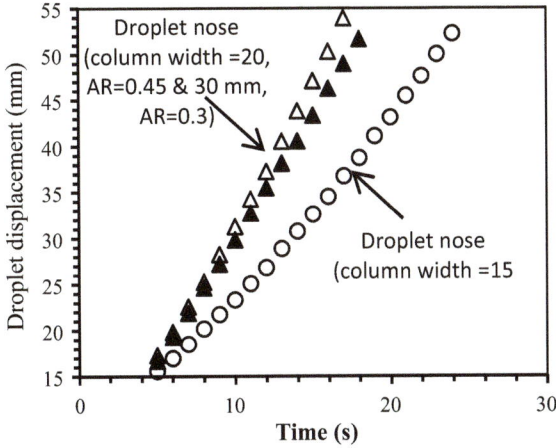

Fig. 2 Compares the *y*-coordinates of 9 mm droplets

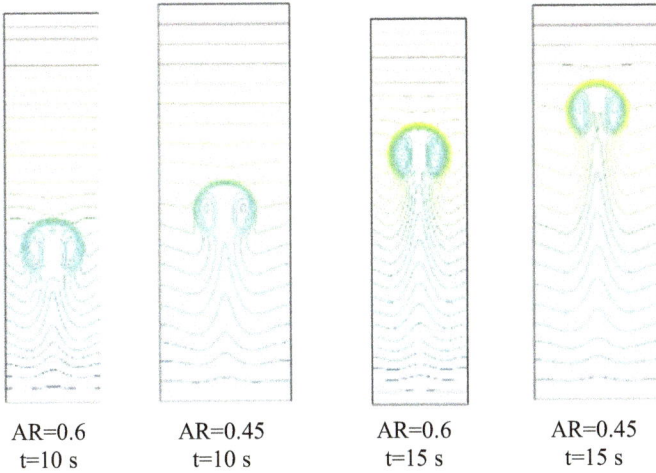

AR=0.6   AR=0.45   AR=0.6   AR=0.45
t=10 s   t=10 s   t=15 s   t=15 s

Fig. 3 2d-axis contours of temperature illustrate the effect of two ARs upon droplet migration at time = 10 s and 15 s

5   Conclusions

Two- and three-dimensional VOF simulations of two-phase (liquid/liquid) transient flow were performed using a multiphase flow algorithm based on the finite-volume method. The current results show conclusive existence of Marangoni droplet flow phenomena in a zero-gravity environment. The present CFD results show that the thermocapillary droplet rise velocity is effected by the column diameter. The results show that when the ratio of the droplet diameter to the column diameter (AR) is less

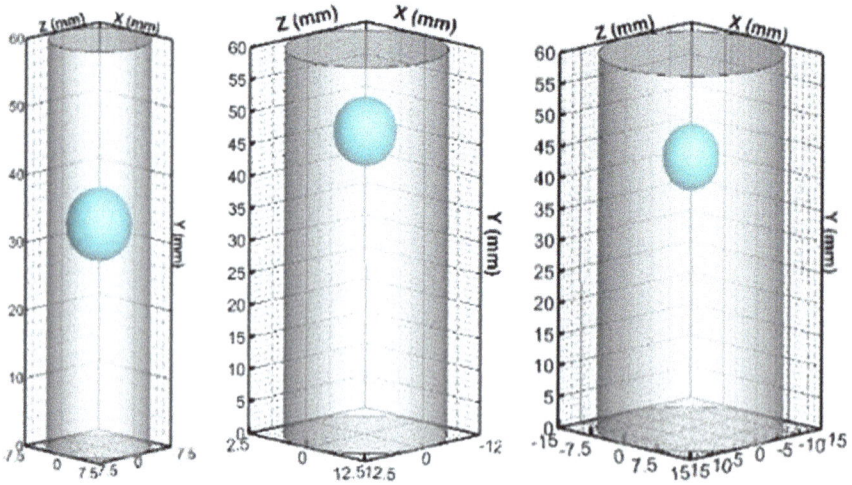

Fig. 4 Three-dimensional droplet migration towards the hotter side of three different column width (15, 20, and 30 mm), at $t = 15$ s

than 0.45, the influence of the column diameter on the ascension velocity is negligible; however, as the AR increases, there is a significant reduction in the droplet's velocity. The VOF model with the UDF was examined properly and results show that the surface tension coefficient was well coded due to the fact that it is based on the geo-reconstruct algorithm, suggesting that it is an appropriate choice to solve thermocapillary and zerogravity problems.

Acknowledgements The principal investigator would like to express his sincere gratitude to Kuwait Foundation for the Advancement of Sciences (KFAS), Kuwait, for supporting and funding this research work.

References

1. J.-C. Xie, H. Lin, P. Zhang, F. Liu, W.-R. Hu, Experimental investigation on thermocapillary drop migration at large Marangoni number in reduced gravity. J. Colloid Interface Sci. **285**, 737–743 (2005)
2. Q. Kang, H.L. Cui, L. Hu, L. Duan, On-board experimental study of bubble thermocapillary migration in a recoverable satellite. Microgravity Sci. Technol. **20**, 67–71 (2008)
3. N.O. Young, J.S. Goldstein, M.J. Block, The motion of bubbles in a vertical temperature gradient. J. Fluid Mech. **6**, 350–356 (1959)
4. ANSYS-FLUENT 2011. Users Guide
5. D.L. Youngs, Time-dependent multi-material flow with large fluid distortion, in *Numerical Methods for Fluid Dynamics* (Academic Press, 1982), pp. 273–285
6. C.W. Hirt, B.D. Nichols, Volume of fluid (Vof) method for the dynamics of free boundaries. J. Comput. Phys. **39**, 201–225 (1981)

7. J.U. Brackbill, D.B. Kothe, C. Zemach, A continuum method for modeling surface tension. J. Comput. Phys. **100**, 335–354 (1992)
8. P.H. Hadland, R. Balasubramaniam, G. Wozniak, R.S. Subramanian, Thermocapillary migration of bubbles and drops at moderate to large Marangoni number and moderate Reynolds number in reduced gravity. Exp. Fluids **26**, 240–248 (1999)

Modelling of Flash Boiling in Two-Phase Geothermal Turbine

S. Rane and L. He

1 Introduction

In a total flow system for geothermal energy extraction, the high pressure and temperature of geofluid are directly passed through the power turbine. During the expansion process, there is a significant drop in fluid pressure and these results into flash boiling. Flash boiling generates steam, which has a very high specific volume and an abrupt increase in fluid volume makes the turbine passage difficult to be designed. The reported study focuses on the computational fluid dynamic model of the two-phase turbine. Flash boiling has been incorporated through the thermal phase change approach that uses energy balance at the liquid-vapour interface considering a two-resistance analogy for heat transfer coefficients of both the phases. The control parameters in this physics have been studied and it was found that the specification of phase Nusselt number, the description of bubble mean diameter and local saturation temperature significantly influence the heat and mass transfer mechanism. In this paper, the influence of Nusselt number based on Ranz–Marshall's and Wolfert's correlation has been reported. The CFD results have been validated with test data available in literature and the heat transfer parameters are tuned in order to get the best fit.

Fabris [1] proposed a turbine (Fig. 1a) that could be used in a total flow system for energy production from geothermal resources. Literature references are available for experimental investigation of such turbines but there is no validated numerical model for prediction of flow field and flashing in the turbines flow passages. The motivation for the reported study was to develop a two-phase model of this turbine. Geometry of the test turbine reported by Date et al. [2] is shown in Fig. 1a, b and shows the computational domain. In the current study, data for tests with average feed water temperature of 117 °C was used.

S. Rane (✉) · L. He
Southwell Laboratory, Department of Engineering Science, University of Oxford, Osney Mead, Oxford OX2 0ES, UK
e-mail: sham.rane@eng.ox.ac.uk

© Springer Nature Singapore Pte Ltd. 2021
C. Wen and Y. Yan (eds.), *Advances in Heat Transfer and Thermal Engineering* ,
https://doi.org/10.1007/978-981-33-4765-6_7

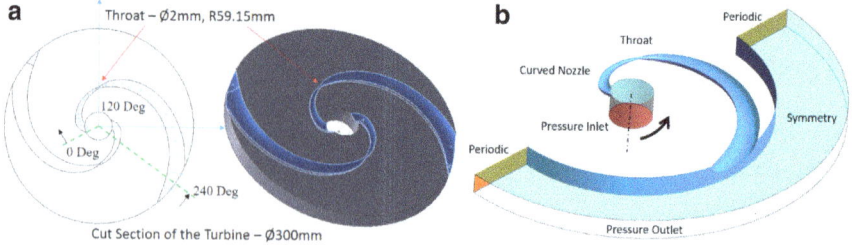

Fig. 1 Base two-phase turbine design tested by Date et al. [2]. **a** Cut section showing the curved nozzle, **b** two-phase CFD model domain with boundaries

2 Methodology

The thermal phase change model is based on thermal non-equilibrium between the phases. Heat, mass and momentum transfer is governed by the local heat balance across the liquid-vapour interface during phase transition. Such a formulation is suitable for modelling of phenomenon such as flash boiling occurring in converging-diverging nozzle flows. At the water-liquid and water-vapour interface, heat balance gives

$$Q_w + Q_v = 0$$

$$h_w A_i (T_{\text{sat}} - T_w) - \dot{m}_{wv} \cdot H_{ws} + h_v A_i (T_{\text{sat}} - T_v) + \dot{m}_{wv} \cdot H_{vs} = 0$$
$$\dot{m}_{wv} = \frac{h_w A_i (T_{\text{sat}} - T_w) + h_v A_i (T_{\text{sat}} - T_v)}{H_{vs} - H_{ws}}$$
$$A_i = 6\frac{\alpha_v}{d_v} = (6\alpha_v)^{2/3} (\pi n_v)^{1/3} \tag{1}$$

The interphase mass transfer rate is obtained from Eq. (1). Equation (2) is used to select the enthalpy for source in energy equation. Sub-script s is selected depending on evaporation or condensation.

If $\dot{m}_{wv} > 0$ (Boiling),

$$H_{ws} = H_w(T_w), \quad H_{vs} = H_v(T_{\text{sat}}) \tag{2}$$

As seen from Eq. (1), heat transfer coefficients h_w, h_v and interfacial area A_i are required to be determined in the thermal phase change model. For the interfacial area, the mean droplet diameter d_v or a bubble number density n_v (BND) is required to be prescribed. In literature [3], computational studies have presumed BND with an optimum value for every operating condition in flows through converging-diverging nozzles or tube depressurization flows. In some studies, bubble number transport equation has been introduced but even this approach requires presumption of source terms in the transport equations derived from nucleation model.

In order to reduce CFD domain, a 180 degree sector of the rotor is modelled with periodic boundaries. Inlet to the nozzle is defined by the feed water pressure of 400 kPa and temperature of 117 °C. The exit of the turbine is open to a flash tank connected to condenser coil which during the start of the experiments is at 6 kPa pressure. CFD calculations were performed over a range of turbine operating speed from 1561 to 4623 rpm using ANSYS CFX solver. Water-liquid and vapour properties were specified using the IAPWS-IF97 definition available within the solver.

3 Results

From literature survey, it was found that broadly the Nusselt number correlations for interfacial heat transfer are based on either Re and Pr or Pe and Ja_T as listed in Table 1. Jakob number Ja_T directly accounts for the degree of liquid superheat. Ranz–Marshall [4] type of correlations has been extensively used in flows with flashing and is based on Reynolds number and Prandtl number. Whereas some other flashing studies [3] have identified Wolfert [5] type of correlations, using Ja_T as superior for vapour bubbles in superheated liquid.

In current study, no suitable Nu correlation was found for flashing flows in rotating channels, hence, both type of formulations were examined for consistency and accuracy. Figure 2a, b presents the comparison of turbine gross shaft power between measurements [2] and CFD model for a presumption of BND $= 5e^{10}$ m^{-3} and $5e^{07}$

Table 1 Nusselt number correlations for interfacial heat transfer—h_w

Interphase heat transfer formulation		Nusselt number correlation
Based on Reynolds number Re and Prandtl number Pr	Ranz–Marshall [4]	$Nu = 2 + 0.6Re^{0.5}Pr^{0.3}$ $0 \leq Re < 200, 0 \leq Pr < 250$
Based on Péclet number Pe and Jakob number Ja_T	Wolfert [5]	$Nu = \left(\frac{12}{\pi}Ja_T + \frac{2}{\sqrt{\pi}}Pe^{0.5} \right)$

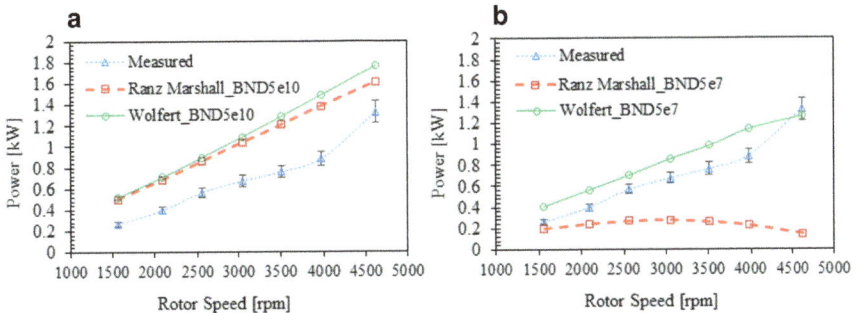

Fig. 2 Comparison of turbine power between measurements and CFD model, **a** BND $(n_v) = 5e^{10}$ m^{-3}, **b** BND $(n_v) = 5e^7$ m^{-3}, with two correlations for Nusselt number

m^{-3}, respectively. In addition, results based on Ranz–Marshall and Wolfert correlation for Nu (Table 1) have been compared. With BND $5e^{07}$ m^{-3}, turbine flow at various operating speed (figure not presented here) was well estimated within 0.5–7.5% deviation from measurements. At BND $5e^{10}$ m^{-3}, turbine flow was underestimated in the range 10–18%. From Fig. 2b, with BND $5e^{07}$ m^{-3}, the CFD model is over predicting the turbine power at operating speeds below 4000 rpm in the range of 30–55%. At 4623 rpm, turbine power deviation in the range of 4.5% was observed. Whereas with BND $5e^{10}$ m^{-3}, the overprediction was high in the range of 20–90%. This suggested that BND $5e^{07}$ m^{-3} or lower was more suitable for the vapour regime formed inside this turbine nozzle after flashing [6].

There was very small difference in flow prediction between the two Nu correlations. With respect to turbine power prediction, however, the response from the two correlations was greatly different. Comparison of Fig. 2a, b shows that the power magnitude drops in case of Ranz–Marshall correlation when the BND was changed from $5e^{10}$ m^{-3} to $5e^{07}$ m^{-3}. Also, the trend of power vs speed variation is largely deviated with BND $5e^{07}$ m^{-3}. In case of Wolfert's correlation, there was a drop in magnitude of power but the trend with speed was consistent. Wolfert's correlation was thus found to be more acceptable choice for further investigation. As reported in literature [3], the specification of BND highly influences the accuracy also observed in this case with the turbine power estimation.

Figure 3 presents the distribution of vapour mass fraction (dryness fraction) in the turbine nozzle. Flashing initiates just after the throat area, and although homogeneous nucleation is set in the CFD model, vapour generation is high adjacent to the throat wall. This effect is due to liquid velocity being high in the core of the flow. Downstream, the flashing is highly non-uniform and vapour mass fraction is high on the low pressure surface of the nozzle. At the nozzle exit, average dryness fraction is 12% which is close to the test data of 12.8%

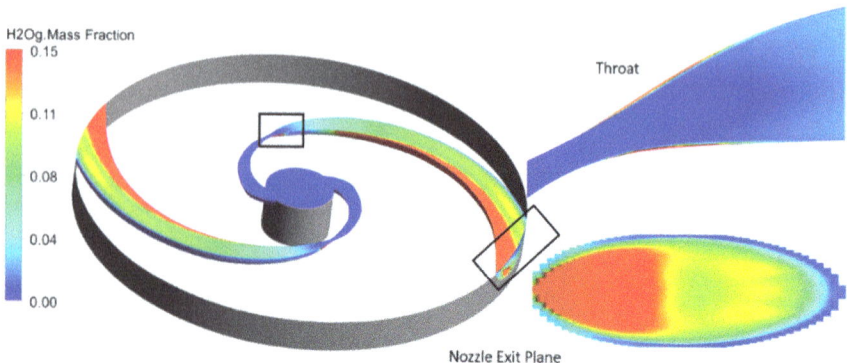

Fig. 3 Vapour mass fraction distribution at 4623 rpm on the curved nozzle mid-surface, throat and exit

4 Conclusions

CFD model of a two-phase turbine has been analysed with respect to the specification of heat transfer parameters of interfacial Nusselt number correlation and vapour bubble number density. Flash boiling has been incorporated through the thermal phase change approach that relies on local thermal non-equilibrium with homogeneous nucleation. The influence of Nusselt number based on Ranz–Marshall's and Wolfert's correlation has been studied. Wolfert's correlation was found to be more consistent when bubble number density was varied. BND $5e^{07}$ m^{-3} or lower was found to be more suitable for the vapour regime formed in the turbine nozzle. Turbine exit vapour dryness fraction, temperature, pressure, feed water flow rate and rotor power have been evaluated for the performance of the test turbine. Test turbine flow at various operating speed was well estimated, within 0.5–7.5% deviation from measurements. Nozzle torque and hence turbine power estimate was within 4.5% at 4623 rpm but was found to be over-estimated at lower speeds. The two-phase model so established will be further used to re-design the turbine passage in order to improve the performance and control flashing. Such a turbine design improvement using the described model has been reported in [6–8].

References

1. G. Fabris, Two-phase Reaction Turbine, United States Patent 5,236,349 (1993)
2. A. Date, S. Vahaji, J. Andrews, A. Akbarzadeh, Experimental performance of a rotating two-phase reaction turbine. Appl. Therm. Eng. **76**, 475–483 (2015)
3. Y. Liao, D. Lucas, Evaluation of interfacial heat transfer models for flashing flow with two-fluid CFD. MDPI-Fluids **3**, 38 (2018)
4. W. Ranz, W. Marshall, Evaporation from drops—Part I. Chem. Eng. Prog. **48**, 141–146 (1952)
5. K. Wolfert, The simulation of blowdown processes with consideration of thermodynamic non-equilibrium phenomena, in *Proceedings of the Specialists Meeting of Transient Two-Phase Flow, OECD/Nuclear Energy Agency, Canada* (1976)
6. S. Rane, L. He, CFD analysis of flashing flow in two-phase geothermal turbine design. J. Comput. Des. Eng. **7**(2), 238–250 (2020). https://doi.org/10.1093/jcde/qwaa020
7. S. Rane, L. He, Numerical analysis of a novel two-phase turbine using thermal non-equilibrium, homogeneous nucleation phase change. Thermal Sci. Eng. Progr., 100827 (2021). ISSN 2451-9049. https://doi.org/10.1016/j.tsep.2020.100827
8. H. Li, S. Rane, Z. Yu, G. Yu, An inverse mean-line design method for optimizing radial outflow two-phase turbines in geothermal systems. Renew. Energy **168**, 463–490 (2021). ISSN 0960-1481. https://doi.org/10.1016/j.renene.2020.12.079

Experimental Investigation on Flash Evaporation of Pure Water at Different Depths with Functional Analysis Method

Siguang Li, Yanjun Li, Longbin Yang, Xiaojin Zhang, and Runzhang Xu

1 Introduction

When a liquid is suddenly exposed to an environment below its saturate pressure, it will become superheat, and the heat surplus is released in form of latent heat of evaporation. This phenomenon is known as flash evaporation. Due to its large steam production and significant temperature drop, flash evaporation is widely used in the industry.

Miyatake et al. [1, 2] studied the variations of the mean liquid temperature in flash evaporation and firstly proposed that the non-equilibrium fraction (*NEF*) is better than temperature to describe the evolution of flash evaporation. The effects of initial pressures and superheats on flash evaporation were studied by Yan et al. [3]. Saury [4] preliminarily reported the temperature stratification during flash evaporation. However, few studies investigated the influences of the temperature stratification on flash evaporation, which was important for the design of steam accumulator and so on.

Therefore, in this study, an experimental system was designed to investigate the temperature variations at different depths during flash evaporation. And the initial water temperatures T_0 were between 65.0 and 84.4 °C, and the superheats ΔT varied from 15.0 to 27.7 K. First, the variations of the water temperature were analyzed by *NEF*. Then, the function analysis method was introduced to study the digital characteristic of *NEF*. The above work would be benefit to understand the heat and mass transfer characteristics of flash evaporation further.

S. Li · Y. Li · L. Yang (✉)
College of Power and Energy Engineering, Harbin Engineering University, Harbin, China
e-mail: yanglb@hrbeu.edu.cn

X. Zhang
China General Nuclear Power Group, Shenzhen, China

R. Xu
College of Mathematical Sciences, Harbin Engineering University, Harbin, China

© Springer Nature Singapore Pte Ltd. 2021
C. Wen and Y. Yan (eds.), *Advances in Heat Transfer and Thermal Engineering* ,
https://doi.org/10.1007/978-981-33-4765-6_8

2 Experimental System

The experimental system for flash evaporation shown in Fig. 1 includes flash chamber, vacuum/pressure tank, data measurement system, steam heating system, and so on, which is basically similar to our former work [5]. The positions of the two thermocouples in the flash chamber are also shown in the left part of Fig. 1.

After obtaining the experimental data of flash evaporation, the corresponding temperature variation is first converted into the variation of *NEF* by the following expression

$$NEF = \frac{T_t - T_e}{T_0 - T_e} \qquad (1)$$

where T_0, T_e, and T_t denote the temperature at initial state, equilibrium state, and time t, respectively. In order to further investigate the digital characteristics of *NEF*, the function analysis method is introduced as follows,

$$I = \int_{t_i}^{t_{i+1}} NEF \, dt \qquad (2)$$

where *NEF* is considered as a function of t and I denotes the integration of *NEF* on the time interval (t_i, t_{i+1}). In this work, both *NEF* and I are used to explore the temperature variation characteristics of flash evaporation.

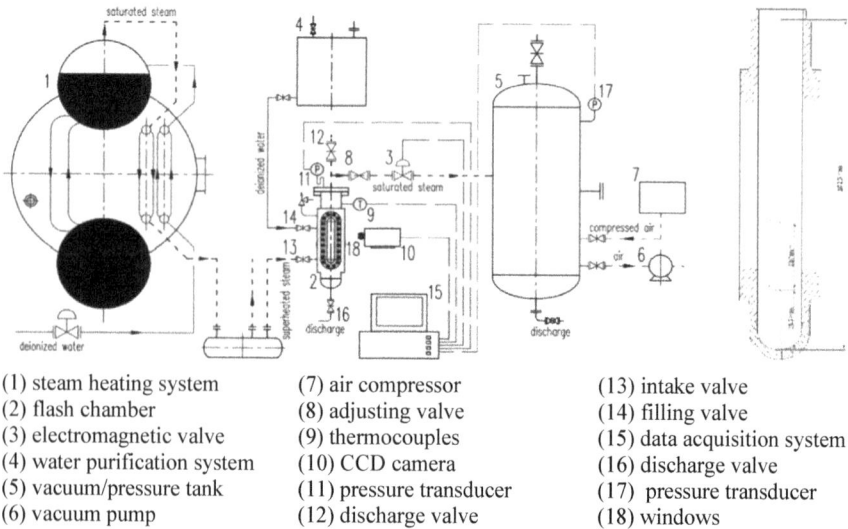

(1) steam heating system	(7) air compressor	(13) intake valve
(2) flash chamber	(8) adjusting valve	(14) filling valve
(3) electromagnetic valve	(9) thermocouples	(15) data acquisition system
(4) water purification system	(10) CCD camera	(16) discharge valve
(5) vacuum/pressure tank	(11) pressure transducer	(17) pressure transducer
(6) vacuum pump	(12) discharge valve	(18) windows

Fig. 1 Schematic diagram of flash evaporation system and the thermocouple positions

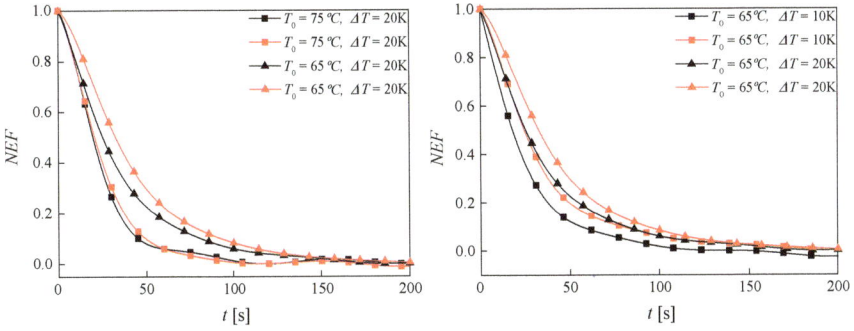

Fig. 2 Evolution of *NEF* with different T_0 and ΔT

3 Results

The evolution of *NEF* with different T_0 and ΔT are shown in Fig. 2. Black and red curves in Fig. 2 represent the variations of *NEF* corresponding to No. 1 and No. 2 thermocouples, respectively. It can be seen that *NEF* decrease fast with the increasing of T_0 when ΔT is fixed. In addition, *NEF* decreases slowly as ΔT increases but reaches the equilibrium state with same time. It indicates that the water with high T_0 will be cooled much faster than that with lower T_0 and increasing ΔT is another way to enhance the decline of water temperature. Meanwhile, the curve of *NEF* for No. 1 thermocouple is always lower than that for No. 2 thermocouple. In other words, the temperature of the shallower water decreases earlier than that of the deeper water, which causes the temperature stratification phenomenon.

In order to capture the feature of temperature stratification phenomenon, the function analysis method is used to convert *NEF* curve into the value of *I*. It can be discovered that the temperature stratification phenomenon could be obviously observed in the first 20 s of flash evaporation. As flash evaporation progress, the difference of the temperature at different depths generally disappears.

4 Conclusions

The conclusions are as follows.

1. The effects of initial temperature and superheat on different depths during flash evaporation are studied.
2. Temperature stratification phenomenon is successfully captured by using the functional analysis method.

Acknowledgements This work was supported by the Fundamental Research Funds for the Central Universities (No. 3072020CFT303) and the program of China Scholarship Council (No. 201906680009) .

References

1. O. Miyatake, K. Murakami, Y. Kawata, T. Fujii, Fundamental experiments with flash evaporation. Heat Transf.-Jpn. Res. **2**, 89–100 (1973)
2. O. Miyatake, T. Fujii, T. Tanaka, T. Nakaoka, Flash evaporation phenomena of pool water. Heat Transf.-Jpn. Res. **6**, 13–24 (1977)
3. J.J. Yan, D. Zhang, D.T. Chong, G.F. Wang, L.N. Liu, Experimental study on static/circulatory flash evaporation. Int. J. Heat Mass Transf. **53**, 5528–5535 (2010)
4. D. Saury, S. Harmand, M. Siroux, Flash evaporation from a water pool: influence of the liquid height and of the depressurization rate. Int. J. Therm. Sci. **44**, 953–965 (2005)
5. Y.X. Shao, Y.J. Li, L.B. Yang, X.J. Zhang, L.J. Yang, H.Y. Wu, R.Z. Xu, New experimental system for high pressure and high temperature flashing evaporation experiments. Appl. Therm. Eng. **66**, 148–155 (2014)

Investigation on the Characteristics of Droplet Evaporation Under Different Heating Conditions Using Lattice Boltzmann Method

Li Wang, Yuying Yan, and Nick Miles

1 Introduction

Evaporation of a liquid drop on a solid substrate is a common phenomenon in everyday life, such as inkjet printing, accidental drippings on a hot surface [1], and the well-known "'coffee-stain" effect. Recently, effects of surface's thermos-physical properties [2], microstructures, roughness, wettability on sessile droplet evaporation characteristics, as well as Marangoni effect during evaporation and droplet evaporation above the Leiden frost temperature were investigated. As a powerful numerical scheme for fluid flow, lattice Boltzmann method (LBM) has been adopted in this paper for its advantages in dealing efficiently with complicated boundaries and interfacial transport dynamics. A LBM model for simulating the droplet evaporation process has been established, and the streamlines and temperature distribution during the evaporating process are investigated. The liquid properties, such as the density and viscosity, as well as the surface wettability, are changed to study their effects on the evaporating process. The gravity is also set to different values to study its effect on the evaporating rate. Moreover, different heat sources are applied, so the heating conditions are changed accordingly.

2 Methodology

The multi-component multiphase (MCMP) LBM model proposed by Zhang el al. [3]. is used in this simulation. Compared to the single component multiphase model,

L. Wang (✉) · Y. Yan
Faculty of Engineering, University of Nottingham, Nottingham NG7 2RD, UK
e-mail: Li.wang@nottingham.ac.uk

N. Miles
Faculty of Science and Engineering, University of Nottingham Ningbo China, Ningbo, China

© Springer Nature Singapore Pte Ltd. 2021
C. Wen and Y. Yan (eds.), *Advances in Heat Transfer and Thermal Engineering* ,
https://doi.org/10.1007/978-981-33-4765-6_9

MCMP model can better give the streamlines of the flow field. Each component is calculated by its own governing equation:

$$f_i^{\sigma}(x + e_i \Delta t, t + \Delta t) - f_i^{\sigma}(x, t) = -\frac{1}{\tau_f^{\sigma}}(f_i^{\sigma}(r, t) - f_i^{\sigma, eq}(r, t))$$

$$\sigma = 1, 2 \tag{1}$$

where σ stands for different components. Pseudopotential model, also known as Shan–Chen model [4, 5], has shown tremendous superiority because of its simplicity, versatility and the distinctive feature of automatic phase separation without any specific techniques for interface capturing or tracking. So, in this paper, Shan-Chen model is adopted to simulating the interaction forces in the streamline studied. The forces include attractive force within component $F_{\sigma,\sigma}$, repulsive force between components $F_{\sigma,\sigma'}$, gravitational force $F_{\sigma,g}$ and solid–fluid interaction $F_{\sigma,s}$. Following reference [6], the energy transport equation in terms of internal energy neglecting viscous heat dissipation is employed in this model, as shown below:

$$\partial_t T + v \cdot \nabla T = \frac{1}{\rho c_\upsilon} \nabla \cdot (\lambda \nabla T) - \frac{p}{\rho c_\upsilon} \nabla \cdot v \tag{2}$$

3   Results

Direction of heat source has an obvious effect on the streamlines in the flow field. When heated from bottom, the buoyance force is larger than the Marangoni force, so the vortex in the middle part of the droplet is mainly affected by the buoyance force. However, when heated from the top, The vortex caused by buoyance force is much smaller and can be neglected compared to that of Marangoni force. That is because when heated from the top, the thermophoretic force is in the opposite direction against the buoyance force, while when heated from bottom, they share the same direction (Fig. 1).

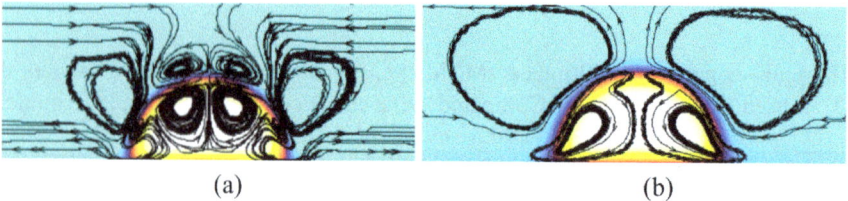

(a)   (b)

Fig. 1 Streamlines in the flow field with different heating source directions. **a** Heated from bottom; **b** heated from above

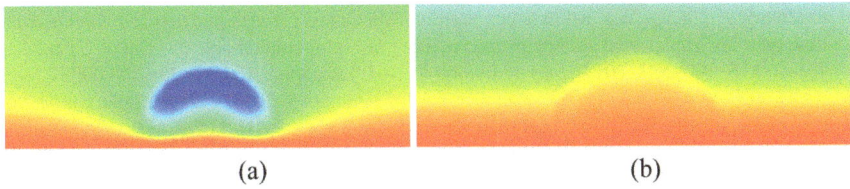

(a)   (b)

Fig. 2 Temperature distributions of different time steps. **a** Time step = 500; **b** time step = 7000

Figure 2 shows the temperature distribution changing with time steps. While water has a larger heat capacity than air, and the thermal diffusivity of water is smaller than air, the temperature inside the droplet corresponds to changes at a slower rate than surrounding air. On the other hand, liquid inside the droplet can better hold the heat it gains from the outside environment, so the droplet appears a higher temperature than the surrounding atmosphere in the later phase. According to the streamline, the contacting point between the droplet and surface has a higher flow speed than other parts in the flow field, so those areas can better exchange heat. Thus, in the beginning, the temperature of the contacting points is lower than the middle of the droplet.

4   Conclusions

The streamlines and temperature distribution have influence on each other, with higher flow rate, and enhances heat transfer, and temperature distribution can change the direction of streamlines. On the contacting points, for the existence of the vortex caused by Marangoni force, the flow rate is relatively highest in the entire droplet, while in the mid part of droplet, the vortexes are caused mainly by the buoyance force. Gravity also has an influence on the evaporation process and that will be investigated in the near future.

Acknowledgements The authors would like to acknowledge the financial support by the scholarship of China Scholarship Council (CSC) for Mr. Li Wang. And this work is also supported by H2020-MSCA-RISE-778104-ThermaSMART.

References

1. V. Bertola, An impact regime map for water drops impacting on heated surfaces. Int. J. Heat Mass Transf. **85**, 430–437 (2015). https://doi.org/10.1016/j.ijheatmasstransfer.2015.01.084
2. Y.H. Chen, W.N. Hu, J. Wang, F.J. Hong, P. Cheng, Transient effects and mass convection in sessile droplet evaporation: the role of liquid and substrate thermophysical properties. Int. J. Heat Mass Transf. **108**, 2072–2087 (2017). https://doi.org/10.1016/j.ijheatmasstransfer.2017.01.050

3. C. Zhang, P. Cheng, W.J. Minkowycz, Lattice Boltzmann simulation of forced condensation flow on a horizontal cold surface in the presence of a non-condensable gas. Int. J. Heat Mass Transf. **115**, 500–512 (2017). https://doi.org/10.1016/j.ijheatmasstransfer.2017.08.005
4. X. Shan, H. Chen, Lattice Boltzmann model for simulating flows with multiple phases and components. Phys. Rev. E **47**, 1815–1819 (1993). https://doi.org/10.1103/PhysRevE.47.1815
5. X. Shan, H. Chen, Simulation of nonideal gases and liquid-gas phase transitions by the lattice Boltzmann equation. Phys. Rev. E **49**, 2941–2948 (1994). https://doi.org/10.1103/PhysRevE.49.2941
6. R. Zhang, H. Chen, Lattice Boltzmann method for simulations of liquid-vapor thermal flows. Phys. Rev. E **67**, 066711 (2003). https://doi.org/10.1103/PhysRevE.67.066711

Evaporating Progress of Hexane Droplet on the Surface of Nacl Solution

Bin Liu, Ruonan Wang, Georges El Achkar, and R. Bennacer

1 Introduction

Understanding the interaction among droplets is one of the most scientific and practical importance in many applications [1]. The interaction of the same droplets on solid (rigid) substrates has been extensively studied for decades. At present, it is very meaningful to study the interaction between different incompatible droplets. In order to better understand the evaporation of hexane on the surface of NaCl solution, we will measure the radius and mass of hexane in different concentrations of NaCl solution, and the experimental results are obtained and discussed in this paper.

2 Methodology

In our evaporative lens experiments, laboratory-produced deionized water (DI) was used with a purity of 99% hexane. Subsequently, we used a salt to change the surface tension between water, hexane and air and form well-defined droplets. The experiment was carried out in the laboratory where the air temperature was maintained at 25 ± 1 °C. The surface tension of hexane is 17.89 mN/m at 25 °C. As for the surface tension of brine (γ_{sa}, mN/m), it is related to the temperature (t, °C) and salinity (s, g/kg), which can be expressed as Eq. (1).

$$\gamma_{sa} = 75.59 + 0.021352 * s - 0.13476 * t - 0.00029529 * s * t \qquad (1)$$

B. Liu (✉) · R. Wang · G. E. Achkar · R. Bennacer
Tiajin Key Lab of Refrigeration Technology, Tianjin University of Commerce, Tianjin 300134, People's Republic of China
e-mail: lbtjcu@tjcu.edu.ccn

R. Bennacer
LMT/ENS-Cachan/CNRS/Paris-Saclay University, 94235 Cachan, France

© Springer Nature Singapore Pte Ltd. 2021
C. Wen and Y. Yan (eds.), *Advances in Heat Transfer and Thermal Engineering* ,
https://doi.org/10.1007/978-981-33-4765-6_10

In our experiments, the room temperature was kept at 25 ± 1 °C, so the surface tensions of the brine with a concentration of 6%, 12%, 18% and 24% (60 g/kg, 120 g/kg, 180 g/kg and 240 g/kg) are 73.3 mN/m, 74.1 mN/m, 75 mN/m and 75.8 mN/m, respectively.

3 Resutls and Analyses

Figure 1 shows the diagram of the lens. When one droplet drops on the liquid surface, the droplet will float on the liquid surface for the balance of the surface tension and the gravity. If the droplet diameter is enough small, the gravity can be neglected, so the tension balance can be shown as Eq. (2).

$$\gamma_{sa} = \gamma_{ha} \cos \theta_1 + \gamma_{sh} \cos \theta_2 \qquad (2)$$

γ_{sa}, γ_{ha} and γ_{sh} indicate the surface tension between NaCl solution and air, the surface tension between hexane droplets and air and the surface tension between hexane droplets and NaCl solution, mN/m, θ_1 and θ_2 are the contact angle for the two different parts, one is over the liquid surface and another is the under the liquid surface. For the surface tension between hexane droplets and NaCl solution, according the Good–Girifalco rule, it can be calculated as Eq. (3).

$$\gamma_{sh} = \gamma_{ha} + \gamma_{sa} - 2\varphi_{sh}\sqrt{\gamma_{ha}\gamma_{sa}} \qquad (3)$$

Figure 2 shows the diameter development of the hexane droplet on the different concentrations of the brine surface. From Fig. 2, we can find that the total process of the evaporation can be divided into three stages, including the expansion stage, the stable evaporation stage and the vanishing stage.

For the expansion stage, it is resulted from the diminution of the surface tension between hexane droplets and NaCl solution. When the droplet of hexane was dripped on the surface, there was an interface between brine and hexane was formed, but it is unstable because of the diffusion of hexane to the brine. With the decrease of the concentration of hexane in the interface, the surface tension between hexane and brine will decrease. From Eq. (2), in order to keep the balance, the contact angle needs

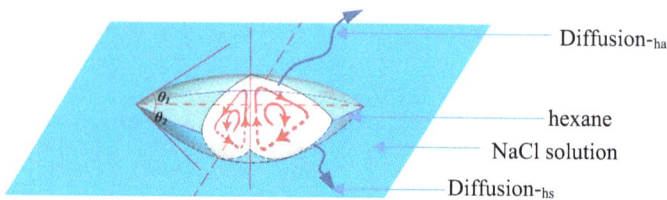

Fig. 1 Process of evaporation of hexane droplets on the surface of NaCl solution

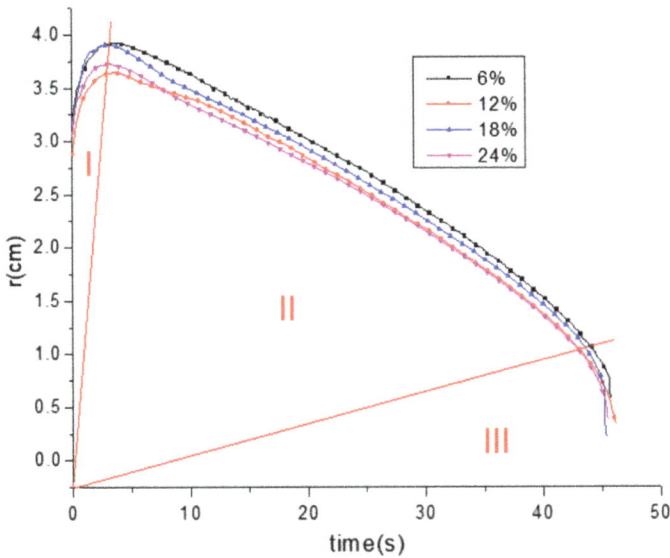

Fig. 2 Hexane-NaCl-Evaporation evaporation process

to be reduced, so the diameter will increase. When the equilibrium state between the brine and the hexane is set up, the evaporation process reaches to a stable evaporation stage. To the end of the evaporation, the contact angle θ_1 almost is zero, and hexane will be a thin film on the brine, so it vanished quickly.

As for the effect of the concentration of the brine, it depends on the surface tension balance among the brine, the air and the hexane. From Eqs. (2) and (3), if the φ_{sh} is a constant coefficient, the θ_1 and θ_2 will decrease with the concentration of the brine. So under a certain volume of hexane, with the decrease of the contact angle θ_1 and θ_2, the diameter of the hexane droplet will increase, so the evaporation process shown the characteristics in Fig. 2.

4   Conclusions

1. When the hexane droplets evaporate on the surface of the NaCl solution, it is mainly divided into three stages. The first stage was the period of the dimeter of the droplet expansion, the second stage was the period of the steady evaporation, and the third stage is the vanishing period of the droplet.
2. At the same temperature, the diameter of the expanded hexane droplets increases with the decrease of the concentration of the NaCl solution.

Reference

1. R. Iqbal, S. Dhiman, A.K. Sen, A.Q. Shen, Dynamics of a water droplet over a sessile oil droplet: compound droplets satisfying a Neumann condition. Lanqmuir **33**(23), 5713–5723 (2017)

Numerical Simulation on Heat Transfer Characteristics of Steam-Water Two-Phase Flow in Smooth Tube and Rifled Tube

Wanze Wu, Baozhi Sun, Jianxin Shi, Xiang Yu, and Zhirui Zhao

1 Introduction

Rifled tubes can enhance boiling heat transfer which is widely used in steam generator. When the working medium flows in the rifled tubes, it rotates under the action of the spiral rising inner threads. With the action of centrifugal force, the droplets entrained by the steam flow keep approaching the tube wall, and finally supple the water film.

Many studies [1–4] researched the two-phase flow heat transfer characteristics of the various working medium in rifled tubes under subcritical and supercritical with experimental methods. The variation of wall temperature trend in post-dryout region with geometrical structure, mass flow rate and pressure was obtained, and the results were compared with those of smooth tubes under the same working conditions. According to these studies, dryout position of rifled pipe is more backward than smooth pipes, and the wall temperature rise is smaller than that of smooth tube. Shen et al. [5] studied the heat transfer characteristics of supercritical water in vertical rifled tubes with numerical simulation, focusing on the influence of structural parameters on the wall temperature changes in the circumference (windward and leeward side). The result shows that because of fluid retention, the wall temperature on the leeward side is higher than that on the windward side.

In this paper, the flow and heat transfer process of steam-water two-phase flow in vertical upward rifled tube are numerically simulated using FLUENT software with a pressure of 7.01 MPa, a mass flow rate of 0.1745 kg/s, a heat flux of 863 kW/m^2 and inlet supercooling is 10 K. Axial and circumferential analysis of the simulation results is carried out to prepare for further research on heat transfer enhancement.

W. Wu · B. Sun (✉) · J. Shi · X. Yu · Z. Zhao
Harbin Engineering University, 145 Nantong Street, Nangang District, Harbin, China
e-mail: sunbaozhi@hrbeu.edu.cn; sunbaozhi@163.com

© Springer Nature Singapore Pte Ltd. 2021
C. Wen and Y. Yan (eds.), *Advances in Heat Transfer and Thermal Engineering* ,
https://doi.org/10.1007/978-981-33-4765-6_11

63

2   Numerical Methods

Geometric structure is drawn by RROE software and unstructured mesh is generated
on the commercial preprocessing software ICEM. Because the total number of grids
is too large, the method of sectional calculation is adopted. The 7 m long pipeline
is divided into five segments, each with 1.7×10^6 cells, which pass through the
verification of grid independence. The mesh structures of wall and inlet cross section
are shown in Fig. 1. Reynolds stress model (RSM) is selected in this study for the
reason of the flow process involving rotation. The three fluid field models of volume
fluid are used to describe continuous steam, continuous liquid phases and droplets
entrained by steam.

3   Results

Figure 2a shows the variation of axial volume fraction of smooth tube and rifled tube.
It can be seen that in the nucleate boiling stage, the volume fraction in the rifled tube
is lower than that in the smooth tube under the same working conditions.

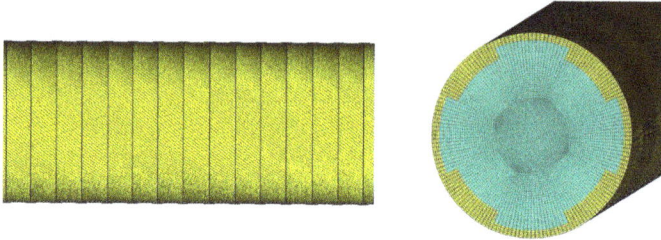

Fig. 1 Mesh structures of wall and inlet cross section

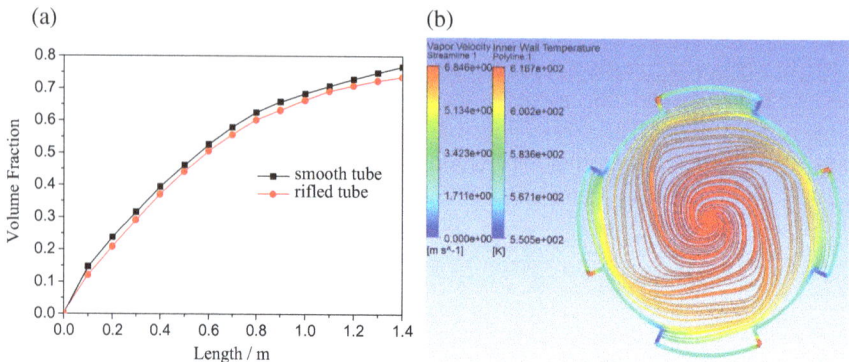

Fig. 2 a Variation of axial volume fraction of smooth tube and rifled tube. **b** Wall temperature and
vapor velocity nephogram at the cross section of 1.4 m

Figure 2b shows the wall temperature and vapor velocity nephogram at the cross section of 1.4 m in which the fluid rotates clockwise. Evidently, the wall temperature distribution of the fluid-solid interface is non-uniform, and the wall temperature on the windward side is lower than that on the leeward side. The reason is that the rotating fluid retains on the leeward side, which reduces the heat transfer efficiency. On the windward side, the heat transfer is enhanced due to the impact of fluid, and the wall temperature is reduced. The nephogram also shows that the axial velocity of vapor phase in the core region is larger, and the closer the vapor phase is to the wall, and the lower the velocity is.

4   Conclusions

The conclusions are as follows.

1. Numerical simulation of boiling heat transfer of vapor-liquid two-phase flow in rifled tube and smooth tube is studied.
2. From the cross-section view, the circumferential wall temperature distribution is non-uniform, and the wall temperature on the leeward side is higher than that on the windward side.
3. Comparisons between calculation results of rifled tube and smooth tube under the same working conditions are carried out. In the nucleation boiling zone, the volume fraction of rifled tube is slightly lower than that of smooth tube on the same working condition.

Acknowledgements This work was sponsored by the National Natural Science Foundation of China (No. 51579048, No. 51709249) which we gratefully acknowledge.

References

1. S.K. Lee, S.H. Chang, Experimental study of post-dryout with R-134a upward flow in smooth tube and rifled tubes. Int. J. Heat Mass Transf. **51**(11), 3153–3163 (2008)
2. J. Pan, D. Yang, Z. Dong, Experimental investigation on heat transfer characteristics of low mass flux rifled tube with upward flow. Int. J. Heat Mass Transf. **54**(13), 2952–2961 (2011)
3. C.H. Kim, I.C. Bang, S.H. Chang, Critical heat flux performance for flow boiling of R-134a in vertical uniformly heated smooth tube and rifled tubes. Int. J. Heat Mass Transf. **48**(14), 2868–2877 (2005)
4. A. Taklifi, P. Hanafizadeh, M.A.A. Behabadi et al., Experimental investigation on heat transfer and pressure drop of supercritical water flows in an inclined rifled tube. J. Supercrit. Fluids **107**, 209–218 (2016)
5. Zhi Shen, Dong Yang, Yaode Li et al., Numerical analysis of heat transfer to water flowing in rifled tubes at supercritical pressures. Appl. Therm. Eng. **133**, 704–712 (2018)

Lubrication Model for Vapor Absorption/Desorption of Hygroscopic Liquid Desiccant Droplets

Zhenying Wang, George Karapetsas, Prashant Valluri, Khellil Sefiane, and Yasuyuki Takata

1 Introduction

The liquid desiccant is a special type of hygroscopic aqueous salt solution, which has been applied in various dehumidification, absorption, and desalination systems. In contact with humid air, the liquid desiccant droplet grows due to vapor absorption, and the droplet motion differs depending on the surface wettability. Previous experimental results from our group [1] show that on hydrophilic substrates, the desiccant droplet gradually spreads with a timescale of 10^3 s as vapor absorption proceeds. For better explaining this phenomenon, a numerical model with moving boundary is required to describe the vapor absorption process into liquid desiccant droplets.

In this study, we develop a lubrication-type model which combines the lubrication theory with the characteristics of hygroscopic ionic solution. The model allows for the free motion of the triple contact line by applying an assumption of precursor film in front, which offsets the vacancy of dynamic models for aqueous solution droplets, and explicitly indicates the mechanisms governing the droplet behaviors during vapor absorption and desorption.

Z. Wang · Y. Takata
Department of Mechanical Engineering, Kyushu University, Fukuoka, Japan

Z. Wang · K. Sefiane · Y. Takata
International Institute for Carbon-Neutral Energy Research (WPI-I²CNER), Kyushu University, Fukuoka, Japan

G. Karapetsas
Department of Chemical Engineering, Aristotle University of Thessaloniki, Thessaloniki, Greece

P. Valluri (✉) · K. Sefiane
Institute for Multiscale Thermofluids, School of Engineering, University of Edinburgh, Edinburgh, UK
e-mail: Prashant.Valluri@ed.ac.uk

© Springer Nature Singapore Pte Ltd. 2021
C. Wen and Y. Yan (eds.), *Advances in Heat Transfer and Thermal Engineering* ,
https://doi.org/10.1007/978-981-33-4765-6_12

2 Formulation of the Model

We take lithium bromide aqueous solution (LiBr-H_2O) droplets as an example for modeling. The droplet is in contact with the gas phase characterized by constant temperature and relative humidity. We consider the droplet to be very thin with an aspect ratio of $\varepsilon = \widehat{H}_0/\widehat{R}_0 \ll 1$. This assumption permits the use of lubrication theory which simplifies the momentum governing equations. Compared with the liquid phase, the density, viscosity, and thermal conductivity of air are significantly smaller and can be neglected, namely, $\hat{\rho}_v \ll \hat{\rho}_l$, $\hat{\mu}_v \ll \hat{\mu}_l$, $\hat{k}_v \ll \hat{k}_l$. Moreover, the vapor absorption is a liquid phase-dominated process. Therefore, here we apply a one-side model which considers only the fluid flow and state evolution of the droplet side. By eliminating the gas phase, the one-side model is efficient and can be solved with modest resources in short time. In simulation, we assume the gas phase as sufficiently large, and the temperature and relative humidity remain unchanged. Along with vapor absorption, the effect of absorptive heating causes a temperature increase at the droplet interface and induces temperature gradient within the droplet. Moreover, we assume the thermal conductivity of the substrate as sufficiently high, and therefore, the substrate temperature remains unchanged. To remove the stress singularity that may arise at the contact line, an ultrathin precursor film is assumed to exist around the periphery of the droplet, which allows for the free motion of the triple contact line.

We take the solutal capillary velocity to characterize the system and apply a weak diffusion approximation to derive the concentration distribution within the droplet. The absorptive mass flux is approximated combining the Hertz–Knusen equation and the thermodynamic equilibrium relationships across the aqueous solution-air interface. In the discretization process, we apply the Galerkin method of weighted residuals and arrive at six independent equations with six independent variables, h, p_0, u, T, χ_{H_2O}, and J. The equation system is then solved using finite element method (FEM), and the temporal variation of the parameters is derived with forward Euler method.

3 Simulation Results

Depending on the initial state of the droplet and the gas phase, vapor absorption or desorption happens. Figure 1 shows the evolution of droplet profile along r direction during vapor desorption and vapor absorption. In the case of vapor desorption, the droplet spreads rapidly at the initial stage driven by the capillary force. Along with droplet spreading, water evaporation ensues, and the evaporation mass flux varies spatially across the droplet profile. At the area near triple contact line, the mass flux reaches the peak due to more efficient heat supply. Due to the hygroscopic property of liquid desiccant, the evaporation near contact line slows down as the water concentration decreases, which means that the evaporation effect will not dry

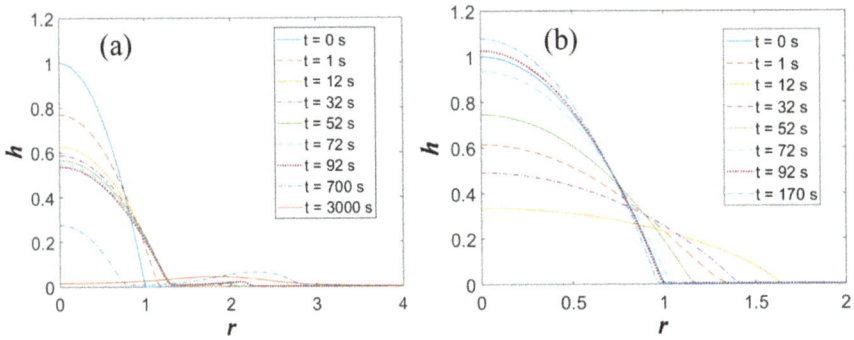

Fig. 1 Evolution of droplet profile (dimensionless droplet height, h) along r direction. **a** Case of vapor desorption, $\chi_{H_2O} = 60\%$, RH $= 30\%$. **b** Case of vapor absorption, $\chi_{H_2O} = 60\%$, RH $= 90\%$

out the liquid near contact line. Consequently, a thin ripple-like film forms as the contact line moves forward. Additionally, the low water concentration at the extended thin film induces solutal capillary flow toward the contact line driven by the surface tension gradient, which further strengthens the spreading of the droplet. The droplet spreads continuously as vapor desorption happens and finally flattens out and reaches equilibrium with the gas phase. The simulation results also provide a reasonable explanation to the experimental results of Shahidzadeh-Bonn et al. [2] and Hadj-Achour et al. [3], where dendritic anhydrous crystals and fractal patterns form at the periphery of drying aqueous solution droplets with hygroscopic properties (Na_2SO_4-H_2O and Ethylene glycol-H_2O), while the patterns are constrained by the triple contact line in cases of normal solutions such as $NaCl$-H_2O and Tetradecane-H_2O.

When the relative humidity of the gas phase is high, vapor absorption happens. The simulation results indicate that the droplet firstly spreads driven by the capillary force, then recedes. As vapor absorption takes place, the heat released due to vapor–water phase change causes a temperature increase at the droplet surface. At the area near contact line, the interfacial temperature keeps low due to more efficient heat removal into the substrate. The non-uniform interfacial temperature causes the preferential distribution of absorptive mass flux near the contact line. As a result, the water concentration near contact line becomes higher than that near droplet center. Consequently, a surface tension gradient is induced and causes the receding of the contact line. However, in our experiments with absorptive droplets, the droplet spreads on hydrophilic glass substrates, and no contact line receding is observed. This is because the practical glass substrates have strong effect of hysteresis, which is not taken into account in the present model. Despite of the inconsistency, the simulation reveals the underneath mechanisms governing the droplet behaviors during vapor absorption. Further modifications will be done to include the effect of hysteresis into the model.

4   Conclusions

This study presents a lubrication-type model to describe the vapor absorption and desorption process into hygroscopic liquid desiccant droplets. The simulation results indicate the non-uniform distribution of mass flux and interfacial parameters across the droplet surface and explains the profile evolution and contact line motion along with vapor desorption/absorption. Further modifications will be done to include the hysteresis effect of the substrate for better predicting the droplet dynamics during vapor absorption.

Acknowledgements The authors acknowledge the supports received from Japanese Society for the Promotion of Science (JSPS), and ThermaSMART project of European Commission (Grant no. EC-H2020-RISE-ThermaSMART-778104).

References

1. Z. Wang, D. Orejon, K. Sefiane, Y. Takata, Water vapor uptake into hygroscopic lithium bromide desiccant droplets: mechanisms of droplet growth and spreading. Phys. Chem. Chem. Phys. **21**(3), 1046–1058 (2019)
2. N. Shahidzadeh-Bonn, S. Rafaï, D. Bonn, G. Wegdam, Salt crystallization during evaporation: impact of interfacial properties. Langmuir **24**(16), 8599–8605 (2008)
3. M. Hadj-Achour, D. Brutin, Fractal pattern formation in nanosuspension sessile droplets via evaporation-spreading on a glass substrate. Colloids Interface Sci. Commun. **1**, 43–46 (2014)

Fixation of Thermocouples and Insulation for Heated Block

Viktor Vajc and Martin Dostál

1 Introduction

In order to calculate heat-transfer coefficient during nucleate pool boiling, one has to use a correlation, e.g., the one created by Rohsenow [1]

$$\mathrm{Nu_b} = \frac{1}{C_{\mathrm{sf}}} \mathrm{Re_b}^{2/3} \mathrm{Pr_L}^{1-m} \tag{1}$$

or Forster and Zuber's [2]

$$\alpha = \frac{0.00122 \, \Delta T^{0.24} \, \Delta p^{0.75} \, c_{p,\mathrm{L}}^{0.45} \, \rho_{\mathrm{L}}^{0.49} \, \lambda_{\mathrm{L}}^{0.79}}{\sigma^{0.5} \, \Delta h_{\mathrm{LG}}^{0.24} \, \mu_{\mathrm{L}}^{0.29} \, \rho_{\mathrm{G}}^{0.24}} \tag{2}$$

However, it might occur that boiling liquid, geometry, or boiling conditions differ from those for which correlations were prescribed. When this happens, there is a non-negligible uncertainty that cast doubts upon the calculated values.

In order to reduce such an uncertainty, we have designed an apparatus for measurement of heat-transfer coefficient during saturated nucleate pool boiling. The purpose of the apparatus is either to compare the measured coefficients with correlations in order to find a suitable one or to let us formulate our own correlation which would correspond with measured data. Furthermore, we want to use the apparatus for various liquids, which are interesting from the perspective of process engineering, but their heat-transfer coefficients were not yet measured.

This paper deals with two problems encountered during our measurements (unwanted horizontal differences in horizontal direction and the impact of forced

V. Vajc (✉) · M. Dostál
Department of Process Engineering, Faculty of Mechanical Engineering, Czech Technical University, Technická 4, 160 00 Prague 6, Czech Republic
e-mail: viktor.vajc@fs.cvut.cz

© Springer Nature Singapore Pte Ltd. 2021
C. Wen and Y. Yan (eds.), *Advances in Heat Transfer and Thermal Engineering ,*
https://doi.org/10.1007/978-981-33-4765-6_13

Fig. 1 Sketch of the
apparatus

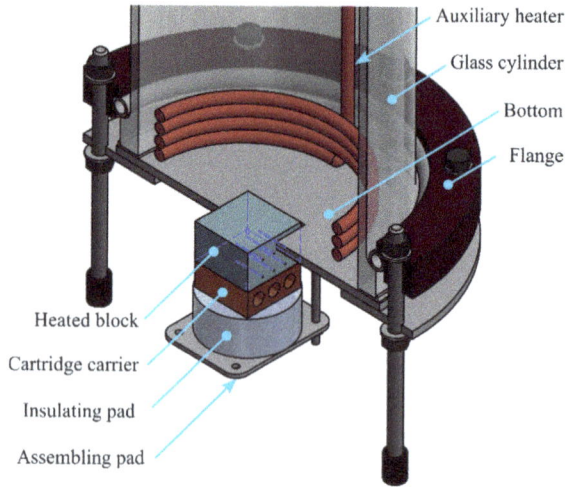

Auxiliary heater
Glass cylinder
Bottom
Flange

Heated block
Cartridge carrier
Insulating pad
Assembling pad

convection of ambient air) and offers appropriate solutions for both of these problems
based on a literary survey and on consultations with various manufacturers.

2 Methodology

The apparatus, see Fig. 1, is quite usual. We indirectly measure the temperature of the
heating surface, i.e., the upper face of a heated stainless-steel block with dimensions
of the surface 48 × 48 mm. For that, we use six sheathed thermocouples placed
inside the block. Thermocouples were simply inserted into six (two rows of three)
1-mm holes with length of 24 mm. The distance between both rows is 13 mm, and
the distance between the upper row and the heating surface is 12 mm. When we
subtract temperature of the boiling liquid (measured with seventh thermocouple)
from the estimated wall temperature, we obtain superheat and calculate the heat-
transfer coefficient with Newton's cooling law.

3 Results

Since the thermocouples inside the block are aligned in two horizontal rows and
the method mentioned above supposes that heat conduction inside the block is one-
dimensional, it is expected that the temperatures in a single row should be equal (in the
ideal case). However, obtained temperatures were usually significantly shifted. For
instance, Fig. 2 shows temperatures obtained on December 19, 2018, after reaching a
stable state during boiling of water at atmospheric pressure. Table 1 presents obtained
temperatures for dataset in Fig. 2 and for two more datasets obtained after completely

Fig. 2 Temperatures measured on 19 December 2018

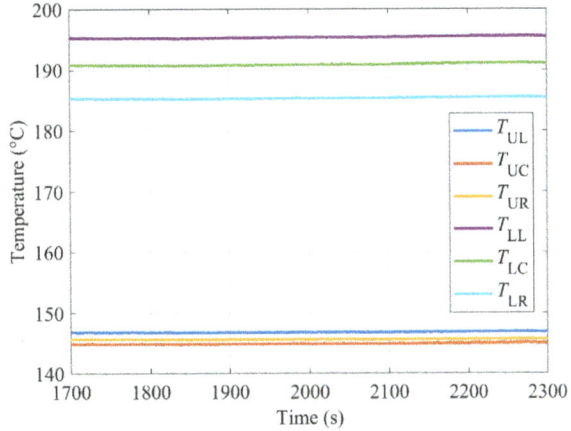

dismantling and reassembling the apparatus. Table 1 also presents average temperatures calculated for both upper and bottom lines, calculated heat flux q and heat-transfer coefficient α. For the first dataset, there is a difference around 10 °C between T_{LL} (lower left position) and T_{LR} (lower right). For the upper row, the difference is smaller, but still around 2 °C between T_{UL} (upper left) and T_{UC} (upper center position). For the second dataset, the discrepancy is somewhat lower—around 4.6 °C for the lower row. The third dataset is perhaps the best result we have ever obtained with maximum difference slightly below 3 °C.

It is necessary to mention that we were unable to predict the magnitude of these differences. It could be as low as several °C or as high as 15 or even 20 °C. Furthermore, we were unable to anticipate positions of maximal and minimal temperatures, because they varied for each experimental run. The fact that in Table 1, regarding the temperatures in the lower row of thermocouples, T_{LL} is always the highest and T_{LR} the lowest is just a mere coincidence.

Lastly, we discovered that the measured temperatures are significantly affected by forced convection of ambient air. We tried to blow air over the uninsulated heated block with a hairdryer and found out that the temperatures measured by embedded thermocouples dropped. The actual decrease depended on the angle from which the air was blown.

4 Embedding of Thermocouples

We suppose that the temperature differences inside the block indicate a problem with temperature measurement—most likely with the way thermocouples were installed into the block. Therefore, we did a literary survey in order to find a suitable method for embedding our thermocouples inside the block. There are basically

Table 1 Measured and calculated values for data in Fig. 2 and two other datasets

Points (–)	T_{UL} (°C)	T_{UC} (°C)	T_{UR} (°C)	T_U (°C)	T_{LL} (°C)	T_{LC} (°C)	T_{LR} (°C)	T_L (°C)	q (kW m^{-2})	α (W m^{-2} K^{-1})
601	146.8	144.9	145.6	145.7	195.3	190.9	185.3	190.5	55.9	6196
291	N/A	179.4	177.9	178.7	259.7	257.3	255.1	257.4	98.4	7329
740	187.1	186.8	187.7	187.2	280.7	278.9	277.9	279.2	115.0	10324

two methods mentioned by various researchers—embedding by melting (soldering, brazing, welding) and embedding with an adhesive.

Embedding by melting is often used for junctions of bare-wire or coated-wire thermocouples, but it might be used for sheathed thermocouples as well. However, we are not convinced that these methods can give a satisfying result considering the 1-mm holes and very narrow space between walls of the holes and thermocouple sheaths.

There are several types of adhesives used for embedment of thermocouples. The most common are various epoxies and cyanoacrylate-based glues. Some special high-temperature adhesives of these types can usually withstand temperatures around 200 or 250 °C. One should thoroughly distinguish between long-term and short-term operating temperatures when comparing several adhesives. Epoxies and cyanoacry-late glues are usually used when boiling of liquid with a lower boiling point or boiling under lower pressure occurs.

Cements, ceramic (alumina) adhesives, or graphite adhesives are usually used for higher-temperature boiling. They can withstand temperatures up to 900 or even 1400 °C. Thermal conductivities of these adhesives are usually just slightly higher than those of epoxies, but they are still quite low (typically around 1 or 2 W m^{-1} K^{-1}).

The crucial criterion for choosing an appropriate method or a suitable adhesive is the temperature inside the heated block. This temperature depends above all on the boiling point of investigated liquid (that includes pressure), range of heat flux which is being investigated, material of the heated block, and positions of thermocouples inside the block. One should also take heed of viscosity, because the adhesive should fill the 1-mm holes without leaving any air gaps. Thermal conductivity is also impor-tant in order to reduce the thermal resistance between the thermocouple sheath and the hole in which the thermocouple is placed.

5   Insulating the Heated Block

Air-blowing over the block proved that the heated block has to be protected from any impact of ambient air. Therefore, we want to enclose the heated block and the cartridge carrier in Fig. 1 with a suitable thermal insulator. In order to find a convenient material, we inquired the scientific papers describing pool boiling apparatuses.

We found out that perhaps the most used materials are various polymers such as Teflon, nylon, polystyrene, bakelite, or PTFE. More heat-resistant often-used polymers are PTFE, PEEK, or PI. Even these materials can be, however, used only at lower temperatures. For instance, the upper working temperature of PI (which is one of the most heat-resistant thermoplastics) lies in the range from 250 to 320 °C. On the other hand, polymers have very low thermal conductivities—typically from 0.1 to 0.5 W m^{-1} K^{-1}.

Another often-mentioned material is ceramic. Such insulations are usually in the form of tablets, blankets made of ceramic fibers, or even special custom-made parts. The advantage of ceramic is an outstanding thermal stability. There are several types of ceramic. Aluminosilicate ceramic can be typically used up to 1500 °C, and some other types such as boron nitride ceramic easily exceed these limits. Thermal conductivity of ceramic varies a lot. More sintered ceramic is characterized with lower resistance against temperature shocks, but higher solidity. It can have conductivities around 7 W m^{-1} K^{-1}. More porous ceramic has higher permeability to moisture, but endure larger temperature shocks and its thermal conductivity is generally lower—typically around 2 W m^{-1} K^{-1}. Special types of ceramic can withstand large temperature shocks and remain impermeable, but conductivities of these types are as high as 15 W m^{-1} K^{-1}.

Third common method is to insulate the block with a wool such as glass or fiberglass wools, mineral, rock, or kaolin wools. Usually, wools offer very low thermal conductivities which are typically lower than 0.1 W m^{-1} K^{-1}. On the other side, they do not perform well in wet environments and their pliability is rather a disadvantage.

We should at least mention some less-common materials such as firebricks, clay bricks, glass, or mica. Of course, several insulation materials might be combined. For example, in [3] copper-heated block was placed on ceramic insulators and the whole assembly was put into housing made of PEEK. Sometimes, vacuum is artificially created in order to insulate a block. It is also possible to leave an air gap between the block and the insulating material. The gap then serves as an insulation, as is mentioned in [4].

The crucial criterion for the right choice of an insulating material is the same as for adhesives, i.e., the operating temperature. Some other important properties to consider are thermal conductivity, machinability, moisture permeability, a possible change of insulating properties when the material becomes moistened, thermal-shock resistance, solidity, or thermal expansion of the material.

After thorough consideration, we have decided to use *Foamglas®Perinsul S* for our apparatus for its remarkably low thermal conductivity (lower than 0.05 W m^{-1} K^{-1}), operational temperature up to 430 °C, sufficient wear hardness and strength, water resistance, and impermeability for steam. We are going to test the modification in the near future.

6   Conclusions

This contribution deals with temperature distribution inside a heated block of an apparatus for research of saturated nucleate pool boiling. It aims at two problems— unwanted temperature differences inside the heated block and influence of ambient air on temperature field inside the block. Remedial actions are briefly described.

Acknowledgements This work was supported by the Ministry of Education, Youth and Sports of the Czech Republic under OP RDE grant number CZ.02.1.01/0.0/0.0/16_019/0000753 "Research centre for low-carbon energy technologies" and by the Grant Agency of the Czech Technical University in Prague, grant No. SGS18/129.

References

1. W.M. Rohsenow, *A Method of Correlating Heat Transfer Data for Surface Boiling of Liquids.* Technical report. Division of Industrial Cooperation, Heat Transfer Laboratory, Cambridge, MA, USA (1952)
2. H.K. Forster, N. Zuber, Dynamics of vapor bubbles and boiling heat transfer. AIChE J. **1**(4), 531–535 (1955)
3. T.Y. Kim, J.A. Weibel, S.V. Garimella, A free-particles-based technique for boiling heat transfer enhancement in a wetting liquid. Int. J. Heat Mass Transf. **71**, 808–817 (2014)
4. S. Petrovic, T. Robinson, R.L. Judd, Marangoni heat transfer in subcooled nucleate pool boiling. Int. J. Heat Mass Transf. **47**(23), 5115–5128 (2004)

Droplet Interplay on Microdecorated Substrates

Veronika Kubyshkina, Khellil Sefiane, Daniel Orejon, and Coinneach Mackenzie Dover

1 Introduction

The study of sessile droplet evaporation has attracted widespread interest due to the plethora of industrial and every day applications ranging from inkjet printing, to DNA micro-array chips, to spray cooling [1, 2]. To date, the bulk of this research has been constrained to single sessile droplet evaporation, which is rarely present in real applications. Hence, to fully reap the benefits of such research and thus optimise the above-mentioned applications, it is crucial to also study evaporation of multiple droplets in close proximity and the interplay between them.

Experimental observation of a pair of evaporating sessile droplets placed in the vicinity of each other has revealed interesting behaviours, such as asymmetric and suppressed evaporation and contact line deformation, in addition to buckling and arching of colloidal droplets [3, 4]. The presence of a neighbouring droplet alters the vapour field around the droplets, thereby disturbing the normal evaporation phenomenon [4]. Droplets placed in close proximity induce a saturation of the vapour field around the liquid–vapour interface, resulting in suppressed evaporation [4]. Moreover, in the case of organic solvent evaporation and/or liquid desiccants, the presence of relative humidity was found to have a strong impact on droplet evaporation or on the droplet growth due to absorption–adsorption and/or condensation onto the droplets [5]. Then, besides the non-uniform distribution of the organic solvent vapour evaporating around the droplets, the distribution of the water vapour field also varies, having a strong influence on the evaporative behaviour.

Due to the complexity of multi-drop systems, current research has been largely focused on the liquid–vapour interactions between neighbouring drops. The present work, however, aims to not only further the understanding of the physical mechanisms

V. Kubyshkina (✉) · K. Sefiane · D. Orejon · C. M. Dover
School of Engineering, University of Edinburgh, Kings Buildings, Edinburgh EH9 3FB, UK
e-mail: veronika.kubyshkina@ed.ac.uk

© Springer Nature Singapore Pte Ltd. 2021
C. Wen and Y. Yan (eds.), *Advances in Heat Transfer and Thermal Engineering* ,
https://doi.org/10.1007/978-981-33-4765-6_14

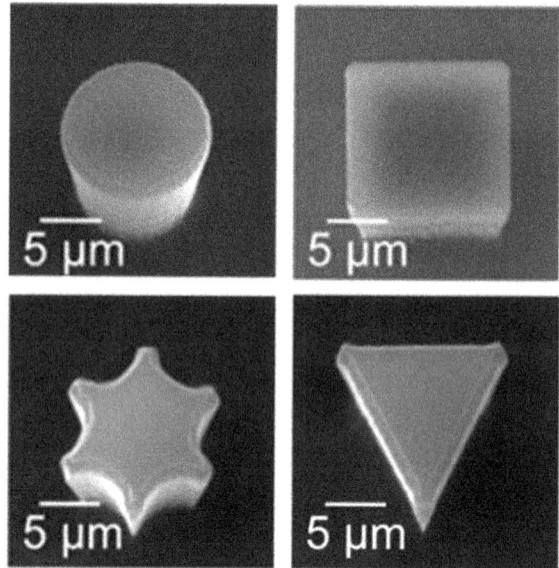

driving the liquid–vapour interactions of such complex droplet systems, but also investigate both the liquid–liquid and solid–liquid interactions of a pair of droplets.

2 Methodology

Double micro-droplets comprising both pure liquids (water and ethanol) and more complex fluid mixtures (binary mixtures of ethanol and water) are tested on microdecorated silicon substrates coated in parylene. Evaporation rate and top view images of the droplets lifetime are recorded. Thin-film interactions are monitored for binary fluid droplets that are able to adopt the hemi-wicking state on such substrates. In the hemi-wicking state, a droplet sits on a composite solid/liquid surface, where a film imbibes the microstructures. Additionally, the effect of surface topography is addressed by varying the pillar shape and interpillar distance. Four pillar shapes will be tested: circular, square, star-shaped and triangular pillars (Fig. 1). An infrared camera will provide insight into the liquid–solid interactions.

3 Results

The obtained results will provide further insight into three important interactions of a pair of neighbouring droplets, namely, liquid–vapour, liquid–liquid and liquid–solid.

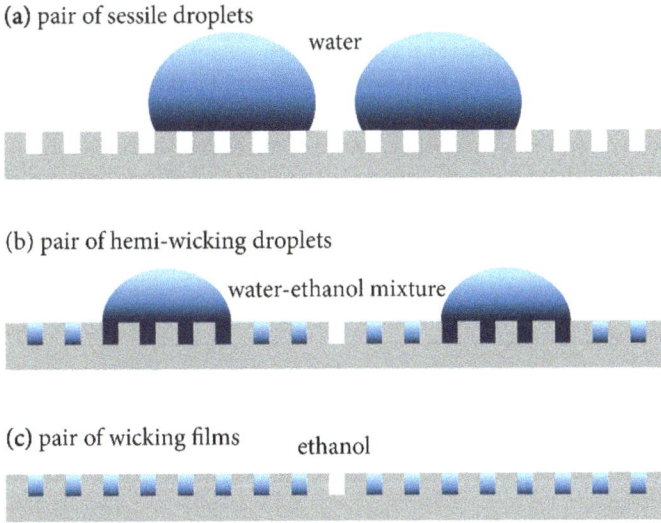

Fig. 2 Schematic diagram of the different configurations of neighbouring droplet pairs: **a** two sessile droplets consisting of pure water, **b** two hemi-wicking droplets consisting of water and ethanol mixture, **c** two wicking films consisting of ethanol

Droplet composition allows for control over the contact angle and as such will provide the means to establish and analyse the interaction of droplets in varying wetting regimes: sessile droplets, hemi-wicking droplets and wicking films. In the case of pure water sessile droplets, the interactions between a pair of droplets take place mainly in the gas phase above the structured substrate, thus suppressing evaporation (Fig. 2a).

In the case of binary droplet mixtures (consisting of ethanol and water) exhibiting the hemi-wicking behaviour, two scenarios are possible (Fig. 2b). For films that are finite in size and that pin at their periphery shortly after deposition, coalescence of neighbouring films does not take place. In this case, it is expected that the saturation of the vapour field would predominantly take place within the microstructures and near the film, with the gas phase interactions between droplets diminishing greatly given their relatively large distance from each other. Experimental observation of single hemi-wicking droplets on the same substrates demonstrated that the wicking films can reach diameters of up to 10 mm. Therefore, to supress the coalescence of two nearby films, the droplets must be relatively distant from each other when compared to the characteristic length of the droplets. Alternatively, when the distance between the two hemi-wicking droplets is reduced, the films will coalesce and mix, thereby inducing a saturation of the vapour field around the liquid–vapour interface. Thus, the experimental data obtained from hemi-wicking droplets will establish the interactions through mass transfer for both the liquid and the vapour phases.

Lastly, in the case of pure ethanol, only the wicking fronts will become saturated and interact with each other (Fig. 2c). In cases where the films are in very close proximity, the pair of wicking films will coalesce and mix.

Additionally, solid–liquid interactions of droplet pairs through heat transfer will be established on surfaces exhibiting a range of microstructures. Variations in surface topography allow for systematic control of surface roughness and wettability, and short time evaporation dynamics are heavily influenced by the surface structure. The influence of pillar shape and pillar-to-pillar spacing on the evaporation of two droplets in close proximity to each other will be determined. Potential application such as evaporation and condensation phase change heat transfer, droplet manipulation and microfluidics will be sought.

4 Conclusions

This research will offer a more encompassing insight into multi-drop evaporation phenomena, as well as the universal features of multi-drop systems, by addressing the vapour, liquid and gas interactions. Emphasis will be placed on the above in the context of custom microdecorated surfaces.

Acknowledgements The present work is supported by the UK Engineering and Physical Sciences Council (EPSRC). The substrates used in the experimental work were fabricated at the Scottish Microelectronics Centre (SMC) at the University of Edinburgh.

References

1. M. Yoshino, T. Matsumura, N. Umehara, Y. Akagami, S. Aravindan, T. Ohno, Engineering surface and development of a new DNA micro array chip. Wear **260**, 274–286 (2006)
2. T. Kokalj, H. Cho, M. Jenko, L.P. Lee, Biologically inspired porous cooling membrane using arrayed-droplets evaporation. Appl. Phys. Lett. **96**, 163703 (2010)
3. L. Chen, J.R.G. Evans, Arched structures created by colloidal droplets as they dry. Langmuir **25**, 11299–11301 (2009)
4. A.J.D. Shaikeea, S. Basu, Insight into the evaporation dynamics of a pair of sessile droplets on a hydrophobic substrate. Langmuir **32**, 1309–1318 (2016)
5. Y. Kita, Y. Okauchi, Y. Fukatani, D. Orejon, M. Kohno, Y. Takata, K. Sefiane, Quantifying vapor transfer into evaporating ethanol drops in a humid atmosphere. Phys. Chem. Chem. Phys. **20**, 19430–19440 (2018)

Effect of Inlet Subcooling on Flow Boiling Behaviour of HFE-7200 in a Microchannel Heat Sink

Vivian Y. S. Lee, Gary Henderson, and Tassos G. Karayiannis

1 Introduction

Miniaturised electronics pose challenging thermal demands, not only at chip-level power dissipation but also at the complete system-level heat rejection, in modern electronic packages. Chip-level power densities are projected to be as high as 4.5 MW/m^2 in computer systems by 2026 [1] and have been reported to exceed 10 MW/m^2 in power modules for defence applications [2]. For instance, microwave power modules used in critical applications such as radar systems and satellites operate at high frequencies to improve dynamic response and reduce component size, albeit at the cost of device efficiency. For a typical efficiency of 20% and input power of 850 W, almost 700 W of waste heat must be rejected from the system into the immediate environment, proving to be particularly challenging in aerospace applications where air-cooling is preferred. Flow boiling in microchannels is regarded as a promising cooling solution [1] for microelectronic systems where surface temperatures are limited to between 85 and 125 °C. The effect of various operational parameters on the performance of the cooling system must be well understood to facilitate successful integration of the developed technology into real-life systems. One of the important parameters in microchannel flow boiling is the degree of subcooling condition at the inlet of the heat sink as it affects the conditions for bubble nucleation, subsequent bubble growth, saturation conditions as well as flow reversal and instabilities in the channel. Deng et al. [3] studied the effect of inlet subcooling at 10 and 40 K of ethanol and found larger flow boiling heat transfer coefficients at higher degree of subcooling and concluded that subcooling had a negligible effect on pressure drop.

V. Y. S. Lee · T. G. Karayiannis (✉)
Department of Mechanical and Aerospace Engineering, Brunel University London, Uxbridge, London UB8 3PH, UK
e-mail: tassos.karayiannis@brunel.ac.uk

G. Henderson
TMD Technologies Limited, Hayes, London UB3 1DQ, UK

© Springer Nature Singapore Pte Ltd. 2021
C. Wen and Y. Yan (eds.), *Advances in Heat Transfer and Thermal Engineering* ,
https://doi.org/10.1007/978-981-33-4765-6_15

On the contrary, in the subcooled boiling study of HFE-7100 in [4], heat transfer coefficient and pressure drop both experienced a slight decrease at higher degree of subcooling. Although considerable effort has been dedicated to the development of microchannel evaporator modules for high heat flux cooling, little attention has been paid to the study of integrated thermal management loops complete with a suitably-sized condenser. The degree of subcooling used will influence the design and size of both the evaporator and the condenser. In this study, the effect of inlet subcooling on the flow boiling characteristics of a microchannel evaporator, developed for the cooling of high heat flux devices using dielectric refrigerant HFE-7200, was investigated.

2   Methodology

The experimental facility is shown in Fig. 1. The main loop consists of a reservoir, a gear pump, Coriolis mass flowmeters, a pre-heater, the test section and a condenser, which is cooled with ambient air aided by two axial fans. Flow visualisation was conducted along the channel at the centre of the heat sink with a high-speed camera. The copper test section has forty-four channels of width 0.36 mm, height 0.7 mm and length 20 mm, milled with a wall thickness of 0.1 mm on a 20×20 mm square area on the top of the block ($D_h = 475$ µm). Five thermocouples were positioned along the channel at the centre of the block to help measure the local heat transfer coefficient at the dimensionless axial positions: $z/L = 0.17, 0.34, 0.5, 0.67$ and 0.83, respectively. The heat flux was based on temperatures recorded by six thermocouples in the vertical direction. The inlet/outlet temperature and heat sink pressure drop are measured at the fluid line in the top plate as illustrated in Fig. 1. Inlet subcooling, ΔT_{sub}, is $T_{sat,p(i)} - T_{in}$, where $T_{sat,p(i)}$ is the fluid saturation temperature evaluated based on the inlet pressure and T_{in} is the inlet temperature. The single-phase length, L_{sub}, was obtained based on the iterative method detailed in [5]. The local heat transfer coefficient, $h_{(z)}$, is calculated as in Eq. 1, where $T_{f(z)}$ is the local fluid temperature,

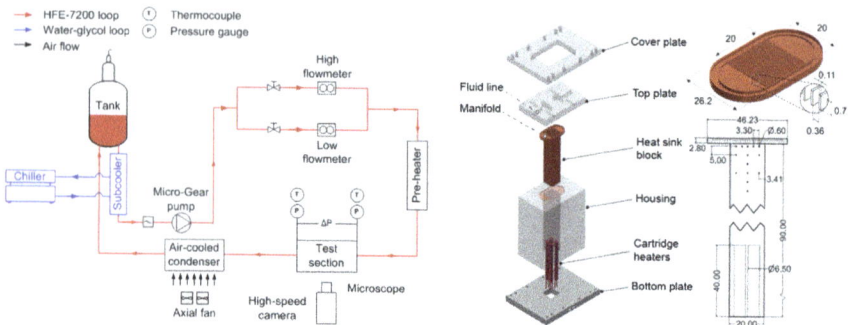

Fig. 1   Experimental facility and details of the microchannel test section

evaluated based on energy balance if in single-phase flow, [5]. In the flow boiling region, $T_{f(z)}$ is the saturation temperature evaluated at the local saturation pressure based on a linear pressure drop assumption. Additionally, $q_b^{"}$ is the base heat flux calculated from the vertical temperature gradient on the copper block and $T_{w(z)}$ is the temperature at the channel bottom wall. N, W, W_{ch}, H_{ch} and η are the heat sink width, channel width, channel height and fin efficiency, respectively. The top plate was assumed to be adiabatic.

$$h_{(z)} = \frac{q_b^{"} W}{\left(T_{w(z)} - T_{f(z)}\right) * N(W_{ch} + 2\eta H_{ch})} \tag{1}$$

The average heat transfer coefficient in this paper, $\overline{h}_{(z)}$, is averaged over the full channel length, covering both the single and two-phase region.

$$\overline{h}_{(z)} = \frac{1}{L} \int_0^L h_{(z)} dz \tag{2}$$

3 Results

Experiments were conducted at a mass flux of 200 kg/m^2 s up to exit vapour qualities of 0.9 at a pressure of 1 bar (evaluated at the inlet). The saturation temperature of HFE-7200 at 1 bar is 75.05 °C. The effect of inlet subcooling was studied at 5 K, 10 K and 20 K for base heat fluxes in the range of 93–692 kW/m^2. Flow patterns including bubbly, slug, churn and annular flow were observed in this study. Flow visualisation showed several differences in the flow behaviour of nucleated bubbles at different inlet temperatures. At $q_b^{"} = 400$ kW/m^2 and inlet subcooling of 5 K, bubbles nucleated from the channel side walls. As the bubbles grow and depart, they coalesce and form larger bubbles and slugs that quickly developed into churn flow near the channel inlet. At the same heat flux condition with 10 K subcooling, bubbles coalesced and formed slugs near the inlet, but only developed into churn flow further downstream. The bubbly flow regime was observed at the channel inlet at 20 K subcooling. Some bubbles departing from the side walls were observed to collapse and re-condense into the flow as they encounter the subcooled bulk fluid.

The single-phase length increased with increasing degree of subcooling and decreased with increasing heat flux. As expected, increasing inlet subcooling delayed the onset of saturated boiling in the channels at a given heat flux and mass flux condition. At $q_b^{"} = 400$ kW/m^2, the single-phase length, L_{sub}, was 1.7 mm and 2.8 mm when $\Delta T_{sub} = 5$ K and 10 K, respectively. The onset of saturated boiling occurred very near the channel inlet (total channel length = 20 mm). As inlet subcooling is increased to 20 K, the single-phase region extended to 5.9 mm and saturated boiling

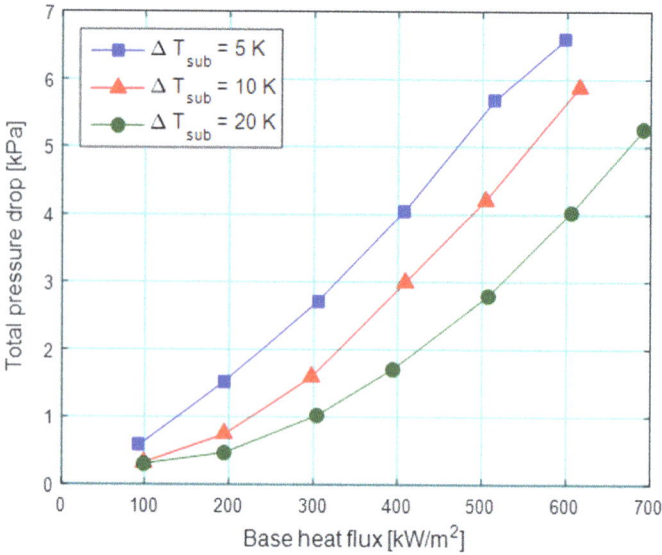

Fig. 2 Pressure drop across the heat sink as a function of base heat flux and degree of subcooling

was triggered further downstream. As mentioned above, departing bubbles near the channel inlet at 20 K subcooling were observed to re-condense into the subcooled fluid. The extended single-phase region, which typically exhibits a lower pressure drop compared to two-phase flow, and the differences in flow patterns observed, may have contributed to the lower total pressure drop measured across the heat sink at higher inlet subcooling conditions, see Fig. 2.

Average heat transfer along the channel also showed a notable dependence on inlet temperature, particularly at low heat fluxes ($q_b^{"} = 100$–$200\,\text{kW/m}^2$) and between inlet subcooling of 5 K and 20 K. Figure 3 shows the average heat transfer coefficient (defined in Eq. 2) as a function of heat flux at different inlet temperatures. For a given heat flux, the average heat transfer coefficient is lower at a higher degree of inlet subcooling, especially at the lowest heat flux condition due to the considerable length of the single-phase region. In fact, at 20 K, the entire channel is in single-phase flow at $q_b^{"} = 100\,\text{kW/m}^2$ and the average heat transfer coefficient is <2000 W/m² K. The heat transfer rates are much lower in the single-phase region and thus resulted in a lower average heat transfer coefficient. The effect of inlet subcooling on heat transfer performance appears to diminish as heat flux is increased. This could be due to the relatively shorter single-phase length at increasing heat fluxes. The low heat transfer coefficients characteristic of the single-phase region seem to have a weaker influence on the average heat transfer coefficient at moderate to high heat fluxes ($q_b^{"} > 300\,\text{kW/m}^2$). Flow pattern development along the channel, highly dependent upon the inlet subcooling degree for a given mass flux and heat flux condition, could also have had an effect on the slightly lower average heat transfer coefficients at moderate to high heat fluxes obtained at higher inlet subcooling conditions.

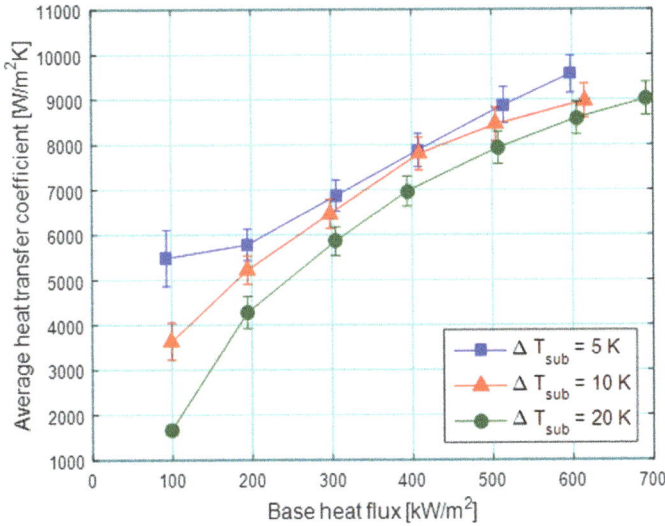

Fig. 3 Average heat transfer coefficient as a function of base heat flux and degree of subcooling

4 Conclusions

Experiments were performed at $G = 200$ kg/m^2 s at a system pressure of 1 bar at three different inlet temperatures to assess the effect of inlet subcooling condition on the flow boiling heat transfer and pressure drop of refrigerant HFE-7200 in a microchannel heat sink. Flow patterns observed in the study were bubbly, slug, churn and annular flow. For a given heat flux and mass flux condition, increasing inlet subcooling extended the single-phase length in the channel. Increasing the applied heat flux reduced the single-phase length, affecting the pressure drop, average heat transfer coefficient and flow pattern development along the channel. The results indicated a notable dependence of pressure drop and heat transfer behaviour on inlet temperature across the range of heat fluxes studied (93–692 kW/m^2). Lower pressure drop values were recorded at higher inlet subcooling conditions due to the characteristically lower single-phase pressure loss and possibly also due to different flow patterns observed at each location in the channels. The average heat transfer coefficient decreased with increased subcooling, especially at low heat fluxes ($q_b^{''}$ < 300 kW/m^2), due to the low heat transfer coefficients in the single-phase region. The effect of inlet subcooling weakened with increasing heat flux, possibly due to the decreasing single-phase length. The difference in flow pattern development could explain the slightly lower average heat transfer coefficients obtained at moderate to high heat fluxes with higher inlet subcooling degrees. The study verified the importance of the degree of subcooling when comparing heat transfer and pressure drop characteristics in microchannel heat sinks and when optimising the design of integrated thermal management systems for high heat flux electronic devices.

Acknowledgements The authors would like to thank TMD Technologies Ltd. for their financial support of the project.

References

1. T.G. Karayiannis, M.M. Mahmoud, Flow boiling in microchannels: fundamentals and applications. Appl. Therm. Eng. **115**, 1372–1397 (2017)
2. I. Mudawar, Assessment of high-heat-flux thermal management schemes. IEEE Trans. Components Packag. Technol. **24**(2), 122–141 (2001)
3. D. Deng, W. Wan, H. Shao, Y. Tang, J. Feng, J. Zeng, Effects of operation parameters on flow boiling characteristics of heat sink cooling systems with reentrant porous microchannels. Energy Convers. Manag. **96**, 340–351 (2015)
4. J. Lee, I. Mudawar, Fluid flow and heat transfer characteristics of low temperature two-phase micro-channel heat sinks—part 2. Subcooled boiling pressure drop and heat transfer. Int. J. Heat Mass Transf. **51**(17–18), 4327–4341 (2008)
5. V.Y.S. Lee, A.H. Al-Zaidi, G. Henderson, T.G. Karayiannis, Flow boiling results of HFE-7200 in a multi-microchannel evaporator and comparison with HFE-7100, in *Proceedings of the 4th World Congress on Momentum, Heat and Mass Transfer (MHMT'19)* (2019)

Heat Transfer Measurements for Condensation of FC72 in Microchannels

Lei Chai, Jiong Hui Liu, Nan Hua, Guang Xu Yu, John W. Rose, and Hua Sheng Wang

1 Introduction

The paper reports heat transfer measurements for condensation of FC72 in ten parallel horizontal microchannels (1.5 mm deep × 1.0 mm wide) in an aluminum test section. The results are compared with annular, laminar flow theory [3], previous measurements [1] and a general correlation [2].

2 Apparatus and Procedure

The closed loop apparatus comprised boiler, superheater, test section, auxiliary condenser and feed pump. The test section (see Fig. 1) comprised two aluminum blocks, the lower block thickness 20.75 mm and the upper block thickness 19.25 mm. Ten equally spaced, parallel, horizontal microchannels were machined in the mating surface of the lower block. The upper and lower blocks each housed 50 thermocouples in 0.6 mm diameter holes spaced through the blocks at 10 locations in the flow direction. The blocks were cooled along the upper and lower surfaces by water in counter flow. The whole of the test section was externally thermally well insulated.

L. Chai · J. H. Liu · N. Hua · J. W. Rose · H. S. Wang (✉)
School of Engineering and Materials Science, Queen Mary University of London, London E1
4NS, UK
e-mail: h.s.wang@qmul.ac.uk

G. X. Yu
DENSO Marston Ltd., Shipley, West Yorkshire BD17 7JR, UK

N. Hua
Faculty of Materials and Energy, Guangdong University of Technology, Guangzhou 510006,
China

© Springer Nature Singapore Pte Ltd. 2021   89
C. Wen and Y. Yan (eds.), *Advances in Heat Transfer and Thermal Engineering* ,
https://doi.org/10.1007/978-981-33-4765-6_16

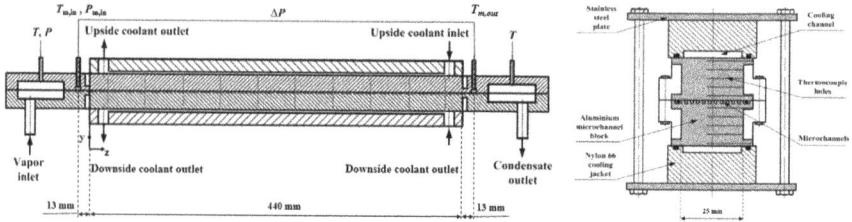

Fig. 1 Cross sections of test section (Note different scales)

Local channel surface temperature and heat flux were determined from the temperature distributions in the blocks and local vapor temperatures found by assuming linear pressure distribution over the two-phase part of the flow [1].

Extreme care was taken to ensure accuracy of all measurements (temperatures ±0.03 K, locations in test block ±0.2 mm) and to avoid effects of non-condensing gas. Tests were conducted for a range of vapor mass fluxes. Under steady-state conditions, temperatures and pressures were measured at the positions indicated in Fig. 1, together with the vapor flow rate (values obtained from the condensate return flowmeter and by energy balance for the boiler agreed to within 2%). The coolant flow rates and inlet temperatures were set to the desired values (same for upper and lower channels). The measured coolant temperature rise agreed with that found by integration using the coolant-side heat flux distribution to within 3%.

3 Results and Discussion

Figure 2 shows typical temperature distributions in the upper and lower blocks at the ten streamwise locations. The channel heat flux and surface temperature were determined from the slopes and intercepts at the interface when the temperatures at each streamwise position were fitted by cubic polynomials. The points differed from the curve fits by less than 0.03 K. Following Kim and Mudawar [1], the local vapor temperature was taken as the saturation value corresponding to a linear pressure variation over the two-phase part of the flow. Based on the difference between values found using the upper and lower block temperatures, uncertainty of the heat transfer coefficients is estimated to be between 15% and 20%. Errors associated with temperature measurement and determination of channel heat flux and surface temperature from the interface values are much smaller.

Figure 3 shows a typical case of dependence of heat transfer coefficient on distance along the channel and local quality together with the prediction of annular, laminar flow theory. As in the case of Kim and Mudawar [1], small systematic dependence on coolant flow rate, and hence on channel surface temperature and heat flux distributions, may be seen. For this comparison, data for all coolant flow rates were taken together. When results for each coolant flow rates were treated separately, the

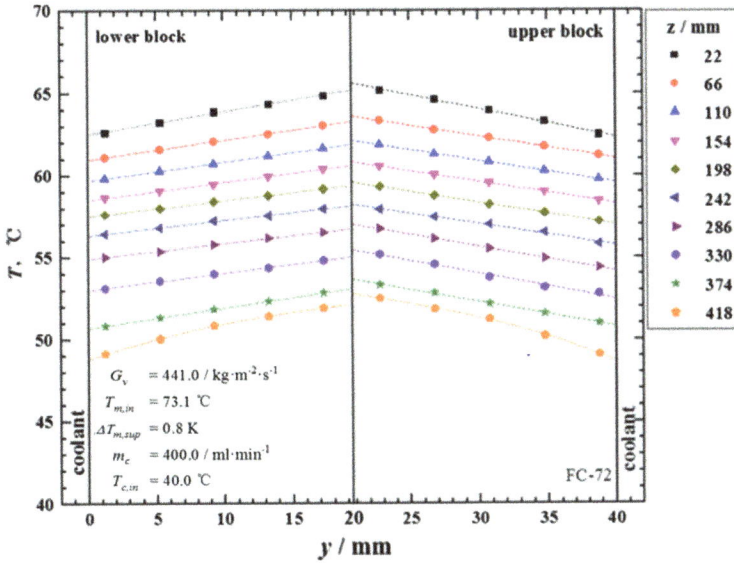

Fig. 2 Typical temperature distributions in test blocks. z denotes streamwise distance, y denotes distance normal to the flow direction

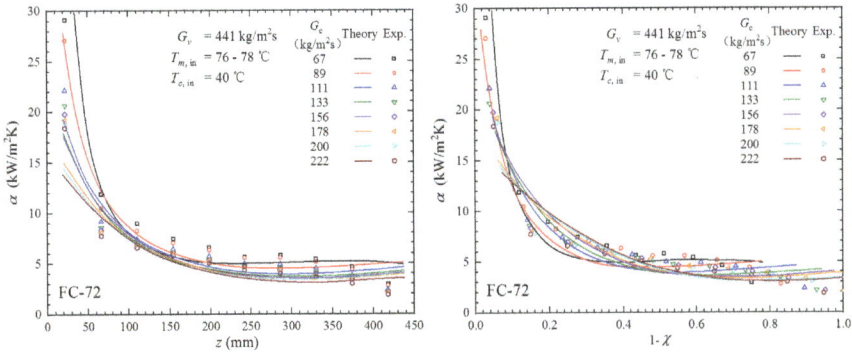

Fig. 3 Typical results. Streamwise dependence of channel heat transfer coefficient on distance along and channel and quality. Comparison of data with laminar, annular flow theory [3]. G_v denotes vapor mass flux, G_c denotes coolant mass flux

predicted dependence on coolant flow rate is the same as that shown by the measurements. The vapor and coolant conditions do not correspond precisely to those used by Kim and Mudawar [1] so direct comparisons cannot be made but the general magnitudes and trends agree very closely. Moreover, agreement of the Kim and Mudawar [1] data with annular laminar flow theory [3] was closely in line with that shown in Fig. 3 (see [4]). As may be seen in Fig. 4, the data are in within 30% bounds of the correlation of Kim and Mudawar [2] which included data for a wide range of

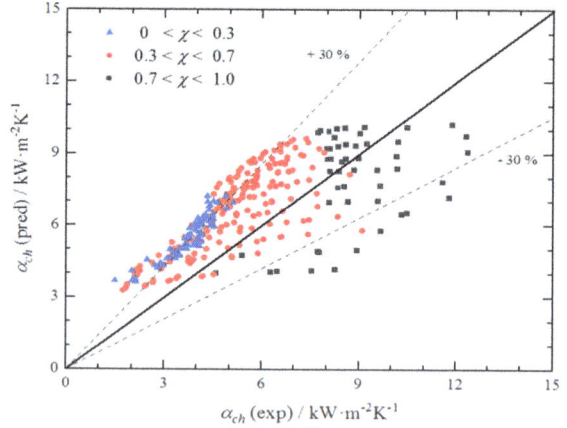

Fig. 4 Comparison of present data with correlation of Kim and Mudawar [2]

fluids. The correlation tends to over-predict the present data at high quality and to under-predict at low quality.

References

1. S.-M. Kim, I. Mudawar, Flow condensation in parallel microchannels—part 2: heat transfer results and correlation technique. Int. J. Heat Mass Transf. **55**, 984–994 (2012)
2. S.-M. Kim, I. Mudawar, Universal approach to predicting heat transfer coefficient for condensing mini/micro-channel flow. Int. J. Heat Mass Transf. **56**, 238–250 (2013)
3. H.S. Wang, J.W. Rose, A theory of film condensation in horizontal noncircular section microchannels. Trans. ASME J. Heat Transf. **127**, 1096–1105 (2005)
4. H.S. Wang, J.W. Rose, Condensation in microchannels: detailed comparisons of annular laminar flow theory with measurements. Trans. ASME J. Heat Transf. **139**, 072403-1 (2017)

Bubble Dynamics and Heat Transfer on Biphilic Surfaces

P. Pontes, R. Cautela, E. Teodori, A. S. Moita, A. Georgoulas, and António L. N. Moreira

1 Introduction

Pool boiling has been proven as a very effective process for heat transfer in cooling applications. Surface wettability plays a vital role in pool boiling heat transfer [1]. Surfaces with customized wetting patterns (hydrophilic surfaces with hydrophobic regions), when properly optimized geometrically, have shown a high potential of enhancing heat transfer [2], by promoting nucleation at lower superheat values and allowing a significant increase in the critical heat flux [3]. However, the scarce number of experimental studies performed so far shows a clear limitation: While authors describe nucleation and bubbles dynamics (and eventually thermal) behavior, with significant detail, very little is known regarding the internal flow inside the bubbles, which nevertheless may strongly affect both dynamic and thermal processes. Hence, a deeper knowledge on the dynamic behavior of both the vapor inside the bubble and the surrounding liquid could provide a deeper insight on the forces acting on the bubble. Such detailed description also allows a clearer analysis of the relation between the geometry and varying wettability patterns of the surface and its performance in terms of an effective heat transfer enhancement. In the previous works, a two-phase, CFD, enhanced numerical model for nucleate boiling that uses an enhanced volume of fluid (VOF)-based method [4] has been developed. In this work, this model will

P. Pontes · R. Cautela · A. S. Moita (✉) · A. L. N. Moreira
IN+ Center for Innovation, Technology and Policy Research, Instituto Superior Técnico, Universidade de Lisboa, Av. Rovisco Pais, 1049-001 Lisboa, Portugal
e-mail: anamoita@tecnico.ulisboa.pt

E. Teodori
ASML Holding N.V, De Run 6501, 5504 DR Veldhoven, The Netherlands

A. Georgoulas
School of Computing, Engineering and Mathematics, Advanced Engineering Centre, University of Brighton, Cockcroft Building, Lewes Road, Brighton BN2 4GJ, UK

© Springer Nature Singapore Pte Ltd. 2021
C. Wen and Y. Yan (eds.), *Advances in Heat Transfer and Thermal Engineering* ,
https://doi.org/10.1007/978-981-33-4765-6_17

be tested in pool boiling applications for biphilic surfaces and validated based on experimental values.

2 Experimental Methdology

The utilized validation data for the proposed numerical model are based on results obtained combining high-speed imaging and timely resolved thermography, following our previous work [5]. The recorded videos are processed using in-house MATLAB routines. Experimental values are averaged from four events.

The experimental setup consisted of a water-filled tank with lateral glass windows for optical access. The water inside the tank is kept at saturation temperature at atmospheric pressure with the help of resistance heaters. Test conditions are controlled by K-type thermocouples and a pressure transducer (OMEGA DYNE INC).

The surface is a 20 μm thick stainless-steel thin foil (AISI304) that is heated by Joule effect, directly fed by DC current, using a HP6274B power supply. For this part of the study, the simple biphilic pattern consisting of a superhydrophobic circular region in the middle of a hydrophilic stainless-steel surface is utilized. A mask with a single circular region was placed on top of the metal sheet and then coated with a superhydrophobic cover for this purpose.

3 Numerical Methodology

For this study, a previously developed, enhanced VOF-based numerical simulation framework, implemented in the general context of OpenFOAM CFD Toolbox, was utilized. To validate the numerical model, the conditions presented in Table 1 were fixed both in the numerical simulation as well as in the laboratory experiments.

Table 1 Parameters of the experiments and numerical simulation

Parameters	Values
Applied heat flux	1.39E3 [W/m^2]
Hydrophilic region wettability characterization:	
Advancing contact angle	85.54 [°]
Receding contact angle	34.37 [°]
Super hydrophobic region wettability characterization:	
Advancing contact angle	160.88 [°]
Receding contact angle	158.98 [°]
Region diameter	1.5 [mm]

In the VOF method, a volume fraction field α identifies the volume of liquid within a cell. The volume of the gaseous phase is therefore given as $(1 - \alpha)$. The value of α is 1 inside the pure liquid cells, 0 in the pure gas cells and between 0 and 1 in the cells containing the interface area. This procedure allows using a single set of continuity and momentum equations for the entire flow domain. A more detailed description of the utilized numerical simulation framework can be found in [4]. For the purposes of the present numerical investigation, 2D axisymmetric numerical simulations were performed. For this purpose, a wedge computational domain representing a 5° section of the actual 3D phenomenon was utilized. A nonuniform, structured computational mesh was constructed consisting of a total number of 59,848 cells. The mash and the vertical and horizontal dimensions of the computational domain were selected after appropriate mesh independency and domain independency studies.

4 Results

Figure 1 depicts a comparison between the experimental results and the model predictions. Figure 1a shows the visual pairing of the experimental and the numerical images at several stages of the considered bubble growth. This first analysis shows a very good agreement between the experimental and the numerical data and a very promising numerical reproduction of the bubble shape, even at later stages of bubble detachment ($t = 2.020$ s and $t = 2.024$ s) where necking is more pronounced. Figure 1b shows the comparison between the temporal variation of the equivalent bubble diameter. This diameter is calculated using the bubble volume and assuming a spherical approximation. In this case, there is a very good agreement between the experimental data and the numerical results. At later stages of bubble detachment, the results for both the volume and maximum diameter are considerably closer than at

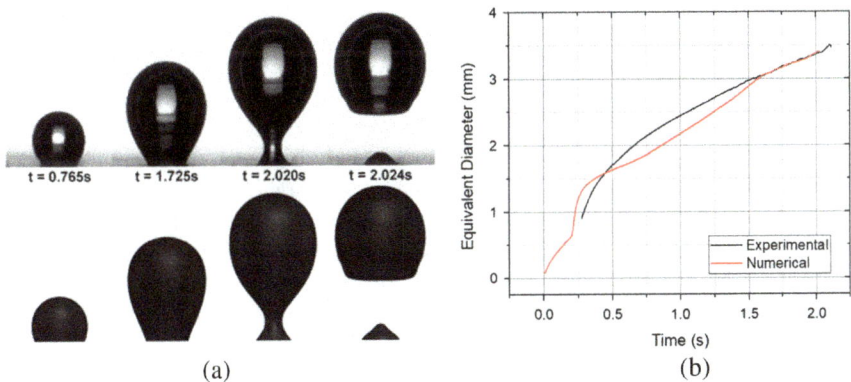

Fig. 1 Comparison between experimental and numerical results to validate the model. **a** Visual comparison of several bubble growth stages. **b** Equivalent diameter comparison

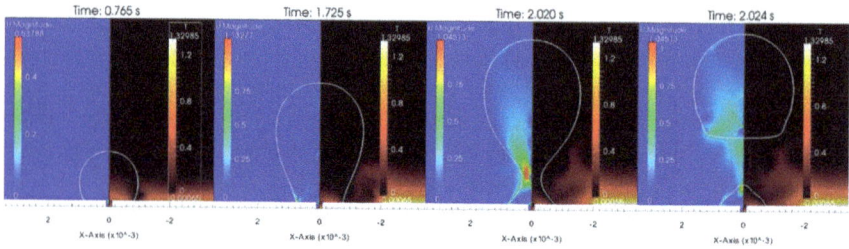

Fig. 2 Velocity magnitude and temperature inside and outside the bubble

earlier stages. This may be due to the fact that, at instant 0, the experiments start with a vapor layer that covers all the superhydrophobic region. This is due to the fact that several bubble cycles have preceded the considered bubble growth and detachment. The numerical model on the other hand starts with no vapor layer, and only one bubble growth and detachment cycle is considered. Regarding the bubble detachment time, the fact that there is no vapor layer at the initial instant for the numerical simulation was considered in the graph by displacing the experimental results to the time when the numerical diameter reached the layer diameter. Hence, the final detachment time observed in both experiments and numerical predictions is the same, showing less than 0.1 s of difference.

Having validated the model, a more in depth study on the velocity and temperature profiles inside the bubble can be made. The extracted colormaps for the studied case are shown in Fig. 2. The temperature results portray a hotter region near the surface and inside the bubble. On the bubble contact line, lower temperatures can be seen. Evaporation occurring in this area can justify this lower temperature values. Regarding the velocity results, the higher velocities can be seen during the detachment, in the neck region. The necking detachment timeframe is very small so these should stand out as the bubble detaches.

5   Conclusions

The dynamic behavior of the bubble was accurately replicated by the model. Data on temperature and velocity profiles that cannot be extracted from the experimental measurements were post-processed from the simulation results. For a more complete validation of the model, further simulations need to be made testing different wettability as well as thermal conditions. In order also to investigate the thermal response of the surface, the conjugate heat transfer version of the utilized numerical simulation framework need to be applied in the future. In conclusion from the overall results of the present investigation, it is obvious that the combination of enhanced direct numerical simulations with high-resolution transient experimental measurements can constitute a promising tool for further and deep understanding of the hydrodynamic and heat transfer characteristics of pool boiling in heated biphilic surfaces.

Acknowledgements The authors from the IN+ team would like to acknowledge Fundação para a Ciência e Tecnologia for partially supporting the research under the framework of the project JICAM/0003/2017 and of project UTAP-EXPL/CTE/0064/2017. A. S. Moita also acknowledges FCT for financing her contract through the IF 2015 recruitment program (IF 00810-2015) and through the exploratory project associated with this contract. Dr. Anastasios Georgoulas would like to acknowledge the financial support from the Engineering and Physical Science Research Council in UK, through the grant EP/P013112/1.

References

1. I. Malavasi, E. Teodori, A.S. Moita, A.L.N. Moreira, M. Marengo, Wettability effect on pool boiling: a review, in *Encyclopaedia of Two-Phase Heat Transfer and Flow III: Macro and Micro Flow Boiling and Numerical Modeling Fundamentals* (A 4-volume set), vol. 4, ed. by J. Thome (World Scientific Publishing Co Pte Ltd., 2018). ISBN: 978-981-3227-31-6
2. A.R. Betz, J. Jenkins, C. Kim, D. Attinger, Boiling heat transfer on superhydrophilic, superhydrophobic, and superbiphilic surfaces. Int. J. Heat Mass Transf. **57**(2), 733–741 (2013)
3. H. Jo, H.S. Park, M.H. Kim, Single bubble dynamics on hydrophobic–hydrophilic mixed surfaces. J. Heat Mass Transf. **93**, 554–565 (2016)
4. A. Georgoulas, M. Andredaki, M. Marengo, An enhanced VOF method coupled with heat transfer and phase change to characterise bubble detachment in saturated pool boiling. Energies **10**(3) (2017)
5. E. Teodori, P. Pontes, A.S. Moita, A.L.N. Moreira, Thermographic analysis of interfacial heat transfer mechanisms on droplet/wall interactions with high temporal and spatial resolution. Exp. Thermal Fluid Sci. **96**, 284–294 (2018)

Molecular Dynamics Simulation of Effect of Temperature Difference on Surface Condensation

J. H. Pu, Q. Sheng, J. Sun, W. Wang, and H. S. Wang

1 Introduction

Vapour condensation is a common and important process in nature and many engineering applications such as the formation of rain and seawater desalination [1]. When vapour is in contact with a cooled solid surface, surface condensation may happen. Condensation mode is conventionally categorized as either dropwise condensation (DWC) or filmwise condensation (FWC). With general understanding, heat transfer with phase change is dramatically higher than that without phase change. Heat transfer of DWC is an order of magnitude higher than that of FWC [2]. Temperature difference and surface wettability are two important factors affecting surface condensation. In this work, we will examine surface condensation mode and dynamical behaviours using molecular dynamics simulation.

2 Methodology

The surface condensation mode under different surface wettabilities and temperature difference (ΔT) conditions is investigated in the present work using molecular dynamics (MD) simulation (see Fig. 1). All the simulations are conducted using LAMMPS software package [3]. The overall size of simulation model measures $l_x \times l_y \times l_z = 471.12 \times 39.26 \times 588.9$ Å. In the x- and y-directions, periodic boundary

J. H. Pu · Q. Sheng · W. Wang · H. S. Wang (✉)
School of Engineering and Materials Science, Queen Mary University of London, London E1 4NS, UK
e-mail: h.s.wang@qmul.ac.uk

J. Sun
School of Chemical Engineering and Technology, Xi'an Jiaotong University, Xi'an 710049, Shaanxi, China

© Springer Nature Singapore Pte Ltd. 2021
C. Wen and Y. Yan (eds.), *Advances in Heat Transfer and Thermal Engineering* ,
https://doi.org/10.1007/978-981-33-4765-6_18

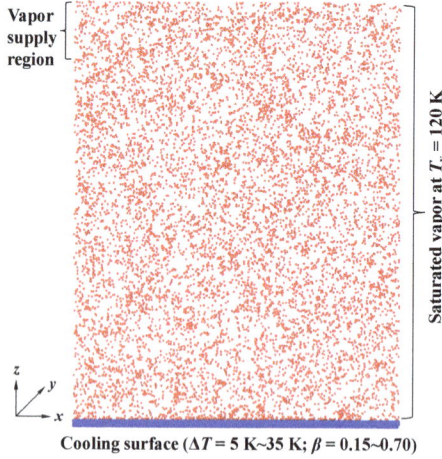

Fig. 1 Schematic of computational model. Solid atoms are in blue, and fluid molecules are in red

conditions are used, while in the z-direction, fixed boundary condition is employed. The solid wall is constructed by three layers of Pt-like atoms arranged as a face-centred cubic (FCC) lattice, and the length of the unit cell is 3.926 Å. Two extra layers of Pt-like atoms are fixed to serve as a frame. The 12-6 L-J potential function is employed for fluid–fluid (Ar–Ar) interaction $\varphi(r) = 4\varepsilon[(\sigma/r)^{12} - (\sigma/r)^6]$, where r is the intermolecular separation, ε and σ are the energy and length characteristic parameters, respectively. The potential function is truncated at the cut-off radius $r_c = 3.5\sigma$. $\varepsilon_{Ar-Ar} = 0.01040$ eV and $\sigma_{Ar-Ar} = 3.405$ Å are used. The interaction between solid atoms is only L-J type with $\varepsilon_{Pt-Pt} = 0.521875$ eV and $\sigma_{Pt-Pt} = 2.475$ Å. The fluid–solid interaction is also governed by L-J potential function but with different energy and length parameters, i.e. $\varepsilon_{Ar-Pt} = \beta\varepsilon_{Ar-Ar}$ and $\sigma_{Ar-Pt} = 0.91\sigma_{Ar-Ar}$, where β is fluid–solid bonding strength parameter indicating the surface free energy, or equivalently the surface wettability.

All the simulations are performed in two stages with a time step of 5 fs. In stage 1 (equilibrium stage), the saturated vapour molecules at $T_v = 120$ K are uniformly arranged. A period of 3 ns guarantees the system to reach the thermal equilibrium state. In stage 2 (condensation/cooling stage), the surface temperature is suddenly reduced to a target temperature and maintained afterwards for a period of 24 ns. Meanwhile, a vapour supply region is arranged on the uppermost region (thickness is $2l_z/15$), where the temperature and density are maintained at fluid saturation state. For the fluid except for those in the supply region, only NVE ensemble is used. All of the temperature control processes are achieved by the Berendsen thermostat. In the present work, we simulated different cases with different β from 0.025 to 0.7 and different ΔT from 5 to 35 K.

3 Results

Based on the simulation results within 24 ns in stage 2, Fig. 2 shows the surface condensation mode map under different β and ΔT. Red, green and blue squares represent no-condensation (NC), dropwise condensation (DWC) and filmwise condensation (FWC), respectively. It can be found that the condensation mode is significantly dependent on β. When ΔT is fixed, the condensation mode transits from NC to DWC and finally to FWC with β increasing. While although ΔT also shows its effect on the condensation mode, the situation is different from β. It can be seen that the boundary between NC and DWC is significantly altered with ΔT increasing, which changes to smaller β (from 0.4 to 0.15 with ΔT increasing from 5 to 35 K). Comparatively, the boundary between DWC and FWC just slightly changes, notwithstanding it also wholly moves to smaller β. It changes from 0.5 to 0.45 with same ΔT increasing from 5 to 35 K. So ΔT is seen significant effect on the transition between NC and DWC compared with its slight effect on the transition between DWC and FWC.

In order to analyse the dynamical behaviours of surface condensation, the number of clusters (cluster size is larger than 50) and the total number of condensed surface molecules (N) are calculated (see Figs. 3 and 4). The Stillinger criterion is used to define the cluster, and two Ar molecules are considered as a cluster when their distance is less than $1.5\sigma_{Ar-Ar}$. A cluster can be regarded as surface cluster only when the distance between the surface and any Ar molecules in the cluster is within that criterion. Two typical cases, $\beta = 0.275$ with different ΔT and $\beta = 0.475$ with different ΔT, are chosen, and just 10 ns results are showed, which is sufficient to show the dynamical evolution. Figure 3 shows the temporal evolution of the number of clusters. It can be seen that for NC situation ($\beta = 0.275$, $\Delta T = 5$ K and 15 K), the number of clusters is 0 in most time. In certain time, one cluster generates but disappears later, which is because its size is less than the critical cluster size. For DWC situation ($\beta = 0.275$ with $\Delta T = 25$ K and 35 K and $\beta = 0.475$ with $\Delta T = 5$ K and 15 K), the appearance time of cluster brings forward with ΔT increasing. $\beta = 0.275$ with $\Delta T = 25$ K is around 2 ns, while $\beta = 0.475$ with $\Delta T = 15$ K is around the initial time. After clusters generate, clusters start to coalescence and new clusters are also formed at the same time, which is why the cluster number fluctuates.

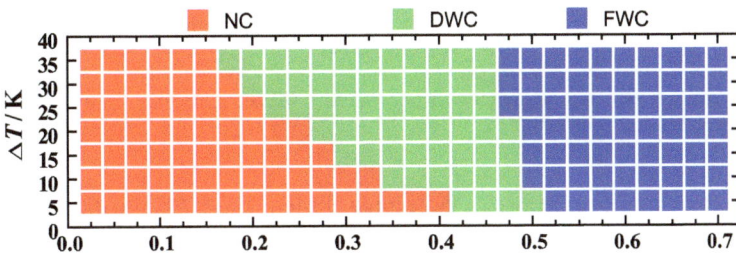

Fig. 2 Surface condensation modes under different β and ΔT

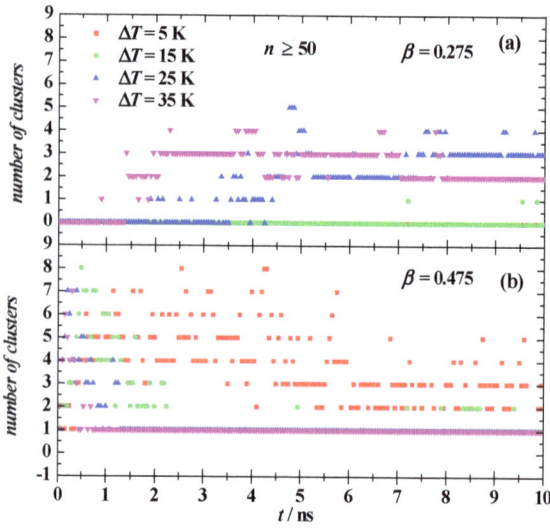

Fig. 3 Temporal evolution of the number of clusters with the size above 50

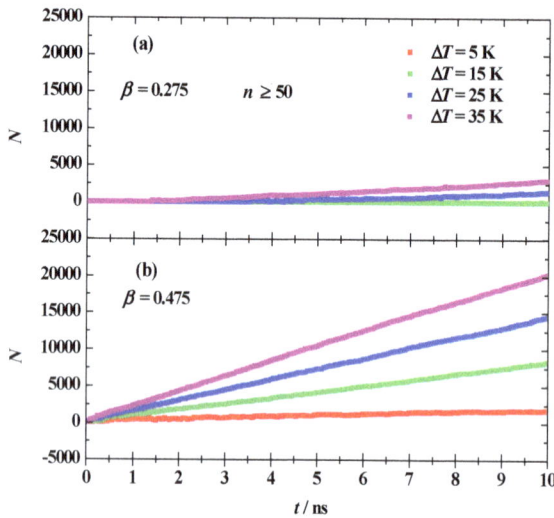

Fig. 4 Temporal evolution of the number of condensed surface molecules (N) that belong to the clusters with the size above 50

Finally, there mostly exist two or three large clusters. For FWC situation ($\beta = 0.475$ with $\Delta T = 25$ K and 35 K), at the initial time, there are some clusters distributing on the surface, but clusters coalescence and generate quickly, which leads to FWC, namely just one cluster. From Fig. 4, it can be found that N increases with time for condensation cases (DWC and FWC). For $\beta = 0.275$, $\Delta T = 5$ K and 15 K, they can also be estimated to NC cases from the change of N because it is nearly equal to 0. It can also found that the condensation intensity increases with ΔT and β. For $\beta = 0.275$, N changes from 0 to around 2500 at 10 ns. For $\beta = 0.275$, it changes from around 2500 to around 20,000 at 10 ns and increases nearly 8 times. Considering we just calculate the onset process of condensation where interfacial thermal resistance takes over the condensation process instead of bulk thermal resistance, so for $\beta = 0.475$, N for FWC ($\Delta T = 25$ K and 35 K) is larger than that for DWC ($\Delta T = 5$ K and 15 K).

4 Conclusions

In the present work, we investigate the surface condensation mode under different β and ΔT using MD simulations. A surface condensation mode map is obtained under different β and ΔT. The results show that β appears significant effect on surface condensation mode, and the surface condensation transits from NC to DWC and finally to FWC with β increasing. ΔT significantly affects the transition between NC and DWC, while ΔT slightly affects the transition between DWC and FWC. During condensation process, the generation, coalescence and growth of different clusters occur simultaneously. For DWC, clusters generate, coalesce and grow into some primary droplets, while for FWC, clusters coalesce into one droplet and occupy the whole surface rapidly, then it growths wholly. On the onset of condensation, the condensation intensity increases with increasing β and ΔT.

Acknowledgements The present work is supported by National Natural Science Foundation of China (51776196), the National Basic Research Program of China (973 Program) (2015CB251505), the Engineering and Physical Sciences Research Council (EPSRC) of the UK (EP/N020472/1, EP/N001236/1) and the Joint Ph.D. Studentship of China Scholarship Council (CSC) and Queen Mary University of London acknowledged. This research utilized Queen Mary's Apocrita HPC facility, supported by QMUL Research-IT.

References

1. S. Gao, W. Liu, Z. Liu, Tuning nanostructured surfaces with hybrid wettability areas to enhance condensation. Nanoscale **11**(2), 459–466 (2019)
2. J. Sun, W. Wang, H.S. Wang, Dependence of nanoconfined liquid behavior on boundary and bulk factors. Phys. Rev. E **87**, 023020 (2013)
3. S. Plimpton, Fast parallel algorithms for short-range molecular dynamics. J. Comput. Phys. **117**, 1 (1995)

The Strong Influence of Thermal Effects on the Lifetime of an Evaporating Droplet

Feargus G. H. Schofield, Stephen K. Wilson, David Pritchard, and Khellil Sefiane

1 Introduction

Determining the lifetime of an evaporating sessile droplet on a solid substrate is an important part of understanding many industrial processes, such as ink-jet printing, coating, and spray cooling, as well as drug delivery systems and chemical spill containment. Consequently, in recent years there has been a rapid growth of experimental and theoretical research into droplet evaporation (see, e.g., the recent reviews by Larson [1], Brutin and Starov [2], and Giorgiutti-Dauphiné and Pauchard [3]). The evolution, and hence the lifetime, of an evaporating sessile droplet depends on both the manner in which it evaporates [4] and the thermal properties of the substrate on which it sits [5]. We have recently considered both effects simultaneously in the special case of a thin droplet on a highly insulating substrate [6]. In the present work, we use the diffusion-limited model for evaporation to calculate the evolution, and hence the lifetime, of droplets with any initial contact angle on a substrate with a wide range of thermal conductivities in both the constant contact radius (CR) and constant contact angle (CA) modes of evaporation. In particular, we show that the lifetime of a droplet decreases with increasing thermal conductivity of the substrate.

F. G. H. Schofield · S. K. Wilson (✉) · D. Pritchard
Department of Mathematics and Statistics, University of Strathclyde, Livingstone Tower, 26 Richmond Street, Glasgow G1 1XH, Scotland, UK
e-mail: s.k.wilson@strath.ac.uk

K. Sefiane
School of Engineering, University of Edinburgh, The King's Buildings, Mayfield Road, Edinburgh EH9 3FB, Scotland, UK

© Springer Nature Singapore Pte Ltd. 2021
C. Wen and Y. Yan (eds.), *Advances in Heat Transfer and Thermal Engineering* ,
https://doi.org/10.1007/978-981-33-4765-6_19

2   Methodology

We adopt the thermally coupled diffusion-limited model for evaporation as used by Ait Saada et al. [7]. In the limit of thin droplets on thin substrates, we recently obtained analytical expressions for the evolution, and hence lifetime, of droplets evaporating in various modes of evaporation [6]. In the general case of non-thin droplets, we use the finite element software package COMSOL Multiphysics to numerically calculate the lifetimes of droplets evaporating in the constant contact radius (CR) and constant contact angle (CA) modes of evaporation.

3   Results

Figure 1 shows the scaled vapour concentration $c/c_{\text{sat}}(T_a)$, where c_{sat} is the vapour saturation value and T_a is the ambient temperature, in the atmosphere for a droplet of water with contact angle $\theta = 0.98\pi$ and contact radius $R = 0.05$ mm evaporating into air with 40% humidity. The substrates considered are (a) highly conducting (aluminium) and (b) poorly conducting (PTFE), each 1 mm thick. Also shown are scaled temperature contours T/T_a. For droplets on poorly conducting substrates, the free-surface temperature is lower than for droplets on highly conducting substrates. This means that the free-surface vapour concentration is lower for poorly conducting substrates than for highly conducting substrates, so the droplet evaporates more slowly on poorly conducting substrates.

Figure 2 shows the scaled lifetimes of droplets in (a) the CR mode, t_{CR}, and (b) the CA mode, t_{CA}, as a function of the scaled initial contact angle θ/π. The lifetimes are scaled with the characteristic timescale of evaporation [4].

$$\frac{\rho}{2D(c_{\text{sat}}(T_a) - c_\infty)}\left(\frac{3V}{2\pi}\right)^{2/3}, \tag{1}$$

where ρ is the density of the droplet, c_∞ is the ambient vapour concentration, D is the diffusivity of vapour through the air and V is the initial volume of the droplet. The solid curves denote the lifetimes obtained by Stauber et al. [4] using the thermally decoupled model, while the circles and the squares are the present numerical results for a droplet of water on the highly conducting substrate (aluminium) and on the poorly conducting substrate (PTFE), respectively, obtained using the thermally coupled model. As for the thermally decoupled model, the lifetimes obtained using the thermally coupled model vary with the initial contact angle and depend on the mode of evaporation. As Fig. 2 illustrates, the lifetimes obtained using the thermally coupled model are always longer than those predicted by the thermally decoupled model. Furthermore, the lifetimes of droplets evaporating on a poorly conducting substrate are always longer than those evaporating on a highly conducting substrate. In the limit of a perfectly conducting droplet on a perfectly conducting substrate, and

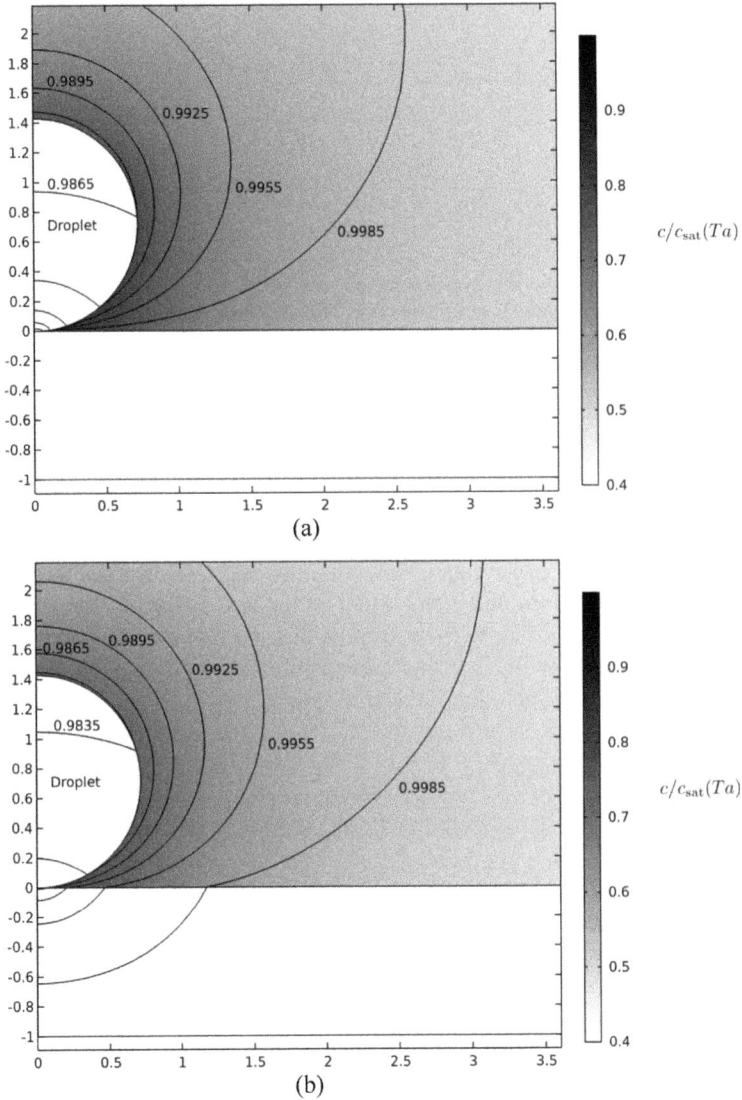

Fig. 1 Scaled vapour concentration $c/c_{\mathrm{sat}}(T_a)$ in the atmosphere for a droplet of water with contact angle $\theta = 0.98\pi$ and contact radius $R = 0.05$ mm evaporating into air with 40% humidity. The substrates considered are **a** highly conducting (aluminium) and **b** poorly conducting (PTFE), each 1 mm thick. Also shown are contours of the scaled temperature T/T_a

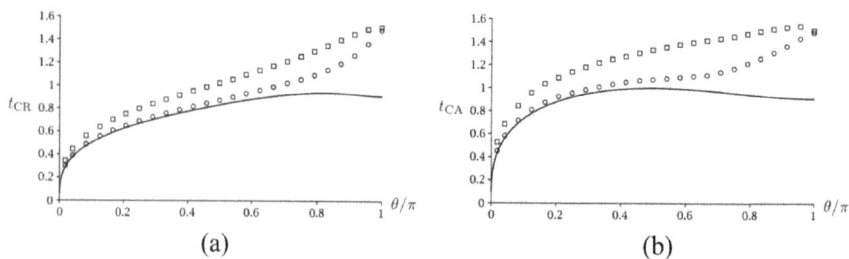

Fig. 2 Scaled lifetimes of droplets evaporating in **a** the CR mode, t_{CR}, and **b** the CA mode, t_{CA}, as a function of the scaled initial contact angle θ/π. The solid curves denote the lifetimes obtained by Stauber et al. [4] using the thermally decoupled model, while the circles and the squares are the present numerical results for a droplet of water on a highly conducting substrate (aluminium) and on a poorly conducting substrate (PTFE), respectively, obtained using the thermally coupled model

in the limit of an infinitely conducting atmosphere, we recover the lifetimes predicted by the thermally decoupled model for all initial contact angles. In the special case of a perfectly hydrophobic substrate, $\theta = \pi$, the CR and CA modes exactly coincide, and so their lifetimes are equal. However, it should also be noted that, for both modes, the lifetimes of droplets evaporating on the two different substrates do *not* exactly coincide when $\theta = \pi$ because, even in this special case, the conductivity of the substrate still has a weak effect on the evaporation of the droplet due to the small but nonzero thermal conductivity of the atmosphere.

4 Conclusions

The thermal properties of the system can have a profound influence on the evolution and hence on the lifetime of an evaporating droplet. In particular, when the initial contact angle is large, the thermally decoupled model for evaporation, which neglects evaporative cooling effects, underpredicts the lifetime of the droplet.

Acknowledgements FGHS acknowledges the financial support of the United Kingdom Engineering and Physical Sciences Research Council (EPSRC) via Doctoral Training Partnership grant EP/N509760/1, the University of Strathclyde, and the University of Edinburgh.

References

1. R.G. Larson, Transport and deposition patterns in drying sessile droplets. AlChE J. **60**, 1538–1571 (2014)
2. D. Brutin, V. Starov, Recent advances in droplet wetting and evaporation. Chem. Soc. Rev. **47**, 558–585 (2018)
3. F. Giorgiutti-Dauphiné, L. Pauchard, Drying drops. Euro. Phys. J. E **41**, 32 (2018)

4. J.M. Stauber, S.K. Wilson, B.R. Duffy, K. Sefiane, On the lifetimes of evaporating droplets. J. Fluid Mech. **744**, R2 (2014)
5. G.J. Dunn, S.K. Wilson, B.R. Duffy, K. Sefiane, The strong influence of substrate conductivity on droplet evaporation. J. Fluid Mech. **623**, 329–351 (2009)
6. F.G.H. Schofield, S.K. Wilson, D. Pritchard, K. Sefiane, The lifetimes of evaporating sessile droplets are significantly extended by strong thermal effects. J. Fluid Mech. **851**, 231–244 (2018)
7. M. Ait Saada, S. Chikh, L. Tadrist, Evaporation of a sessile drop with pinned or receding contact line on a substrate with different thermophysical properties. Int. J. Heat Mass Transf. **58**, 197–208 (2013)

Coupled Water and Ethanol Vapour Transfer to and from Volatile Ethanol Drops in Humid Air: Diffusion Model Revisited

Yutaku Kita, Daniel Orejon, Yasuyuki Takata, and Khellil Sefiane

1 Introduction

Drop evaporation of pure fluids and binary mixtures has been extensively studied, revealing influential factors such as substrate and fluid properties and surrounding atmosphere. Although many experimental and theoretical works focus mainly on vapour transfer from the drop surface towards surroundings, very little is known about mass transfer of a second component present in the atmosphere onto the drop. Indications of absorption–adsorption and/or condensation of water vapour onto evaporating organic solvents have been observed by means of time-resolved infrared spectroscopy [1] and temperature measurement near the drop interface using infrared (IR) thermography [2]. However, none of these studies reported accurate and direct quantification of water intake during the evaporation, which could provide a more detailed description of the coupling heat and mass transfer during organic solvent drop evaporation. In the present study, we performed a systematic analysis of water intake during ethanol drop evaporation in humid air at different ambient temperatures (T_{amb}) and relative humidity (RH). Combining drop profile measurement with gas chromatography (GC) allowed to directly quantify the drop composition in time. We observed that higher RH led to faster evaporation of ethanol as well as a larger amount of water intake. The physical mechanisms governing the increase in ethanol evaporation and that in water absorption–adsorption and/or condensation were thoroughly discussed. Based on the experiments, we proposed an empirical correlation

Y. Kita (✉) · Y. Takata
Department of Mechanical Engineering, Kyushu University, Fukuoka, Japan
e-mail: kita@mech.kyushu-u.ac.jp

Y. Kita · D. Orejon · Y. Takata
International Institute for Carbon-Neutral Energy Research (WPI-I^{2}CNER), Kyushu University, Fukuoka, Japan

D. Orejon · K. Sefiane
School of Engineering, University of Edinburgh, Edinburgh EH9 3JL, United Kingdom

© Springer Nature Singapore Pte Ltd. 2021
C. Wen and Y. Yan (eds.), *Advances in Heat Transfer and Thermal Engineering* ,
https://doi.org/10.1007/978-981-33-4765-6_20

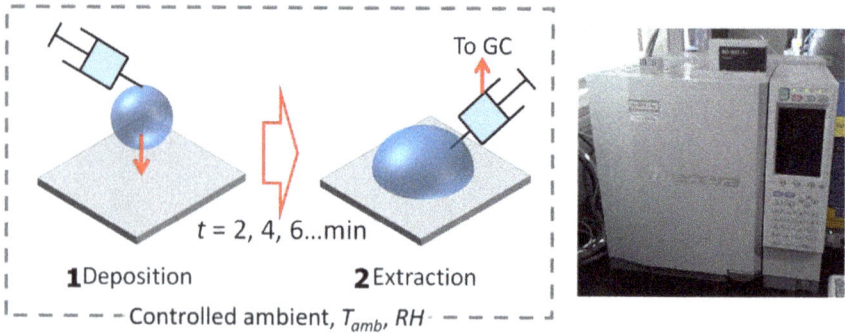

Fig. 1 (Right) Schematic description of drop deposition (stage 1) and liquid extraction (stage 2), and (left) a picture of the GC used in the present study

of the water intake combined with the diffusion model for drop evaporation, which agreed remarkably with the experimental observations.

2 Methodology

Experiments were carried out in an environmental chamber (ESPEC Corp. PR-3KT), where T_{amb} and RH were controlled. T_{amb} and RH tested in the present study were 30 and 40 °C, and 35, 50, 65 and 90%, respectively. An ethanol drop of *ca.* 7 μL was gently placed on a CYTOP-coated copper substrate ($9.5 \times 9.5 \times 1.0$ mm^3). A CCD camera (Sentech STC-MC152USB) and an IR camera (FLIR SC4000) were used to follow the evolution of the drop profile and the drop interfacial temperature. The drop composition was analysed using a gas chromatography (Shimadzu GC-2010 Plus). Using a Hamilton syringe, we extracted the liquid of 0.5 μL from the drop at different evaporating times and injected into the GC as described in Fig. 1. The GC typically showed two distinctive peaks which correspond to water (earlier one; *ca.* 4 min) and ethanol (later one; *ca.* 9 min), respectively. The ethanol concentration within the injected liquid was determined based on the ratio of the area of ethanol peak to the total area of ethanol and water peaks.

3 Results

From the GC analysis of the drop composition at different times t, an approximately linear decrease in the ethanol concentration $x_{ethanol}$ was observed at every T_{amb} and RH investigated (see Fig. 2a for $T_{amb} = 40$ °C and $RH = 90\%$). The amounts of ethanol and water within the drop were calculated by multiplying the drop volume V by their concentrations (Fig. 2b). Increases in the rate of ethanol evaporation and

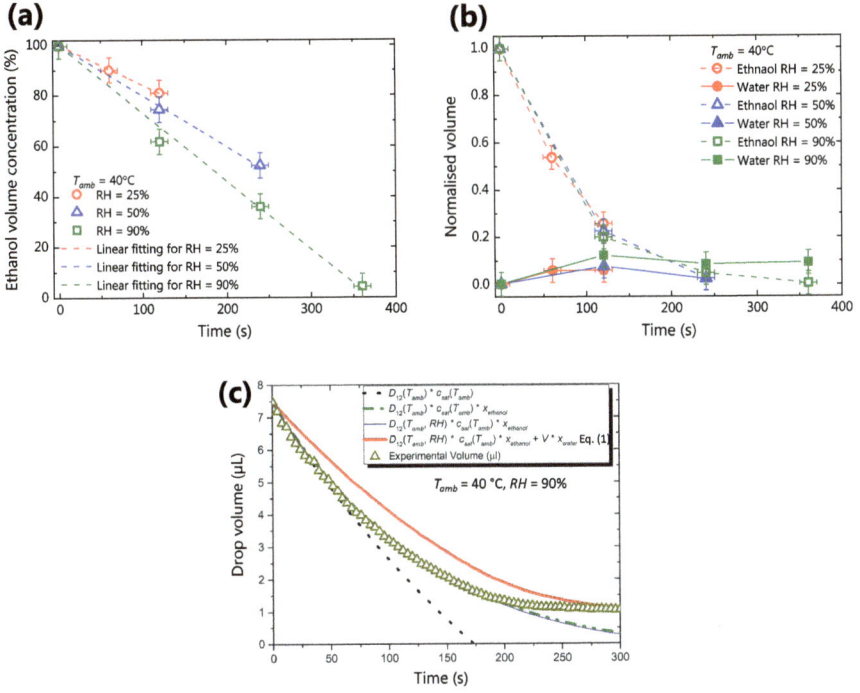

Fig. 2 Evolution of **a** ethanol concentration and **b** normalized volumes of ethanol and water within the drop. **c** Drop volume evolution obtained by the experiments (triangles) and predicted by the model proposed (lines)

that of water intake were found at higher *RH*. The increase in ethanol evaporation at higher *RH* was attributed to the larger diffusion coefficient D_{12} of ethanol into moist air than dry air. Moreover, the drop surface temperature was found to fall below the dew point due to the evaporative cooling, leading to more water condensation onto the drop [2].

Based on the coupling mechanisms of ethanol evaporation and water absorption–adsorption and/or condensation, we propose an empirical model to capture the experimental observations. The rate of drop evaporation can be estimated using a mass diffusion model from the work of Stauber et al. [3]. We then added to the model terms that account for the effect of T_{amb} and *RH* on the diffusion coefficient, the decrease in ethanol concentration within the drop and the water intake as

$$\frac{dV}{dt} = -\frac{\pi D_{12}(T_{amb},RH)c_{sat}(T_{amb})x_{ethanol}}{\rho} \frac{Rg(\theta)}{(1+\cos\theta)^2} + \frac{dx_{water}}{dt} V \qquad (1)$$

where R and θ are the drop radius and the contact angle changing in time, ρ and c_{sat} are the density and the saturated concentration of ethanol vapour, x_{water} is the volume concentrations of water. $g(\theta)$ is a function that accounts for the non-uniform evaporative flux along the drop interface. Experimental drop volume (triangles) and predictions by our model (lines) are plotted as a function of time in Fig. 2c. In addition to the values predicted by Eq. (1) (red line), we also plotted Popov's model (dashed line), the models that account for the effect of $x_{ethanol}$ (green dashed line) and the effect of RH on D_{12} (blue line).

4   Conclusions

We conclude that our revisited drop evaporation model with the empirical correlation, accounting for the change in ethanol concentration, the effect of RH on D_{12} and the water intake in time, remarkably captures both qualitatively and quantitatively the dynamic evolution of drop volume. Further insights into the microscopic and dynamical origin of coupled water and ethanol vapour transfer will be sought in future work. Nonetheless, we stress the importance of modelling the influence of humidity present in the atmosphere for the accurate prediction of organic solvent evaporation [4].

Acknowledgements The authors acknowledge the support received from the ThermaSMART network (EU Horizon 2020 research and innovation programme under the Marie Sklodowska–Curie grant agreement No. 778104).

References

1. P. Innocenzi, L. Malfatti, S. Costacurta, T. Kidchob, M. Piccinini, A. Marcellil, Evaporation of ethanol and ethanol-water mixtures studied by time-resolved infrared spectroscopy. J. Phys. Chem. A **112**, 6512–6516 (2008)
2. Y. Fukatani, D. Orejon, Y. Kita, Y. Takata, J. Kim, K. Sefiane, Effect of ambient temperature and relative humidity on interfacial temperature during early stages of drop evaporation. Phys. Rev. E **93**, 043103 (2016)
3. J.M. Stauber, S.K. Wilson, B.R. Duffy, K. Sefiane, On the lifetimes of evaporating droplets with related initial and receding contact angles. Phys. Fluids **27**, 122101 (2015)
4. Y. Kita, Y. Okauchi, Y. Fukatani, D. Orejon, M. Kohno, Y. Takata, K. Sefiane, Quantifying vapor transfer into evaporating ethanol drops in a humid atmosphere. Phys. Chem. Chem. Phys. **20**, 19430–19440 (2018)

Significantly Enhanced Pool Boiling Heat Transfer on Multi-walled Carbon Nanotubes Self-assembly Surface

Lan Mao, Wenbin Zhou, Xuegong Hu, Yu He, and Rong Fu

1 Introduction (Details for Submitting Paper)

Traditionally, it's thought that pool boiling is the most efficient technique to solve the heat dissipation in many thermal management systems due to the huge latent heat of evaporation. Employing nanofluids to enhance pool boiling heat transfer has been widely investigated for the last decade. And among the various nanomaterials, carbon nanotubes (CNTs) have been referred to the most promising nanomaterials because of their superior thermal properties [1].

In general, it's considered that the pool boiling heat transfer enhancement with nanofluids is attributed to the modified microstructure and topography of the heater surface by nanoparticles depositing during the boiling process [1, 2]. However, the redundant deposition of nanoparticles during nucleate boiling process causes nucleation sites filling, and higher thermal resistance thus deteriorates the boiling heat transfer performance. So chemical and mechanical methods are widely used for depositing nanoparticles on the plate substrate. But those conventional methods such as chemical vapour deposition (CVD) require complicated manufacturing equipment and process. Excitingly, the nanoparticles self-assembly under nucleate boiling condition can address the issues, which involves a single-step, low temperature and low cost [3].

In this study, MWCNTs' coating on the flat copper substrate was fabricated by MWCNTs' self-assembly under nucleate pool boiling condition. And the MWCNTs' coating surface was studied for the enhancement of the saturated pool boiling heat

L. Mao · W. Zhou (✉) · X. Hu (✉) · Y. He · R. Fu
Institute of Engineering Thermophysics, Chinese Academy of Sciences, Beijing, China
e-mail: zhouwenbin@iet.cn

X. Hu
e-mail: xuegonghu@iet.cn

L. Mao · X. Hu · Y. He
University of Chinese Academy of Sciences, Beijing, China

© Springer Nature Singapore Pte Ltd. 2021
C. Wen and Y. Yan (eds.), *Advances in Heat Transfer and Thermal Engineering* ,
https://doi.org/10.1007/978-981-33-4765-6_21

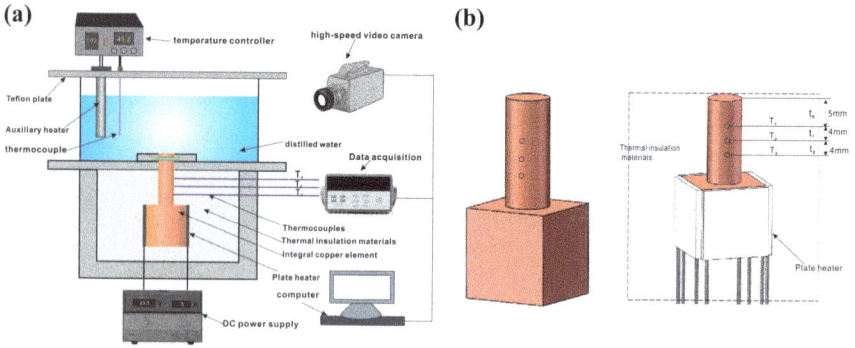

Fig. 1 Schematic diagram of **a** experimental apparatus, **b** the integral copper heating element

transfer using distilled water as working fluid at atmospheric pressure. Furthermore, the bubble growth characteristics were visualized with a high-speed digital camera.

2 Methodology (Length and Layout)

2.1 Experimental Apparatus

As illustrated in Fig. 1a, the experimental apparatus mainly consists of a cylindrical glass test vessel, the integral copper heating element, DC power supply, measurement, and data acquisition device. The test surface with diameter of 10 mm is the top surface of the integral copper element, which is heated by plate heaters connected to the DC power supply (150 V/20 A). Three K-type thermocouples are embedded in three holes with diameter of 0.5 mm in the centreline of the cylinder, as illustrated in Fig. 1b. For the sufficient thermal insulation by ceramic fibre blanket, one-dimensional linear temperature distribution can be assumed in axial direction of the cylinder. The surface temperature and heat flux are calculated by measuring temperatures with three thermocouples according to Fourier's law.

2.2 Experimental Procedure

Before the experiment, distilled water was heated to saturated state for an hour via the auxiliary heater for degassing. Then, 0.5 ml MWCNTs' nanofluids with a concentration of 2 mg/ml were added in the distilled water. By altering the voltage and current, a heat flux of 90 W/cm^2 was applied to the test surface for 2 h boiling. Finally, the MWCNTs' coating surface was obtained, and the pool boiling experiment was followed. The working fluids were replaced with distilled water and heated for an hour for degassing. Afterwards, the heat flux was applied step-wise until the CHF occurred. Heat transfer data under each heat flux could be considered acceptable when its fluctuations are within 0.2 K, and the steady-state temperature datum was time-averaged for data plots. Bubble dynamics were also observed by the high-speed video camera at each heat flux. The temperature of working fluids was maintained saturated in the whole experiment.

3 Results

Figure 2a depicts scanning electron microscopy (SEM) image of MWCNTs' coating surface. It can be clearly seen that SWCNTs are randomly deposited and form nanoporous network structures on the flat copper substrate. As in Fig. 2b, the MWCNTs' coating surface shows an augmentation of heat transfer performance when compared with flat copper surface. The superheat at the onset of nucleate boiling (ONB), critical heat flux (CHF) and maximum heat transfer coefficient (HTC) of flat copper surface is 8.48 K, 98.14 W/cm^2 and 5.23 W/(cm^2 K), respectively. While they are 5.35 K, 182.96 W/cm^2 and 8.96 W/(cm^2 K) of MWCNTs' coating surface, which translate a decrease of 36.91% in superheat at the ONB, an enhancement of 86.43% is in CHF and 71.32% is in maximum HTC.

In general, bubble nucleation sites are activated; then, the bubbles begin to grow as the heat flux increases during pool boiling. With heat flux further increasing, bubbles coalesce to form mushrooming bubbles, and the depletion of liquid under mushrooming bubbles leads to local hot/dry spots. Then, CHF occurs with many local hot/dry spots overheating, which results in a dramatic decrease in the heat transfer and physical breakdown of the heating surface. This should be avoided in thermal management systems. During pool boiling, the MWCNTs' deposited coating with high thermal conductivity can effectively delay the formation of hot/dry spots

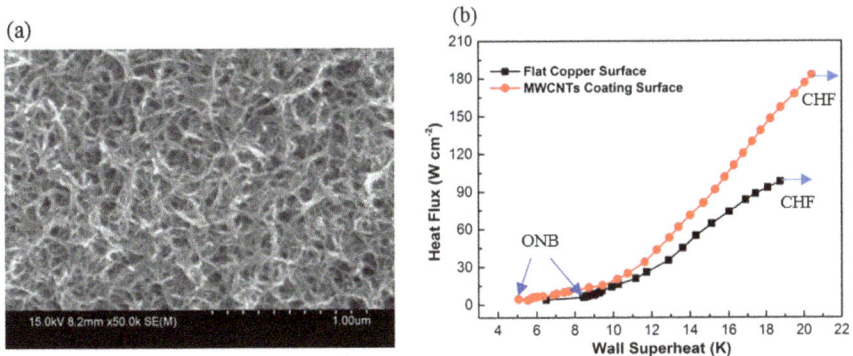

Fig. 2 **a** SEM image of MWCNTs' coating surface. **b** Boiling curves of flat copper surface and MWCNTs' coating surface

and thus improving CHF and heat transfer performance. What's more, nanoporous network structures of MWCNTs' coating result in capillary wicking, ensuring good liquid rewetting by continuous liquid supply at high heat flux conditions, and this good liquid rewetting delays local hot/dry spots formation and enhances the CHF [4].

In addition, the nanoporous network structures of MWCNTs' coating act as the original nucleation sites for initial formation of vapour bubbles [5]. Then, the bubble nucleation sites of MWCNTs' coating surface are highly increased than the original copper substrate. And the increased bubble nucleation sites cause more and quicker heat dissipation through liquid–vapor phase change, resulting in higher HTC of MWCNTs' coating surface. This is verified by the bubble dynamic visualization research. It's also observed decreased bubble departure diameter when compared with the flat copper surface during visualization research. Overall, the pool boiling heat transfer enhancement of MWCNTs' coating surface is attributed to improved surface thermal conductivity of MWCNTs' coating and MWCNTs' deposited nanoporous network structures.

4 Conclusions

MWCNTs' coating on flat copper substrate was fabricated by MWCNTs' self-assembly under nucleate pool boiling condition, and the pool boiling heat transfer performance was investigated in this work. MWCNTs' coating surface shows a better pool boiling heat transfer performance than the flat copper surface. The enhancement of CHF and maximum HTC are 86.43% and 71.32%, respectively. Furthermore, MWCNTs' coating surface exhibits a 36.91% reduced superheat at ONB when compared with flat copper surface. The heat transfer enhancement could be contributed to the delay in local hot/dry spots formation via high thermal conductivity of MWCNTs' coating and capillary wicking and the increased bubble nucleation sites via MWCNTs' deposited nanoporous network structures. The MWCNTs' coating surface is really an attractive candidate to be employed in thermal management systems where high-power density heating devices are needed.

Acknowledgements The authors are profoundly grateful to the financial supports of the National Natural Science Foundation of China (Grant No. 51706225).

References

1. M. Xing, J. Yu, R. Wang, Effects of surface modification on the pool boiling heat transfer of MWNTs/water nanofluids. Int. J. Heat Mass Transf. **103**, 914–919 (2016)
2. A. Amiri, M. Shanbedi, H. Amiri, S.Z. Heris, S.N. Kazi, B.T. Chew, H. Eshghi, Pool boiling heat transfer of CNT/water nanofluids. Appl. Therm. Eng. **71**(1), 450–459 (2014)

3. H.S. Ahn, J.W. Jang, M. Seol, J.M. Kim, D.J. Yun, C. Park, H. Kim, D.H. Youn, J.Y. Kim, G. Park, S.C. Park, J.M. Kim, D.I. Yu, K. Yong, M.H. Kim, J.S. Lee, Self-assembled foam-like graphene networks formed through nucleate boiling. Sci. Rep. **3**, 1396 (2013)
4. G. Udaya Kumar, K. Soni, S. Suresh, K. Ghosh, M.R. Thansekhar, P. Dinesh Babu, Modified surfaces using seamless graphene/carbon nanotubes based nanostructures for enhancing pool boiling heat transfer. Exp. Therm. Fluid Sci. **96**, 493–506 (2018)
5. S. Lee, G.H. Seo, S. Lee, U. Jeong, S.J. Lee, S.J. Kim, W. Choi, Layer-by-layer carbon nanotube coatings for enhanced pool boiling heat transfer on metal surfaces. Carbon **107**, 607–618 (2016)

Heat Transfer Analysis of a Tubular Solar Still Considering the Non-uniform Temperature Distribution on the Condensing Shell

Tiantong Yan, Guo Xie, and Licheng Sun

1 Introduction

Solar distillation is one of the most promising solutions for freshwater shortage in arid areas with abundant solar energy. Several types of solar distillers have been designed [1]. Tubular solar still (TSS) is a typical one characterized by the compactness and ease of fabrication. Various techniques for enhancing performance of the TSS have been proposed and experimentally testified [2], such as vacuum operation or integrations with different types of solar concentrator. Modelling and design work has also been carried out to improve the productivity of the TSS. Ahsan and Fukuhara [3] developed a model to predict freshwater productivity considering the properties of the humid air for the first time. Xie et al. [4] focused on the effect of vacuum pressure on the performance of a TSS and built a model based on the mass diffusion of water vapour rather than natural convection.

The shell temperature in the models mentioned above was considered as uniform, which is not exactly true, as observation shows in many experiments [4]. When vapour is condensed on the shell, the condensation mainly occurs on the upper part of the shell [5], and the other part plays insignificant role. However, this non-uniform heat transfer process was rarely considered in previous researches.

In the present work, non-uniform temperature distribution and its effect on the heat transfer of TSS were discussed in detail. Experiments were firstly carried out to investigate the temperature distribution along the outer shell of TSS. A modified model is then proposed considering the non-uniform temperature distribution on the shell. The model was verified against experimental data and was compared with the previous one [4] in respects of freshwater yield prediction.

T. Yan · G. Xie (✉) · L. Sun
State Key Laboratory of Hydraulics and Mountain River Engineering, College of Water Resource and Hydropower, Sichuan University, Chengdu, China
e-mail: 2008xieguo@scu.edu.cn

© Springer Nature Singapore Pte Ltd. 2021
C. Wen and Y. Yan (eds.), *Advances in Heat Transfer and Thermal Engineering* ,
https://doi.org/10.1007/978-981-33-4765-6_22

2 Methodology

A transient model is built based on the heat transportation in a TSS shown in Fig. 1. The heat balance equations for the water and the shell can be formulized as follows:

$$Q_h = Q_{ew} + Q_{rw} + Q_{cw} + \rho_w \cdot c_{p\_w} \cdot V_w \cdot \frac{dt_w}{d\tau} \tag{1}$$

$$Q_{ew} + Q_{cw} + Q_{rw} = Q_{cs} + Q_{rs} + Q_{fin} + \frac{m_s}{2} \cdot c_{p\_s} \cdot \frac{dt_s}{d\tau} \tag{2}$$

where Q_h is the heating power; Q_{ew}, Q_{rw} and Q_{cw} are the heat transfer rates due to natural convection, evaporation and radiation, respectively; Q_{cs}, Q_{rs} are the convective and radiative heat transfer rates between the upper half shell and the ambient; t_w and t_s are the temperature of the water and the shell, respectively; m_s is the mass of the shell. During the distillation process, heat released from the water in the trough is mainly absorbed by the upper part of the shell. Therefore, the lower half part is seen as a straight fin attached to the upper half, and Q_{fin} is the heat transfer rate from it to the ambient. The modified model in this study is based on that proposed in [4].

Indoor experiments were carried out at different operating pressures, where the water temperature was fixed at 50, 60, 70 and 80 °C, respectively. The operating pressure ranged from 20 to 101 kPa, and it was maintained by a vacuum pump in operation. The freshwater yield was weighed every half an hour. The temperatures of the shell were measured by five resistance thermometers circumferentially attached on the outer surface at 30 mm intervals.

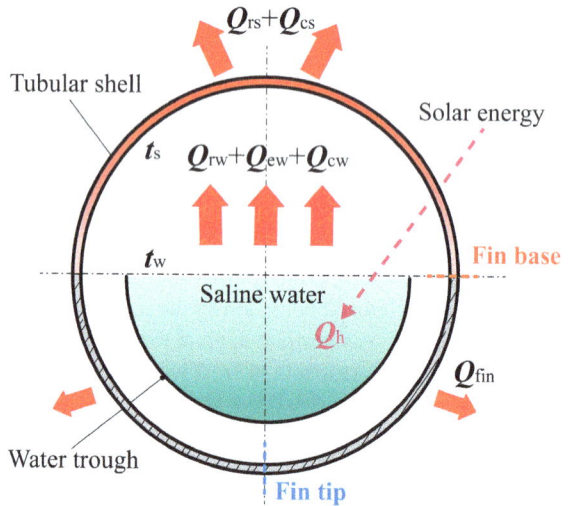

Fig. 1 Schematic diagram of heat transfer in TSS

3   Results

Figure 2 shows the dimensionless excess temperature distribution along the tubular shell at $t_w = 70\,°C$. The dimensionless excess temperature at a specified position on the shell was defined by

$$\Theta = \frac{t - t_a}{t_{s1} - t_a} \tag{3}$$

where t_a is ambient temperature and t_{s1} is the temperature of point 1 in Fig. 2. It is obvious that the surface temperature decreases along the shell from top to bottom. Θ at positions 2, 3 and 4 are greater than 0.9, while is relatively low at the position 5. This agrees with the simulation results in [5]. Higher temperature at upper position indicates that more vapour was condensed on the top region of the shell. The influence of operating pressure is also shown in Fig. 2. With the decrease in operating pressure, the temperatures at positions of 2–5 increase and the temperature distribution tends to be more uniform. A likely reason is that more vapour is generated at vacuum, and thus, more area of the shell is activated for condensation. Equation (4) gives the fitting correlation of Θ obtained based on the experimental data.

$$\Theta = \frac{0.006p}{101}(t_w - 100) \cdot \cos^2 \alpha$$
$$- \frac{0.002p}{101}(t_w - 100) \cdot \cos \alpha + 1, \quad 0 \le \alpha \le \frac{\pi}{2} \tag{4}$$

where p is the operating pressure, t_w is the water temperature, and α is the central angle of a specified position on the shell.

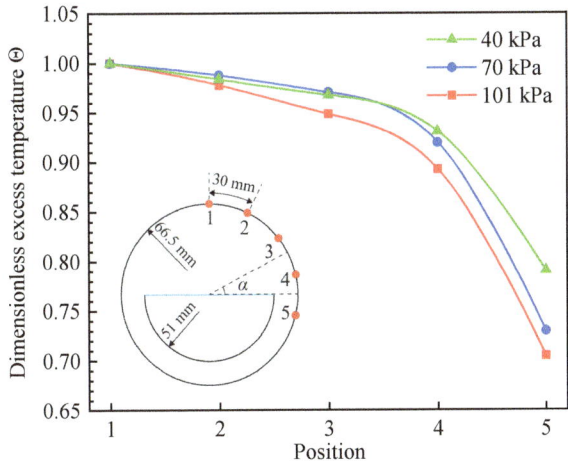

Fig. 2 Dimensionless excess temperature distribution along the tubular shell at $t_w = 70\,°C$

The modified model is validated against experiment data at constant heating power of 100 W. The relative deviations of the predicted freshwater yield rate at 101, 60 and 40 kPa are 13%, 1% and 0.8%, respectively. t_w is well predicted by the present model, while the predicted t_{s1} is 10% higher than the measured values. The temperature at the bottom of the shell is also predicted with a deviation of 9.3%. The present model was proved to be more accurate than the previous one in [4]. In the present model, the lower half of the shell is considered as a fin attached to the upper half, and water vapour is considered to condense only on the upper half part of the shell. This assumption is closer to the actual working condition than the previous one in [4]. The prediction for productivity is hence improved due to more reasonable prediction of the condensing temperature.

4 Conclusions

An indoor experiment at different operating pressures was carried out to investigate the temperature distribution on the tubular shell of TSS. A modified model that considers non-uniform temperature distribution of the shell was then developed. It has a better accuracy than the previous model, and the maximum deviations in predicting freshwater yield rate are 13%, 1% and 0.8% at 101, 60 and 40 kPa, respectively.

Acknowledgements The authors are profoundly grateful to the financial supports of the National Natural Science Foundation of China (Grant No.51606130).

References

1. G. Xiao, X. Wang, M. Ni, F. Wang, W. Zhu, Z. Luo, K. Cen, A review on solar stills for brine desalination. Appl. Energy **103**, 642–652 (2013)
2. S.W. Sharshir, Y.M. Ellakany, A.M. Algazzar, A.H. Elsheikh, M.R. Elkadeem, E.M.A. Edreis, A.S. Waly, R. Sathyamurthy, H. Panchal, M.S. Elashry, A mini review of techniques used to improve the tubular solar still performance for solar water desalination. Process Saf. Environ. Prot. **124**, 204–212 (2019)
3. A. Ahsan, T. Fukuhara, Mass and heat transfer model of tubular solar still. Sol. Energy **84**, 1147–1156 (2010)
4. G. Xie, L. Sun, T. Yan, J. Tang, J. Bao, M. Du, Model development and experimental verification for tubular solar still operating under vacuum condition. Energy **157**, 115–130 (2018)
5. N. Rahbar, J.A. Esfahani, E. Fotouhi-Bafghi, Estimation of convective heat transfer coefficient and water-productivity in a tubular solar still—CFD simulation and theoretical analysis. Sol. Energy **113**, 313–323 (2015)

Droplet Deposition Pattern Affected by Heating Directions

Zeyu Liu, Yuying Yan, Xinyong Chen, Li Wang, and Xin Wang

1 Introduction

'Coffee ring' effect is a stain pattern deposited by non-volatile solute of liquid after evaporation. However, due to unexpected deposition patterns especially happened in industry, this phenomenon is not welcomed in many manufacturing processes, such as inkjet printing, DNA microarrays and nanotechnology. Therefore, the drying of suspension droplets has not only encouraged many researchers to study the mechanism of coffee ring effect, but also to explore effective ways to actively control it.

Many scholars make great contributions to controlling deposition patterns in different fields. In 2016, Kim et al. [1] described key parameters to achieve uniform particles coating from binary solutions. The experimental results showed that Marangoni effect driven by surfactant makes significant effect on uniform distribution on the final deposit. In 2010, Bhardwaj et al. [2] investigated that the pH values of solution influence the dried deposit pattern, which can be ring-like or more uniform. In this research, they found a critical point, pH value is 5.8, which affects deposition patterns by influencing Van der Waals interaction. In addition, Brutin [3] reported that depending on the concentration, two different patterns are observed: an O-ring pattern and a continuous nanoparticle flower pattern.

Z. Liu · Y. Yan (✉) · L. Wang · X. Wang
Fluids and Thermal Engineering Research Group, Faculty of Engineering, University of Nottingham, Nottingham, UK
e-mail: yuying.yan@nottingham.ac.uk

X. Chen
Faculty of Science, School of Pharmacy, University of Nottingham, Nottingham, UK

Y. Yan
Fluids and Thermal Engineering Research Centre, University of Nottingham Ningbo, Ningbo, China

© Springer Nature Singapore Pte Ltd. 2021
C. Wen and Y. Yan (eds.), *Advances in Heat Transfer and Thermal Engineering* ,
https://doi.org/10.1007/978-981-33-4765-6_23

Droplet evaporation can be typically divided into two stages, constant contact line (CCL) mode and constant contact angle (CCA) mode. The interests in droplet deposition pattern research were inspired after O-ring pattern was observed and explained by Deegan [4]. If heated uniformly from substrate, the evaporation flux at the edge is evidently higher than other positions, forcing particles inside to transport from centre towards to periphery for compensating the solvent loss. Eventually, evaporation flow carries the central particles to the edge, resulting in most solute deposited at the edge, forming a ring-like pattern. Ring-like pattern is normally unexpected, and scholars have made much effort on developing methods for suppressing the formation. And also, many researchers found that Marangoni flow and capillary flow are two significant factors influencing deposition pattern and tried to subject formation through balancing them both.

Most strategies use chemicals to control droplet deposition patterns. Our aim is to explore an effective method to control droplet deposition patterns by changing heating directions without chemicals. We investigated deposition patterns on both hydrophilic and hydrophobic materials by heating from apex and substrate. 500 mW laser light is applied on test section to supply heat from apex side. The whole process of droplet evaporation is recorded. Physical mechanisms of such phenomenon in different conditions are also analysed.

2 Experimental Set-up

To apply heat flux from different directions, a 500 mW laser system is applied on the apex of saline with the concentration of 2 wt%. The volume of saline is 2 μL, which is controlled by pipette for each test. A phantom v640 CCD camera is used to record the whole process of droplet evaporation. LEDs array illuminators are used for enough exposure. In the experiment, the substrate surface is coated by hydrophobic and hydrophilic, respectively. In addition, ambient conditions were controlled by control unit.

3 Results

In the experiment, evaporation process has been investigated on both hydrophilic and hydrophilic surfaces. The whole process can be divided into two stages, constant contact line (CCL) mode and constant contact angle (CCA) mode. In CCL stage, the contact line is pinned, and contact angle decreases during drying process. CCA stage comes when contact angle reaches the receding contact angle. In addition, progressions are standardized to its relative time, T/T_0, where T_0 is for complete evaporation. In the evaporation process, CCL mode occupies most time, of which T/T_0 is at around 0.8 as shown in Fig. 1a, c.

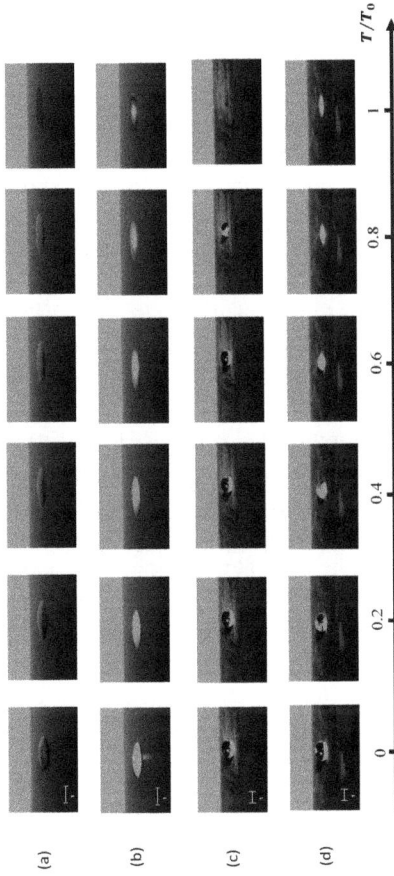

Fig. 1 Modes of saline droplet evaporation from hot plate heating to localized heating on both hydrophilic and hydrophobic surfaces. **a** 2 μL saline droplet evaporation process heated from substrate on hydrophilic surface. **b** 2 μL saline droplet evaporation process heated by laser system on hydrophilic surface. **c** 2 μL saline droplet evaporation process heated from substrate on hydrophobic surface. **d** 2 μL saline droplet evaporation process heated by laser system on hydrophilic surface

In substrate heating, saline deposition patterns exhibited similar phenomena, ring-like structure along contact line when droplet evaporates completely. However, when it comes to heat flux supplied from apex, different wettability shows different phenomena. For hydrophilic surface, most saline particles assembled from contact line to central area to form dot-like deposition pattern, but few saline particles are deposited between the edges and the central region. In terms of hydrophobic surface, particles on the boundary also tend to gather at central area. However, many particles deposit between boundary and centre, which makes deposition pattern more uniform.

When heat flux is supplied from substrate, droplet apex is the coolest position, while surface temperature at the boundary is the highest, leading to more solvent losing at the droplet boundary than that at central area. CCL stage occupies most of the evaporating time shown in Fig. 1, of which T/T_0 is at around 0.8. Hence, solvent must flow from droplet centre towards boundary in order to compensate losing solute at the edge. When evaporating completely, most solvent has been transported and deposited at the pinned contact line, forming a ring-like deposition pattern. In addition, rush hour behaviour also makes a great contribution based on mass balance. The evaporation flow inside droplet is driven by the evaporation from the droplet surface, bringing out that evaporation flow rate inside is approximately constant over time. Since the boundary needs to be replenished from the centre, a continuous volume flow towards contact line is generated inside the drop. However, the drop height is decreasing during evaporation, which can be characterized by a contact angle. Therefore, the same amount of solvent has to be squeezed through an area which is vanishing, inducing a diverging radial velocity. Hence, rush hour behaviour is also regarded as having greatly contributed to coffee ring effect.

When heated from apex shown in Fig. 2, temperature at the edge became the coolest positions, while apex is the highest point, resulting in more solvent losing at the apex than other positions. More solvent must flow from periphery to apex to compensate losing solute at apex area. Therefore, particles are hanging around apex. With the decrease of droplet, hanging particles drop down, and then most particles gathered at central area which form a dot-like formation. In this way, saline droplet flow direction is changed through altering heating directions. Moreover, surface tension also needs to be considered when it comes to hydrophobic surface. This is because surface tension is the tendency of fluid surfaces to shrink into the minimum

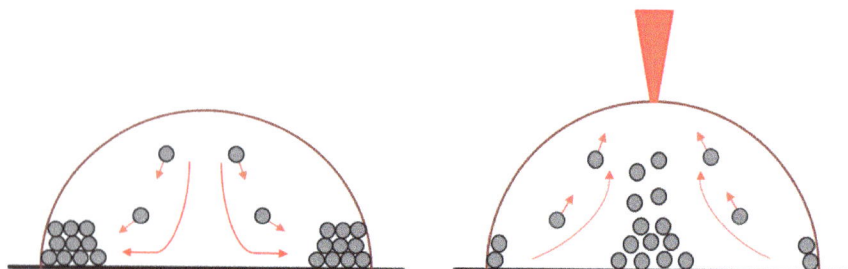

Fig. 2 Mechanism of coffee ring effect and reversing internal flow by localized heating

surface area [5]. With the same volume, the area of droplet exposed to the air on the hydrophobic surface is smaller than that on the hydrophilic surface. This means that at the gas–liquid interface, the attraction between molecules on hydrophobic surface is much greater than that on hydrophilic one. Meanwhile, surface tension gradients also affect the motion of liquid–gas interface, which is sometimes referred to as the Marangoni effect. In the field of heat transfer, the surface tension gradient mostly results from a temperature gradient. The surface tension of liquids decreases as temperature increases. When laser system is on, fluids naturally flow from apex point to the boundary, which is a tendency to impede the internal flow of saline droplet during evaporation.

To some extent, on the hydrophobic surface, it is very likely that the surface tension overcomes the internal flow, so that some of the saline particles are finally dispersed between the edge and the centre, making the deposition pattern more uniform than that of hydrophilic surface.

4   Conclusion

Experiments of saline droplet evaporating on both hydrophobic and hydrophilic surfaces have been conducted. All experiments are started from CCL mode and then switched to CCA mode. Results show that the CCL mode occupies most evaporation process. It is found that heating from the substrate, both hydrophobic and hydrophilic surfaces, exhibits ring-like deposition patterns along contact lines. In addition, when it comes to hydrophobic surface, it is feasible to change deposition pattern from ring-like to dot-like structure by reversing inner flow. However, Marangoni effect also needs to be considered when it comes to hydrophobic surface. To some extent, the surface tension can balance the internal flow induced by laser system from apex. This has shown not only the potential for overcoming internal flow, but also the opportunity of making deposition pattern more uniform as well.

Acknowledgements The authors would like to acknowledge the financial support by the scholarship of China Scholarship Council (CSC) for Mr. Zeyu Liu. And this work is also supported by H2020-MSCA-RISE-778104-ThermaSMART.

References

1. H. Kim et al., Controlled uniform coating from the interplay of Marangoni flows and surface-adsorbed macromolecules. Phys. Rev. Lett. **116**(12), 124501 (2016)
2. R. Bhardwaj et al., Self-assembly of colloidal particles from evaporating droplets: role of DLVO interactions and proposition of a phase diagram. Langmuir **26**(11), 7833–7842 (2010)
3. D. Brutin, Influence of relative humidity and nano-particle concentration on pattern formation and evaporation rate of pinned drying drops of nanofluids. Colloids Surf. A **429**, 112–120 (2013)
4. R.D. Deegan, Pattern formation in drying drops. Phys. Rev. E **61**(1), 475 (2000)

5. M.F.D.A. Silva, *Impacto de diferentes métodos de esterilização nas propriedades físico-químicas de polímeros e na degradação da bacitracina de zinco utilizados em bandagens adesivas* (2016)

Effect of Vapour Pressure on Power Output of a Leidenfrost Heat Engine

Prashant Agrawal, Gary G. Wells, Rodrigo Ledesma-Aguilar, Glen McHale, Anthony Buchoux, Khellil Sefiane, Adam Stokes, Anthony J. Walton, and Jonathan G. Terry

1 Introduction

In the Leidenfrost effect, when a liquid droplet comes in contact with a surface heated to temperatures significantly above the liquid's boiling point, a vapour layer forms instantly due to which the droplet levitates on the surface [1]. The vapour layer also acts as a thermal insulator that reduces the evaporation rate of the liquid. The same effect is observed for sublimating solids, such as dry ice [2]. Due to the reduced friction, the levitating liquids/solids are highly mobile and can self-propel [2–4] on asymmetrically textured surfaces such as herringbone patterns [5] or ratchet grooves [4, 6]. The substrate asymmetry rectifies the escaping vapour in a preferential direction, which produces a viscous drag on the levitating component, propelling it in a specific direction. This low friction propulsion can also be used to rotate liquid droplets and sublimating dry ice discs on asymmetrically textured surfaces [7] or asymmetrically arranged levitating components [8]. We have previously shown that on turbine-like patterned substrates, these rotating droplets can continuously transfer torque to additional surface tension coupled non-volatile solids [9]. This continuous thermal to mechanical energy conversion establishes the concept of a closed cycle Leidenfrost heat engine.

P. Agrawal (✉) · G. G. Wells · R. Ledesma-Aguilar · G. McHale
Smart Materials and Surfaces Laboratory, Faculty of Engineering and Environment, Northumbria University, Newcastle upon Tyne, UK
e-mail: prashant.agrawal@northumbria.ac.uk

A. Buchoux · K. Sefiane
School of Engineering, Institute for Multiscale Thermofluids, The University of Edinburgh, Edinburgh, UK

A. Stokes · A. J. Walton · J. G. Terry
School of Engineering, Institute for Integrated Micro and Nano Systems, The University of Edinburgh, Edinburgh, UK

© Springer Nature Singapore Pte Ltd. 2021
C. Wen and Y. Yan (eds.), *Advances in Heat Transfer and Thermal Engineering* ,
https://doi.org/10.1007/978-981-33-4765-6_24

In this Leidenfrost engine, the power output depends on the pressure in the vapour layer that supports the weight of the levitating rotors [3, 9]. Above the Leidenfrost temperature, these rotors demonstrate an invariance in the power output with temperature, thereby hampering its efficiency at increased temperatures [9]. In this work, we demonstrate a method to control this output power by mechanically altering the pressure in the vapour layer. This is achieved by supporting the weight of the evaporating liquid and a coupled solid plate using a bearing assembly and mechanically altering the gap between the solid plate and the heated substrate. Although we introduce a bearing friction by removing levitation, we observe an increase in the terminal speed of rotation, which gives an indication of the increase in power output.

By controlling its power output and understanding the dynamics of rotation of these Leidenfrost rotors, these Leidenfrost engines open possibilities of developing heat engines for power generation at millimetre and sub-millimetre scales. At microscales, the virtually frictionless vapour bearing can be advantageous for energy harvesting [10], while at macroscales, operation in microgravity conditions, such as for space and planetary exploration, can be a potential application. These engines can provide transportation ease by using alternate naturally occurring substances such as ices of H_2O, CO_2 and CH_4 in extreme temperature and pressure conditions for thermal energy harvesting [11, 12].

2 Methodology

The experimental set-up, shown in Fig. 1, comprises an aluminium solid component connected to a Z-axis motion stage through a bearing assembly, to allow for free rotation above the turbine-like substrate. The solid component comprises a shaft and a plate which is coupled via surface tension to an evaporating liquid. The substrate is heated to temperatures above 280 °C to ensure a thin-film boiling regime. Liquid is continuously supplied to the assembly via a syringe pump. As the assembly is lowered onto the substrate, the liquid comes in contact with the hot substrate and undergoes thin-film boiling. The escaping vapour is rectified which is observed as rotation of the solid plate. The angular speed of the plate is tracked over time for

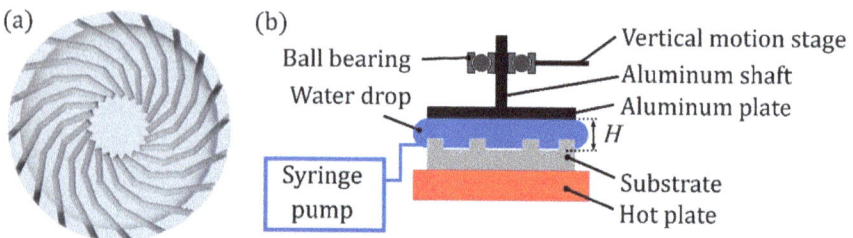

Fig. 1 a Depiction of the turbine-like substrate. The groove depth is 100 μm. b Experimental set-up

different gaps (H) to ascertain the final constant angular speed (termed as the terminal angular speed).

3 Results

Figure 2a shows the variation of angular speed of the plate with time as it accelerates from rest to an eventual constant speed. The rotation of the coupled liquid–solid components is driven by the torque from the vapour flow over the substrate and is resisted by an inertial resistance due to liquid deformation over the substrate [5, 9]. The equation of motion of this coupled liquid–solid rotor can be written as follows [9]:

$$I\dot{\omega} = \Gamma_v - c_i\omega^2, \tag{1}$$

where ω is the angular speed of the rotor, I is the inertia of the rotor, Γ_v is the driving torque from the vapour flow, and c_i is the coefficient of inertial resistance due to liquid deformation. The solution to Eq. (1) is obtained as $\omega = \omega_t\tanh(t/\tau)$, where ω_t is the terminal velocity as $t \to \infty$ and τ is the relaxation time which is a measure of the rotor acceleration. By fitting the Eq. (1) on the experimental data, the starting torque on the rotor can be calculated by $\Gamma_v = I\omega_t/\tau$. However, as the rotation speed increases, the centrifugal force on the liquid forces droplets to eject from the gap. This droplet ejection releases some built-up pressure in the liquid, which leads to a momentary drop in the speed, as indicated in Fig. 2a. As a result, due to this random droplet ejection event, the rotor configuration changes frequently during acceleration that does not provide a reliable fit for torque estimation. Therefore, we rely only on the terminal angular speed as a measure of power output from rotation.

Figure 2b shows an increase in the terminal angular speed with decrease in the gap between the plate and the substrate above the Leidenfrost temperature. For a

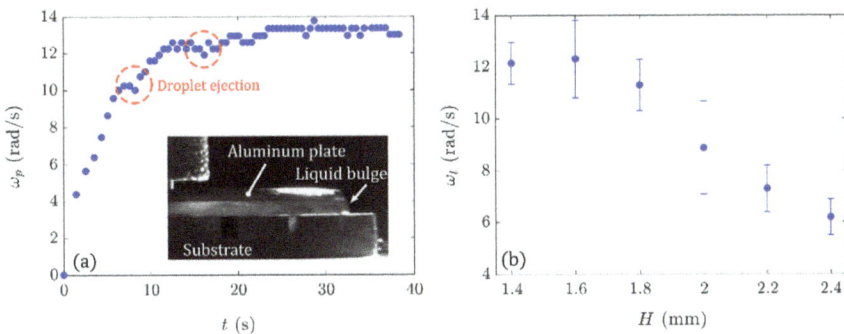

Fig. 2 **a** Plate rotation speed over time. **b** Comparison of terminal speed for different gap thickness H

given volume of liquid between the plate and the substrate, a change in the gap H alters the liquid distribution over the substrate which can be observed from the extent by which the liquid bulges out of the assembly, as depicted in Fig. 1b and seen in the inset of Fig. 2a. For instance, for a decreasing H, the curvature of the liquid bulge will increase. This liquid bulge indicates an increase in the pressure in the liquid. Considering that the radius of the substrate is an order of magnitude higher than H, the capillary pressure in the liquid can be approximated as $P_c \approx 2\gamma/H$. Therefore, as H decreases, the pressure in the liquid increases, which corresponds to an increase in the vapour layer pressure. An increase in the vapour layer pressure results in a higher torque and, by extension, terminal speed [3], which agrees with our observations in Fig. 2b. Due to the high pressure at the centre of this Leidenfrost pool, a vapour bubble develops [13], which is aided by the centrifugal force due to rotation. As a result of this bubble, the liquid distribution over the substrate assumes the shape of a ring. This bubble formation is suppressed by the hydrostatic pressure from the liquid pool. Therefore, at lower values of H, and at higher temperatures, the bubble formation is enhanced and the radius of the liquid ring increases [9]. As a result, a smaller area of the liquid levitates over the substrate and therefore reduces the generated torque. This reduction in the area covered by the liquid ring is observed as a saturation in the terminal angular speed, despite the increased pressure due to decreasing H. Nevertheless, despite the added friction from the bearings, the speeds obtained here are higher than that obtained in the case of pure levitation (~ 10 rad/s) [9].

4   Conclusions

In the present work, we have demonstrated a method to control the power output (via the angular speed) of a conceptual Leidenfrost heat engine. Although we remove the low friction aspect of levitation by supporting the rotors with a friction-included bearing, we observe an increased angular speed of rotation by decreasing the gap between the substrate and the plate. Therefore, this configuration of a Leidenfrost rotor allows us to control and increase the output power beyond that obtained in the case of levitation only. These design principles can be extrapolated to alternative liquid and solids to develop engines utilizing thin-film boiling for eliminating friction. Potential applications of such engines can be in extreme environments with naturally occurring low pressures and high-temperature differences, such as in space or for planetary exploration. Additionally, their virtually frictionless operation can be used for developing microscale engines for thermal energy harvesting.

Acknowledgements The authors acknowledge funding from EPSRC (EP/P005896/1 and EP/P005705/1) for supporting this work.

References

1. J.G. Leidenfrost, On the fixation of water in diverse fire (trans: C. Wares). Int. J. Heat Mass Transf. **9**, 1153–1166 (1966)
2. T. Baier, G. Dupeux, S. Herbert, S. Hardt, D. Quéré, Propulsion mechanisms for Leidenfrost solids on ratchets. Phys. Rev. E Stat. Nonlinear Soft Matter Phys. **87**, 3–6 (2013)
3. G. Lagubeau, M. Le-Merrer, C. Clanet, D. Quéré, Leidenfrost on a ratchet. Nat. Phys. **7**, 395–398 (2011)
4. H. Linke, B.J. Alemán, L.D. Melling, M.J. Taormina, M.J. Francis, C.C. Dow-Hygelund, V. Narayanan, R.P. Taylor, A. Stout, Self-propelled Leidenfrost droplets. Phys. Rev. Lett. **96**, 2–5 (2006)
5. D. Soto, G. Lagubeau, C. Clanet, D. Quere, Surfing on a herringbone. Phys. Rev. Fluids 013902, 2–3 (2016)
6. T.R. Cousins, R.E. Goldstein, J.W. Jaworski, A.I. Pesci, A ratchet trap for Leidenfrost drops. J. Fluid Mech. **696**, 215–227 (2012)
7. G. Wells, R. Ledesma-Aguilar, G. McHale, K.A. Sefiane, A sublimation heat engine. Nat. Commun. **6**, 6390 (2015)
8. G. Dupeux, T. Baier, V. Bacot, S. Hardt, C. Clanet, D. Quéré, Self-propelling uneven Leidenfrost solids. Phys. Fluids **25**, 1–7 (2013)
9. P. Agrawal, G.G. Wells, R. Ledesma-Aguilar, G. McHale, A. Buchoux, A. Stokes, K.A. Sefiane, Leidenfrost heat engine: sustained rotation of levitating rotors on turbine-inspired substrates. Appl. Energy **240**, 399–408 (2019)
10. M. McCarthy, C.M. Waits, R. Ghodssi, Dynamic friction and wear in a planar-contact encapsulated microball bearing using an integrated microturbine. J. Microelectromech. Syst. **18**, 263–273 (2009)
11. R.P. Mueller, L. Sibille, J. Mantovani, G.B. Sanders, C.A. Jones, Opportunities and strategies for testing and infusion of ISRU in the evolvable mars campaign, in *AIAA SPACE 2015 Conference and Exposition* (2015)
12. B. Palaszewski, Solar system exploration augmented by in-situ resource utilization: human planetary base issues for mercury and saturn, in *10th Symposium on Space Resource Utilization* (2017)
13. A.L. Biance, C. Clanet, D. Quéré, Leidenfrost drops. Phys. Fluids **15**, 1632–1637 (2003)

Three-Dimensional Computer Simulation of Heat Flux-Controlled Pool Boiling of Water-Based Nanofluids by the Coupled Map Lattice Method

Asheesh Kumar and Partha Ghoshdastidar

1 Introduction

The paper presents a three-dimensional coupled map lattice (CML) simulation of heat flux-controlled atmospheric saturated pool boiling of copper–water, silver–water and zirconia–water nanofluids on a 0.2-mm-thick horizontal stainless steel flat plate. The size of the plate is 4 mm × 4 mm. The pool height is 0.7 mm. The experimental study of Kathiravan et al. [1] revealed that copper nanoparticles got deposited on the heater surface during boiling, and hence, there was a decrease in cavity diameters. Due to the decrease in the cavity diameter, the wall superheat increased, and therefore, the critical heat flux (CHF) increased as compared to pure water. The same trend is seen for the case of boiling of zirconia–water [2]. On the contrary, Kathiravan et al. [3] experimentally demonstrated that silver nanoparticles showed no such deposition on the heater surface; rather there was oxidation of the surface. Hence, the cavity diameters increased resulting in decrease in wall superheat, and hence, decrease in CHF as compared to pure water. The objective of the present study is to reproduce the aforementioned trends of nanofluids boiling by a computer simulation based on a 3D CML model [4].

2 Methodology

The coupled map lattice (CML) method is used to model spatiotemporal chaotic phenomena such as boiling [5]. A CML is a dynamical system with discrete time, discrete space and continuous states. In the present CML model, it is assumed that the

A. Kumar · P. Ghoshdastidar (✉)

Department of Mechanical Engineering, Indian Institute of Technology Kanpur, Kanpur, U.P. 208016, India

e-mail: psg@iitk.ac.in

© Springer Nature Singapore Pte Ltd. 2021

C. Wen and Y. Yan (eds.), *Advances in Heat Transfer and Thermal Engineering* ,

https://doi.org/10.1007/978-981-33-4765-6_25

boiling is governed by: (a) nucleation from cavities on a heated surface; (b) thermal conduction; (c) bubble rising motion and associated convection and (d) phase change in the bulk of the fluid. Thus, there are three maps here, namely *thermal conduction map*, *convection due to bubble rising motion map* and *phase change in the bulk map*. The thermal conduction in the heater plate is also taken into account. The bubble stirring action is modelled by increasing the fluid thermal diffusivity by an enhancement factor. The basic advantage of using CML to model boiling is that individual bubbles are not tracked, yet the effects of bubbles in the flow are reflected qualitatively in the final solution.

The computational domain has a 3D grid or lattice with 12 grid points in the liquid pool, five grid points in the heater plate in the y-direction and 18 grid points each in x- and z-directions. The aforementioned number of grid points is arrived at after conducting a grid independence test based on the pool boiling curve. At the centre of each sub-lattice or control volume, there is a grid point. The field variable is temperature, T, and the flag function, $F_{i,j,k} = 0$ when a sub-lattice is completely filled with liquid and $F_{i,j,k} = 1$ when a sub-lattice is full of vapour. The state of a sub-lattice can be either 0 or 1. The nucleation is formulated as follows. If $F_{i,j,k} = 0$ and $T_{i,j,k} > T_{act}$, then $F_{i,e,k} = 1$, where a grid point (i, j, k) is located at the centre of a sub-lattice or control volume. The subscript $j = e$, indicates the location at the solid–liquid interface. T_{act} is the nucleation activation temperature obtained from Eq. (1) given below.

$$\Delta T_{act} = T_{act} - T_{sat} = \frac{4\sigma T_{sat}}{\rho_v h_{fg} D_c} \tag{1}$$

Note that in Eq. (1), T_{sat} is in kelvin, σ is the liquid–vapour surface tension (N/m), ρ_v is the saturation density of vapour (kg/m^3), and h_{fg} is the latent heat of vapourization (J/kg). D_c in Eq. (1) is in metre.

D_c, the diameter of the largest nucleating cavity, is given by the following expression.

$$D_c(i, e, k) = D_m + \beta R(i, k) \tag{2}$$

D_m is the diameter of the smallest nucleating cavity on the heated surface (taken as 0.8 µm), $\beta = 0.99$ µm, and $R(i, k)$ is a random number between 0 and 1 assigned at each (i, k) surface sub-lattice.

Thermal Conduction. 3D transient heat conduction equations in the plate and the liquid are solved subject to the compatibility conditions at the plate–liquid interface, periodic boundary conditions on lateral sides, the top surface $(j = n)$ being at T_{sat}. The bottom surface $(j = 1)$ is exposed to heat flux, q''. The explicit finite difference scheme is used with $\Delta t = 10^{-6}$ s after considering the stability criteria. The initial conditions are as follows: $T_{i,j,k} = T_{sat}$ and $F_{i,j,k} = 0$. The temperature at each grid point at the end of the time step, Δt, based on the solution of the conduction problem is $T'_{i,j,k}$.

Bubble Rising Motion and Associated Thermal Convection. The following equation is used to map T' to T'' in the same time step.

$$T''_{i,j,k} = T'_{i,j,k} + \frac{s_{i,j,k}}{2}\left(\rho_{i,j+1,k} - \rho_{i,j-1,k}\right)T'_{i,j,k} \tag{3}$$

where $\rho = 1$ for liquid sub-lattice and $\rho = 0$ for vapour sub-lattice. It may be noted that ρ is the dimensionless density of the liquid or vapour normalized by the saturation density of the liquid. The parameter, $s_{i,j,k}$, is the representative of the velocity of the bubble rising motion and strength of thermal convection.

Phase Change in the Bulk of the Fluid. The phase change mapping is formulated as follows.

$$\text{If } F_{i,j,k} = 0 \text{ and } T''_{i,j,k} \geq T_{c(i,j,k)} \text{ then } T^{p+1}_{n(i,j,k)} = T''_{n(i,j,k)} - \eta \tag{4}$$

$$\text{If } F_{i,j,k} = 1 \text{ and } T''_{i,j,k} \leq T_{c(i,j,k)} \text{ then } T^{p+1}_{n(i,j,k)} = T''_{n(i,j,k)} + \eta \tag{5}$$

The suffix, $n(i, j, k)$, represents the nearest neighbouring six sub-lattices and $T_{c(i,j,k)}$ is the phase change temperature, which is T_{act} if a liquid sub-lattice is adjacent to the heating surface, homogeneous nucleation temperature if a liquid sub-lattice is surrounded by liquid sub-lattices, T_{sat} if a liquid or a vapour sub-lattice is adjacent to a vapour or liquid sub-lattice, respectively. The parameter, η, is related to enthalpy of vapourization or condensation and has a unit of degree celsius. $\eta = 2$ if the sub-lattice is on the heater surface, otherwise $\eta = 1$.

Time Advancement and Program Termination Criterion. The time is advanced by repeating a set of mapping the aforementioned dynamic processes till steady state is reached. The starting value of heat flux is 1000 W/m², and it is incremented by 1000 W/m². For each heat flux, the steady-state average wall temperature is calculated. The computations stop when the overall vapour fraction is 1 in the pool accompanied by a sudden rise in average wall temperature as compared to that corresponding to previous heat flux, indicating that CHF has been reached. Thus, a pool boiling curve can be drawn.

Modelling of Nanofluids. Modification of Properties: The nanofluids are treated as homogeneous, and their properties are calculated using available correlations based on nanoparticle concentration.

Surface Modifications: Since the boiling of nanofluids used in the present study results in modification of cavity diameters as mentioned in Sect. 1, the minimum cavity diameter is modified using the following expression.

$$(D_m)_{new} = a(\text{wt.fraction}) + (D_m)_{old} \tag{6}$$

'a' is −0.65 for copper–water nanofluid, −9.47 zirconia–water nanofluid and +
0.6 for silver–water nanofluid. The value of 'a' in each case has been obtained from
the corresponding experimental data reported in literature [1–3].

3   Results

Figure 1 shows the comparison of CML-predicted atmospheric saturated pool boiling
curves for pure water and copper–water nanofluid for different weight fractions of
nanoparticles with the same obtained in the experiment of Kathiravan et al. [1]. It is
evident that there is a good qualitative match between the two as both show shifting
of the curves to the right and rise in CHF with increase in nanoparticle concentration.

The present CML simulation predicts 6%, 13% and 31% increase in CHF with
respect to pure water for Cu–water nanofluid having nanoparticle concentration of
0.25%, 0.5% and 1% by weight, respectively, as compared to 21%, 36% and 43%
in experiment [1]. In the case of ZrO_2–water nanofluid, an increase of 12%, 31%
and 34% in CHF is predicted for nanoparticle concentration of 0.02%, 0.05% and
0.07% by volume, respectively, as compared to 32%, 48% and 39% in experiment
[2]. On the contrary, a decrease of 33%, 37% and 41% in CHF for Ag–water nanofluid
corresponding to nanoparticle concentration of 0.25%, 0.5% and 0.75% by weight,
respectively, is shown by the CML simulation as compared to 27%, 35% and 47% in

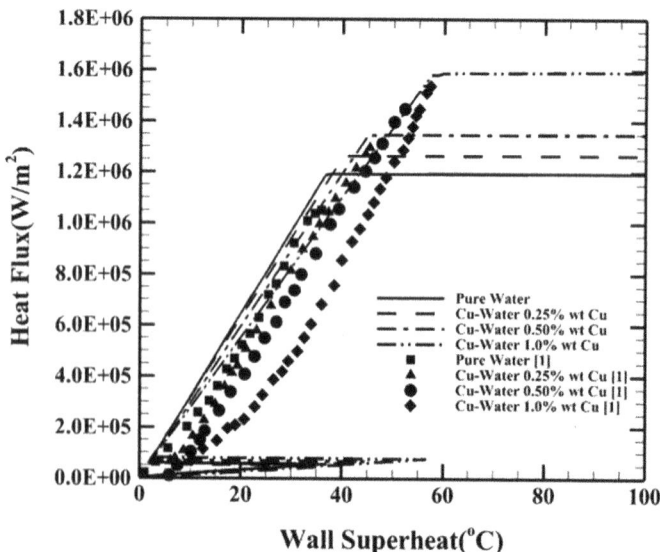

Fig. 1 Comparison of CML-predicted saturated pool boiling curves for pure water and Cu–water
nanofluid at 1 atm for different nanoparticle concentrations with the corresponding experimental
results of Kathiravan et al. [1]

experiment [3]. The quantitative accuracy of CML-prediction of CHF for pure water is excellent.

4   Conclusions

A 3D coupled map lattice-based simulation of atmospheric saturated pool boiling of copper–water, silver–water and zirconia–water has been reported in this work. The results show that CML has been able to reproduce the experimental pool boiling curves realistically for all three nanofluids.

References

1. R. Kathiravan, R. Kumar, A. Gupta, R. Chandra, Preparation and pool boiling characteristics of copper nanofluids over a flat plate heater. Int. J. Heat Mass Transf. **53**, 1673–1681 (2010)
2. M. Chopkar, A.K. Das, I. Manna, P.K. Das, Pool boiling heat transfer characteristics of ZrO_2-Water nanofluids from a flat surface in a pool. Heat Mass Transf. **44**, 999–1004 (2008)
3. R. Kathiravan, R. Kumar, A. Gupta, R. Chandra, Preparation and pool boiling characteristics of silver nanofluids over a flat plate heater. Heat Transf. Eng. **33**, 69–78 (2012)
4. A. Gupta, P.S. Ghoshdastidar, A three-dimensional numerical modeling of atmospheric pool boiling by the Coupled Map Lattice method. ASME J. Heat Transf. **128**, 1149–1158 (2006)
5. K. Kaneko, *Theory and Applications of Coupled Map Lattices* (Wiley, Chichester, 1993).

Heat Transfer of Condensing Saturated and Non-saturated Hydrocarbons Inside Horizontal Tubes

S. Fries and A. Luke

1 Introduction

Saturated and unsaturated hydrocarbons such as propane and propylene are widely used in the process industry as refrigerant. Additionally, those fluids are considered as components in natural gas liquefaction plants, which have high-energy requirements. In these applications, tube bundle heat exchangers are applied as evaporators and condensers due to the high pressures. Nevertheless, engineering data for the design of the heat exchanger are scarce.

Few experimental data are available [1, 2] for plain tubes in a small range of industrial-sized tubes considering condensation heat transfer at higher pressures. The main reason of the lack of experimental data is the light flammability and explosion potential especially at high pressures, denoted by the CLP regulation [3] with H 220 and H 280 at the one hand side and at the other side the geometry of the tubes.

However, the newly proposed correlation by Shah [4] already considers the experimental results in industrial-sized tubes by Garimella and Macdonald [1], Fries et al. [2] and Lee and Son [3]. The experimental investigation of [3] is carried out at the same saturation temperature for different refrigerants, whereas the saturation pressure is varied as experimental parameter by the other authors [1, 2].

2 Experimental Procedure

The experiments are carried out in different sized plain steel and copper tubes with propane and propylene as fluids. The heat transfer measurements cover the state of

S. Fries (✉) · A. Luke
Department of Technical Thermodynamics, University of Kassel, 34125 Kassel, Germany
e-mail: ttk@uni-kassel.de

© Springer Nature Singapore Pte Ltd. 2021
C. Wen and Y. Yan (eds.), *Advances in Heat Transfer and Thermal Engineering* ,
https://doi.org/10.1007/978-981-33-4765-6_26

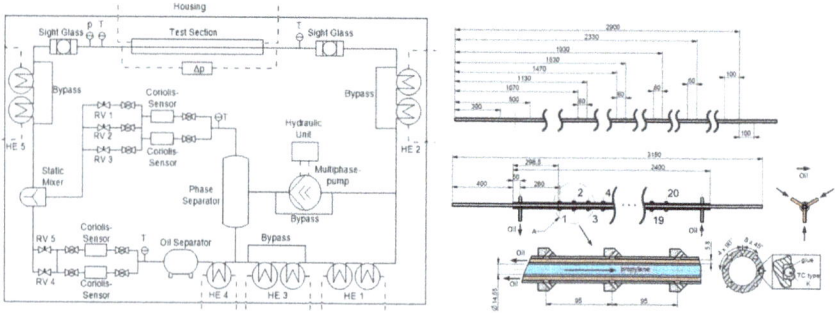

Fig. 1 PID Schema of the test facility (left) and design of the test tube (right)

subcooled liquid, saturated fluid (condensation) and superheated vapour. The experimental set-up is reported in detail in [2]. A multiphase pump conveys the single or multiphase test fluid; see Fig. 1. The mass flux is varied in a large range. The vapour quality is adjusted by a static mixer. Extensive precaution is provided for safety reasons as, e.g. a drag pressure system and special active coal filter system (see [5]).

The test section itself consists of a shell and tube heat exchanger. The test fluid flows within the test tube, where the cooling fluid flows in counterflow with the annulus; see Fig. 1 right. The temperatures are measured in high resolution in the wall by thermocouples as well as in the annulus by resistance thermometers. The heat flux and the heat transfer are determined by the local temperature field as described by Fries et al. [2].

3 Results

The experimental results for the single-phase and two-phase heat transfer and pressure drop are shown for propane and propylene and are shown for varying pressure. In the case of condensation, the heat transfer coefficient calculated by the correlation of Shah [4] is in overall good agreement to the experimental results for a reduced pressure of 0.36 and different mass fluxes; see Fig. 2. The increase of the heat transfer coefficient with the vapour quality and mass flux is underpredicted for higher vapour qualities by Shah [4].

The heat transfer coefficients are very similar for both fluids for all mass fluxes and vapour qualities with slight differences being in the experimental uncertainty range; see Fig. 2. This is expected, since the thermophysical properties of fluids are nearly the same at the same reduced pressure. Hence, the comparison of the fluids should be conducted at the same reduced pressure but not at the same saturation temperature as done by [3] in order to consider the influence of the inclination of the vapour–pressure curve.

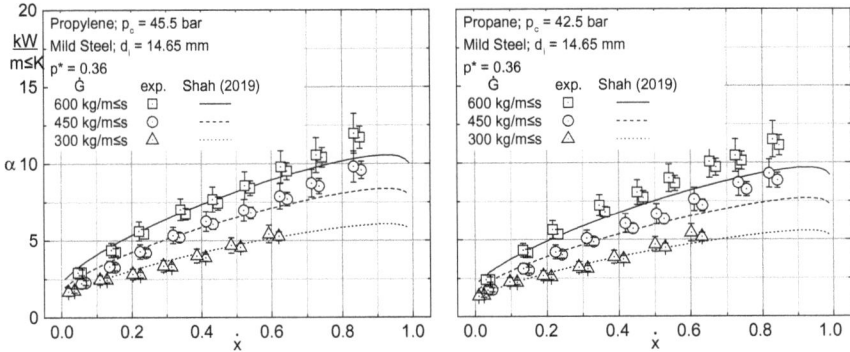

Fig. 2 Heat transfer coefficient as function of the vapour quality of condensing propane and propylene at $p^* = 0.36$ for different mass flux densities

4 Conclusions

Experimental heat transfer investigations are carried out for propane and propylene in a wide range of temperature and pressure. The results for condensation compared to the correlation of Shah [4] are in good agreement. The deviation increases with higher vapour qualities ($x > 0.7$) to a maximum difference of 24.8%.

In order to consider the influence of the vapour–pressure curve, the heat transfer coefficients have to be compared at the same reduced pressure.

References

1. S. Garimella, M. Macdonald, Hydrocarbon condensation in horizontal tubes. Part I Meas. Int. J. Heat Mass Transf. **93**, 75–85 (2016)
2. S. Fries, S. Skusa, A. Luke, Heat transfer and pressure drop of condensation of hydrocarbons in tubes. J. Heat Mass Trans. **55**, 33–40 (2019)
3. H.-S. Lee, C.-H. Son, Condensation heat transfer and pressure drop characteristics of R-290, R-600a, R-134a and R22 in horizontal tubes. J. Heat Mass Transf. **46**, 571–584 (2010)
4. M.M. Shah, Improved correlation for heat transfer during condensation in conventional and mini/micro channels. Int. J. Refrig. **98**, 222–237 (2019)
5. Regulation (EC) No 1272/2008 of the European Parliament and of the Council of 16 December 2008 on classification, labelling and packaging of substances and mixtures, amending and repealing Directives 67/548/EEC and 1999/45/EC, and amending Regulation (EC) No 1907/2006

Analysis of the Influence of Thermophysical Properties on the Coupled Heat and Mass Transfer in Pool Boiling

Niklas Buchholz and Andrea Luke

1 Introduction

Pool boiling is mainly used in flooded evaporators for heat pumps, refrigeration and air-conditioning systems as well as within the process industry, e.g. in the liquefaction of natural gas. Although many theoretical and experimental investigations on the integral heat transfer of various boiling liquids are presented in the literature, however the secondary conditions for the experimental investigations are often not reported. Heat transfer in pool boiling is mainly influenced by saturation pressure, heat flux, as well as micro- and macrostructure of the heating wall and thermophysical properties of the boiling liquid.

Current approaches try to separate these influencing parameters for the calculation of the heat transfer coefficient, e.g. the VDI correlation by Gorenflo [1]. This method is justified when the impact of certain influencing parameters is to be assessed. Nevertheless, the influencing parameters are interdependent, and a more accurate description of physical interconnections in pool boiling is gained by examining this interdependence. As thermophysical properties of liquids have not yet been investigated in the literature, this study focuses on their influence during pool boiling and their interaction with other parameters, especially the surface roughness.

2 Modelling

It is essential to consider heterogeneous nucleation in cavities on a solid surface to describe the influence of thermophysical properties on two-phase heat transfer, as it is critical for the integral heat transfer coefficient. The influence of surface

N. Buchholz (✉) · A. Luke
Department of Technical Thermodynamics, University of Kassel, Kassel, Germany
e-mail: ttk@uni-kassel.de

© Springer Nature Singapore Pte Ltd. 2021
C. Wen and Y. Yan (eds.), *Advances in Heat Transfer and Thermal Engineering* ,
https://doi.org/10.1007/978-981-33-4765-6_27

roughness is included in the VDI correlation [1] through the standard arithmetic mean roughness value. Activation of vapour bubbles, their growth and departure diameter and frequency is highly dependent on the shape of the nucleation site or cavity. This integral parameter is insufficient, as it neither includes the shape of the cavity nor the wetting of the surface.

Therefore, it is necessary to identify a parameter that considers cavity shape, surface wetting and fluid properties. Several cavity parameters have been defined which are obtained by the envelope method by Schömann [2]. Luke [3] modifies the cavity parameters by Schömann [2]. Thus, a characteristic parameter

$$P_5^* = \sqrt[3]{V_K} \qquad (1)$$

is given for a potential nucleation site with the volume V_K of a sphere with the diameter d as the maximal distance between the surface profile and the envelope. The pressure dependence of active nucleation sites and thereby the heat transfer is given with the choice of the spherical radius R_K used in the envelope method. In addition, the wetting of the heating surface by the boiling liquid influences R_K resp. the cumulative distribution of P_5^*. A characteristic size of P_5^* must be chosen from the cumulative distribution according to

$$P_5^* = f\left(p^*, \Theta\right) \qquad (2)$$

with the reduced pressure p^* and the contact angle Θ. The critical nucleation radius can be calculated for low saturation temperatures using the Thomson and the Clausius–Clapeyron equations. A new parameter is defined to include this interaction between heating surface and boiling liquid for the calculation of the heat transfer for different fluids on the basis of the VDI correlation by Gorenflo [1]. In this factor, the bond number Bo, with the mean value or function of P_5^* as the characteristic length, and the fluid parameter

$$P_f = (\mathrm{d}p/\mathrm{d}T)_{\mathrm{sat}}/\sigma \qquad (3)$$

defined by Kotthoff and Gorenflo [4] include the interaction between the heating surface and the boiling liquid.

3   Results

Heat transfer measurements are performed in a standard boiling apparatus according to Goetz [5] to fit Eq. (2) and the function for the characteristic length used in Bo with experimental data and develop the parameter that includes Bo and P_f in the VDI correlation by Gorenflo [1]. The product of Bo and P_f is shown in Fig. 1 for different values of P_5^* as function of the reduced pressure.

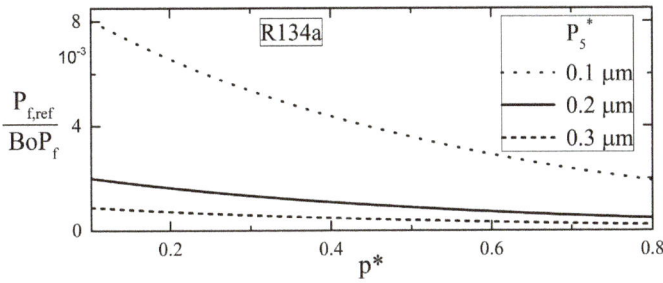

Fig. 1 Fluid parameter P_f and bond number as function of the reduced pressure

The critical nucleation radius is reduced at higher reduced pressure, and therefore smaller nucleation sites are activated at increased pressure for the same superheat. This trend is represented by the factor with Bo, P_f and a reference value of P_f in Fig. 1.

A stochastic distribution of cavities is required to ensure that nucleation sites are evenly distributed on the surface of the heating tube to ascribe active nucleation sites to a characteristic cavity. Accordingly, the surface is fine sandblasted with $P_a = 0.5$ μm. Roughness measurements are performed with a mechanical stylus system and an ultrasonic stylus to determine the cavity parameters. To avoid damage to the heating surface, contactless measuring methods are chosen. The distribution of P_5^* for the sandblasted surface is calculated by comparing the surface topography with the envelope areas in Fig. 2.

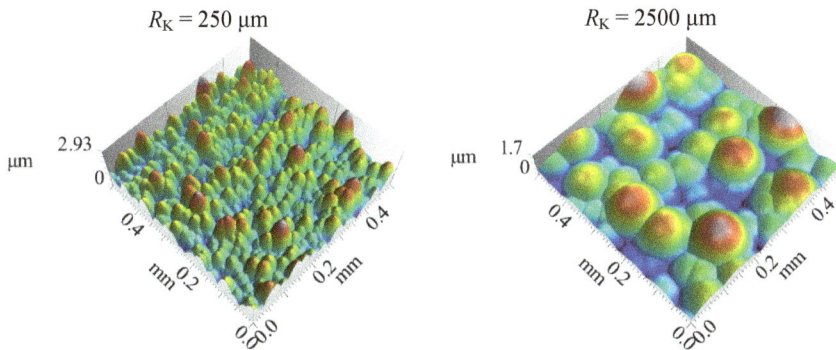

Fig. 2 Envelope areas calculated for the sandblasted heating surface for two different values of R_K

4 Conclusions

Experimental heat transfer measurements are carried out to find a mathematical description for interaction between boiling fluid and microstructure of the heating surface in pool boiling regime. The bond number, the fluid parameter P_f and the cavity parameter P_5^* are identified as important influences to create this parameter. In future, the heat transfer measurements are used to fit the new parameter to the experimental data.

References

1. D. Gorenflo, H2 pool boiling, in *VDI Heat Atlas*, 11th edn. (Springer, Berlin, 2010)
2. H. Schömann, Beitrag zum Einfluss der Heizflächenrauhigkeit auf den Wärmeübergang beim Blasensieden. Dissertation, University of Paderborn (1994)
3. A. Luke, Thermo- and fluid dynamic in boiling—connection between surface roughness, bubble formation and heat transfer, in *Proceedings of 5th International Conference on Boiling Heat Transfer*, Montego Bay, Jamaica (2003)
4. S. Kotthoff, D. Gorenflo, Heat transfer and bubble formation on horizontal copper tubes with different diameters and roughness structures. Heat Mass Transf. **45**, 893–908 (2009)
5. J. Goetz, Entwicklung und Erprobungeiner Normapparatur zur Messung des Wärmeübergangs beim Blasensieden. Dissertation, University of Karlsruhe (1980)

Experimental Investigation of Various Influence Parameters on the Onset of Nucleate Boiling

M. E. Newton, H. Margraf, J. Addy, and A. Luke

1 Introduction

Conventional single-phase cooling systems are no longer able to meet the cooling demands of modern electronics. Decreasing size and increasing processing power result in increasing heat flux densities [1]. Multiphase systems are therefore proposed due to their capability of transporting high heat fluxes at small temperature gradients. Heat pipes are an example of such multiphase systems with the added benefit of offering a wide operation and application range [2]. Operation limits nevertheless need to be considered, especially the boiling limitation in the case of higher superheats represented by the onset of nucleate boiling.

Nucleate boiling is a highly efficient variant of two-phase heat transfer. However, the nucleate boiling regime must be avoided in a heat pipe, because the fluid flow within the capillary structure is obstructed or even blocked by vapour bubbles due to nucleation [3]. This not only reduces the efficiency of the heat pipe but also results in damaging it in special cases. The first viable bubble occurs within a cavity of the heated wall or capillary structure, respectively, which is influenced by saturation pressure, superheat, thermophysical properties of the boiling liquid and the material of the heated wall as well as the microstructure and geometry of the heated wall.

2 Heat Transfer Measurements

Measurement of the onset of nucleate boiling to determine the boiling limitation cannot be performed directly with heat pipes. Reaching the boiling limitation would only be observable by a sudden increase in temperature in the evaporator section

M. E. Newton (✉) · H. Margraf · J. Addy · A. Luke
Department of Technical Thermodynamics, University of Kassel, Kassel, Germany
e-mail: ttk@uni-kassel.de

© Springer Nature Singapore Pte Ltd. 2021
C. Wen and Y. Yan (eds.), *Advances in Heat Transfer and Thermal Engineering* ,
https://doi.org/10.1007/978-981-33-4765-6_28

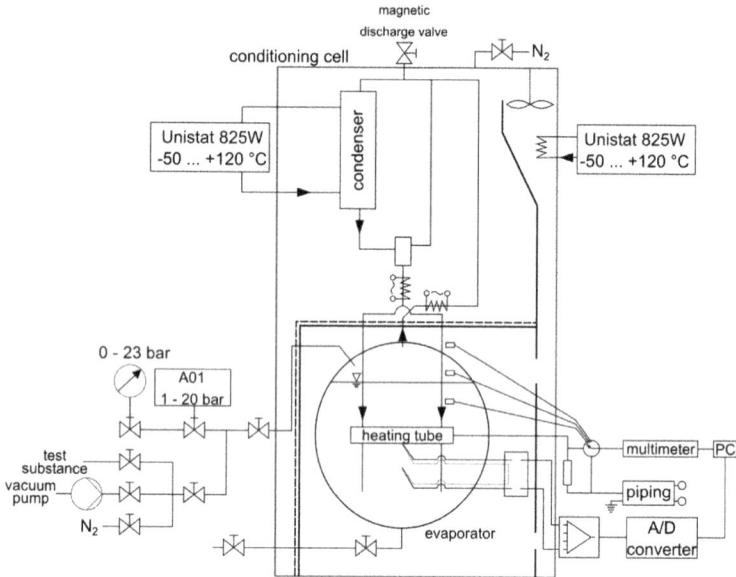

Fig. 1 Schematic of the standard boiling apparatus with natural circulation, including thermal circuit, conditioning cell and measuring devices

of the heat pipe due to dry out. As this is not precise, instead the onset of nucleate boiling is determined by pool boiling experiments. A standard boiling apparatus with modifications by Luke and Bujok [4] is used to perform heat transfer measurements and determine the onset of nucleate boiling, as shown in Fig. 1.

Built into a conditioning cell, the apparatus features natural circulation between evaporator and condenser. An electrically heated horizontal test tube serves as the heating surface in the evaporator. The surface of the test tube is treated by various means, e.g. with corundum to achieve fine and rough sandblasted surfaces in the first step. In the second step, surface structures as those of the capillary structure in heat pipes will be investigated. Nucleation on the surface of the heating tube is observed and recorded through a sight glass. Superheat ΔT for the determination of the heat transfer coefficient is calculated from the temperature difference between thermocouples fixed within the test tube and liquid bulk of the evaporator. The measurements are carried out within a large range of saturation pressure and heat flux in various boiling liquids. The boiling regimes of nucleate and of convective boiling are investigated with a focus of the transition between them.

3   Results

Measurements with the test rig are performed with propane and a mild still tube for a wide pressure range and varying surface roughness by Luke [5]. A comparison of results is shown on the left-hand side of Fig. 2 in the form of a double logarithmic plot of the heat transfer coefficient α as function of the heat flux $\dot{q}$ for reduced pressures of 0.1, 0.2 and 0.4. Measurements with a fine and rough sandblasted surface with a roughness of 0.16 μm and 11.32 μm, respectively, are also compared. The onset of nucleate boiling is marked by the sudden change of the inclination of the slopes of the graphs and an enhanced increase of the heat transfer coefficient. Additionally, the onset of nucleate boiling is shifted to smaller superheats with increased surface roughness and saturation pressure due to the activation of smaller cavities within the structure, as shown in Fig. 2. The influence of saturation pressure on the onset of nucleate boiling is also depicted in the left-hand side of Fig. 2. As saturation pressure and saturation temperature are coupled by the vapour pressure curve, an increase in pressure results in an increase of temperature. Many of the thermophysical properties influencing the onset of nucleate boiling are dependant of temperature and decrease with increasing temperature, e.g. surface tension, viscosity and density. As the values of these parameters decrease, so does the required superheat for formation of viable vapour bubbles. An increase in pressure therefore results in a decrease of the heat flux at the onset of nucleate boiling.

On the right-hand side of Fig. 2, the merit number for propane along the entire vapour pressure curve is depicted with the investigated pressures marked. The merit number is used to compare different working fluids but is also used to determine the efficiency of a working fluid at different operating conditions [2, 3]. As a consequence of increased pressure, the merit number decreases. In combination with the results

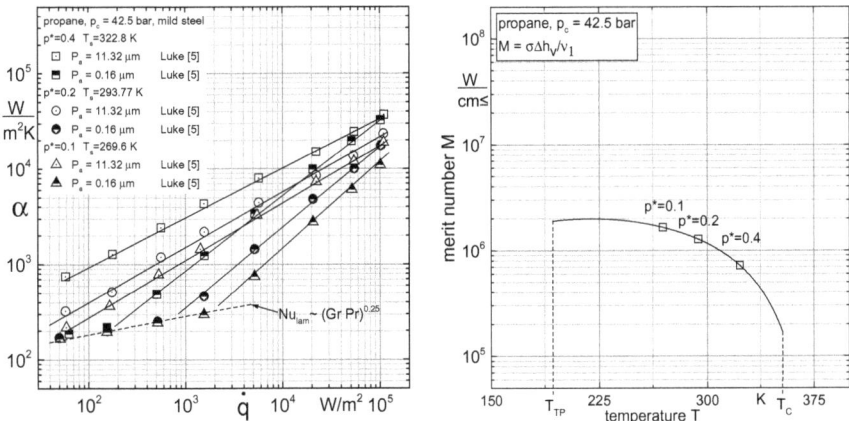

Fig. 2 Heat transfer coefficient versus heat flux for varying surface roughness and saturation pressure at a mild steel tube [5] (left) and the merit number for propane calculated along the vapour pressure (right) at a reduced pressure of 0.1, 0.2 and 0.4

in the left-hand side of Fig. 2, this indicates that a low pressure is not only preferable due to the delay of the onset of nucleate boiling but also due to an increased merit number.

The activation criteria of the first viable bubble are still not known due to the many parameters influencing nucleate boiling. A new, semi-empirical correlation is therefore proposed by Addy [6] to calculate the onset of nucleate boiling which is based on the VDI correlation by Gorenflo [7]. This correlation considers all the parameters influencing the onset of nucleate boiling such as surface roughness, operating conditions and thermophysical properties of both the boiling fluid and the heating surface. Several aspects of heat conduction and heat convection near the growing bubble within its cavity are introduced in the model.

4 Conclusions

The boiling limitation of heat pipes has to be determined. Therefore, the incipience of nucleate boiling on similar surfaces as the capillary structure in heat pipes with varying surface roughness has been experimentally investigated in a pool boiling test rig in a wide range of saturation pressure. Experimental results show that the onset of nucleate boiling is shifted to a smaller superheat with increasing surface roughness and saturation pressure. As nucleate boiling is to be avoided in heat pipes during operation, it is recommended to avoid high surface roughness at the inner wall and capillary structure of heat pipes. This may prove difficult for mesh wick heat pipes, as surface roughness is not a priority in production of the mesh. Nevertheless, regarding grooved heat pipes, the consideration of the influence of surface roughness might prove useful for cases in which the boiling limitation is the limiting factor. As for the influence of pressure, results indicate that operation at low pressure is preferable with propane as working fluid. Not only is the onset of nucleate boiling moved to higher heat fluxes but the merit number is also increased at low operating pressure.

References

1. F. Crößmann, Untersuchung der Verdampfung aus strukturierten Oberflächen in Reinstoffatmosphäre. Dissertation, Technical University Darmstadt (2016)
2. D. Reay, P. Kew (eds.), *Heat Pipes—Theory, Design and Applications* (Butterworth-Heinemann, Oxford, 2005)
3. P. Stephan, N5 heat pipes, in *VDI Heat Atlas*, 11th edn. (Springer, Berlin, 2010)
4. A. Luke, P. Bujok, Influence of fin geometry on heat transfer in boiling pure refrigerants and their mixtures, in *Proceedings of the 4th IIR Conference on Thermophysical Properties and Transfer Processes of Refrigerants*, Delft, The Netherlands (2013)
5. A. Luke, Thermo- and fluid dynamic in boiling—connection between surface roughness, bubble formation and heat transfer, in *Proceedings of 5th International Conference on Boiling Heat Transfer*, Montego Bay, Jamaica (2003)

6. J. Addy, Contribution to Onset of nucleate boiling on structured tubes. Dissertation (in preparation), University of Kassel (2019)
7. D. Gorenflo, H2 pool boiling, in *VDI Heat Atlas*, 11th edn. (Springer, Berlin, 2010)

Study on Morphological Development of Fuel Droplets After Impacted on Metal Surfaces with Different Wettability

Liang Guo, Gaofei Chen, Wanchen Sun, Degang Li, Yuheng Gao, and Peng Cheng

1 Introduction

In order to meet increasingly stringent emission regulations and fuel economy requirements, the high efficient clean combustion and emission control technologies for internal combustion engine were developed rapidly in recent years. Among them, the development of basic combustion science and new combustion modes such as homogeneous compression combustion (HCCI), premix compression combustion (PCCI), low-temperature combustion (LTC), and reactive controlled compression combustion (RCCI) plays a vital role in improving the combustion and emission performance of engines. Injection-related technologies such as variable injection timing, multiple injection, and high-pressure injection are always used to for HCCI, PCCI, and other premixed compression ignition modes to improve the atomization and combustion quality, but they are more likely to result in wall-wetting phenomenon, comparing to the traditional injection methods. It is well known that wall wetting is one of the most important factors that cause high HC and CO emissions. In order to reduce HC and CO emissions caused by wall wetting, it is necessary to conduct an in-depth study on detailed fluid–solid coupling development process between the fuel droplets and the surface of the combustion chamber after wall wetting, to determine the controllability of the fuel spray diffusion and evaporation process [1, 2]. To understand the influential factors of spreading and evaporation of the wall-wetting fuel droplets, the development process of single droplets after impacting mental walls is studied. The development characteristics of wall-wetting droplets can be affected by many factors such as surface roughness, temperature, wettability and many other factors. The surface contacting characteristics between the droplet

L. Guo · G. Chen · W. Sun (✉) · D. Li · Y. Gao · P. Cheng
State Key Laboratory of Automotive Simulation and Control, Jilin University, Changchun 130022, China
e-mail: sunwc@jlu.edu.cn

© Springer Nature Singapore Pte Ltd. 2021
C. Wen and Y. Yan (eds.), *Advances in Heat Transfer and Thermal Engineering* ,
https://doi.org/10.1007/978-981-33-4765-6_29

and the surface affect its morphological development process [3] and heat transfer performance and these will further influence the combustion and emission performance of internal combustion engines [4]. In this study, the microstructure of surfaces made of aluminum alloy was reconstructed with physical and chemical processing, and substrate surface with diffident wettability was obtained. The spreading and rebounding process of fuel droplets upon surfaces with different wettabilities was studied at various boundary conditions. The results are expected to provide a theoretical basis for understanding the interaction mechanism between fuel spray droplets and the wall of the combustion chamber for guiding future design of combustion chamber-related components.

2   Methodology

To obtain different surface wettabilities, the micropattern of the tested surfaces made of aluminum alloy was constructed with physical and chemical processing, including

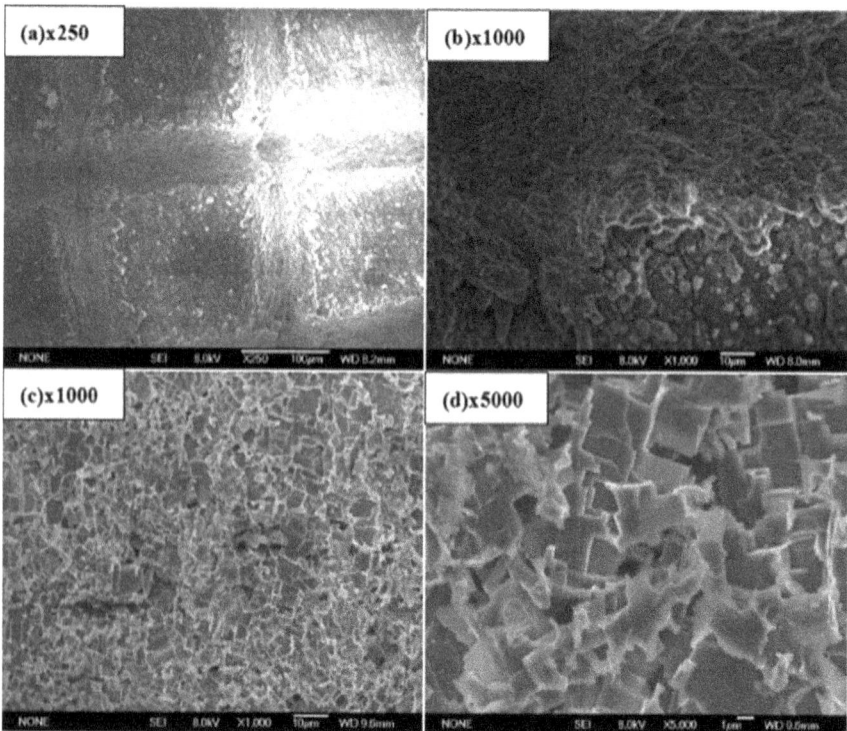

Fig. 1 SEM photographs of aluminum alloy substrate surface, **a** laser-etched surfaces, 250×, **b** laser-etched surfaces, 1000×, **c** chemically treated surface, 1000×, **d** chemically treated surface, 5000×

laser etching and chemical etching. Figure 1a, b shows the micro-morphology of a laser-etched surface with roughness of 2.8 μm, taken with scanning electron microscope (SEM) under magnifications of 250 and 1000 times, respectively. Image (c) and (d) in Fig. 1 is the SEM photographs of the chemically treated surface, taken under magnifications of 1000 and 5000 times, respectively.

On the surface of the laser-etched aluminum alloy, the static contact angles of all the tested fuel droplets, including diesel and n-butanol and dimethyl carbonate (DMC) were measured as less than 2°. And on the chemically treated surface, the static contact angles of diesel, n-butanol and DMC droplets were increased to 85°, 65° and 15°, respectively. The morphological developing process of the above fuel droplets after impacting onto different walls was studied under various boundary conditions. The wettability of the fuels against different surfaces at different temperature was obtained through contact angle measurement.

3   Results

Figure 2 shows the results of spreading/retraction factor of diesel droplets upon surfaces with different wettabilities, obtained at different surface temperatures. Two surfaces, a laser-etched surface with a roughness of 2.8 μm and a chemically treated surface, are tested.

As shown in Fig. 2, upon the laser-etched surface with a same surface roughness value, the dynamic contact angle of diesel is the larger than the other two fuels. The dynamic contact angle of the DMC is the smallest among all the fuels. This is because the falling inertial force of all droplets is similar to each other, but the viscous force and surface tension of the diesel droplet are the largest, leading to a

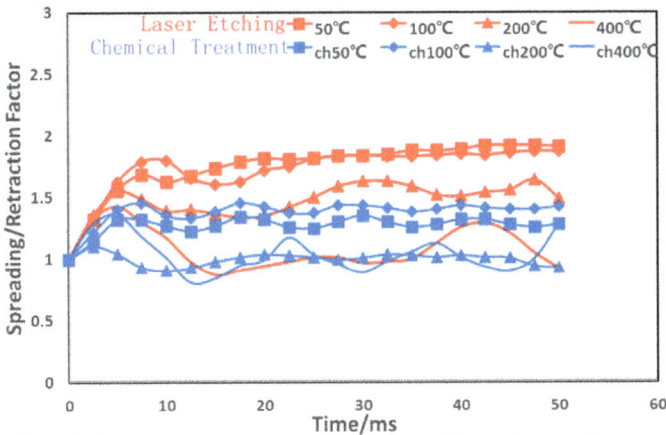

Fig. 2 Effect of wall wettability on the spreading/retraction factor of diesel droplets at different temperatures

poorest spreading performance and wettability upon the surface. The surface tension and viscous force of DMC fuel are the smallest among all three fuels, so that it can infiltrate the surface of the substrate more easily, and therefore a better spreading performance. The dynamic contact angle of the fuel droplets upon the chemically treated aluminum alloy surface is greater than the contact angle of the other three laser-etched aluminum alloy surfaces. According to the Cassie and Baxter models, because the chemically treated substrate surface has a certain oleophobic surface structure, the pits on the surface can hardly be filled with the liquid fuel, so that a combined solid–liquid–gas three-phase boundary is formed between the fuel droplet and the wall. And the contact angle of all fuel droplets upon the chemically treated surface is larger than that upon the laser-etched surface. The contact angle of the diesel fuel on the chemically etched surface is about 90° and the contact angle of n-butanol is about 65°. The surface tension and viscous force of DMC droplet are the smallest among all the three fuels, leading to a smallest contact angle of about 15°.

For the effect of surface temperature, it can be seen from Fig. 2 that at a lower temperature level, like 50 and 100 °C, the spreading/retraction factor of the fuel on the chemically treated surface is clearly smaller than that on the laser-etched surface, indicating that the spreading/retraction level is mainly influenced by the wettability of the surfaces. However, at a high surface temperature condition, e.g., when the wall temperature reaches 400 °C, the difference of spreading/retraction factor between the two surfaces tends to be smaller. This is because that at such a high temperature, the fuel droplet enters a nuclear boiling state. Due to the lifting effect of the steam film generated between the droplet and the surface, it is difficult for the droplet to contact with the surface directly. And the effect of wall properties on the spreading process spreading/retraction factor is greatly diminished.

4   Conclusions

When the diesel droplets impact on the chemically etched surface, a solid–liquid–gas three-phase boundary is formed between the liquid and solid surface, and the high-pressure air layer hinders the spreading process of the droplets. The oleophobicity of the wall reduces the moving speed of the three-phase contact line, so that the friction loss and the viscous dissipation energy within the fuel molecules decrease, the spreading/retraction factor of the fuel droplet is reduced. When the diesel droplets are in convective heat transfer and nuclear boiling states, the wettability of the wall surface has a great influence on the spreading/retraction factor of the fuel droplets; when the diesel droplets are in a film boiling state, the effect of surface wettability on the morphological development process of the fuel droplets turns weak, at such a heat transfer states, the spreading /retraction factor of droplets of all the fuels upon all the tested surfaces tend to be consistent and the influence of temperature occupies a dominant position.

Acknowledgements The authors gratefully acknowledge financial support from the National Natural Science Foundation of China (51676084, 51776086); Industrial Innovation Special Funding Project of Jilin Province (2019C058-3); and Natural Science Foundation of Jilin Province, China, (20180101059JC).

References

1. N. Fukushima, M. Katayama, Y. Naka, Combustion regime classification of HCCI/PCCI combustion using Lagrangian fluid particle tracking. Proc. Combust. Inst. **35**(3), 3009–3017 (2015)
2. A. Montanaro, L. Allocca, M. Lazzaro, G. Meccariello, Impinging jets of fuel on a heated surface: effects of wall temperature and injection conditions.SAE Technical Paper, 2016-01-0863 (2016)
3. L. Dongwei, N. Zhi, Lü. Ming, Study on the rebound and breakup of the droplet after impacting on the super-hydrophobic wall. Chin. J. Comput. Mech. **33**(1), 106–112 (2016)
4. A.L.N. Moreira, A.S. Moita, M.R. Panão, Advances and challenges in explaining fuel spray impingement: how much of single droplet impact research is useful? Prog. Energy Combust. Sci. **36**(5), 554–580 (2010)

Optimizing the Design of Micro-evaporators via Numerical Simulations

Mirco Magnini and Omar K. Matar

1 Introduction

Flow boiling and heat transfer in microchannels are recognized as one of the most efficient cooling solutions for high-power density electronic devices [1]. The high thermal transfer efficiency of two-phase flow allows the removal of heat fluxes on the order of MW/m^2 from heat exchanger areas of ~1 cm^2. This is possible thanks to the use of multi-microchannel evaporators, where several parallel microchannels are etched on a wafer substrate, which is directly bonded on the surface to be refrigerated (e.g., a CPU). Several different designs of the microchannels cross-sections have been proposed in the recent years, with the most traditional configuration being square or rectangular cross-sections of aspect ratios $\epsilon = H/W = 1 - 20$. However, despite the large number of experimental studies conducted on boiling heat transfer in microchannels, aimed to characterize different micro-evaporators performances, there is still general disagreement on the best configuration maximizing heat transfer. One of the possible reasons is that the global analysis of a micro-evaporator, with its experimental uncertainties, makes it difficult to draw clear conclusions about which geometry provides the best performance given the operating conditions. This has resulted in a lack of established thermal design guidelines for micro-heat sinks. On the other hand, numerical simulation methods have advanced rapidly in the last decades and, if computational resources are available, the direct numerical simulation of boiling flows with interface-tracking methods may provide valuable information on fluid dynamics and governing heat transfer mechanisms [2]. The objective of the present work is to study, via numerical simulations, the impact of the flow conditions

M. Magnini (✉)
Department of Mechanical, Materials and Manufacturing Engineering, University of Nottingham, Nottingham NG7 2RD, UK
e-mail: mirco.magnini@nottingham.ac.uk

O. K. Matar
Chemical Engineering Department, Imperial College London, London SW7 2AZ, UK

© Springer Nature Singapore Pte Ltd. 2021
C. Wen and Y. Yan (eds.), *Advances in Heat Transfer and Thermal Engineering* ,
https://doi.org/10.1007/978-981-33-4765-6_30

163

and microchannel geometry on the heat transfer performance, analyzing different cross-section shapes under flow conditions pertinent to practical applications. We utilize a customized version of the open-source software OpenFOAM, based on a volume-of-fluid (VOF) method to capture the interface motion, which implements a second-order level-set-based surface tension algorithm and a non-equilibrium evaporation model. In previous works [3, 4], the numerical tool has been validated versus several benchmark cases and it showed potential to accurately simulate flow boiling in microchannels of non-circular cross-sections. This study shows that the heat transfer performance of flow boiling in non-circular microchannels is tightly related to the distribution of the liquid film along the perimeter of the channel. Square channels exhibit the highest heat transfer rates at lower flow rates, but are more at risk of film dryout, whereas larger aspect ratios may be beneficial at larger flow rates as they promote the formation of an extended liquid film covering a large fraction of the cross-section perimeter.

2   Methodology

The numerical solver is based on the solution of the flow equations discretized with a finite-volume method. The interface is captured by a VOF algorithm. The volume fraction, mass, momentum, and energy equations are solved as:

$$\frac{\partial \alpha}{\partial t} + \nabla \cdot (\alpha \boldsymbol{u}) = \frac{1}{\rho_v} m_i'' |\nabla \alpha| \tag{1}$$

$$\nabla \cdot \boldsymbol{u} = \left(\frac{1}{\rho_v} - \frac{1}{\rho_l} \right) m_i'' |\nabla \alpha| \tag{2}$$

$$\frac{\partial (\rho \boldsymbol{u})}{\partial t} + \nabla \cdot (\rho \boldsymbol{u} \boldsymbol{u}) = -\nabla p + \nabla \cdot \left[\mu \left(\nabla \boldsymbol{u} + \nabla \boldsymbol{u}^{\mathrm{T}} \right) \right] + \boldsymbol{F}_\sigma \tag{3}$$

$$\frac{\partial (\rho c_p T)}{\partial t} + \nabla \cdot (\rho c_p \boldsymbol{u} T) = \nabla \cdot (\lambda \nabla T) - m_i'' \left[h_{lv} - \left(c_{p,v} - c_{p,l} \right) T \right] |\nabla \alpha| \tag{4}$$

where α indicates the VOF fraction ($\alpha = 1$ in the vapor), t the time, $\boldsymbol{u}$ the fluid velocity vector, ρ the density, m_i'' the evaporation mass transfer, p the pressure, μ the dynamic viscosity, $\boldsymbol{F}_\sigma$ the surface tension force vector, c_p the specific heat, T the temperature, λ the thermal conductivity, and h_{lv} the latent heat. The subscripts l and v refer to the liquid and vapor-specific properties, respectively. The surface tension force is evaluated as $\boldsymbol{F}_\sigma = \sigma \kappa \nabla \alpha$, where σ is the (constant) surface tension coefficient, and κ is the interface curvature, estimated according to a self-implemented

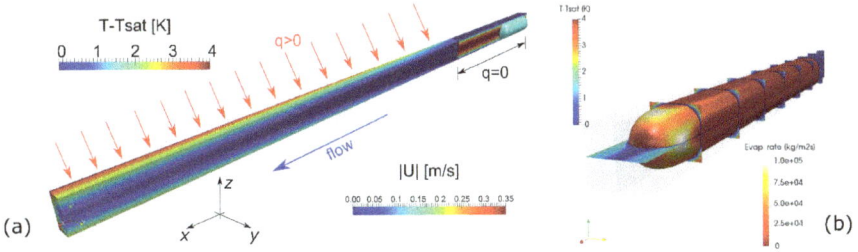

Fig. 1 **a** Setup of the numerical simulation with a bullet-like bubble (cyan color) starting at the channel upstream; **b** snapshot of the elongated bubble near the end of the channel colored by the local evaporation rate

algebraic coupled level set and VOF method (flexCLV) [3]; note that thermally-induced Marangoni stresses will not be considered in the present work. The evaporation mass transfer is calculated based on the local temperature field according to the Hertz–Knudsen–Schrage relationship [4].

The simulation setup is illustrated in Fig. 1a. We simulate square/rectangular channels of aspect ratios $\epsilon = H/W = 1 - 8$, maintaining a constant hydraulic diameter $D_h = 2HW/(H + W)$. A bubble of fixed volume is initialized at the upstream of the channel and an initial adiabatic section allows the flow to reach steady state. This is followed by a section heated with a constant heat flux q, uniformly distributed along the channel perimeter. The bubble is transported by liquid flow that enters the channel with a fully developed laminar velocity profile of average speed U, at saturated conditions $T = T_{sat}$. No-slip conditions are set at the channel walls, while zero-gradient conditions are imposed at the outlet. The initial velocity and temperature fields are obtained from a preliminary liquid-only steady-state simulation run under the same flow conditions.

The adiabatic flow of long bubbles in capillary channels is determined by the capillary number, $Ca = \mu_l U/\sigma$, the Reynolds number $Re = \rho_l U D_h/\mu_l$, the liquid/gas density and viscosity ratios; heat transfer and phase change bring in the Prandtl number $Pr = \mu c_p/\lambda$, the boiling number $Bl = q/(\rho_l U h_{lv})$, the liquid/gas specific heat and thermal conductivity ratios. In this work, we study the flow conditions $Ca = 0.005 - 0.05$, $Re = 100$, and $Bl = 10^{-4}$, which are relevant to the flow of refrigerants in capillary channels ($D_h \sim 0.1$ mm) at low heat flux ($q \sim 10$ kW/m^2), while the Prandtl number and properties ratios are set to model the fluid R245fa. Under the low heat flux conditions simulated, the liquid film does not dry out and therefore no contact angle modeling is required. The minimum thickness of the liquid film achieved in this work is on the order of $\delta/D_h = 0.01$ and below; mesh independence is ensured by discretizing the film with at least 10 computational cells using near-wall refined cells.

3 Results

At onset, the bubble flows downstream and its shape/velocity achieves a steady shape
in the adiabatic section of the domain. When the liquid–gas interface meets the
superheated thermal boundary layer, evaporation begins, and the bubble elongates
and accelerates downstream under the confining effect of the channel walls. Film
evaporation cools the liquid trapped between the liquid/gas interface and the channel
wall. Interestingly, Fig. 1b shows that the highest evaporation rate is detected in
the vicinity of the bubble nose, where the liquid temperature is highest, and at the
channel corners along the bubble, where the thick liquid film does not allow efficient
cooling, and the temperature remains high.

The bubble dynamics and liquid film perimetric distribution differ substantially
depending on the capillary number and aspect ratio. When $Ca = 0.005$, all the
cross-sectional bubble profiles exhibit a dimple, with the minimum film thickness
value located midway between the channel symmetry axis and the corners. When
$Ca \geq 0.01$, the region with the thinnest film shifts to the channel symmetry axis
when the channel is square, see Fig. 2 (left). The distance between the channel
symmetry axis and the location where the minimum film is detected increases with ϵ
for $\epsilon > 1$ (Fig. 2 (left)). Importantly, while for $\epsilon = 1$, the film thickness profiles are
symmetric with respect to the diagonal to the cross-section, when $\epsilon > 1$ a significant
asymmetry develops, with a thin liquid film covering a large fraction of the longest
side (as observed in Fig. 2 (left)) and a thick film deposited along the shortest wall.

The heat transfer trends follow closely those of the film thickness, as the heat
transfer across the thin film region is governed by cross-stream heat conduction.
Therefore, the highest heat transfer rates are measured in the proximity of the
minimum film thickness regions, see the perimetric profiles of the Nusselt number
reported in Fig. 2 (right); here, $Nu = hD_h/\lambda_l$, with $h = q/(T_{wall} - T_{sat})$. Note

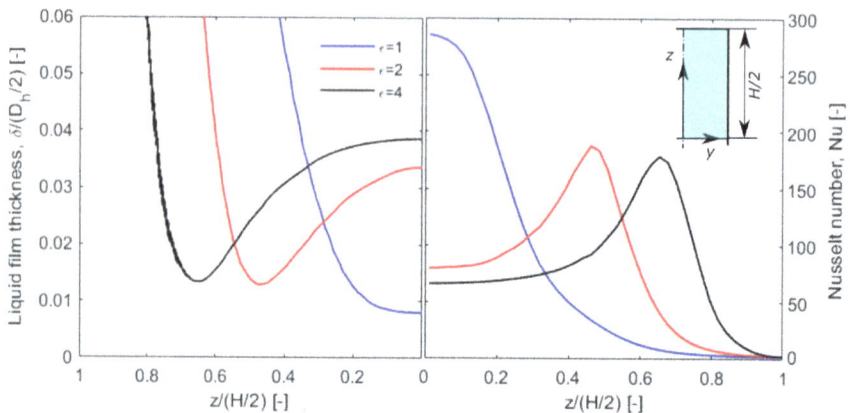

Fig. 2 Local profiles of liquid film thickness (left) and Nusselt number (right) along the vertical
wall for aspect ratios from 1 to 4, $Ca = 0.01$

from Fig. 2 that, since heat transfer in the thin film is heat conduction-dominated, Nu ≈ D_h/δ. Relatively large heat transfer rates are detected along the longest channel wall at all values of ϵ, due to the significant extension of the liquid film region. However, the spatial average of Nu along the shortest wall drops considerably as $\epsilon > 1$ due to the significant increase of the film thickness.

The overall results of this study suggest that the channel aspect ratio that maximizes the heat transfer performance depends strongly on Ca, due to distinctive film thickness patterns. When Ca = 0.005 (lower flow rates), the square channel exhibits the largest average Nu due to a very thin film localized at the interface dimple; performance drops as ϵ is increased, due to the thickening of the film along the shortest channel wall, while heat transfer rises again as $\epsilon > 4$, due to the formation of an extended thin film covering the longest channel wall, and it reaches an asymptotic limit as ϵ approaches 8. Note that the very thin film observed for $\epsilon = 1$ may be at serious risk of dryout. When Ca ≥ 0.01 (larger flow rates), the bubble profile in the square channel becomes axisymmetric and therefore a thick liquid film is formed, with a consequent drop of the heat transfer performance. In these conditions, Nu increases monotonically with ϵ as the fraction of the heated area covered with a thin liquid film becomes larger.

4 Conclusions

This work shows that the heat transfer performance of flow boiling in non-circular microchannels is tightly related to the distribution of the liquid film along the perimeter of the channel, which in turn depends on the flow conditions (here Ca and Re), and channel aspect ratio. Square channels may be more appropriate at lower flow rates, but are more at risk of film dryout. Larger aspect ratios may be beneficial at larger flow rates. A strong asymmetry in the heat transfer performance is observed along the vertical and horizontal walls as $\epsilon > 1$, with Nu being up to one order of magnitude larger along the longest channel wall compared to the value calculated along the shortest wall. The latter represents an important observation, because if the microchannel is unevenly heated (as usually happens with the evaporator sitting upon the device to be refrigerated), very different heat transfer rates are achieved depending on the orientation of the microchannels. The present study is limited to the slug flow regime and low heat fluxes (no dryout occurs). Extension to larger heat fluxes would be of interest, in order to approach conditions, which are typical in the thermal management of high-power electronic devices.

References

1. T.G. Karayiannis, M.M. Mahmoud, Flow boiling in microchannels: fundamentals and applications. Appl. Therm. Eng. **115**, 1372–1397 (2017)

2. M. Magnini, B. Pulvirenti, J.R. Thome, Numerical investigation of hydrodynamics and heat transfer of elongated bubbles during flow boiling in a microchannel. Int. J. Heat Mass Transf. **59**, 451–471 (2013)
3. A. Ferrari, M. Magnini, J.R. Thome, A flexible coupled level set and volume of fluid (flexCLV) method to simulate microscale two-phase flow in non-uniform and unstructured meshes. Int. J. Multiphase Flow **91**, 276–295 (2017)
4. A. Ferrari, M. Magnini, J.R. Thome, Numerical analysis of slug flow boiling in square microchannels. Int. J. Heat Mass Transf. **123**, 928–944 (2018)

Simultaneous Laser- and Infrared-Based Measurements of the Life Cycle of a Vapour Bubble During Pool Boiling

Victor Voulgaropoulos, Gustavo M. Aguiar, Matteo Bucci, and Christos N. Markides

1 Introduction

Nucleate boiling is one of the most effective heat removal modes and has found use in a wide range of cooling applications, from the scale of state-of-the-art densely packed integrated circuits to the majority of current nuclear reactors. While a substantial amount of research has been performed over the years on both pool and flow boiling, this has predominantly focused on qualitative visualisation, often high-speed, aimed at observing the complex and multiphase transport phenomena involved in nucleate boiling, and the development of empirical methods to try to quantify global quantities of interest, such as heat transfer coefficients and pressure drops. These methods have allowed us to make significant progress in these areas, however, recent developments in mechanistic modelling methods and advanced computational capabilities [1, 2] for the prediction of relevant flows have resulted in a need for detailed, spatiotemporally resolved (instantaneous, local) measurements in the liquid phase and on the solid-wall surface, for validation and model development purposes, and for providing further insight into the physics of these phenomena.

This work has a fundamental nature, focusing our attention on low superheat levels of pool boiling of water at saturated temperature and ambient pressure, where isolated vapour bubbles form at an artificial predefined nucleation site on a heater. We develop and apply a series of simultaneous high-speed laser-based [two-colour laser-induced fluorescence (2cLIF) and particle image/tracking velocimetry (PIV/PTV)] and infrared (IR) thermometry diagnostic tools to extract spatiotemporally resolved

V. Voulgaropoulos · C. N. Markides (✉)
Clean Energy Processes (CEP) Laboratory, Department of Chemical Engineering, Imperial College London, London SW7 2AZ, UK
e-mail: c.markides@imperial.ac.uk

G. M. Aguiar · M. Bucci
Department of Nuclear Science and Engineering, Massachusetts Institute of Technology, Cambridge MA02139, USA

© Springer Nature Singapore Pte Ltd. 2021
C. Wen and Y. Yan (eds.), *Advances in Heat Transfer and Thermal Engineering* ,
https://doi.org/10.1007/978-981-33-4765-6_31

information on the velocity and temperature of the water and heater during the life cycle of a nucleating bubble. Our ultimate goal is to obtain a detailed understanding of the complex and interacting heat transfer mechanisms and fluid flow phenomena during the boiling process, in order to shift away from the empiricism that has dominated the field. This work focuses on the technical description of the experimental methodology followed to perform simultaneous 2cLIF, PIV/PTV, and IR measurements during pool boiling and acts as the precursor to a study with the emphasis placed on interpreting the phenomena and physical mechanisms.

2 Methodology

The experimental system used for the pool boiling experiments is illustrated in Fig. 1a. The cell was filled with DI water, which was brought to its saturation temperature via a circulating heating bath. Current was then supplied to a thin-film indium tin oxide (ITO) heater coated on an infrared transparent sapphire glass. Nucleation of single vapour bubbles was achieved at a predefined PTFE-coated spot at the centre of the heater for heat fluxes ranging from 30 to 60 kW m^{-2}. For more details of the apparatus, the reader is directed to Ref. [3].

Figure 1b illustrates the normalised excitation and emission spectra of the temperature-sensitive fluorescein disodium (FL) dye, the temperature-insensitive sulforhodamine 101 (SR) dye, and the fluorescent tracer PIV particles (mean size 10 μm) added in the water at concentrations of 4×10^{-5} M, 1.4×10^{-7} M, and <0.01 M, respectively. The effect of those additives on the nucleating characteristics has been found negligible [3]. Two high-speed (HS) cameras were connected to a pellicle beamsplitter (R45%:T55%) and a Nikon 105 mm lens to visualise the emission of each dye individually and allow for two-colour laser-induced fluorescence

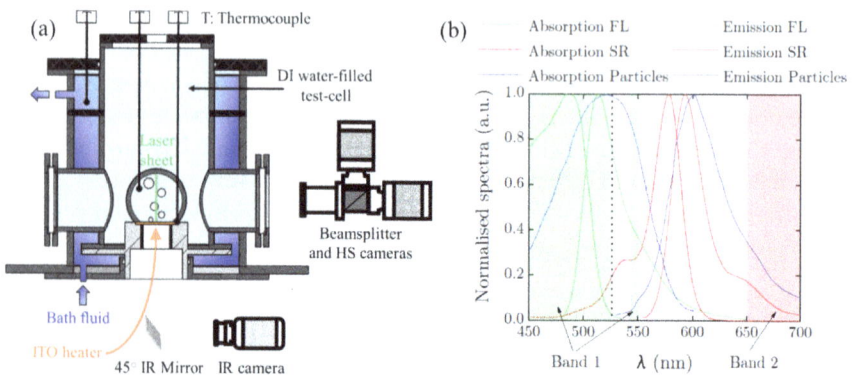

Fig. 1 Illustration of the arrangement used for optical measurements: **a** drawing of the boiling cell and arrangement of the cameras and optics; **b** absorption and emission spectra of the two fluorescent dyes and particles

(2cLIF) measurements using a ratiometric method [4]. The spectral bands recorded with each camera were achieved with the use of long-pass, short-pass, and notch filter combinations and are illustrated with the shaded rectangles in Fig. 1b. The dotted black line denotes the second harmonic wavelength ($\lambda = 527$ nm) of a high-speed Nd:YLF laser light, with a sheet thickness of ~200 μm, which was synced with the cameras and was firing in a single-cavity mode at the same frequency of 2.5 kHz.

The images from the two HS cameras were spatially calibrated and corrections were applied to minimise rotation or magnification differences by using a native spatial calibration algorithm of DaVis 8.4. For the 2cLIF measurements, the images were then corrected for the thermal degradation of the SR dye at elevated temperatures, by quantifying the degradation rates at various temperatures. Additionally, both dyes were found to photo-degrade, so we minimised the overall number of a laser pulses fired, resulting in degradations below 1% and 3% for the FL and SR dyes, respectively. Finally, the particles were removed from the images and the remaining 'holes' were interpolated. The resulting images were used for the ratiometric 2cLIF method to obtain the water temperature field close to the bubble surface. For the PIV measurements, we first applied an algorithmic mask to separate the water from the vapour phase and we then used DaVis PIV cross-correlation algorithm in a three-pass iterative mode to obtain the water velocity field.

The wall temperature measurements were performed with an IR camera, which recorded the thermal radiation emitted by the ITO surface during the boiling experiments. The heater temperature and the heat flux distribution were calculated from the IR images by solving a coupled three-dimensional heat conduction and two-dimensional radiation inverse problem, following the methodology developed by Bucci et al. [5].

3 Results

Figure 2 presents the phase, temperature, and velocity fields for four time-steps during the bubble nucleation, from birth (a) at $t^* = 0$ ms (first time-step of nucleation) to detachment (d) at $t^* = 20$ ms, for a typical nucleating bubble at saturation temperature and for a heat flux of 30 kW m^{-2}. The blue regions denote the vapour phase. The phase boundaries are overshot uniformly by 20% to account for the LIF-measured interface bias. For more details on the bubble size LIF errors, the interested reader is directed to Ref. [3]. The water phase is plotted with temperature contours, which were measured using this ratiometric 2cLIF method. Our measurements have a low signal to noise ratio and sensitivity with the current arrangement, and thus the resulting temperature fields are smoothed with a Gaussian filter to better illustrate the trends. Superheated liquid of about 2 °C was found present close to the nucleation site and the bubble interface, which was displaced away from the thermal boundary layer of the heater during the formation and departure of the bubbles. The temperature gradients measured in the bulk and away from the bubble can be attributed to the natural circulation pattern inside the cell. We are not yet able to capture the thermal

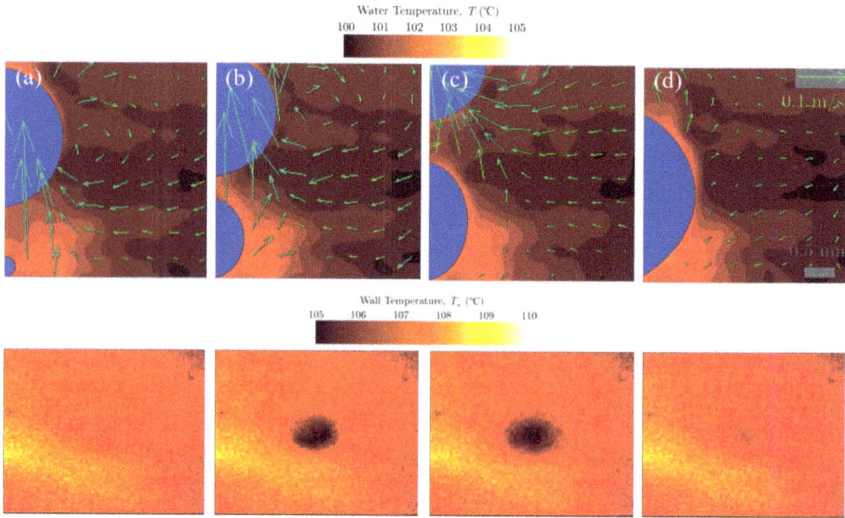

Fig. 2 Top row: temperature (from 2cLIF) and velocity (from PIV) fields around a vapour bubble from nucleation to departure under a nominal heat flux of 30 kW/m^2 and $T_{sat} = 100$ °C for various time-steps: **a** $t^* = 0$ ms; **b** $t^* = 2$ ms; **c** $t^* = 6$ ms; **d** $t^* = 20$ ms. The blue region represents the vapour phase and each temperature contour level corresponds to 0.5 °C. Bottom row: temperature of the heater (from IR) for the same time-steps. The IR images are portraying the whole heater area of 10 mm^2

boundary layer of the heater, primarily due to resolution limitations; we expect the thermal boundary layer to be approximately a few pixels thick for this heat flux.

The velocity fields, calculated with PIV, are also shown in Fig. 2a–d for the same time-steps during the bubble nucleation, growth, and departure. An upwards motion of high velocity magnitude appears during the initial time-steps of the bubble nucleation, which is the contributing factor for the upwards displacement of the superheated liquid from the thermal boundary layer of the heater. This upwards motion decreases with time as the bubble growth rate reduces. Two counteracting vortices form on each side of the bubble after its departure, with their centres propagating upwards and always remaining at the bubble sides. We can also further calculate the velocity gradients (based on the two spatial and velocity components), and information on the mixing patterns formed can be obtained.

The temperature distribution over the heater was recorded with the IR camera during the nucleation and departure of the vapour bubbles. We observe a sharp decrease in the wall temperature below the nucleating bubble. The bubble grows due to the microlayer evaporation and heat transferred from the superheated liquid layer adjacent to the heater, rendering the drop in the temperature to become more prominent, as is also shown in Fig. 2b, c. After the bubble departs and the vapour layer breaks, a liquid film rewets the nucleation spot of the heater and the temperature starts increasing again, to the initial wall superheat level and until the next bubble is generated.

4 Concluding Remarks

This paper describes the technical challenges encountered in, and selected information from, an experimental campaign aimed at measuring single-bubble nucleate boiling with simultaneous laser-based diagnostic (in the liquid) and IR (at the wall) methods. From the resulting temperature fields around departing bubbles, we observed the presence of superheated liquid being transferred to the surrounding water. Vortices were formed on either side of rising bubbles, which contributed to mixing close to the heater. The infrared measurements supported these findings, showing low temperatures at the nucleation sites during bubble formation, which were accompanied by high local heat fluxes. The results show the intrinsic coupled nature of the flow and thermal fields during this important phenomenon. An on-going effort is being made to extract information closer to the thermal boundary layer for low heat fluxes to enable a detailed analysis of the heat flux partitioning and for model validation purposes.

Acknowledgements This work was supported by the MIT International Science and Technology Initiatives (MISTI) Global Seed Funds, the Department for International Development (DFID) through the Royal Society-DFID Africa Capacity Building Initiative, the Imperial College London Faculty of Engineering Dame Julia Higgins Postdoc Collaboration Fund and Chevron Corporation. Data supporting this publication can be obtained on request from cep-lab@imperial.ac.uk.

References

1. M. Magnini, B. Pulvirenti, J.R. Thome, Numerical investigation of hydrodynamics and heat transfer of elongated bubbles during flow boiling in a microchannel. Int. J. Heat Mass Transf. **59**, 451–471 (2013)
2. E.R. Smith, P.E. Theodorakis, R.V. Craster, O.K. Matar, Moving contact lines: linking molecular dynamics and continuum-scale modeling. Langmuir **34**, 12501–12518 (2018)
3. V. Voulgaropoulos, G.M. Aguiar, O.K. Matar, M. Bucci, C.N. Markides, Temperature and velocity field measurements of pool boiling using two-colour laser-induced fluorescence, infrared thermometry and particle image velocimetry, in *10th International Conference on Multiphase Flow (ICMF)*, Rio de Janeiro, Brazil (2019)
4. W. Chaze, O. Caballina, G. Castanet, F. Lemoine, Spatially and temporally resolved measurements of the temperature inside droplets impinging on a hot solid surface. Exp. Fluids **58**, 96 (2017)
5. M. Bucci, A. Richenderfer, G.Y. Su, T. McKrell, J. Buongiorno, A mechanistic IR calibration technique for boiling heat transfer investigations. Int. J. Multiph. Flow **83**, 115–127 (2016)

Experimental Observations of Flow Boiling in Horizontal Tubes for Direct Steam Generation in Concentrating Solar Power Plants

Hannah R. Moran, Victor Voulgaropoulos, Dimitri Zogg, Omar K. Matar, and Christos N. Markides

1 Introduction

Direct steam generation (DSG) for concentrating solar power (CSP) is an emerging technology that can unlock new avenues for efficient and affordable energy utilisation and expand the current capabilities of CSP. However, the direct evaporation of steam in parabolic trough solar collectors presents control and operational challenges due to the inherently complex two-phase nature and inherent unsteadiness of the boiling flow inside the tubes. Thus, a fundamental understanding of the hydrodynamic and heat transfer characteristics of this two-phase flow is required.

Many studies have investigated flow boiling by measuring integral quantities such as pressure gradients [1] and heat transfer coefficients [2]. The flow boiling of refrigerants, in particular, has been widely studied, but the literature lacks detailed spatiotemporally resolved information on the velocity and phase-distribution fields, e.g., the thickness and frequency characteristics of films in annular flow. Laser-diagnostic techniques based on particle image velocimetry (PIV) and laser-induced fluorescence (LIF) have been used in two-phase flows [3–6] to provide such data, give insight into thermal and hydrodynamic interactions, and promote our understanding of the phenomena involved in these flows.

2 Methodology

A new experimental facility has been constructed to investigate the flow boiling of refrigerant R245fa in horizontal pipes using laser-diagnostic techniques. The flow loop, as shown in Fig. 1, consists of a vertical receiver from which liquid is pumped

H. R. Moran (✉) · V. Voulgaropoulos · D. Zogg · O. K. Matar · C. N. Markides
Department of Chemical Engineering, Imperial College London, London SW7 2AZ, UK
e-mail: hannah.moran11@imperial.ac.uk

© Springer Nature Singapore Pte Ltd. 2021
C. Wen and Y. Yan (eds.), *Advances in Heat Transfer and Thermal Engineering* ,
https://doi.org/10.1007/978-981-33-4765-6_32

175

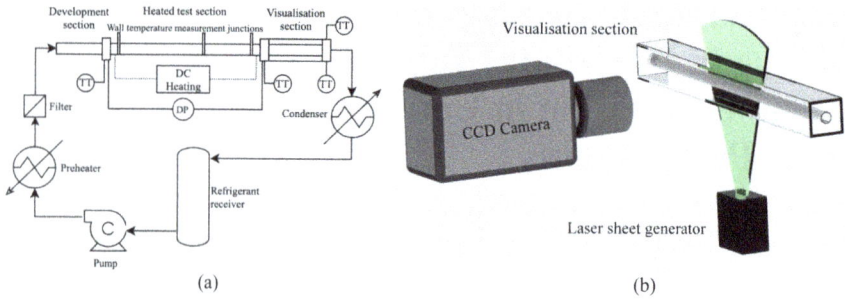

Fig. 1 **a** Flow diagram of the experimental facility. **b** Illustration of the arrangement for the optical measurements

to a preheater, followed by a filter and a turbine flow metre. The inlet to the heated test section consists of a 60-cm long, 12.6-mm inside diameter flow straightening section to establish a developed flow. A 2-m long, 12.6-mm inside diameter stainless-steel test section is heated using direct current with a maximum supply of 135 kW/m^2 of uniform heating along the pipe. The two-phase flow generated in the heated test section enters a visualisation section consisting of a FEP pipe enclosed in a Perspex correction box filled with deionised water to minimise distortions due to the curvature of the pipe. Downstream of the visualisation section, the flow is cooled and condensed before being returned to the receiver.

In-flow temperature measurements are taken at the inlet and outlet of each section, and wall temperature measurements are taken at three junctions along the heated test section. To obtain measurements in the visualisation section, PIV is employed. The flow is seeded with silver-coated tracer particles of 10 μm size and the test section is illuminated by a laser sheet. The scattered particle light is recorded by a high-speed CCD camera, positioned perpendicularly to the laser light sheet, as shown in Fig. 1b. The camera was also used in a volumetric illumination arrangement, employing a backlight to qualitatively identify and record images of the different flow patterns.

3 Results

The operation of the facility during single-phase water and single-phase refrigerant flow has been compared to well-established pressure drop and heat transfer coefficient correlations and prediction methods. Preliminary refrigerant two-phase flow experiments have been performed, and high-speed imaging was employed to identify the flow patterns. It has been shown that a full range of flow patterns can be achieved using the experimental apparatus described in Sect. 2, and a selection of representative images is displayed in Fig. 2.

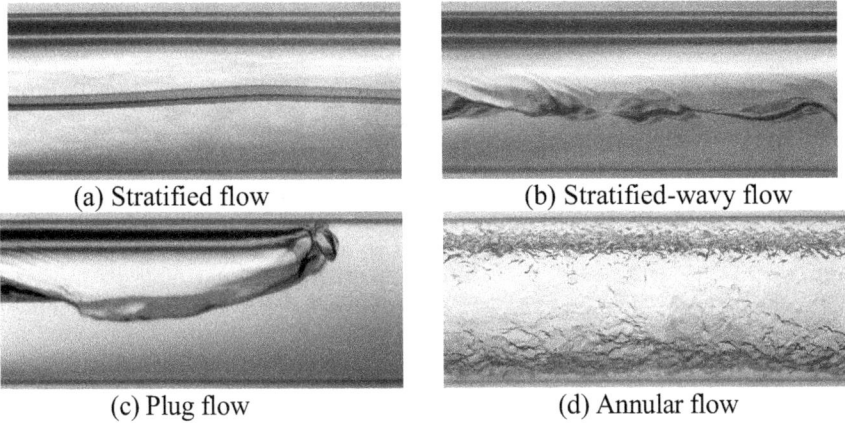

(a) Stratified flow   (b) Stratified-wavy flow

(c) Plug flow   (d) Annular flow

Fig. 2 High-speed images of various flow patterns

Preliminary measurements demonstrated thermal and hydrodynamic behaviours in agreement with those reported in the available literature, e.g., heat transfer coefficients were higher for annular flow than other flow patterns, with a subsequent drop-off indicating the occurrence of dryout. A proof-of-concept for the novel application of PIV to horizontal boiling flows has also been performed, simultaneously recording the interface location and the velocity field primarily in stratified and plug flow configurations.

4 Conclusions

A new experimental facility for the investigation of flow boiling of R245fa in horizontal pipes using laser-diagnostic techniques has been constructed and validated in single-phase flow conditions. Preliminary two-phase flow experiments show agreement with existing literature. The applicability of PIV to this boiling flow has been established, and work is on-going to refine this technique and produce simultaneous phase and velocity field data for a wide range of flow conditions, in particular, with the use of planar LIF (PLIF) methods.

Acknowledgements This work was supported by the UK Engineering and Physical Sciences Research Council (EPSRC) [EP/P004709/1], and the Department for International Development (DFID) through the Royal Society-DFID Africa Capacity Building Initiative. Data supporting this publication can be obtained on request from cep-lab@imperial.ac.uk. Thanks are also extended to Marit Mohn for her Ph.D. scholarship support.

References

1. O. Turgut, M. Coban, M. Asker, Comparison of flow boiling pressure drop correlation for smooth macrotubes. Heat Transf. Eng. **37**, 487–506 (2016)
2. X. Fang, F. Zhuang, C. Chen, Q. Wu, Y. Chen, Y. Chen, Y. He, Saturated flow boiling heat transfer: review and assessment of prediction methods. Heat Mass Transf. **55**, 197–222 (2018)
3. A. Charogiannis, J. An, C.N. Markides, A simultaneous planar laser-induced fluorescence, particle image velocimetry and particle tracking velocimetry technique for the investigation of thin liquid-film flows. Exp. Thermal Fluid Sci. **68**, 516–536 (2015)
4. I. Zadrazil, O.K. Matar, C.N. Markides, An experimental characterisation of downwards annular flow by laser-induced fluorescence: flow regimes and film statistics. Int. J. Multiph. Flow **67**, 42–53 (2014)
5. A. Charogiannis, C.N. Markides, Spatiotemporally resolved heat transfer measurements in falling liquid-films by simultaneous application of planar laser-induced fluorescence (PLIF), particle tracking velocimetry (PTV) and infrared (IR) thermography. Exp. Thermal Fluid Sci. (2018) (in press)
6. A. Charogiannis, I. Zadrazil, C.N. Markides, Thermographic particle velocimetry for simultaneous interfacial temperature and velocity measurements. Int. J. Heat Mass Transf. **97**, 589–595 (2016)

Multiscale Investigation of Nucleate Boiling and Interfaces

E. R. Smith, M. Magnini, and V. Voulgaropoulos

1 Introduction

Nucleate boiling is still poorly understood, as it involves a multitude of scales, from the bubble formation at the scale of individual molecules, to its detachment as a bubble spanning millimetres or more. The state-of-the-art research on nucleate boiling is based on three methods, each with limitations: (i) experiments can track bubble growth but it is challenging to capture and understand the nanoscopic inception of the bubble. (ii) Computational fluid dynamics (CFD) provide insight into the bubble growth and its dynamics, but the modelling process necessitates empirical or simplified descriptions of nucleation and phase change. (iii) Molecular dynamics (MD) provides fundamental information on the essential nature of nucleation, but it is not yet clear if simulations at such small spatial and temporal scales are relevant to engineering problems. In this work, we combine the three techniques in order to gain unique insights into the process of boiling. Molecular dynamics is applied to capture the dynamics of the initial nucleation of a bubble, which can be combined with CFD to model the bubble growth, rise and eventually its detachment. This coupled modelling approach allows the similarities and differences between the MD and CFD approaches to be explored, guided by comparison to experimental results.

E. R. Smith (✉)
Aerospace and Mechanical Engineering, Brunel University London, Kingston Lane, Uxbridge UB8 3PH, UK
e-mail: edward.smith@brunel.ac.uk

M. Magnini
Faculty of Engineering, University of Nottingham, Nottingham NG7 2RD, UK

V. Voulgaropoulos
Chemical Engineering, Imperial College London, South Kensington Campus, London SW7 9AZ, UK

© Springer Nature Singapore Pte Ltd. 2021
C. Wen and Y. Yan (eds.), *Advances in Heat Transfer and Thermal Engineering* ,
https://doi.org/10.1007/978-981-33-4765-6_33

2 Methodology

For the experiments (i), a series of synchronous high-speed laser- and infrared-based measurements were performed in a pool boiling cell filled with deionised water at saturation temperature, as shown in Fig. 1. Nucleation was achieved by supplying current to a thin-film indium tin oxide heater coated on an infrared transparent sapphire glass covering heat fluxes in the range of 30 to 60 kW m^{-2}. The generation of isolated vapour bubbles was achieved by coating a small area of the heater with a hydrophobic (PTFE) material. Two-colour laser-induced fluorescence measurements were performed by adding a pair of dyes, and by having their individual fluorescence monitored by two cameras, which were connected to a pellicle beamsplitter and a 105 mm lens. A ratiometric method was applied to obtain spatiotemporally resolved information on the temperature of the water close to the heater and bubble surfaces [1]. Particle image velocimetry was employed to obtain the water velocity and velocity-gradient fields. The temperature distribution over the heater surface was obtained with an IR camera and by solving a coupled three-dimensional heat conduction and two-dimensional radiation inverse problem.

The CFD model (ii) is based on a minimal scripted solution of the Navier–Stokes and energy equation for a two-phase flow; it utilises a front-tracking method to model the sharp liquid–gas interface, and thus have a representation of interfacial effects such as surface tension and phase change [2]. The phase change is modelled by means of the Hertz–Knudsen–Schrage equation and therefore the local rate of evaporation is calculated proportional to the local superheating of the interface above the saturation temperature. All coefficients including surface tension, viscosity and heat flux are

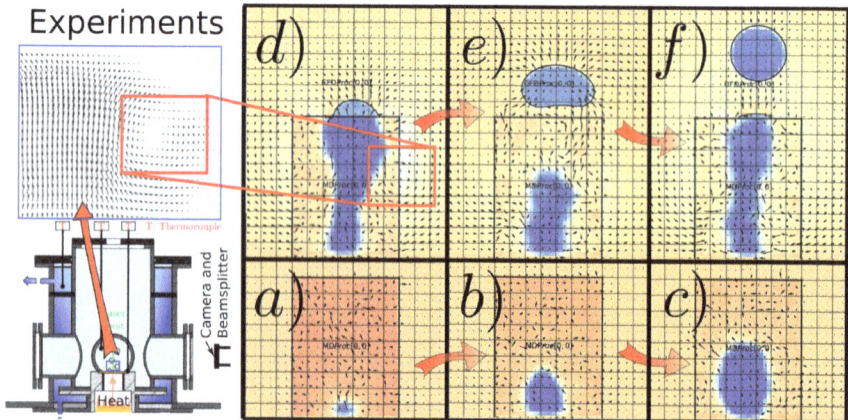

Fig. 1 Experimental schematic and results (left) compared to coupled MD-CFD boiling (right) with density shown by colour (blue low, yellow high) and vectors showing flow fields for **a** nucleation of a bubble in MD, **b** and **c** its growth, **d** the start of the transfer to CFD by boundary conditions with interface tracking, **e** complete transfer and rising in CFD and **f** start of the transfer of a second bubble. The red box highlights one similarity to the experimentally measured flow field

obtained from a-priori MD simulations. The key details of the nucleation of the initial bubble and the movement of the contract line cannot be easily modelled by CFD simulation and so are delegated to the MD solver.

The MD (iii) solves Newton's law with inter-molecular interactions based on electrostatic potentials. As a result, it is the most fundamental classical model for both solids and fluids. Far from the oft-stated picture of random particle collisions, liquids are governed by the deformation and evolution of a lattice structure. It is this structure that allows viscosity and surface tension to be measured from an MD simulation and nucleation to be reproduced. A parallelised in-house solver, Flowmol, is used to simulate boiling with MD, using a simple Lennard-Jones model, with variable wetting between the liquid and thermostatted tethered walls [3].

The MD simulations is coupled to CFD via the boundary conditions, i.e. nucleation occurs in the MD and the density, velocity and temperature are used to set time-dependent boundary values in part of the CFD domain, a technique established in the literature for single phase flows and fixed boundaries [4] and extended here for arbitrary regions, interface tracking and phase change.

3   Results

The results from the coupled MD-CFD simulation are presented in Fig. 1; the box at the bottom of the domain is the time-evolving MD results used to specify mass, momentum and temperature in a section of the CFD domain. The nucleation of the bubble occurs naturally in the MD and eventually creates an interface as it crosses into the CFD domain with a self-contained bubble eventually passed completely to the CFD model.

Fluid features such as a recirculation vortex next to the rising bubble are compared to experimental results, along with temperature and density fields. The flow field around the rising bubble shows good agreement to experimental measurements, despite crossing the interface between the continuum and molecular solvers. In addition to the notable similarities, there are differences between the experimental and model results. The separation of the bubble in the MD occurs slower, growing like a plume, which eventually pinches off due to thinning by the recirculation vortex. The experiments exhibit a much cleaner bubble departure. Such differences are not at all surprising given the discrepancy in length scales between the two systems, where the assumption in comparing nano-scale coupled models to experiments is the validity of dimensional similitude at the atomic scale. Through comparisons between the simulation and experiments, we gain insights into the nucleation process and validate the modelling approach developed herein.

4 Conclusions

The molecular simulation captures the nanoscopic origins of nucleation in a range of geometries. A nucleated bubble is then passed to a minimal CFD solver, which models the evolution and growth of this bubble. The coupled system is compared to the results from experiments. This approach allows a range of interesting questions to be addressed. How much complexity in the CFD model is required to match the MD behaviour? Do the predicted results quantitatively and qualitatively match the experiments, and if not, why not? Is any of the fluid dynamics observed in the MD system incompatible with the CFD model? These questions are of fundamental importance; if we can show that the modelling approach describes the experimental results, we can apply this to the construction of surface features, modelling the uniform heating of a random fractal wall, and the essential interplay of surface and liquid can be studied at the first moment of a bubble lifetime. This allows detailed insight into exactly how surface design effects the onset of bubble nucleation, as well as an unprecedented insight into the science of nucleation. In this work, we start to answer these questions and highlight the potential for simulations of this type.

Acknowledgements VV is grateful to Prof. Christos N. Markides and Prof. Matteo Bucci for the support with the experiments. ES and MM acknowledge support from Prof. Omar Matar and the Matar fluids group. All authors acknowledge funding from the Julia Higgins postdoctoral collaboration fund at Imperial College, London.

References

1. V. Voulgaropoulos, G.M. Aguiar, O.K. Matar, M. Bucci, C.N. Markides, Temperature and velocity field measurements of pool boiling using two-colour laser-induced fluorescence, infrared thermometry and particle image velocimetry, in *10th International Conference on Multiphase Flow (ICMF)*, Rio de Janeiro, Brazil (2019)
2. A. Prosperetti, G. Tryggvason, *Computational Methods for Multiphase Flow* (Cambridge University Press, Cambridge, 2009)
3. E.R. Smith, Ph.D. thesis, Imperial College London (2014)
4. S. O'Connell, P. Thompson, Molecular dynamics continuum hybrid computations. Phys. Rev. E **52**, 6 (1995)

Effect of Surface Wettability on Nanoscale Boiling Heat Transfer

Longyan Zhang, Jinliang Xu, Junpeng Lei, and Guanglin Liu

1 Introduction

In recent years, rapid development of microelectronic manufacturing technologies has attracted growing interest in the thermal management of nanoelectronics. Nucleate boiling of thin liquid films confined in a nanochannel plays a key role in electronic cooling. Classical theory indicates that high superheating is required for bubble nucleation on hydrophobic surfaces [1]. However, previous studies provided insights into the characteristics of phase change in nanoscale thermal systems [2], which was opposite to the prediction by classical nucleation theory and the observation in pool boiling experiments. Hence, a deeper understanding of nanoscale nucleation mechanisms is necessary to improve the thermal management efficiency in engineering processes.

Molecular dynamics simulation is a useful tool for investigating the mechanism of nucleation boiling at the nanoscale [2–5]. Nagayama et al. [3] found that solid–liquid interfacial interaction has a significant influence on bubble nucleation behavior. In particular, homogeneous nucleation occurred in the case of a hydrophilic surface, where a bubble was generated at the center of a nanochannel, while heterogeneous nucleation appeared in the case of a hydrophobic surface, where a bubble was directly formed on the solid wall. In addition, a non-wetting surface was also favorable to the formation of vapor films. Hens et al. [2] simulated the phase change process of liquid argon films with a 2 nm thickness on smooth platinum substrates. Their results showed that at a constant wall temperature of 250 K, explosive boiling occurs on the hydrophilic surface, while only evaporation appears on the hydrophobic surface. Apparently, a higher wall temperature is required to trigger explosive boiling on the hydrophobic surface. Wang et al. [4] indicated that a high surface wettability reduced

L. Zhang · J. Xu (✉) · J. Lei · G. Liu
Beijing Key Laboratory of Multiphase Flow and Heat Transfer, North China Electric Power University, Beijing, China
e-mail: xjl@ncepu.edu.cn

© Springer Nature Singapore Pte Ltd. 2021
C. Wen and Y. Yan (eds.), *Advances in Heat Transfer and Thermal Engineering* ,
https://doi.org/10.1007/978-981-33-4765-6_34

the onset temperature. She et al. [5] studied bubble formation on a solid wall that contained a cavity. Their simulation results indicated that the cavity was conducive to the growth of the bubble. Compared to the liquid atoms near the plane surface, those in the cavity experienced much larger repulsive forces and repulsive force gradients along the direction perpendicular to the solid walls, leading to an increase in both the density gradients and potential energy.

Although previous studies provided insights into the properties of phase changes on substrates in nanoscale thermal systems, many of them focused on evaporation and explosive boiling phenomena. The effect of wall wettability on the nucleation boiling of liquid films has rarely been studied despite their extensive applications in engineering and science with the advances in nanotechnology. In this study, molecular dynamics simulations are conducted to investigate the bubble nucleation behavior of liquid films on solid surfaces under different wettability and temperature. The main goal of this study is to investigate how does the wall wettability influences the onset temperature of nucleation boiling and the nucleation time.

2 Methodology

In this paper, molecular dynamics method is utilized to investigate the phase change behavior of an argon film placed on a cavity substrate. Figure 1 shows the simulation system. The computation domain has a size of $L_x = 58.87$ nm, $L_y = 1.96$ nm, and $L_z = 157.04$ nm. Periodic boundary conditions are applied along x- and y-directions. The simulation system is composed of solid platinum atoms, liquid argon atoms, and vapor argon atoms. The vapor layer is sandwiched between the liquid layers in the z-direction to maintain the supersaturated state of the liquid surrounding the bubble. A total of 95,035 liquid atoms are divided into two parts to be placed on the top and bottom walls with thicknesses of $H_2 = 3.92$ nm and $H_1 = 39.26$ nm, respectively. The top wall consists of 15,600 solid atoms arranged as an FCC lattice structure. Its $\langle 1\ 0\ 0 \rangle$ surface is in contact with the argon atoms. Similarly, the bottom wall with the same structure consists of 13,500 solid atoms. The nanostructure depth is $h = 5.89$ nm, and the width is $w = 6.28$ nm. The Langevin thermostat technique is applied to control the wall temperature. The governing equation is as follows [6]:

$$\frac{\mathrm{d}p_i}{\mathrm{d}t} = -\xi p_i + f(t) + F(t) \tag{1}$$

where p_i is the momentum vector of the ith solid atom, $\xi = 77.92$/ps is the damping constant, and $f(t)$ represents the interaction force vector between atoms. $F(t)$ represents the exciting force vector, whose components are randomly sampled from Gaussian distribution with zero mean average value and standard deviation of $\sigma_G = \sqrt{2\xi k_B T_w / \delta_t}$, where $k_B = 1.38 \times 10^{-23}$ J/K is the Boltzmann constant and $\delta_t = 5$ fs is the time step of the simulation.

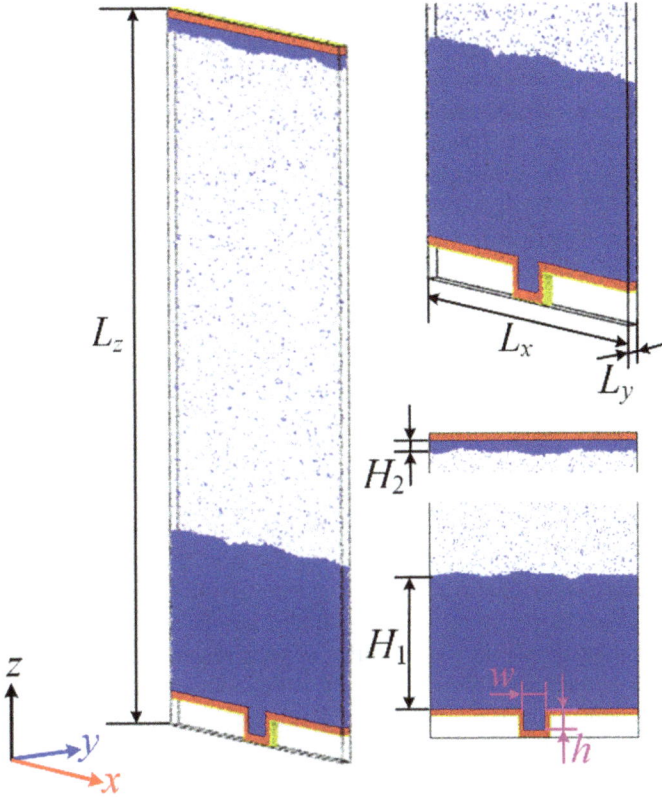

Fig. 1 Initial configuration of the simulation system

The pair interaction uses the Lennard-Jones (L-J) potential [7]

$$\phi(r) = 4\alpha\varepsilon_{sl}\left[\left(\frac{\sigma_{sl}}{r}\right)^{12} - \beta\left(\frac{\sigma_{sl}}{r}\right)^6\right] \tag{2}$$

where ε is the energy scale, σ is the length scale, α is the potential energy factor indicating the strength of hydrophilic interaction, and β is the potential energy factor indicating the attraction for hydrophobic interaction. In this paper, we fix $\alpha = 0.14$ but β is changed from case to case to show the wall wettability effect, where the contact angles of 91.3° and 0° correspond to $\beta = 0.5$ and 1.0, respectively. Reference [3] shows that the wall wettability increases with the increase of parameter β. For solid–solid and liquid–liquid computations, the parameters are set as $\alpha = 1$ and $\beta = 1$. The parameters are $\varepsilon_1 = 1.67 \times 10^{-21}$ J and $\sigma_1 = 3.405 \times 10^{-10}$ m for argon, and they are $\varepsilon_s = 83.5 \times 10^{-21}$ J and $\sigma_s = 2.475 \times 10^{-10}$ m for solid. For solid–liquid interaction, $\varepsilon_{sl} = (\varepsilon_1 \cdot \varepsilon_s)^{0.5}$ and $\sigma_{sl} = 0.5(\sigma_1 + \sigma_s)$ are applied, which is called the Lorentz-Berthelot combining rule [7].

The Velocity-Verlet method is used for integrating the momentum equation. The simulation includes two stages, the equilibrium stage and the bubble nucleation stage. In the equilibrium stage, the simulation is performed in canonical ensemble (NVT) for 3 ns. After achieving the equilibrium state, the system temperature and pressure are kept constant at $T = 100$ K and $P = 0.324$ MPa, respectively. The following period of 30 ns is the phase change stage. The temperature of the bottom heating wall rises to $T = 157.32$ K, while that of the top wall is changed to $T = 87$ K, where argon atoms are in the liquid phase. Simulations were carried out using the open-source large-scale atomic/molecular massively parallel simulator (LAMMPS) [8] and data visualization was performed using the software OVITO [9].

3 Results

The evolution of phase change on surfaces with various wettability was successfully observed after the temperature of the bottom substrate rose to $T = 157.32$ K. First, to verify bubble nucleation conditions at the nanoscale, the relationship between the liquid temperature and the density of the nucleation region was examined to compare them with the coexistence curve and the spinodal line. Two cases with $\beta = 0.5$ and 1.0 were selected to demonstrate the reliability of the simulation results. The nucleation region, which is the region of the biggest bubble, is obtained based on the snapshots of the phase change. The liquid density of the nucleation region is compared with the coexistence curve and the spinodal line [10] in Fig. 2. The results show that in case of $\beta = 1.0$, the liquid density gradually decreases and finally exceeds the spinodal line, indicating that the bubble nucleation occurs. While in case of $\beta = 0.5$, the liquid density curve only rises around the saturated liquid line, indicating that the

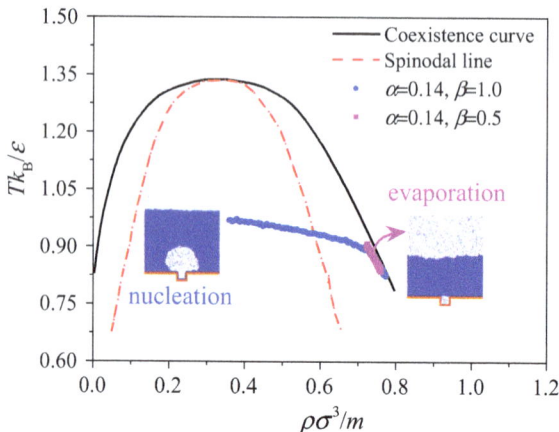

Fig. 2 Simulation points and phase diagram with coexistence curve and spinodal line

evaporation phenomenon occurs. Thus, the liquid film in the simulations satisfies the condition of bubble nucleation.

Figure 3 illustrates the snapshots of phase change for the liquid film on surfaces with $\beta = 0.5$ and 1.0. After the temperature of the bottom wall rises, the heat transfer occurs between the liquid argon and the solid substrate and then the bubble nucleus appears where the heat energy is sufficiently high. In the case of $\beta = 1.0$, a small bubble nucleus is formed in the nanostructure at 0.8 ns, indicating the start of nucleation boiling (Fig. 3a). With the increasing heating time, the bubble nucleus grows upward and sideward. Another small bubble nucleus appears on the two sides of the bottom substrate at 3.8 ns. Then, the bubbles are merged into a bigger bubble at 4.25 ns. Eventually, the coalescence happens due to the period boundary conditions used in the directions parallel to the bottom wall. Nucleation boiling changes to the film boiling mode. Compared to the surface with strong wettability, phase change behavior is significantly different on the surface with weak wettability. As shown in Fig. 3b, in case of $\beta = 0.5$, bubble nucleus appears in the nanostructure, owing to the strong repulsive force of surface. While the bubble nucleus keeps unchanged, and only evaporation occurs at the liquid–vapor interface. This behavior indicates that surface wettability has a significant influence on the bubble nucleation of liquid films in nanoscale. The strong wettability surface facilitates local heat accumulation and thus strengthens local atomic collisions. The results indicate that bubble nucleation prefers the hydrophilic surface.

Figure 4a shows the time evolution of bubble equivalent radiuses. It is observed that the bubble growth trend under various wettability conditions is similar. In the initial heating stage, the equivalent radius of the bubble keeps almost 0, due to only evaporation of the liquid film. As heating time goes on, the bubble generates in the cavity when the heat energy is sufficiently high. Then the bubble grows and the equivalent radius increases with the heating time. During the bubble growth stage, the bubble growth rate increases with the increase of the wall wettability. Further heating

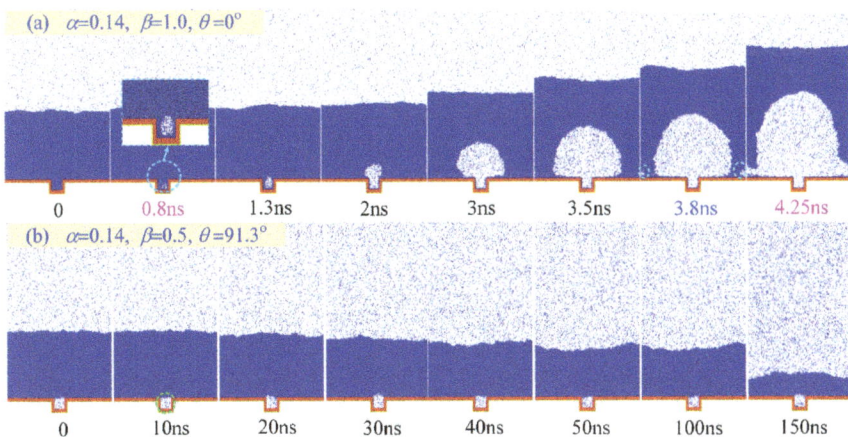

Fig. 3 Snapshots of liquid argon nucleation boiling process **a** $\beta = 1.0$ and **b** $\beta = 0.5$

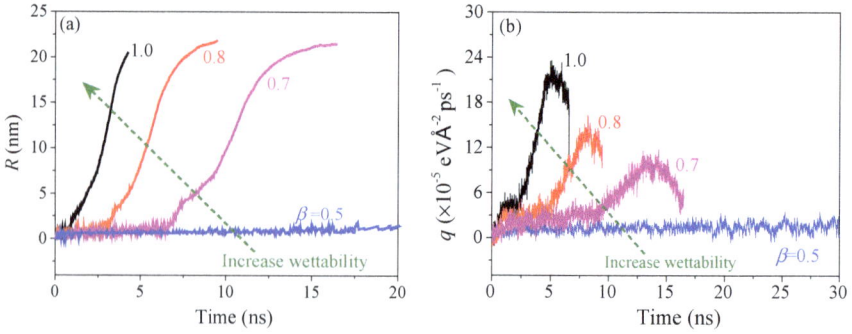

Fig. 4 Time evolution of **a** bubble equivalent radius and **b** the absorbed heat flux of argon on various surfaces

the liquid, the nucleation boiling mode turns into the film boiling mode and the vapor volume in nucleation region keep unchanged. It is worth noting that for the solid–liquid potential energy parameter of $\beta = 0.5$, the volume of embryo formed in the nanocavity keeps unchanged, indicating that the nucleation boiling does not occurs and only the evaporation phenomenon exists. Therefore, the hydrophilic surface promotes the nucleation boiling, indicating that the onset temperature of nucleation boiling increases with the increase of the surface wettability.

To further compare the heat transfer efficiency of the surfaces with various wettability, heat flux through the liquid argon film is calculated, as shown in Fig. 4b. The heat flux curves have four stages, initial slow increase, smooth transition, the rapid increase and the rapid decrease stage, corresponding to initial heating, stable evaporation, nucleation boiling and film boiling periods. For potential parameters of $\beta = 1.0$, heat flux rapidly increases, nearly without a smooth transition. For potential parameters of $\beta = 0.7$ and $\beta = 0.8$, the heat flux curves are almost synchronized at the initial stage until nucleation boiling occurs. In general, heat transfer between the hot substrate and the liquid atoms is determined by solid–liquid atomic collisions; more liquid atoms that participate in the collision with the surface lead to a stronger heat transfer efficiency. Hence, the strong wettability surface that holds more liquid atoms will shorten the time of the heat transfer to a greater extent than the surface with weak solid–liquid interaction. The turning points of the heat flux curves correspond to the incipient nucleation time and the heat flux plunges after bubble departure, resulting from the large thermal resistance of the vapor film. Thus, the surface with stronger solid–liquid interaction provide favorable conditions for bubble nucleation and the formation of vapor films, which shorten the incipient nucleation time and reduce the onset temperature of nucleation boiling because of the strong heat transfer efficiency.

4 Conclusions

Molecular dynamics simulations were conducted to investigate the effect of surface wettability on bubble nucleation behavior. The simulation results show that the onset of nucleate boiling and bubble nucleation rate is strongly dependent on surface wettability. The onset temperature of nucleate boiling decreases significantly with increases of surface wettability because of the strong heat transfer efficiency. The bubble embryo is found to generate rapidly and bubble grows faster on surface with strong liquid–solid interaction, which does not support the classical bubble nucleation theory regarding the wettability effect. As the heating time goes on, when the bubble volume attains a critical value, the boiling mode is switched from nucleate boiling to Leidenfrost vapor film on the wall, corresponding to the increase stage and decrease stage of dynamic heat fluxes, respectively. The findings enhance the understanding of nanoscale boiling heat transfer and provide the guideline for the development of micro/nano-thermal management devices.

Acknowledgements The authors appreciate financial support provided by the National Natural Science Foundation of China (51436004).

References

1. C.H. Wang, V.K. Dhir, Effect of surface wettability on active nucleation site density during pool boiling of water on a vertical surface. J. Heat Transf. **115**, 659–669 (1993)
2. A. Hens, R. Agarwal, G. Biswas, Nanoscale study of boiling and evaporation in a liquid Ar film on a Pt heater using molecular dynamics simulation. Int. J. Heat Mass Transf. **71**, 303–312 (2014)
3. G. Nagayama, T. Tsuruta, P. Cheng, Molecular dynamics simulation on bubble formation in a nanochannel. Int. J. Heat Mass Transf. **49**, 4437–4443 (2006)
4. Y.H. Wang, S.Y. Wang, G. Lu, X.D. Wang, Explosive boiling of nano-liquid argon films on high temperature platinum walls: effects of surface wettability and film thickness. Int. J. Therm. Sci. **132**, 610–617 (2018)
5. X. She, T.A. Shedd, B. Lindeman, Y. Yin, X. Zhang, Bubble formation on solid surface with a cavity based on molecular dynamics simulation. Int. J. Heat Mass Transf. **95**, 278–287 (2016)
6. P. Yi, D. Poulikakos, J. Walther, G. Yadigaroglu, Molecular dynamics simulation of vaporization of an ultra-thin liquid argon layer on a surface. Int. J. Heat Mass Transf. **45**, 2087–2100 (2002)
7. J. Delhommelle, P. Millié, Inadequacy of the Lorentz-Berthelot combining rules for accurate predictions of equilibrium properties by molecular simulation. Mol. Phys. **99**, 619–625 (2001)
8. S. Plimpton, Fast parallel algorithms for short-range molecular dynamics. J. Comput. Phys. **117**, 1–19 (1993)
9. A. Stukowski, Visualization and analysis of atomistic simulation data with ovito-the open visualization tool. Modell. Simul. Mater. Sci. Eng. **18**, 2154–2162 (2010)
10. T. Kinjo, M. Matsumoto, Cavitation processes and negative pressure. Fluid Phase Equilib. **144**, 343–350 (1998)

Effect of Bulk Temperature on the Evaporation of a Nitrogen Drop in Different Immiscible Liquids

Neville Rebelo, Huayong Zhao, François Nadal, Colin Garner, and Andrew Williams

1 Introduction

Cryogenic fluids such as liquid nitrogen can be used to produce power with negligible tailpipe emissions. Since the 1990s, researchers have extracted energy from liquid nitrogen by expansion in various thermal cycles; however, these cycles produced limited power and had low efficiencies due to the low evaporation rate in the proposed indirect heat exchange system [1]. To overcome this limitation of low evaporation rate, a novel approach of directly injecting liquid nitrogen (LN_2) into a heat exchange fluid was proposed. The achievable heat flux during direct contact heat exchange was estimated to be between 0.9 and 5.5 MW/m^2, but these values have not been verified experimentally [2, 3].

In this work, quantitative evaluations of evaporation rates of a nitrogen droplet immersed into different immiscible liquids maintained at different bulk temperatures were carried out by accurately tracking the three-dimensional evaporation process using a high-speed backlight imaging system. A comparison of the evaporation rates measured by the surrounding bubble growth and heat transfer rates at different bulk temperatures has been made for a droplet evaporating in 2-propanol, methanol, n-pentane and n-hexane.

In summary, earlier works estimated the heat transfer rates during nitrogen evaporation in a heat exchange fluid based on the measured pressure traces while obviating the measurement of the interfacial area that is essential for an accurate measurement of the heat transfer rate, this study provides a more accurate quantification of the evaporation rate which is crucial for power generation in cryogenic energy systems.

N. Rebelo (✉) · H. Zhao · F. Nadal · C. Garner
Wolfson School of Mechanical, Electrical and Manufacturing Engineering, Loughborough University, Loughborough LE11 3TU, Leicestershire, UK
e-mail: N.J.Rebelo@lboro.ac.uk

A. Williams
Department of Mechanical Engineering, University of Chester, Chester CH1 4BJ, UK

© Springer Nature Singapore Pte Ltd. 2021
C. Wen and Y. Yan (eds.), *Advances in Heat Transfer and Thermal Engineering* ,
https://doi.org/10.1007/978-981-33-4765-6_35

2 Methodology

An experimental set-up for visualizing a LN_2 droplet evaporation process and quantifying the heat transfer rate was built. The set-up consisted of an LN_2 drop injector built in-house, a test section (Hellma) and an optical imaging system. Two identical backlight imaging systems, each consisting of a LED light source (GSVITEC), a condensing lens, a diffuser, a focussing lens and a camera (VEO710L, Phantom), placed orthogonally to each other were used to visualize the droplet evaporation process. The calibrated images had pixel size of 14.0845 μm/pixel, field of view (FoV) of 16.8 × 11.27 mm and a depth of field (DoF) of approximately 5 mm.

The captured images were then post-processed and analysed using a programme built in MATLAB that processed each pair of images from the two cameras using either the Sobel or Canny edge detection functions [4] to binarise the images. An ellipse closest to the shape the nitrogen droplet/vapour bubble was then estimated using the randomized Hough transform. A 3D point cloud was finally generated from the two views, with a volume computation using the alpha shape function in MATLAB. The MATLAB programme was validated using a precision grade 3 ceramic ball captured at different image contrasts controlled by the camera exposure time. In the worst case, this led to a ±3.86% error in the ball diameter, ±5.45% in the surface area, and ±6.68% in the ball volume.

A LN_2 droplet having a density $\rho_d = 807$ kg m^3, viscosity $\mu_d = 161 \times 10^{-3}$ Pa s, surface tension $\sigma_d = 11$ mN m^{-1}) was injected into 2-propanol, methanol, n-hexane and n-pentane maintained at different bulk temperatures. The evaporation process was recorded by the imaging system for post-processing to quantify the evaporation rate.

The experimental data obtained was scaled using the classical D^2-law [5] that accurately predicts the diffusion-controlled evaporation of a droplet in air at normal temperature pressure (i.e. 1 bar, 294 K), to obtain a relation of normalized bubble growth $\left(\frac{V_b - V_0}{v_0}\right)$ against the dimensionless time $\left(\tau = \frac{\alpha t}{r_0^2}\right)$ as shown in Eq. (1), where V_b is the volume of the bubble at time t, V_0 is the initial volume of the bubble, v_0 is the initial droplet volume, r_0 is the initial droplet radius and α is the diffusion parameter. The constant of proportionality κ is computed from the nitrogen bubble density ρ_b, mean temperature within the bubble $\overline{T}$, atmospheric pressure P_a assumed within the bubble, the ideal gas constant $\mathfrak{R}$ and the molecular weight of nitrogen $\mathcal{M}_{N_2}$.

$$v_0^{-1} \frac{dV_b}{d\tau} = \kappa, \text{ where } \kappa = \frac{3}{2} \frac{\rho_b \mathcal{R} \overline{T}}{\mathcal{M}_{N_2} P_a} \tag{1}$$

3 Results

A LN_2 droplet boiling in different bulk liquids (Fig. 1) shows an increase in the surrounding bubble growth rate with an increase in the bulk liquid temperature when the initial droplet size is similar. The initial droplet size has a stronger influence on the bubble growth, possibly due to the convection effects of the vapour within the bubble and the complex motion of the droplet within the bubble, which have been ignored in the scaling analysis.

Figure 2 shows that Eq. (1) with a correction, i.e. $\kappa' = 4\kappa$ scales, the data very well for a droplet evaporating in 2-propanol, methanol and n-hexane. The value of κ increases by only 2.5% when the bulk temperature T_b is increased by 10 K, therefore only a single curve of κ is shown in Fig. 2. The rescaled plot shown in Fig. 2 indicates that the evaporation rate of a LN_2 droplet in n-pentane is the highest when compared with the other bulk liquids. A nitrogen droplet evaporating in n-pentane at 294 K has a larger bubble growth rate than a nitrogen droplet evaporating in the other three bulk liquids at 313 K. The different bubble growth behaviour could be due to n-pentane having the lowest surface tension among all the bulk liquids (16 mN m^{-1}).

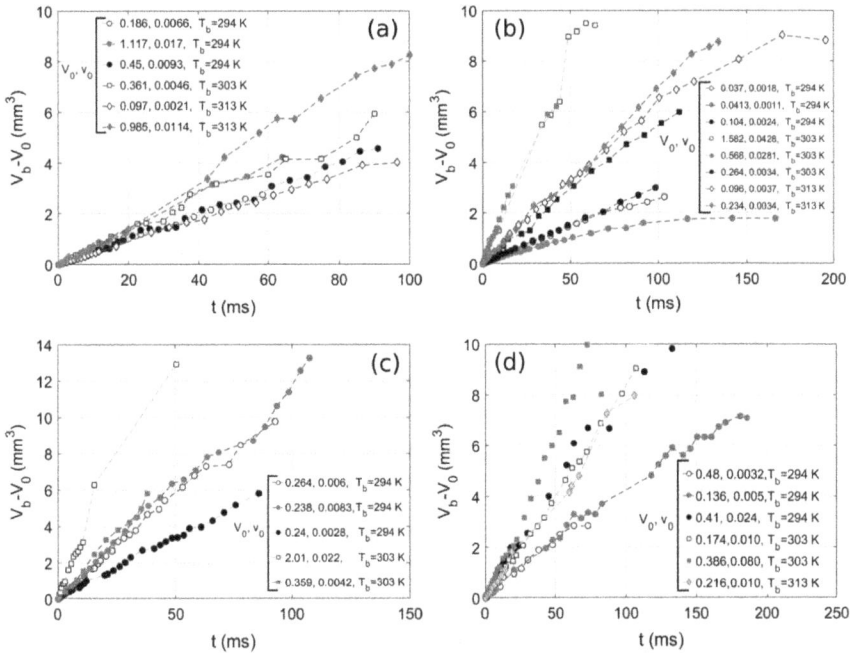

Fig. 1 Bubble growth rate for a nitrogen droplet evaporating in **a** 2-propanol, **b** methanol, **c** n-pentane and **d** n-hexane at different bulk temperatures. For each data set, initial bubble volume V_0 (mm^3) and the initial droplet volume v_0 (mm^3) along with the bulk temperature T_b are given in the legend. Dotted lines with similar colour compare data based on bulk temperature and initial droplet sizes

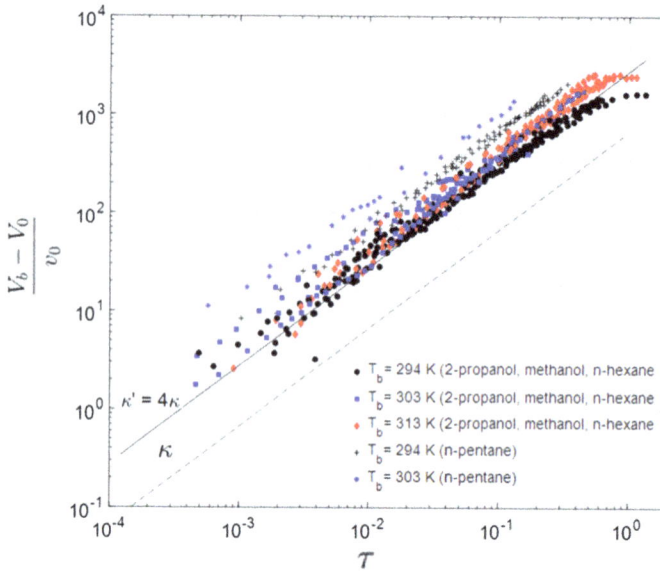

Fig. 2 Experimental data rescaled using the relationship in Eq. (1). For 2-propanol, methanol and *n*-hexane all the points corresponding to the bulk liquid temperature T_b collapse well on a single curve

The average value of heat flux to the droplet (based on the droplet surface area) obtained in all the current experiments was between 25 and 69.5 W cm^{-2}, which is higher than the heat flux (based on the heating surface area) measured during pool boiling of liquid nitrogen.

4   Conclusions

The experimental results show that liquid nitrogen droplets of similar initial size are evaporating faster in liquids maintained at higher temperatures. However, the impact of the initial droplet size on the evaporation rate was higher than the increased bulk fluid temperature up to 20 K. Therefore, in applications that require a faster evaporation rate of cryogenic droplets, changing the size of the drops might be more effective than increasing the bulk fluid temperature slightly, e.g. <20 K.

A scaling analysis of the bubble growth rate with dimensionless time scaled the data well for 2-propanol, methanol and *n*-hexane but tends to underestimate the bubble growth rate in *n*-pentane, possibly due to the lower surface tension of the *n*-pentane.

The average values of heat flux obtained were higher than reported values of heat flux during the pool boiling of liquid nitrogen.

Acknowledgements The work presented has been carried out with funding from a Ph.D. studentship by Loughborough University, UK. The financial support is gratefully acknowledged by the authors'.

References

1. C. Ordonnez, Liquid nitrogen fueled, closed Brayton cycle cryogenic heat engine. Energy Convers. Manage. **41**, 331–341 (2000)
2. D. Wen, H. Chen, Y. Ding, P. Dearman, Liquid nitrogen injection into water: pressure build-up and heat transfer. Cryogenics **46**, 740–748 (2006)
3. H. Clarke, A. Martinez-Herasme, R. Crookes, D.S. Wen, Experimental study of jet structure and pressurisation upon liquid nitrogen injection into water. Int. J. Multiph. Flow **36**, 940–949 (2010)
4. W. Burger, M. Burge, *Digital Image Processing an Algorithmic Introduction Using Java* (Springer, Berlin, 2008)
5. D. Spalding, *Combustion and Mass Transfer: a Textbook with Multiple-Choice Exercises for Engineering Students* (Pergamon Press, Oxford, 1979)

CFD Modelling of Gas-Turbine Fuel Droplet Heating, Evaporation and Combustion

Mansour Al Qubeissi, Geng Wang, Nawar Al-Esawi, Oyuna Rybdylova, and Sergei S. Sazhin

1 Introduction

The modelling of droplet heating, evaporation and combustion processes is crucial to the design and advancement of combustors [1, 2], and essential to the assessment of suitability of fuel [3]. In this study, we have conducted a detailed analysis of kerosene fuel droplet heating and evaporation, using the previously developed discrete component model (DCM). Kerosene fuel composition (approximated by 44 components of the full composition reported in [4]) is replaced with 2 surrogate components to reduce the computational time. In contrast to the classical industrial analyses of aviation fuel (e.g. the distillation curve method [5]), the DCM takes into account gradients of temperatures and species mass fractions in droplets. Our application of this model is based on the analytical solutions to the heat transfer and species diffusion equations subject to appropriate boundary and initial conditions [6]. Numerical codes using these solutions were extensively verified and validated [7–9]. The effective thermal conductivity and effective diffusivity approaches to modelling of moving droplets are used in the model.

The DCM was implemented in the commercial CFD software ANSYS-Fluent which was applied to study the processes in a can combustor. A polyhedral mesh was used, as shown in Fig. 1. This opened up opportunities for the simulation of the full combustion cycle. The influence of droplet evaporation on the combustion process was investigated.

The computational domain and polyhedral mesh used for the hydrodynamic model are shown in Fig. 1. The model features the diffusion flamelet generated manifold

M. Al Qubeissi (✉) · G. Wang · N. Al-Esawi
Faculty of Engineering, Environment and Computing, Coventry University, Coventry, UK
e-mail: ac1028@coventry.ac.uk

O. Rybdylova · S. S. Sazhin
School of Computing, Environment and Mathematics, Advanced Engineering Centre, University of Brighton, Brighton, UK

© Springer Nature Singapore Pte Ltd. 2021
C. Wen and Y. Yan (eds.), *Advances in Heat Transfer and Thermal Engineering* ,
https://doi.org/10.1007/978-981-33-4765-6_36

197

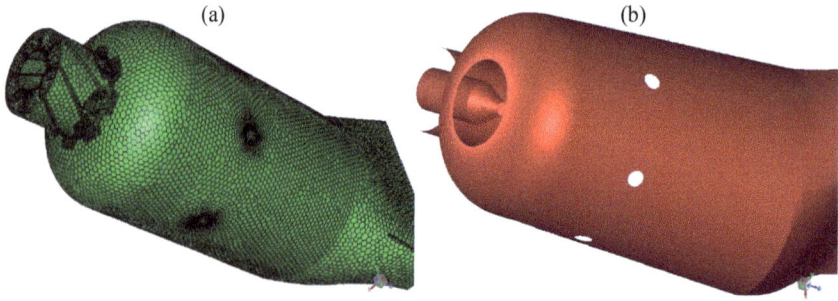

Fig. 1 Polyhedral mesh used for the can combustor simulation. The figure presents **a** the domain polyhedral mesh and **b** the internal walls of the system. The cell volume range is 0.0057647–470 mm^3, the face cell area range is 0.014–8 mm^2 and the total number of cells is 262,255

(FGM) method to simulate partially premixed gaseous combustion. This opened up new opportunities for the simulation of full combustion cycle. The influence of droplet evaporation on the combustion process was investigated.

2 Results

The analysis was applied to a balanced mixture of kerosene and diesel fuels, represented by decane ($C_{10}H_{22}$) and cyclododecane ($C_{12}H_{24}$), respectively. The initial droplet diameter and temperature were 100 μm and 375 K, respectively. The ambient gas temperature and pressure were 800 K and 4 bar, respectively. A co-axial air-blast atomizer was used with primary and secondary air and fuel mass flow rates of 0.15, 0.025 and 0.003 kg/s, respectively, with an injection speed of 1 m/s. Figure 2 shows the evolution of droplet radii with time predicted by three approaches: (1) by standard ANSYS-Fluent software using constant properties; (2) by ANSYS-Fluent with the transient properties of fuel using the user defined function (UDF), but without the DCM; and (3) by ANSYS-Fluent with full implementation of the DCM and transient thermodynamic and transport properties. In the latter case, the impact of thermal swelling on droplet heating and evaporation was taken into account.

As can be seen from Fig. 2, incorporating the DCM into the ANSYS-Fluent leads to up to 10.4% increase in predicted evaporation time compared to the case when a standard ANSYS-Fluent model is used. Also, our results indicate that the fuel composition and temperature gradient inside droplets, which are ignored in the standard ANSYS-Fluent model, can lead to noticeable impact on the spray formation and combustion processes. The new results have been compared with those reported in the literature [10] (see Fig. 3) for kerosene droplets. The droplets of 1.8 mm Sauter mean diameter and 298 K initial temperature were injected at ambient pressure and temperature of 1 bar and 673 K, respectively. The average relative velocity between liquid and air was negligible, and the droplets were assumed stationary.

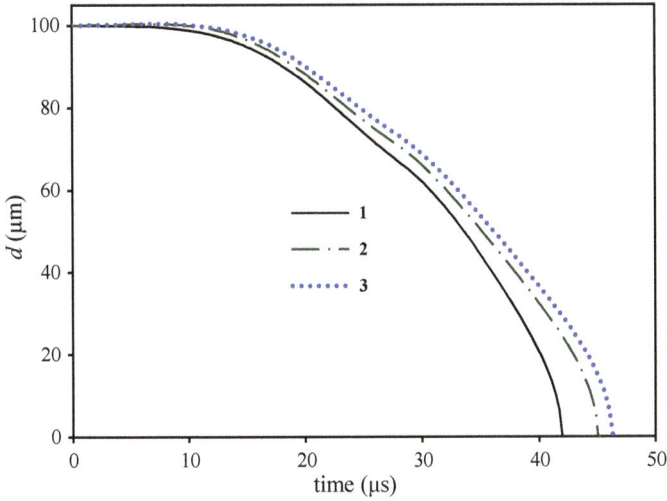

Fig. 2 Evolutions of droplet diameter using the three modelling approaches: 1 refers to Standard ANSYS-Fluent results, with constant properties, 2 refers to ANSYS-Fluent results, with in-house properties using UDF and 3 refers to ANSYS-Fluent results with the in-house DCM using UDF

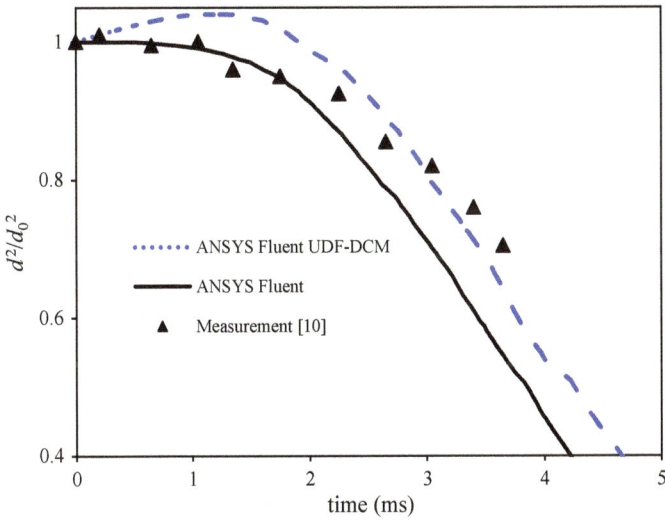

Fig. 3 Validation of the models for the normalised droplet diameters squared predicted by the standard ANSYS-Fluent (solid curve), and ANSYS-Fluent with the DCM (dotted curve). The modelling results are compared with data reported in [10] (bold triangles) for kerosene fuel droplets

As can be seen from Fig. 3, general agreement between the numerical results and experimental data was found. In our analyses, we considered the impact of thermal swelling on droplet evaporation. Finally, the combustion of the blended fuel droplets was simulated, and the influence of fuel evaporation and species diffusion on flame properties was investigated. A realizable κ-ε turbulence model is used for the hydrodynamic region with enhanced wall treatment. The species prediction is based on partially premixed combustion model with FGM state relation of diffusion flamelet for a non-adiabatic system. The kinetic reaction pathways (Chemkin mechanisms) have been imported from the NIST Chemical Kinetics Database (https://kinetics.nist.gov/kinetics/index.jsp). The domain pressure and velocity are coupled in a quasi-transient manner.

3 Conclusion

It is shown that a customized version of ANSYS-Fluent with the DCM leads to predictions that are closer to experimental data than the predictions of the conventional ANSYS-Fluent code. The combustion of the blended fuel droplets was simulated, and the influence of fuel evaporation and species diffusion in droplets on flame properties was investigated.

Acknowledgements The authors are grateful to Coventry University (Grant No. ECR019), and the EPSRC (Grant No. MR/T043326/1) for their financial support of this project.

References

1. M. Al Qubeissi, Predictions of droplet heating and evaporation: an application to biodiesel, diesel, gasoline and blended fuels. Appl. Therm. Eng. **136**(C), 260–267 (2018)
2. S.S. Sazhin, *Droplets and Sprays* (Springer, London, 2014).
3. E.G. Jones, L.M. Balster, Impact of additives on the autoxidation of a thermally stable aviation fuel. Energy Fuels **11**(3), 610–614 (1997)
4. K. Lissitsyna, S. Huertas, L.C. Quintero, L.M. Polo, PIONA analysis of kerosene by comprehensive two-dimensional gas chromatography coupled to time of flight mass spectrometry. Fuel **116**, 716–722 (2014)
5. T.M. Lovestead, T.J. Bruno, Application of the advanced distillation curve method to the aviation fuel avgas 100LL. Energy Fuels **23**(4), 2176–2183 (2009)
6. S.S. Sazhin, Modelling of fuel droplet heating and evaporation: recent results and unsolved problems. Fuel **196**, 69–101 (2017)
7. S.S. Sazhin, A.E. Elwardany, P.A. Krutitskii, V. Deprédurand, G. Castanet, F. Lemoine, E.M. Sazhina, M.R. Heikal, Multi-component droplet heating and evaporation: numerical simulation versus experimental data. Int. J. Therm. Sci. **50**(7), 1164–1180 (2011)
8. M. Al Qubeissi, N. Al-Esawi, S.S. Sazhin, M. Ghaleeh, Ethanol/gasoline droplet heating and evaporation: effects of fuel blends and ambient conditions. Energy Fuels **32**(6), 6498–6506 (2018)

9. N. Al-Esawi, M. Al Qubeissi, R. Whitaker, S.S. Sazhin, Blended E85–diesel fuel droplet heating and evaporation. Energy Fuels **33**(3), 2477–2488 (2019)
10. F. Wang, R. Liu, M. Li, J. Yao, J. Jin, Kerosene evaporation rate in high temperature air stationary and convective environment. Fuel **211**, 582–590 (2018)

A Study of Nucleate Boiling Conjugate Heat Transfer

Robin Kamencky, Michael Frank, Dimitris Drikakis, and Konstantinos Ritos

1 Introduction

There has been a limited number of studies attempting to numerically estimate fluid flow and heat transfer phenomena taking part in the course of the complete quenching process. During this process, a metal part undergoes cooling from a temperature around 1000 K to a temperature around 330 K. The majority of studies is devoted to one of the heat transfer regimes or to one specific aspect of boiling.

AVL-FIRE, which utilises the Eulerian–Eulerian multi-fluid approach with the mixture energy equation, is the most commonly used software for such applications. Due to the need to accurately estimate the heat transfer coefficient, various boiling heat transfer regimes need to be taken into account. Film boiling and transition boiling regimes have been considered to be of significant importance. On the other hand, the developers of AVL-FIRE claim that nucleate boiling only plays a minor role, because the temperature range in which it occurs is much smaller than the temperature ranges of the other two regimes. The claim is questionable and might be reasonable to quantify its impact. AVL-FIRE's capability to accurately predict temperatures in the solid varies depending on conditions such as coolant subcooling, submerging direction of the solid and other factors [1].

Apart from AVL-FIRE, other softwares have also been used in quenching research. For example, Fluent has been used with its mixture model, and the volume of fluid (VOF) solver in OpenFOAM. However, all of them, including the quenching targeted AVL-FIRE, have limited capability to fully describe with high accuracy the challenging and complex problem of fluid flow and heat transfer during quenching.

R. Kamencky (✉) · M. Frank · K. Ritos
Department of Mechanical and Aerospace Engineering, University of Strathclyde, Glasgow, UK
e-mail: robin.kamencky@strath.ac.uk

D. Drikakis
University of Nicosia, Nicosia, Cyprus

© Springer Nature Singapore Pte Ltd. 2021
C. Wen and Y. Yan (eds.), *Advances in Heat Transfer and Thermal Engineering* ,
https://doi.org/10.1007/978-981-33-4765-6_37

This paper presents the results on nucleate boiling and conjugate heat transfer and aims to investigate the physics behind those processes in order to develop a code capable of modelling accurately the complete process of immersion quenching. An Eulerian–Eulerian framework is used in conjunction with the nucleate boiling by Kurul & Podowski. In addition to the fluid flow governing equations, an energy equation for the solid region is solved. The present results are compared with numerical and experimental results from the literature for three different cases. Firstly, we validate the conjugate heat transfer and solid–liquid interface parts of the solver against a backward-facing step flow benchmark case [2]. Secondly, we also compare the numerical results for a subcooled boiling flow in a vertical annular pipe against experimental data [3]. Finally, the proposed model for the nucleate boiling is validated against a pool boiling conjugate heat transfer experiment [4].

2 Methodology

The code is capable of solving two or more regions which can be either of solid or fluid type. The fluid is allowed to undergo a phase change; hence, both liquid and gas can be present with mass transfer between them. Each fluid phase is governed by the Navier–Stokes equations. The solid region is treated using only an energy equation.

The regions' interface must allow heat transfer. Our approach is based on the Open-FOAM method which estimates the temperature values for the interfacial boundary condition of each region using a harmonic mean. For our purposes of two-phase flow, the volume fractions need to be taken into account for accurate temperature estimation.

A precise evaluation of nucleate boiling phenomena requires not only the usage of the nucleate boiling model but also an appropriate choice of auxiliary closure models. The foundations, that nucleate boiling is based on, are the nucleation site density, the bubble departure diameter and the bubble detachment frequency models. In the course of our investigation, various combinations of models have been tested.

Further challenges arise when interactions between the phases are also modelled. Precise predictions of interfacial forces and heat transfer are of considerable interest because it can alter the flow behaviour significantly. Our study has taken into account lift, drag, turbulent dispersion, wall lubrication and virtual mass effects.

Bubble behaviour is another important aspect demanding attention. Initially, we use Anglart and Nylund model [5], which predicts the bubble diameter as a function of liquid subcooling. The model, however, does not account for phenomena which a bubble can undergo (coalescence, agglomeration and bubble breakage). As a consequence, we also employ the alternative approach of population bubble balance. The population balance approach takes into account the aforementioned phenomena together with grow and condensation. We use the discrete class method, where a number of classes describe the bubbles size distribution. Bubbles are classified by their size into the classes which are governed by a scalar transport equation.

3 Results

The developed code has been validated against three different problems. A representative example of the validation process is described in Fig. 1 and Table1, where we model conjugated heat transfer without boiling. In the figure, the fluid velocity field with two recirculation regions and the solid region showing dimensionless temperature field are depicted. The fluid velocity field is shown using colourful scheme in comparison with the solid temperature field being shown in gray scale. The fluid flows into the domain from the left side over the backward-facing step and separates creating two recirculations (grey, solid lines in Fig. 1). The first recirculation is found directly behind the inlet on the bottom, and second recirculation is located on the upper side slightly behind the middle of the fluid domain. The outflow is on the right side of the domain. The reattachment of the bottom recirculation is compared with the original work by Ramsak [2] in the first column of Table1. The fluid interacts with the heated solid region located on the bottom of the depicted geometry. The effect can be seen at the dimensionless temperature field of the solid region. The largest heat flux into the liquid is observed at the reattachment point of the bottom recirculation, where the Nu number reaches its peak value.

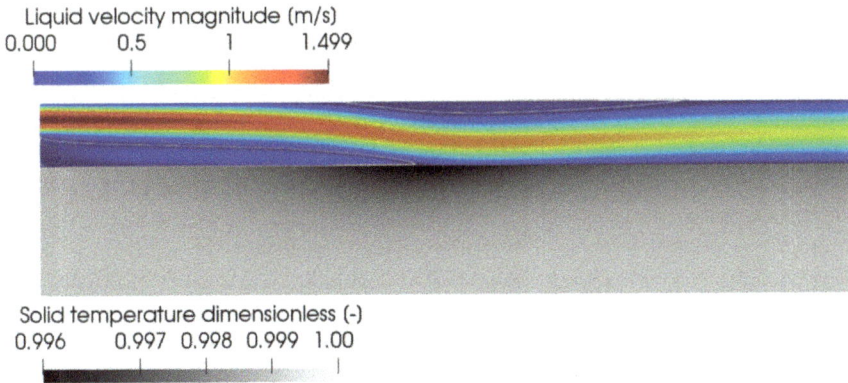

Liquid velocity magnitude (m/s)
0.000   0.5   1   1.499

Solid temperature dimensionless (-)
0.996   0.997 0.998 0.999 1.00

Fig. 1 Fluid velocity magnitude and solid dimensionless temperature field

Table 1 Comparison of hydrodynamic and conjugate heat transfer values

	$\times$ 1/h	Err (%)	$k = 1$		$k = 1000$	
			Nu'	Err (%)	Nu'	Err (%)
Ramšak (0.025)	12.315		0.425		2.719	
Ramšak (0.0125)	12.222		0.425		2.721	
OF (0.025)	11.933	3.103	0.421	0.796	2.692	0.983
OF (0.0125)	12.109	0.926	0.425	−0.0003	2.701	0.727

Additional results for two various conductivity ratios ($k = k_{SOLID}/k_{FLUID}$) are provided in Table 1. These results are for the conductivity rations, $k = 1$ and $k = 1000$. We have chosen two extreme cases to demonstrate the capabilities of the solver. The first and second columns show the position of the first recirculation reattachment, which allow us to validate the hydrodynamic solution. The third and fourth columns give an average Nusselt number and its percentage error for $k = 1$ and the last two columns tabulate the same variable but for $k = 1000$. The positions of the recirculations are independent of the conductivity ratio.

The comparison of the nucleate boiling model has been performed with the experiment of subcooled boiling flow in an annulus [3]. The focus has been placed on velocity profiles of vapour and liquid phases, vapour volume fraction and bubble Suater mean diameter in radial direction at the predefined location.

Finally, our solver has been compared against experimental and numerical results of a pool boiling problem [4]. In this case, the computational domain consists of two parallel pipes submerged at various directions into a water tank. The comparison focuses on wall superheat and heat transfer coefficient estimated under various heat fluxes and submerging directions.

4 Conclusions

Our newly developed solver has been compared against three different cases, found in the literature, that involve phase change and conjugate heat transfer. We have individually validated the nucleate boiling and conjugate heat transfer parts of the solver. The final comparison has been performed with a more complete problem, where heat is transferred from a solid to a boiling fluid. Further work is required in order to extend the capabilities of the solver to deal with the complex problem of quenching. Necessary future steps include extending the current models to deal with phenomena beyond the critical heat flux, definition of the Leidenfrost point and usage of models for transient and vapour film boiling stages.

References

1. R. Kopun, L. Škerget, M. Hriberšek, D. Zhang, B. Stauder, D. Greif, Numerical simulation of immersion quenching process for cast aluminium part at different pool temperatures. Appl. Thermal Eng., 74–78 (2014)
2. M. Ramsak, Conjugate heat transfer of backward-facing step flow: a benchmark problem revisited. Int. J. Heat Mass Transf. **84**, 791–799 (2015)
3. T.H. Lee, G.C. Park, D.J. Lee, Local flow characteristics of sub-cooled boiling flow of water in a vertical concentric annulus. Int. J. Multiphase Flow, 1351–1368 (2002)
4. Norri Rahim Abadi, S.M.A., Ahmadpour, A., Mayer, J.P., Numerical simulation of pool boiling on smooth, vertically aligned tandem tubes. Int. J. Thermal Sci., 628–644 (2018)
5. H. Anglart, O. Nylund, N. Kurul, M.Z. Podowski, CFD prediction of flow and phase distribution in fuel assemblies with spacers. Nuclear Eng. Des., 215–228 (1997)

Flow Boiling of Water in a Square Metallic Microchannel

S. Korniliou, F. Coletti, and T. G. Karayiannis

1 Introduction

Recent advancements in the electronics industry led to smaller and more powerful systems. Therefore, efficient heat dissipation for cooling of microelectronics systems, integrated circuit chips, power semiconductor devices such as IGBTs and laser diodes is required. In such systems, heat fluxes in the order of MW/m^2 need to be removed from small spaces, while maintaining the temperature below a certain design limit [1]. Flow boiling in microchannels is one of the most promising methods for achieving these high cooling demands because it can dissipate large heat fluxes over a small surface area by utilizing the latent heat of the coolant [2]. Literature review indicates that there are still disagreements on the prevailing flow patterns, heat transfer rates, and pressure drop trends; see Mahmoud and Karayiannis [3]. The main objective of the present work was to investigate the flow boiling flow patterns, heat transfer coefficient, and pressure drop characteristics in a square metallic microchannel at different inlet subcooling, mass flux and heat flux conditions, using water as the working fluid.

2 Methodology

The experimental facility consists of a reservoir, a subcooler, a magnetic drive gear pump, a Coriolis flow meter, a preheater, an inline filter of 90 μm placed after the preheater, a sight glass, test section, a glycol–water circulation chiller, and a reflux condenser. After degassing in the reservoir, DI water was circulated in the

S. Korniliou (✉) · F. Coletti · T. G. Karayiannis
Department of Mechanical and Aerospace Engineering, Brunel University London, Uxbridge UB8 3PH, UK
e-mail: sofia.korniliou@brunel.ac.uk

© Springer Nature Singapore Pte Ltd. 2021
C. Wen and Y. Yan (eds.), *Advances in Heat Transfer and Thermal Engineering*,
https://doi.org/10.1007/978-981-33-4765-6_38

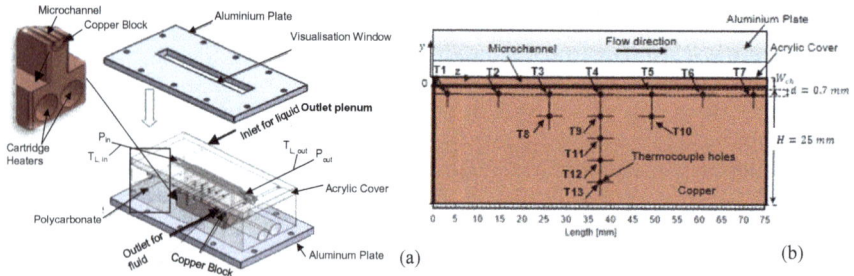

Fig. 1 a Test section assembly depicting the two aluminum plates, the copper block with microchannel, the polycarbonate block and acrylic cover and **b** thermocouples along the channel and copper block

flow loop system at the desired flow rate and inlet temperature. System pressure was maintained constant at 1 bar by controlling the reservoir temperature. A single rectangular microchannel 1 mm high, 1 mm wide, and 75 mm long was cut in the top surface of a copper block using a high-speed micro-milling machine (Kern HSPC 2216); see Fig. 1a. The average surface roughness (Ra) of the microchannel was found to be 0.252 ± 0.08 μm. The average was produced from five locations equally spaced along the channel. The inlet and outlet plenums of the test section were fabricated in polycarbonate in order to reduce the heat transfer losses. Seven K-Type thermocouples (accuracy of ± 0.12 K) were inserted into the copper block at a distance of 0.7 mm below the channel bottom to obtain surface temperatures along the channel. The thermocouple locations are shown in Fig. 1b. Readings from thermocouples T4 and T9-T13 where used to evaluate the heat flux distribution in the copper block. Two cartridge heaters (total power of 800 W) were placed horizontally, parallel to the channel inside the copper block. An acrylic transparent plate on the top was supported in place with the copper block by an aluminum cover and bottom plate. The inlet pressure was measured at the inlet plenum using an absolute pressure transducer. The pressure drop between the microchannel inlet and outlet was measured using a differential pressure transducer. Inlet and outlet liquid temperatures were measured using two K-type thermocouples. The camera used to obtain the flow patterns was a High-Speed MicroLAB110 Phantom. The flow patterns were obtained at a frame rate of 4,609 fps and resolution of 768×480. Flow visualization was carried out at the inlet, middle, and outlet of the microchannel. The flow boiling experiments were carried out at different subcooling conditions of $\Delta T_{\text{sub}} = 5$, 15, and 50 K for heat fluxes between 200 and 1,130 kW/m^2 at three constant mass fluxes of 200, 400, and 600 kg/m^{2}s and inlet pressure of 1 bar. The experimental setup was validated with single-phase experiments. The relationships for the theoretical friction factor for fully developed and developing flow given by Shah [4] predicted the experimental friction factor data well for all test sections with a mean absolute error (MAE) of 2.8–14.5%. The single-phase Nusselt number in the test section was predicted well with the correlation of Choi et al. [5], for the range of 200–1580 kg/m^{2}s and for Re < 2000, i.e., MAE of 2.03–8.50%.

3   Results

The flow patterns observed in this study where bubbly, confined bubble, slug, churn, and annular flow. Churn flow was observed to occur only at a very small section of the channel or not observed at all in some cases due to fast transition from bubbly and slug to the annular flow regime, which is usually observed to occur only in small tubes [3] and fluids like water. Figure 2 shows the observed flow patterns along the entire channel length for the medium mass flux values of $G = 400$ kg/m²s at a constant heat flux of $q'' = 380$ kW/m² and inlet subcooling of $\Delta T_{sub} = 5$ K. (Vertical lines indicate different photos that were put together.) The flow patterns changed along the channel as a result of channel confinement, bubble coalescence and increasing vapor quality. Figure 2 shows that for these conditions, single phase occurred at the first 2 mm and bubbly flow occurred at the next 4 mm of the channel. This was followed by confined bubble-slug flow at the next 10 mm, churn flow (~5 mm) and then annular flow, which occupied a big section of the channel length (~ 47 mm).

As expected, increasing the inlet subcooling, resulted in increasing entry length (single-phase). The annular flow regime was established almost along the whole channel length for low degree of subcooling ($\Delta T_{sub} = 5, 15$ K). For low mass flux of $G = 200$ kg/m²s at $\Delta T_{sub} = 5$ K, periodic flow reversal was observed to occur, which was caused by bubbles that expanded both upstream and downstream. It is worth mentioning that bubble nucleation was observed to occur in the liquid thin film during the annular flow regime for all mass fluxes with increasing heat flux. Figure 3 shows part of the channel where bubble nucleation in the thin film occurred

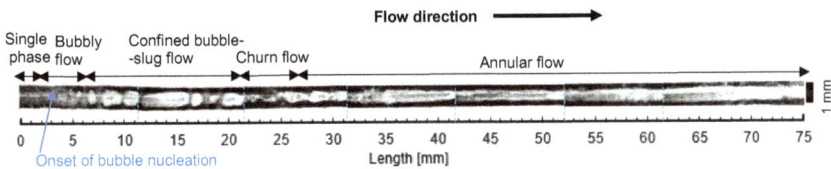

Fig. 2   Flow patterns along the microchannel length for the mass flux of $G = 400$ kg/m²s, inlet subcooling of $\Delta T_{sub} = 5$ K and heat flux of $q'' = 380$ kW/m²

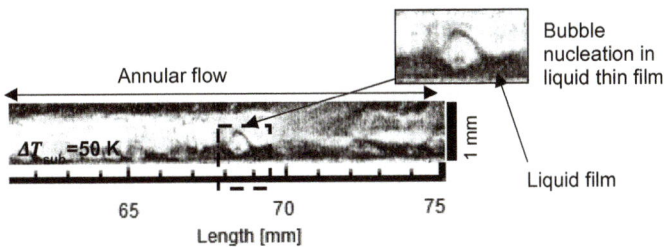

Fig. 3   Annular flow with bubble nucleation in thin liquid film observed at $G = 600$ kg/m²s, inlet subcooling of $\Delta T_{sub} = 5$ K and heat flux of $q'' = 680$ kW/m²

for the conditions of $G = 600$ kg/m^{2}s, $q'' = 680$ kW/m^2 and $\Delta T_{\text{sub}} = 50$ K. This is in agreement with a recent study by Ali et al. [4], where nucleating bubbles were visualized in the liquid film in the slug and annular flow regime with increasing heat flux in rectangular multi-channels of 0.46 mm hydraulic diameter using HFE-7100. As the heat flux increased and with the thinning of the liquid film, bubbles disappeared.

Thinning of the liquid film was observed to occur along the channel length during the annular flow regime that can cause local dryout at high heat fluxes ($q'' > 680$ kW/m^2). Local dryout can account for reduced heat transfer coefficient as vapor quality increases with channel length.

4 Conclusions

Flow boiling experiments were carried out in a copper microchannel test section having 1 mm hydraulic diameter. Flow patterns were recorded along the whole channel length for mass flux from 200 to 600 kg/m^{2}s and a pressure of 1 bar for three different inlet degree of subcooling, namely 5 K, 15 K, and 50 K and heat fluxes up to 1130 kW/m^2. Bubbly, confined bubble flow-slug flow, churn flow and annular flow were observed to occur in the channel. Churn flow occurred for a very short length of the channel due to early transition to annular flow. Annular flow with bubble nucleation in thin liquid film was observed to occur for all mass fluxes. This was not sustained at high heat fluxes where the liquid film was very thin.

A detailed comparison of the flow regimes observed in this study with flow pattern maps presented in the literature, e.g., see [5] will be presented. The effect of inlet subcooling, heat flux and mass flux on two-phase pressure drop and local and average two-phase heat transfer coefficient will be also presented. In addition, heat transfer rates and pressure drop results will be compared with correlations available in the literature and presented at the meeting.

References

1. T.G. Karayiannis, M.M. Mahmoud, Flow boiling in microchannels: fundamentals and applications. Appl. Therm. Eng. **115**, 1372–1397 (2017)
2. S. Wang et al., Cooling design and evaluation for photovoltaic cells within constrained space in a CPV/CSP hybrid solar system. Appl. Therm. Eng. **110**, 369–381 (2017)
3. M.M. Mahmoud, T.G. Karayiannis, Chapter 4: Flow boiling in mini to microdiameter channels, in *Encyclopedia of Two-Phase Heat Transfer and Flow IV*, pp. 233–301 (2018)
4. A.H. Al-Zaidi, M.M. Mahmood, T.G. Karayiannis, Flow boiling of HFE-7100 in multi-microchannels: aspect ratio effect, in *6th International Micro and Nano Flows Conference*, Atlanta (2018)
5. M.M. Mahmoud, T.G. Karayiannis, Flow pattern transition models and correlations for flow boiling in mini-tubes. Exp. Therm. Fluid Sci., 270–282 (2016)

Experimental Study on Flow Boiling Heat Transfer of Refrigerant R1233zd in Microchannels

Xin Yu You, Jiong Hui Liu, Nan Hua, Ji Wang, Rongx Hui Xu, Guang Xu Yu, and Hua Sheng Wang

1 Introduction

A new test rig has recently been built to implement an inverse method [1] to measure local heat transfer in microchannels. Hua et al. [2] report the measurement results for condensation heat transfer of refrigerant R1233zd in microchannels. This paper reports measurement results for flow boiling heat transfer of refrigerant R1233zd in microchannels.

2 Experimental Setup

Figure 1 shows the schematic of the apparatus. Figure 2 shows the photograph of the apparatus. The refrigerant is circulated in the refrigerant loop. The liquid refrigerant is heated a desired temperature and enters the test section where flow boiling occurs.

X. Y. You · J. H. Liu · N. Hua · R. H. Xu · H. S. Wang (✉)
School of Engineering and Materials Science, Queen Mary University of London, London E1 4NS, UK
e-mail: h.s.wang@qmul.ac.uk

G. X. Yu
DENSO Marston Ltd., Shipley BD17 7JR, West Yorkshire, UK

N. Hua
Faculty of Materials and Energy, Guangdong University of Technology, Guangzhou 510006, China

R. H. Xu
Beijing University of Civil Engineering and Architecture, Beijing 100044, China

J. Wang
College of Mechanical and Transportation Engineering, China University of Petroleum, Beijing 102249, China

© Springer Nature Singapore Pte Ltd. 2021
C. Wen and Y. Yan (eds.), *Advances in Heat Transfer and Thermal Engineering* ,
https://doi.org/10.1007/978-981-33-4765-6_39

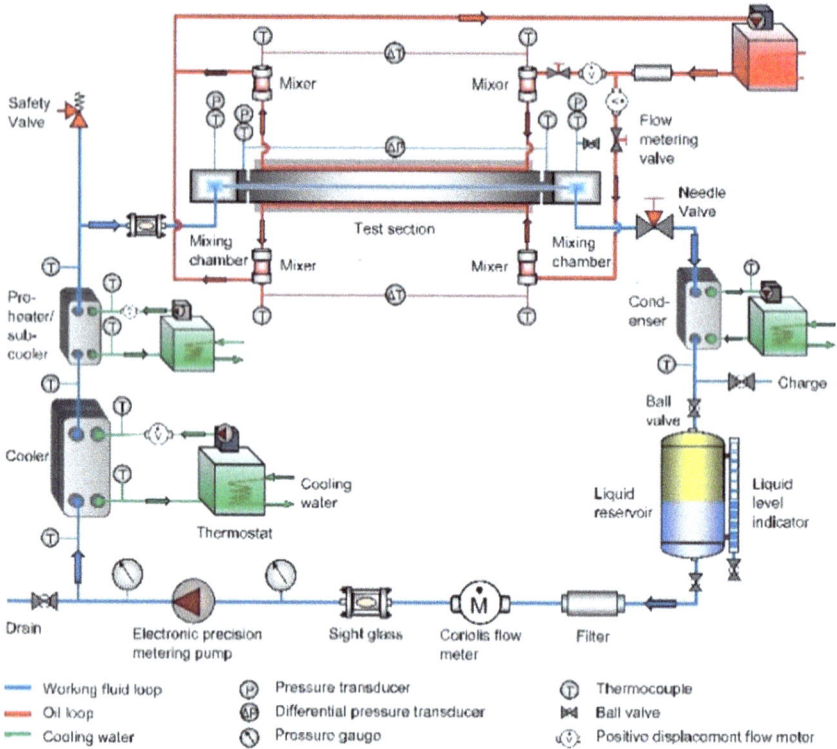

Fig. 1 Schematic of the apparatus

The vapor is completely condensed in the condenser before entering into the liquid reservoir. The heating water (deionized), supplied by a thermostat, flows counter-currently in the upper and lower heating jackets. The refrigerant flow rate is measured by a Coriolis flow meter. Four mixers are installed to accurately measure the bulk temperatures at the inlet and outlet of the heating water flowing through the upper and lower heating jackets, respectively, while the temperature drops of the heating water are measured using two thermopiles composed of ten thermocouples, respectively. Sufficient immersion of the thermocouples and thermopiles in the mixers and tubes is ensured. The temperature and pressure at the mixing chambers at the inlet and outlet of the test section are measured by thermocouples and pressure transducers. The temperature and pressure at the measuring slots at the inlet and outlet of the microchannels are also measured using thermocouples and pressure transducers. The pressure drop between the measuring slots at the inlet and outlet of the microchannels is measured by a differential pressure transducer.

Figures 3 and 4 show the cross-sectional views of the test section. The test section is made of two aluminum blocks. Ten parallel horizontal microchannels are machined in the mating surface of the lower block. The channel is 1.0 mm in width, 1.5 mm in

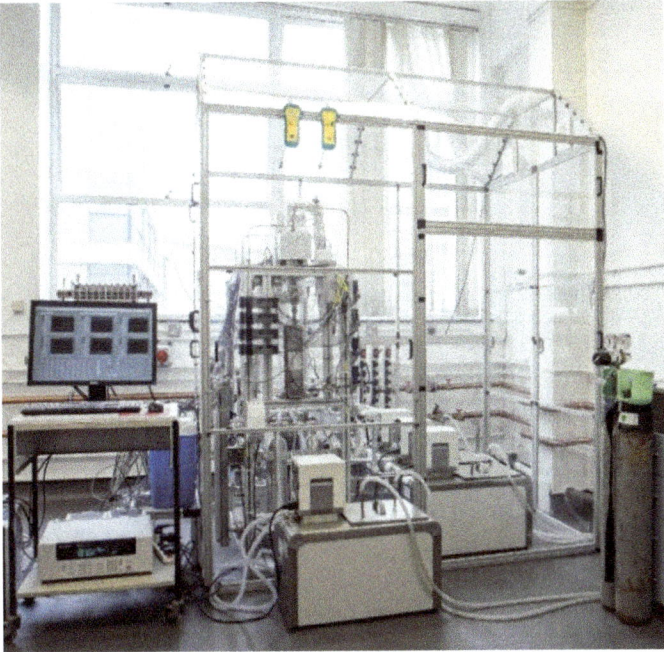

Fig. 2 Photograph of the apparatus

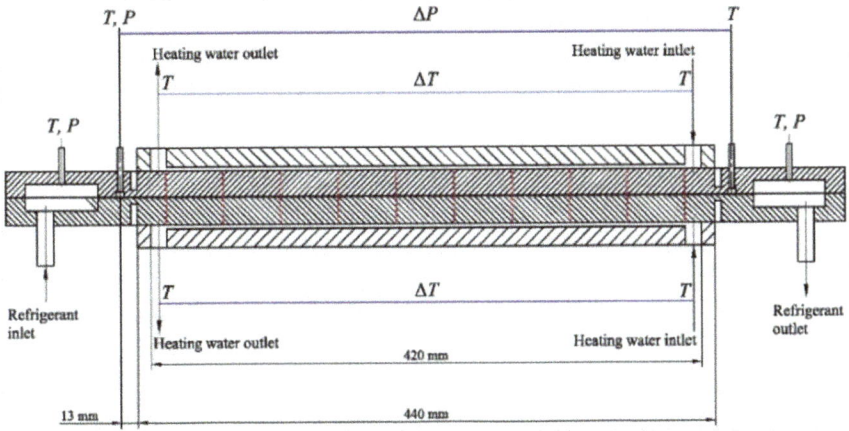

Fig. 3 Cross-sectional view of the test section

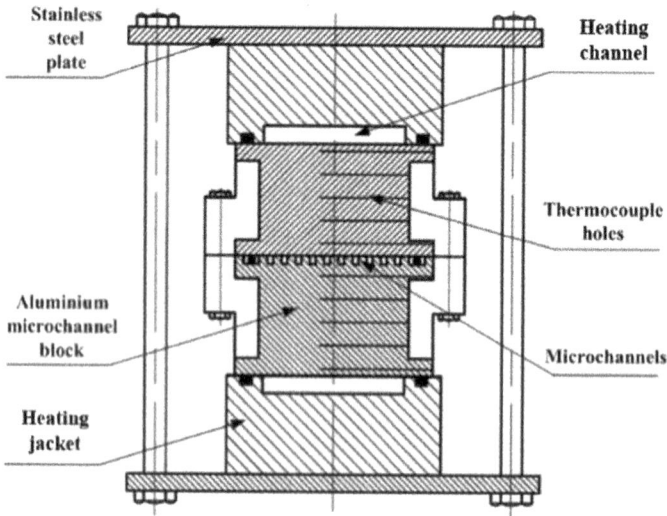

Fig. 4 Cross-sectional view of the test section

height, and 440 mm in length. 50 thermocouples holes with 0.6 mm in diameter and 20 mm in depth are drilled in the upper and lower blocks, respectively. Two thermal barriers are machined in the upper and lower blocks, respectively, to reduce heat conduction to the mixing chambers at the two ends of the test section. O-rings are placed in a round slot machined on the lower aluminum block and on the upper and lower nylon cooling jackets to seal. Figure 5 shows the photograph of the test section after thermocouples inserted and insulation. A desired pressure of the refrigerant loop is adjusted and maintained by controlling the temperature of refrigerant in the condenser. A data acquisition system (Keysight 34980A) is employed to set the experiment conditions and to record the temperatures, pressures, and flow rates.

Fig. 5 Test section after thermocouples inserted and insulation

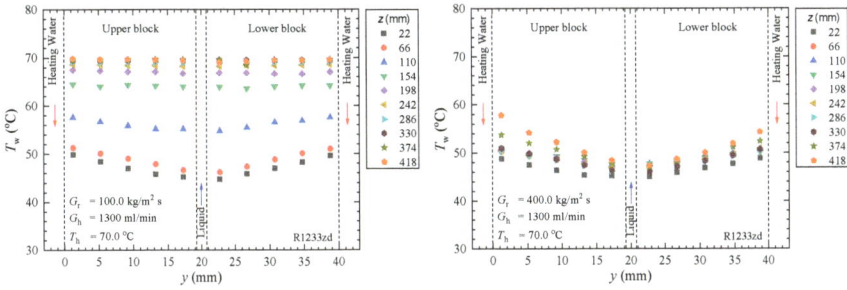

Fig. 6 Samples of typical temperatures measured in the upper and lower test blocks

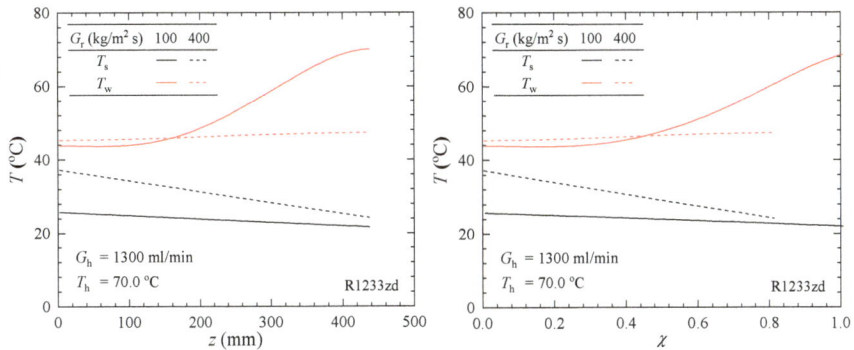

Fig. 7 Variations of local saturation and channel surface temperature with distance and vapor quality along the channel

3 Results and Discussion

Figure 6 shows samples of typical measured temperatures in the test blocks for refrigerant mass fluxes of 200 and 400 kg/m² s. Figure 7 shows the distributions of channel surface and saturation temperatures along the channel, using the inverse method, based on the measured temperatures shown in Fig. 6. Figure 8 shows the local heat flux and local heat transfer coefficient along the channel for refrigerant mass fluxes of 200, 250, 300, 350 and 400 kg/m² s.

4 Conclusions

Measurements have been conducted under different conditions. Local channel surface temperature, heat flux, and heat transfer coefficient are obtained, using the inverse method, based on accurately measured 100 temperatures in the test blocks.

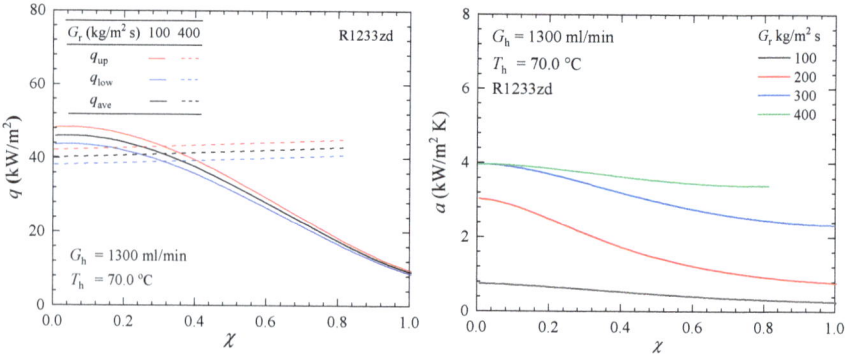

Fig. 8 Variations of local heat flux (left) and heat transfer coefficient (right) along the channel

The results show successful implement of the inverse method to measure local heat transfer during flow boiling of refrigerant R1233zd in microchannels.

Acknowledgements Financial supports from EPSRC of the UK (EP/L001233/1, EP/N001236/1), DENSO Marston UK, the Science and Technology Program of Guangzhou, China (201704030108), and the Joint PhD Studentship of China Scholarship Council (CSC) and Queen Mary University of London are acknowledged. We would also like to thank Mario Ciaffarafe, Andrew Diakiw, and Andrew Roberts of DENSO for their kind support.

References

1. G.X. Yu, J. Sun, H.S. Wang, P.H. Wen, J.W. Rose, Meshless inverse method to determine temperature and heat flux at boundaries for 2D steady-state heat conduction problems. Exp. Therm. Fluid Sci. **52**, 156–163 (2014)
2. N. Hua, J.H. Liu, X.Y. You, J. Wang, R.J. Xu, G. X. Yu, H.S. Wang, Experimental investigation of condensation heat transfer of refrigerant R1233zd in microchannels, in *ICR2019*, 25–31 August 2019, Montreal, Canada

Study on the Pool Boiling Bubble Departure Diameter and Frequency from Porous Graphite Foam Structures

I. Pranoto, K. C. Leong, A. A. Rofiq, H. M. Arroisi, and M. A. Rahman

1 Introduction

As elaborated by Mudawar [1], cooling technologies for electronic devices have shifted from natural convection to single-phase forced convection and then to phase-change cooling systems. The heat transfer performance limitations of natural and single-phase forced convection have driven the development of two-phase cooling or boiling heat transfer. Hence, two-phase cooling is generally considered to be one of the promising techniques for high heat flux electronic devices in the future. The bubble departure diameter (D_b) and frequency (f_d) are important parameters in bubble dynamics that directly affect boiling heat transfer performance. Surface heating, liquid heating, nucleation, bubble growth and departure occur continuously and repeatedly during the boiling process such that the cycle is normally known as an ebullition cycle. During the boiling process, a bubble is generated from an activated cavity on the boiling surface. The generated bubble grows during the bubble growth time and then departs from the nucleation cavity. The bubble diameter at the time of departure from the nucleation site is called the bubble departure diameter. The bubble departure frequency indicates how fast the bubble grows and departs from the cavity. It is affected directly by the bubble departure diameter. At the same heat flux, a smaller bubble departure diameter will result in a higher bubble departure frequency.

This study focuses on the experimental study to determine the bubble departure diameter and frequency from porous graphite foam structures with dielectric

I. Pranoto (✉) · A. A. Rofiq · H. M. Arroisi · M. A. Rahman
Department of Mechanical and Industrial Engineering, Faculty of Engineering, Gadjah Mada University, Jl. Grafika No. 2 Kampus UGM, Yogyakarta 55281, Indonesia
e-mail: indro.pranoto@ugm.ac.id

K. C. Leong
School of Mechanical and Aerospace Engineering, Nanyang Technological University, 50 Nanyang Avenue, Singapore 639798, Singapore

© Springer Nature Singapore Pte Ltd. 2021
C. Wen and Y. Yan (eds.), *Advances in Heat Transfer and Thermal Engineering* ,
https://doi.org/10.1007/978-981-33-4765-6_40

coolants. During the previous decade, studies on the D_b and f_d measurement and predictions were conducted by many researchers. The bubble growth mechanism and parameters from the smooth boiling surfaces have been studied experimentally and analytically by Forster and Zuber [2], Han and Griffith [3] and Cole and Rohsenow [4]. Bubble dynamics from structured surfaces were also studied by Nakayama et al. [5], and the bubble departure diameter and frequency predictions were proposed from their works. Chien and Webb [6] conducted a visualisation and analytical study on the bubble dynamics from the structured surface with a circular fin base. Their structure surface possessed surface pores and sub-surface tunnels to enhance the boiling surface area and the bubble escaping process. Dhir et al. [7] studied the bubble departure and frequency by simulations and experiments of pool and flow boiling. The simulations were based on the solution of the conservation equations of mass, momentum and energy for both phases. Furberg [8] conducted bubble dynamics investigation and proposed a model for D_b and f_d on a dendiritic and micro-porous structure. The bubble frequency was determined based on the heat and mass balance during the nucleation process while the bubble diameter was determined by balancing the forces acting on the micro-structure surface.

The results of this study would yield better understanding of the bubble dynamics from the porous graphite foam structures that have been proven to enhance boiling heat performance and are becoming a promising material for cooling electronic components in the near future as reported by White et al. [9], Williams and Roux [10], Jin et al. [11] and Pranoto et al. [12].

2 Methodology

To study the boiling heat transfer performance, bubble departure diameter and frequency, a compact pool boiling experimental setup was developed. The experimental facility consists of three main parts: heating base, evaporator and air-cooled condenser as shown in Fig. 1. A customised cartridge heater is inserted at the centre of a Teflon heating base to generate heat. The ceramic insulation layer and Teflon base effectively minimised the heat loss during the experiments. To adjust the level of generated heat, a power controller was used. The maximum heat flux achievable is 1.5×10^3 kW/m^2. The graphite foam evaporator insert was bonded on a copper plate by using highly thermal conductive epoxy "OMEGABOND 101". The designed copper plate was clamped tightly on the heater surface and properly sealed with the adhesive. The heater was fastened to provide good contact at the interface between the top of the heater and the copper plate. A rubber O-ring was employed at the interface between the copper plate and Teflon surface to improve liquid tightness.

The bubble departure diameter from the graphite foam structure was measured from the captured boiling images using a high speed camera and "Image Pro Software". The isolated bubbles from three different locations (i.e., nucleation site) at the graphite foam were used to calculate the bubble diameter. For each location, there are ten (10) bubbles of different captured frames were measured and analysed by

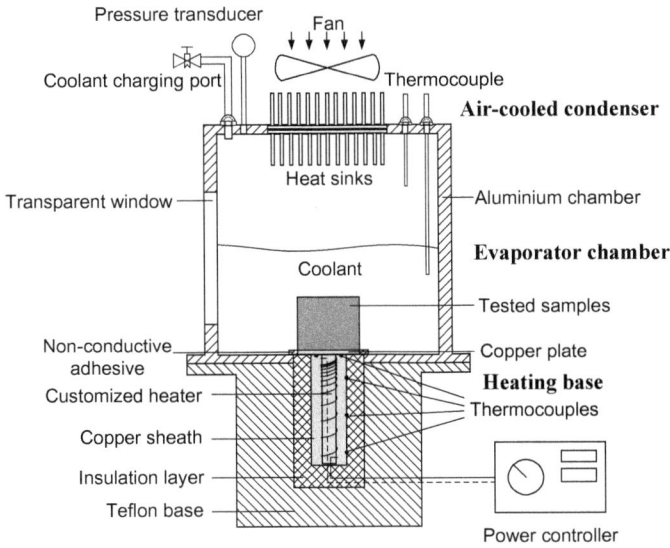

Fig. 1 Schematic diagram of the pool boiling facility

using the software. Hence, the number of measurements are $N = 30$ for each graphite foam evaporator.

The average bubble departure diameter $(\overline{D_B})$ and the standard deviation (σ_B) are calculated by

$$\overline{D_b} = \frac{1}{N} \sum_{i=1}^{N} D_{b_i} \tag{1}$$

$$\sigma_{D_b} = \sqrt{\frac{\sum \left(D_{b_i} - \overline{D_b}\right)^2}{N - 1}} \tag{2}$$

To determine the bubble departure frequency, the bubble growth and departure processes were recorded by using a high speed camera at 2005 frames per second (fps). The captured images were analysed frame by frame to determine the bubble growth and departure phenomena. The bubble departure frequency calculation method is portrayed in Fig. 2 which shows that a period of bubble growth and departure can be divided into waiting and bubble departure times. Once these times are determined, the bubble departure frequency can be calculated by $f_d = 1/(t_w + t_d)$, where t_w and t_d are waiting and departure times, respectively. At the frame rate of 2005 fps, the number of frames in the waiting and departure periods are M and N, respectively. Therefore, t_w and t_d can be calculated as $t_w = (M \text{ frames}/2005 \text{ fps})$ and $t_d = (N \text{ frames}/2005 \text{ fps})$, respectively.

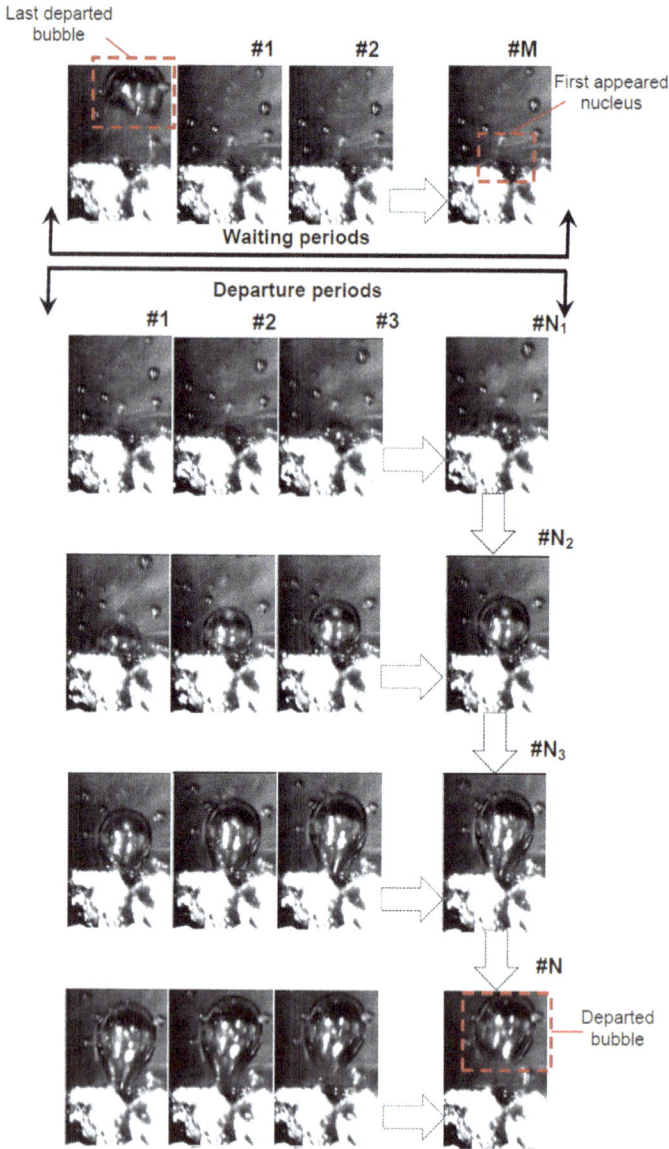

Fig. 2 Captured images of bubble growth and departure process, diameter, and frequency determination

3 Results

From the captured boiling process, bubble departure diameters from different graphite foams and dielectric coolants at heat flux of $q'' = 56.34$ W/cm^2 are presented in Fig. 3a. It is noted that at $q'' = 56.34$ W/cm^2, the isolated bubbles from all the graphite foams and working fluids can be observed clearly by using the high speed camera. The average bubble departure diameters and their standard deviations are given in Table 1. The measurement results show that "Pocofoam" of 61% had produced the smallest bubble departure diameter and bubble growth time compared to other tested graphite foams. The results also show that the measured bubble diameters are much larger as compared to the respective graphite foam pore diameters given in Table 1. This finding can be used to analyse the boiling heat transfer performance between different graphite foams and the enhancement mechanism from the porous graphite foam structures. The results will also useful to optimise the size in the two-phase cooling system.

By using the method described in the previous section, the bubble departure frequencies from different graphite foams were determined. The results of the bubble departure frequency from the different graphite foams are shown in Fig. 3b. It shows that at a heat flux of $q'' = 56.34$ W/cm^2, the measured f_d from the "Pocofoam" 61%, "Pocofoam" 75%, "Kfoam" 78% and "Kfoam" 72% are 173, 156, 163 and 149 Hz, respectively. It can be found that "Pocofoam" of 61% produced the highest bubble departure diameter as compared to the other tested graphite foams. The bubble

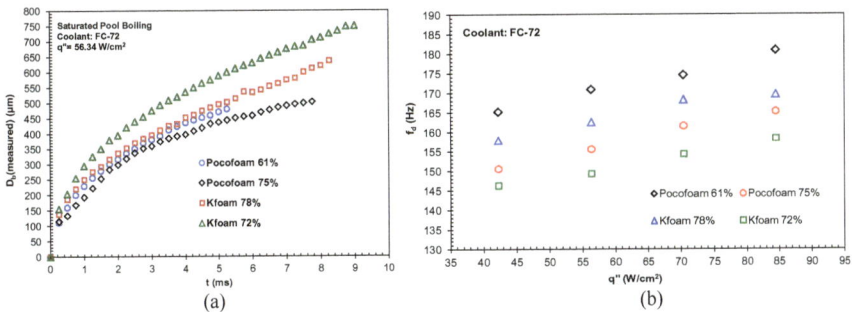

Fig. 3 Measured **a** bubble departure diameter and **b** frequency from the porous graphite foams

Table 1 Average and standard deviation of bubble departure diameter at $q'' = 56.34$ W/cm^2

Graphite foam	$\overline{D_B}$ (μm)	σ_{D_B} (μm)
"Pocofoam" 61%	480.8	9.5
"Pocofoam" 75%	502.4	10.2
"Kfoam" 78%	635.8	6.5
"Kfoam" 72%	750.1	8.3

departure frequency values were found to be increased with the increase of heat flux level. It is also shown that "Kfoam" of 78% porosity have generated higher bubble departure frequency compared to "Pocofoam" of 75% porosity. It is noted from the thermophysical properties of the graphite foams, "Pocofoam" of 75%, porosity possesses higher effective thermal conductivity which is about 2.5 times that of "Kfoam" of 78% porosity. Hence, "Pocofoam" 75% should be more efficient in conducting heat from the heater section to the fluid–solid contact surface where bubble nucleation occurred. Hence, higher bubble frequency should be generated from the "Pocofoam" of 75% porosity. However, the experimental results show that the use of "Kfoam" of 78% porosity resulted in higher bubble departure frequency and boiling performance compared to "Pocofoam" of 75% porosity. With similar pore structure, the main morphological difference between "Pocofoam" 75% and "Kfoam" 78% is the pore diameter of the foam d_P. The pore diameter would affect the bubble departure mechanism and boiling heat transfer significantly.

4 Conclusions

The bubble departure diameter and frequency from porous graphite foam structures with FC-72 were studied under saturated pool boiling condition. From this study, the following important findings can be concluded:

(1) The thermophysical properties of the graphite foams have affected significantly the bubble departure diameter and frequency. It was also found that the bubble departure frequency values increased linearly with the increase of heat flux level.
(2) "Pocofoam" of 61% porosity have produced the smallest bubble departure diameter and highest bubble departure frequency compared to the other graphite foams.
(3) The measured bubble departure diameter of the graphite foams was found to be 480.8–750.1 μm.
(4) Bubble departure frequencies of up to 149–173 Hz were produced by the porous graphite foam structures with FC-72 coolant.

Acknowledgements The authors acknowledge the financial supports of the research funding from Department of Mechanical and Industrial Engineering, Faculty of Engineering, Universitas Gadjah Mada Year 2019.

References

1. I. Mudawar, Assessment of high heat flux thermal management schemes. IEEE Trans. Compon. Packaging Technol. **24**, 122–141 (2001)
2. H.K. Forster, N. Zuber, Dynamics of vapor bubbles and boiling heat transfer. AIChE J. **1**, 531–535 (1955)

3. C.-H. Han, P. Griffith, The mechanism of heat transfer in nucleate pool boiling-Part I. Int. J. Heat Mass Transfer **8**, 887–904 (1965)
4. R. Cole, W.M. Rohsenow, Correlation of bubble departure diameters for boiling of saturated liquids. Chem. Eng. Progr. Symp. Ser. **65**, 211–221 (1968)
5. W. Nakayama, T. Daikoku, H. Kuwahara, T. Nakajima, Dynamic model of enhanced boiling heat transfer on porous surfaces Part II: analytical modelling. J. Heat Transfer **102**, 451–456 (1980)
6. L.-H. Chien, R.L. Webb, Measurement of bubble dynamics on an enhanced boiling surfaces. Exp. Thermal Fluid Sci. **16**, 177–186 (1998)
7. V.K. Dhir, H.S. Abarajith, D. Li, Bubble dynamics and heat transfer during pool and flow boiling. Heat Transfer Eng. **28**, 608–624 (2007)
8. R. Furberg, Enhanced boiling heat transfer on a dendritic and micro-porous copper structure. Ph.D. Thesis, KTH Industrial Engineering and Management (2011)
9. S.B. White, N.C. Gallego, D.D. Johnson, K. Pipe, A.J. Shih, E. Jih, Graphite foam for cooling of automotive power electronics, in *Proceedings of IEEE Power Electronics in Transportation Conference, USA*, pp. 61–65 (2004)
10. Z.A. Williams, J.A. Roux, Graphite foam thermal management of a high packing density array of power amplifiers. ASME J. Electron. Packag. **128**, 456–465 (2001)
11. L.W. Jin, K.C. Leong, I. Pranoto, Saturated pool boiling heat transfer from highly conductive graphite foams. Appl. Therm. Eng. **31**, 2685–2693 (2011)
12. I. Pranoto, K.C. Leong, L.W. Jin, The role of graphite foam pore structure on saturated pool boiling enhancement. Appl. Therm. Eng. **42**, 163–172 (2012)

Investigation of Droplet Evaporation on Copper Substrate with Different Roughness

Xin Wang, Zeyu Liu, Li Wang, and Yuying Yan

1 Introduction

Droplet evaporation is one of physical phenomena omnipresent in nature. This phenomena was attracted much attention recently, because it itself has rich fundamental phenomenon and relates to number of applied aspects.

Coffee-ring effect is one ubiquitous phenomenon, which is deposited by non-volatile solute of liquid after evaporation. However, this effect is unexpected result in adverse effect during manufacturing process involving inkjet printing, DNA microarrays, fabrication of ordered structures, and nanotechnology. Therefore, many studies have been conducted to understand and controlling the process of solute deposition in the presence of coffee-ring effect. Most of studies investigated the regulation of coffee-ring effect by changing the of the droplet composition. However, a high-purity vapor-deposited film is hard to producing by this method [1]. So it is necessary to find alternative method to adjust the droplet evaporation. Zhang el al. presented that the pattern formation of droplet has strong relevant to the property of substrates [2].

Currently, the understanding of droplet evaporation on substrates with different roughness was poor. Thus, the evaporation and the patterns of saline droplets on a smooth copper substrate with different roughness was investigated in my research. The relationship between the roughness and droplet evaporation reveals further.

X. Wang (✉) · Z. Liu · L. Wang · Y. Yan
Faculty of Engineering, University of Nottingham, Nottingham NG7 2RD, UK
e-mail: ezxxw2@exmail.nottingham.ac.uk

© Springer Nature Singapore Pte Ltd. 2021
C. Wen and Y. Yan (eds.), *Advances in Heat Transfer and Thermal Engineering* ,
https://doi.org/10.1007/978-981-33-4765-6_41

2 Methodology

Droplets of the saline solution were deposited by a pipette. The volume and concentration of the droplet are 0.2 μL and 1.75%, respectively. A high speed camera was placed vertical to the saline droplet and recorded the whole process of the droplet evaporation on substrates with different roughness (400 pp, 1200 pp, 0, 2 μm). In order to achieve the substrates with different roughness, the copper substrate was polished by abrasive paper with the help of grinding and polishing equipment. The grits of abrasive paper is 400 pp and 1200 pp, respectively. Additional, a high polish copper substrate was polished by hemp wheel (0.2 μm). In the whole experiment, the substrate temperature is set to 40 °C which is higher than normal temperature and controlled by a water bath.

3 Results

Figure 1a, b represent hydrophobic surface which apparent contact angle is larger than 90°. Meanwhile, the apparent contact angle of hydrophobic surface increases as the substrates roughness increase. The other two show the saline droplet on hydrophilic surface (contact angle < 90°), and the contact angle decreases as the substrates roughness increases. The particle distribution is difference between hydrophobic surface and hydrophilic surface. The particle distribution of hydrophobic surface is more uniform than hydrophilic surface. For the hydrophilic surface, the particles more close to boundary and tendency more obvious. The ring shape thickness of hydrophilic larger than hydrophobic surface as the roughness decrease.

The thickness boundary of droplet is the thinnest part relative to whole droplet on the hydrophilic. Besides, the surface temperature of droplet is the highest. Thus, the evaporation rate of the boundary is the fastest than other part, which results in more saline evaporated at the droplet boundary than center of droplet. Therefore, the saline move from the center to the boundary of droplet to offset the losing part. When the whole process of evaporation is completed, the concentration of boundary is higher than center of droplet due to the transport and deposit of the saline. However, the boundary thickness of droplet is not the thinnest part on the hydrophobic surface. Hence, there is less saline particle that moves to boundary of the droplet and distributes in the center of the droplet on the hydrophobic surface, which leads to the difference distribution of particles between hydrophobic surface and hydrophilic surface.

To sum up, the hydrophilic surface is more conducive to coffee-ring formation. In other word, the coffee-ring effect has strong relationship to roughness of substrate.

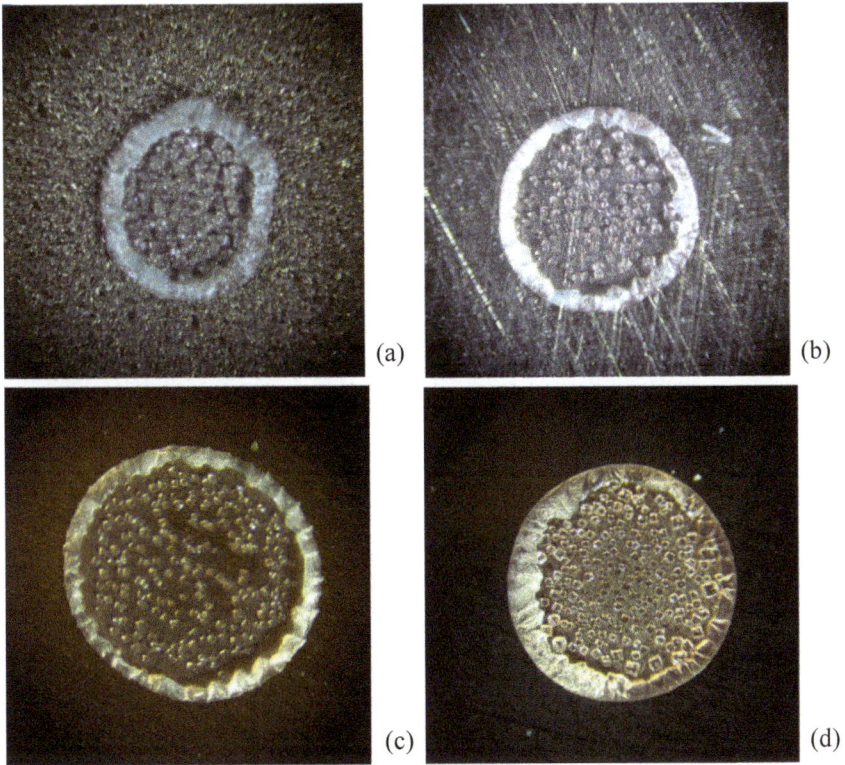

Fig. 1 Evaporation patterns of saline droplets on copper substrates with different roughness: **a** substrate without polishing, **b** 400 pp, **c** 1200 pp, **d** high polishing (0.2 μm)

4 Conclusions

The experiment of saline droplet deposition pattern on the copper substrate with different roughness has been conducted. We found that the particles distribution is different between the hydrophobic surface and hydrophilic surface. The particle distribution of hydrophobic surface is more uniform than hydrophilic surface. For the hydrophilic surface, more particles aggregate at the boundary of the droplet and the boundary thickness is thicker than the hydrophobic surface. Because the roughness affect the contact angle of droplet results in the variation of evaporation rate at the boundary of droplet, more saline evaporated and deposited at the boundary of droplet. Thus, the coffee-ring effect has strong relevant to the roughness of substrate. In the future, the experiment of nano-fluids on different roughness substrate will be carried on, which is based on the finding of this research, in order to enhance the understanding of the mechanism of coffee-ring effect with nano-fluids.

References

1. Y. Zhang, X. Chen, F. Liu, L. Li, J. Dai, T. Liu, Enhanced coffee-ring effect via substrate roughness in evaporation of colloidal droplets. Adv. Condensed Matter Phys. **2018**, Article ID 9795654, 9 p (2018). https://doi.org/10.1155/2018/9795654
2. Y. Zhang, Y. Qian, Z. Liu, Z. Li, D. Zang, Surface wringkling and cracking dynamics in the drying of colloidal droplets. Eur. Phys. J. E **37**(9), article 84, 7 p (2014)

Enhanced Heat Transfer

Enhance Heat Transfer and Mass Transfer in the Falling Film Absorption Process by Adding Nanoparticles

Hongtao Gao, Fei Mao, Yuchao Song, Jiaju Hong, and Yuying Yan

Nomenclature

C	Concentration
c_p	Specific heat at constant pressure, $J\,kg^{-1}\,K^{-1}$
D	Diffusion coefficient, $m^2\,s^{-1}$
g	Gravitational acceleration, $m\,s^{-2}$
H_{abs}	Heat of absorption, $kJ\,kg^{-1}$
k	Thermal conductivity, $W\,m^{-1}\,K^{-1}$
L	Falling film length, m
m	Mass transfer flux, $kg\,m^{-2}\,s^{-1}$
M	Mass transfer rate, $kg\,m^{-1}\,s^{-1}$
P	Pressure, kPa
q	Heat transfer flux, $kW\,m^{-2}$
Q	Heat transfer rate, $W\,m^{-1}$
T	Temperature, K
u	Velocity in x-direction, $m\,s^{-1}$
v	Velocity in y-direction, $m\,s^{-1}$
Re	Reynolds number
δ	Liquid film thickness, m
μ	Dynamic viscosity, Pa s
ρ	Density, $kg\,m^{-3}$
Γ	Film flow rate, $kg{\cdot}m^{-1}\,s^{-1}$

H. Gao (✉) · F. Mao · Y. Song · J. Hong
Institute of Refrigeration and Cryogenics Engineering, Dalian Maritime University, Dalian 116026, China
e-mail: gaohongtao@dlmu.edu.cn

Y. Yan
Fluids and Thermal Engineering Research Group, Faculty of Engineering, University of Nottingham, University Park, Nottingham NG7 2RD, UK

© Springer Nature Singapore Pte Ltd. 2021
C. Wen and Y. Yan (eds.), *Advances in Heat Transfer and Thermal Engineering* ,
https://doi.org/10.1007/978-981-33-4765-6_42

Φ   Volume fraction of nanoparticles

Subscripts

nf   Nanofluid
c   Cooling water
p   Nanoparticle
f   Fluid
abs   Absorption
in   Inlet
out   Outlet
w   Wall
i   Liquid–vapor interface

1 Introduction

The research of absorption refrigeration technology has been paid attention to by many scholars. The absorber is one of the most important components in the absorption refrigeration system, and its mass transfer and heat transfer performance directly affect the performance of the entire unit. Therefore, improving mass transfer and heat transfer in the absorber is one of the most effective means to improve the performance of the whole refrigeration unit. The addition of nanoparticles is one of the means to effectively improve the heat transfer and mass transfer performance of the absorber [1–3].

Many scholars have studied the effects of nanofluids on mass transfer and heat transfer performance in absorbers through experimental methods. Under the condition of constant temperature wall, the effects of adding copper oxide and alumina nanoparticles on the heat transfer properties of laminar falling film absorption in a circular tube were studied by Heris et al. [4]. The experimental results show that for both nanofluids, the heat transfer coefficient increases with the increase of nanoparticle concentration. The influence of nanofluids on heat transfer performance mainly depends on thermal conductivity, chaotic movements, fluctuations, and interactions. Kang et al. [5] studied the heat transfer and mass transfer in falling film absorption of binary nanofluids with iron nanoparticles and carbon nanotubes. The results show that the mass transfer enhancement of carbon nanotubes nanoparticles is higher than that of iron nanoparticles, and the mass transfer enhancement is more significant than heat transfer enhancement in binary nanofluids with iron nanoparticles and carbon nanotubes nanoparticles. Kim et al. [6] experimentally studied the effect of nanoparticles on bubble absorption. The results show that the absorption performance of nanoparticles increases by 3.21 times, and the absorption rate increases with the increase of the concentration of nanoparticles. For the solution with lower absorption capacity, the enhancement effect of nanoparticles is more obvious. Liu et al. [7] experimentally studied the effects of Fe_2O_3 and $ZnFe_2O_4$ nanoparticles on the mass transfer and heat transfer properties of falling film absorption of ammonia–water. The results show that the effective absorption of Fe_2O_3 and $ZnFe_2O_4$ nanofluids increases by 70% and 50%, respectively, when the initial ammonia content is 15%. It is concluded that the enhancement of absorption is due to the enhancement of heat transfer and the decrease of viscosity of nanofluids. Li et al. [8] presented a falling film generating test bench for testing ammonia vapor generation rate with/without nanoparticles, in which ammonia–water was used as working fluid. The enhancement mechanism was analyzed from the micro-motion, interface effect, Marangoni effect, and physical properties of nanofluids. The results show that the micro-motion of nanofluids and the physical properties of nanofluids are the two main factors of the enhancement of ammonia falling film by nanofluids.

Some scholars have studied the effect of nanoparticles on mass transfer and heat transfer performance by numerical simulation. Armou et al. [9] established a mathematical and physical model for falling film absorption of nanofluids in laminar flow. The numerical results show that the mass transfer flux can be increased by adding Ag nanoparticles. The mass transfer and heat transfer performance of binary nanofluids are stronger than that of pure lithium bromide solution. Wang et al. [10] established a mathematical model for the falling film absorption of nanofluids on sloping plates. The results show that the nanoparticles can significantly increase the water vapor absorption rate. Also, the water vapor absorption rate increases as the amount of the nanoparticles added increases. When the flow rate of nanofluids with 0.05% and 0.1% nanoparticles is 1.0L min^{-1}, the mass transfer coefficients increase by 1.28 and 1.41 times, respectively. Zhang et al. [11] established a mathematical model for falling film absorption of nanofluids, which consisted of Fe_3O_4 nanoparticles and lithium bromide solution. The results show that the enhanced heat transfer and mass transfer of nanofluids is related to the concentration and size of nanoparticles. The higher the concentration of nanoparticles, the stronger the heat and mass transfer of falling film, and the smaller the size of nanoparticles, the greater the heat and mass transfer of falling film.

Some scholars have found other ways to improve the heat and mass transfer performance of falling film absorption. Gao et al. [12] added alcoholic surfactants with different carbon atom numbers into lithium bromide solution. It was found that the lower the surface tension of the solution, the better the mass transfer ability of lithium bromide solution. Zhu et al. [13] added alcoholic surfactants into water or lithium bromide aqueous solution and found that surfactant molecules adsorbed at the gas–liquid interface, hydrophobic groups pointed to the gas phase, and hydrophilic groups pointed to the dominant orientation of the liquid phase. Niu et al. [14] established a mathematical model of falling film absorption of ammonia–water in magnetic field. The effect of magnetic field on the absorption process was studied. The results show that the magnetic field has a significant effect on the falling film absorption of ammonia–water. When the magnetic induction intensity of solution inlet is 3 T (Tesla), the concentration of ammonia solution at outlet increases by 1.3% and the refrigeration coefficient of ammonia–water absorption refrigeration system increases by 4.73%.

In this paper, COMSOL Multiphysics simulation software based on finite element method is used to simulate the vertical falling film absorption process of lithium bromide solution. In order to make the model closer to the actual situation, the convection on the cooling water side was considered and the temperature of the cooling water changed linearly along its flow direction. The effect of adding copper oxide nanoparticles on the mass/heat transfer of falling film absorption was investigated.

2 Mathematical Physics Model

2.1 Physical Model

The flow pattern of the falling film is shown in Fig. 1. The x-direction is the flow direction of the falling film of lithium bromide, and the y-direction is the direction of the thickness of the liquid film. At the inlet, the solution is sprayed onto the wall. Due to gravity, the solution flows down the inner wall of the absorber and begins to absorb water vapor, while releasing a large amount of absorbed heat, which causes the temperature of the solution to rise. On the outside of the wall, there is a countercurrent upward cooling water to cool the solution, ensuring the absorption process go continuously. In order to establish a physical mathematical model that reflects both the actual absorption process and the analysis and comparison, the model assumes the following:

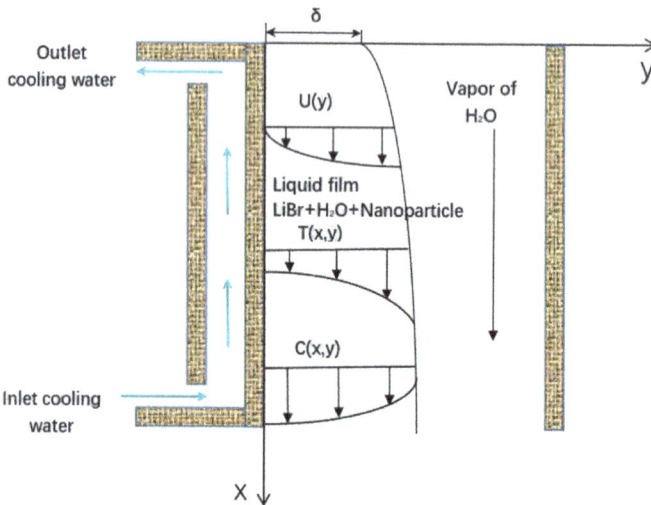

Fig. 1 2D model of the absorption process

(1) The physical properties of the solution are constant and are incompressible Newtonian fluids.
(2) Falling film flow is a fully developed laminar flow state.
(3) The nanoparticles are uniformly dispersed in the liquid.
(4) The liquid film thickness is considered constant.
(5) The gas–liquid interface is in a phase equilibrium state.
(6) It was assumed that the heat transfer to the water vapor phase is negligible.

2.2 Governing Equations

According to the basic assumptions mentioned above, in the vertical falling film absorption process, the continuous equation, the energy equation, the momentum equation, and the concentration equation can be given in the following form:

$$\frac{\partial}{\partial x}(\rho_{nf}u) + \frac{\partial}{\partial y}(\rho_{nf}v) = 0 \tag{1}$$

$$\frac{\partial(\rho_{nf}uu)}{\partial x} + \frac{\partial(\rho_{nf}vu)}{\partial y} = \frac{\partial}{\partial x}\left(\mu_{nf}\frac{\partial u}{\partial x}\right) + \frac{\partial}{\partial y}\left(\mu_{nf}\frac{\partial u}{\partial y}\right) + \rho_{nf}g \tag{2}$$

$$\frac{\partial(\rho_{nf}uv)}{\partial x} + \frac{\partial(\rho_{nf}vv)}{\partial y} = \frac{\partial}{\partial x}\left(\mu_{nf}\frac{\partial v}{\partial x}\right) + \frac{\partial}{\partial y}\left(\mu_{nf}\frac{\partial v}{\partial y}\right) \tag{3}$$

$$\left(\rho c_p\right)_{nf}\left(u\frac{\partial T}{\partial x} + v\frac{\partial T}{\partial y}\right) = k_{nf}\left(\frac{\partial^2 T}{\partial x^2} + \frac{\partial^2 T}{\partial y^2}\right) + \frac{\partial}{\partial y}(m H_{abs}) \tag{4}$$

$$\frac{\partial(\rho_{nf}uC)}{\partial x} + \frac{\partial(\rho_{nf}vC)}{\partial y} = \frac{\partial}{\partial x}\left(\rho_{nf}D\frac{\partial C}{\partial x}\right) + \frac{\partial}{\partial y}\left(\rho_{nf}D\frac{\partial C}{\partial y}\right) \tag{5}$$

2.3 Related Parameters

The film Reynolds number is defined as [15]:

$$Re = \frac{4\Gamma}{\mu_{nf}} \tag{6}$$

The film thickness is determined by Nusselt's analysis as follows:

$$\delta = \left(\frac{3\Gamma\mu_{nf}}{\rho_{nf}^2 g}\right)^{\frac{1}{3}} \tag{7}$$

According to the Nusselt theory, the downstream velocity profile u is given as follows:

$$u = \frac{3}{2}u_0\left[2\frac{y}{\delta} - \left(\frac{y}{\delta}\right)^2\right]$$

(8)

where $u_0 = \frac{\Gamma}{\rho_{nf}\delta}$.

2.4 Boundary Conditions

(1) At the solution inlet, it can be assumed that the liquid film temperature and concentration distribution are uniform.

$$x = 0; \quad T = T_{in}; \quad C = C_{in}$$

(9)

(2) At plate wall, the wall is smooth and impermeable. The wall temperature can be determined as follows:

$$u = v = 0; \quad \frac{\partial C}{\partial y} = 0; \quad T_w = T_{c,in} + \left(\frac{L-x}{L}\right) \times \left(T_{c,out} - T_{c,in}\right)$$

(10)

(3) At solution outlet, the flow of liquid film is fully developed state:

$$x = L; \quad \frac{\partial T}{\partial x} = 0; \quad \frac{\partial C}{\partial x} = 0$$

(11)

(4) The liquid–vapor interface is in thermodynamic equilibrium state. The mass transfer flux is given as follows:

$$C_i = C_i(P, T)$$

(12)

$$m = -\rho D\left(\frac{\partial C}{\partial y}\right)_{y=\delta_0}$$

(13)

2.5 Nanofluid Properties

The density, the dynamic viscosity, the specific heat, and the thermal conductivity of the nanofluid are, respectively, determined by the following [16–18]:

$$\rho_{nf} = (1 - \Phi)\rho_f + \Phi\rho_n$$

(14)

Table 1 Grid verification

The number of meshes	1000×20	1500×20	2000×20	1500×20	1500×30	1500×40
$m_{avg} \times 10^{-3}$	1.8012	1.8016	1.8018	1.8016	1.8096	1.8130

$$\mu_{nf} = \frac{\mu_f}{(1 - \Phi)^{2.5}} \tag{15}$$

$$(\rho c_P)_{nf} = (1 - \Phi)(\rho c_P)_f + \Phi(\rho c_P)_n \tag{16}$$

$$\frac{k_{nf}}{k_f} = \frac{k_n + 2k_f - 2\Phi(k_f - k_n)}{k_n + 2k_f + \Phi(k_f - k_n)} \tag{17}$$

3 Numerical Procedure

3.1 Solution Method and Meshing

The laminar flow, heat transfer in fluids, and transport of concentrated species physics in COMSOL Multiphysics are used to establish a two-dimensional falling film absorption model of lithium bromide solution with a vertical wall length of 150 mm. Steady-state and PAEDISO solver are used, and a preordering algorithm of nested dissection multithreaded is adopted. The model uses a structured quadrilateral mesh that is compatible with the computational region. The maximum deviation is 0.63% (Table 1). After grid-independence verification, the number of meshes in the x-direction and y-direction is determined to be 1500 × 20.

3.2 Model Validation

Under the same operating conditions (Table 2) and physical properties (Table 3), the calculated results of this model are in good agreement with the analytical solutions of Kawae et al. [19]. The maximum deviation is 0.4%, as shown in Fig. 2.

Table 2 Operating conditions

Parameters	Value
Inlet concentration/$C_{s,in}$	60%
Inlet temperature/$T_{s,in}$	319.65 K
System pressure/P	1 kPa
Length of falling film/L	150 mm
Cooling water inlet temperature/$T_{c,in}$	305.15 K
Cooling water outlet temperature/$T_{c,out}$	309.15 K

Table 3 Physical properties of LiBr solution

Physical properties	Value
Density/ρ	1699.7 kg m^{-3}
Diffusion coefficient/D	1.6175×10^{-9} m^2 s^{-1}
Heat of absorption/H	2772.7 kJ kg^{-1}
Thermal conductivity/k	0.431 W m^{-1} K^{-1}
Specific heat/C_P	1567.5 J kg^{-1} K^{-1}
Dynamic viscosity/μ	5.35×10^{-3} Pa s

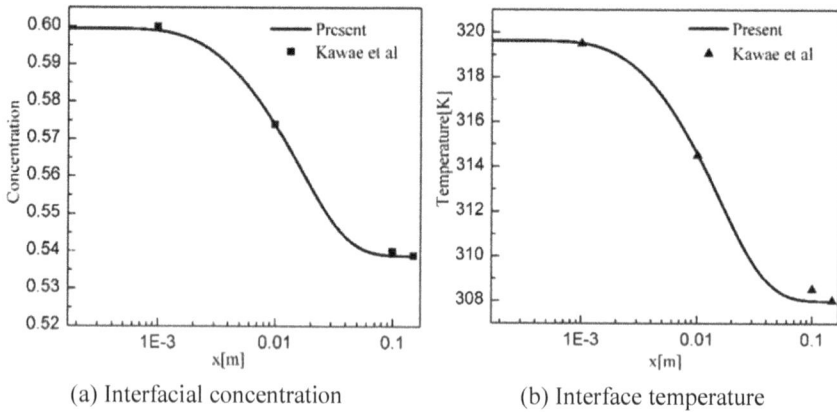

(a) Interfacial concentration   (b) Interface temperature

Fig. 2 Comparison of interface concentration and temperature of present study with previous investigations

4 Results and Discussion

4.1 Mass Transfer Flux at Interface

The effect of the volume fraction of copper oxide nanoparticles on the absorption properties was mainly studied. Figure 3 shows the effect of nanoparticles with increasing volume fraction from $\Phi = 0$ to $\Phi = 0.001$ on mass transfer flux. When no

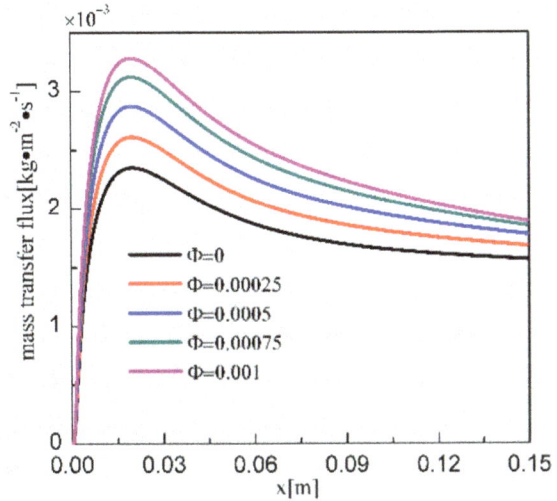

Fig. 3 Variation of mass transfer flux with volume fraction of nanoparticles

nanoparticles are added ($\Phi = 0$), the mass transfer flux rises rapidly within 20 mm of the inlet and reaches the maximum value ($m_{i,max} = 2.32 \times 10^{-3}$ kg m^{-2} s^{-1}), then began to decline, declining rapidly from 20 to 60 mm, but slowly from 80 mm. The interface liquid–vapor solution at the inlet is in phase equilibrium. When the solution enters the tube, the cooling effect quickly reaches the interface, reduces the temperature of the liquid film interface, and reduces the saturated vapor pressure of the lithium bromide aqueous solution, which increases the mass transfer driving force and therefore increases the mass transfer flux at the interface. However, as the absorbed water vapor gradually increases, the latent heat released gradually increases, which causes the temperature of the interface to increase, the saturation pressure of the solution to increase, the mass transfer driving force to decrease, and the mass transfer flux to decrease slowly. At the same time, with the increase of falling film distance, lithium bromide solution has absorbed more water vapor, which leads to the decrease of water vapor absorption capacity of solution, so the mass transfer flux decreases slowly.

In this paper, it is assumed that nanoparticles are uniformly and stably dispersed in solution. Therefore, the current numerical studies only consider the low concentration of nanoparticles. When the volume fraction of nanoparticles increases from $\Phi = 0$ to $\Phi = 0.001$, the mass transfer flux is improved, and the trend of the five curves is consistent. The effect of the nanoparticles is not obvious near the inlet and near the outlet. When falling film distance is 20 mm, the average mass transfer flux of lithium bromide solution with 0.1% nanoparticles is 1.4 times higher than that of pure lithium bromide solution. This is because the irregular Brownian motion of nanoparticles makes the mass diffusion coefficient increase. At the same time, the density of lithium bromide aqueous solution with copper oxide nanoparticles is slightly higher than that of pure lithium bromide aqueous solution. Therefore, the mass transfer flux increases with the increase of the volume fraction of nanoparticles.

4.2 Variation of Mass Transfer Coefficient and Mass Transfer Rate in Liquid Phase

The liquid mass transfer coefficient is defined as follows:

$$h_m(x) = \frac{m(x)}{\rho(C_w(x) - C_i(x))} \tag{18}$$

Figure 4 shows the change in liquid phase mass transfer coefficient with volume fraction of copper oxide nanoparticles. When the volume fraction of nanoparticles is $\Phi = 0$, the liquid mass transfer coefficient reaches the maximum near the entrance $(h_{m,max} = 1.47 \times 10^{-4} \text{ m s}^{-1})$, but with the increase of falling film length, the liquid mass transfer coefficient decreases rapidly and then becomes stable. Near the entrance, due to wall cooling, the absorption process begins and the concentration of the interface decreases rapidly, so the concentration difference between the interface and the wall increases rapidly, which makes the mass transfer coefficient of the liquid phase decrease rapidly. However, with the increase of falling film distance, the absorption capacity of the solution decreases, which makes the concentration difference change slowly and makes the liquid mass transfer coefficient change smoothly.

It can be seen from Fig. 4 that a small amount of nanoparticles has a positive effect on the liquid phase mass transfer coefficient. Compared with pure lithium bromide solution, the liquid phase mass transfer coefficient of lithium bromide solution with 0.1% nanoparticles at the outlet is 1.62 times higher than that of pure lithium bromide solution. This is because the perturbation of the nanoparticles makes the mass transfer enhanced, and the difference in concentration between the vapor–liquid interface

Fig. 4 Variation of mass transfer coefficient in liquid phase with volume fraction of nanoparticles

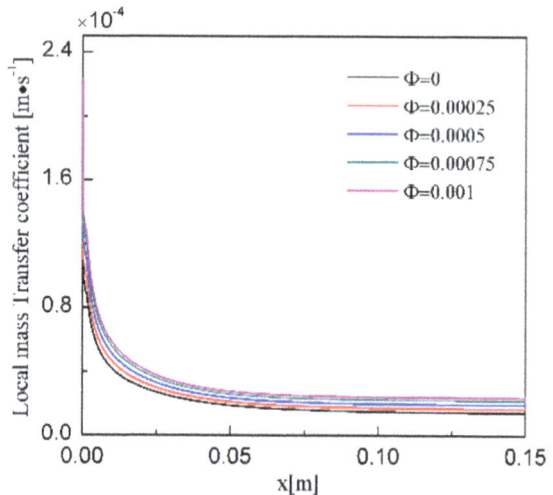

Fig. 5 Variation of mass
transfer rate with volume
fraction of nanoparticles

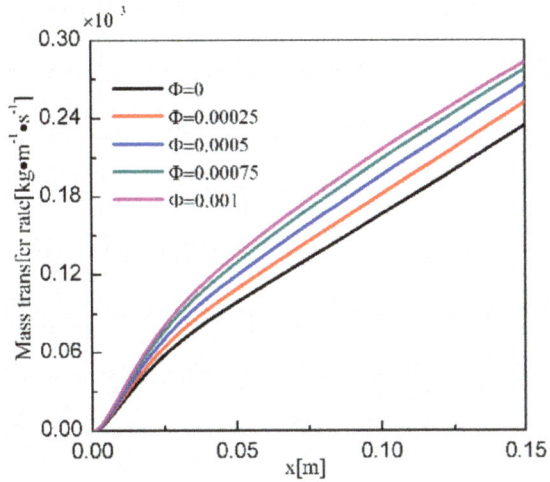

and the wall surface decreases. Therefore, the liquid phase mass transfer coefficient increases as the volume fraction of the nanoparticles increases.

Figure 5 shows the change in mass transfer rate. It can be concluded from the figure that the mass transfer rate increases with the falling film distance. As the volume fraction of nanoparticles increases, the mass transfer rate increases, but near the inlet, the effect of the nanoparticles is not significant.

4.3   Effect of Film Flow Rate

The mass transfer enhancement factor is defined as the ratio of the mass transfer flux of the lithium bromide aqueous solution to which the copper oxide nanoparticles are added under the same conditions to the mass transfer flux of the pure lithium bromide aqueous solution. This coefficient can be expressed as:

$$R = \frac{m_{nf}}{m_f} \tag{19}$$

Figure 6a shows the effect of the volume fraction of copper oxide nanoparticles on the average mass transfer flux as a function of film flow rate. When the nanoparticle volume fraction is $\Phi = 0$, the average mass transfer flux first increases and then decreases with the increase of the film flow rate. When the spray density is 0.05 $kg\,m^{-2}\,s^{-1}$, the average mass transfer flux reaches the maximum ($m_{avg} = 2.28 \times 10^{-3}\ kg\,m^{-2}\,s^{-1}$)). This is because as the film flow rate increases, the lithium bromide aqueous solution is renewed faster, the absorption capacity is enhanced, and more water vapor can be absorbed, so that the average mass transfer flux is increased. As the film flow rate continues to increase, the liquid film renews

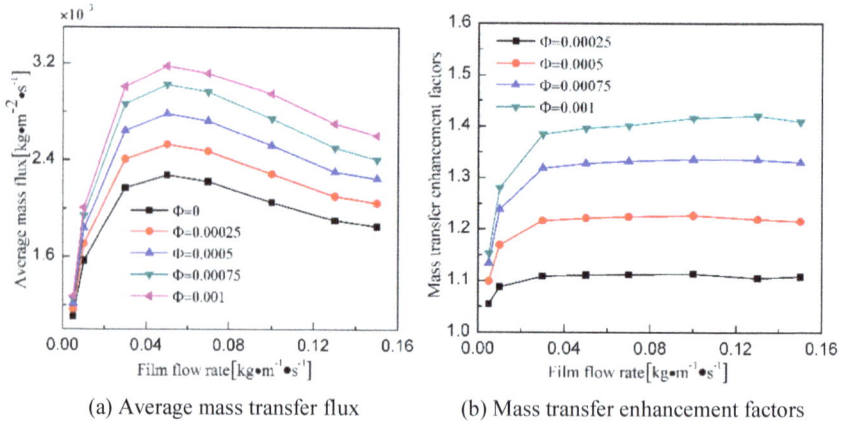

(a) Average mass transfer flux   (b) Mass transfer enhancement factors

Fig. 6 Film flow rate and different volume fractions of CuO nanoparticles

too quickly, resulting in a shorter time for the water vapor to contact the liquid film, resulting in a decrease in the amount of water vapor absorbed and a decrease in the average mass transfer flux. Therefore, the average mass transfer flux does not always increase with the increase of the film flow rate, and the excessive or small film flow rate is not conducive to the absorption process.

When the volume fraction of copper oxide nanoparticles increases from $\Phi = 0$ to $\Phi = 0.001$, the average mass transfer flux is improved. To analyze in more detail, the effect of nanoparticle addition on mass transfer flux, Fig. 6b shows the change in mass transfer enhancement factor with film flow rate and nanoparticle volume fraction. When the film flow rate is very small, the increase of average mass transfer flux by nanoparticles is not obvious. When the film flow rate was $0.01 \text{ kg m}^{-2} \text{ s}^{-1}$, the average mass transfer flux increased by 1.28 times with the addition of 0.1% nanoparticles. With the increase of film flow rate, the increase of average mass transfer flux of nanoparticles tends to be stable. When the film flow rate is $0.05 \text{ kg m}^{-2} \text{ s}^{-1}$, the average mass transfer flux is 1.4 times higher than that of the pure lithium bromide solution with 0.1% of the nanoparticles.

4.4 Effect of Solution Inlet Temperature

Figure 7a shows the effect of the volume fraction of nanoparticles on the average mass transfer flux with the increase of the solution inlet temperature. When the volume fraction of copper oxide nanoparticles is $\Phi = 0$, the increase of the inlet temperature is not conducive to the mass transfer process. When $\Phi = 0$, the inlet temperature increases from 313.15 to 333.15 K, the average mass transfer flux decreases by 9.4%. This is because the water vapor pressure of lithium bromide solution increases with

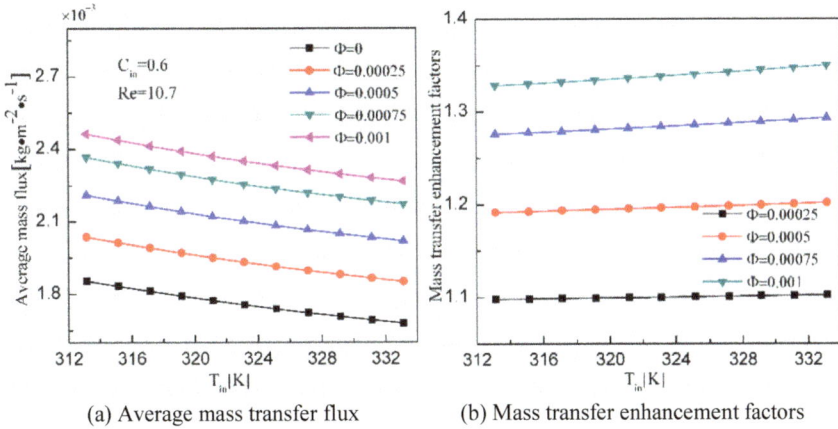

(a) Average mass transfer flux

(b) Mass transfer enhancement factors

Fig. 7 Solution inlet temperature and different volume fractions of CuO nanoparticles

the increase of solution inlet temperature, which leads to the decrease of mass transfer driving force and the decrease of average mass transfer flux.

With the increase of volume fraction of copper oxide nanoparticles, the average mass transfer flux increases, and the trend of the five curves is consistent. Figure 7b shows the effect of nanoparticles on mass transfer enhancement factors as the inlet temperature increases. When the inlet temperature is 313.15 K and 333.14 K, the average mass transfer flux of lithium bromide solution with 0.1% nanoparticles is 1.32 times and 1.35 times higher than that of pure lithium bromide solution, respectively. When the inlet temperature is higher, the effect of nanoparticles is more obvious.

4.5 Effect of Solution Inlet Concentration

Figure 8a, b show the effect of volume fraction of nanoparticles on mass transfer flux and mass transfer enhancement factor, respectively, as the inlet concentration of the solution increases. Increasing the inlet concentration of the solution helps to increase the mass transfer flux (Fig. 8a). When $\Phi = 0$, the average mass transfer flux increases to 2.97 times of the original when the inlet concentration increases from 60 to 70%. The higher the inlet concentration, the lower the vapor pressure of lithium bromide solution, the greater the mass transfer driving force, thus increasing the mass transfer flux.

When the volume fraction of nanoparticles increases from $\Phi = 0$ to $\Phi = 0.001$, the average mass transfer flux increases. When the inlet concentration was 60% and 70%, the average mass transfer flux was increased to 1.33 times and 1.31 times, respectively, compared with the lithium bromide solution added with 0.1% of the nanoparticles. When the inlet concentration is low, the effect of the nanoparticles is more pronounced.

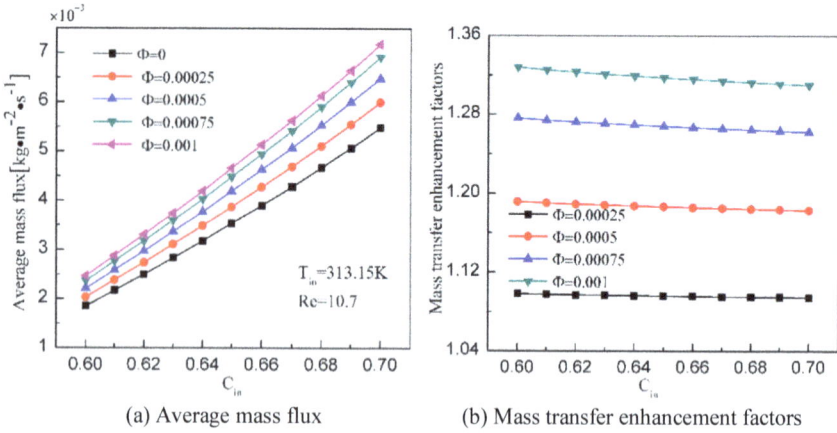

(a) Average mass flux (b) Mass transfer enhancement factors

Fig. 8 Solution inlet concentration and different volume fractions of CuO nanoparticles

4.6 Effect of Cooling Water Inlet Temperature

Figure 9a, b show the effect of copper oxide nanoparticles on mass transfer flux and mass transfer enhancement factor, respectively, as the inlet temperature of the cooling water increases. It can be seen from Fig. 9a that lowering the temperature of cooling water is beneficial to the mass transfer process. When $\Phi = 0$, the cooling water inlet concentration decreased from 307.15 to 301.15 K and the average mass transfer flux increased to 1.65 times. This is because as the temperature of the inlet cooling water decreases, the heat transfer rate increases, resulting in a decrease in the partial pressure of the lithium bromide water vapor and an increase in the mass transfer driving force. Therefore, the mass transfer flux increases.

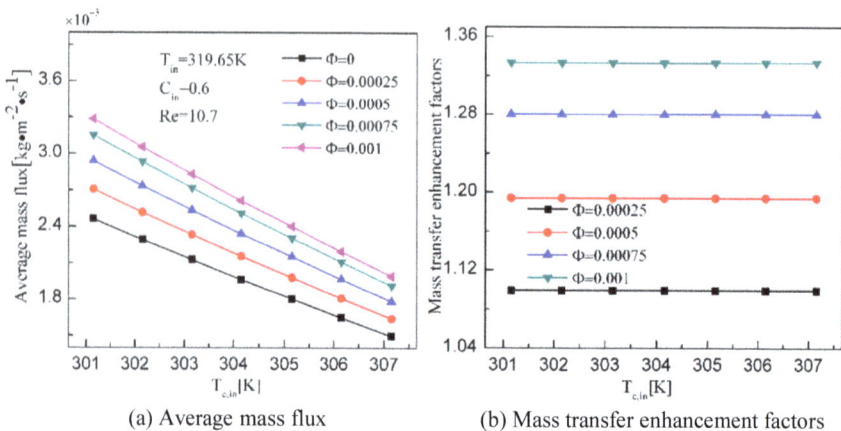

(a) Average mass flux (b) Mass transfer enhancement factors

Fig. 9 Cooling water inlet temperature and different volume fractions of CuO nanoparticles

When the inlet temperature of cooling water is constant, the average mass transfer flux increases with the increase of the volume fraction of nanoparticles. When the cooling water inlet temperature was 301.15 K and 307.15 K, the average mass transfer flux was increased to 1.33 times compared with the pure lithium bromide solution with 0.1% of the nanoparticles. Figure 9b shows that with the increase of cooling water inlet temperature, the size of mass transfer enhancement factor hardly changes when the volume fraction of nanoparticles is constant. Therefore, the effect of cooling water inlet temperature on mass transfer enhancement factor is not obvious.

5 Conclusions

In this paper, the finite element method is used to establish a thermal-mass coupling model for lithium bromide falling film absorption. The effect of adding nanoparticles and film flow rate on the heat/mass transfer in lithium bromide falling film absorption was investigated. The main conclusions are as follows:

(1) For the falling film absorption of pure lithium bromide solution, as the film flow rate increases, the average mass flux increases first and then decreases. When the film flow rate is too large or too small, it is not conducive to the absorption process, and there is an optimum film flow rate.
(2) Reducing the solution inlet temperature, increasing the solution inlet concentration, or decreasing the cooling water inlet temperature are beneficial to the mass transfer process.
(3) The mass transfer flux at the interface increases rapidly near the inlet and reaches the maximum value, then decreases slowly; the mass transfer coefficient of liquid phase decreases rapidly after reaching the maximum near the inlet and tends to be stable with the increase of falling film distance.
(4) After adding copper oxide nanoparticles into lithium bromide solution, the mass transfer flux, liquid phase mass transfer coefficient, and mass transfer rate at the interface are increased. Therefore, adding copper oxide nanoparticles is conducive to improve the absorption performance.
(5) When the solution inlet temperature is high and the solution inlet concentration is low, the effect of the nanoparticles is more significant.

Acknowledgements This work was financially supported by research funds of the Maritime Safety Administration of the People's Republic of China (2012_27) and the Fundamental Research Funds for the Central Universities (3132019305).

References

1. S.S. Ashrafmansouri, E.M. Nasr, Mass transfer in nanofluids: a review. Int. J. Therm. Sci. **82**, 84–99 (2014)
2. H.-T. Gao, M.-Y. He, Numerical modelling and performance analysis of falling film absorption under rolling conditions. J. Dalian Maritime Univ. **42**, 79–83 (2016)
3. S. Krishnamurthy, P. Bhattacharya, P.E. Phelan et al., Enhanced mass transport in nanofluids. Nano Lett. **6**, 419–423 (2006)
4. S.Z. Heris, S.G. Etemad, M.N. Esfahany, Experimental investigation of oxide nanofluids laminar flow convective heat transfer. Int. Commun. Heat Mass Transfer **33**, 529–535 (2006)
5. Y.T. Kang, H.J. Kim, K.I. Lee, Heat and mass transfer enhancement of binary nanofluids for H_2O/LiBr falling film absorption process. Int. J. Refrig **31**, 850–856 (2008)
6. J.K. Kim, J.Y. Jung, Y.T. Kang, The effect of nano-particles on the bubble absorption performance in a binary nanofluid. Int. J. Refrig **29**, 22–29 (2006)
7. L. Yang, K. Du, X.F. Niu et al., Experimental study on enhancement of ammonia–water falling film absorption by adding nano-particles. Int. J. Refrig **34**, 640–647 (2011)
8. Y.-J. Li, Du. Kai, W.-X. Jiang, Mechanism analysis on performance enhancement of ammonia-water falling film generation by nanofluid. Refrigeration **44**, 51–57 (2016)
9. S. Armou, R. Mir, Y. El Hammami et al., Heat and mass transfer enhancement in absorption of vapor in laminar liquid film by adding nano-particles. J. Appl. Fluid Mech. **10**, 1711–1720 (2017)
10. G. Wang, Q. Zhang, M. Zeng et al., Investigation on mass transfer characteristics of the falling film absorption of LiBr aqueous solution added with nanoparticles. Int. J. Refrig. **89**, 149–158 (2018)
11. L.Y. Zhang, Y. Li, Y. Wang, et al., Effect of nanoparticles on H_2O/LiBr falling film absorption process, in *ASME 2016 5th International Conference on Micro/Nanoscale Heat and Mass Transfer* (American Society of Mechanical Engineers, 2016), pp. 1–11.
12. H.-T. Gao, H. Eiji, Surface tension of LiBr aqueous solution with heat mass transfer enhancement additives. J. Refrig. **3**, 5–8 (2004)
13. B.-B. Zhu, H.-T. Gao, Molecular modeling at liquid-vapor interface of lithium bromide aqueous solutions with alcohols surfactants. J. Dalian Maritime Univ. **34**, 29–33 (2008)
14. N. Xiaofeng, D. Kai, D. Shunxiang, Numerical analysis of falling film absorption with ammonia–water in magnetic field. Appl. Therm. Eng. **27**, 2059–2065 (2007)
15. H.C. Chang, E.A. Demekhin, *Complex Wave Dynamics on Thin Films* (Elsevier Press, Netherlands, 2002)
16. H.C. Brinkman, The viscosity of concentrated suspensions and solutions. J. Chem. Phys. **20**, 571–581 (1952)
17. Y. Xuan, W. Roetzel, Conceptions for heat transfer correlation of nanofluids. Int. J. Heat Mass Transfer **43**, 3701–3707 (2000)
18. C.P. Bock, I.C. Young, Hydrodynamic and heat transfer study of dispersed fluids with submicron metallic oxyde particles. Experim. Heat Transfer **11**, 151–170 (1998)
19. N. Kawae, T. Shigechi, K. Kanemaru, et al. Water vapor evaporation into laminar film flow of a lithium bromide-water solution (influence of variable properties and inlet film thickness on absorption mass transfer rate). Heat Transfer-Japanese Research, 18, 58–70 (1989)

Pulsating Heat Stripes: A Composite Polymer Sheet with Enhanced Thermal Conductivity

Oguzhan Der, Marco Marengo, and Volfango Bertola

1 Introduction

The use of polymeric materials to replace metallic parts is the obvious choice to address weight and cost constraints in a large number of devices and applications, including space, aircraft and portable electronics applications. While polymeric materials offer excellent features of mechanical flexibility, resistance to fatigue, low weight and low cost in comparison with metallic materials, they exhibit poor heat transfer performance due to their low-thermal conductivity. Recently, there were several attempts to increase the thermal conductivity of polymers by means of high-thermal conductivity additives and fillers, such as minerals, fibres and metals [1].Commonly used fillers include particles [2, 3], fibres [4], metal powders or particles [5, 6] and carbon nanotubes [7–9].

In the present work, it is proposed to enhance the thermal conductivity of polymer sheets by embedding a self-driven liquid–vapour mixture, which transfers heat from an evaporator to a condenser region of the material according to the well-known working principle of pulsating (or oscillating) heat pipes (PHP) [10]. The heat transfer fluid circulates in a closed-loop serpentine channel, which is cut out in a polypropylene sheet and sandwiched between two transparent polypropylene sheets, bonded together by selective transmission laser welding. The resulting channel has a rectangular cross-section characterized by a large aspect ratio, hence the denomination *pulsating heat stripes* (PHS).

O. Der · V. Bertola (✉)
Laboratory of Technical Physics, School of Engineering, University of Liverpool, Liverpool L69 3GH, UK
e-mail: Volfango.Bertola@liverpool.ac.uk

M. Marengo
School of Computing, Engineering and Mathematics, University of Brighton, Brighton BN2 4GJ, UK

© Springer Nature Singapore Pte Ltd. 2021
C. Wen and Y. Yan (eds.), *Advances in Heat Transfer and Thermal Engineering* ,
https://doi.org/10.1007/978-981-33-4765-6_43

The thermal performances of composite polypropylene sheets with different designs of the serpentine channel and containing FC-72 as heat transfer fluid were tested by applying to the evaporator an ascending/descending power ramp ranging between 2 and 35 W and measuring the temperatures on the sheet surface for different orientations (vertical, inclined at 45°, horizontal). At the maximum heat supply, the equivalent thermal conductance of the PHS in vertical position exhibits a five-fold increase with respect to the composite polypropylene sheet without working fluid.

2 Methodology

Prototype composite sheets consisted of one black polypropylene layer (0.7 mm thick) sandwiched between two transparent polypropylene layers (0.4 mm thick), all with a length of 250 mm and a width of 100 mm. The transparent polypropylene layers were bonded on the two sides of the channel by selective transmission laser welding [11], as shown schematically in Fig. 1a. A serpentine-shaped channel with either five or seven turns was cut out in the central black layer, as illustrated in Figs. 1b, c. The channel had a width of 5 mm, which was determined so that the hydraulic diameter, $D_H = 1.1$ mm, satisfies the design criterion given in Eq. (1), which ensures surface forces prevail on gravity [12]:

$$0.7\sqrt{\frac{\sigma}{g(\rho_L - \rho_G)}} \leq D_H \leq 1.8\sqrt{\frac{\sigma}{g(\rho_L - \rho_G)}} \tag{1}$$

According to Eq. (1), the hydraulic diameter depends on the fluid properties; in particular, for the refrigerant fluid FC-72 ($\rho = 1680$ kg/m^3; $\sigma = 10$ mN/m), the

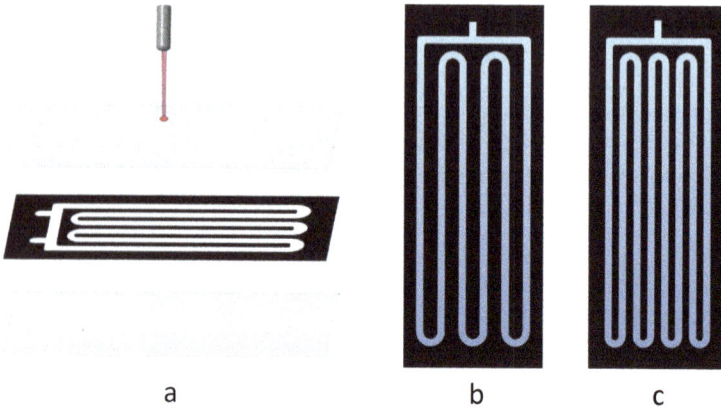

a   b   c

Fig. 1 Schematic of the manufacturing process (**a**) and top views of the PHS channels with five turns (**b**) and seven turns (**c**)

criterion becomes 0.54 mm $\leq D_H \leq$ 1.4 mm. The five-turn channel (Fig. 1b) had
a total volume of 8.4 ml, while the seven-turn channel (Fig. 1c) had a total volume
of 11 ml. A polypropylene fitting was used to connect a pressure transducer and a
micro-metering valve used for introducing the heat transfer fluid.

Experiments were conducted by applying to the evaporator section an
ascending/descending stepped heating power ramp ranging approximately between 2
and 35 W, and measuring the temperatures on the composite polymer sheet surface in
the evaporator and in the condenser regions. For each power step, the heat supply was
kept constant until a pseudo-steady-state regime was attained. Tests were interrupted
earlier in case any point of the material reached a temperature of 110 °C.

The equivalent thermal resistance of the engineered composite polypropylene
sheet in the longitudinal direction was calculated as:

$$R_{eq} = \frac{T_{ev} - T_{cond}}{\dot{Q}} \qquad (2)$$

where T_{ev} and T_{cond} are the averages of the four thermocouples temperature measurements of the evaporator and condenser sections, respectively, and $\dot{Q}$ is the heating
power supply.

3 Results

Temperatures measured in the evaporator and in the condenser zones of the composite
polymer sheet during the ascending/descending heating power supply ramp are
displayed in Fig. 2. Due to the relatively small thermal conductivity of polypropylene, and to the intrinsically unstable gas–vapour flow, the composite sheet exhibits
significant thermal inertia and reaches a pseudo-steady state in about 40 min after
each step change in the heating power supply. In all cases, the maximum performance
is limited by the temperature of the composite sheet in the evaporator region, which
is limited by the maximum continuous service temperature of the material used.

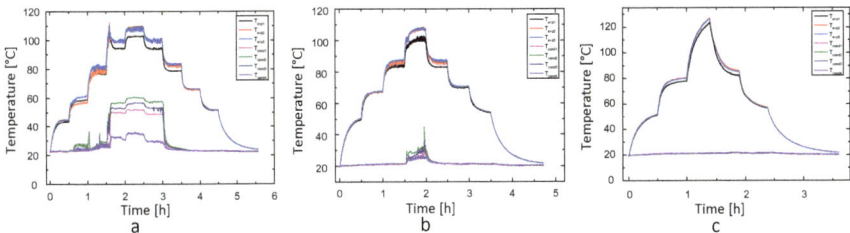

Fig. 2 Temperatures measured in the evaporator and condenser sections of the seven-turn PHS in
vertical position (**a**), with 45° inclination (**b**) and horizontal (**c**)

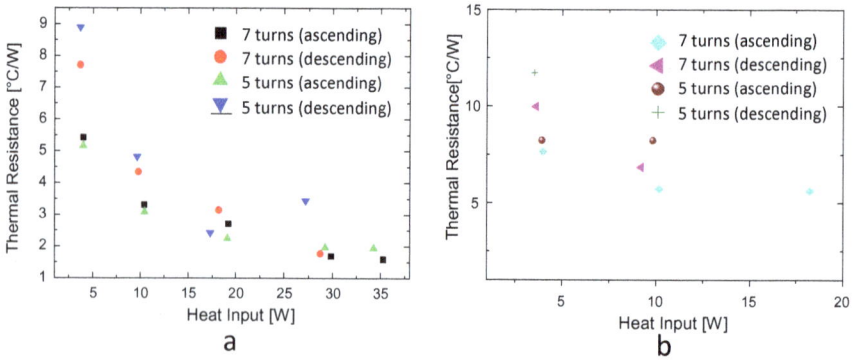

Fig. 3 Equivalent thermal resistances of the five-turn and seven-turn PHS in vertical position (**a**) and horizontal position (**b**)

The overall thermal performances of the composite polymer sheet in vertical and horizontal arrangement are shown, respectively, in Figs. 3a, b, which display the equivalent thermal resistance of the material between a hot end and a cold end, calculated according to Eq. (2), as a function of the heating power supply.

4   Conclusions

A novel concept of engineered composite polymer sheet was designed and manufactured using three polypropylene sheets, bonded together by selective laser welding, where the central sheet contains a serpentine channel filled with a heat transfer fluid (FC-72). The thermal response was evaluated for two different geometries of the channel, for different values of the heat input at the evaporator. Preliminary results indicate a 500% increase of the equivalent thermal conductance for the seven-turn channel in vertical position, while the increase achieved in the horizontal position is significantly smaller. The proposed technology represents a promising route to produce composite polymeric materials with enhanced thermal characteristics.

Acknowledgements O. Der gratefully acknowledges a YLSY doctoral studentship from the Republic of Turkey, Ministry of National Education.

References

1. S.K. Mazumdar, *Composites Manufacturing: Materials, Product, and Process Engineering* (CRC Press, 2001)
2. D. Kumlutas, I.H. Tavman, A numerical and experimental study on thermal conductivity of particle filled polymer composites. J. Thermoplast. Compos. Mater. **19**, 441–455 (2006)

3. I. Krupa, A. Boudenne, L. Ibos, Thermophysical properties of polyethylene filled with metal coated polyamide articles. Eur. Polymer J. **43**, 2443–2452 (2007)
4. I.H. Tavman, H. Akinci, Transverse thermal conductivity of fiber reinforced polymer composites. Int. Commun. Heat Mass Transfer **27**, 253–261 (2000)
5. S.N. Maiti, K. Ghosh, Thermal characteristics of silver powder-filled polypropylene composites. J. Appl. Polym. Sci. **52**, 1091–1103 (1994)
6. D.W. Chae, S.S. Hwang, S.M. Hong, S.P. Hong, B.G. Cho, B.C. Kim, Influence oh height contents of silver nanoparticles on the physical properties of poly(vinylidene fluoride). Mol. Cryst. Liq. Cryst. **464**, 233–241 (2007)
7. M. Biercuk, M. Llaguno, M. Radosavljevic, J. Hyun, A. Johnson, J. Fischer, Carbon nanotube composites for thermal management. Appl. Phys. Lett. **80**, 2767–2769 (2000)
8. R. Haggenmueller, C. Guthy, J.R. Lukes, J.E. Fischer, K.I. Winey, Single wall carbon nanotube/polyethylene nanocomposites: thermal and electrical conductivity. Macromolecules **40**, 2417–2421 (2007)
9. K.I. Winey, T. Kashiwagi, M. Mu, Improving electrical conductivity and thermal properties of polymers by the addition of carbon nanotubes as fillers. MRS Bull. **32**, 348–353 (2007)
10. H. Ma, *Oscillating Heat Pipes* (Springer, 2015)
11. B. Acherjee, A. Kuar, S. Mitra, D. Misra, Laser transmission welding of polycarbonates: experiments, modeling, and sensitivity analysis. Int. J. Adv. Manuf. Technol. **78**, 853–861 (2015)
12. S.B. Paudel, G.J. Michna, Effect of inclination angle on pulsating heat pipe performance, in *Proceedings of the ASME 2014 12th International Conference on Nanochannels, Microchannels and Minichannels*, Chicago, Illinois (2014)

Study of Heat Transfer Enhancement by Pulsating Flow in a Rectangular Mini Channel

Parth S. Kumavat, Richard Blythman, Darina B. Murray, and Seamus M. O'Shaughnessy

1 Introduction

Pulsating flow is viewed as a promising mechanism for augmenting heat transfer in single-phase cooling systems and could help to better regulate temperatures in applications involving high heat flux densities and/or those applications where fluid flow is restricted to confined spaces. Although the phenomenon is commonly observed in biological processes such as arterial blood flow, pulsating flow may be particularly suited to areas such as electronics cooling and performance enhancement of heat exchangers. Due to the complexity of the coupled pulsating fluid flow and heat transfer and the wide range of pulsation formats that could be employed, a comprehensive parametric analysis of the phenomenon is considered to be of great interest.

The current study experimentally investigates the heat transfer associated with unsteady pulsating flows in a bottom-heated rectangular minichannel. The objective is to ascertain a more effective heat removal method (in relation to comparable steady flows) without compromising the net mass transport. This is achieved by determining the local time-dependent variation of wall temperature in response to different pulsation formats. The dimensionless numbers used to describe pulsating flows are the Reynolds number ($\mathrm{Re} = \rho u D_h/\mu$), which represents the ratio of convective and diffusive time scales, and the Womersley number ($\mathrm{Wo} = (D_h/2)(\sqrt{\omega/\nu})$), which quantifies the ratio of pulsating to diffusive time scales. Additionally, a pulsation amplitude may be defined as Q_A/Q_0 which represents the ratio of the oscillating flow rate to the steady flow rate.

Previous pulsating flow experimental studies have characterized the associated bulk temperature modulations to determine the heat transfer enhancement. Shi et al.

P. S. Kumavat (✉) · R. Blythman · D. B. Murray · S. M. O'Shaughnessy
Department of Mechanical and Manufacturing Engineering, Trinity College Dublin, University of Dublin, Dublin-2, Ireland
e-mail: kumavatp@tcd.ie

© Springer Nature Singapore Pte Ltd. 2021
C. Wen and Y. Yan (eds.), *Advances in Heat Transfer and Thermal Engineering*,
https://doi.org/10.1007/978-981-33-4765-6_44

[1] visually investigated the unsteady heat transfer processes using laser-induced fluorescence in an acoustically excited compressible flow, with acetone employed as the working fluid. Mamoru and Akira [2] numerically observed sinusoidally varying axial temperature gradients at the wall in their single local experiment at a frequencies corresponding to Wo = 4.4 and 12.2. These results were later confirmed experimentally. Blythman et al. [3] used infrared thermography to achieve high temporal and spatial resolution of the non-uniform convective heat flux at the heated wall of a rectangular minichannel and determined good agreement between analytical and experimental wall temperature profiles at moderate frequencies (Wo = 1.4, 3.1 and 7.0). 'Annular' effects were observed in the oscillating temperature profiles which are characterized by near-wall overshoots and flatter profiles near the mid-channel. These phenomena were first characterized by Richardson and Tyler [4] who linked the thermal and local hydrodynamic profiles.

2 Methodology

The experimental setup consists of multiple plate profiles clamped rigidly together, as represented in Fig. 1. The base section is the heater support plate which is made from an *Ertacetal* plastic material. This plate houses two pairs of copper electrodes that help to supply a constant current of 42 A to an Inconel 600 foil of 12.5 μm thickness. The 350 mm long and 90 mm wide foil is laterally tensioned using a spring-loaded bolt system and forms the bottom surface of the channel which is approximated as a constant heat flux boundary. A rectangular channel of dimensions 360 mm $\times$ 22 mm $\times$ 1.4 mm (length $\times$ width $\times$ height) is machined into an acrylic plate which is placed on the top of the foil. The small thickness of foil approximates an isoflux wall and fluid boundary condition with one heated long wall, denoted here as H2(1L) and in accordance with the work of Blythman et al. [3]. The resultant maximum heat flux generation is 4.5 kW/m^2, measured at the outer contact points of copper electrodes. De-ionized water is used as a coolant and is supplied in a closed flow loop that connects the inlet and outlet of the minichannel. Heat removed from the foil in the minichannel is subsequently withdrawn from the coolant by a fan-driven heat exchanger before returning to the channel inlet. A membrane contactor and vacuum pump are used to remove air bubbles and other impurities prior to testing.

The flow is driven by a McLennan 34HSX-108 stepper motor controlled by a ST5-Q-NN applied motion drive, operates for 0.02 Hz and 2 Hz as presented in this study. A National Instruments NI-DAQ 9269 is configured using LABVIEW to modulate the amplitude and frequency of a continuous sine wave pulsation delivered to the motor. An Atrato 710 series ultrasonic flow meter measures the coolant flow rate non-intrusively. Two calibrated T-type thermocouples are located at the minichannel inlet and outlet.

A small section of the underside of foil is exposed to enable infrared temperature measurements. This section of the foil is sprayed with a thin layer of matte black to

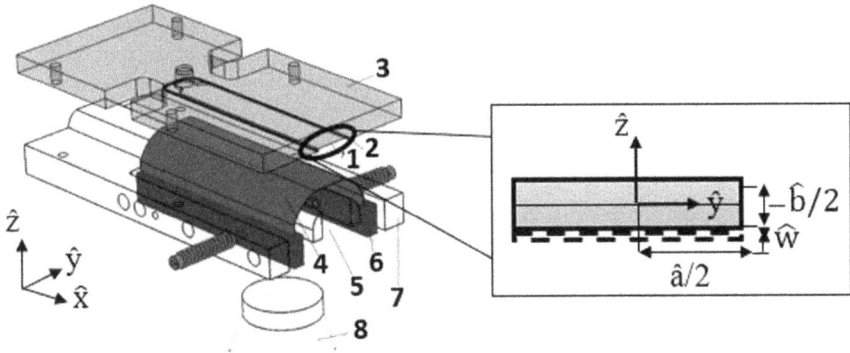

Fig. 1 Exploded sectional view showing: (1) channel profile, (2) o-ring, (3) channel plate, (4) foil, (5) air cavity, (6) electrodes, (7) heater plate, (8) camera. A channel inset view is shown with dimensional notations. Dashed lines mark the foil boundary

enhance its surface emissivity. To prevent heat loss from the foil via natural convection, a 1 mm thick sapphire glass is placed in a window approximately 13.5 mm beneath the foil. The glass has high IR transmission and forms an air cavity beneath the foil. The air temperature in this cavity is recorded in two locations, and an approximation of 1D conduction heat transfer as described by Bejan [5] is used to estimate thermal losses in this region. The local time-dependent variations of the foil temperature are recorded using a FLIR SC6000 high speed, high resolution, infrared camera. It has a maximum 640 × 512-pixel focal plane array with sensitivity of 2.5–5.1 μm range. Prior to testing, a two-point non-uniformity correction (NUC) is performed with a black body calibrator and an in situ temperature counts calibration is performed. The camera is triggered from LABVIEW to facilitate phase-locked measurements. IR video files are exported to MATLAB for postprocessing and analysis.

The experimental setup ensures that the flow is hydrodynamically and thermally developed at the region where the IR measurements are recorded which is 220 mm downstream of the channel inlet. The thermal boundary layer development length is determined in accordance with the work of Durst et al. [6] and is determined at the maximum Reynolds number (Re_{max}) used in this study, where Re_{max} equals the sum of the steady (Re_0) and oscillating (Re_A) Reynolds numbers, i.e. $Re_{max} = Re_0 + Re_A$ = 81. For the configurations tested, the maximum predicted steady entry length is $L_e = 9.72$ mm, and the thermally developed flow calculated by $Pr \cdot L_e = 54.43$ mm.

3   Results

Figure 2 plots the temperature profiles along the heated wall in the transverse ($\hat{y}$) direction at different phase angles. The x-axis (T''_w) represents the difference between

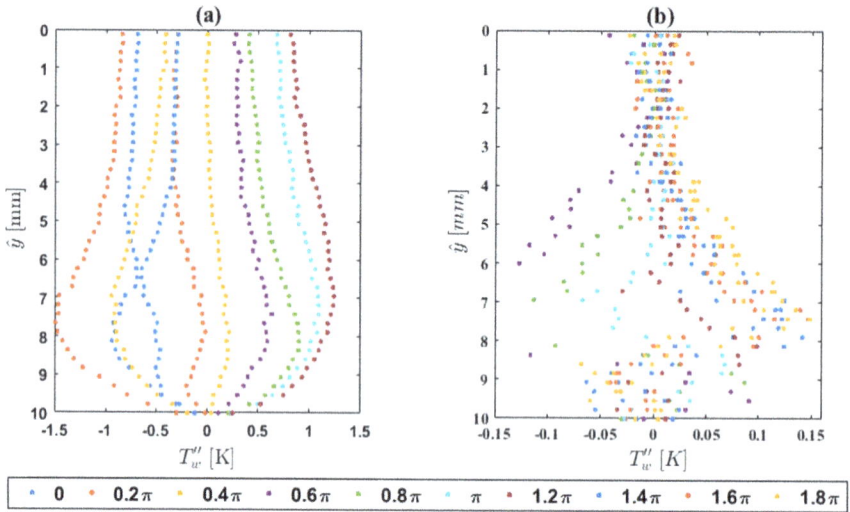

Fig. 2 Experimental oscillating wall temperature profiles in the transverse dimension, for Wo = 1.41 (0.02 Hz) and 14.1 (2 Hz) and Pr = 5.6, at the heated surface ($\hat{z} = -(\hat{b}/2 + \hat{w})$). Colours indicate successive phase angles from 0 in increments of 0.2π

the phase-averaged pulsating flow temperature and the steady temperature, where the phase-averaged data represents the mean value recorded over 20 pulse cycles. Figure 2a, b corresponds respectively to pulsation frequencies of 0.02 Hz and 2 Hz for a constant flow rate amplitude $Q_A/Q_0 = 0.9$. For the lower frequency pulsation, larger near-wall temperature fluctuations are observed close to the channel corners (i.e. $\hat{y} > 5$ mm). The temperature profiles suggest improved cooling in the phases 0, $0.2\pi, 1.4\pi, 1.6\pi, 1.8\pi$ and reduced cooling in the phases $0.4\pi, 0.6\pi, 0.8\pi, \pi, 1.2\pi$. This exhibits similar behaviour to the annular effects observed by Blythman et al. [3] and suggests that the temperature profile is formed as a result of similar velocity and displacement profiles. By increasing the pulsation frequency and lowering the excitation amplitude to maintain Q_A/Q_0 constant, the temperature profiles are observed to show only minimum deviations from the steady flow case. There appears to be no significant transverse thermal diffusion. The time scale of the pulsations is short compared to the time scale of thermal diffusion, which may explain why the transverse rate of spreading of heat is small. As the excitation frequency is quick enough, the temperature field represents smaller amplitudes and does not readily adjust, since the fluid elements are displaced over small axial distance.

4 Conclusions

The scope of this study is to examine the transverse temperature profiles at the heated wall of a rectangular minichannel in response to pulsating flow for very low flow rates corresponding to Womersley numbers of 1.41 and 14.1. A constant flow rate amplitude is set for all experiments, and the excitation frequency is varied between cases to investigate thelocal thermal behaviour, which is achieved through infrared imaging of the heated surface. The nature of the temperature fluctuations observed near the side walls of the minichannel agrees with the concept of annular effects hypothesized by other researchers. The variation in pulsation frequency leads to a significant variation in temperature amplitudes which signifies that the local heat flux can be altered by flow pulsations. Increasing the frequency for a fixed flow rate amplitude appears to reduce heat transfer gains observed at lower frequencies. The aim of future studies is to explore the heat transfer enhancement associated with different complex pulsating waveforms and to validate the experimental findings through computational simulations which will also offer deeper insight into the local flow dynamics.

Acknowledgements The research was supported by the Trinity College Dublin, School of Engineering Postgraduate Scholarship.

References

1. L. Shi, X. Mao, A.J. Jaworski, Application of planar laser-induced fluorescence measurement techniques to study the heat transfer characteristics of parallel-plate heat exchangers in thermoacoustic devices. Meas. Sci. Technol. **21**(11) (2010)
2. O. Mamoru, K. Akira, Lumped-parameter modeling of heat transfer enhanced by sinusoidal motion of fluid. Int. J. Heat Mass Transf. **34**(12), 3083–3095 (1991)
3. R. Blythman, T. Persoons, N. Jeffers, D.B. Murray, Effect of oscillation frequency on wall shear stress and pressure drop in a rectangular channel for heat transfer applications. J. Phys. Conf. Ser. **745**(3) (2016)
4. E.G. Richardson, E. Tyler, The transverse velocity gradient near the mouths of pipes in which an alternating or continuous flow of air is established. Proc. Phys. Soc. **42**(1), 1–15 (1929)
5. A. Bejan, *Convection Heat Transfer* (Wiley, 2013)
6. F. Durst, S. Ray, B. Ünsal, O.A. Bayoumi, The development lengths of laminar pipe and channel flows. J. Fluids Eng. **127**(6), 1154 (2005)

The Potential Use of Graphene to Intensify the Heat Transfer in Adsorption Beds

Ahmed Rezk, Laura J. Leslie, and Rees Davenport

1 Introduction

Adsorption system for cooling, heat pumps, heat storing and water desalination has gained increasing attention in the last few decades. Adsorption systems have the advantage of utilizing low-grade heat (below 200 °C), available by solar energy, wasted from industrial processes or other potential domestic waste heat sources. Adsorption bed is the core component of such systems, where the solid adsorbent material is put in contact with a heat exchanger. Granular-packed adsorption bed is the most used design in the commercially available adsorption equipment because of its high permeability, but nonetheless it has poor heat transfer performance. Many other approaches were investigated to enhance the heat transfer performance of the adsorbent bed, blending the adsorbent granules with metal additives to enhance the overall thermal conductivity of the adsorbent bed, coating the adsorbent bed surface with thin layers of adsorbent to reduce the contact thermal resistance, adsorbent deposition over metallic foam and other consolidation approaches [1]. On the one hand, blending the adsorbent granules with metal additives has shown great potential to improve the heat transfer performance without reducing the adsorbent bed permeability; on the other hand, using metals has not significantly enhanced the weight of the systems [2].

Graphene is a single layer of carbon atoms which are arranged in hexagonal shape. A single layer of graphene exhibits thermal conductivity of up to 5300 W/mK [3]. However, the thermal conductivity of graphene-based materials can be reduced by increasing the number of carbon atomic layers. Reduced graphene platelets of 30–45 layers showed thermal conductivity close to the bulk thermal conductivity of graphite [4]. Graphite is a 3D structured graphene of relatively high bulk thermal conductivity (1950 W/mK) and is commonly used in adsorption equipment. Whilst

A. Rezk (✉) · L. J. Leslie · R. Davenport
Aston Institute of Materials Research, Aston University, Birmingham B4 7ET, UK
e-mail: a.rezk@aston.ac.uk

© Springer Nature Singapore Pte Ltd. 2021
C. Wen and Y. Yan (eds.), *Advances in Heat Transfer and Thermal Engineering* ,
https://doi.org/10.1007/978-981-33-4765-6_45

some research has been carried out by the International Institute for Carbon–Neutral Energy Research in early 2019 on developing consolidated adsorbent composites using graphene nano-platelets, no studies have been found which investigated the bulk thermal performance of granular-packed bed using graphene-based material additives. Also, there is a little understanding of the molecular characteristics of the used graphene nano-platelets (i.e., the number of carbon layers). This research attempts to cover these knowledge gaps by investigating the effect of using graphene-based materials of identified molecular characteristics as additives to enhance the overall heat transfer performance of granular-packed adsorbent beds. It is the most practical approach to intensify the heat transfer of the adsorbent bed for rapid commercialization of adsorption systems.

2 Methodology

The bulk heat transfer performance of granular-packed silica gel as a baseline adsorbent blended with new species of additives dubbed graphene-based additives was investigated. Two types of graphene-based additives were investigated: graphene oxide of 1–3 atomic layers and graphene nano-platelets of 1–5 atomic layers. The heat transfer performance of silica gel adsorbent granules blended with two reference additives, aluminum and graphite, was investigated in order to quantify the enhancement of heat transfer performance as a result of using graphene-based materials of few atomic layers.

The bulk thermal diffusivity of granular-packed adsorbent beds enhanced by graphene-based additives was experimentally measured using the apparatus shown in Fig. 1. Similar experimental approach was first developed by Demir to measure

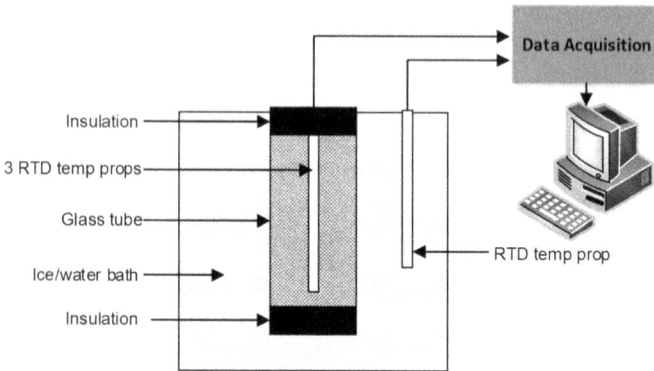

Fig. 1 Schematic diagram of the test apparatus

the effect of using different metal additives on the thermal performance of granular-packed beds [5]. It consists of a glass cylinder, ice/water bath and resistance temperature detector (RTD) props. Three RTD props, 30 mm apart, measured the centerline temperature of the packed cylinder. The thermal insulation at both ends of the cylinder eliminated the heat transfer in the axial direction, which reduced the heat transfer case to 1D in the radial direction. Silica gel was chosen as a baseline adsorbent granular. The cylindrical container was packed with pure silica gel granules, pure additives or homogeneous silica gel/additive mixtures at different volume percentages: 10%, 20% and 30%. Prior the tests, each sample was dried under vacuum for 24 h at 100 °C to remove any adsorbed gas including water vapor. The packed cylinder was heated up to 45 °C and then placed in the ice/water bath at 0 °C. Since the thermal diffusivity is inversely proportional to the rate of temperature change, the time elapsed to cool down the packed cylinder was employed to quantify the enhancement of the thermal diffusivity.

Equation (1) presents the thermal diffusivity, which is the ratio of the thermal conductivity and the heat capacity of a substance. Equation (2) presents the change of the thermal diffusivity of adsorbent/additive-packed cylinder with respect to the thermal diffusivity of purely packed silica gel as a function of the time elapsed to cool down the packed volume from 45 to 0 °C.

$$\alpha = \frac{k}{\rho c} \tag{1}$$

$$\frac{\alpha_{silica+additive} - \alpha_{silica}}{\alpha_{silica}}\% = \left[\frac{t_{silica}}{t_{silica+additive}} - 1 \right]\% \tag{2}$$

where α is the effective thermal diffusivity in m^2/s, k is the effective thermal conductivity in W/mK, ρ is bulk density of the packed granules in kg/m^3, c is the bulk specific heat in J/kgK and t is the elapsed time to cool the packed volume from the initial to final temperature.

3 Results

Figure 2 presents the percentage enhancement of the bulk thermal diffusivity of the silica gel/additive-packed cylinder with respect to purely packed silica gel. Since the bulk density of the used additives is less than that of silica gel, the weight of the silica gel/additive-packed cylinder was generally less than the purely packed silica gel. Aluminum flake additive showed the highest enhancement of the thermal diffusivity, and graphene nano-platelet additives showed the second highest enhancement. Although graphene nano-platelets showed lower enhancement of the thermal diffusivity compared to aluminum flakes, it provided more reduction in the packed volume weight.

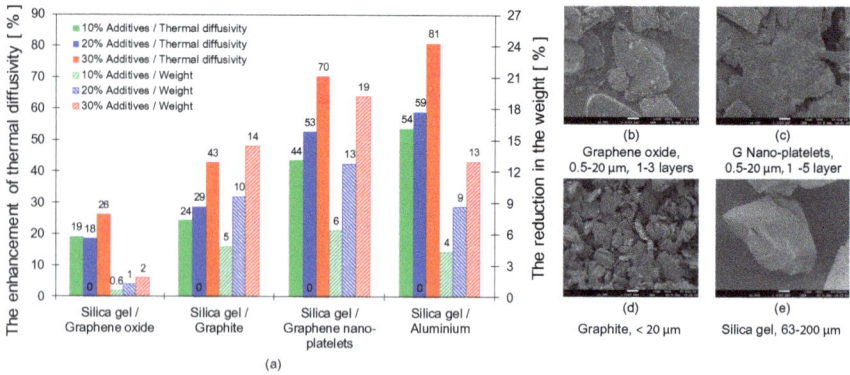

Fig. 2 **a** Effect of graphene-based and aluminum additives on the overall thermal diffusivity of granular-packed cylinder and **b–e** SEM images of the tested materials

The thermal behavior of graphene oxide depends on the percentage of intercalated oxygen and carbon interlayer spacing [4]. The percentage of intercalated oxygen is difficult to control in the bulk application and component level, which is revealed in this investigation. Despite graphene oxide consisting of fewer carbon layers, it showed the least enhancement of the thermal diffusivity. Using 20 vol% and 30 vol% graphene oxide additive, the enhancement of the thermal diffusivities was slightly equal: 18.4% and 19.2%, respectively. The bulk density of graphene oxide is slightly lower than silica gel adsorbent and showed negligible reduction in the weight of the packed volume.

Graphite is a major feedstock of graphene derivatives and widely used for enhancing the heat transfer performance of thermal equipment. In this investigation, graphite flakes showed less enhancement of the thermal diffusivity in comparison with graphene nano-platelets and aluminum flakes. Using graphite additives reduced the weight of the packed cylinder by 14.4%, 9.6% and 4.8% at 30 vol%, 20 vol% and 10 vol%, respectively. The reduction in weight is higher than graphene oxide and aluminum flakes but less than graphene nano-platelets.

Based on the thermal performance merit, aluminum might still the best option to enhance the thermal performance of the packed adsorbent beds; however, considering the bulkiness and the weight of the adsorption systems might lead to choosing graphene nano-platelets as an optimal additive. The research findings confirm the association between the thermal performance and the number of atomic layers of graphene derivatives as well as the adverse effect of any intercalated substance into graphene derivatives. Nevertheless, further research is recommended to determine the phenomenological relation between the atomic structure and the bulk thermal performance of graphene derivative.

4 Conclusions

The main goal of this work was to investigate through experiments the potential use of graphene-based materials of few atomic layers to enhance the heat transfer performance of adsorbent beds in comparison with two reference materials: graphite and aluminum. Results showed that graphene nano-platelets provide the greatest increase of the thermal diffusivity aside from the aluminum flakes. Graphene nano-platelets provided 19.2% improvement of the packed bad weight compared to 13% improvement in aluminum-enhanced packed bed. The second major finding was that graphene oxide did not provide a significant improvement in the heat transfer performance or the weight of the adsorption bed, compared to other additives. The results of bulk thermal performance of the graphene oxide contradict the general consensus of relating the number of atomic layers of graphene derivative and its thermal performance primarily due to the presence of intercalated oxygen. However, from the literature, there is a potential to modulate the heat transfer performance of graphene oxide by controlling the oxygen content and the interlayer spacing. Graphite showed an improvement of 42.8% in the heat transfer performance and reduced the bed weight by 14.4% at 30 vol%.

The most obvious finding to emerge from this work is that graphene derivatives at few atomic layers behaved differently at the component level. Using graphene additives will make noteworthy contributions to the heat transfer performance of granular-packed adsorbent bed, which is the most utilized design in real-life application of adsorption systems of different applications. The future work will focus on modulating graphene-based materials to provide further improvement of the heat transfer in adsorption equipment and reveal the phenomenological relation between the atomic structure and the bulk thermal performance of such materials.

Acknowledgements The authors would like to thank Graphitene Ltd. for supporting the project.

References

1. A. Rezk, R.K. Aldadah, S. Mahmoud, A. Elsayed, Effects of contact resistance and metal additives in finned-tube adsorbent beds on the performance of silica gel/water adsorption chiller. Appl. Therm. Eng. **53**, 278–284 (2013)
2. A.A. Askalany, S.K. Henninger, M. Ghazy, B.B. Saha, Effect of improving thermal conductivity of the adsorbent on performance of adsorption cooling system. Appl. Therm. Eng. **110**, 695–702 (2017)
3. M. Anju, N.K. Renuka, Graphene-dye hybrid optical sensors. Nano-Struct. Nano-Objects **17**, 194–217 (2019)
4. N.K. Mahanta, A.R. Abramson, Thermal Conductivity of graphene oxide nanoplatelets, in *13th IEEE Intersociety Conference on Thermal and Thermomechanical Phenomena in Electronic Systems (ITherm)* (2012)
5. H. Demir, M. Mobedi, S. Ulku, The use of metal additives to enhance heat transfer rate through an unconsolidated adsorbent bed. Int. J. Refrig. **33**, 714–720 (2010)

A Parametric Study into a Passively Enhanced Heat Separation System

Chidiebere Ihekwaba and Mansour Al Qubeissi

1 Background and Model

Traditional thermal separation technologies require a type of work to either heat up or cool down the fluid in question. One of the most promising approaches, in comparison with the more commercially popular Ranque–Hilsch vortex tube, is the Leontiev tube which is the primary advantage of decreased full pressure loss at the expense of reduced thermal separation [1, 2]. It is known that lowering the adiabatic wall temperature in the supersonic flow channel will increase the overall heat transfer between the supersonic and subsonic flow channels, allowing for increased temperature separation and device efficiency. When air passes through a nozzle designed to induce supersonic flow, the fluid expands and cools down and the kinetic energy dissipates. As a result, the stagnating supersonic fluid flow increases the fluid temperature near the wall boundaries (i.e. viscous heating).

Several studies have been made in the literature to explain the "temperature separation" effect. Xue et al. [3] presented a critical review of explanations on the working concept of a vortex tube. They discussed hypotheses of temperature, pressure, turbulence, viscosity, acoustic streaming and secondary circulation. Based on the observed velocity, turbulence intensity, temperature and pressure distributions, Xue et al. [4] proposed multi-circulation as the main reason for thermal separation. In our analysis, a numerical model validation is performed using experimental data conducted by Leontiev et al. [2]. In order to reduce the computational time, with detailed simulation of the fundamental physics, a series of simplifications are made to the original geometry (see Fig. 1 for the modified domain). The subsonic flow entry and exits were altered to flow parallel to the supersonic channel rather than perpendicular. Also, the

C. Ihekwaba · M. Al Qubeissi (✉)
Faculty of Engineering, Environment and Computing, Coventry University, Coventry CV1 2JH, UK
e-mail: ac1028@coventry.ac.uk

© Springer Nature Singapore Pte Ltd. 2021
C. Wen and Y. Yan (eds.), *Advances in Heat Transfer and Thermal Engineering* ,
https://doi.org/10.1007/978-981-33-4765-6_46

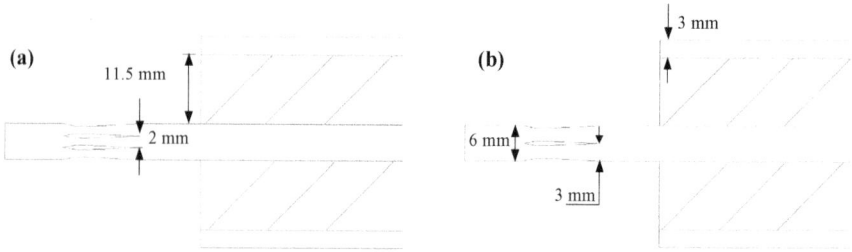

Fig. 1 Geometry dimensions for **a** triple- and **b** dual-nozzle designs

supersonic flow channel was modified to maintain a consistent geometry through the device with no expansion. The model has been conducted as a 2D problem.

A transition SST model was used for its high accuracy in modelling supersonic shocks and conjugate heat transfer. The turbulent intensity was taken as 5%, and coupled interface was selected for any regions of the contact between the subsonic/supersonic fluid and the heat conducting walls. The model was solved using a density-based, numerical least square-based method with an implicit ROE-FDS for its convergence capability with supersonic flows. The mesh profiles were maintained as quads throughout the domain within the range of 480,000–510,000 elements. Third-order MUSCL was used for its increased accuracy in modelling compressible supersonic flows [5].

For the investigation into the effects of evaporating water droplets in the supersonic channel to lower the adiabatic wall temperature, a discrete phase model was used injecting 4-μm-diameter water droplets at a pulse rate of 1 million droplets at 6 source locations every 0.1 s with a Leontiev tube model's initial temperature of 400 K to ensure the water vapour did not undergo condensation. Total mass fraction of water within the flow was maintained to be below 1%.

2 Results

The system outlet temperatures have been investigated for different shapes versus different fractions of mass flowrate (see Fig. 2). The Mach number range used for this analysis is $M = 1.22 - 1.3$. The inlet air temperature and pressure are 299 K and 10.5 bar, respectively. The outlet air pressure for supersonic channels is 7.39–7.42 bar for the single-nozzle, 7.06–7.08 bar for the dual-nozzle and 6.90 bar for the triple-nozzle designs. Pressure losses within the subsonic channel are comparatively very minor. As can be seen from this figure, the introduced simplification to the geometry in the validation model leads to up to 2 °C warmer subsonic channel outlet temperatures at high subsonic mass flowrate ratios, but a much steeper decline from those initial temperatures at the lower subsonic mass flowrate ratios. This difference can be attributed to the differences in the supersonic channel geometry between the

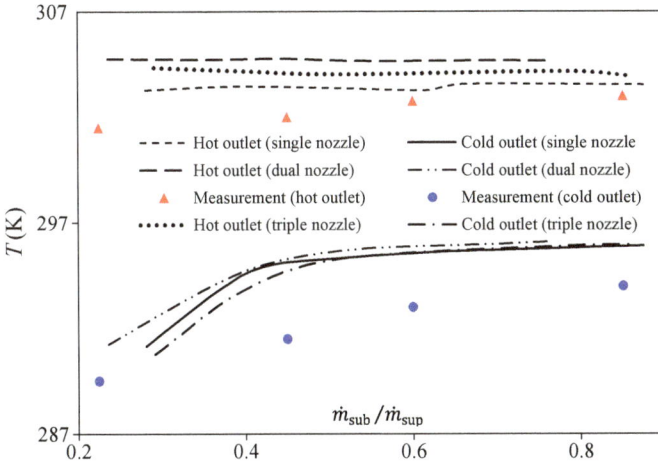

Fig. 2 System heating and cooling temperatures versus subsonic/supersonic mass flowrate fractions ($\dot{m}_{sub}/\dot{m}_{sup}$) for single-, dual- and triple-nozzle designs, in comparison with the experimental data [3]

model and experiment due to the expansion within the supersonic channel. This leads to a high-speed flow, reaching $M = 1.2$ in our model, compared to the $M = 1.6$ reached during the experiment.

The modification of the single-nozzle design (SND) to dual-nozzle design (DND) has enhanced the heating effect at the supersonic region, but it is less effective in cooling the subsonic region with up to 1 °C less cooling compared to SND. The triple-nozzle design has shown average heating temperature in the supersonic region, but it is generally better than both, SND and DND, in terms of thermal separation for all subsonic mass flowrate ratios. All models show higher temperature and adiabatic efficiencies, compared to the experimental design due to the sensitivity of results to Mach number. In terms of thermal efficiency (η) and C.O.P, the triple-nozzle design is observed to be the least efficient ($\eta = 0.110 – 0.244$ and C.O.P $= 0.041 – 0.055$) due to its high pressure loss within the supersonic channel. The dual design has moderate efficiency range of $\eta = 0.119 – 0.243$ and C.O.P $= 0.0364 – 0.04650$. And the single nozzle is the most efficient design with $\eta = 0.143 – 0.275$ and C.O.P $= 0.0425 – 0.0603$. Overall, it observed peak thermal separation of 28 °C when the supersonic flow reaches its steady state. Maintaining this significant thermal separation and reducing the losses is a potential avenue for increasing the system thermal efficiency.

For the investigation of the effects of water droplets' evaporation on the thermal separation, these are found to improve the cooling effects in the range of 2–8 °C along the thermal profile (see Table 1). Nozzle geometries and droplet sizes that promote evaporation near the channel walls prove most effective. Further cooling of the subsonic channel can also be achieved through this method at the expense of total thermal separation with higher mass flowrate of water droplets.

Table 1 Impact of water droplet mass flowrate on the separation temperatures (K) at the exit cold and hot streams[a]

$\dot{m}_{sub}/\dot{m}_{sup}$	No droplets		Small droplet injection $\left(\dot{m} = 0.005 \frac{kg}{s}\right)$		Large droplet injection $\left(\dot{m} = 0.025 \frac{kg}{s}\right)$	
	T_c	T_h	T_c	T_h	T_c	T_h
0.347	390.137	407.668	388.479	404.326	389.258	406.535
0.409	391.308	407.644	389.913	404.115	390.557	406.366
0.528	394.355	407.783	393.482	403.996	393.898	406.035
0.724	395.280	407.832	394.562	404.213	394.905	406.093

[a]The subsonic mass flowrates $\left(\dot{m}_{sub}/\dot{m}_{sup}\right)$ are approximated to the nearest values. These might differ with up to 0.1% of the displayed figures

3 Conclusions

A new approach for passive thermal separation has been investigated. Two methods of reducing the adiabatic wall temperature have been proposed. Overall, it was shown for the specified geometry that:

- Triple- and dual-nozzle geometries for a Leontiev tube are more effective for thermal separation than a single-nozzle design, but at the expense of thermal efficiency.
- Water droplets' injection into the supersonic stream can lead to significant impact on the total thermal separation. This approach is more effective with nozzle geometries that promote droplet evaporation near channel walls.
- Peak thermal separation of 28 K was observed before the steady-state flow condition.

Further investigation is suggested to be made for enhancing the system thermal separation, including adding water droplets to the subsonic channel with the inclusion of baffles, so that droplet flow stream does not interfere with the thermal conduction through the channel. An investigation into the impacts of various supersonic channels and appropriate nozzle geometries, with a study of the system thermal efficiency, is suggested at the next stage of research.

References

1. G.M. Azanov, A.N. Osiptsov, The efficiency of one method of machineless gas dynamic temperature stratification in a gas flow. Int. J. Heat Mass Transfer **106**, 1125–1133 (2017)
2. A. Leontiev, A. Zditovets, Y. Vinogradov, M. Strongin, N. Kiselev, Experimental investigation of the machine-free method of temperature separation of air flows based on the energy separation effect in a compressible boundary layer. Exp. Thermal Fluid Sci. **88**, 202–219 (2017)

3. Y. Xue, M. Arjomandi, R. Kelso, A critical review of temperature separation in a vortex tube. Exp. Therm. Fluid Sci. **34**(8), 1367–1374 (2010)
4. Y. Xue, M. Arjomandi, R. Kelso, Experimental study of the thermal separation in a vortex tube. Exp. Therm. Fluid Sci. **46**, 175–182 (2013)
5. P. Cyklis, P. Młynarczyk, The influence of the spatial discretization methods on the nozzle impulse flow simulation results. Procedia Eng. **157**, 396–403 (2016)

Heat Transfer During Pulsating Liquid Jet Impingement onto a Vertical Wall

J. Wassenberg, P. Stephan, and T. Gambaryan-Roisman

1 Introduction

Liquid jet impingement is used in industry when surfaces must be cooled or cleaned, since this process is characterized by high heat or mass transport rates. One of the methods suggested for heat transfer optimization is using pulsating jets. Zumbrunnen and Aziz investigated pulsating water jets impinging on a heated horizontal wall. The pulsation with frequencies up to 140 Hz leads to an enhancement of the heat transfer in the stagnation flow area of up to 100% [1]. The influence of pulsation on heat transfer by jet impingement onto a vertical wall has not been studied so far. The impingement of an initially circular liquid jet onto a vertical wall is—as for the horizontal wall—accompanied by a stagnation flow and the development of a radial thin liquid film bounded by a hydraulic jump, but the latter is not circular as in the case of normal impingement onto horizontal surfaces, but has an arc shape. The impingement of a pulsating jet (consisting of successive jet sections) leads to development of complex periodic wall flow patterns, which typically can be subdivided into several phases: (i) the spreading phase directly after impingement, eventually leading into (ii) the quasi-stationary phase lasting until the end of the jet section impingement phase, followed by (iii) the drainage of the wall film. The latter is dominated by the drainage of water accumulated in the arc-shaped hydraulic jump during the stationary phase. This flow contributes to convective heat transport without ongoing inflow of cooling liquid, which leads to improvement of resource efficiency.

In the present work, the heat transfer between a horizontally impinging pulsating liquid jet and a vertical wall is investigated to discover if or to what extent jet pulsation increases heat transfer efficiency. Experiments were performed at jet Reynolds numbers $Re = 35000 - 75000$ and pulsation frequencies $f = 1 - 5$ Hz with a pipe nozzle with diameter of $d_N = 4$ mm, and nozzle-to-wall distances from 3 to

J. Wassenberg (✉) · P. Stephan · T. Gambaryan-Roisman
Institute for Technical Thermodynamics, Technische Universität Darmstadt, Darmstadt, Germany
e-mail: wassenberg@ttd.tu-darmstadt.de

© Springer Nature Singapore Pte Ltd. 2021
C. Wen and Y. Yan (eds.), *Advances in Heat Transfer and Thermal Engineering*,
https://doi.org/10.1007/978-981-33-4765-6_47

250 nozzle diameters. Additionally, it has been found that the jet impingement in this range of parameters is accompanied by a high rate of liquid splattering, which affects both hydrodynamics and heat transfer.

2 Experimental Method

The main part of the test setup used in this work is the open water circuit pictured in Fig. 1a. The water used in the test stand is a mixture of tab water provided by the city of Darmstadt and deionized water (1:3). During the measurement, water is pumped from the tank (1) by a multistage centrifugal pump (2, IN-VB 2-140, Speck) passing the Coriolis mass flow meter (TME 5, Heinrichs Messtechnik GmbH). The water temperature is kept at 20 ± 0.5 °C using a counterflow heat exchanger and a bath thermostat (not depicted). The sectional valves (3, EV210B, Danfoss) are used to create the pulsation by either directing water to the tank or to the nozzle (4) where the liquid jet is formed. This leads to an intermittent pulsation form. The jet Reynolds number is then defined as follows:

$$Re = \frac{4\dot{M}}{\pi d_N \mu DC} \tag{1}$$

In Eq. (1), DC is the duty cycle of the pulsation, which indicates the ratio of mass flow through the nozzle to the total mass flow before the valves, $\dot{M}$ the mass flow rate through the nozzle and μ the dynamic viscosity of water. This definition of Re has the effect that liquid jets with different pulsation duty cycles, but with equal velocity at nozzle exit, have equal Reynolds numbers.

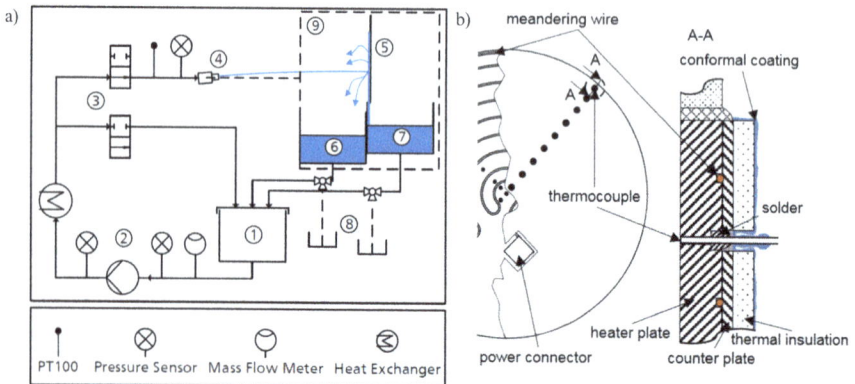

Fig. 1 **a** Schematic of the test stand used in this work: tank (1), centrifugal pump (2), sectional valves (3), nozzle (4), wall with integrated heater (5), collecting tubs (6) and (7), sampling position (8) and wet cell (9). **b** Drawing of heater and detailed cross section at a thermocouple position

The jet impinges on the center of a plate heated by the meandering wire heater (5, Fig. 1b) inside the wet cell (9), having passed the nozzle distance L. After impingement, a part of liquid with the mass flow rate $\dot{M}_s$ is splattered, and the rest of the liquid with the mass flow rate $\dot{M}_r$ remains on the wall and cools the heater. Both mass flows are collected by separate tubs (6 and 7) and can be withdrawn from the circuit at the sampling point (8) into two separate containers. The nozzle is placed on a sled to be able to vary the distance L. In order to guarantee a horizontal impact, the nozzle's orientation is slightly deviating from horizontal. The angle between the nozzle axis and horizontal direction has been calculated using the parabolic throwing, neglecting drag.

The heater plate is manufactured from nickel and has a cylindrical shape with a thickness of 8 mm and a radius of $R = 67$ mm. The meandering wire on the back side is connected to a 1.5 kW power supply (Delta Electronica 1500) and countered by a 2-mm-thick aluminum plate to ensure a thermal contact. Additional thermal and electric insulation is applied according to Fig. 1b. Thermocouples soldered in at surface level are used to measure the wall temperature $T_w(r)$ along the radial coordinate from the point of impingement (POI). The heater can be rotated around its axis to allow measurements along different directions from the POI. The water temperature at the nozzle inlet T_l is measured as well. The heat transfer at a radial distance r from POI has been quantified using the modified Stanton number. This number relates the heat transferred from the wall to the liquid in this region, $\dot{Q}(r/R)^2$, to the maximum heat $\dot{Q}_{max}$ that could be transferred, if the water temperature at the position r were uniform and equal to the wall temperature:

$$St = \frac{A_H(r) \cdot h}{\dot{M} \cdot c} = \frac{\dot{Q} \cdot (r/R)^2}{|T_w(r) - T_l| \cdot \dot{M} \cdot c} = \frac{\dot{Q}_{1D}(r)}{\dot{Q}_{max}(r)} \qquad (2)$$

In Eq. (2), h denotes the heat transfer coefficient, A_H the area of the heater surface and c the specific heat. $\dot{Q}_{1D}(r) = \dot{Q} \cdot (r/R)^2$ is the heat transported from the circular area with a center at POI and radius r to the liquid, evaluated under an assumption that radial heat conduction within the heater plate is negligible. The temperatures T_w and T_l have been averaged over a time interval of 30 s.

3 Results

Figure 2a shows the measured temperature differences $\Delta T = T_w(r) - T_l$ for Re = 75000, for continuous jets ($f = 0$ Hz), and different pulsation frequencies (DC = 0.5). The wall temperatures are measured above and below the POI, and the nozzle-to-target distance is $L/d_N = 33$. It can be seen that in case of the continuous jet ΔT close to the POI ($r \leq 24$ mm) is independent of the orientation.

As the impinging mass flow is reduced by 50% for the pulsating jet, ΔT increases. The heat transfer strongly depends on the pulsation frequency and, in contrast to

a)

b)

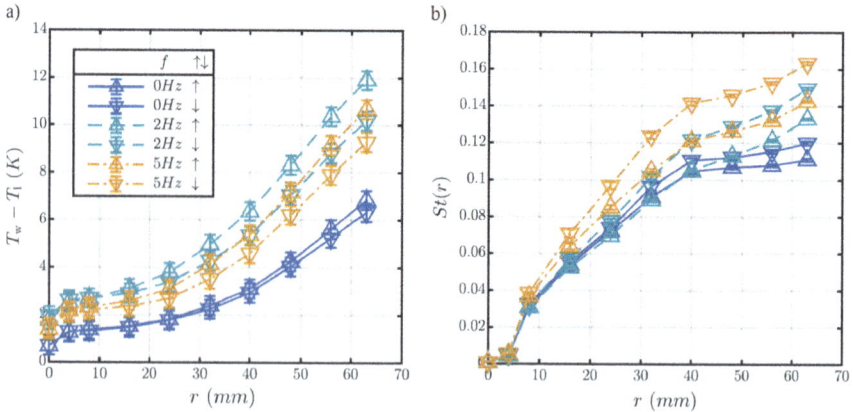

Fig. 2 **a** Mean temperature difference between heated wall and liquid at the nozzle inlet, **b** modified Stanton number for Re = 75000 and different pulsation frequencies (DC = 0.5). $L/d_N = 33$, $\vartheta_l = 20\,°C$. The upward directed arrow denotes the data collected by thermocouples directly over POI, and the downward directed arrow denotes the data collected by thermocouples located under POI

the continuous jets, is significantly different for the regions above and below POI. This is due to the difference in duration of the phases (i)–(iii) of the wall flow (see the "Introduction" section). For example, the hydraulic jump reaches the position $r \approx 50$ mm above the POI during the phase (iii) for $f = 5$ Hz, but passes the POI for $f = 2$ Hz. Figure 2b shows the Stanton number calculated based on the temperature values presented in Fig. 2a. It is shown that pulsation leads to a higher Stanton number compared to continuous jets at the same Reynolds numbers and thereby to a higher heat transfer efficiency. It can be suggested that increasing heat transfer efficiency by pulsation is governed by the complex wall flow pattern, which is determined by the pulsation frequency.

4 Conclusions

In this work, pulsating water jets impinging horizontally on a vertical wall have been investigated regarding their heat transfer efficiency. It is shown that the heat transfer efficiency can be increased using pulsating jets, while the heat transport itself is—due to the reduced amount of cooling liquid—weakened. The heat transfer efficiency strongly depends on the pulsation frequency. For pulsating jets, in contrast to continuous jets, the heat transfer is significantly different for the regions above and below POI.

Reference

1. D.A. Zumbrunnen, M. Aziz, Convective heat transfer enhancement due to intermittency in an impinging jet. J. Heat Transfer **115**, 91–98 (1993)

Icing of Water Droplet Enhanced with the Electrostatic Field

Qiyuan Deng, Hong Wang, Xun Zhu, Qiang Liao, Rong Chen, and Yudong Ding

1 Introduction

Freezing of liquid benefited a lot in human activities, such as food preservation [1], biological material cryopreservation [2], ice energy storage and so on. For food and biological material cryopreservation, it should be solidified as fast as possible. The crystalline should be as fine as possible to make cells completeness. Icing energy storage technology (IEST), considered as one of the effective ways to reduce energy consumption in refrigeration systems, was an important technical measure to transfer peak power to develop off-peak power, and it has positive effects on both of optimization of resource allocation and protection of ecological environment. During facing plenty of water solidification in the storing ice for energy storage, a practical problem urgently needs to be solved, that is, how to accelerate the icing process and save the cost of energy. Therefore, it is necessary to find a way to promote water freezing. In the present study, a water droplet solidification was investigated with the action of electrostatic field by using a visual experiment. The effects of EF strength on freezing behavior of water droplet were considered.

H. Wang (✉) · X. Zhu · Q. Liao · R. Chen · Y. Ding
Key Laboratory of Low-Grade Energy Utilization Technologies and Systems, Chongqing University, Chongqing, China
e-mail: hongwang@cqu.edu.cn

Q. Deng · H. Wang · X. Zhu · Q. Liao · R. Chen · Y. Ding
Institute of Engineering Thermophysics, Chongqing University, Chongqing, China

© Springer Nature Singapore Pte Ltd. 2021
C. Wen and Y. Yan (eds.), *Advances in Heat Transfer and Thermal Engineering* ,
https://doi.org/10.1007/978-981-33-4765-6_48

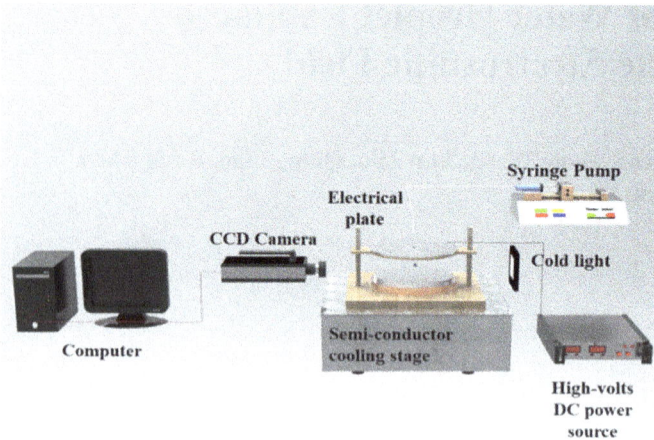

Fig. 1 Experimental system

2 Material and Method

Experimental setup is shown in Fig. 1. It included a semiconductor thermoelectric cooler system, an EF-produced system, a data acquisition system, an image acquisition system and a test unit.

To achieve electrostatic shielding and avoid the frosting happening on the test surface due to high humidity, a closed chamber made of the stainless steel with 4 organic glass windows was adopted to keep the humidity less than $20\% \pm 1\%$ by the nitrogen ventilation. The temperature of the test surface was cooled from the ambient temperature to $-15°C \pm 0.2 °C$. Distilled water with 10 μL was placed on the surface by the syringe pump. Aluminous electrodes are paralleled to each other of 3 cm. The bottom one was superhydrophobic with the contact angle (CA) 153.1°. EF was set as 0, 2, 3.3, 5, 6 (kV/cm) via a direct current (DC) power, respectively. The experimental process was recorded by data acquisition system and visual capture system.

In the present experiments, T_d was defined as the time of delay of the freezing, and T_i was defined as the time of the freezing. Each experiment was repeated more than 10 times to avoid the random effects.

3 Results and Discussion

Figure 2a shows the variation of T_d at the different EFs. It decreased sharply when droplet suffered the EF with 2 kV/cm. Obviously, EF promoted droplet freezing. With EF strength rising, the T_d continually decreased, but the descent tendency of T_d became gently with the increasing EF. Figure 2b presents the variations of T_i

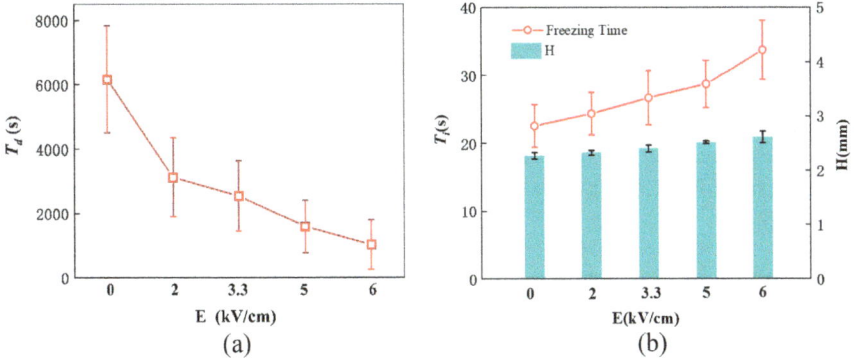

Fig.2 **a** Freezing-delay time changed with field strength and **b** freezing time changed with field strength

and droplet height H with field strength. Both of them increased with EF strength slightly.

Solidification of water droplet included two stages, nucleation and ice growth. According to the experimental results, it is the fact that EF promoted the nucleation. And the effect increased with the increment of EF strength. Water molecule is a kind of polar molecule, whose electric dipole moment ($\mu = 6.127 \times 10^{-30}$ cm). Thus, they were easily polarized in the EF. Hydrogen and oxygen atoms suffered from EF force toward the opposite direction. The EF moment made molecules rotate until the molecule along the EF line. The rearrangement of water molecules was hexagonal cluster which was similar to one kind of crystalline structures [3, 4]. The electrical dipole moment of ordered molecule clusters is larger than single molecule. This would attract surrounding water molecules. Once the diameter of hexagonal clusters was larger than critical nucleation radius, crystalline structure of the cooled water spontaneously grew up and freezing happened.

Another potential reason was that the interfacial flow enhanced between the liquid and the air, because the charges gathered at the interfacial surface. The EF made the surface flow accelerate and promoted the internal flow of the water droplet, which increased the destabilization in water droplet. This kind of destabilization provided the energy for nucleation.

For ice growth, T_i was increasing with EF strength. When EF was applied, droplet is charged with electrons on the bottom electrode. Therefore, charged droplet suffered from Coulomb force toward the upper one. Owing to pulling, droplet became higher. It meant the heat resistance of droplet increased under EF. Thus, T_i became longer comparing with no EF.

4 Conclusions

The solidification of droplet was promoted significantly under the EF. Freezing-delay time was reduced sharply at a relatively high strength. It is analyzed that water molecules were polarized under EF. Polarization made water cluster more crystalline-like. Meanwhile, freezing time was prolonged due to heat resistance enlargement vertically.

Acknowledgements The authors would like to acknowledge the research grant from National Natural Science Foundation of China (No. 51676022), the Fundamental Research Funds for the Central University (2018CDXYDL0001) and Venture & Innovation Support Program for Chongqing Overseas Returnees (cx2018053).

References

1. G. Jia, X. He, S. Nirasawa, E. Tatsumi, H. Liu, H. Liu, Effects of high-voltage electrostatic field on the freezing behavior and quality of pork tenderloin. J. Food Eng. **204**, 18–26 (2017)
2. S. Wei, X. Xiaobin, Z. Hong, X.J.C. Chuanxiang, Effects of dipole polarization of water molecules on ice formation under an electrostatic field. Cryobiology **56**(1), 93–99 (2008)
3. P.V. Acharya, V. Bahadur, Fundamental interfacial mechanisms underlying electrofreezing. Adv. Colloid Interface Sci. **251**, 26–43 (2018)
4. J.Y. Yan, G. N. Patey, Heterogeneous ice nucleation induced by electric fields. J. Phys. Chem. Lett. **2**(20), 2555–2559 (2017)

Micro-Nano Heat Transfer

Performance Enhancement of Vapour-Compression Refrigeration Systems Using Nanoparticles: An Experimental Study

V. La Rocca, M. Morale, A. Ferrante, and A. La Rocca

1 Introduction

In recent years, the development and characterization of new refrigerants with higher energetic efficiency have gained considerable interest [1], with a particular emphasis on the energetic performance of suitable replacements. Nanofluids have been proposed as possible alternative due to their role in enhancing the thermophysical proprieties of traditional refrigerants [2]. The comprehensive review in [2] suggests that thermal conductivity of nanofluid increases with temperature and volume concentration, but it is dependent on nanoparticle size distribution; specific heat is also increased if nanoparticles are added to the refrigerant. As consequence, the improved heat transfer coefficients lead to COP improvements for the same cooling capacity. Recently, the addition of nanoparticles to the compressor oil from a refrigerator has been found to reduce the friction coefficient up to 90% [3]; this suggests that nanoparticles migrating to the lubricant oil can also contribute to improve efficiency and reliability of the system. Findings from the literature are however contradictory at times [2]; it is therefore necessary to further explore the benefits of using nanoparticles dispersed in refrigerants. The present work focuses on a quantitative analysis of the performance enhancement for a vapour-compression refrigeration system due to addition of aluminium oxide (Al_2O_3) nanoparticle to a commercially available refrigerant, tetrafluoroethane (R134a).

V. La Rocca (✉) · M. Morale
Dipartimento di Ingegneria, Università degli Studi di Palermo, Palermo, Italy
e-mail: vincenzo.larocca@unipa.it

A. Ferrante · A. La Rocca
Department of Mechanical Materials and Manufacturing Engineering, The University of Nottingham, Nottingham, UK

© Springer Nature Singapore Pte Ltd. 2021
C. Wen and Y. Yan (eds.), *Advances in Heat Transfer and Thermal Engineering* ,
https://doi.org/10.1007/978-981-33-4765-6_49

2 Methodology

The experimental apparatus was assembled and instrumented to experimentally measure the heat transfer coefficients at the evaporator and the coefficient of performance (COP) of the system. The components are: a hermetic compressor, an evaporator consisting of a shell with a straight copper pipe centrally mounted, a thermal expansion valve with adjustable throttle, a commercial condenser and a subcooler.

R134a was used as the refrigerant working fluid, the evaporator exchanged heat with water flowing over the copper tube and through the shell, while condenser was cooled by forced convection with air. Flow metres, pressure transducers and type T copper–constantan thermocouples were used to monitor performance at different points of the refrigeration systems. The evaporator has a length of 1800 mm and an external diameter of 9 mm; it is instrumented with 14 thermocouples located axially, seven on the top and seven at the bottom, to monitor phase change across its length as shown in Fig. 1. Preparation of nanofluid R134a + Al_2O_3 was carried out with a bespoke sealed vessel. Nanoparticles were firstly introduced, and vacuum was produced; refrigerant in liquid state was then added. The results obtained for the R134a were used as baseline to compare the refrigeration performance when aluminium oxide (Al_2O_3) nanoparticles were added to the working fluid. Four different mass concentrations were investigated: $0.005, 0.01, 0.015$ and 0.02%. Transmission electron microscopy (TEM) was used to characterize morphology of the Al_2O_3 nanoparticles and samples were dispersed onto TEM support films, while dynamic light scattering (DLS) was used to check size of nanoparticles deposited into the lubrication oil during operation of the refrigeration system.

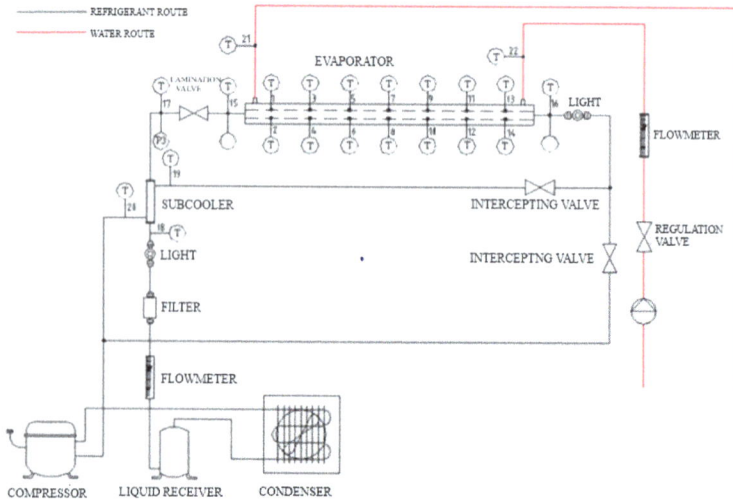

Fig. 1 Schematic representation of the test rig setup

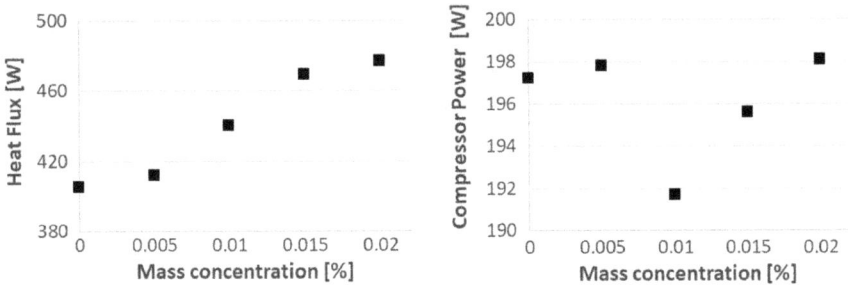

Fig. 2 Heat flux exchanged at the evaporator (left) and power measured at the compressor (right) as function of nanoparticle mass concentration in the refrigerant. Zero mass concentration refers to R134a with no nanoparticles

3 Results

The total heat transfer to the wall all along the pipe was calculated from the measurements of mass flow rate and inlet/outlet temperatures of the counterflow exchanger. For each of the mass concentrations investigated, the heat flux value is given in Fig. 2. Increasing the concentration of Al_2O_3 in the refrigerant results in a higher heat flux. A 0.01% concentration leads to a 8.7% increase; the heat flux doubles when the concentration of nanoparticles is doubled to 0.02%. This increase is mainly related to the increased heat transfer coefficient, which increases from 2985 to 4495 W/m^2 K increasing the mass concentration of aluminium oxide from 0% (R134a only) to 0.01%.

TEM analysis of the Al_2O_3 sample shows fractal agglomerates of 1000 nm in size were composed by spherical-like particles of 50–100 nm in diameter. The DLS analysis of oil sampled from the compressor at the end of the experimental campaign shows the presence of particles with an average size of 800 nm. Figure 2 shows a 3% reduction in power to the compressor when a 0.01% concentration is added to the refrigerant. In agreement with [3], nanoparticles transferred to the compressor oil might be responsible for a decrease in friction. This is corroborated by the DLS results.

4 Conclusions

Aluminium oxide nanoparticles were successfully added to R134a in mass concentration up to 0.02%. The nanofluid was used as the refrigerant in a vapour-compression system. The experimental results show an increase in the heat flux exchanged at the evaporator; in particular, there is a monotonic increase in heat flux which is linked to the increase in nanoparticle concentration, reaching a plateau at around 0.015%. A maximum increase of 18% in heat flux was measured for a concentration of 0.02%. The heat transfer coefficient from the refrigerant side almost doubles when a 0.01% Al_2O_3 is added to the tetrafluoroethane. DLS analysis of lubricant shows that nanoparticles migrate to the lubricant oil; a 0.01% mass concentration of Al_2O_3 in the refrigerant was found to reduce the power consumption of the compressor by 3% compared to the R134a case. However, increasing the concentration of particles in the oil does not necessarily lead to a further reduction in friction.

References

1. A. La Rocca, V. La Rocca, A. Messineo, D. Panno, Use of HFC fluids as suitable replacements in low-temperature refrigeration plants ARPN J. Eng. Appl. Sci. **9**(1), 74–79 (2014)
2. A. Bhattad, J. Sarkar, P. Ghosh, Improving the performance of refrigeration systems by using nanofluids: a comprehensive review. Renew. Sustain. Energy Rev. **82**(Part 3), 3656–3669 (2018). https://doi.org/10.1016/j.rser.2017.10.097
3. K. Lee, Y. Hwang, S. Cheong, L. Kwon, S. Kim, J. Lee, Performance evaluation of nano-lubricants of fullerene nanoparticles in refrigeration mineral oil. Curr. Appl. Phys. **9**(2, Supplement), e128–e131 (2009). https://doi.org/10.1016/j.cap.2008.12.054

Thermal Conductivity Correlation for Microscale Porous Media by Using OpenPNM

Ángel Encalada, Mayken Espinoza-Andaluz, and Martin Andersson

1 Introduction

Over time, efforts have been intensified to study transport phenomena in porous media due to its multiple applications in various areas, such as physics, chemistry, biology, and geology [1]. Within the area of energy sciences, porous media have a leading role since it is the fundamental base in the generation of electrical energy inside fuel cells (FCs) [2, 3]. Gas diffusion layers (GDLs) or microporous membranes are used in FCs for the flow of reagents and waste disposal during the energy conversion process. Therefore, it is important to understand how the different transport phenomena—electron, energy, mass—occur through the porous media, to improve their properties and therefore enhance the performance.

A porous medium is a solid material composition with empty space, distributed uniformly or randomly, depending on the medium. In the solid material, the electron and energy transport take place, while in the empty space the flow of fluid is allowed, i.e., momentum and mass transport are analyzed. However, representing in some way a porous medium, due to its randomness, geometric and morphological complexity, has continuously been considered a challenge. At the computational level, this task becomes less exhaustive if the porous medium geometry is approximated through the representation of spheres as pores and cylindrical bodies as connection paths

Á. Encalada · M. Espinoza-Andaluz (✉)
Escuela Superior Politécnica del Litoral, ESPOL, Facultad de Ingeniería Mecánica y Ciencias de la Producción, Centro de Energías Renovables y Alternativas, Campus Gustavo Galindo km. 30.5 Vía Perimetral, P.O. Box 09-01-5863, Guayaquil, Ecuador
e-mail: masespin@espol.edu.ec

M. Andersson
School of Materials and Energy, University of Electronic Science and Technology of China, 2006 Xiyuan Ave, West Hi-Tech Zone, Chengdu, Sichuan, China

Department of Energy Sciences, Faculty of Engineering, Lund University, P.O. Box 118, Lund, Sweden

© Springer Nature Singapore Pte Ltd. 2021
C. Wen and Y. Yan (eds.), *Advances in Heat Transfer and Thermal Engineering* ,
https://doi.org/10.1007/978-981-33-4765-6_50

between pores. In this way, knowing the dimensions of both the spheres (pores) and the cylinders (throats), it becomes easier to construct a porous medium considering parameters of its morphology such as porosity (degree of empty space contained in the medium).

For this work, the effective thermal conductivity of the porous media is analyzed. The challenge is to propose a correlation considering the morphology of a digitally recreated porous medium. Various studies investigating effective thermal conductivity have been carried out on highly complex porous media of different nature. In an investigation on (porous) metallic foams, it was possible to successfully correlate the thermal conductivity and the effect of the porosity in the medium. It was considered a highly complex geometry of the medium, approximating the pores as regular or irregular polyhedrons [4]. As these transport phenomena occurs on the solid material, it was possible to prove that effectively a more porous medium is less capable of performing heat transfer in contrast to a minimally porous medium [5].

Unlike these mentioned studies, our contribution is directed with the recreation of a porous medium, by means of a less complex geometric representation but equally efficient in the computation of transport phenomena. For the computational modeling of our porous media, a Python library named Open Porous Media Network (OpenPNM) has been used. It is a powerful tool to calculate and analyze transport phenomena inside porous media, both in solid phase and vacuum phase. OpenPNM makes use of mathematical resources, matrix algebra, Voronoi triangulation, and Delaunay tessellation to reduce the computational load in the design of these highly complex and exhaustive media.

2   Methodology

The following is a description of the methodology that was followed in our research work. The first section describes the process of generating the porous media by applying algorithms such as Voronoi triangulation and Delaunay tessellation, the simulation parameters used, and some additional considerations. The second section explains the parameters regarding the morphology and structure of the media, which are set prior to the simulation and computation of the transport phenomena.

2.1   Porous Media Generation

The porous medium is created as a cube, with dimensions of 200 μm per side. The pores have a diameter corresponding to 1/50 of the size of the porous medium, i.e., 4 μm, while the connection throats have a diameter corresponding to 1/2 of the diameter of the pore, i.e., 2 μm. The location of the pores and the definition of throats are stochastically determined; however, a seed number of pores must be considered to start the generation of the porous medium. As the number of pores distributed in the medium defines its porosity, this must correspond to a useful and commercial range, i.e., the medium must correspond to a porosity that can be attributed to a real medium such as clay or metallic foam. Therefore, seed values of pores from 1400 to 2000 pores are set, varying in step of 100 and representing a porosity range between 50 and 85%. For each seed value, the porosity of the medium and the effective thermal conductivity are computed, bringing together a total of seven records for the two involved variables. For each porous media, in which the porosity is set, 15 different pore positions were generated. This number of iterations is justified by the computational cost that this process represents.

2.2 Initial and Boundary Conditions

The digitally created porous medium can be associated with a carbon fiber membrane which is the material of the GDLs. A medium of these characteristics has a thermal conductance along its main flow direction of 1.7 W/mK according to the commercial datasheet of GDLs [6]. To compute the thermal conductivity of a determined material, a gradient temperature between the ends of the generated volume is applied. In the current study, at the inlet a 50 °C has been set as boundary condition while at the other surface 40 °C was established. For the computation of the effective thermal conductivity, the differential equation represented by Fourier's Law is solved:

$$q = -k\nabla T \tag{1}$$

where q is the heat flux density, k is the thermal conductivity, and ∇T is the gradient temperature applied to the sample material.

3 Results

From the digitally generated porous medium, keeping the pore diameter and throat diameter constant, the number of pores present in the medium was varied, and each time the porosity and the effective thermal conductivity were computed. For each variation, as indicated above, 15 different samples were generated, obtaining a certain amount of data that ensured the stochastically stability of the values. At the end of the simulation, seven porosities and their corresponding thermal conductivity were determined. Figure 1 shows the computed thermal conductivity values versus the porosity.

According to the obtained results, the thermal conductivity decreases as the porosity of the medium increases, with a nonlinear trend. It is also clear that, due to the randomness of the medium, there are porosity values in which the deviation error of the thermal conductivity is greater than others. The effective thermal conductivity for this medium decreases in a range between 4.85×10^4 and 2.92×10^4 W/mK.

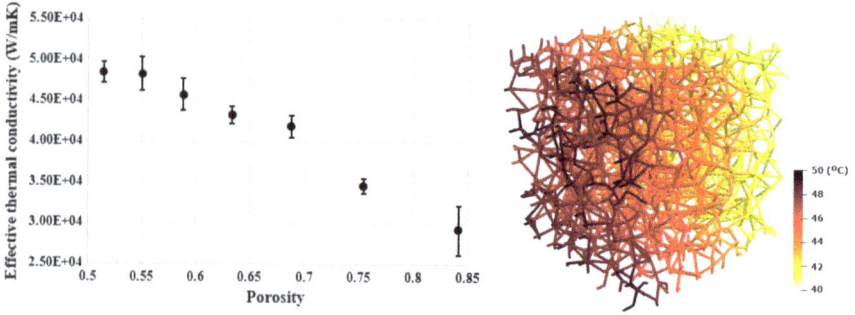

Fig. 1 Left: Computed thermal conductivity versus porosity. Right: temperature distribution

	Function type	Correlation	R-square
Table 1 Empirical correlations to determine the effective thermal conductivity as a function of porosity (ε)	Power	$k_{\text{eff}}(\varepsilon) = 2.71 \times 10^4 \varepsilon^{-095}$	0.9117
	Linear	$k_{\text{eff}}(\varepsilon) = -6.11 \times 10^4 \varepsilon + 8.15 \times 10^4$	0.9678
	Exponential	$k_{\text{eff}}(\varepsilon) = 1.08 \times 10^5 \, e^{-1.48\varepsilon}$	0.9441

In order to propose a correlation between the two variables, Table 1 shows three correlations with the best R-square value, a statistical parameter that allows us to evaluate how efficient is the prediction of the proposed correlations.

The three correlations describe an efficient way to predict effective thermal conductivity as a function of porosity. However, according to Table 1, the most efficient prediction function is the linear followed by the exponential one. These correlations can accurately predict about 97% of the computed data within a 95% confidence interval.

4   Conclusions

OpenPNM library has proven to be a powerful computational modeling tool for porous media. The characteristics of a porous medium can be managed according to the researcher's criteria, such as maintaining the diameter of pores and throats while varying the amount of pores arranged inside the porous medium. Simulations of porous media were carried out, maintaining between 1400 and 2000 pores. For each generated medium, the effective thermal conductivity and porosity were computed. According to the results obtained, as the porosity increases, the thermal conductivity of the medium decreases. Considering the three proposed correlations, the linear type presents the best prediction with an R-square of 0.9678 and within a 95% confidence interval.

Acknowledgements The authors kindly acknowledge the financial support from G8-DI-2014 and FIMCP-CERA-05-2017 projects. Computational and physical resources provided by ESPOL are also very grateful.

References

1. W.C. Tan, L.H. Saw, H. San Thiam, J. Xuan, Z. Cai, M.C. Yew, Overview of porous media/metal foam application in fuel cells and solar power systems. Renew. Sustain. Energy Rev. **96**, 181–197 (2018)
2. M. Espinoza-Andaluz, M. Andersson, B. Sundén, Comparing through-plane diffusibility correlations in PEFC gas diffusion layers using the lattice Boltzmann method. Int. J. Hydrogen Energy **42**(16), 11689–11698 (2017)
3. M. Espinoza, B. Sundén, M. Andersson, Pore-scale analysis of diffusion transport parameters in digitally reconstructed SOFC anodes with gradient porosity in the main flow direction. ECS Trans. **78**(1), 2785–2796 (2017)
4. H.Q. Jin, X.L. Yao, L.W. Fan, X. Xu, Z.T. Yu, Experimental determination and fractal modeling of the effective thermal conductivity of autoclaved aerated concrete: effects of moisture content. Int. J. Heat Mass Transf. **92**, 589–602 (2016)
5. R.A. Mahdi, H.A. Mohammed, K.M. Munisamy, N.H. Saeid, Review of convection heat transfer and fluid flow in porous media with nanofluid. Renew. Sustain. Energy Rev. **41**, 715–734 (2015)
6. https://www.fuelcellsetc.com/store/DS/Toray-Paper-TGP-H-Datasheet.pdf. Web page retrieved in November 2018

Ethylene Glycol and Propanol: Understanding the Influence of an Extra Hydroxyl Group on the Mechanisms of Thermal Conductivity

Likhith Manjunatha, Hiroshi Takamatsu, and James J. Cannon

1 Introduction

Propanol and ethylene glycol are two similar alcohols that find wide application in industry. Propanol is made up of a chain of three carbon atoms with a hydroxyl group attached, while glycol is highly similar in structure, except with the end carbon being replaced by a second hydroxyl group (Fig. 1). While this change in structure is very subtle, the resulting change in thermal conductivity is not, with 0.15 W/(m K) for propanol and 0.26 W/(m K) for glycol [1, 2].

It is of interest to understand the mechanisms of this change of thermal conductivity. Molecular dynamics simulation can provide detailed information regarding the mechanisms of thermal transport, and there have been a number of studies examining thermal conductivity of alcohols, focusing, for example, on the role of hydroxyl groups in the thermal transport process [3, 4] and the effect of molecular conformation [5]. In this manuscript, we utilize this simulation technique to give an insight into the large sensitivity of thermal conductivity to the small structural difference between propanol and glycol.

2 Methodology

The Green-Kubo (GK) method is employed to enable calculation of thermal conductivity under equilibrium conditions. The GK method works by relating the thermal conductivity of the medium to the time correlation of the heat flux (Eq. 1). The heat flux can in turn be expressed as a linear combination of diffusive heat flux and heat

L. Manjunatha · H. Takamatsu · J. J. Cannon (✉)

Department of Mechanical Engineering, Kyushu University, 744 Motooka, Nishi-Ku, Fukuoka, Japan

e-mail: cannon@mech.kyushu-u.ac.jp

© Springer Nature Singapore Pte Ltd. 2021

C. Wen and Y. Yan (eds.), *Advances in Heat Transfer and Thermal Engineering* ,

https://doi.org/10.1007/978-981-33-4765-6_51

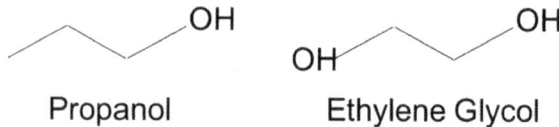

Fig. 1 Schematic of the two molecules. The only difference between the two molecules is the hydroxyl group at the end of the glycol molecule in place of the carbon atom at the end of the propanol molecule

flux arising through atomic interaction (Eq. 2).

$$k = \frac{V}{3k_B T^2} \int_0^\infty \langle J(t).J(0) \rangle dt \qquad (1)$$

$$J = J_{\text{diffusion}} + J_{\text{interaction}} \qquad (2)$$

The virial term can further be broken down into flux arising from interactions internal to the molecule (e.g., atoms bonded to each other) and interactions external to the molecule (i.e., flux between molecules) (Eqs. 3 and 4). This ability to drill deep down into the heat flux pathways is a unique benefit of this approach.

$$J_{\text{interaction}} = J_{\text{internal}} + J_{\text{external}} \qquad (3)$$

$$J_{\text{external}} = J_{OO} + J_{OH} + J_{CC} + J_{CO} + \cdots \qquad (4)$$

3   Results

In order to validate the accuracy of the simulations, first the calculated thermal conductivity is compared to experiment. Values of 0.20 W/(m K) and 0.37 W/(m K) were obtained for propanol and glycol, respectively. This is an over-estimation compared to experiment and is typical of this simulation method, however the higher and lower order of conductivity is clearly preserved.

In order to understand the origins of the difference in conductivity, the relative contribution of internal and external correlations to the total conductivity can be observed (Fig. 1, left). Since the conductivity arises from correlations, there is a cross-correlation term too, although its influence is very small. Interestingly, although the conductivity of glycol is much larger than propanol, the relative contribution from internal and external conduction is roughly unchanged.

Next, the external contributions to conductivity can be broken down into contributions from hydroxyl groups and carbon atoms. The influence of the change from

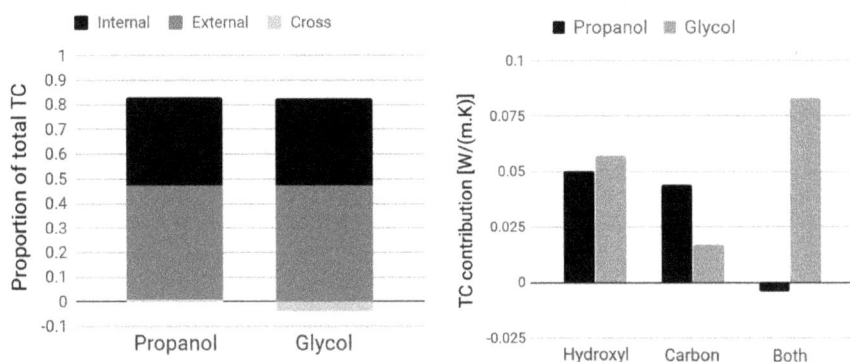

Fig. 2 (Left) *Relative* contribution to thermal conductivity of internal and external contributions to thermal conductivity for each molecule. Note that the sum is not unity because the diffusion term is not included here; only the interaction term (Eq. 2). Despite different conductivities, the relative contributions remain largely the same. (Right) The contribution to external thermal conductivity (Eq. 4) by atomic type. The interaction of hydroxyl groups with carbon atoms ("both") is seen to contribute significantly to the higher thermal conductivity displayed by glycol

the carbon atom to a hydroxyl group can be observed (Fig. 2). It is interesting to note that addition of the second hydroxyl group does not significantly increase the conduction between hydroxyl groups. Instead, it leads to a significant increase in conduction between the hydroxyl groups and carbons atoms. This comes partially at the expense of inter-carbon conduction, but is more than sufficient to compensate for this loss and ultimately helps result in a higher net thermal conductivity of glycol compared to propanol.

4 Conclusions

The influence of a single atomic change to the structure of propanol on the mechanisms of thermal conductivity has been investigated using molecular dynamics simulations. By breaking down the thermal conductivity of propanol and ethylene glycol, it has been shown how the relative importance of internal and external conduction does not change, and a large factor in the higher conduction between glycol molecules is the increase in conduction between hydroxyl groups and carbon atoms.

References

1. K. Ogiwara, Y. Arai, S. Saito, Thermal conductivities of liquid alcohols and their binary mixtures. J. Chem. Eng. Jpn. **15**(5), 335–342 (1982)
2. D. Bohne, S. Fischer, E. Obermeier, Thermal conductivity, density, viscosity, and Prandtl-numbers of ethylene glycol-water mixtures. Phys. Chem. **88**, 739–742 (1984)

3. H. Matsubara, G. Kikugawa, T. Bessho, S. Yamashita, T. Ohara, Molecular dynamics study on the role of hydroxyl groups in heat conduction in liquid alcohols. Int. J. Heat Mass Transf. **108**, 749–759 (2017)
4. L. Manjunatha, H. Takamatsu, J.J. Cannon, An investigation into application of the Green-Kubo method in molecular simulation to help understand the mechanisms of thermal conductivity of alcohols, in *JSME Thermal Engineering Conference* (2018)
5. Y. Lin, P.-Y. Hsiao, C.-C. Chieng, Constructing a force interaction model for thermal conductivity computation using molecular dynamics simulation: ethylene glycol as an example. J. Chem. Phys. **134**, 154509 (2011)

Experimental Investigation of Thermal Performance and Liquid Wetting of a Vertical Open Rectangular Microgrooves Heat Sink with Fe_3O_4–Water Nanofluids

R. Fu, W. B. Zhou, J. H. Wang, and X. G. Hu

1 Introduction

Open microgrooves heat sink (OMHS) represents one of the best cooling solutions for high-power electronic devices. As shown in Fig. 1, it features a bunch of open capillary microgrooves on the surface. Liquid is sucked into the groove and forms an extended meniscus in the cross section, which is typically divided into three regions: the intrinsic meniscus region, the evaporating thin film region, and the adsorbed layer region. The second region enables intensive evaporation on the side wall of the groove. At higher heat flux, besides such evaporation, nucleate boiling occurs within the intrinsic meniscus region at the same time. This combined phase change leads to a very efficient heat transfer. It is found that the heat transfer performance of an OMHS associates intimately with liquid wetting in the grooves. Once the driving capillary forces are not capable to overcome the friction forces and gravitational forces of liquid, the axial wetting length (AWL, Fig. 1c) will decrease, leading to a heated section non-cooled by the evaporation or boiling. The heat flux applied to this section can cause sharp temperature rise and shifting of some of the heat load by conduction toward the cooler region, resulting in a further fluid recession [1]. At present, besides optimization of groove dimensions, enhancing wetting in capillary microchannels can be achieved by hydrophilic surface treatment, applications of electrohydrodynamics and ferrohydrodynamics, and reduction of friction losses with bionic structure.

Nanofluids are colloidal suspensions of nanoscaled solid particles. Evidence shows that nanofluids are able to promote thermal performances of capillary-driven

R. Fu · W. B. Zhou · J. H. Wang · X. G. Hu (✉)
Institute of Engineering Thermophysics, Chinese Academy of Sciences, Beijing 100190, China
e-mail: xuegonghu@iet.cn

J. H. Wang
University of Chinese Academy of Sciences, Beijing 100049, China

© Springer Nature Singapore Pte Ltd. 2021
C. Wen and Y. Yan (eds.), *Advances in Heat Transfer and Thermal Engineering* ,
https://doi.org/10.1007/978-981-33-4765-6_52

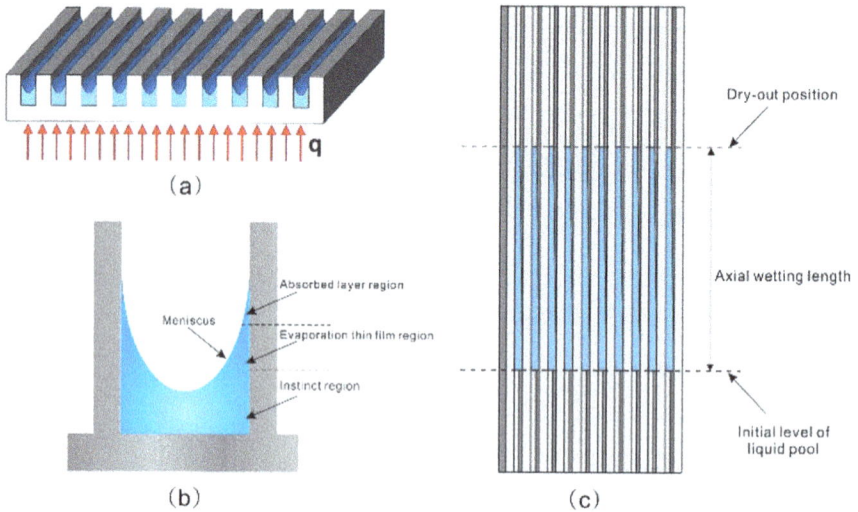

Fig. 1 Schematic of **a** the open capillary microgrooves heat sink, **b** the extended meniscus in the cross section of the groove, and **c** the OCMHS placed vertically in a liquid pool

heat transfer systems [2]. The enhancements are often attributed to the particle depositions that improve nucleate boiling and capillary performance of the wick structure. The depositions alter surface roughness and wettability, which may increase the density of active nucleation sites depending on particles' properties and concentration. Moreover, the particle deposition in the wick structure can supply an extra pumping effect and reduces the contact angle between the fluid and the surface, resulting in a stronger capillary force. More importantly, the particles dispersed in the liquid can further enhance the wetting behavior of an evaporating meniscus. They are ordered layered in the evaporating thin-film region, inducing an additional disjoining pressure, known as structural disjoining pressure. This structural disjoining pressure promotes the wetting by reducing the equilibrium contact angle and stabilizing the thin film.

This work aims to experimentally investigate the thermal performance of a vertical OMHS with rectangular grooves when Fe_3O_4–water nanofluids are used as the working fluids. The thermal performance was estimated by the wall superheat. At the same time as the heat transfer test, the AWL was characterized with infrared(IR) thermal imaging, which has been proven to be an accurate method to locate the dry-out point of the liquid in the capillary-driven system [3]. It was found that thermal performance and wicking ability of the OMHS were significantly enhanced by using the nanofluids.

2   Methodology

Figure 2 illustrates the schematic diagram of the experimental apparatus, which mainly consists of a visualization unit, a test section, a liquid pool, and a data acquisition system. The visualization unit includes a high-speed camera (Vision Research Inc., Phantom 5.0, maximum shooting rate: 100,000 frames per second), an IR thermal imager (Testo 885-2, measuring range: 0 to 350 °C), and a LED lighting device. No heating facility was utilized to heat the liquid in the pool. All the experiments were carried out at an ambient temperature, which is maintained between 23 and 25 °C. The relative humidity was around 20%.

The OMHS made of borosilicate glass was fixed onto the PTFE support and meanwhile attached to the heating component (40 mm × 20 mm), which was formed by sandwiching a ceramic heater with a copper plate and a piece of closed-cell foam. The copper plate possessed four K-type thermocouples to measure the temperature of the back of the heat sink T_b. The boiling surface temperature (T_w) was calculated with the Fourier law based on the T_b, the thermal conductivity of borosilicate glass, and the distance between the back of the heat sink and the bottom of the grooves. The heater was powered by a constant voltage DC power supply (0–60 V, 0–5 A, EVERFINE WY605). Another thermal couple was put into the liquid pool to monitor the temperature of the liquid. All the thermocouples were pre-calibrated, and their signals were collected by a data acquisition instrument (AGILENT 34970A) during each experiment.

A two-step method was applied in this work to prepare Fe_3O_4–water nanofluids. Fe_3O_4 nanoparticles were prepared by co-precipitation of Fe^{3+} and Fe^{2+} ions in basic solution. To improve the colloidal stability, Fe_3O_4 nanoparticles were modified with

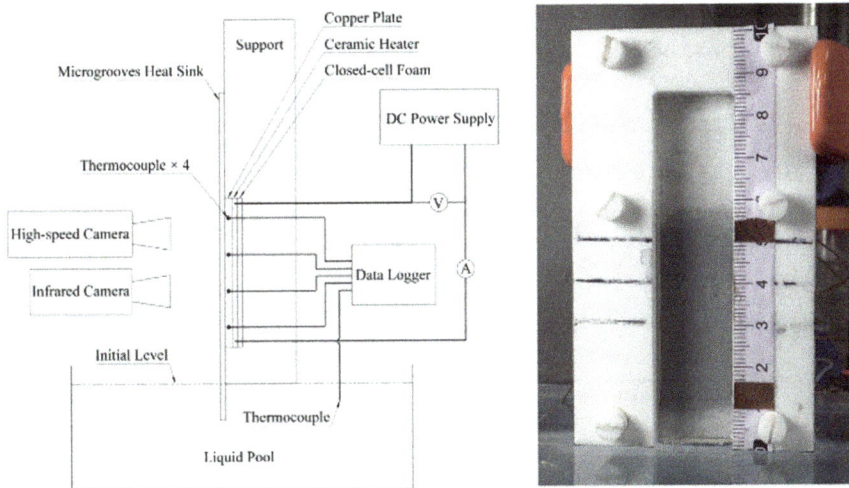

Fig. 2   Schematic of the experimental apparatus and photograph of the test section

citric acid through dissolving citric acid to the water suspension of the particles followed by three hours of heating. The modified particles were washed with diluted HCl solution to remove the residue citric acid. Finally, the particles were dispersed to de-ionized water to generate the Fe_3O_4–water nanofluids.

The AWL of liquid in the OMHS was determined with the reported method based on IR thermal imaging. When a heat flux was applied to the heat sink, there was an apparent temperature difference between the sections cooled and uncooled by liquid evaporation and/or boiling. In an IR image of the OMHS, it is easy to locate the dry-out points, which are the tips of the strip-like isothermal regions. Pixel counting work was conducted with MATLAB software.

3 Results

Figure 3 displays the influences of the Fe_3O_4–water nanofluids on the thermal performance and wicking ability of the OCMHS at different particle concentrations. The heat flux was changed from 16 to 55 kW/m^2. All the nanofluids enhanced the performance and the wicking, which is evidenced by the significantly reduced ΔT_w (Fig. 3a) and improvements of AWL (Fig. 3b) in comparison with those for BL. There exists an optimum concentration for the enhancement. The insert of Fig. 3a gives the decrease percentage of the wall superheat ΔT_w for each concentration. The ΔT_w was reduced by 41% and HTC was improved by 21% when the nanofluid of 0.00375 vol.% was used. Nonetheless, the enhancement became weaker regardless of whether the concentration was increased or reduced. For the concentrations of 0.0025, 0.005, and 0.0075 vol.%, the ΔT_w was reduced by 30%, 37%, and 24%, respectively.

Thermal performance of the heat sink strongly depended on the AWL. By sorting the nanofluids with the percentage increase of the AWL from the highest to the lowest, the order is the same with that generated by sorting them with the decrease percentage

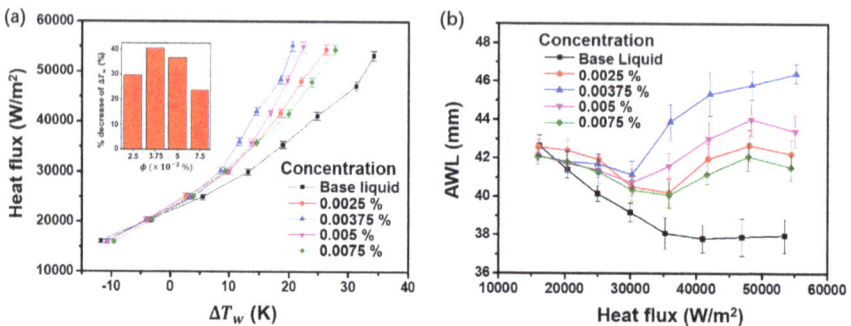

Fig. 3 **a** Plots of heat flux against the wall superheat and **b** the dependences of the AWL on the heat flux for various particle concentration. The insert shows the decrease percentage of the wall superheat for each concentration

of ΔT_w from largest to lowest. The average increase of AWL over the concentrations was 15%. The optimum concentration was also 0.00375 vol.%, which led to a 22% increase in the AWL. Moreover, Fig. 3a shows that the nanofluids started to enhance thermal performance since the heat flux of 25 kW/m^2. It is where the differences of AWLs between the nanofluids and the BL became obvious. And the thermal performance with the nanofluid of 0.00375 vol.% kept superior to the others since 36 kW/m^2, corresponding to where the AWL at this concentration begun to being the highest.

Otherwise, the AWLs of all the nanofluids were almost the same as that of the BL at the beginning of increasing the heat flux. However, they decreased apparently slower than the BL did and turned to rise at heat fluxes of 30–36 kW/m^2. With using the high-speed camera, the phenomena of the boiling started to be detected around 35 kW/m^2 for all the concentrations. The mechanism of the effect of the nanofluids will be discussed based on the study on the roles of the particle deposition and the particles dispersed in the liquid.

4 Conclusions

This work investigated the effects of Fe_3O_4–water nanofluids on the axial wetting length and thermal performance of a vertical OCMHS with rectangular microgrooves being applied with heat flux. Four particle concentrations were selected, including 0.0025, 0.00375, 0.005, and 0.0075 vol.%. The AWL was characterized by IR thermal imaging at the same as the heat transfer test. Thermal performance and wicking ability of the OCMHS were both significantly enhanced with the nanofluids, and the former strongly depends on the later. By sorting the nanofluids with the percentage increase of the AWL from the highest to the lowest, the order is the same with that generated by sorting them with the decrease percentage of the wall superheat from largest to lowest. There existed an optimum concentration, which led to a 41% decrease in the wall superheat and a 22% increase in the AWL. Either increasing or decreasing the concentration reduced the percentage improvements.

Acknowledgements This work was financially supported by the National Key R&D Program of China (Grant No. 2017YFB0403200).

References

1. R.H. Nilson, S.W. Tchikanda, S.K. Griffiths, M.J. Martinez, Steady evaporating flow in rectangular microchannels. Int. J. Heat Mass Transf. **49**, 1603–1618 (2006)
2. M.H. Buschmann, Nanofluids in thermosyphons and heat pipes: overview of recent experiments and modelling approaches. Int. J. Therm. Sci. **72**, 1–17 (2013)

3. Y. Tang, D. Deng, L. Lu, M. Pan, Q. Wang, Experimental investigation on capillary force of composite wick structure by IR thermal imaging camera. Exp. Therm. Fluid Sci. **34**, 190–196 (2010)

On the Measurement Error of Temperature in Nanocomposite Thermal Insulation by Thermocouples

Chao Fan, Xiao-Chen Zhang, Chuang Sun, and Xin-Lin Xia

1 Introduction

The use of nanocomposite thermal insulation (NTI) has been extensively studied due to the wide range of potential engineering applications, such as thermal protection of hypersonic vehicle, energy storage, solar thermal utilization and energy conservation of buildings. Accurately obtaining the temperature distribution within the NTI is fundamental for characterizing the thermal performance. According to the contacting schemes, the techniques used to measure temperature can be divided into three categories: invasive, semi-invasive and noninvasive [1]. Thermocouples, one of the invasive temperature measuring devices, are widely used in many systems for their easy operation and fast response. However, it has been reported that there are errors between the readings of the thermocouple and the actual temperature [2]. This uncertainty greatly impedes the understanding of mechanism of those NTI as well as the developments of material science. In order to accurately get the thermal property of NTI, it is critical to analyze the heat transfer process of the thermocouple.

As a typical invasive measurement technique, thermocouples have gained a reputation for its accurate temperature measurements of various fields, such as internal temperature, surface temperature. However, it has been reported that the transient temperature measurement is artificial lower than the real value [3]. Some scholars emphasized that the heat conduction through the leads and thermal inertia of thermocouple may affect the measurements. Sun et al. [4] investigated the performances of the thermocouples with different junctions for measuring transient surface temperature, their study shows that the isothermal junction has the least measurement errors and degree of deviation, whereas the dynamic property of the separate junction is the best. However, due to the complex heat transfer mechanism of the thermocouple and

C. Fan · X.-C. Zhang · C. Sun · X.-L. Xia (✉)
Key Laboratory of Aerospace Thermophysics of MIIT, School of Energy Science and Engineering, Harbin Institute of Technology, Harbin 150001, China
e-mail: xiaxl@hit.edu.cn

© Springer Nature Singapore Pte Ltd. 2021
C. Wen and Y. Yan (eds.), *Advances in Heat Transfer and Thermal Engineering* ,
https://doi.org/10.1007/978-981-33-4765-6_53

some other factors, such as the diameter of the thermocouple, the heating temperature remains to be further studied.

In this study, the transient thermal response of thermocouple for the internal temperature measurement of NTI is obtained with a detailed simulation of the heat transfer process of thermocouple. The energy equation is discretized by the finite volume method (FVM). Meanwhile, the radiative source term is calculated based on radiation distribution factor (*RD*), which is loaded to grid. Then, an energy balance model is built for the thermocouple. The thermocouple temperature is predicted and compared with the medium grid where the thermal couple node located. This work can facilitate the understanding of measurement error by thermocouples. Such errors can be corrected by this model, thus providing a more accurate temperature value.

2 Methodology (Length and Layout)

2.1 Problem Description

Figure 1 illustrates the geometry and heat transfer model of NTI specimen, thermocouple. The geometrical sizes of the setup are highlighted in Table 1. At the initial time, the specimen and the thermocouple are kept at 300 K. The heating surface of the specimen is 600 K. The opposite side of the specimen is irradiated with the environment, where the emissivity of the specimen is 0.8, and other surfaces are considered as adiabatic walls. The exposed part of the thermocouple is irradiated with the environment, where the emissivity of the thermocouple is ε_c.

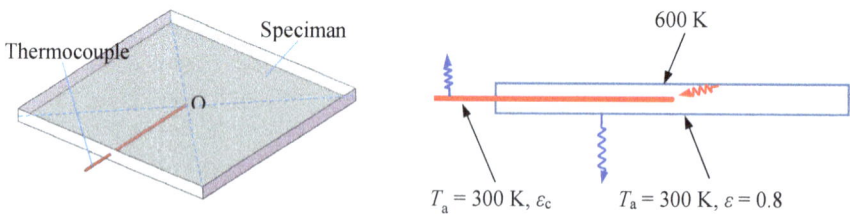

Fig. 1 Geometry and heat transfer model of the specimen, thermocouple

Table 1 Geometrical size of each part of the model (mm)	Parts	Size
	Insulation	$300 \times 300 \times 20$
	Thermocouple	$\Phi D \times 200$

3 Results

Based on the three-dimensional heat conduction differential equation, the influence of the heating slab temperature, the emissivity of the thermocouple and the diameter of the thermocouple on the temperature measurement errors are calculated, respectively. The emissivity of the surface of the cavity and the heating slab is set to 0.75 in all simulations. The initial temperature of the specimen and the thermocouple is set to 300 K. The environment temperature T_∞ is set to 300 K. Thermal-physical properties of the specimen are selected as: $\rho_s = 320$ kg/m^3, $c_s = 870$ J/(kg K), $k_s = 0.05$ W/(m K), $\varepsilon_s = 0.8$. The thermal properties of the thermocouple are $\rho_t = 8665$ kg/m^3, $c_t = 486$ J/(kg K), $k_t = 25$ W/(m K), where subscript s and t are, respectively, specimen and thermocouple.

3.1 Thermal Response of Thermocouple

Figure 2 shows the results at $\varepsilon_c = 0.5$ and $D = 3.0$ mm for the transient measuring process. The measurement errors of thermocouple $\delta = |T_t - T_s|/T_s \times 100\%$ are introduced here, where T_t and T_s, respectively, represent the temperature of the thermocouple and the specimen. It can be seen that the errors increase rapidly at first and then decrease gradually. Due to the thermal diffusivity of the specimen is far smaller than that of the thermocouple, thus the diffusion rate of heat in the thermocouple is faster than that in the specimen. Consequently, the temperature of the thermocouple is lower than that of the surrounding specimen. In addition, as time passes by, the temperature within the specimen and the thermocouple tends to balance gradually, the errors decrease gradually.

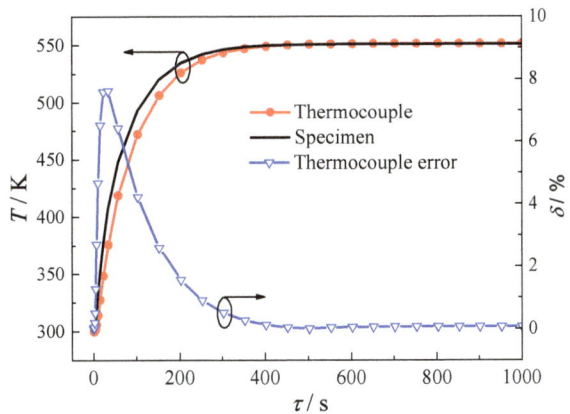

Fig. 2 Temperature response of thermocouple and internal of the specimen

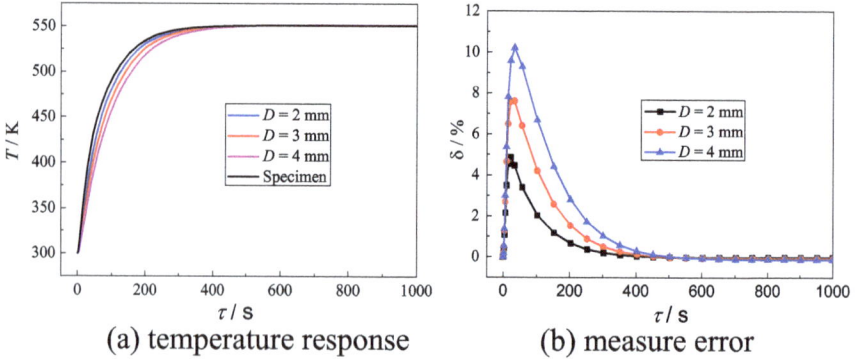

Fig. 3 Temperature response and measure error of the thermocouple with different thermocouple diameter **a** Temperature response, **b** measure error

3.2 Influence of Thermocouple Diameter on Measurement Error

Figures 2 and 3 illustrate the effects of thermocouple diameters ($D = 2, 3, 4$ mm) on measurement error, while other parameters are kept unchanged from those adopted in Fig. 2. It can be seen that the temperature field of the thermocouple changes more quickly with the increases of the thermocouple diameter, and the measurement error increases with the increase of the thermocouple diameter. As expected, the response of thermocouple junction to heat-up becomes quicker as the diameter gets smaller. The effect of thermal inertia becomes obvious as the diameter increases.

4 Conclusions

In this study, a numerical study has been performed to analyze thermocouple error in temperature measurement of nanocompiste thermal insulation. A combining Monte Carlo ray tracing method and finite volume method are used to solve the transient temperature field of the thermocouple and the specimen, and the measurement error is obtained. The impact of the diameter of the thermocouple on temperature response of thermocouple and measurement error is calculated. The following conclusions are obtained:

(1) The temperature response of the thermocouple lags behind the specimen. With the increase of the measuring time, the measurement errors increase rapidly and then decrease.

(2) Decreasing the diameter of the thermocouple can reduce the delay of the thermo-couple and reduce the measurement errors. The measurement errors during the

measurement time are mostly smaller using a smaller diameter, higher emissivity thermocouple.

Acknowledgements This research was supported by the National Natural Science Foundation of China (NO. 51536001 and NO. 51776053).

References

1. P.R.N. Childs, J.R. Greenwood, C.A. Long, Review of temperature measurement. Rev. Sci. Instrum. **71**(8), 2959–2978 (2000)
2. S. Yang, W. Tao, *Heat Transfer*, vol. 61–62, 4th edn. (Higher Education Publishing Company, Beijing, 2006) p. 433–434
3. A. Ishihara, Temperature measurement of a burning surface by a thermocouple. J. Propul. Power **20**(3), 455–459 (2004)
4. Z.Q. Sun, Y. Chen, X. Chen, Measurement of transient surface temperature of conductive solid using thermocouples with different junctions. Sens. Transducers **148**(1), 22–27 (2013)

Thermal Effect on Breakup Dynamics of Double Emulsion Flowing Through Constricted Microchannel

Yong Ren, Yuning Huang, Yuying Yan, and Jing Wang

1 Introduction

Double emulsion templated microcapsules fabricated using microfluidics have drawn dramatically increased attention nowadays because of high monodispersity of size and intrinsic core/shell structure, giving rise to versatile control over drug release kinetics [1]. Double emulsions have been widely used in chemical and biological areas as they can protect substance inside which can be released in a controllable way. Despite of the latest advances, the thermal effect in drug release especially when drug loaded microcapsules deliver in microcapillaries is not fully understood or sufficiently studied. In this paper, numerical model has been established to investigate the thermal effect on breakup dynamics of double emulsion when it flows through constricted microchannel subject to heat input. Two different cases with inflow velocity of 0.5 m/s were studied: water/oil/water double emulsion with the viscosity independent of temperature; and water/oil/nanofluids where single-wall carbon nanotubes (SWCN) were dispersed in ethylene glycol (EG) solution with volume fraction of 0.05% [2], and viscosity changing versus temperature induced

Y. Ren (✉) · Y. Huang
Department of Mechanical, Materials and Manufacturing Engineering, University of Nottingham Ningbo China, Ningbo, China
e-mail: yong.ren@nottingham.edu.cn

Y. Ren · Y. Yan
Research Group for Fluids and Thermal Engineering, University of Nottingham Ningbo China, Ningbo, China

Y. Yan
Research Group for Fluids and Thermal Engineering, University of Nottingham, Nottingham NG7 2RD, UK

J. Wang
Department of Electrical and Electronic Engineering, University of Nottingham Ningbo China, Ningbo, China

© Springer Nature Singapore Pte Ltd. 2021
C. Wen and Y. Yan (eds.), *Advances in Heat Transfer and Thermal Engineering* ,
https://doi.org/10.1007/978-981-33-4765-6_54

by local heat transfer, the temperature applied at narrow region of microchannel was controlled in a range from 30 to 60 °C.

2 Methodology

Volume of fluid (VOF) multiphase model [3] was used in the numerical simulation for two cases. For the first case, the fluid field can be divided into three phases including pure water as outer phase denoted by 'o', the silicone oil as middle phase denoted by 'm', and pure water as inner phase denoted by 'i'; while for the second case, the three phases including EG-SWCN as outer phase, the silicone oil as middle phase, and pure water as inner phase were modeled. The viscosity of EG-SWCN nanofluid is affected by temperature. The governing equations for incompressible three-phase fluids can be described by advection equation of each phase (Eq. 1), continuity equation (Eq. 2), and Navier–Stokes equation (Eq. 3), the cell-averaged density and dynamic viscosity can be obtained from Eqs. 4 to 5, respectively [4], the interfacial tension force f is modeled by Eq. 6, and the enducergy equation is shown in Eq. 7 [5].

$$\frac{\partial \alpha_{ph}}{\partial t} + \nabla \cdot \left(v \cdot \alpha_{ph}\right) = 0 \tag{1}$$

$$\nabla \cdot v = 0 \tag{2}$$

$$\frac{\partial v}{\partial t} + \nabla \cdot (VV) = -\frac{P}{\rho} + \nabla \cdot \frac{\mu}{\rho}\left(\nabla v + \nabla v^T\right) + f \tag{3}$$

$$\rho = \sum \alpha_{ph}\rho_{ph} \tag{4}$$

$$\mu = \sum \alpha_{ph}\mu_{ph} \tag{5}$$

$$f = \sigma K \hat{n} \delta_s \tag{6}$$

$$\rho C_p\left(\frac{\partial T}{\partial t} + v \cdot \nabla T\right) = \nabla \cdot k \nabla T \tag{7}$$

where α is the volume fraction of one fluid phase denoted by subscript 'ph' in a cell of the computational domain (ph = i, m or o), the volume fraction of all phases in each cell sum up to unity. v is the flow velocity governed by continuity equation (Eq. 2), t is the time, P is the pressure, ρ is the volume averaged density, μ is the volume averaged dynamic viscosity, and σ is the interfacial tension between two phases, K denotes the mean curvature of the interface, $\hat{n}$ is the unit normal to the interface, δ_s is the interface delta function, C_p is specific heat at constant pressure, T is temperature, and k is thermal conductivity of a fluid. No-slip condition was applied at the solid boundaries of the walls of the microchannel. Zero gauge pressure was applied at the outlet of the microdevice. The flow velocity was specified at each inlet of the microdevice. The numerical data were analyzed by the Ansys CFX-Post Processor 15.0.

3 Results

Figure 1 shows the overall microchannel configuration, and the narrow region is located at the middle of channel. The left surface of channel represents the flow inlet, while the right surface represents the flow outlet. The channel has 0.15 × 0.15 mm in cross section and is 2 mm long. The constricted region has height of 0.05 mm. The mesh size sensitivity study was performed first to validate the accuracy of the numerical model. Three models with the same geometry and the same boundary conditions yet various mesh sizes were used: the coarse meshing with total element number 339,231, the normal meshing with total element number 465,834, and the refined meshing with total element number 766,613. The velocity results of fluid flow in the channel show small discrepancy when compared with the normal and refined meshing model. Therefore, the normal meshing model was chosen for further computational study due to the less computational load. Figure 2b shows the shear stress along the vertical line in the center position of constricted passage of microchannel. As the temperature on the surface of constrained channel increases, the shear stress value will decrease, which confirms that the viscosity varies in reverse proportion to the change of temperature for EG-SWCN solution [2], implying low temperature can facilitate the breakup double emulsion when nanofluids are used. At the outer surface of the constricted channel, the temperature effect is obvious but when it approaches to the central position of the channel, the difference among the four cases at various wall temperatures for the shear stress becomes less pronounced. The total range of shear stress in the first case is one order of magnitude less than that when nanofluid was used in the second case.

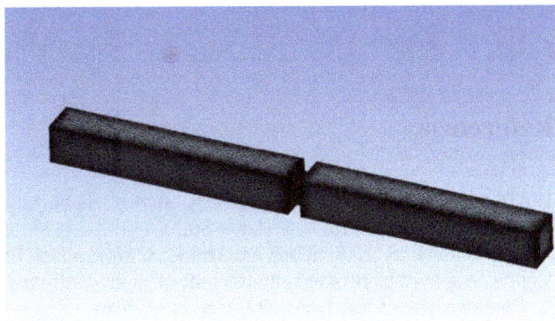

Fig. 1 Schematic of microchannel

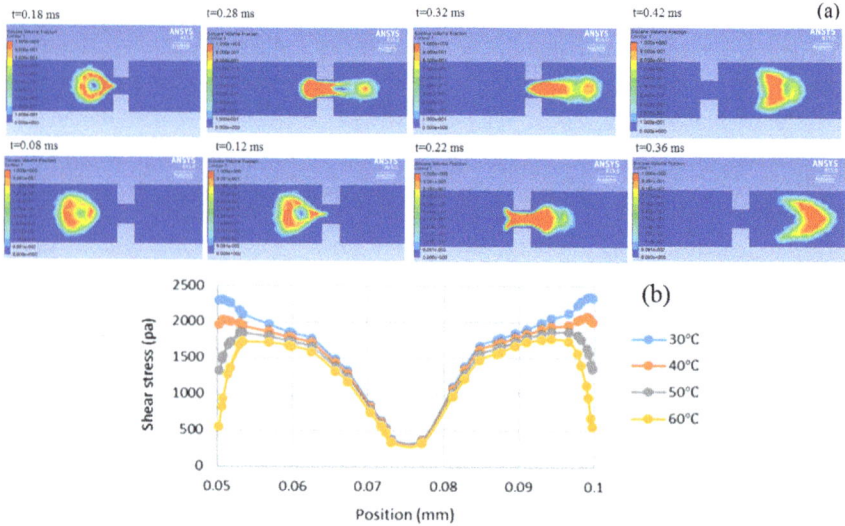

Fig. 2 **a** Time lapse images of silicone oil of middle phase distribution in constricted microchannel for the first case at top row, and the second case at bottom row; **b** the shear force distribution in microchannel when EG-SWCN is used

4 Conclusions

The shear stress distribution affects the break mechanism of emulsion. High shear stress will facilitate the breakup of double emulsion when flowing through constricted channel passage. The double emulsion breaks up when passing through the constricted microchannel at velocity of 0.5 m/s. The research will lead to new insights to understanding of thermal induced drug release kinetics of microcapsules flowing in microcapillaries after being administrated.

Acknowledgements This research was supported by Zhejiang Provincial Natural Science Foundation of China under Grant No. LY19E060001, National Natural Science Foundation of China under Grant No. NSFC51506103/E0605, Ningbo Science and Technology Bureau under the project code 2019F1030, as well as Research Seed and Supplementary Support Fund from Nottingham China Health Institute.

References

1. J. Sander, L. Isa, P. Rhs, P. Fischer, A. Studart, Stabilization mechanism of double emulsions made by microfluidics. Soft Matter **8**(45), 11471–11477 (2012)
2. M. Baratpour, A. Karimipour, M. Afrand, S. Wongwises, Effects of temperature and concentration on the viscosity of nanofluids made of single-wall carbon nanotubes in ethylene glycol. Int. Commun. Heat Mass Transf. **74**, 108–113 (2016)

3. R. Davarnejad, M. Jamshidzadeh, CFD modeling of heat transfer performance of MgO-water nanofluid under turbulent flow. Eng. Sci. Technol., Int. J. **18**(4), 536–542 (2015)
4. Y. Chen, L. Wu, L. Zhang, Dynamic behaviors of double emulsion formation in a flow-focusing device. Int. J. Heat Mass Transf. **82**, 42–50 (2015)
5. M. Fischer, D. Juric, D. Poulikakos, Large convective heat transfer enhancement in microchannels with a train of coflowing immiscible or colloidal droplets. J. Heat Transf. **132**(11), 112402 (2010)

Design and Analysis of Synthetic Jet for Micro-channel Cooling

Ashish Mishra, Akshoy Ranjan Paul, Anuj Jain, and Firoz Alam

1 Introduction

Miniaturization of electronic component has made the heat dissipation more momentous problem than ever because of which failure due to heating has increased by 50% [1]. Various methods have been developed for electronics cooling such as liquid cooling, fans, active and micro-channel cooling. The present study focuses on new technology like sinusoidal synthetic jet employed for the micro-channel cooling (Fig. 1). Synthetic jet works on a principle of zero-net-mass-flow, i.e. it does not introduce any new mass in the systems, and it just recirculates the mass of the system in the system. It works by inhaling and exhaling the surrounding air by vibrating the diaphragm. Initially, its application was limited to the flow control [2], increasing the turbulence in boundary layer [3]. Later, its use in cooling of micro-channels came into light. The difference it makes from the conventional fans is that during its exhaling cycle it creates as series of vortices in the surroundings which helps in creating turbulence hence increasing the heat transfer coefficient. Ming and Zhou [4] studied the flow and heat transfer characteristics under synthetic jets impingement driven by piezoelectric actuator. Zhang et al. [5] used synthetic jet at low frequency along with the cross-flow known as conjugate flow. Fanning et al. [6] studied the heat transfer characteristics of pair of synthetic jet over a flat plate. Since no works on the tapered synthetic jet for cooling purpose are available in open literature, this work may provide an insight on synthetic jet for cooling.

A. Mishra · A. R. Paul (✉) · A. Jain
Department of Applied Mechanics, M.N. National Institute of Technology Allahabad, Prayagraj, India
e-mail: arpaul@mnnit.ac.in

F. Alam
School of Aerospace, Mechanical and Manufacturing Engineering, RMIT University, Melbourne, Australia

© Springer Nature Singapore Pte Ltd. 2021
C. Wen and Y. Yan (eds.), *Advances in Heat Transfer and Thermal Engineering* ,
https://doi.org/10.1007/978-981-33-4765-6_55

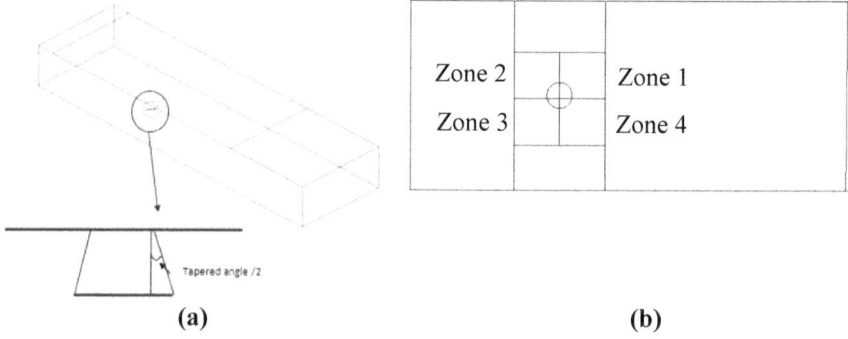

Fig. 1 **a** Computational domain, **b** different zones of a processor chip

2   Modelling and Analysis

For present work, three-dimensional geometry consisting a heated plate at one end and synthetic jet on the opposite was created in CAD as shown in Fig. 1. The computational domain is extended from $8d$ in upstream to $20d$ in downstream, $8d$ in vertical and $4d$ in spanwise direction. Taper angle of the jet as shown in Fig. 1a is varied. In order to demonstrate the effectiveness of synthetic jet for cooling of heated processor chip, the later is placed on the top of the heated plate and is divided into four different zones (Fig. 1b).

The computational domain is divided into structured computational grids using multizone method. Grid independency test is carried out, and the final grid having 6.98×10^5 cells is shown in Fig. 2. Unsteady CFD simulations are performed using a finite volume method-based ANSYS-Fluent (v.15) solver. Momentum and continuity coupling are achieved by pressure correction scheme. Because of periodic nature of synthetic jet, turbulence closure is achieved by RNG k–ε turbulence model. Energy equation is used to capture the changes in the convective heat transfer coefficient h. Time step size is different for different frequency, and convergence criteria of 10^{-6}

Fig. 2 Longitudinal view of grid generation

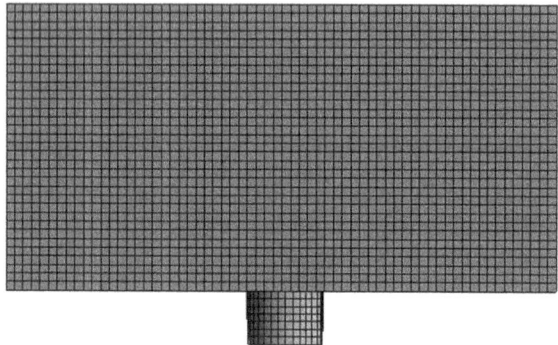

Fig. 3 Validation

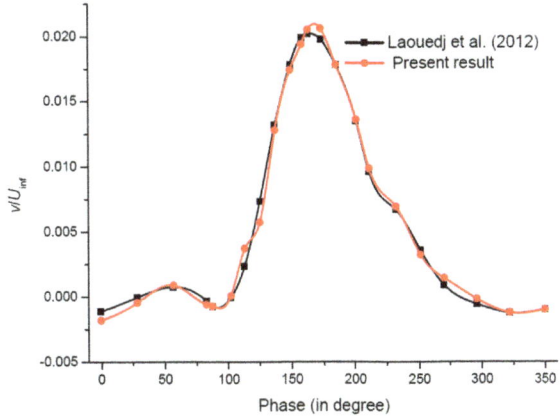

for each step are used. The CFD results are validated with the experimental data reported in Laouedj et al. [7] and are shown in Fig. 3.

3 Results and Discussion

Discussion on results is divided into two sections, in first one, the optimized configuration of jet is narrowed down by aerodynamic and heat transfer analysis, and in the second section, the optimized configuration is used to cool the processor chip.

Aerodynamic effect of synthetic jet in the domain: For finding out the optimum jet configuration, the plate was kept at a temperature of 350 K in presence of synthetic jet. It is observed that vortices start generating at the jet exit and hit the plate before the actual jet, trapping the air in the centre, thereby increasing the effective area of the jet. Vortices generation is found independent of the plate distance, but the disturbance caused by it depends on the same, thus making the velocity as the function of plate distance.

Heat transfer analysis: Heat transfer analysis of the heated plate is done on the basis of calculation of heat transfer coefficients and dissipated heat value.

Effects of frequency variation: For finding an optimum frequency, conjugate flow with frequency variation of jet (100–200 Hz) is initialized with an airflow of 34.6 m/s imposed at the inlet of flow domain. The heated plate is kept at 350 K for the optimization study. Outlet of domain is kept at atmospheric conditions. Frequency of synthetic jet is varied from 100 to 200 Hz for different z/d ratios. Amplitude for all the frequencies is kept same, so the frequency is sole responsible. Heat transfer coefficient is maximum for 150 Hz frequency. Thus, it is considered as the optimum frequency, and all the results discussed further are analysed on this frequency only.

Effect of plate distance: Heated plate to jet exit distance is varied from 10 to 25 mm. The maximum heat transfer in the case occurs at 15 mm ($z/d = 2.14$). Thus,

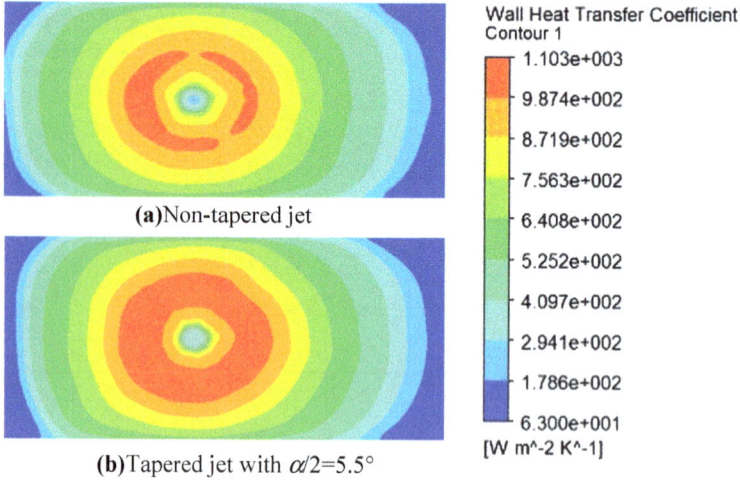

(a)Non-tapered jet

(b)Tapered jet with $\alpha/2=5.5°$

Wall Heat Transfer Coefficient
Contour 1

1.103e+003

9.874e+002

8.719e+002

7.563e+002

6.408e+002

5.252e+002

4.097e+002

2.941e+002

1.786e+002

6.300e+001
[W m^-2 K^-1]

Fig. 4 Wall heat transfer coefficient contours for constant centerline velocity at $z/d = 2.14$

for the used jet diameter, optimum frequency and amplitude, this plate distance offers the best heat transfer results.

Effect of jet taper angle: Taper angle of the jet (α) is varied from 5° to 11°. Average dissipated heat value increases with the taper angle. But the increase in taper angle is limited because of space constraint, as taper angle increases the area of jet inlet.

Comparison of optimized jet configuration with non-tapered jet: The maximum and average velocities are kept same. On comparing wall heat transfer coefficient shown in Fig. 4, it is seen that the intense red region for tapered jet is more than the non-tapered jet signifying the effect of tapered shape in increasing heat transfer coefficient.

A 47.89% increase in average heat transfer coefficient is observed for optimized jet configuration (frequency = 150 Hz, jet taper angle = 11°, plate distance from jet = 15 mm and jet length = 4 mm) as compared to that of a conventional jet.

Cooling of processor chip: For practical implementation of the present study, the heated plate is considered as a computer motherboard, and Intel i7 chip is placed over it. Since heat generation is not constant for the processor chip, chip was initialized with heat flux of 10,000 W/m^2 each at all four zones (not shown here for brevity) or at zones 2 and 4, while the rest of motherboard is kept at 2000 W/m^2 (Fig. 5) for both cross-flow and non cross-flow conditions. For each case, optimized synthetic jet is applied for cooling of the processor chip. Heat transfer coefficient using synthetic jet is found higher as compared to the conventional fan-based cooling (cross-flow) system.

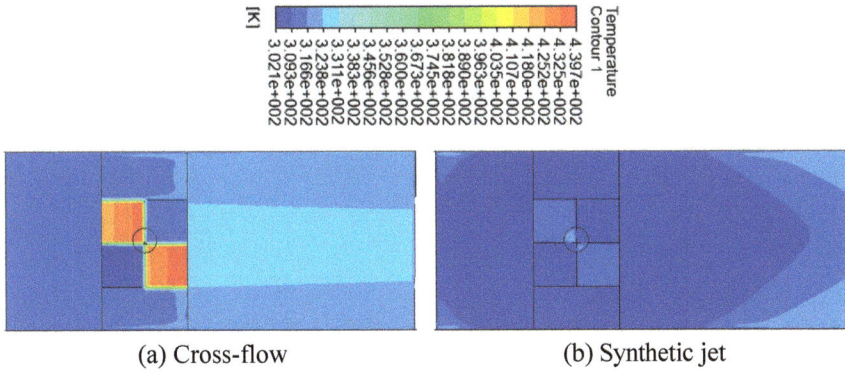

(a) Cross-flow   (b) Synthetic jet

Fig. 5 Temperature contours when zones 2 and 4 are heated

4 Conclusions

The synthetic jet assisted cooling is also found effective in removing hot spots (localized heating) on the processor chip. To depict this computationally, the chip region is divided into four zones, and the constant heat flux value is given for different combinations. The results of synthetic jet are compared with the cross-flow results at same heating configuration.

References

1. J.D. Narayanaswamy, T.T. Chandratilleke, Analysis of a synthetic jet-based electronic cooling module. Numer. Heat Transf., Part A: Appl. **56**(2018), 211–229 (2009)
2. B. Smith, A. Glezer, The formation and evaluation of synhetic jets. Phys. Fluids **10**(9), 2281–2297 (1998)
3. A.A. Hassan, R.D. Jannakiram, Effects of zero-mass synthetic jets on the aerodynamics of the NACA0012 airfoil. J. Amer. Helicopter SOC. **43**, 303–311 (1998)
4. T. Ming, Z. Zhou, Flow and heat transfer charecterictics under synthetic jet impingement driven by piezoelectric actuator. Int. J. Heat Mass Transf. **48**, 321–330 (2013)
5. Z. Zhang, G. Shan, T. Ming, Convective heat transfer on a flat plate subjected to normally synthetic jet and horizontally forced flow. Int. J. Heat Mass Tranf. **57**, 321–330 (2013)
6. E. Fanning, T. Persoons, D. Murray, Heat transferd flow characteristics of a pair of adjacent impinging synthetic jets. Int. J. Heat Mass Transf. **54**, 153–166 (2015)
7. S. Laouedj, A. Azzi, A. Benazza, New analysis in 3D synthetic jets within numerical investigation using CFD validation code: moving boundary techniques. Energy Procedia **19**, 226–238 (2012)

Thermal Bioeffect of Hybrid Microfluidic System Used for Particle and Cell Separation

Ali Mohammad Yazdani, Hossein Alijani, Arzu Özbey, Mehrdad Karimzadehkhouei, Ali Koşar, Alper Şişman, Emre Alpman, and Rana Altay

1 Introduction

Separation of circulating tumor cells (CTCs) is crucial for diagnosis and therapeutic purposes since metastasis is leading cause of cancer deaths. Many separation methods are proposed in the literature to purify the CTCs. The study analyzes the thermal bioeffect of hybrid system combining the passive and active cell separation system on human cells during separation process. The separation speed and sensitivity are optimized using the hybrid structure (Fig. 1). The passive and active systems are tested through simulation studies, and the design is optimized. The simulation studies and the results are reported in the paper. The fabrication method and characterization results are also presented by the study.

A. M. Yazdani (✉) · E. Alpman
Mechanical Engineering Program, Faculty of Engineering, Marmara University, Kadıköy, Istanbul 34722, Turkey
e-mail: aliyazdani20@gmail.com

H. Alijani · A. Özbey · M. Karimzadehkhouei · A. Koşar · R. Altay
Mechatronics Engineering Program, Faculty of Engineering and Natural Sciences, Sabanci University, Tuzla, Istanbul 34956, Turkey

A. Koşar
Center of Excellence for Functional Surfaces and Interfaces for Nano-Diagnostics (EFSUN), Sabanci University, Tuzla, Istanbul 34956, Turkey

A. Koşar · R. Altay
Sabanci University Nanotechnology and Applications Center (SUNUM), Tuzla, Istanbul 34956, Turkey

A. Şişman
Electrical and Electronics Engineering Program, Faculty of Engineering, Marmara University, Kadıköy, Istanbul 34722, Turkey

© Springer Nature Singapore Pte Ltd. 2021
C. Wen and Y. Yan (eds.), *Advances in Heat Transfer and Thermal Engineering* ,
https://doi.org/10.1007/978-981-33-4765-6_56

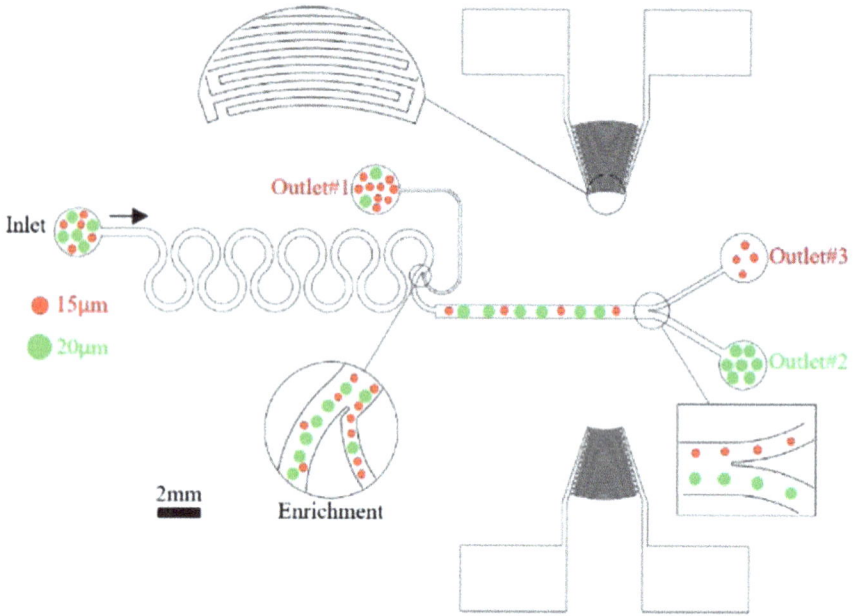

Fig. 1 Illustration of hybrid separation system

2 Methodology

Passive systems do not require any external field and use the internal flow dynamics like flow field and channel structure. These systems can sort particles using hydrodynamic forces, inertial forces, transient adhesion or filtration. The benefit of the passive systems is relatively high throughput, while they have poor separation sensitivity [1]. The passive system in our study increases the efficiency as a first stage to enrich the particles that are separated. It has a curvilinear structure (Fig. 1) that creates a velocity profile affecting the 3D position of the microparticles in each turn and accordingly the direction and magnitude of the forces acting on the microparticles, such as shear gradient lift force and Dean drag force.

Label-free active separation systems have four main methods, electrokinetics, optics, acoustophoresis and magnetophoresis. Magnetophoresis or electrokinetics separates the particle according to its magnetic or electrical properties, while acoustophoresis uses volume or density [2]. Acoustophoresis was used as the active system since volume and the shape are the characteristic property of the CTCs. Focused IDT electrodes generate a standing wave to create the nodes around the focal area. FIDT structure decreases the power requirement of the system and increases the speed. One main concern of using FIDT structure active cell separator is the local temperature increase in the microfluidic channel caused by parasitic power absorption and viscous damping which may lead to a thermal bioeffect [3].

The structure used in the experiments and simulations has eight nodes to observe the particle behavior. The microchannel was fabricated out of PDMS by SU-8 molding. The PDMS microchannel was bonded to the quartz wafer using plasma cleaner. Channel width and height are 400 μm × 200 μm, respectively. The system structure is simulated using COMSOL 5.3 Multiphysics. Passive structure simulation is performed by solving Navier–stokes equations under the uncompressible flow and single-phase assumptions. The channel is simulated as a PDMS channel wall and fluid flowing inside.

3 Results

The passive system experimental results show that the best flow rate for separation is 1900 μL/min which is the input stream flow rate of the active system. Since passive part has no external field acting on the system, there is no concern about temperature variations which could affect cells. The main concern of thermal side effect on cells is in active part. In Fig. 2, a temperature variation between $\pm 3 \times 10^{-4}$ K can be observed. Since the tolerable temperature for human cells is 33–39 °C [4], the temperature variation could be simply neglected, and there is no concern about thermal bioeffect of surface acoustic wave on separating cells.

It was observed from the simulations that the particles alignment duration is 5 s after SAW is on. Therefore, the separation capability of the overall system is demonstrated through numerical simulations. The fabrication of the active system and the electrical characterization is performed. The results show that, the operating frequency is 15.02 MHz as expected, and about 10 dB response is observed.

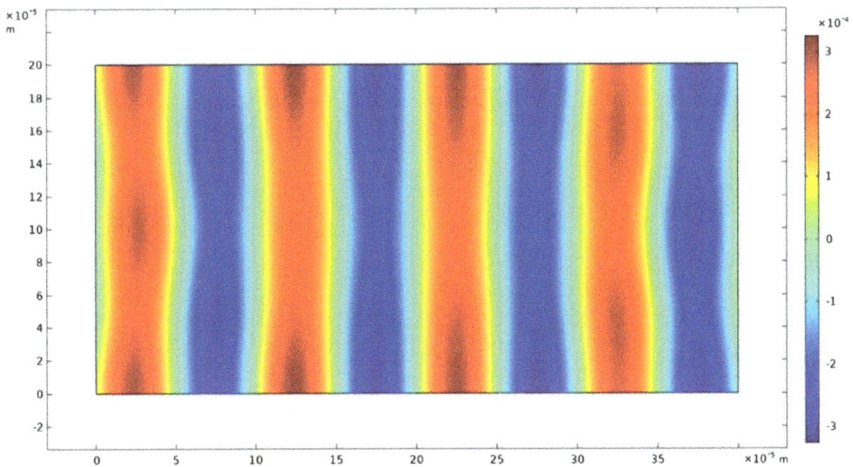

Fig. 2 Acoustic temperature profile in microfluidic channel

4   Conclusions

As a conclusion, the study proved that there is no concern about the thermal bioeffect of the proposed hybrid cell separation system and local temperature increase in the microfluidic channel caused by parasitic power absorption and viscous damping can be simply neglected in our system.

References

1. A. Özbey, M. Karimzadehkhouei, S. Akgönül, D. Gozuacik, A. Koşar, Inertial focusing of microparticles in curvilinear microchannels. Sci. Rep. **6**, 38809 (2016)
2. M. Toner, D. Irimia, Blood on a chip. Annu. Rev. Biomed. Eng. **7**, 77–103 (2005)
3. A. Winkler, R. Brünig, C. Faust, R. Weser, H. Schmidt, Towards efficient surface acoustic wave (SAW)-based microfluidic actuators. Sens. Actuators, A: Phys. **247**, 259–268 (2016)
4. M. Wiklund, Acoustofluidics 12: biocompatibility and cell viability in microfluidic acoustic resonators. Lab Chip **12**(11), 2018–2028 (2012)

Computational Heat Transfer

Numerical Simulation of Natural Convection in Solar Chimney

Hichem Boulechfar and Hadjer Bahache

1 Introduction

The solar chimney system is one of the applications that interest several countries it is already implemented and has shown success in the field. But the increase in the efficiency of the solar chimney has always been the subject of several studies [1]. Presented a numerical analysis on the performance of a solar chimney power plant using steady state Navier–Stokes and energy equations in cylindrical coordinate system. The fluid flow inside the chimney is assumed to be turbulent and simulated with the k–ε turbulent model, using the FLUENT software package [2]. Reports in their paper an experimental work to investigate effects of number of solar chimneys, chimney height, air gap width and chimney orientation on natural ventilation in a space [3]. Presented a numerical study of the natural convection in a solar chimney containing the air as the fluid which is considered as a Newtonian and incompressible and by using the Boussinesq approximation, the governing equations are taken to be in the vorticity-stream function formulation in hyperbolic coordinates [4]. Analyses the performance of a solar chimney power plant expected to provide the remote villages located in Algerian southwestern region with electric power.

Our work concerns the study of the natural convection inside a solar chimney where the fluid conveyed is the air. We have examined the effect of the temperature difference between the soil and the collector on the heat transfer inside the chimney. We performed a numerical simulation using CFD calculation code which is computing software based on the finites volumes method. The dimensional equations model is written in the Cartesian coordinates with consideration of some simplifying hypotheses. The general equations system which governs the phenomenon of natural convection is represented by continuity, momentum and heat equations.

H. Boulechfar (✉) · H. Bahache
Faculty of Sciences, Department of Physics, University of M'sila, Med BOUDIAF-BP 166, M'sila 28000, Algeria
e-mail: hichem.boulechfar@univ-msila.dz

© Springer Nature Singapore Pte Ltd. 2021
C. Wen and Y. Yan (eds.), *Advances in Heat Transfer and Thermal Engineering* ,
https://doi.org/10.1007/978-981-33-4765-6_57

2 Methodology

We considered a natural convection in a solar chimney, and the geometry is represented in Fig. 1. In order to simplify our problem, we have retained some simplifying hypotheses. The fluid is Newtonian and incompressible; we assume that the flow is two-dimensional, permanent and laminar. The viscous dissipation and the work of the forces of pressure are negligible in the equation of the heat. The density of the fluid varies linearly with the temperature according to the approximation of Boussinesq, and we consider the no exchange by radiation within the fluid.

The mathematical model equations are written in Cartesian coordinates with consideration of the previous hypotheses. The general model which governs the phenomenon of natural convection is represented by the following equations:

$$\frac{\partial u}{\partial x} + \frac{\partial v}{\partial y} = 0 \tag{1}$$

$$u\frac{\partial u}{\partial x} + v\frac{\partial u}{\partial y} = -\frac{1}{\rho_0}\frac{\partial p}{\partial x} + \upsilon\left(\frac{\partial^2 u}{\partial x^2} + \frac{\partial^2 u}{\partial y^2}\right) u\frac{\partial u}{\partial x} + v\frac{\partial u}{\partial y}$$

$$= -\frac{1}{\rho_0}\frac{\partial p}{\partial x} + \upsilon\left(\frac{\partial^2 u}{\partial x^2} + \frac{\partial^2 u}{\partial y^2}\right) \tag{2}$$

$$u\frac{\partial v}{\partial x} + v\frac{\partial v}{\partial y} = -\frac{1}{\rho_0}\frac{\partial p}{\partial y} + g\beta(T - T_0) + \upsilon\left(\frac{\partial^2 v}{\partial x^2} + \frac{\partial^2 v}{\partial y^2}\right) \tag{3}$$

$$u\frac{\partial T}{\partial x} + v\frac{\partial T}{\partial y} = \frac{\lambda}{\rho c_p}\left(\frac{\partial^2 T}{\partial x^2} + \frac{\partial^2 T}{\partial y^2}\right) \tag{4}$$

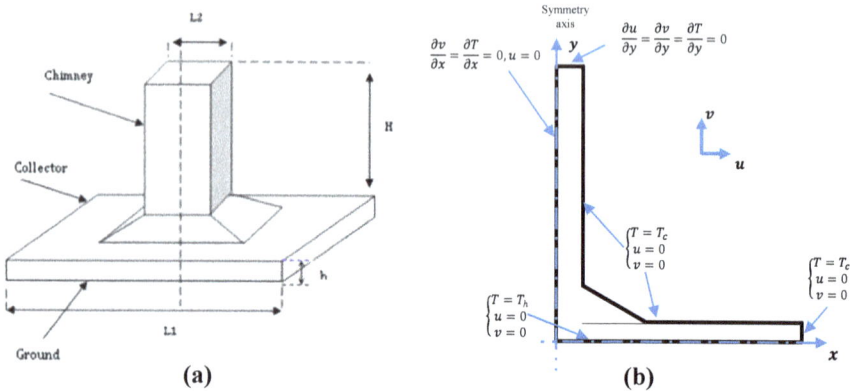

Fig. 1 a Geometry configuration of solar chimney, **b** boundary conditions for two dimensions geometry

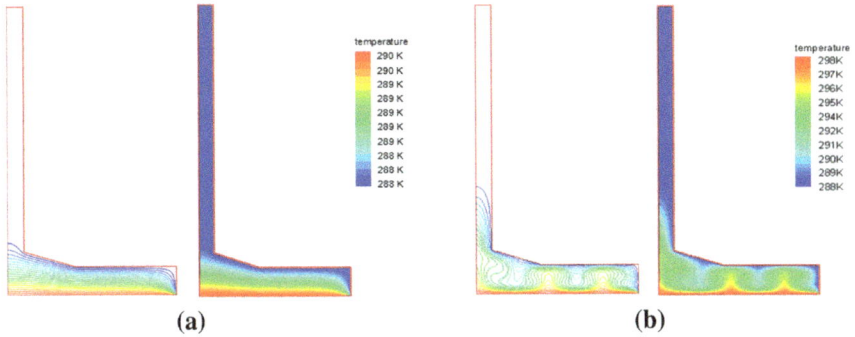

Fig. 2 Isotherms and temperature distribution in the solar chimney for two different gradients **a** ΔT = 5 K, **b** ΔT = 15 K

3   Results

We present the isotherms for different values of the temperature difference which corresponds to different values of the Rayleigh number which is the dimensionless number characterizing natural convection. We see in Fig. 2 that the isotherms are parallel lines and follow the profile of the walls; the temperature is simply decreasing from the ground which has the hot temperature towards the cold wall which is the roof of the collector. In this case, the heat transfer within the solar chimney is mainly dominated by pseudo-conduction. At the entrance of to the chimney, we notice that the isotherms slightly deform announcing a presence of very low intensity of a natural convection.

In the same figure, the increase of the thermal gradient generates an intensification of the natural convection which is interpreted by the distortion of the isotherms in all the space of the solar chimney where we can observe a phenomenon of Rayleigh–Bénard happening between the ground and the roof of the collector. This region has become an area of instability due to the extent of natural convection which intensifies with the increase of the temperature gradient in the solar chimney.

Our results are very similar in comparison with results of references [3] and [5] but for different geometries. In both cases, the authors used the finite volume method with direct programming using FORTRAN language. Their numerical results show that the variations of the isotherms are in perfect similarity to our results. Qualitatively both results are in very good agreement.

4   Conclusions

The two-dimensional laminar and permanent natural thermal convection in a solar chimney has been studied numerically using the CFD calculation code. The effect of the Rayleigh number, which is characterized by a temperature gradient on heat transfer within the solar chimney, was analyzed. For this, we have a fluid conveyed inside which is Newtonian and the flow is two-dimensional, laminar and permanent. We have chosen the Boussinesq approximation for the variation of the density of the fluid. The mathematical model is represented by the equations of continuity, momentum and the equation of heat. The finite volume method was used for the discretization of the equations performed by the simulation tool. Simulation results show that for low temperature gradients that are lower than $\Delta T \leq 5K$ which correspond to low Rayleigh number values, the isotherms represent parallel lines, and the temperature is simply decreasing from the ground to the collector roof where the heat within the solar chimney is mainly transferred by pseudo-conduction. For temperature gradients $5K \prec \Delta T \prec 10K$, the results show the simultaneous presence of two modes of heat transfer. In the right part of the collector, the heat transfer is dominated by pseudo-conduction. On the other hand, at the entrance of the vertical chimney, the natural convection intensifies but remains relatively weak. As the temperature gradient increases furthermore $\Delta T \succ 10K$, the natural convection increases significantly, and the rise of warm air in the chimney becomes obvious, the isotherms deform further throughout the solar chimney space, and the heat transfer is mainly by natural convection.

References

1. A. Dhahri et al., Numerical study of a solar chimney power plant. Res. J. Appl. Sci., Eng. Technol. **8**(18), 1953–1965 (2014)
2. S.A. Hassanein et al., Improvement of natural ventilation in building using multi solar chimneys at different directions. J. Eng. Sci., Assiut Univ. **40**(6), 1661–1677 (2012)
3. T. Tayebi, Laminar natural convection process in a solar chimney. Magister thesis of physics and renewable energies in 2010 at the University of Mentouri, Constantine, Algeria
4. S. Larbi et al., Performance analysis of a solar chimney power plant in the southwestern region of Algeria. Renew. Sustain. Energy Rev. **14**, 470–477 (2010)
5. T. Chergui, Flow modelling in the solar chimneys. Magister thesis in mechanical engineering in 2007 at National Polytechnic School, Algeria

Numerical Investigation on Heat Transfer Performance of Triply Periodic Minimal Surface Structures for Supercritical CO₂ Cycles

Weihong Li, Guopeng Yu, and Zhibin Yu

1 Introduction

The supercritical carbon dioxide (sCO$_2$)-based Brayton cycle is a proposed as an alternative to conventional power cycles due to high cycle efficiency, compact turbomachinery and compact heat exchangers. In the sCO$_2$ cycle, large portion of heat transfer (approximately 60–70% of total cycle heat transfer) occurs in the regenerator [1]. The present study proposes to utilize triply periodic minimal surface (TPMS) structures [2, 3] (gyroid and Schwarz D surface) as heat exchangers for increasing cycle efficiency. TPMS is a class of structures composed of two distinct inter-penetrating volume domains separated by an area-minimizing wall, which have been observed as biological membranes and co-polymer phases. Computational fluid dynamics modelling is performed to identify effects of geometrical properties, e.g. structure shape and dividing wall thickness, on the conjugate heat transfer performance and pressure drop. The results are further compared with the widely used printed circuit heat exchanger (PCHE) [4] and observed to have better overall thermohydraulic performance, which shows the potential to utilize TPMS for enhancing thermal efficiency of sCO$_2$ cycle (Fig. 1).

2 Methodology

The Reynolds-averaged continuity, momentum and temperature equations are given as follows.

$$\frac{\partial U_i}{\partial x_i} = 0 \tag{1}$$

W. Li · G. Yu · Z. Yu (✉)
School of Engineering, University of Glasgow, Glasgow G12 8QQ, UK
e-mail: Zhibin.Yu@glasgow.ac.uk

© Springer Nature Singapore Pte Ltd. 2021
C. Wen and Y. Yan (eds.), *Advances in Heat Transfer and Thermal Engineering* ,
https://doi.org/10.1007/978-981-33-4765-6_58

(a)   (b)

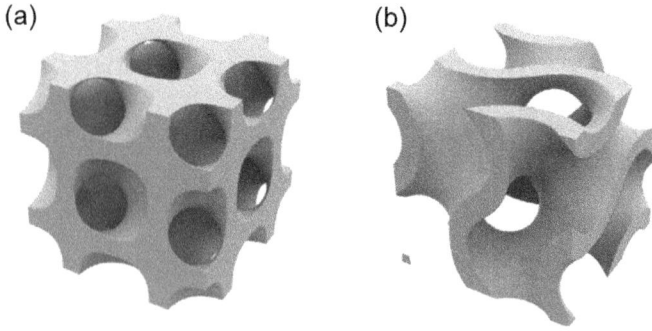

Fig. 1 Triply periodic minimal surface (TPMS) structures: **a** Gyroid channel and **b** Schwarz-D channel

$$\frac{\partial U_i U_j}{\partial x_j} = -\frac{1}{\rho}\frac{\partial P}{\partial x_i} + \frac{\partial}{\partial x_j}\left[\nu\left(\frac{\partial U_i}{\partial x_j} + \frac{\partial U_j}{\partial x_i}\right) - \overline{u_i u_j}\right] \tag{2}$$

$$\frac{\partial T U_j}{\partial x_j} = \frac{\partial}{\partial x_j}\left[\frac{\nu}{\mathrm{Pr}}\frac{\partial T}{\partial x_j} - \overline{u_{j\theta}}\right] \tag{3}$$

Employing the concepts of eddy viscosity and the Boussinesq assumption, the Reynolds stress and heat flux are approximated using equations which are given by

$$-\overline{u_i u_j} = \nu_t\left(\frac{\partial U_i}{\partial x_j} + \frac{\partial U_j}{\partial x_i}\right) - \frac{2}{3}\delta_{ij}k \tag{4}$$

$$-\overline{u_j\theta} = \frac{\nu_t}{\mathrm{Pr}_t}\frac{\partial T}{\partial x_j} \tag{5}$$

Within these equations, ν_t and Pr_t are the turbulent eddy viscosity and turbulent Prandtl number, respectively. Effects of turbulence are included by utilizing the Reynolds stress $\overline{u_i u_j}$, which is computed by means of transport equations in turbulence models. Different closure models, which are developed to close the RANS equations (and to compute the Reynolds stresses), often utilize different transport equations for fluctuating quantities. The overall goal of any such model is a more physically realistic flow description.

The commercial computer code ANSYS CFX 18 is employed for the study. The SST k–ω turbulence model is employed. The solvers for both codes employ finite volume solution methods for the momentum, energy and turbulence transport equations. High resolution is employed for both the advection scheme and for the turbulence numerical computations.

3   Results

Figure 2 shows the heat transfer coefficient and pressure drop variations for three heat exchangers with different Reynolds numbers. Apparently, the heat transfer coefficient increases with the increase of Reynolds number. The heat transfer coefficients for Gyroid channel and Schwarz-D channel are significantly higher than that of PCHE channel. At low Reynolds number, the Schwarz-D channel shows slightly higher heat transfer performance than the Schwarz-D channel. As the Reynolds number increases above around 34,000, the Gyroid channel produces higher heat transfer coefficients. For friction number along the channel, the Schwarz-D channel produces the highest friction number for all the investigated Reynolds number. And, the PCHE shows the lowest friction value for all the investigated Reynolds number.

To provide a clearer comparison between different channels, Fig. 3 shows the performance evaluation coefficient for hot channel (left) and cold channel (right) on the basis of PCHE. The performance evaluation coefficient is defined as $\frac{Nu/Nu(\text{PCHE})}{(f/f(\text{PCHE}))^{1/3}}$. For both hot and cold channel, the PEC values are generally higher than 1.2. In particular, the PEC values are higher than 1.2 and up to 1.8 at Reynolds number lower than 20,000. Meantime, it is also observed that Schwarz-D channel performs

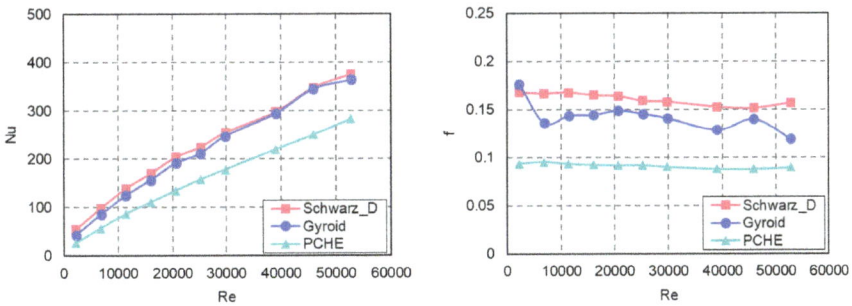

Fig. 2   Nusselt number and friction number in hot channel with varying Reynolds numbers

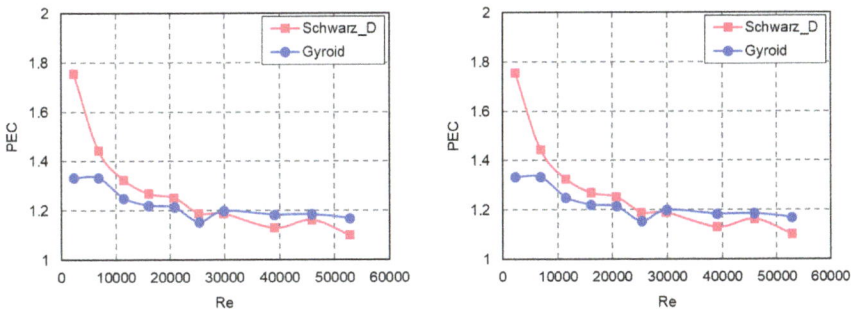

Fig. 3   Performance evaluation coefficient for hot channel (left) and cold channel (right) for Gyroid channel and Schwarz-D channel on basis of PCHE

better than Gyroid channel when Reynolds number is lower than 25,000, while Gyroid channel has superior performance at high Reynolds number.

4 Conclusions

The present study proposes two new heat exchangers, triply periodic minimal surface (TPMS) structures (gyroid and Schwarz D surface) for enhancing heat exchanger efficiency and decreasing pressure drop of sCO$_2$ cycle. The simulation results indicate that the performance evaluation coefficient can be increased by 20–80% compared with the PCHE. The complicated TPMS structures can be fabricated with additive manufacturing and utilized for enhancing thermal efficiency of sCO$_2$ cycle.

Acknowledgments This research is funded by EPSRC (EP/P028829/1) in United Kingdom.

References

1. A. Meshram, A.K. Jaiswal, S.D. Khivsara, J.D. Ortega, C. Ho, R. Bapat, P. Dutta, Modeling and analysis of a printed circuit heat exchanger for supercritical CO$_2$ power cycle applications. Appl. Therm. Eng. **109**, 861–870 (2016)
2. T. Femmer, A.J. Kuehne, M. Wessling, Estimation of the structure dependent performance of 3-D rapid prototyped membranes. Chem. Eng. J. **273**, 438–445 (2015)
3. N. Sreedhar, N. Thomas, O. Al-Ketan, R. Rowshan, H.H. Hernandez, R.K.A. Al-Rub, H.A. Arafat, Mass transfer analysis of ultrafiltration using spacers based on triply periodic minimal surfaces: effects of spacer design, directionality and voidage. J. Membr. Sci. **561**, 89–98 (2018)
4. D.E. Kim, M.H. Kim, J.E. Cha, S.O. Kim, Numerical investigation on thermal–hydraulic performance of new printed circuit heat exchanger model. Nucl. Eng. Des. **238**(12), 3269–3276 (2008)

Temperature-Dependent Conductances to Improve the Accuracy of the Dynamic Model of an Electric Oven

Michael Lucchi, Nicola Suzzi, and Marco Lorenzini

1 Introduction

Electric ovens, with a usual energy efficiency between 10 and 12%, rank among low-efficiency household appliances. The energy classification of such appliances is regulated by the EN 60350-1 European standard [1] which mandates proper control of the oven centre temperature; the design of suitable, advanced strategies thus represents a central field of investigation. In this work, a low-order dynamic model suitable for control design was developed to simulate the transient behaviour of an electric domestic oven in natural convective operation. When dealing with thermal problems, many numerical approaches can be used, spanning from the most simplified with a single lumped parameter, e.g. [2] to the more complex with high level of information in the spatial domain like those using the finite-elements method, [3]. A number of works aiming at defining the transient behaviour of household ovens, spanning from experimental, [2] to numerical, [4] with various degrees of complexity, yet most of them are either too cumbersome to be suitable for control design purposes, or give oversimplified descriptions. In this work, a lumped-parameter approach based on the thermoelectric analogy was used to keep computational cost low whilst granting an acceptable level of discretization. The main temperatures within the cavity were identified, in particular those of the oven centre, T_{OC}, and of the Pt500 sensor, which is used by the oven control board as the only temperature feedback during operation. A 11th order model including the cavity walls, the glass door, the heaters, the oven centre, and the Pt500 sensor was obtained, introducing nonlinearized terms to model the radiative heat exchanges within the cavity. In order to evaluate the effects

M. Lucchi (✉) · M. Lorenzini
Department of Industrial Engineering, University of Bologna, via Fontanelle 40, Forlì, Italy
e-mail: michael.lucchi2@unibo.it

N. Suzzi
DPIA-Dipartimento Politecnico di Ingegneria e Architettura, University of Udine, Via delle Scienze, Udine, Italy

© Springer Nature Singapore Pte Ltd. 2021
C. Wen and Y. Yan (eds.), *Advances in Heat Transfer and Thermal Engineering* ,
https://doi.org/10.1007/978-981-33-4765-6_59

on model accuracy of the temperature dependence of the heat transfer coefficients in natural convection, both constant and nonlinear conductances were determined through an optimization procedure based on experimental data, leading to two semi-physical grey-box models, whose predictive capabilities were tested under different temperature set-points and compared.

2 Methodology

An experimental campaign aimed at measuring temperature and electric power data for the design of a "grey-box" model was conducted on a commercial household oven with a 72 dm^3 cavity. The oven has three heaters, but only two, a 2300 W top heater and a 1000 W bottom heater, are activated in the heating mode investigated. Thermocouples were used to get the temperatures of the main elements (Fig. 1) and of the air at the geometric centre of the cavity. The temperature sensed by the Pt500 probe was also recorded. The power absorbed by the heaters was metered (one metre for each heater and one for the overall consumption). Three temperature set-points T_{set} were investigated, as per prescription of [1], namely 160, 200, and 240 °C, keeping temperature in the laboratory at 23 °C.

The model is based on a lumped-parameter approach which uses the thermoelectric analogy to simulate the heat transfer among the main elements of the cavity.

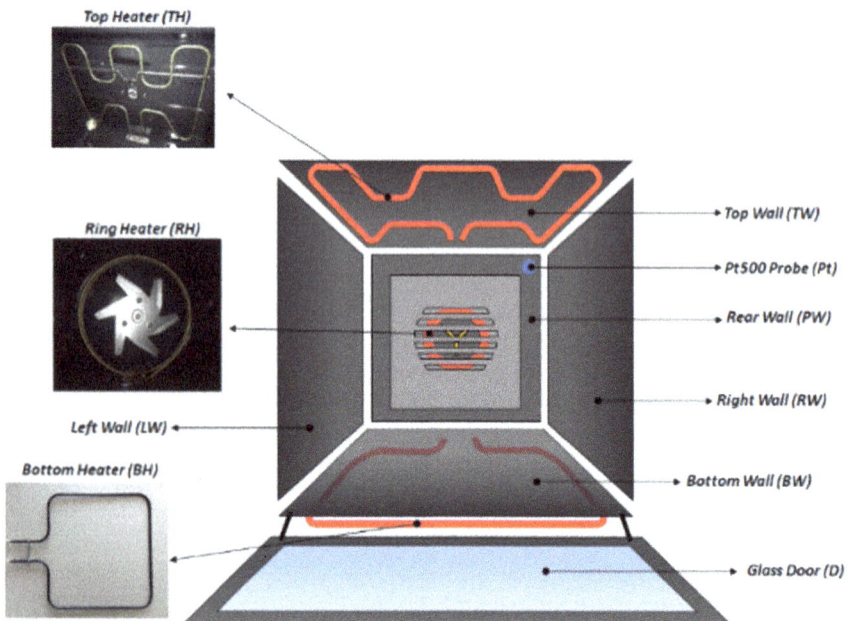

Fig. 1 Sketch of the oven with the main elements of the cavity

Equation (1) reports the energy balance for the generic ith node, where C_i is the thermal capacity, $\dot{Q}_{ij}$ and $\dot{Q}_{ij}^{\text{rad}}$ are the convective–conductive and radiative heat exchange between two generic nodes i and j, and $W_{\text{el}\,i}$ is the electric power absorbed (nonzero for the heaters only).

$$C_i \frac{dT_i}{d\tau} = \sum_{j=1}^{n} (\dot{Q}_{ij} + \dot{Q}_{ij}^{\text{rad}}) + W_{\text{el}\,i} \tag{1}$$

Detailed information on the connections between the main elements can be found in [5]. The convective–conductive terms can be expressed as $\dot{Q}_{ij} = G_{ij}(T_j - T_i)$, where G_{ij} is the thermal conductance between nodes i and j; the radiative heat exchange is calculated as $\dot{Q}_{ij}^{\text{rad}} = G_{ij}^{\text{rad}}\sigma\left(T_j^4 - T_i^4\right)$, where T_i and T_j are the node temperatures in kelvin. G_{ij}^{rad} is the radiative thermal conductance.

As explained in detail in [5], thermal capacities and conductances were determined through an optimization procedure based on the experimental data acquired at $T_{\text{set}} = 200\,°C$. In this work, also the influence of temperature on the convective heat transfer in the cavity was investigated. In particular, for the connections between air and the walls and the heaters, G_{ij} was written as reported in Eq. (2), where G_{0ij} and a_{ij} are optimized constants.

$$G_{ij} = G_{0ij} + a_{ij}\frac{T_i + T_j}{2} \tag{2}$$

3 Results

One example of the outcome of the optimization procedure is shown in Fig. 2a, for the case of temperature-independent conductances. Air temperature at cavity centre, T_{OC}, is predicted within 1% when steady state is reached. During transient operation, the difference is within 2%, except at the very start, owing to much lower values of air temperature. Predictive use of the model for T_{OC} is shown in Fig. 2b, where the maximum difference with the experimental data is 10%. As the green, dashed-line plot shows, no significant influence of temperature on thermal conductances was found, as it almost perfectly superposes with the black dashed line, which refer to temperature-independent conductances. This would seem to indicate that no significant overall variation of the convective heat transfer coefficient with temperature occurs for the range of T_{set} investigated.

The model showed good prediction capabilities also for the other elements of the thermoelectric grid. In Fig. 3a, the results obtained at $T_{\text{set}} = 240\,°C$ for the Pt500 sensor are reported, showing how the percentage deviation between numerical and experimental data always lies under 10%. The low influence of temperature-dependent conductances is confirmed. The capability of the model in predicting the

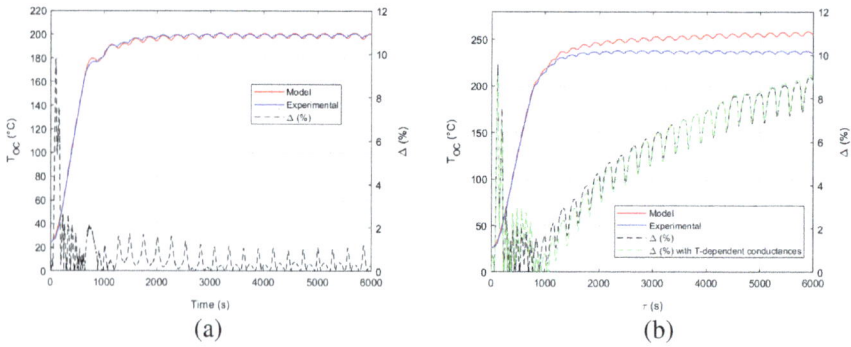

Fig. 2 **a** Comparison between experiment and numerical data during the optimization at T_{set} = 200 °C for T_{OC}. **b** Prediction for the same location at T_{set} = 240 °C and influence of T on conductances

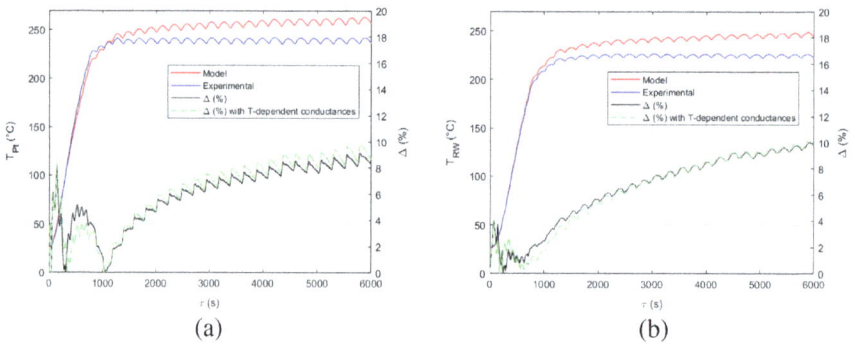

Fig. 3 **a** Comparison between experiment and numerical data at T_{set} = 240 °C for T_{Pt} and influence of T on conductances. **b** Prediction under the same test conditions for the right wall (RW) and effect of T on conductances

temperature of the oven centre and of the Pt500, which is the only temperature feedback during real operation, ensures its usefulness in control design. Figure 3b shows the numerical and experimental trend of the mean temperature of the right wall under the same temperature set-point. In this case too, the percentage deviation is lower than 10% and is not changed appreciably by temperature-dependent conductances, confirming that the convective heat transfer coefficients do not vary conspicuously with the operating temperatures of oven.

The prediction capability of the model was also investigated at temperature set-points lower than 200 °C, at which the parameters identification was carried out. Figure 4a, b shows the results obtained for the oven centre and the Pt500 sensor at T_{set} = 160 °C. Differently from the test at 240 °C, the model underestimates the temperatures of the two nodes of the grid. Excluding the peaks at the beginning of the simulation, the percentage deviation is always lower than 10% for the oven centre and lower than 8% for the Pt500 sensor.

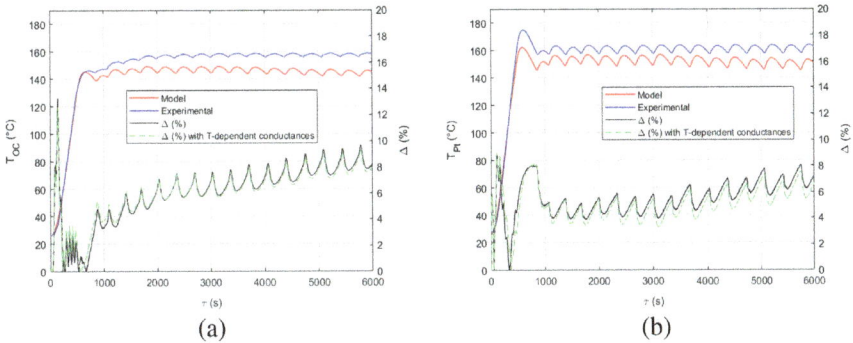

Fig. 4 **a** Comparison between experiment and numerical data at $T_{set} = 160$ °C for T_{OC} and influence of T on conductances. **b** Prediction under the same test conditions for the Pt500 sensor and effect of T on conductances

4 Conclusions

One grey-box, lumped-parameter model of a household oven was obtained, based on experimental data and conductance optimization. The predictive capabilities of the model were proved, but no significant influence of temperature on conductances was found.

References

1. CENELEC 2013 EN 60350-1 Household electric cooking appliances Part 1: Ranges, ovens, steam ovens and grills. Methods for measuring performance (latest version including all amendments)
2. J. Abraham, E. Sparrow, A simple model and validating experiments for predicting the heat transfer to a load situated in an electrically heated oven. J. Food Eng. **62**, 409–415 (2004)
3. M. Lucchi, M. Lorenzini, Effects of pipe angular velocity and oven configuration on tube temperature distribution in the radiative heating of PVC pipes. Int. J. Energy Environ. Eng. **9**(2), 123–134 (2018)
4. H. Mistry, S. Ganapathisubbu, H. Dey, P. Bishnoi, J. Castillo, Modeling of transient natural convection heat transfer in electric ovens. Appl. Therm. Eng. **26**, 2448–2456 (2006)
5. M. Lucchi, M. Lorenzini, Control-oriented low-order models for the transient analysis of a domestic electric oven in natural convective mode. Appl. Therm. Eng. **147**, 438–449 (2019)

Numerical Simulation on Combustion Simulation Based on Soft-Measuring Technique

Yingai Jin, Mingyu Quan, Shijuan Yan, Yuying Yan, and Jiatong Guo

1 Introduction

Power plants have their own corresponding designed coal types, however, in practice, typically in China; they normally cannot burn a single type of coal for long term as the cost of transportation and limited reserves of coal needed to be considered. Changing coal type or quality during the operation period of a boiler will likely decrease boiler combustion thermal efficiency and result in a series of problems, such as combustion instability, ash accumulation and slagging, tube explosion.

In this paper, a soft-measuring model of coal quality was developed by first finding the corresponding relationship between flue gas composition and coal quality and then combining heat balance equations with coal quality elements, which can obtain real-time data of coal composition. In addition, the software program for online measurement of coal quality was created. Input the percentage of CO_2, H_2O, O_2, SO_2 in flue gas and other physical parameters such as atmospheric density and air humidity into the interface, then the coal type used in the boiler at this time can be determined.

According to the volatile content and low calorific value of different types of coal, targeted boiler air distribution was carried out. Numerical simulation software was used to conduct simulation analysis on different combustion conditions of the boiler,

Y. Jin · M. Quan · S. Yan (✉)
State Key Laboratory of Automotive Simulation and Control, Jilin University, Changchun 130022, China
e-mail: 1779157087@qq.com

College of Automotive Engineering, Jilin University, Changchun 130022, China

Y. Yan
Faculty of Engineering, University of Nottingham, Nottingham NG7 2RD, UK
e-mail: Yuying.Yan@nottingham.ac.uk

J. Guo
Changchun HechengXingye Energy Technology Co., Ltd., Changchun 130021, China

© Springer Nature Singapore Pte Ltd. 2021
C. Wen and Y. Yan (eds.), *Advances in Heat Transfer and Thermal Engineering* ,
https://doi.org/10.1007/978-981-33-4765-6_60

and the boiler thermal efficiency and NO_X emissions before and after air distribution were compared to judge the optimization effect of air distribution.

2 Methdology

As shown in Fig. 1, input the flue gas composition, atmospheric density, moisture content and other parameters in the left column, click the calculate button, then the coal quality information of the coals burning at this time in the boiler can be output on the right column, compare the volatile matter and low calorific value of coal with the designed coal and determine the air distribution scheme.

When burning Hesgwula coal, because its volatile matter is higher than that of designed coal, the primary air velocity should be increased to ensure the oxygen required for pulverized coal combustion. While the calorific value is low, the coal consumption should be increased, the secondary air velocity should be increased, the flue gas re-circulation should be increased, so that the heat loss of combustion can be reduced. Similarly, for Shaertala coal, the volatile matter is lower than that of designed coal, the combustion is unstable, and fire is easily extinguished so the primary air velocity should be reduced. Because of the low calorific value of Shaertala coal, the volume of high speed secondary air should be increased to mix strongly with the burning pulverized coal flow, increase oxygen concentration and intensify combustion. For Gepu coal in Shanxi Province, the primary air velocity should be reduced as the volatile content is lower than that of the designed coal, and

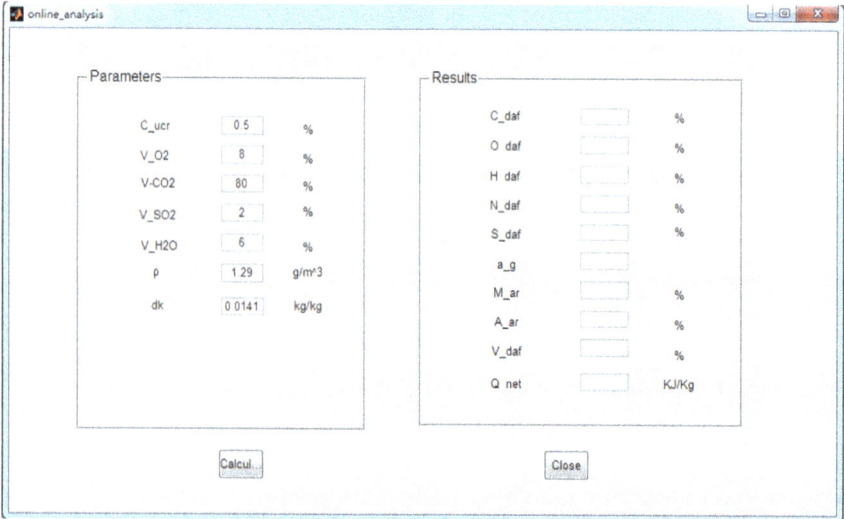

Fig. 1 Coal quality online analysis program interface

the secondary air velocity should be reduced as the low calorific value is higher than that of the designed coal.

3 Results

Five kinds of coal commonly used in power plants were selected to compare the actual and calculated values of volatile matter and low calorific value. The results are shown in Fig. 2. According to Fig. 2, the relative error of V_{daf} is less than 3%, and the relative error of $Q_{\mathrm{net,ar}}$ is less than 5%. That is to say, the mathematical model of online calculation of coal quality used in this paper can be applied in industry.

Table 1 shows the thermal efficiency and NO_x emission of the boiler before and after air distribution when the boiler burns Hesgwula coal, Shaertala coal and Gepu coal in Shanxi Province. According to Table 1, after air distribution, the boiler efficiency can be increased by at least 0.5%, and the NO_x emission can be reduced by at least 5.9%.

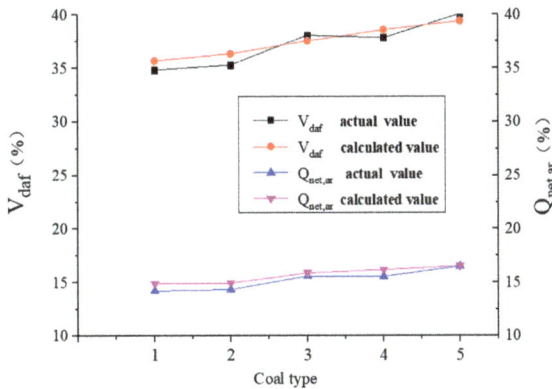

Fig. 2 Relative error of V_{daf}, $Q_{\mathrm{net,ar}}$

Table 1 Results of wind distribution optimization

Type of coal	Hesgwula		Shaertala		Gepu coal in Shanxi Province	
	Before	After	Before	After	Before	After
Thermal efficiency (%)	87.74	88.34	88.16	88.67	87.98	88.74
Emission of NO_x (mg/m^3)	672	628	591	556	589	540

4 Conclusions

Comparing with the actual data, the error of coal quality information calculated by the soft-sensing model in this paper is about 5%. Adjusting the air distribution according to different coal qualities can improve the efficiency of the boiler by at least 0.5% and reduce the emission of NO_x by at least 5.9%. Determining the coal type through online analysis and adjusting air distribution contrapuntally can achieve efficient, economical and clean combustion of boilers.

Acknowledgements 1. This work was supported by the Royal Academy of Engineering under the UK-China Industry Academia Partnership Programme scheme (UK-CIAPP\201).

2. This work was supported by Jilin Province Science and Technology Development Plan Project (No. 20180414021GH).

Bibliography

1. X.C. Xu, H.F. Lv, H. Zhang, *Combustion Theory and Combustion Equipment* (Science Press, Beijing, 2011).
2. G.F. Liu, W.D. Hao, X.G. Han, Y.Q. Guo, Model of monitoring coal grade for utility boiler basing on flue gas compositional measurement. J. Combustion Sci. Technol. (05), 441–445 (2002)
3. D.W. Pershing, J.O.L. Wendt, Pulverized coal combustion: the influence of flame temperature and coal composition on therm and fuel NO_X. Symp. Combustion **16**(1), 389–399 (1997)
4. W.P. Jin, Formation mechanism and control measures of fuel-type NOx. China Sci. Technol. Inf. (22), 20+29 (2005)

Numerical Investigation of Mass Redistribution in Supercritical Water-Cooled Reactor Flow Channels

Zhirui Zhao, Yitung Chen, Baozhi Sun, Jianxin Shi, Xiang Yu, and Wanze Wu

1 Introduction

Supercritical water-cooled reactor (SCWR), as the only water-cooled reactor of six Generation-IV reactor systems, has obtained a significant attention. SCWR is designed as a once-through water-cooled reactor system because water as the working fluid is considered as a single-phase fluid under the supercritical pressure. Due to the different types sub-channel in the fuel assembly and the dramatic changes of the physical properties, the circumferential non-uniformity occurs. Gang et al. [1] carried out an experimental study on the heat transfer characteristics of supercritical water in vertical annular channels and summarized the effect of heat flux, mass flux, operating pressure and geometry parameters on wall temperature and heat transfer coefficient. Chen et al. [2] investigated the circumferential heat transfer behavior in a tight hexagonal 19-rod bundle by using supercritical R134a as the working fluid. The circumferential non-uniform distribution of wall temperature was found to be strongly related to the variation of the local hydraulic diameter which was defined to represent the heterogeneity of channels in the tight bundle.

Many other studies [3–5] are mentioned about the circumferential non-uniformity of the flow and heat transfer process in the rod fuel bundle. Some of them considered that the mass redistribution is the main reason that causes the non-uniformity. In this study, two adjacent sub-channel is selected as the physical model to investigate the mass flux distribution along the rod in each sub-channel, and the effect of mass flux, heat flux and operation pressure on mass redistribution is discussed, in order to provide the reference for the design and operation of SCWR.

Z. Zhao · B. Sun (✉) · J. Shi · X. Yu · W. Wu
Harbin Engineering University, 145 Nantong Street, Nangang District, Harbin, China
e-mail: sunbaozhi@163.com

Y. Chen
Department of Mechanical Engineering, University of Nevada Las Vegas, 4505 S. Maryland Parkway, Las Vegas, USA

© Springer Nature Singapore Pte Ltd. 2021
C. Wen and Y. Yan (eds.), *Advances in Heat Transfer and Thermal Engineering* ,
https://doi.org/10.1007/978-981-33-4765-6_61

2 Methodology

The flow and heat transfer of supercritical water in two adjacent different type sub-channels is investigated in this study. Supercritical water is considered as a single-phase fluid, and the flow process is regarded as a three-dimensional flow of compressible Newtonian fluid, which involves the coupling process of flow and heat transfer. Considering the anisotropy of turbulence and solving Reynolds stress directly, Reynolds stress model (RSM) is selected in this study. Meanwhile, enhanced wall treatment with high requirement for near-wall mesh was selected to ensure Y^+ value is around 1. Based on the finite volume method, the convection term, the diffusion term and the pressure gradient term in the governing equation are discretized by the second-order upwind style, the central difference scheme and the discrete form based on the least squares method. The coupling between pressure and speed is performed by the simple algorithm.

3 Results

Figure 1 shows the dimensionless mass flux distribution of interior sub-channel and side sub-channel under different mass flux. As can be seen, the mass flux of interior sub-channel decreases slightly, then increases to a peak point and decreases again until the exit. The mass flux changes opposite in the side sub-channel. Besides, as the mass flux increases, the magnitude of mass redistribution decreases.

Figure 2 shows the mass flux distribution of each sub-channel under different heat flux. It can be seen from the figure that since the inlet mass flow rate is the same, the maximum and minimum values of the mass flow rate of each channel under different heat flux densities are unchanged, but the position appears as the heat flux increases, so the heat flux only changes the speed of mass redistribution. When the heat flux density reaches $800 \, \text{kW} \cdot \text{m}^{-2}$, it can be found that the fluid starts to flow from the side

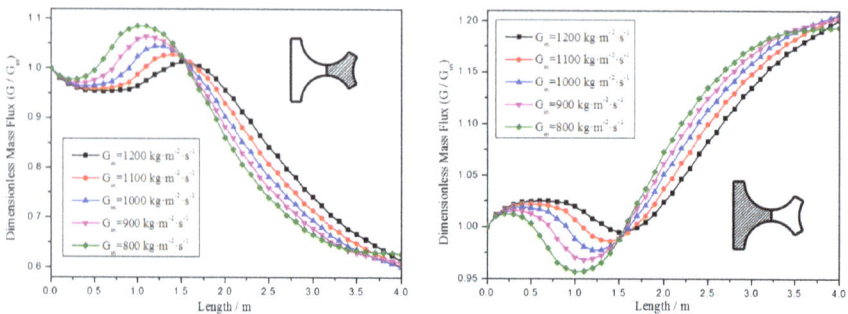

Fig. 1 Axial distribution of mass flux with different inlet mass flux in interior sub-channel and side sub-channel

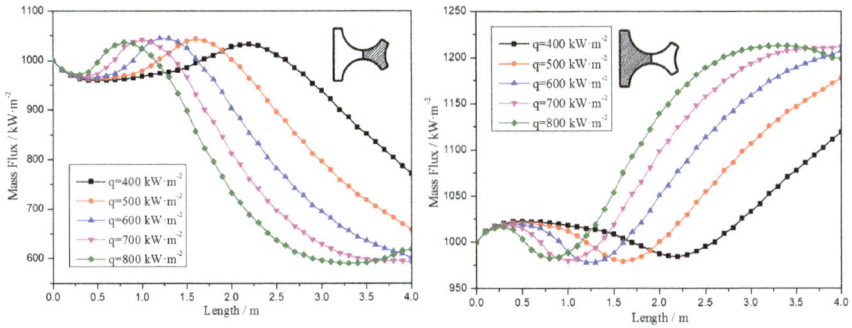

Fig. 2 Axial distribution of mass flux with different heat flux in interior sub-channel and side sub-channel

sub-channel to the interior sub-channel near the outlet, because at this time the fluid temperature of the side channel reaches the pseudo-critical temperature, the volume begins to expand rapidly, and the fluid temperature of the interior sub-channel has been out of the large specific heat area.

4 Conclusions

The conclusions are as follows.

(1) A numerical investigation of mass redistribution in two adjacent sub-channels is carried out.
(2) Effects of mass flux, heat flux and operation pressure on the mass redistribution are studied.
(3) The influence of mass redistribution on circumferential non-uniformity along the flow direction is discussed.

Acknowledgements This work was sponsored by the National Natural Science Foundation of China (No. 51579048, No. 51709249) which we gratefully acknowledge.

References

1. W. Gang, Q. Bi, Z. Yang et al., Experimental investigation of heat transfer for supercritical pressure water flowing in vertical annular channels. Nucl. Eng. Des. **241**, 4045–4054 (2011)
2. J. Chen, H. Gu, Z. Xiong et al., Experimental investigation on heat transfer behavior in a tight 19 rod bundle cooled with supercritical R134a. Ann. Nucl. Energy **115**, 393–402 (2018)
3. J. Yang, Y. Oka, Y. Ishiwatari et al., Numerical investigation of heat transfer in upward flows of supercritical water in circular tubes and tight fuel rod bundles. Nucl. Eng. Des. **237**, 420–430 (2007)

4. Z.X. Hu, H.B. Li, J.Q. Tao et al., Experimental study on heat transfer of supercritical water flowing upward and downward in 2 × 2 rod bundle with wrapped wire. Ann. Nucl. Energy **111**, 50–58 (2018)
5. Z.X. Hu, H.Y. Gu, Heat transfer of supercritical water in annuli with spacers. Int. J. Heat Mass Transf. **120**, 411–421 (2018)

Numerical Study of the Impacts of Forced Vibration on Thermocapillary Bubble Migration in a Rotating Cylinder

Fatima Alhendal and Yousuf Alhendal

1 Introduction

In a non-uniform temperature gradient fluid medium, the surface tension varies according to the local temperature conditions. The surface tension in the cold fluid region is greater than that in the hot region. Therefore, in a zero-gravity stagnant fluid of a non-uniform temperature field, this difference in surface tension creates a net force acting on the fluid particles, which leads to a general fluid motion from the hot region to the cold region. This phenomenon is known as Marangoni or thermocapillary migration phenomenon. It is necessary to carry out an appropriate numerical simulation on the behavior of bubbles under the influence of forced vibration and rotation in a microgravity fluid medium in order to achieve a better understanding of this complicated phenomenon. Clearly, an in-depth understanding of the flow patterns, including the shape and the area of the varying complex interfaces, is of vital importance to the understanding and prediction of the physics behind these flow systems and as the flow patterns of some regimes remain largely undiscovered; consequently, accurate predictions of flow patterns are highly desirable. This is the goal of the present study, wherein this microgravity phenomenon is numerically investigated in a three-dimensional domain (3D).

F. Alhendal (✉)
American University of the Middle East (AUM), State of Kuwait, Kuwait, Kuwait
e-mail: fa_alhendal@gmail.com

Y. Alhendal
College of Technological Studies (CTS), Public Authority for Applied Education and Training (PAAET), Kuwait, Kuwait

© Springer Nature Singapore Pte Ltd. 2021
C. Wen and Y. Yan (eds.), *Advances in Heat Transfer and Thermal Engineering* ,
https://doi.org/10.1007/978-981-33-4765-6_62

2 Methodology

The thermocapillary motion of a bubble was first examined experimentally by Young, Block, and Goldstein [1], when Reynolds number (Re) and Marangoni number (Ma) are small, which means that both convective momentum and energy transport are negligible, who also found an analytical expression for its terminal velocity in the creeping flow:

$$V_{YGB} = \frac{2|d\sigma/dT|r_b\lambda dT/dx}{(2\mu + 3\mu')(2\lambda + \lambda')} \tag{1}$$

commonly called the YGB model, which is suitable for small Reynolds and Marangoni numbers:

$$Re_T = r_b V_T / v \tag{2}$$

$$Ma_T = r_b V_T / \alpha = Re_T \cdot Pr \tag{3}$$

where Prandtl number is the ratio of kinematic viscosity to thermal diffusivity:

$$Pr = v/\alpha \tag{4}$$

and v is the kinematic viscosity in m^2/s: $v = \mu/\rho$.

The velocity V_T derived from the tangential stress balance at the free surface is used for scaling the migration velocity (m/s) in Eqs. (2) and (3):

$$V_T = \frac{(d\sigma/dT) \cdot (dT/dx) \cdot r_b}{\mu} \tag{5}$$

where μ and μ', λ and λ' are the dynamic viscosity and thermal conductivity of continuous phase and bubble, respectively. ρ is the density and r_d is the radius of the bubble. The constant $d\sigma/dT$ or σ_T is the rate of change of interfacial tension, and dT/dx is the temperature gradient imposed in the continuous phase fluid.

3 VOF Model and Computational Procedure

The governing continuum conservation equations for two-phase flow were solved using the Ansys Fluent commercial software package [2], and the volume of fluid (VOF) method was used to track the liquid/gas interface. This method deals with completely separated phases with no diffusion. The geometric reconstruction scheme, based on the piece-wise linear interface calculation (PLIC) method of Young's [3] in Ansys Fluent, was chosen for the current investigation. Geo-reconstruction is an added module to the already existing VOF scheme that allows for a more accurate definition of the free surface [4].The movement of the gas–liquid interface is tracked based on the distribution of the volume fraction of the gas, i.e.,α_G, in a computational cell, where the value of α_G is 0 for the liquid phase and 1 for the gas phase. Therefore, the gas–liquid interface exists in the cell where α_G lies between 0 and 1. A single momentum equation, which is solved throughout the domain and shared by all the phases, is given by:

$$\frac{\partial}{\partial t}(\rho\vec{v}) + \nabla \cdot (\rho\vec{v}\vec{v}) = -\nabla p + \nabla \cdot [\mu(\nabla\vec{v} + \nabla\vec{v}^{T})] + \vec{F} \tag{6}$$

In Eq. 6, $\vec{F}$ represents volumetric forces at the interface, resulting from the surface tension force per unit volume. The continuum surface force (CSF) model proposed by Brackbill et al. [5] is used to compute the surface tension force for the cells containing the gas–liquid interface:

$$\vec{F} = \sigma \left(\rho k \hat{n} \bigg/ \frac{1}{2}(\rho_L + \rho_G) \right) \tag{7}$$

where σ is the coefficient of surface tension,

$$\sigma = \sigma_0 + \sigma_T(T_0 - T) \tag{8}$$

and σ_0 is the surface tension at a reference temperature T_0, and T is the liquid temperature.

4 Results

A constant temperature gradient of 0.208 K/mm was prescribed for each simulation, and the corresponding thermal Reynolds (Re_T) and Marangoni (Ma_T) numbers for all cases were set to 257 and 4188, respectively. The thermocapillary flow pattern and trajectory of a single gas bubble in a stationary, vibrated and rotating cylinder are illustrated in the figures below. These two cylinders have the same vibration amplitude and frequency, $A_P = 0.02$ m/s^2, $f = 0.2$ Hz. Thus, the bubble path in the non-rotary container adhered to the wall and then went up toward the higher temperature. In a vibrating rotating cylinder, which vibrates with the same vibration capacity, the bubble migrates toward the hotter side and rotates around the axis of rotation due to the angular speed of the container. By comparing the two forms, we can found that there are three major forces that affect the flow of the bubble, namely vibration, rotation and thermocapillary forces. More detailed on the effect of these three forces on the bubble movement in the lack of gravity will be addressed in the next part.

Fig. 1 Bubble oscillations around the X-coordinate, $d_b = 9$ mm

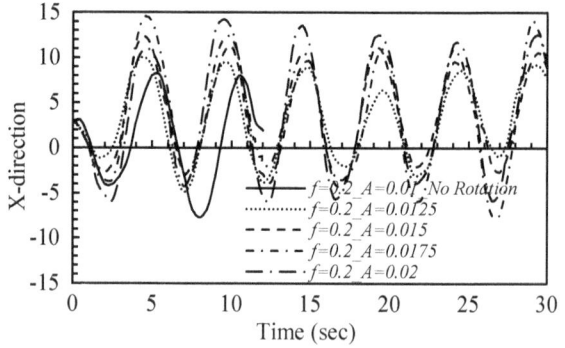

Fig. 2 Bubble migration toward the hotter side, $d_b = 9$ mm

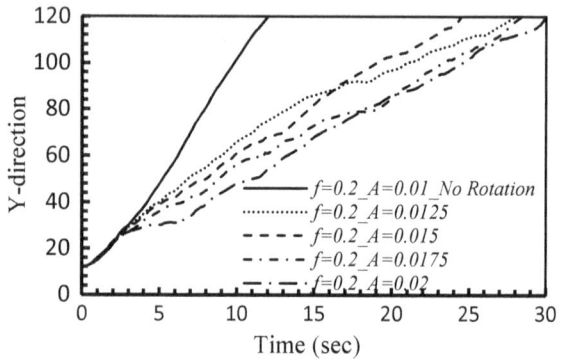

As amplitude gets larger, bubble goes from small to longer oscillations around the axis of rotation and toward the hotter side demonstrating the effect of the three forces on the bubble shape and behavior. Figures 1 and 2 illustrate that at the vibration amplitude selected and beyond, can cause a major reduction in the bubble velocity toward the hotter side and becomes considerable at higher vibration amplitude, and affects the forces acting on and consequently the translational motion of the bubbles. Figures 3 and 4 show the results of the numerical simulations of stationary and vibrating/rotating cylinder. In other words, increasing the speed of rotation attracts the bubble in the direction of the axis of rotation, while increasing the vibration amplitude, it distancing it from the axis of rotation and vice versa. In the case of vibration with rotation, the bubble remains rotating around the axis of rotation and does not center the cylinder.

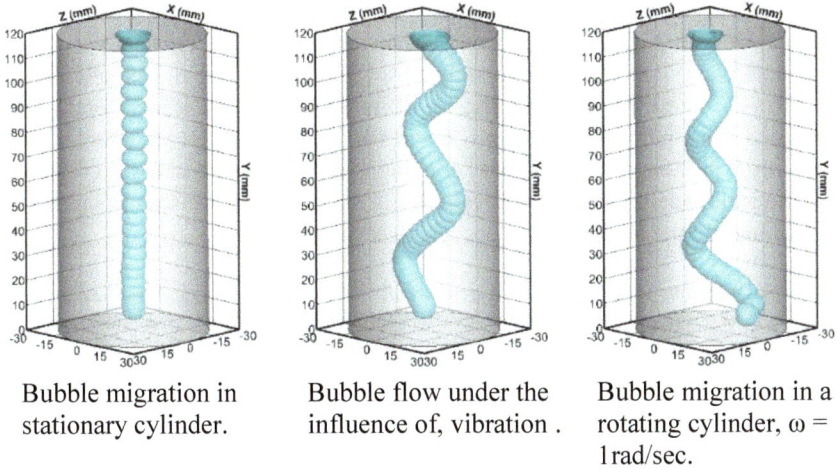

| Bubble migration in stationary cylinder. | Bubble flow under the influence of, vibration . | Bubble migration in a rotating cylinder, ω = 1rad/sec. |

Fig. 3 Bubble dynamics in stationary, vibrating and rotating cylinder $d_b = 9$ mm, $\nabla T = 0.208$ K/mm

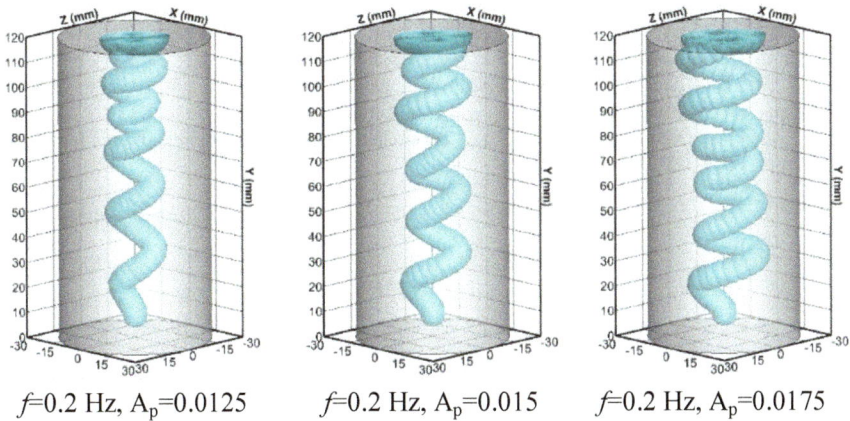

| f=0.2 Hz, A_p=0.0125 | f=0.2 Hz, A_p=0.015 | f=0.2 Hz, A_p=0.0175 |

Fig. 4 Bubble dynamics in a vibrating/rotating cylinder, $d = 9$ mm, $\omega = 1$ rads^{-1}

5 Conclusions

The thermocapillary flow of isolated bubble in a vibrating/rotating container was studied for single bubble migrating with different amplitudes. Small vibrations/rotations aboard space platforms can have significant effects on bubble movement. Conducting future experiments on this topic aboard the space station will be of interest to us since better understanding of the vibration effects will help us better understand the onset of oscillation mechanisms and ways to design g-jitter

resistant thermocapillary bubble migration in the future. Since zero gravity is difficult to achieve in a laboratory setting, one can demonstrate the relevant phenomena
using numerical simulations. Simulating these phenomena also allowed one to study
the effects of altering the sensitivities of different parameters. It therefore may be
concluded that computer simulations proving their worth as a valuable tool to study
the complex problems in zero-gravity conditions and one can observe the credentials
of numerical modeling to simulate realistic 3D Marangoni cases. This has important
implications on the Marangoni flow characteristics in space.

Acknowledgements The principal investigator would like to express his sincere gratitude to Kuwait
Foundation for the Advancement of Sciences (KFAS), Kuwait, for supporting and funding this
research work.

References

1. N.O. Young, J.S. Goldstein, M.J. Block, The motion of bubbles in a vertical temperature gradient.
   J. Fluid Mech. **6**, 350–356 (1959)
2. ANSYS-FLUENT 2011. Users Guide
3. D.L. Youngs, Time-dependent multi-material flow with large fluid distortion, in *Numerical
   Methods for Fluid Dynamics*. Academic Press, pp. 273–285 (1982)
4. C.W. Hirt, B.D. Nichols, Volume of fluid (Vof) method for the dynamics of free boundaries. J.
   Comput. Phys. **39**, 201–225 (1981)
5. J.U. Brackbill, D.B. Kothe, C. Zemach, A continuum method for modeling surface tension. J.
   Comput. Phys. **100**, 335–354 (1992)

Numerical Study on the Freezing Point of Methane Hydrate Dissociation by Depressurization

Qun Zhang, Longbin Yang, Yazhou Shao, Shidong Wang, and Runzhang Xu

1 Introduction

Natural gas hydrate is an ice-like non-stoichiometric crystal compound which shows a series of advantages, such as large reserves and clean combustion. It is widely distributed in the permafrost and deep ocean sediments. As a substitute for traditional fossil energy, it is popular among the energy science workers. Various decomposition methods include depressurization, thermal stimulation, inhibitor injection, and replacement [1]. The depressurization is considered as the most economical and effective method. A number of numerical simulations and experiments investigation into the hydrate production behaviors by depressurization have been studied. However, the behavior of hydrate dissociation at low pressures has rarely studied.

Li et al. [2] investigated ice formation/melting on the dissociation of methane hydrate in micro-porous media channels by depressurization. The results showed the change in heat as one of the key factors with the ice formation/melting. Zhao et al. [3] analyzed the dissociation behavior for different depressurization modes via magnetic resonance imaging (MRI) visualization in situ. The results indicated that obvious hydrate reformation and ice generation can be effectively avoided using piecewise and continuous depressurization methods.

Ice formation can easily happen under low production pressures, due to insufficient heat transfer in the deposit. When ice is present in the porous media, it will affect the decomposition process of hydrate. Therefore, it is important to study the ice formation during the dissociation of hydrates.

In this paper, we used the TOUGH + HYDRATE_v1.5 to simulate the decomposition process of hydrates under icing conditions. It mainly focuses on investigating

Q. Zhang · L. Yang (✉) · Y. Shao · S. Wang
College of Power and Energy Engineering, Harbin Engineering University, Harbin, China
e-mail: ylbhrb@163.com; yanglb@hrbeu.edu.cn

R. Xu
College of Mathematical Sciences, Harbin Engineering University, Harbin, China

© Springer Nature Singapore Pte Ltd. 2021
C. Wen and Y. Yan (eds.), *Advances in Heat Transfer and Thermal Engineering* ,
https://doi.org/10.1007/978-981-33-4765-6_63

the evolution of local pressure and temperature, which are caused by the flow of water and methane. It can reflect the heat and mass transfer in the pore space.

2 Methodology

In this work, we use the TOUGH + HYDRATE_v1.5 and choose equilibrium model. The model includes four phases (aqueous A, hydrate H, gas G, and ice I), two components (water and methane), and considers non-permeability boundaries ($k_z = k_r = 0$). According to the mass and energy balance equation, the multi-phase flow, and seepage during the hydrate decomposition process, the following reasonable assumptions are made:

(1) Darcy's law is valid in the simulated domain under the conditions of the study.
(2) The gas produced by the hydrate only methane and $N_H = 6$;
(3) Uniform and isotropic distribution of porous media in reservoirs;
(4) The compressibility and thermal expansivity of hydrate are the same as those of ice.

Four-phase saturations:

$$\sum S_\kappa = 1 \, , \ \kappa = \text{I,H,G,A,} \tag{1}$$

Mass and energy conservation equation:

$$\frac{d}{dt} \int_{V_n} M^\kappa dV = \int_{\Gamma_n} F^\kappa \cdot n \, dA + \int_{V_n} q^\kappa \, dV \tag{2}$$

where V and V_n are volume and volume of subdomain n (m^3); M^κ is mass accumulation term of component κ (kg m^{-3}); A and Γ_n are surface area and surface area of subdomain n (m^2); F^κ is Darcy flux vector of component κ (kg m^{-2} s^{-1}); q_κ is source/sink term of component κ (kg m^{-3} s^{-1}).

The physical model of 1L cylindrical reactor is established, as shown in Fig. 1. The vessel (ϕ 120 mm × 100 mm) is discretized into a two-dimensional grid because of symmetry. The domain consists of 22 × 50 + 23 × 6 = 1238 grid blocks in (r, z), of which 937 are active (each parameter changes with the reaction). The active areas include hydrate reservoir and well. The well wall and boundaries are inactive; it means non-permeability ($k_z = k_r = 0$). Discretization along the radial direction is nonuniform from well to boundary (with Δr increasing from 1 to 4 mm), a total of 22 grids. Discretization along the z-axis grid is constant; the uppermost and lowermost boundaries are $\Delta z = 4$ mm, and the rest are $\Delta z = 1.75$ mm and $\Delta z = 2$ mm, a total of 54 grids. The perforated interval is located in the mid of the center vertical well and has 12 mm in length, with the well radius of $r_w = 3$ mm.

The five measuring points (1 #, 2 #, 3 #, 4 # and 5 #) are located on the horizontal line of $z = -55$ mm, and the distance is 9 mm, 22 mm, 36 mm, 48 mm, and 60 mm from the well.

Based on the experimental parameters, the cases are set to wellhead pressure (2.0 MPa, 1.5 MPa, 1.0 MPa, 0.5 MPa). The production well is a vertical well in the center, and porosity $\phi = 1.0$, permeability $K = 5000$ Darcies. The initial parameters of hydrate reservoir are given in Table 1.

3 Results

Figure 2a shows the cumulative volumes of produced gas at the well (V_P), gas released from hydrate dissociation (V_R), and the free gas remained in the reservoir (V_F) under different pressures. The results indicate that in the hydrate decomposition process, the V_P and V_R gradually increased and V_F gradually reduced. The gas production at the wellhead in 0.5 MPa is much lower than in 1.0 MPa over time. This is because the dissociation rate is too fast and the amount of ice is large in the reactor when P_w = 0.5 MPa. However, the wellhead accumulative gas production in $P_w = 0.5$ MPa is much higher than in others. This is because the driving force is the highest in four cases.

Fig. 1 Physical model and grid used in the numerical simulation

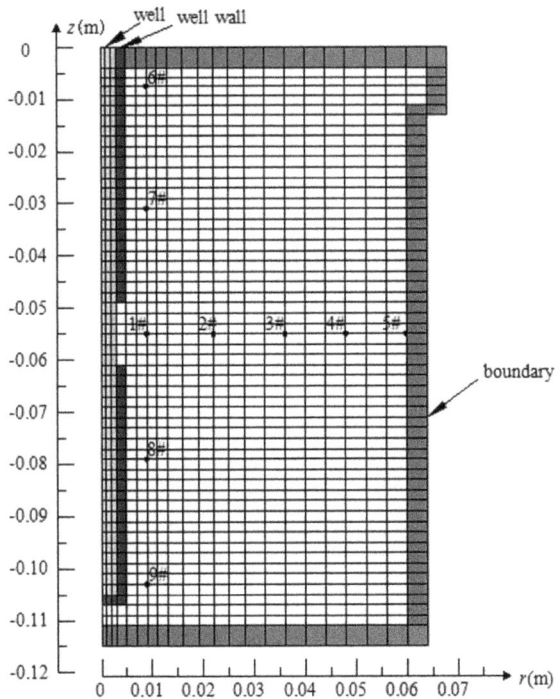

Figure 2b shows the gas production rate of wellhead (Q_G) with time during hydrate decomposition. In all cases, the rate of hydrate dissociation decreases significantly and appear fluctuation. The process can be divided into three stages: In stage I ($0 < t < 6$ min), it shows production rate decreases, which indicates that the decomposition absorbs a large of heat; in stage II (6 min $< t < 15$ min), the gas production rate appears a slightly rise. This is because that the ice melts due to boundary heat transfer, and the temperature increases; in stage III (15 min $< t <$ end), the production rate decreases again until it is zero because of the remaining hydrate mass drops. In the stage I, the decomposition rate in 0.5 MPa drops rapidly and is soon below in 1.5 MPa, 1.0 MPa. This phenomenon is because a large amount of ice inhibits the dissociation of hydrate. In all decomposition process, the time required for complete decomposition of hydrate is the shortest in 1.0 MPa.

Table 1 Initial conditions and physical conditions in the reactor

Parameter	Value
Initial pressure p_0/ MPa	3.70
Initial temperature T_0/°C	4
Initial hydrate saturation S_{H0}	0.20
Initial water saturation S_{A0}	0.19
Initial gas saturation S_{G0}	0.61
Gas composition	100%CH_4
Sand porosity	0.39
Density of the quartz sand/kg m^{-3}	2600
Specific heat/J kg^{-1} K^{-1}	1000
Intrinsic permeability/m^2	5.0×10^{-11}
Thermal conductivity/W m^{-1} K^{-1}	1.0

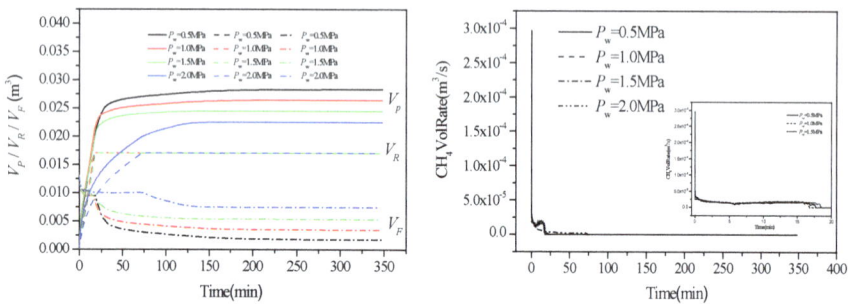

Fig. 2 Evolution of **a** V_R, V_P, and V_F; **b** Q_G during the production process under different wellhead pressures

4 Conclusions

The influence of pressure was studied by TOUGH + HYDRATE numerical simulation software. The following conclusions can be drawn: (1)The icing area is mainly concentrated near the well and is distributed in a ring shape from the perforated area; (2)when the pressure is very low in a 1L reactor, a large amount of ice hinders fluid flow and inhibits hydrate dissociation.

References

1. X.S. Li, C.G. Xu, Y. Zhang, X.K. Ruan, Y. Wang, Investigation into gas production from natural gas hydrate: a review. Appl. Energ. **172**, 286–322 (2016)
2. X. Wang, B. Dong, F. Wang, W.Z. Li, Y.C. Song, Pore-scale investigations on the effects of ice formation/melting on methane hydrate dissociation using depressurization. Int. J. Heat Mass Transf. **131**, 737–749 (2019)
3. B. Wang, Z. Fan, P.F. Wang, Y. Liu, J.F. Zhao, Y.C. Song, Analysis of depressurization mode on gas recovery from methane hydrate deposits and the concomitant ice generation. Appl. Energ. **227**, 624–633 (2018)

Multi-bubble Coalescence Simulations with Large Density Ratio Using Improved Lattice Boltzmann Method

Hongtao Gao, Xiupeng Ji, Jiaju Hong, Yuchao Song, and Yuying Yan

1 Introduction

Although the conventional absorption refrigeration system uses low-grade waste heat [1] as the heating source, such as exhaust gas and waste water, it requires solution pump to provide circulating power for the entire system. Solution pump requires electrical energy to drive, and absorption refrigeration system still consumes a significant amount of high-quality energy. The emergence of the bubble pump absorption refrigeration system solves this problem. The bubble pump uses low-quality waste heat as driving heat source to provide circulating power for the system. The entire system is completely driven by low-quality heat source, and there is no pollutant emission. Early research on bubble pump mainly focused on the structural sizes and operating parameters of the bubble pump. Different structural sizes and operation parameters affect the flow pattern of the bubble pump, and the flow pattern characteristics determine the performance parameters of the bubble pump, such as the solution lift. Therefore, more and more scholars have carried out simulation and experimental researches on the flow pattern of the bubble pump. As shown in Fig. 1, the flow patterns in the lifting pipe experience bubble flow, slug flow, plug flow, churn flow, and annular flow; totally five different flow patterns [2].

When the bubble pump is in operation, the flow pattern in the lifting pipe is gas–liquid two-phase flow. Therefore, the research on the characteristics of bubble motion is very important. From the beginning operation of the bubble pump to the stable operation, the flow patterns in the lifting pipe are changing. At different point

H. Gao (✉) · X. Ji · J. Hong · Y. Song
Institute of Refrigeration & Cryogenics Engineering, Dalian Maritime University, Dalian 116026, China
e-mail: gaohongtao@dlmu.edu.cn

Y. Yan
Fluids and Thermal Engineering Research Group, Faculty of Engineering, University of Nottingham, University Park, Nottingham NG7 2RD, UK

© Springer Nature Singapore Pte Ltd. 2021
C. Wen and Y. Yan (eds.), *Advances in Heat Transfer and Thermal Engineering* ,
https://doi.org/10.1007/978-981-33-4765-6_64

(a) bubble flow   (b) slug flow   (c) plug flow   (d) churn flow   (e) annular flow

Fig. 1 Two phase flow patterns in lifting pipe

of flow patterns, the number, size, shape, speed of the bubbles, and the character-istics of the bubble coalescence are different. Studying the bubble motion in the lifting pipe can help to better analyze the flow characteristics of the bubble pump and promote the performance improvement of the bubble pump and its theoretical research. The lattice Boltzmann method is a new computational fluid dynamics simu-lation method developed in the 1980s. Unlike the macroscopic theoretical basis of the traditional CFD simulation method, the lattice Boltzmann method is based on the molecular dynamic theory [3]. Since the mesoscopic properties of the lattice Boltzmann method can accurately describe and calculate the gas–liquid interface, effective simulation can be performed in the field of multiphase flow. In 1988, McNa-mara and Zanetti [4] used f_i to represent the local particle distribution function, and the lattice Boltzmann equation to replace the evolution equation of the lattice pneu-matic machine. This is the earliest lattice Boltzmann model. Higuera and Jimenez [5] proposed a simplified model, introducing an equilibrium distribution function f_i^{eq} for the model, and linearizing the collision operator. The model replaces the original collision operator with the collision matrix. The values in the matrix satisfy the conservation of momentum and mass. Compared with the multi-particle colli-sion model, the collision matrix ignores the collision details of each particle, so it is easier to construct. Later, Higuera et al. [6] proposed a method of strengthening the collision operator to increase the numerical stability of the model. At the same time, the matrix model reduces the complexity of the model, and the storage and the calculation are greatly reduced. In 1991, Chen and Qian et al. [7–9] proposed a single relaxation time, which further simplified the collision operator. Chen and Qian et al. used the time relaxation coefficient to control the speed at which different particles reach their equilibrium state. This method is called lattice BGK model. The essence of the lattice BGK model is to simplify the process of particle collision. Even if the particle type and collision operator change, the model cannot become complicated. By selecting the appropriate f_i^{eq}, the state equation is independent of the speed of particles. The early lattice Boltzmann model can only be used for the simulation of isothermal incompressible flow field, and a variety of isothermal incompressible models have been developed [10]. Later, the lattice Boltzmann model is successfully coupled with the thermal model[11], which can simulate the flow field with variable temperature, and increase the application of the lattice Boltzmann model. Then, the double distribution functions and multiple distribution functions appear. The double distribution function successfully couples the density distribution function and the temperature distribution function, respectively, derives the velocity field by the density distribution function, and obtains the temperature field by the temperature distribution function. The double distribution function has the same numerical stability as the isothermal incompressible model.

The field of multiphase flow has always been a difficult and hot topic of research. Due to the large difference in physical properties between different phases, the interface is difficult to capture and unstable. The traditional computational fluid dynamics method has many limitations in simulating two-phase flow. The lattice Boltzmann method can accurately describe and calculate the interface between different phases due to its mesoscopic characteristics. More and more scholars use the lattice Boltzmann method to study the multiphase flow field. In 1997, Hou, Shan et al. [12] used lattice Boltzmann method to simulate mutually incompatible fluids and multi-component non-ideal gases, comparing different physical parameters of multiphase flow. At the same time, static bubbles were simulated. Surface tension, interface thickness, Laplace's theorem, numerical stability, and other issues were studied. He et al. [13] proposed a lattice Boltzmann model for incompressible two-phase flow, which improves the stability of numerical simulation by reducing the numerical errors in the calculation of molecular interactions. At the same time, an exponential function is introduced to track the interface between different phases. During the simulation process, the interface thickness is maintained at 3–4 lattice lengths, and no manual interference is required. He and Doolen [14] analyzed the advantages and disadvantages of the previous lattice Boltzmann model, and introduced the thermal model for lattice Boltzmann multiphase flow model. The model satisfies the conservation equations of mass, momentum, and energy. The thermal parameters change continuously during the simulation process and have good stability. Premnath and Abraham [15] proposed a three-dimensional multi-relaxation time (MRT) lattice Boltzmann model, which significantly improved the numerical stability of lattice Boltzmann method in simulating low-viscosity fluids and reduced the influence of compressibility of flow field on simulation. In view of the advantages of MRT lattice Boltzmann model in improving numerical stability, increasing the application of the model and simulating anisotropic flow field, Kuzmin and Mohamad [16] also studied MRT lattice Boltzmann model. Compared with single relaxation time lattice BGK model, Kuzmin and Mohamad quantitatively analyze the advantages of the MRT model in improving numerical stability, and theoretically analyze how the MRT model improves numerical stability. Gong and Cheng [17] improved the lattice Boltzmann model to simulate the process of phase transition and multiphase flow. The droplet motion and coalescence process were analyzed numerically. The velocity field distribution of the flow field was elaborated in detail. The improved model was more accurate and stable, and the pseudopotential flow could be significantly reduced.

With the development of lattice Boltzmann method in the field of multiphase flow, the simulated flow field is getting closer to the actual flow field, and the contradiction between the larger gas–liquid density ratio and numerical stability is more prominent. For example, the density ratio of water to air is nearly 1000, and the density ratio of water to steam is more than 1000. For the bubble pump, lithium bromide absorption refrigeration system studied in this paper, and the density ratio of lithium bromide solution to water vapor is 2778. Under the condition of high density, numerical instability limits the further development and application of lattice Boltzmann method. Therefore, more and more scholars begin to study lattice Boltzmann model with high density ratio. Teng et al. [18] introduced TVD/AC method into lattice Boltzmann multiphase flow model in order to improve the numerical stability of simulation when the density ratio reached 100. The improved model can significantly improve the numerical stability of simulation under the condition of high density ratio. Excessive density ratio will result in difficulties in interface capture, unclear interface, long operation time, unstable calculation, and easy divergence. In reality, some two-phase flow has high gas–liquid density ratio, even up to 1000–2000. Therefore, a lattice Boltzmann free energy model with high density ratio has emerged. Inamuro et al. [19] used lattice Boltzmann free energy model with high density ratio to simulate the process of bubble coalescence, and good results were obtained when the density ratio reached 1000. Inamuro et al. simulated the rising process of 24 bubbles in a square pipe. It was found that every bubble deformed, and the bubbles coalesce with each other. With the increase of simulation time, the degree of bubble deformation became larger and larger. The simulation results show that when the density ratio reaches 1000, the multi-bubble motion in the complex flow field can still be simulated stably. Gao et al. [20] used lattice Boltzmann free energy model with high density ratio to simulate the characteristics of bubble motion in the lifting pipe of bubble pump lithium bromide absorption refrigeration system. The density ratio is as high as 2778. Excessive density ratio made it difficult to maintain numerical stability in simulation, so the number of simulated bubbles is relatively small. In order to solve the contradiction between the excessive density ratio and the numerical stability, the lattice Boltzmann free energy model with large density ratio is improved, and dozens of bubble motions are simulated by the improved model. The simulation results are analyzed from different angles. The density ratio of the model is 2778. Then, the improved model is coupled with the thermal model [21] to analyze the effect of the process of multi-bubble coalescence on the temperature distribution of the flow field.

2 Methodology

The lattice Boltzmann free energy model with high density ratio proposed by Inamuro in 2004 [19] is used as the basic model in this paper. However, the density ratio can only reach 1000. Numerical instability occurs when the density ratio continues to increase. The density ratio of lithium bromide solution to water vapor is as high as 2778. Therefore, the original model is locally modified to improve the numerical stability of multi-bubble simulation under large density ratio conditions. At the same time, in order to study the effect of multi-bubble motion on the temperature distribution of flow field under large density ratio, a non-isothermal lattice Boltzmann model is established by coupling the improved model with the thermal model.

The original model uses three distribution functions, namely density distribution function, velocity distribution function, and pressure distribution function, which are used to solve density field, velocity field, and pressure field in turn. In this paper, the unit is dimensionless lattice unit, and the three-dimensional D3Q15 model is used to discretize the velocity in 15 directions. The stereoscopic diagram is shown in Fig. 2.

In this paper, the distribution function f_i is used to solve the order parameters ϕ, which τ_f is dimensionless relaxation time, Δt is the time required for a particle to move a lattice unit, and the f_i^{eq} is equilibrium distribution function. The distribution function is as follows:

$$f_i(\boldsymbol{x} + c_i \Delta t, t + \Delta t) = f_i(\boldsymbol{x}, t) - \frac{1}{\tau_f} \left[f_i(\boldsymbol{x}, t) - f_i^{eq}(\boldsymbol{x}, t) \right] \tag{1}$$

The equilibrium distribution function f_i^{eq} is given by:

Fig. 2 D3Q15 model

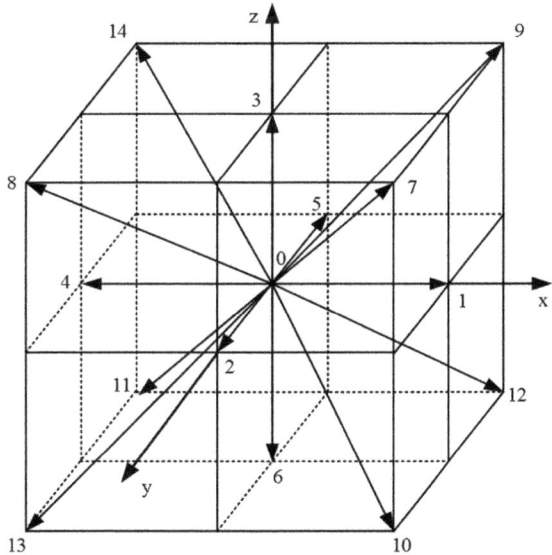

$$f_i^{eq} = H_i\phi + F_i \left[p_0 - k_f\phi \frac{\partial^2 \phi}{\partial x_\alpha^2} - \frac{k_f}{6} \left(\frac{\partial \phi}{\partial x_\alpha} \right)^2 \right]$$

$$+ 3E_i\phi c_{i\alpha}u_\alpha + E_i k_f G_{\alpha\beta}(\phi)c_{i\alpha}c_{i\beta} \tag{2}$$

Among them, k_f is a constant parameter to determine the width of gas–liquid interface, α, β, γ represent the Cartier coordinate system x, y, z. The order parameter ϕ is given by:

$$\phi = \sum_{i=1}^{15} f_i \tag{3}$$

Density is calculated by order parameter and the function is given by:

$$\rho = \begin{cases} \rho_G, & \phi = \phi_{min} \\ \frac{\Delta\rho}{2}\left[\sin\left(\frac{\phi-\bar{\phi}}{\Delta\phi}\pi\right) + 1\right] + \rho_G, & \phi_{min} < \phi < \phi_{max} \\ \rho_L, & \phi = \phi_{max} \end{cases} \tag{4}$$

The velocity distribution function is used to solve the velocity distribution of the flow field. The distribution function g_i is given by:

$$g_i(x + c_i\Delta t, t + \Delta t) = g_i(x, t) - \frac{1}{\tau_g}\left[g_i(x, t) - g_i^{eq}(x, t)\right]$$

$$+ 3E_i c_{i\alpha}\frac{1}{\rho}\left\{ \frac{\partial}{\partial x_\beta}\left[\mu\left(\frac{\partial u_\beta}{\partial x_\alpha} + \frac{\partial u_\alpha}{\partial x_\beta} \right) \right] \right\}\Delta x \tag{5}$$

Equilibrium distribution Function g_i^{eq} is given by:

$$g_i^{eq} = E_i\left[1 + 3c_{i\alpha}u_\alpha - \frac{3}{2}u_\alpha u_\alpha + \frac{9}{2}c_{i\alpha}c_{i\beta}u_\alpha u_\beta \right.$$

$$+ \frac{3}{2}\left(\tau_g - \frac{1}{2} \right)\delta_x\left(\frac{\partial u_\beta}{\partial x_\alpha} + \frac{\partial u_\alpha}{\partial x_\beta} \right)c_{i\alpha}c_{i\beta}\right]$$

$$+ E_i\frac{k_g}{\rho}G_{\alpha\beta}(\rho)c_{i\alpha}c_{i\beta} - \frac{2}{3}F_i\frac{k_g}{\rho}\left(\frac{\partial \rho}{\partial x_\alpha} \right)^2 \tag{6}$$

The preliminary macro speed is u^* as follows:

$$u^* = \sum_{i=1}^{15} c_i g_i \tag{7}$$

Equation (7) does not consider the influence of the pressure of flow field on the velocity distribution, so it is necessary to correct the velocity by pressure of flow field. Firstly, the pressure distribution needs to be solved by the pressure distribution function, then the Poisson equation associated with pressure and velocity is established. The velocity is corrected to obtain the real velocity of flow field. The pressure distribution function h_i is given by:

$$h_i^{n+1}(x + c_i \Delta t, t + \Delta t) = h_i^n(x, t) - \frac{1}{\tau_h}\left[h_i^n(x) - h_i^{eq}(x)\right]$$
$$- \frac{1}{3}E_i \frac{\partial u_\alpha^*}{\partial x_\alpha}\Delta x \tag{8}$$

Pressure of flow field is given by:

$$p = \sum_{i=1}^{15} h_i \tag{9}$$

Poisson's equation is used to modify the velocity, and the real velocity a of the flow field is obtained as follows:

$$u = u^* - \frac{\Delta t}{\rho \cdot Sh}\nabla p \tag{10}$$

In view of the limitation of the original model in studying the motion of multiple bubbles with high density ratio, the calculation method of pressure and velocity field in lattice Boltzmann free energy model is improved. The improved model uses the same distribution function to calculate the pressure field and velocity field, simplifies the program algorithm, and eliminates the pressure distribution function introduced by the original model for calculating the pressure field. At the same time, the original model introduces complex Poisson equation for velocity correction, which greatly increases the complexity of program calculation and easily causes instability of gas–liquid interface, and finally leads to program divergence and calculation failure. The improved model eliminates the velocity correction method of Poisson equation, introduces pressure gradient, and simplifies the model algorithm without affecting the calculation accuracy. It can shorten the calculation time and improve the stability of numerical calculation. The improved model can simulate the motion of two-phase flow with large density ratio and multiple bubbles in good stability.

The improved equation for calculating pressure is as follows:

$$p = \frac{1}{3}\sum_{i=1}^{15}\left[\Delta P_i(x, t) + g_i^{eq}(x - c_i \Delta t, t)\right] \tag{11}$$

Velocity equation is given by:

$$u = \sum_{i=1}^{15} c_i \left[\Delta P_i(x, t) + g_i^{\text{eq}}(x - c_i \Delta t, t) \right] \tag{12}$$

Among them, ΔP_i is the pressure gradient of flow field and g_i^{eq} is the equilibrium distribution function. The functions are as follows:

$$\Delta P_i(x, t) = \frac{3}{2} E_i \left[\frac{1}{\rho(x - c_i \Delta t, t)} + \frac{1}{\rho(x, t)} \right] \\ \times \left[p(x - c_i \Delta t, t) - p(x, t) \right] \tag{13}$$

$$g_i^{\text{eq}} = E_i \left[3 c_{i\alpha} u_\alpha - \frac{3}{2} u_\alpha u_\alpha + \frac{9}{2} c_{i\alpha} c_{i\beta} u_\alpha u_\beta \right. \\ + \frac{3}{4} \Delta x \left(\frac{\partial u_\beta}{\partial x_\alpha} + \frac{\partial u_\alpha}{\partial x_\beta} \right) c_{i\alpha} c_{i\beta} \\ + 3 c_{i\alpha} \frac{1}{\rho} \frac{\partial}{\partial x_\beta} \left\{ \mu \left(\frac{\partial u_\beta}{\partial x_\alpha} + \frac{\partial u_\alpha}{\partial x_\beta} \right) \right\} \Delta x \\ \left. + \frac{k_g}{\rho} c_{i\alpha} c_{i\beta} G_{\alpha\beta}(\rho) - \frac{2}{3} \frac{F_i}{E_i} \frac{k_g}{\rho} \left(\frac{\partial \rho}{\partial x_\alpha} \right)^2 \right] \tag{14}$$

3 Results

When the bubble pump works, the flow field in the lifting pipe is two-phase flow filled with a large number of bubbles. If only a small number of bubbles such as single or double bubbles are simulated, it is quite different from the actual situation. And the results are difficult to reflect the real process of bubble motion. In this paper, the coalescence motion of multi-bubbles in the lifting pipe with large density ratio of two-phase flow is researched. The number of bubbles is set to 20, and the bubbles are randomly distributed. For gas–liquid two-phase flow in lifting pipe, the size and location of bubbles are different. The bubbles coalesce with each other, and may burst at any time. The process is complex, so it is difficult to simulate directly. Therefore, the size of 20 bubbles simulated in this paper is tentatively the same. With the development of simulation and calculation method, the simulation will get closer to the real flow field step by step. Firstly, the coalescence motions of 20 bubbles are analyzed by the figure of density distribution. Based on the better simulation results of 20 bubbles, the number of bubbles is increased to 40 and 60. In this paper, the influence of bubble coalescence motion on velocity, pressure, and temperature distribution of two-phase flow field in lifting pipe is analyzed in detail, and the process of flow pattern changing from bubble flow to slug flow is simulated.

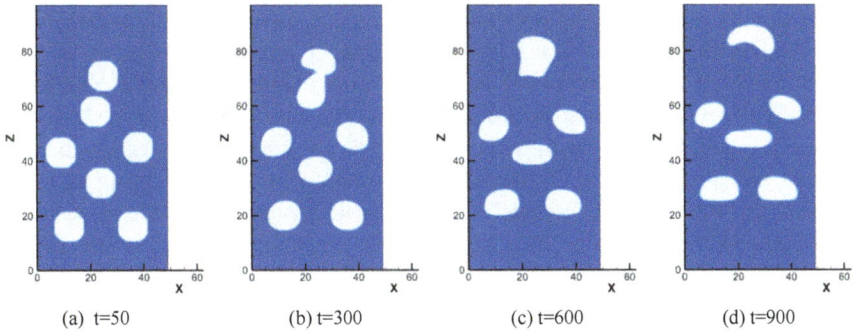

(a) t=50   (b) t=300   (c) t=600   (d) t=900

Fig. 3 Distribution figures of density field of 20 bubbles coalescence

Figure 3 shows the density distribution of multi-bubble coalescence. The number of bubbles is 20. Seven bubbles can be observed in the longitudinal section. Because the number of bubbles is relatively small and the distance between bubbles is relatively far, many bubbles do not coalesce and rise independently in the flow field. As can be seen from Fig. 3, only two bubbles close to each other coalesce, while the other five bubbles do not coalesce. However, due to the influence of disturbance of flow field, the shape of bubbles is different and irregular.When the calculation runs to 300 time steps, the two bubbles at the top of the flow field begin to contact and coalesce. As the simulation continues, the two bubbles eventually coalesce into one bubble, which is similar to a curved moon. The other five bubbles rise independently because they are far away from each other.

On the basis of the result that the model can better simulate the motion of 20 bubbles, the number of bubbles will continue to increase, and the number of bubbles will reach 40. Figure 4 is distribution figures of density field of multi-bubble simulations at different time steps, and 14 bubbles can be observed on the longitudinal section. The distributions of bubbles are random and contain a variety of phenomena of bubble coalescence, such as the independent rising process of a single bubble,

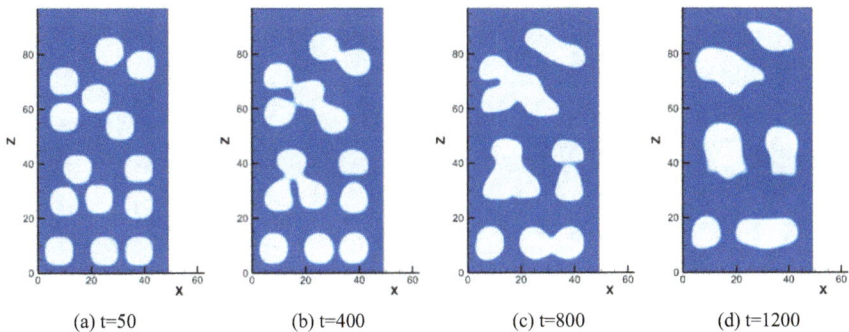

(a) t=50   (b) t=400   (c) t=800   (d) t=1200

Fig. 4 Distribution figures of density field of 40 bubbles coalescence

the coalescence of the closer two bubbles, and the coalescence of three bubbles or four bubbles to larger bubbles. It is found that when one bubble is far away from other bubbles, the rising process of the bubble is less affected by other bubbles, so the bubble is difficult to coalesce with other bubbles. There is one bubble in the lower left corner of Fig. 4, which is far from the right bubble and above, and does not coalesce with the two bubbles. However, the shape of the bubble is distorted during the rising process, and the bubble is obviously affected by the disturbance of the surrounding flow field. At the same time, two closer bubbles coalesce with each other. Due to the influence of disturbance of flow field caused by the bubble motions, the process of bubble coalescence is no longer regular, and finally the two bubbles coalesce to a spherical bubble with irregular shape. If multiple bubbles are close to each other in the lifting pipe, such as the coalescence of three bubbles or even four bubbles shown in Fig. 4, the multiple bubbles will coalesce with each other to form a larger bubble. The bubble is unstable. The unpredictability of broken bubbles is also the main reason why the program cannot continue to work.

In order to further explore the motion of multi-bubble coalescence and simulate the process of flow pattern transition from bubble flow to slug flow in lifting pipe, the number of bubbles is increased to 60. Similarly, a longitudinal profile cloud image is used to analyze the process of multi-bubble coalescence into large bubbles, and 20 bubbles can be observed in the density distribution figure. As shown in Fig. 5, the bubble shape is set to a spherical shape at the initial time, the positions of bubbles in the flow field are random, and the distances between bubbles are different. When the time step is $t = 50$, it is found that there are 10 bubbles at the bottom of the flow field, and these bubbles are close to each other and easy to coalesce. In the middle and upper part of the flow field, there are several groups of bubbles with different numbers, which are far away from each other and not easy to coalesce. With the development of bubble motion, when the time step is $t = 400$, it will be found that bubbles close to each other and coalesce within the group of bubbles. The flow pattern in the lifting pipe begins to change from bubble flow to slug flow. The motion of bubble coalescence keeps on developing. When the simulation reaches $t = 800$, it is found that the bubbles in each bubble group are connected with each other, forming

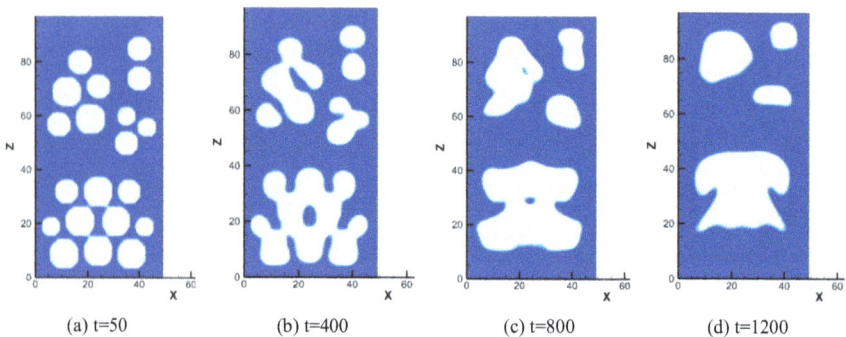

(a) t=50   (b) t=400   (c) t=800   (d) t=1200

Fig. 5 Distribution figures of density field of 60 bubbles coalescence

several large irregular bubbles. These bubbles rise under the action of buoyancy, and the gas–liquid interface shrinks gradually under the action of surface tension, and the curvature becomes smooth. Under the combined action of surface tension and flow resistance, the top of the bubbles protrudes like the head of a bullet. Finally, in the flow field, multiple bubbles coalesce into several bullet-like large bubbles, and the flow pattern in the lifting pipe completes the transition process from bubble flow to slug flow.

Figure 6 is velocity distribution, which reveals the velocity magnitude and velocity vector direction in the flow field. At the beginning of the simulation, bubbles begin to rise under buoyancy, and the velocity of bubbles is obviously higher than surrounding liquids, forming high-speed centers near bubbles. When the calculation runs to 400 steps, bubbles begin to contact and coalesce with each other. The increase of bubbles velocity leads to the increase of the velocity of liquid flow around bubbles. The high-speed center in the flow field expands continuously, forming several regions with larger velocities. When the calculation runs to 800 steps, it can be concluded from the color change of Fig. 6 that the velocity of the entire flow field continues to increase and the direction of the velocity is generally upward. The velocity is small near the wall of the simulated flow field because of friction resistance; the maximum velocity in the flow field occurs where the bubble group composed of 10 bubbles is located. With the continuation of the simulation, 60 bubbles eventually coalesce into several larger bubbles. Because the shape of the top of the bubbles is similar to that of bullets, the flow pattern is called slug flow, in which the slug bubble at the bottom of the flow field is the most representative.The maximum velocity in the flow field occurs at the location of the slug bubble, and the high-speed area in the bullet bubble appears as an inverted capital V word.

Figure 7 shows the pressure distribution of the flow field. The law of pressure distribution is similar to that of 20 bubbles and 40 bubbles. At the beginning of the simulation, the pressure near the bubbles is higher than that of the surrounding liquid, and many high pressure centers are formed near the bubbles. Their distribution is almost the same as that of the bubbles and the pressure at the top of the bubbles is obviously higher than that at the bottom of the bubbles.It is found that the pressure of these high pressure centers decreases with the bubble coalescence, and the pressure of the whole simulated flow field decreases. Therefore, it can be concluded that in the process of multi-bubble coalescence, there are high-pressure centers near the bubbles, and the high-pressure center coalesce with the bubble coalescence, but the pressure in the flow field decreases as a whole with the increase of velocity.

The motion of bubble coalescence will affect the temperature distribution in the flow field. The figure of temperature distribution is shown in Fig. 8. At the initial time, due to the limitation of the accuracy of the model, there may be some errors in the temperature distribution. As the simulation proceeds, the temperature distribution can more accurately reveal the law of temperature change in the flow field. In the process of bubbles rising and coalescence, the velocity of bubbles is increasing, and the temperature of bubbles is decreasing. Especially in the region of bubbles coalescence, the temperature decrease is more obvious.The lowest temperature in the flow field occurs in the region where multiple bubbles coalesce. The regions

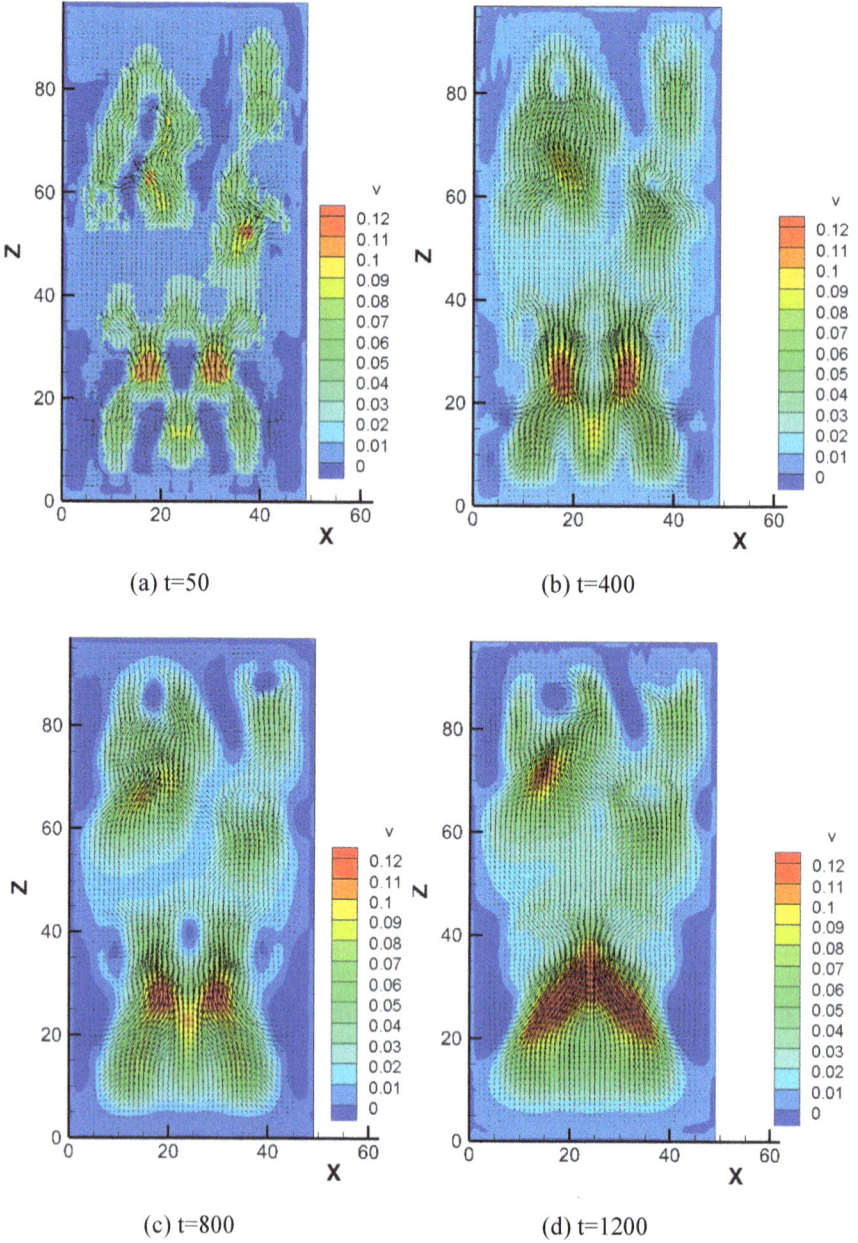

(a) t=50

(b) t=400

(c) t=800

(d) t=1200

Fig. 6 Distribution figures of velocity field of 60 bubbles coalescence

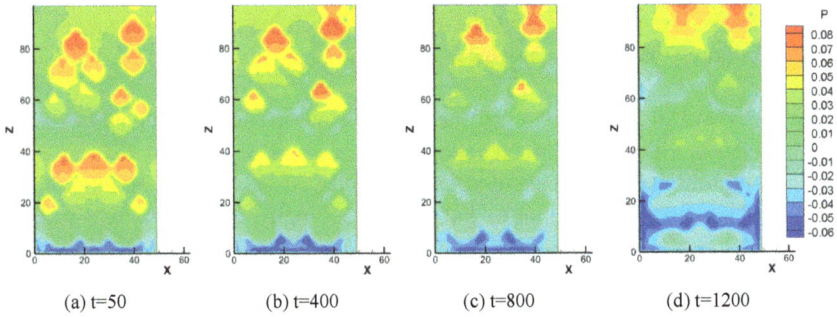

Fig. 7 Distribution figures of pressure field of 60 bubbles coalescence

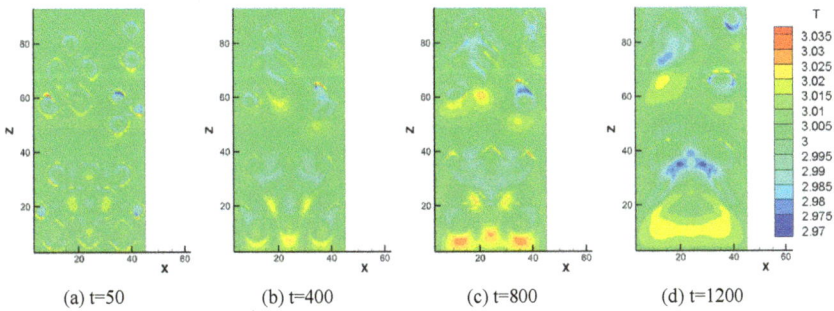

Fig. 8 Distribution figures of temperature field of 60 bubbles coalescence

with lower temperature in the flow field coincide with the regions where bubbles coalesce with each other. Therefore, it can be concluded that the process of bubble coalescence is a process of temperature decreasing. Bubble coalescence is a fast and intense process.

4 Conclusions

The conclusions are as follow:

(1) The improved lattice Boltzmann free energy model with large density ratio eliminates the Poisson equation and introduces the pressure gradient, which simplifies the model algorithm without affecting the calculation accuracy and improves the numerical stability. The improved lattice Boltzmann method can be well applied to the investigation of multi-bubble motions in the field of two-phase flows with large density ratio.

(2) As the process of bubble coalescence is affected by the disturbance of the surrounding flow field, the shape of the bubble will be distorted and vary in various forms.
(3) When there are enough bubbles, the bubbles close to each other coalesce into big bubbles. Under the combined action of surface tension and flow resistance, the gas–liquid interface shrinks gradually, the curvature becomes smooth, and the top of the bubbles is similar to that of bullets. Finally, in the flow field, multiple bubbles coalesce into several bullet-like large bubbles, and the flow pattern in the lifting pipe completes the transition process from bubble flow to slug flow.
(4) Considering the density distribution, velocity distribution, pressure distribution, and temperature distribution of multi-bubble coalescence, it is found that they are not isolated, but have a certain connection. In the region of multi-bubble coalescence, the velocity increases, and the pressure and temperature decrease, thereby ensuring energy conservation of the entire flow field.

Acknowledgements This work was financially supported by National NaturalScience Foundation of China (No. 50976015), Liaoning S&T Project (No. 2010224002), and the Fundamental Research Funds for theCentral Universities (3132019305).

References

1. Y. Ammar, S. Joyce, R. Norman, Y. Wang, A.P. Roskilly, Low grade thermal energy sources and uses from theprocess industry in the UK. Appl. Energy **89**(1), 3–20, (2012)
2. Y. Taitel, D. Bornea, A.E. Dukler, Modelling flow pattern transitions for steady upward gas-liquid flow in vertical tubes. AIChE J. **26**(3), 345–354 (1980)
3. M. Born, H.S. Green, A general kinetic theory of liquids. I. The molecular distribution functions. Proc. Roy. Soc. Londn. **188**(1012), 10–18 (1946)
4. G.R. MeNamara, G. Zanetti, Use of the Boltzmann equation to simulate lattice automata. Phys. Rev. Lett. **61**(20), 2332–2335 (1988)
5. F.J. Higuera, J. Jiménez, Boltzmann approach to lattice gas simulations. Europhys. Lett. 9(7), 663–668 (1989)
6. F.J. Higuera, S. Succi, R. Benzi, Lattice gas dynamics with enhanced collisions. Europhys. Lett. **9**(4), 345–349 (1989)
7. S. Chen, H. Chen, D. Martnez et al., Lattice Boltzmann model for simulation of magnetohy-drodynamics. Phys. Rev. Lett. **67**(27), 3776–3779 (1991)
8. Y.H. Qian, D. d'Humières, P. Lallemand, Lattice BGK models for Navier-Stokes equation. Europhys. Lett. **17**(6), 479–484 (1992)
9. P.L. Bhatnagar, E.P. Gross, M. Krook, A model for collision processes in gases. I. Small amplitude processes in charged and neutral one-component systems. Phys. Rev. **94**(3), 511–525 (1954)
10. Q. Zou, S. Hou, S. Chen et al., A improved incompressible lattice Boltzmann model for time-independent flows. J. Stat. Phys. **81**(1–2), 35–48 (1995)
11. Y. Chen, H. Ohashi, M. Akiyama, Thermal lattice Bhatnagar-Gross-Krook model without nonlinear deviations in macrodynamic equations. Phys. Rev. E **50**(4), 2776–2783 (1994)
12. S. Hou, X. Shan, Q. Zou et al., Evaluation of two lattice Boltzmann models for multiphase flows. J. Comput. Phys. **138**(2), 695–713 (1997)

13. X. He, S. Chen, R. Zhang, A lattice Boltzmann scheme for incompressible multiphase flow and its application in simulation of Rayleigh-Taylor instability. J. Comput. Phys. **152**(2), 642–663 (1999)
14. X. He, G.D. Doolen, Thermodynamic foundations of kinetic theory and lattice Boltzmann models for multiphase flows. J. Stat. Phys. **107**(1–2), 309–328 (2002)
15. K.N. Premnath, J. Abraham, Three-dimensional multi-relaxation time (MRT) lattice Boltzmann models for multiphase flow. J. Comput. Phys. **224**(2), 539–559 (2007)
16. A. Kuzmin, A.A. Mohamad, S. Succi, Multi-relaxation time lattice Boltzmann model for multiphase flows. Int. J. Mod. Phys. C **19**(06), 875–902 (2008)
17. S. Gong, P. Cheng, Numerical investigation of droplet motion and coalescence by an improved lattice Boltzmann model for phase transitions and multiphase flows. Comput. Fluids **53**, 93–104 (2012)
18. S. Teng, Y. Chen, H. Ohashi, Lattice Boltzmann simulation of multiphase fluid flows through the total variation diminishing with artificial compression scheme. Int. J. Heat Fluid Flow **21**(1), 112–121 (2000)
19. T. Inamuro, T. Yokoyama, K. Tanaka et al., An improved lattice Boltzmann method for incompressible two-phase flows with large density differences. J. Comput. Phys. **198**(2), 628–644 (2004)
20. H. Gao, B. Liu, Y. Yan, Numerical simulation of bubbles motion in lifting pipe of bubble pump for lithium bromide absorption chillers. Appl. Therm. Eng. **115**, 1398–1406 (2017)
21. T. Inamuro, Lattice Boltzmann methods for viscous fluid flows and for two-phase fluid flows. Fluid Dyn. Res. **38**(9), 641–659 (2006)

A New Approach to Inverse Boundary Design in Radiation Heat Transfer

Mehran Yarahmadi, J. Robert Mahan, and Farshad Kowsary

1 Introduction

Unlike direct heat transfer problems, inverse boundary design problems in radiation heat transfer have an ill-posed characteristic [1]. According to Daun et al. [2], there exist two types of inverse boundary design problems, categorized as regularization and optimization types. Regularization methods include methods such as singular value decomposition (SVD) and the conjugate gradients linear solution and have been reviewed comprehensively by Erturk et al. [3] and Daun et al. [2]. In optimization methods, an appropriate least-square type function is defined and its value is minimized iteratively. The main idea of these methods is to define an objective function so that its minimum point corresponds to the ideal design outcome. The optimization methods replace the ill-posed problem with a well-posed one that must be solved repetitively through a systematic method to obtain an optimum solution.

Optimization methods are classified into two general groups: gradient-based methods and metaheuristic search methods. Unlike the metaheuristic search methods, the gradient-based methods need gradient information of the objective function. Furthermore, in gradient-based methods, unknown design variables must be continuous. Thus, metaheuristic search methods are superior to gradient-based methods because of the lack of dependency on gradient information and the ability to work with non-continuous variables. Among metaheuristic search methods, the genetic algorithms (GAs) have received attention for their flexibility, global applicability, robustness, and simplicity. This method has been widely applied to various optimization problems. Safavinejad et al. [4] implement a micro-genetic algorithm to

M. Yarahmadi · J. R. Mahan (✉)
Virginia Polytechnic Institute and State University, Blacksburg, VA 24061, USA
e-mail: jrmahan@vt.edu

F. Kowsary
University of Tehran, Tehran, Iran

© Springer Nature Singapore Pte Ltd. 2021
C. Wen and Y. Yan (eds.), *Advances in Heat Transfer and Thermal Engineering* ,
https://doi.org/10.1007/978-981-33-4765-6_65

377

optimize the number and location of the heaters in two-dimensional radiant enclosures composed of specular and diffuse surfaces. They also use this method to solve
the inverse boundary design problem in two-dimensional radiant enclosures with
absorbing-emitting media [5]. Darvishvand et al. [6] use a GA for the design and
optimization of three-dimensional radiant enclosures to satisfy the uniform thermal
conditions on the surfaces of the irregularshaped design body.

2   Formulation of the Direct Problem

In the MCRT method [7, 8], a distribution factor matrix is populated by tracing a large
number of rays as they navigate within an enclosure subject to the rules of geometrical
optics, which treat the surface absorptivities and reflectivities as probabilities. Use
of distribution factors in radiation analysis requires that the enclosure is subdivided
into a sufficiently large number n of surface elements to assure the desired spatial
resolution. For the case of a diffuse gray enclosure in which the surface net heat fluxes
are specified on the first N surfaces and the surface temperatures are specified on the
remaining n–N surfaces, the distribution factors D_{ij} are related to the temperatures
and net heat fluxes from the surfaces according to

$$q_i + \varepsilon_i \sum_{j=N+1}^{n} \sigma T_j^4 D_{ij} = \varepsilon_i \sum_{j=1}^{N} \sigma T_j^4 \left(\delta_{ij} - D_{ij} \right), 1 \leq i \leq N \qquad (1)$$

where δ_{ij} is the familiar Kronecker delta. The algorithm for calculating radiation
distribution factors using the Monte Carlo Ray-Trace method is described in Mahan
[7, 8] and elsewhere.

3   Inverse Boundary Design

We consider the design problem illustrated in Fig. 1 involving a two-dimensional
oven containing a structure, the workpiece, whose surfaces, the design surfaces,
constitute a box-like geometry. This enclosure is a benchmark in the literature [9,
10]. All surfaces are considered to be gray and diffuse with an emissivity of 0.8,
and the interior medium is assumed to be non-participating. It is desired to maintain
a uniform net heat flux $q_d = q_{d,\text{emitted}} - q_{d,\text{absorbed}}$ and a uniform temperature T_d
on the design surfaces, as indicated in the figure. The problem is to prescribe the
temperature of the heaters required to achieve the design conditions. The surfaces
on the floor of the oven are assumed to be insulated.

The total number of surface elements for the design problem described above is
$n = N_R + N_H + N_D$, where N_H is the number of heater surface elements, N_D is
the number of design surface elements, and N_R is the number of insulated surface

Fig. 1 Cross section of the two-dimensional oven

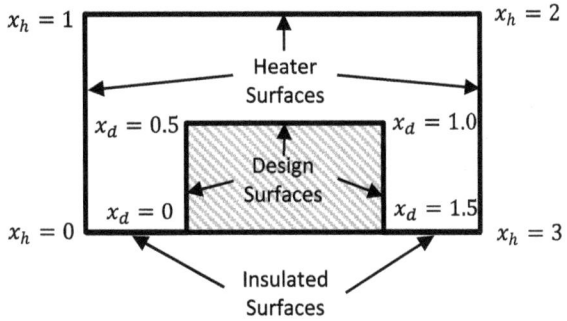

elements. Then, for $n = 100$ total surface elements, we consider the case where $N_R = 10$, $N_H = 60$ and $N_D = 30$. The governing equations to be solved are Eqs. (1) with $q_d = -2000\,\text{W/m}^2$ and $T_d = 300\,\text{K}$, subject to the specified boundary conditions $q_i = 0$, $i \in N_R$, and $T_i = T_d$, $i \in N_D$.

We solve this problem using two approaches. In the traditional approach, a full genetic algorithm (GA) is employed to establish a reference solution against which the solution obtained using the new approach can be compared. The new approach also uses a GA, but while the GA used in the traditional approach requires a population of 2000, 60-gene chromosomes, and 5000 generations to obtain acceptable accuracy, the much-abridged GA in the new approach requires only 100, one-gene chromosomes, and only 64 generations to obtain a significantly more accurate solution. Therefore, the reduction in execution time afforded by the new method is dramatic. In the new approach, we assume a cosine Fourier series of order K for the heater temperature distribution, i.e.,

$$T^4(x) = \sum_{k=0}^{K} a_k \cos(k\omega x) \tag{2}$$

where x is the position of the center of the heater surface elements. A cosine series is dictated by the inherent symmetry of the solution space. Direct solution of the resulting equation for the Fourier series coefficients, a_k, is straightforward using standard matrix operations assuming prior knowledge of the fundamental angular frequency ω. The value of ω is generally unknown.

The difference between the estimated net heat flux distribution on the design surface,q_i, and desired net heat flux,q_d, is well described by the root mean square of relative error,

$$E_{\text{rms}} = \frac{1}{N_D} \sqrt{\sum_{i=1}^{N_D} \left[\frac{q_d - q_i}{q_d}\right]^2} \times 100 \tag{3}$$

Equation (3) is used as the fitness criteria in a GA for the selection of Fourier series fundamental angular frequency, ω.

4 Results

Fourier series with orders K ranging from 1 to 30 were considered. The results for E_{rms} are plotted as a function of the Fourier series order in Fig. 2. It is clear that increasing the order of the Fourier series beyond 6 provides negligible improvement in E_{rms}. Therefore, a Fourier series of order 6 seems to be a reasonable choice for defining the temperature profile of the heater surface. Results obtained here for orders of 6 or more compare well with those reported in the previous efforts [9, 10].

The variation of the fundamental angular frequency with Fourier series order, illustrated in Fig. 3, is intriguing. For a given order, meaningful solutions are obtained only for specific values of ω; that is, ω is an eigenvalue of the problem. The fundamental angular frequency for the current problem peaks at the order, 6, which provides the optimum solution. This suggests the existence of an underlying principle whose judicious exploitation might reveal the eigenvalue without need for iteration. However, realization of this possibility remains elusive.

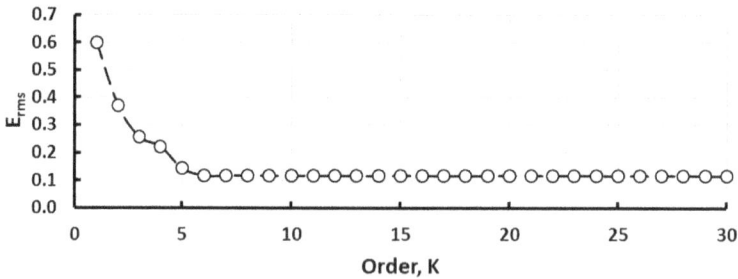

Fig. 2 Variation of the penalty function with the order of the Fourier series

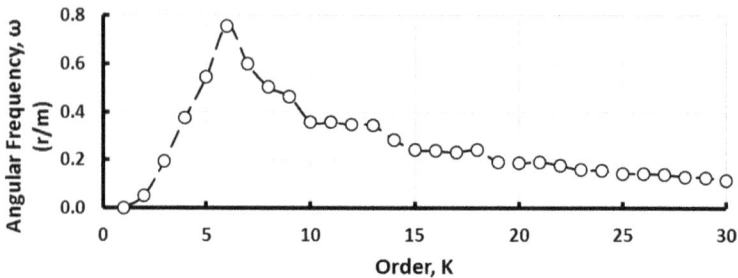

Fig. 3 Results for the fundamental angular frequency versus the order of the Fourier series

Figure 4 displays the temperature profile on the heater surface and the estimated net heat flux profile over the design surface for a Fourier series of order 6. The value of E_{rms} in this case is 0.1161. The horizontal line at 2000 W/m² in Fig. 2b represents the design goal. We see that the net heat flux estimate is within about 0.5 percent of this value over most of the width of the workpiece, with a maximum variation of only about 1–1/2 percent near the ends.

Figure 5 is a direct comparison of the heater temperature profiles obtained using the two approaches. The classical method, based on a genetic algorithm requiring a population of 2000, 60-gene chromosomes, and 5000 generations, yields an E_{rms}

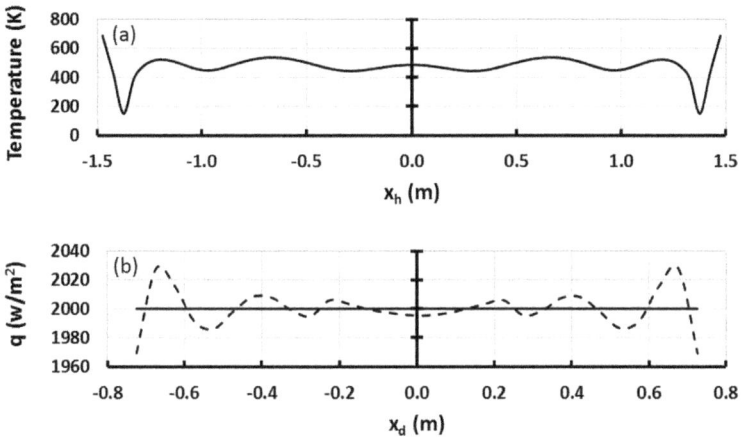

Fig. 4 a Temperature distribution on heater surfaces, and **b** comparison of design (solid line) and calculated (dashed line) net heat flux over design surfaces

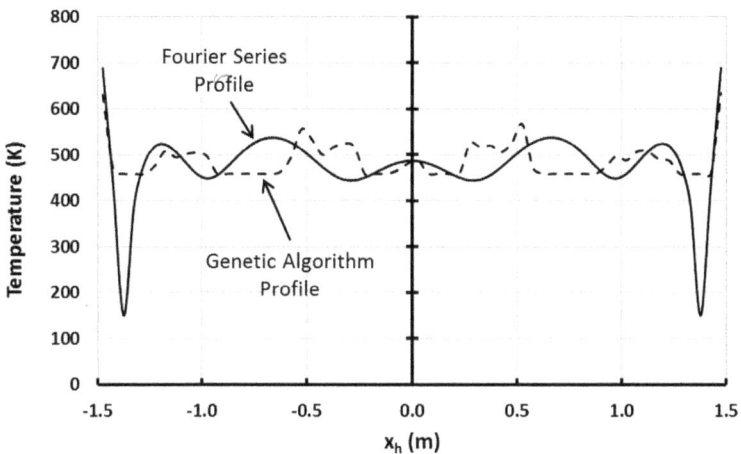

Fig. 5 Comparison of the heater temperature results obtained using the two approaches

value of 0.14514; while the new method introduced here requires only 100, one-gene chromosomes, and 64 generations to produce an E_{rms} value of 0.1162 for a Fourier series of order 6.

5 Conclusions

A new approach is presented for solving inverse boundary design problems in radiation heat transfer. The temperature profile along the heater surface is represented as a Fourier cosine series whose fundamental angular frequency is unknown. The $K + 1$ coefficients of the Fourier series are computed directly using standard matrix methods, with the unknown fundamental angular frequency being determined using a genetic algorithm. The results are compared with those obtained using only a classical genetic algorithm. It is established that the new approach provides more accurate results in a fraction of the execution time.

Acknowledgements The authors gratefully acknowledge NASA's Langley Research Center for its partial financial support for this effort under NASA Contract NNL16AA05C with Science Systems and Applications, Inc., and Subcontract No. 21606-16-036, Task Assignment M.001C (CERES) with Virginia Tech.

References

1. J. Beck, B. Blackwell, C. Clair, *Inverse Heat Conduction* (Wiley, New York, 1985).
2. K. Daun, F. França, M. Larsen, G. Leduc, J. Howell, Comparison of methods for inverse design of radiant enclosures. J. Heat Transfer **128**, 269 (2006). https://doi.org/10.1115/1.2151198
3. H. Ertürk, O. Ezekoye, J. Howell, Comparison of three regularized solution techniques in a three-dimensional inverse radiation problem. J. Quant. Spectrosc. Radiat. Transfer **73**, 307–316 (2002). https://doi.org/10.1016/s0022-4073(01)00212-6
4. A. Safavinejad, S. Mansouri, A. Sakurai, S. Maruyama, Optimal number and location of heaters in 2-D radiant enclosures composed of specular and diffuse surfaces using micro-genetic algorithm. Appl. Therm. Eng. **29**, 1075–1085 (2009). https://doi.org/10.1016/j.applthermaleng.2008.05.025
5. A. Safavinejad, S. Mansouri, S. Sarvari, Inverse boundary design of two-dimensional radiant enclosures with absorbing—emitting media using micro-genetic algorithm. Proc. Inst. Mech. Eng. Part C: J. Mech. Eng. Sci. **221**, 945–948 (2007). https://doi.org/10.1243/09544062jmes154
6. L. Darvishvand, F. Kowsary, P. Jafari, Optimization of 3-D radiant enclosures with the objective of uniform thermal conditions on 3-D design bodies. Heat Transfer Eng. **37**, 1–12 (2015). https://doi.org/10.1080/01457632.2015.1025002
7. J.R. Mahan, *Radiation Heat Transfer: A Statistical Approach*, Wiley (2002)
8. J.R. Mahan, *The Monte Carlo Ray-Trace Method in Radiation Heat Transfer and Applied Optics*, Wiley-ASME Press (2019)

9. M. Mosavati, F. Kowsary, B. Mosavati, A. Novel, Noniterative inverse boundary design regularized solution technique using the backward Monte Carlo method. J. Heat Transfer **135**, 042701 (2013). https://doi.org/10.1115/1.4022994
10. H. Ertürk, J.R. Howell, O. Ezekoye, Inverse solution of radiative heat transfer in two-dimensional irregularly shaped enclosures, in *Proceedings of 2000 IMECE* (2000)

Numerical Investigation of Thermal Dynamic Response in Porous Media—A Pore-Scale Study

Rabeeah Habib, Bijan Yadollahi, and Nader Karimi

1 Introduction

Investigating fluid flow in porous media has been of considerable interest for researchers over the past few decades. Countless applications, such as fluid flow and heat transfer in compact heat exchangers, packed beds, aerosol transport and blood flow in vessels, are all reliant on the behaviours of the flow in the porous media [1].

Several areas of applied science and engineering technologies have come across the interlinked phenomena of fluid flow and heat transfer in porous media. Porous media are widely available in ceramic foams with an open pore structure, having a random dodecahedral-like geometry. Recent availability of porous ceramic foams and discrete ceramic matrices has accelerated research in this field. However, their durability suffers from cracking caused by thermal stress during pore-scale heat transfer [2]. Thus, experimental analyses of fluid flow and heat transfer in porous media are still proven to be a challenge due to the inherent space constraints within the materials, even with the introduction of new experimental techniques and advancements in technology [3]. Also, for several decades, only a handful of experimental studies were undertaken to understand flow physics involving heat transfer. Hence, computational approaches are preferred particularly when the problem at hand has not been previously studied. However, numerical modelling can be thought-provoking because of the restricted knowledge of the fundamentals of thermal, radiative and fluid mechanical processes within porous media [2].

The existing studies on forced convection in porous media have focused predominately on the steady phenomena. As a result, there is a gap on understanding the dynamic response of heat transfer in porous media to time varying flows. Yet, such

R. Habib (✉) · B. Yadollahi · N. Karimi
School of Engineering, University of Glasgow, Glasgow G12 8QQ, UK
e-mail: r.habib.1@research.gla.ac.uk

© Springer Nature Singapore Pte Ltd. 2021   385
C. Wen and Y. Yan (eds.), *Advances in Heat Transfer and Thermal Engineering* ,
https://doi.org/10.1007/978-981-33-4765-6_66

understanding is central to the design of several practical systems including porous burners.

2 Methodology

An unsteady, three-dimensional computational fluid dynamic pore-scale model was developed (Fig. 1) to investigate the dynamic response of the porous medium to disturbances in the inlet flow. A grid independency check was carried out on the computational model, and it was validated against the existing experimental studies conducted on a single obstacle.

Figure 1 shows the system under investigation, which is a pore-scale structure, subject to isothermal boundary conditions and an inlet flow with sinusoidal velocity fluctuations. The working fluids which include air, hydrogen (H_2), carbon dioxide (CO_2) and the particle elements had a constant temperature of 700 K. Time-dependent Nusselt number was calculated over each elements of the porous structure to evaluate the thermal response of the system. A parametric study was subsequently conducted in which the Reynolds number, amplitude and frequency of the inlet velocity disturbances and the porosity of the structure were varied systematically.

Two different fundamental approaches were then undertaken to analyse the dynamics of the system. First, the system was assumed to act as a classical linear, dynamic system, and thus, a transfer function was devised between the inputs (fluctuating flow) and outputs (Nusselt number over obstacles). Second, the system was assumed to be totally nonlinear, and the inlet and outlet relations were studied on a phase portrait. Next, the validity of linear assumption was assessed through introduction of a measure of nonlinearity, defined as

$$\mu = \frac{x - y}{x} \tag{1}$$

where μ is the measure of nonlinearity, x, Euclidean distance of the normalized Nusselt number recorded at each particle and y, discrete Fourier transform single-sided amplitude spectrum of the normalized Nusselt number at each particle.

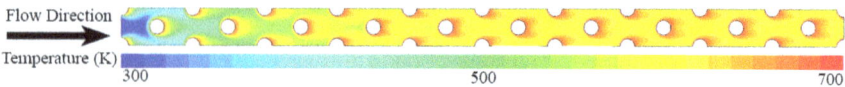

Fig. 1 Schematic view of the developed porous model

3   Results

Figure 2 shows the transfer function between the average Nusselt number of different cylinders and the fluctuation in the inlet velocity. In this figure, the frequency of velocity fluctuation has been expressed as a Strouhal number on the basis of bulk flow velocity and diameter of the cylinders. It is clear from Fig. 2a that the amplitude of the transfer function closely resembles a low-pass filter. This shows that the system has a high tendency to respond to low frequencies but strongly damps the higher frequencies. In reality, the flow disturbances are non-sinusoidal and can have any arbitrary temporal form. Nonetheless, they can be always viewed as an ensemble of harmonic disturbances with different amplitudes. The observed low-pass filter implies that those disturbances with characteristic time scales shorter than the cut-off frequency of the filter do not have any influence upon the thermal response of the system. Figure 2b displays the lag of transfer function of the first five obstacles (C1–C5), of the porous system, operating on air with Reynolds number of 50 and a porosity of 0.874. In this figure, the phase difference between the input (velocity fluctuation) and output (Nusselt number fluctuations) has been converted to the time delay between these two signals. Such delay was then non-dimensionalized by dividing with the average convective time. The overall trend in Fig. 2 indicates a typical convective lag. Nonetheless, there appears to be a change in the phase speed of the disturbance as it travels through the first few pores of the system.

Figure 3 displays a surface plot of the maximum measure of nonlinearity with the changes in porosity and Reynolds number for the three-dimensional porous model of air across all ten obstacles and six frequencies. The largest measure of nonlinearity occurs at point E with value of around 0.75, and the smallest at point G with a numerical value close to zero. Importantly, this shows that the extent of nonlinearity does not feature a simple relation with the Reynolds number and varies with porosity as

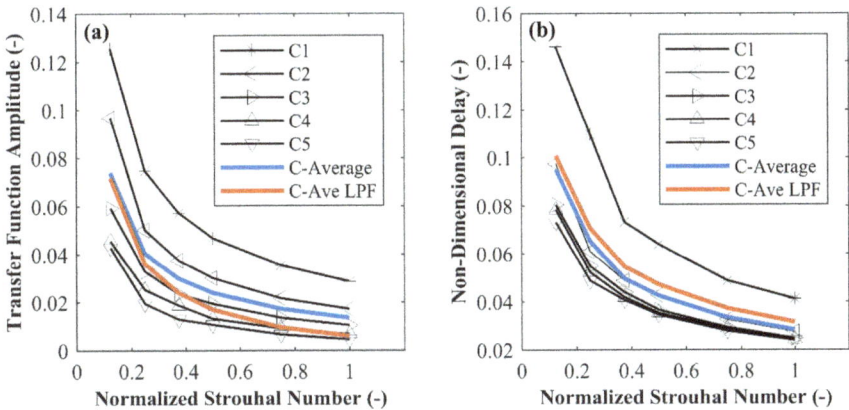

Fig. 2 a Air_410_050_V—Amplitude of transfer function on cylinders 1–5, **b** Air_410_050_V—phase of transfer function on cylinders 1–5

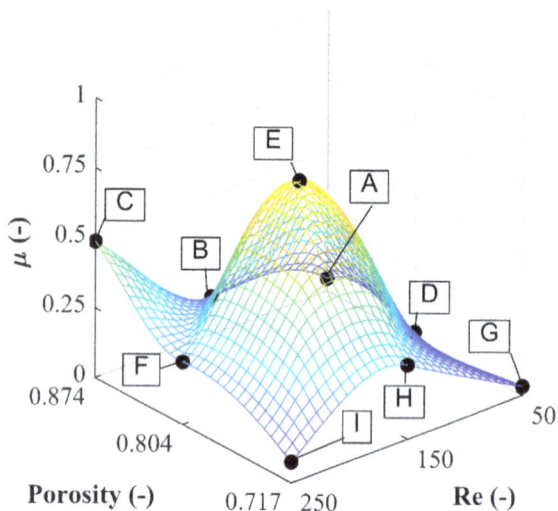

Fig. 3 Max nonlinearity factor in Nusselt number in air—3D. **a** St = 0.125, C9 **b** St = 0.125, C7 **c** St = 0.125, C2 **d** St = 0.125, C2 **e** St = 0.125, C8 **f** St = 0.125, C2 **g** St = 1, C2 **h** St = 0.125, C6 **i** St = 0.25, C2

well. Although not shown in here, changes in type of fluid also affect the deviation of the system response from linearity with CO_2 also showing the maximum nonlinearity to be at point E. An interesting result shown by Fig. 3 is the non-monotonic dependency of measure nonlinearity upon Reynolds number. Also, it is can be seen, seven out of nine points for maximum nonlinearity occur in the lowest frequency. This trend is consistent and visible across all fluids, reaffirming that the system in general responds to lower frequencies but strongly damps higher frequencies. In general, the system can either diverge from or approach a linear response by increasing the Reynolds number or increasing the frequency.

4   Conclusions

A parametric study was carried out with varying input parameters and working fluids inside a three-dimensional fluid dynamic pore-scale model. The system was evaluated with two central approaches. First, a transfer function was developed assuming that the system behaves linearly with the fluid flow as the input and the Nusselt number as the output. This demonstrated that the porous model acts as a low-pass filter; responding strongly to lower frequencies but strongly damping higher frequencies. Although when the system is assessed through the introduction of a measure of nonlinearity, the thermal response of the system in some instances includes significant nonlinearities. It is concluded that the use of inherently linear models such as the Darcy equation, although might be acceptable under steady condition, can be highly questionable under dynamic forced convection.

Acknowledgements The authors acknowledge the partial support of EPSRC through grant number EP/N020472/1.

References

1. M. Torabi, G.P. Peterson, M. Torabi, N. Karimi, A thermodynamic analysis of forced convection through porous media using pore scale modeling. Int. J. Heat Mass Transfer **99**, 303–316 (2016)
2. J.R. Howell, M.J. Hall, J.L. Ellzey, Combustion of hydrocarbon fuels within porous inert media. Prog. Energy Combust. Sci. **22**, 121–145 (1996)
3. X. Chu, G. Yang, S. Pandey, B. Weighard, Direct numerical simulation of convective heat transfer in porous media. Int. J. Heat Mass Transfer **133**, 11–20 (2019)

Numerical Simulation of Thermal–Hydraulic Performance of a Round Tube-Fin Condenser with Liquid–Vapour Separation

Nan Hua, Ying Chen, and Hua Sheng Wang

1 Introduction

A lumped-parameter model of round tube-fin condensers with (RTFCs-LS) and without (RTFCs) liquid–vapour separation was proposed by Hua et al. [1]. This model appears accurate, simplified and low computational cost and was used to optimize the tube-pass strategy of the condenser. Zhong et al. [2–4] implemented the same model to simulate and evaluate the performance of single/duo-slab parallel flow microchannel condensers (PFMCs) with liquid–vapour separation. Luo et al. [5, 6] applied the model for the optimization of the MPFC-LS in organic Rankine circle. This paper reports a distributed parameter model, using the effectiveness and number of transfer units (ε-NTUs) method, to numerically simulate heat transfer performance of RTFCs-LS and RTFCs.

2 Description of the Condenser with Liquid–Vapour Separation

As seen in Fig. 1, the RTFC-LS has multi-pass parallel flow configuration and uses round tubes. In the tube side, tubes are divided into several passes separated by baffles in both headers. Compared with conventional multi-pass parallel flow condensers, the baffles separate condensate from the main stream. The baffles have purposely

N. Hua · H. S. Wang (✉)
School of Engineering and Materials Science, Queen Mary University of London, London E1 4NS, UK
e-mail: h.s.wang@qmul.ac.uk

Y. Chen
Faculty of Materials and Energy, Guangdong University of Technology, Guangdong 510006, China

© Springer Nature Singapore Pte Ltd. 2021   391
C. Wen and Y. Yan (eds.), *Advances in Heat Transfer and Thermal Engineering* ,
https://doi.org/10.1007/978-981-33-4765-6_67

Fig. 1 Schematic diagram
of the RTFCs-LS

designed uniform diameter orifices of approximately 0.5 mm to 2.0 mm. When two-phase refrigerant from the previous tube-pass flows into the intermediate header, the condensate of the stream accumulates on the baffle surface and forms a liquid film. The liquid film only allows the condensate to pass through the orifices due to the capillary force and pressure difference between the two sides of the baffle. The two-phase main stream with increased vapour quality flows into the next flow pass. This is the working principle of the liquid–vapour separation. The baffles serve as the liquid–vapour separators.

3   Description and Validation of Model

A two-dimensional model is developed in the present work. The entire condenser is divided into three levels: (1) the condenser is divided into tube-passes separated by the baffles in the headers; (2) a tube-pass part consists of several branches, i.e. tubes connecting to the headers at their two ends; (3) each branch consists of three types of elements including a dividing T-junction, a combining T-junction and tube segments. Each finite tube segment is a tube-centred element, which can be treated as an independent cross-flow arrangement between air flow outside the tube and refrigerant flow inside the tube.

The ratios of predicted and measured heat transfer rates, Q_{pre}/Q_{exp}, and pressure drops, $\Delta P_{pre}/\Delta P_{exp}$, are plotted as functions of χ_{ave} in Fig. 2. The experimental data using R134a include the data of the χ_{ave} at the range of 0.27–0.72 and 0.30–0.66 for heat transfer rate of 1125 W and 1525 W, respectively, under the fixed conditions that the inlet pressure is 1160 kPa and the refrigerant mass flux is 533 kg/m^2 s. The root-mean-square errors (r.m.s) of predicted and measured heat transfer rate and pressure drop for the RTFC-LS are within ±7.5% and ±20.6%, respectively.

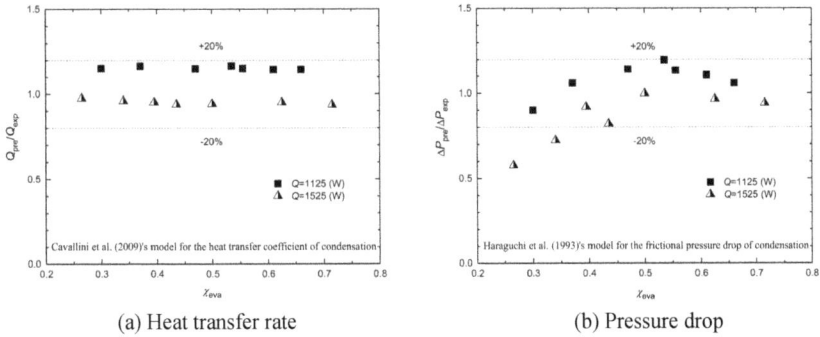

(a) Heat transfer rate   (b) Pressure drop

Fig. 2 Comparison of predicted and measured results with vapour quality

4 Results and Discussion

For the experimental conditions of refrigerant R134a, inlet mass flux 533.0 kg/m^2 s, saturation temperature 45.0 °C, average vapour quality 0.6, average heat flux 5.0 kW/m^2 and of air, inlet temperature 35.0 °C, the simulation results, as shown in Fig. 3, indicate that the average *HTC* on the refrigerant side of the RTFC-LS is 4548.0 W/m^2 K, which is 34.6% higher than that of 3378.0 W/m^2 K of the RTFC. The pressure drop on the refrigerant side of the RTFC-LS is 6.8 kPa, which is 74.2% lower than that of 26.3 kPa of the RTFC. These improvements in the RTFC-LS are caused by the increased average vapour qualities and the decreased mass fluxes inside the tube-passes downstream due to the liquid–vapour separation.

The mechanisms of the improvements are also demonstrated in the distributions of the wall temperatures for the RTFC-LS and RTFC as shown in Fig. 4. The inlet air temperatures for the two condensers are set as identical, higher wall temperatures distributed of the RTFC-LS mean larger temperature difference driving the heat transfer in the air side, which could lead to lower air velocity required on the condition of fixed heat transfer rate. This can lower the fan power consumption and noise pollution.

5 Conclusions

A distributed parameter model was proposed. The predictions of this model are in good agreement with the experimental data. The r.m.s. values of predicted and measured heat transfer capacities and pressure drops are within ±7.5% and ±20.6%, respectively. There are some notable findings from the simulation analysis:

(1) The average *HTC* and pressure drop of the RTFC-LS are 34.6% higher and 74.2% lower than those of the RTFC, respectively.
(2) Higher temperatures distributed in the downstream of the RTFC-LS are depicted.

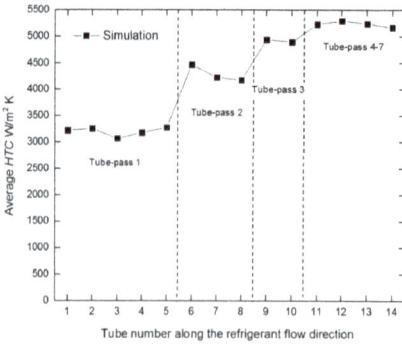
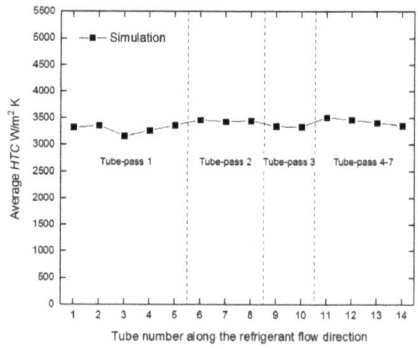
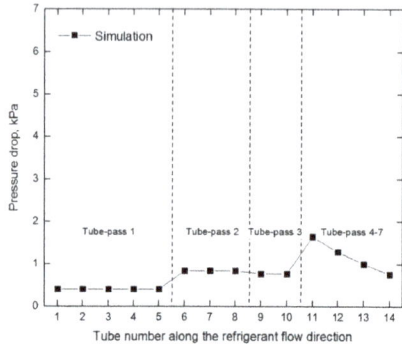
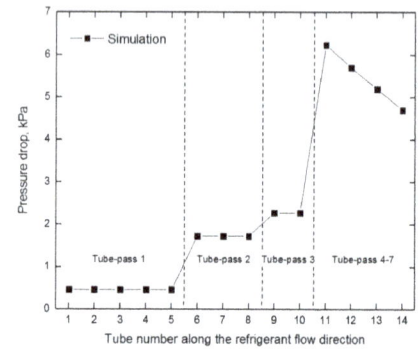

(a) RTFC-LS   (b) RTFC

Fig. 3 Average heat transfer coefficient (*HTC*) and pressure drop of the tubes along the refrigerant flow direction

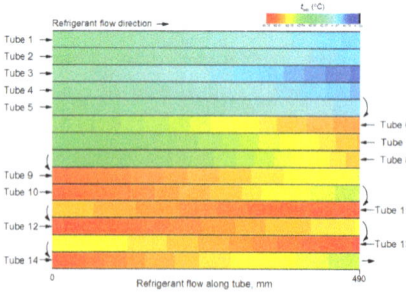
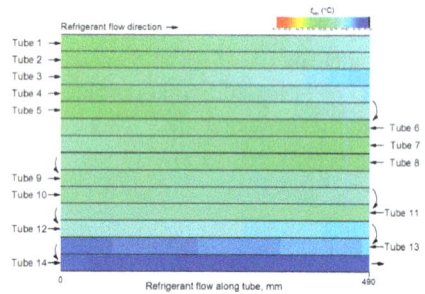

(a) RTFC-LS   (b) RTFC

Fig. 4 Wall temperature distribution

Acknowledgements The authors gratefully acknowledge the financial supports from the Engineering and Physical Sciences Research Council (EPSRC) of the UK (EP/N020472/1), the Royal Society of IEC\NSFC\170543-International Exchanges 2017 Cost Share (China) and the National Natural Science Foundation (NSFC) of China (51736005).

References

1. N. Hua, Y. Chen, E. Chen, L. Deng, W. Zheng, Z. Yang, Prediction and verification of the thermodynamic performance of vapour–liquid separation condenser. Energy **58**, 384–397 (2013)
2. T. Zhong, Y. Chen, N. Hua, W. Zheng, X. Luo, S. Mo, In-tube performance evaluation of an air-cooled condenser with liquid–vapour separator. Appl. Energy **136**, 968–978 (2014)
3. T. Zhong, Y. Chen, W. Zheng, N. Hua, X. Luo, Experimental investigation on microchannel condensers with and without liquid–vapour separation headers. Appl. Therm. Eng. **73**, 1510–1518 (2014)
4. T. Zhong, Y. Chen, Q. Yang, S. Mo, X. Luo, Experimental investigation on the thermodynamic performance of double-row liquid–vapour separation microchannel condenser. Int. J. Refrig. **67**, 373–382 (2016)
5. X. Luo, J. Xu, Y. Chen, S. Mo, Mathematical optimization of the liquid separation condenser used in the organic Rankine-cycle. Energy Proc. **75**, 3127–3132 (2015)
6. X. Luo, Z. Yi, B. Zhang, S. Mo, C. Wang, M. Song, Y. Chen, Mathematical modelling and optimization of the liquid separation condenser used in organic Rankine cycle. Appl. Energy **185**, 1309–1323 (2017)

Numerical Modelling of Wet Steam Flows in Turbine Blades

Chuang Wen, Xiaowei Zhu, Hongbing Ding, and Yan Yang

1 Introduction

Steam turbine is widely used in fossil-fired power plants. During the expansion of steam in the transonic flow inside the turbine cascade, the state path may cross the saturation line which induces the homogeneous condensation. The produced liquid existing in the fluid flow results in energy losses, which leads to the decrease of turbine efficiency [1]. A series of studies has been carried out both numerically [2] and experimentally [3, 4] to improve energy efficiencies by considering two-phase flows. However, the steam flow is still not fully understood due to the complicated interactions of transonic flow, phase transition, boundary layers, and shock waves. In this study, a wet steam model is developed to predict steam condensations in transonic flows within turbine blade cascades. The condensation parameter is analysed including the static pressure, subcooling, nucleation rate, wetness, and droplet number.

2 Numerical Model

The Eulerian-Eulerian approach is employed to model wet steam flows in the turbine blade considering the homogeneous condensation process. Two-dimensional

C. Wen
Faculty of Engineering, University of Nottingham, Nottingham NG7 2RD, UK

X. Zhu · Y. Yang (✉)
Department of Mechanical Engineering, Technical University of Denmark, Nils Koppels Allé, 2800 Kgs. Lyngby, Denmark
e-mail: yanyang2021@outlook.com

H. Ding
School of Electrical and Information Engineering, Tianjin University, Tianjin 300072, China

© Springer Nature Singapore Pte Ltd. 2021
C. Wen and Y. Yan (eds.), *Advances in Heat Transfer and Thermal Engineering* ,
https://doi.org/10.1007/978-981-33-4765-6_68

compressible Navier–Stokes equations are adopted to govern the fluid flow in transonic flows. To describe the phase transition in supersonic flows, two transport equations are utilised including the liquid fraction (ζ) and droplet numbers per unit volume (η) [5]. Due to the tiny size of the condensed droplet, the slip velocity is ignored between vapour and liquid phases in this numerical study [6]. The governing equations are expressed as:

$$\frac{\partial(Q)}{\partial t} + \frac{\partial(F_x)}{\partial x} + \frac{\partial\left(F_y\right)}{\partial y} = \frac{\partial(G_x)}{\partial x} + \frac{\partial\left(G_y\right)}{\partial y} + S \tag{1}$$

where

$$Q = \begin{bmatrix} \rho \\ \rho u \\ \rho v \\ \rho E \\ \rho\zeta \\ \rho\eta \end{bmatrix}, \quad F_x = \begin{bmatrix} \rho u \\ \rho u u + p \\ \rho u v \\ \rho u(E + p) \\ \rho u\zeta \\ \rho u\eta \end{bmatrix}, \quad F_y = \begin{bmatrix} \rho v \\ \rho u v \\ \rho v v + p \\ \rho v(E + p) \\ \rho v\zeta \\ \rho v\eta \end{bmatrix} \tag{2}$$

$$G_x = \begin{bmatrix} 0 \\ \tau_{xx} \\ \tau_{xy} \\ q_x \\ 0 \\ 0 \end{bmatrix}, \quad G_y = \begin{bmatrix} 0 \\ \tau_{xy} \\ \tau_{yy} \\ q_y \\ 0 \\ 0 \end{bmatrix}, \quad S = \begin{bmatrix} -\Gamma \\ -u\Gamma \\ -v\Gamma \\ -(h_v - h_{fg})\Gamma \\ \Gamma \\ \rho J \end{bmatrix} \tag{3}$$

where Q represents conservation variables, F and G are inviscid and viscid fluxes, S is the source term. ρ and p are density and pressure, respectively. E is the total energy, u and v are the velocity components. J is the nucleation rate. Γ is the condensation mass per unit vapour volume per unit time:

$$\Gamma = \frac{4\pi r_c^3}{3}\rho_l J + 4\pi r^2 \rho_l \eta \frac{dr}{dt} \tag{4}$$

$$J = \frac{q_c}{1 + \phi}\frac{\rho_v^2}{\rho_l}\sqrt{\frac{2\sigma}{\pi m_v^3}}\exp\left(-\frac{4\pi\sigma}{3k_B T_v}r_c^2\right) \tag{5}$$

$$\frac{dr}{dt} = \frac{\lambda_v(T_s - T_v)}{\rho_l h_{lv} r}\frac{\left(1 - r_c/r\right)}{\left(\frac{1}{1+2\beta\mathrm{Kn}} + 3.78(1 - v)\frac{\mathrm{Kn}}{\mathrm{Pr}}\right)} \tag{6}$$

where ρ_l is the droplet density, r is the droplet radius. dr/dt is the growth rate of droplets [7], and r_c is critical droplet radius. q_c is the condensation coefficient, σ is the liquid surface tension, m_v is the mass of a vapour molecule, k_B is the Boltzmann's constant. T_v is the vapour temperature. ϕ is a correction factor. T_s is the saturated temperature, Pr is the Prandtl number, Kn is the Knudsen number, and v is the modelling correction coefficient.

The numerical simulation is performed based on the commercial platform ANSYS FLUENT 18.2. The User-Defined-Scalar (UDS) and User-Defined-Function (UDF) interfaces are employed to solve two scalar equations and source terms. The pressure inlet and pressure outlet conditions are assigned for the turbine blade with an assumption of no-slip and adiabatic walls [8].

3   Results

The geometry is adopted from Dykas et al. [3] experiments with linear blade cascades. In the experimental test, the inlet total pressure and temperature were $89{,}000 \pm 250$ Pa(a) and $100 \pm 0.25\,°\mathrm{C}$, respectively. The mean static back pressure measured in a distance of 80 mm downstream from the cascade outlet was $39{,}000 \pm 250$ Pa(a) [3]. The detailed boundary conditions are shown in Fig. 1 including one turbine blade in the simulation with periodic boundary conditions. Figure 2 compares pressure distributions at the pressure and suction sides of a turbine blade. The comparison shows that the numerical simulation agrees well with experimental data. This demonstrates that the wet steam model captures accurately the non-equilibrium condensation inside turbine blades.

Figure 3 describes the contours of subcooling, nucleation rate, and wetness inside turbine blades. Steam expands in the turbine blade and reaches maximum subcooling

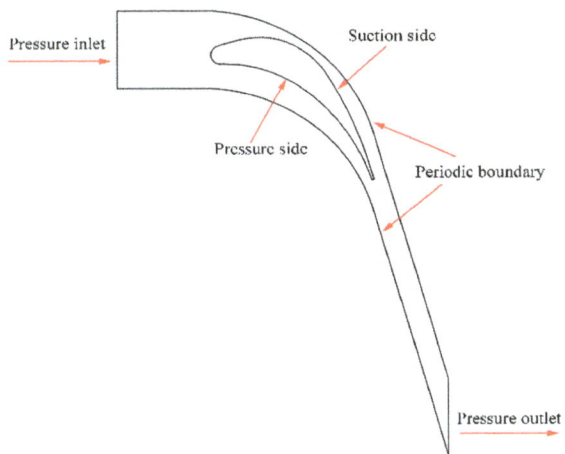

Fig. 1 Geometry and boundary conditions

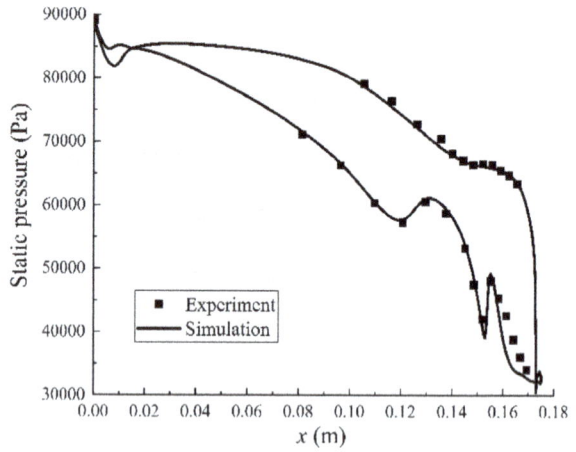

Fig. 2 Pressure distributions at turbine blades [3]

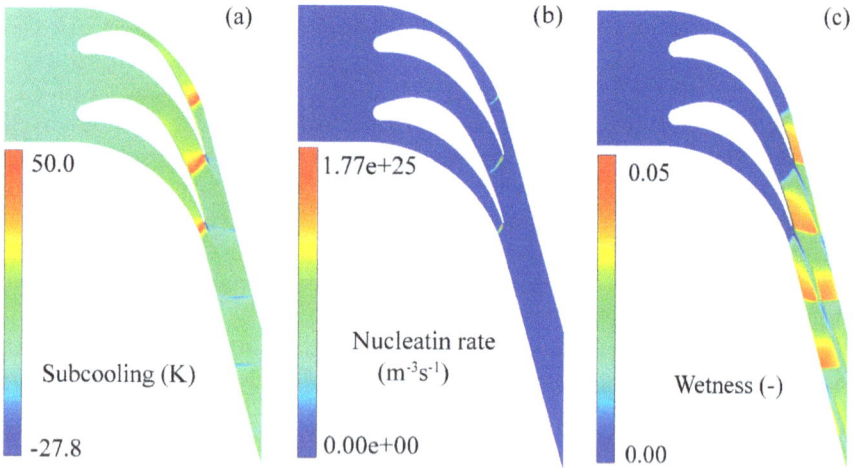

Fig. 3 Contours of subcooling, nucleation rate and wetness inside turbine blades

value around 50.0 around the trailing edge of the blade. This indicates that the steam represents an extremely non-equilibrium state which induces the nucleation process. The maximum value of the nucleation rate is approximately 1.77×10^{25} m^{-3} s^{-1}, and the location where the nucleation occurs is accordant with the subcooling of steam. Following the nucleation process, the liquid wetness is produced and increases along the flow direction in the turbine blade downstream. The liquid wetness can reach 5% of the total mass of steam, leading to energy losses in turbine blades and reduces the turbomachinery efficiency.

4 Conclusions

Wet steam flow is investigated within turbine blades based on the single-fluid model. The numerical simulation shows that the non-equilibrium condensation of steam occurs due to the steam expansion in transonic flows inside turbine cascades. Approximately, 5% liquid wetness of the total mass of steam is produced during the homogeneous condensation. The optimisation of the turbine blade will be performed to reduce the wetness and improve turbomachinery efficiency in future studies.

References

1. E.Y. Rad, M.R. Mahpeykar, Studying the effect of convergence parameter of CUSP's scheme in 2D modeling of novel combination of two schemes in nucleating steam flow in cascade blades. Numer. Heat Transfer Part B: Fundamentals **72**(4), 325–347 (2017)
2. G. Zhang, F. Wang, D. Wang, T. Wu, X. Qin, Z. Jin, Numerical study of the dehumidification structure optimization based on the modified model. Energy Convers. Manage. **181**, 159–177 (2019)
3. S. Dykas, M. Majkut, M. Strozik, K. Smołka, Experimental study of condensing steam flow in nozzles and linear blade cascade. Int. J. Heat Mass Transf. **80**, 50–57 (2015)
4. S. Dykas, M. Majkut, K. Smołka, M. Strozik, Study of the wet steam flow in the blade tip rotor linear blade cascade. Int. J. Heat Mass Transf. **120**, 9–17 (2018)
5. C. Wen, N. Karvounis, J.H. Walther, Y. Yan, Y. Feng, Y. Yang, An efficient approach to separate CO_2 using supersonic flows for carbon capture and storage. Appl. Energy **238**, 311–319 (2019)
6. Y. Yang, X. Zhu, Y. Yan, H. Ding, C. Wen, Performance of supersonic steam ejectors considering the nonequilibrium condensation phenomenon for efficient energy utilisation. Appl. Energy **242**, 157–167 (2019)
7. J. Young, The spontaneous condensation of steam in supersonic nozzle. Physico Chem. Hydrodyn. **3**(1), 57–82 (1982)
8. C. Wen, N. Karvounis, J.H. Walther, H. Ding, Y. Yang, Non-equilibrium condensation of water vapour in supersonic flows with shock waves. Int. J. Heat Mass Transf. **149**, 119109 (2020)

Evaluation of Neck Tissue Heat Transfer in Case of Stenosis in the Carotid Artery

Ashish Saxena, Vedabit Saha, and E. Y. K. Ng

1 Introduction

The convective heat exchange through the blood flow in the carotid artery can be modelled as one of the major heat sources to the neck tissue bio-heat conjugate heat transfer [1]. For the given ambient thermal conditions, this conjugate heat transfer will result in external neck skin surface temperature map. Any abnormality in the carotid blood flow shall affect the conjugate heat transfer to the neck tissue, and hence, will be reflected in the temperature map of the neck skin [2]. Carotid stenosis, a disease where thrombosis formation reduces the blood flow passage as a result of endothelial cell dysfunction, may lead to debilitating stroke [3]. In the past, many studies on the stenosis altered carotid artery hemodynamics were done; however, its effect on the neck skin temperature map is not found in the literature [4]. In the present study, patient-specific flow and stenosis geometry characteristics are numerically modeled in a fictitious tissue encapsulated three-dimensional (3D) carotid artery geometry.

2 Methodology

Using the angiography geometry data available in the literature [5, 6], an idealized 3D carotid artery accompanied by the jugular vein, encapsulated in the neck tissue, was reconstructed (Fig. 1a). Eight patients (63 ± 7 years old) were recruited under an approved ethical study (CIRB Ref No.: 2017/2119). On each patient, Duplex

A. Saxena · E. Y. K. Ng (✉)
School of Mechanical and Aerospace Engineering, Nanyang Technological University, Singapore, Singapore
e-mail: mykng@ntu.edu.sg

V. Saha
Department of Mechanical Engineering, Manipal University Jaipur, Jaipur, India

© Springer Nature Singapore Pte Ltd. 2021 403
C. Wen and Y. Yan (eds.), *Advances in Heat Transfer and Thermal Engineering* ,
https://doi.org/10.1007/978-981-33-4765-6_69

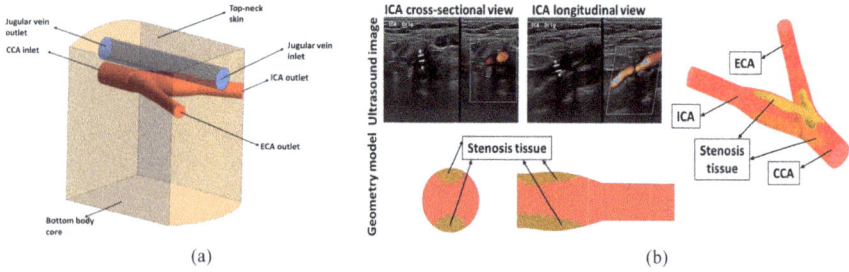

Fig. 1 a 3D model, **b** Carotid model incorporated with the patient-specific stenosis details from the ultrasound

Ultrasound examination was performed to evaluate the extent of stenosis in both the carotid arteries (left and right side), and hence the two groups of patients were defined as normal and diseased. While in the normal group, both the carotid arteries were normal (0% stenosis, N_N=8), in the diseased group, stenosis in the range of 20% to 100% ($43 \pm 17\%$) was found at various locations of the arteries on either side of the neck (N_D=8). For the diseased group, the measured geometrical dimensions of the stenosis were incorporated in the carotid artery model (Fig. 1b). The common carotid artery (CCA) peak systolic velocity inlet (fully developed flow at 37 °C) was calculated based on the patient-specific Duplex Ultrasound measured Reynolds number (Normal: 2158 ± 276 versus Diseased: 1884 ± 268, p-value = 0.03), while both the internal (ICA) and external (ECA) carotid artery outlets were given a zero gauge pressure boundary condition.

Laminar time-independent Navier–Stokes (blood flow) and energy (conjugate heat transfer) equations were solved using finite volume discretization, with Semi-Implicit Pressure Linked Equations (SIMPLE) algorithm, available in the FLUENT software (version 19.1) by ANSYS Inc. The side (left, right, front, and back), bottom (body core), and top walls (neck skin) of the neck encapsulation were given an adiabatic, isothermal (37 °C), and ambient natural convective (T_{amb} =21 °C, $h = 10 \, \text{m}^{-2} \, \text{C}^{-1}$) heat transfer boundary conditions, respectively. The jugular vein (same diameter as CCA), in all the simulations, was given a fully developed inlet velocity of 0.22 m/s [1], temperature of 35.6 °C [7], and zero gauge pressure outlet. The neck tissue encapsulation is thermally modelled using Pennes bio-heat equation [8]; a tissue perfusion rate of 0.4333 kg/ms and a volumetric heat generation of 33 W/m³ were used [9, 10]. A convergence criterion of 10^{-10}, for velocity and temperature approximation residual, was applied. Following a mesh independent study, all the geometry models ($N_N + N_D = 16$) were meshed with an approximate overall mesh size of 5×10^6 cells.

3 Results

For the given inflow conditions at CCA, the flow division between ICA:ECA was found to be 64:36 for the normal cases, while this ratio was found to be 67:33 for the diseased cases (excluding the 100% blocked ICA case). Applying a statistical student t-test to this ratio shows that the two groups of patients are significantly different in terms of flow division (Table 1). Given the blockage in ECA was higher as compared to ICA, in most of the cases, the flow division ratio is higher for diseased cases as compared to the normal.

Figure 2 shows a sample velocity streamlines and top-neck skin temperature contour from the two groups of patients. For the diseased patient results in Fig. 2, as diagnosed in the Duplex Ultrasound examination, the CCA, ICA, and ECA are 30%, 50%, and 55% blocked, respectively. This has a direct effect on the velocity streamlines, higher velocity region before the bifurcation due to reduction in the cross-section area. The presence of stenosis is accompanied with the formation of recirculation zones in the vicinity of the stenosis [11]. Given the low volumetric flow and formation of recirculation zones, the heat transfer to the top-neck skin surface is expected to be negatively affected in the carotid artery with stenosis cases. As observed in Fig. 2, the top-neck skin, in the diseased case, indicates an obvious low temperature contours when compared with the normal case. For the normal cases, the average temperature on the top-neck skin was found to be slightly higher but significantly different as compared to the diseased cases (Table 1).

To further quantify the low temperature contour in the top-neck skin temperature map, a ratio (R_n) of number of low temperature nodes (with temperature less than the threshold temperature, T_{th}) divided by the total number of nodes can be evaluated. The value of T_{th} can be selected as the maximum of minimum temperature occurring in all the cases (normal and diseased) under consideration. In the present study, the value of T_{th} was found to be 33.21 °C. As compared to the normal cases, the value of R_n was found to be around 5.5 times higher in the diseased cases (Table 1).

Table 1 Simulation result parameters for normal and diseased group of patients

Parameters	Normal ($N_N = 8$)	Diseased ($N_D = 8$)	p-value (student t-test)
ICA: ECA mass flow ratio	1.74 ± 0.01	2.09 ± 0.32	0.03
Top-neck skin average temperature (°C)	33.33 ± 0.03	33.24 ± 0.03	5.9×10^{-6}
R_n	0.10 ± 0.15	0.53 ± 0.07	1.9×10^{-5}

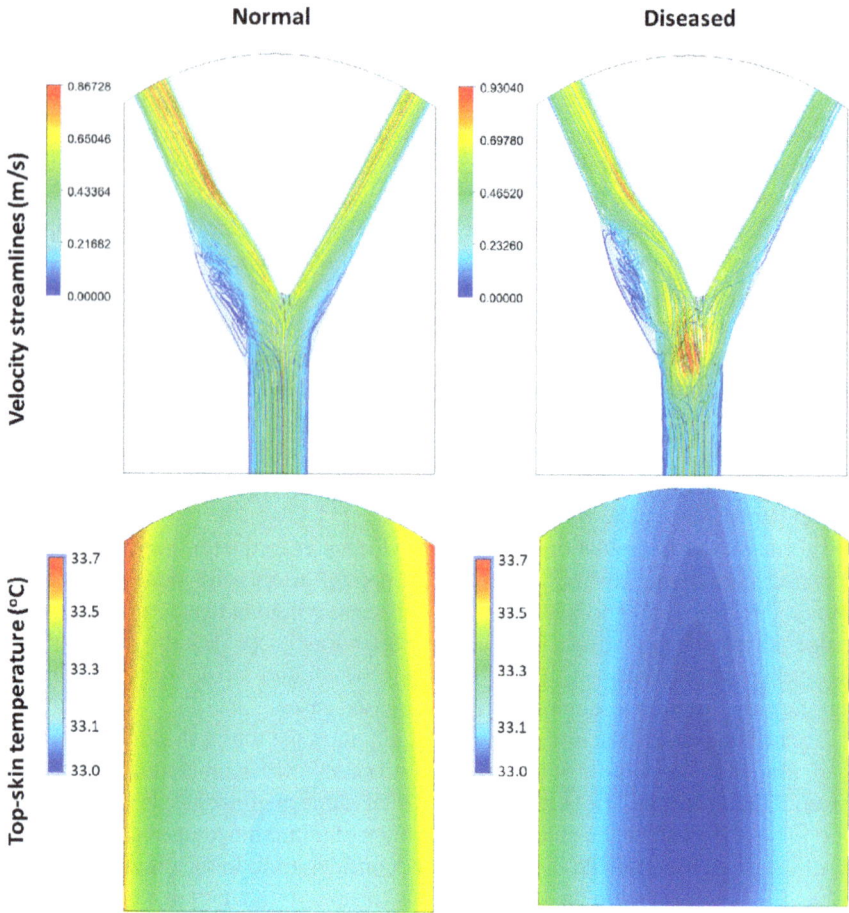

Fig. 2 Streamlines and top-neck skin temperature contour for one sample each from the normal and diseased group of patients

4 Conclusions

First of its kind, this study incorporates patient-specific flow and geometry conditions of the carotid artery (with and without stenosis) to computationally analyze conjugate bioheat transfer phenomenon. The effect of carotid artery hemodynamics, in the presence of stenosis, on the top-neck skin temperature contour was investigated. Significant difference in the top-neck skin temperature contour was found. The results show the potential of neck skin surface temperature map in evaluating the presence of stenosis in the carotid artery.

Acknowledgements The authors would like to acknowledge the SingHealth-NTU collaborative research grant (Grant number: SHS-NTU/014/2016) for the funding support to carry out this project.

The authors would also like to acknowledge the NTU-India Connect program that availed the authors to collaborate and complete the study.

References

1. L. Zhu, Theoretical evaluation of contributions of heat conduction and countercurrent heat exchange in selective brain cooling in humans. Ann. Biomed. Eng. **28**, 269–277 (2000)
2. Y. Yang, J. Liu, Detection of atherosclerosis through mapping skin temperature variation caused by carotid atherosclerosis plaques. J. Therm. Sci. Eng. Appl. **3**, 1–9 (2011)
3. J. Sanz, Z. Fayad, Imaging of atherosclerotic cardiovascular disease. Nature **451**(7181), 953–957 (2008)
4. A. Saxena, E.Y.K. Ng, S.T. Lim, Imaging modalities to diagnose carotid artery stenosis: progress and prospect. Biomed. Eng. Online **18**(1), 66 (2019)
5. B.K. Bharadvaj, R.F. Mabon, D.P. Giddens, Steady flow in a model of the human carotid bifurcation. Part I—flow visualization. J. Biomech. **15**(5), 349–362 (1982)
6. J. Breeze, A. West, J. Clasper, Anthropometric assessment of cervical neurovascular structures using CTA to determine zone-specific vulnerability to penetrating fragmentation injuries. Clin. Radiol. **68**(1), 34–38 (2013)
7. C.M. Crowder, R. Tempelhoff, M.A. Theard, M.A. Cheng, A. Todorov, R.G. Dacey, Jugular bulb temperature: comparison with brain surface and core temperatures in neurosurgical patients during mild hypothermia. J. Neurosurg. **85**(1), 98–103 (1996)
8. H.H. Pennes, Analysis of tissue and arterial blood temperature in the resting human forearm. J. Appl. Phisiology **1**, 93–122 (1948)
9. A. Saxena, V. Raman, E.Y.K. Ng, Single image reconstruction in active dynamic thermography: a novel approach. Infrared Phys. Technol. **93**, 53–58 (2018)
10. A. Saxena, E.Y.K. Ng, V. Raman, Thermographic venous blood flow characterization with external cooling stimulation. Infrared Phys. Technol. **90**, 8–19 (2018)
11. A. Saxena, E.Y.K. Ng, Steady and pulsating flow past a heated rectangular cylinder(s) in a channel. J. Thermophys. Heat Tr. **8722**, 1–13 (2017)

CFD-Enabled Optimization of Polymerase Chain Reaction Thermal Flow Systems

Hazim S. Hamad, N. Kapur, Z. Khatir, Osvaldo Querin, H. M. Thompson, and M. C. T. Wilson

1 Introduction

The polymerase chain reaction (PCR) is a significant technique used to make thousands to millions of copies of particular DNA, and involves a process of heating and cooling called thermal cycling which can be conducted by a PCR device. The PCR process requires exposing the sample to different temperature zones which are denaturation (95 °C), annealing (54 °C), and extension (72 °C) [5]. These three temperature levels are essential in the PCR reaction, and this process is considered to have failed in the absence of one of these cycles. The conventional PCR that is commercially available consists of 96 wells containing the reagent mixture and DNA sample [6]. The thermal cycle of this type of PCR is controlled by liquid or air circulation to heat up and cool down the overall chamber in order to achieve the PCR steps, so this results in large thermal mass and is time-consuming (about $1-2$ h) and requires high power consumption [6, 7]. Recently, with advances in the field of miniaturization and microelectromechanical systems (MEMS) technology, which has been used to fabricate microfluidic devices [8], the trend has changed from macro-PCR to microfluidic-based PCR, which offers several advantages including smaller sample volumes, total reaction times, and power and material consumption [6, 8]. Micro-PCR based on a stationary reaction chamber made of silicon was first developed in 1993 by Northrup [9], in which a fast and efficient PCR compared to conventional PCR was

H. S. Hamad (✉) · N. Kapur · O. Querin · H. M. Thompson · M. C. T. Wilson
School of Mechanical Engineering, University of Leeds, Leeds, UK
e-mail: mnhsh@leeds.ac.uk

H. S. Hamad
Midland Refineries Company, Ministry of Oil, Baghdad, Iraq

Z. Khatir
School of Engineering and the Built Environment, Birmingham City University, Birmingham B4 7XG, UK

© Springer Nature Singapore Pte Ltd. 2021
C. Wen and Y. Yan (eds.), *Advances in Heat Transfer and Thermal Engineering* ,
https://doi.org/10.1007/978-981-33-4765-6_70

obtained. The design of micro-PCR was improved to perform continuous flow polymerase chain reaction (CFPCR) [10], which dramatically shortened PCR process times [11]. The CFPCR process is achieved by cycling the PCR mixture through three PCR zones to achieve the desired temperature instead of heating and cooling all the entire chip. Therefore, the heating and energy consumption are reduced and the thermal cycles could be achieved faster than in a stationary PCR-based chamber [12]. A CFPCR channel was first presented by Kopp et al. [12] in 1998 who designed serpentine cyclic identical channels etched on a glass chip and equipped with an independent copper block to support temperature uniformity. In addition, a temperature gradient technique was employed to simplify the process of creating the desired temperature zones. Crews et al. [13] presented channels with thirty and forty serpentine PCR cycles, made of glass and heated and cooled by two thermal aluminum strips placed underneath. A comparison between DNA amplification results with commercial macroscale devices was achieved and the amplification time of control samples was 10 min for the 30-cycle PCR and 13 min for the 40-cycle PCR. Moschou et al. [14] presented thirty serpentine cycle channels made of polyimide with relative time ratios of 1:1:2 for denaturation, annealing, and extension, respectively, and three integrated resistive copper heaters as heating elements. They achieved efficient DNA amplification within 5 min. Schneegaß et al. [15] fabricated a 25 cycle PCR device made of silicon and glass. The device consisted of reaction chambers etched into a glass chip and covered by a silicon chip and equipped with heaters. They obtained residence times of a sample volume for the 25 cycle device of approximately 25 min, and the energy consumption was 0.012 kWh for a 35 min of the PCR process. Hashimoto et al. [16] presented a 20-cycle microchannel in a spiral shape made of polycarbonate to study and evaluate the effect of high velocity on thermal and biochemical characteristics in CFPCR. Zhang et al. [17] used ANSYS Fluent to optimize the effect of different material on CFPCR chip design, but the flow effect was neglected. Chen et al. [18] presented an analytical study to investigate the influence of geometrical chip parameters and various chip materials on temperature uniformity with no flow condition.

Previous studies have focused on comparatively small fluidic PCR channels in order to achieve the desired temperature by decreasing the effect of convection, albeit with larger pressure drop. A number of questions regarding the relationships among residence time, heating power, and pressure drop and temperature uniformity still need to be addressed. Since there is a wide design space in terms of geometrical, flow, and thermal parameters, and important constraints on the process, research in microfluidic PCR systems could greatly benefit from the multi-objective optimization studies. With this motivation, this paper aims to use CFD-based optimization to investigate the effect of microfluidic geometry and operating conditions on the effectiveness of PCR design within each of the three temperature zones.

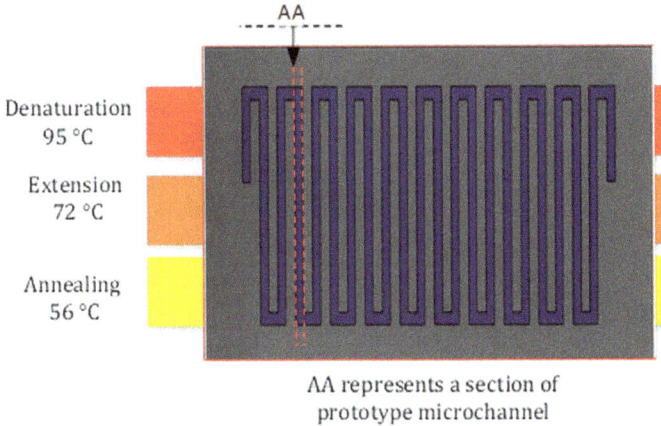

Denaturation
95 °C

Extension
72 °C

Annealing
56 °C

AA represents a section of
prototype microchannel

Fig. 1 PCR process and design

2 PCR Design

The CFPCR chip under consideration has been designed based on a single microflu-idic channel etched on glass for its low thermal conductivity, bounded by a PMMA cover for its low cost and good optical access and three copper blocks with cartridge heaters placed underneath the glass chip. The microfluidic channel passes through isothermal temperature regions as shown in Fig. 1. Water has been chosen as the working fluid, the inlet temperature is 20 °C, and the Reynolds number is fixed at 0.7 and the outside wall is assumed to be insulated. Moreover, the bottom of the copper heaters was set at 95 °C, 72 °C, and 56 °C, respectively, and the distance between heaters S is kept constant 1.5 mm, the length of heater (L_h) at each zone is 9 mm, and the height of the heater H_h (100 μm). The schematic diagram of the channel geometry is described in Fig. 2 where W_c(μm), H_c (μm), W_w (μm), H_b (μm), L (mm) are, respectively, the microchannel width, height, the wall thickness W_w μm, the bottom height, and the length.

3 Modeling

Steady state, three-dimensional single-phase water flow and heat transfer in the PCR microchannel is modelled using the governing Eqs. (1)–(4), comprising the continuity and Navier-Stokes equations together with the energy equations for the liquid and solid:

$$\nabla \cdot (\rho \boldsymbol{u}) = 0 \ (\text{continuity equation}) \tag{1}$$

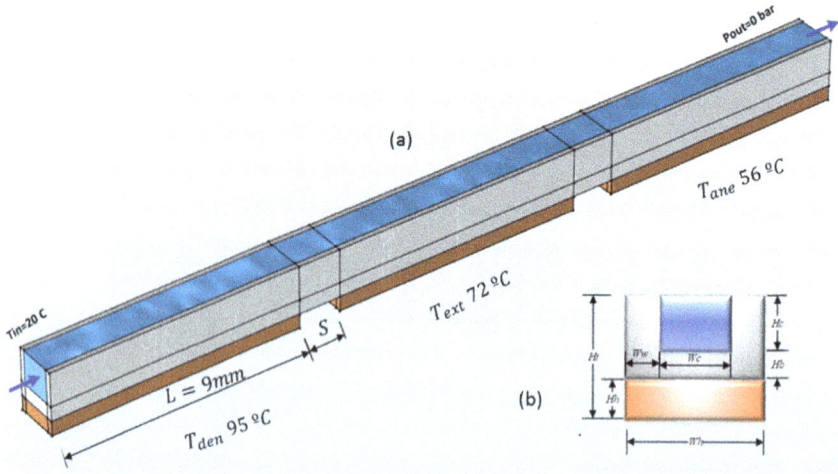

Fig. 2 Schematic diagram of microfluidic channel

$$\rho_f(\mathbf{u} \cdot \nabla \mathbf{u}) = -\nabla p + \mu \nabla^2 \mathbf{u} \text{ (momentum equation)} \tag{2}$$

$$\rho_f C_{pf}(\mathbf{u} \cdot \nabla T) = k_f \nabla^2 T \text{ (Heat transfer (energy) equation for the liquid)} \tag{3}$$

$$k_s \nabla^2 T_s = 0 \text{ (Heat transfer (energy) equation for the solid)} \tag{4}$$

where u and p are the fluid velocity vector [m/s] and the fluid pressure (Pa), respectively, and C_{pf}, ρ_f, k_f, and k_s represent the specific heat, density, and thermal conductivity of the fluid and thermal conductivity of the solid, respectively. COMSOL Multiphysics® version 5.4 is used to solve this conjugate heat transfer model by assuming the fluid flow to be incompressible, Newtonian, and laminar. Radiative heat transfer is neglected, and there are no internal heat sources.

4 Numerical Validation

The numerical model has been validated by comparison with the numerical results of Toh et al. [19] and experimental results of Tuckerman [20]. For the purpose of this comparison, the constant-temperature boundary condition and Reynolds number were replaced with a heat flux $q\left(\text{W/cm}^2\right)$ and a volumetric flow rate $\dot{V}\left(\text{cm}^3/\text{s}\right)$ and it showed a good agreement with available numerical and experimental results as listed in Table 1.

Table 1 Thermal resistances validation of the present work against Tuckerman [20] experimental work, and numerical data of Toh et al. [19]

Case	$q\left(\text{W/cm}^2\right)$	$\dot{V}\left(\text{cm}^3/\text{s}\right)$	$R_{\text{th}}\left(\text{cm}^2 \text{ K/W}\right)$			
			Tuckerman [20]	Toh et al. [19]	Present numerical results	Error (%)
1	181	4.7	0.110	0.157	0.149	0.35
2	277	6.5	0.133	0.128	0.115	0.13
3	790	8.6	0.090	0.105	0.0958	0.06

5 Optimization

The optimization of PCR to minimize the temperature deviation (T_{dev}) and pressure drop Δp has been conducted to explore the influence of two design variables W_c and H_c,. The optimization problem is defined as:

$$\text{Objective function } \min(\text{STD}) \text{ and } \min(\Delta P)$$
$$\text{Subject to: } 150\,\mu\text{m} < W_c < 500\,\mu\text{m} \; ; 50\,\mu\text{m} < H_c < 150\,\mu\text{m}$$

After choosing the range of the design variables, eighty design of experiments (DOE) points were generated and distributed within design space using D-Optimal design technique. Then, these DOE points were used to provide input parameters for the CFD model implemented in COMSOL Multiphysics® 5.4 which used to determine the corresponding ΔP and, T_{dev} values. MATLAB code was used to create metamodels surfaces based on these CFD results and to build surrogate models using cubic Radial Basis Functions (RBF) throughout the design space. In the present study, PCR is designed according to the conflicting requirements of minimising temperature deviation T_{dev} and pressure drop ΔP. The T_{dev} with respect to target temperatures at each PCR zones was determined according to the volumetric integral:

$$T_{\text{dev}} = \left(\iiint \left(T_{fi,j,k} - T_{\text{target}}\right)^2 \mathrm{d}V \right) / \iiint \mathrm{d}V. \tag{5}$$

The second objective is the Δp which is calculated based on COMSOL results as:

$$\Delta P = P_{\text{in}} - P_{\text{out}} \tag{6}$$

where P_{in} and P_{out} represent the pressure at channel inlet and outlet, respectively.

6 Results

The surrogate models are investigated for each objective individually. The meta-models of ΔP and T_{dev} are depicted in Fig. 3. These show that each objective has a simple dependency on design variables in that that optima lie on design space boundaries. These dependencies are shown more clearly in Fig. 4. The optimum solution values for each objective are depicted in Table 2. This reveals the conflict between the objectives and it can be noted from Figs. 3 and 4 that it is not possible to

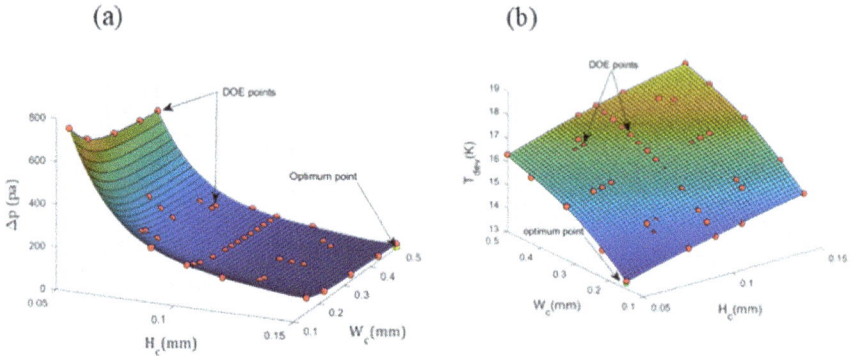

Fig. 3 Response surfaces of Δp(pa) and T_{dev}(K) in terms of W_{c} and H_{c} for the PCR

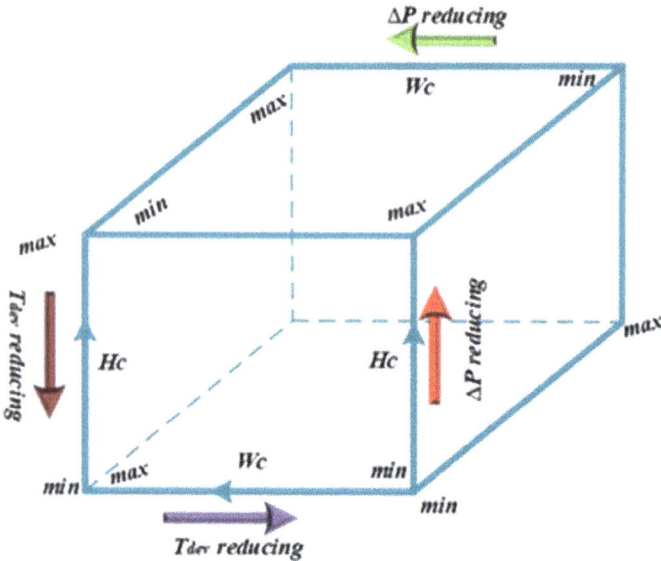

Fig. 4 Global optimization design trend obtained from metamodels

Table 2 Global optimization

Parameter	Optimum value	Corresponding values			
		Δp(pa)	T_{dev}(K)	W_{c}(mm)	H_{c} (mm)
$T_{\text{dev}}(K)$	13.27	708	–	0.15	0.05
Δp(pa)	23.3	–	17.9	0.5	0.15

move along the design points on the meta-surface to minimize any of the objective functions without increasing the other objective function.

6.1 Single Objective Optimization

The surrogate models are investigated for each objective individually in order to optimize each objective and explorer the competence between them. Figure 3a–b shows the surrogate model surface between ΔP and T_{dev} against W_{c} and H_{c}. Fig. 3a shows that minimizing ΔP demand biggest channel size (W_{c} and H_{c}) whereas Fig. 3b illustrates that minimizing the T_{dev} requires smallest channel size (W_{c} and H_{c}). The optimum solution values at each objective are depicted in Table 2 reveals the conflict between the objectives and it will be interesting in future work to examine the trade-off between the objectives through a multi-objective analysis.

7 Conclusion

In this paper, a three-dimensional CFD mathematical model has been solved using COMSOL Multiphysics® 5.4 and coupled with MATLAB optimization code to achieve optimization of PCR systems. This work used optimisation to improve the thermal performance of CFPCR in terms of temperature deviation, pressure drop, and heating power in microfluidic channels. It can be noted from Fig. 2a–b that when W_{c} increases, the pressure drop decreases, and when H_{c} increases, the pressure drop decreases. Also, the temperature deviation T_{dev} increases with an increase of W_{c} and H_{c} which means that the temperature uniformity decreases. This first optimization study using accurate meta-modelling and stochastic optimization methods will now be taken forward to multi-objective optimization that determines the compromises that can be struck between the competing multiple objectives in PCR systems.

Acknowledgements The authors are grateful to the Ministry of Oil, Iraq, and British petroleum (BP), Rumaila Oil field Iraq, for sponsoring the work on this project. This work was undertaken on ARC4, part of the High Performance Computing facilities at the University of Leeds, UK.

References

1. A.F. Al-Neama et al., An experimental and numerical investigation of chevron fin structures in serpentine minichannel heat sinks. **120**, 1213–1228 (2018)
2. A.D. Stroock, et al., Chaotic mixer for microchannels **295**(5555), 647–651 (2002)
3. M.D. Tarn, et al., The study of atmospheric ice-nucleating particles via microfluidically generated droplets **22**, 1–25 (2018)
4. J. Chen, K.J.M. Li, Analysis of PCR kinetics inside a microfluidic DNA amplification system **9**(2), 48 (2018)
5. Y.S. Shin, et al., PDMS-based micro PCR chip with parylene coating **13**(5), 768 (2003)
6. C. Zhang, D. Xing, Y.J. B.A. Li, Micropumps, microvalves, and micromixers within PCR microfluidic chips: advances and trends **25**(5), 483–514 (2007)
7. Z. Chunsun et al., Continuous-flow polymerase chain reaction microfluidics based on polytetrafluoethylene capillary. Chin. J. Anal. Chem. **34**(8), 1197–1203 (2006)
8. C.D. Ahrberg, A. Manz, B.G. Chung, Polymerase chain reaction in microfluidic devices. Lab Chip **16**(20), 3866–3884 (2016)
9. M.A. Northrup, DNA amplification with a microfabricated reaction chamber, in *Technical Digest of 7th International Conference on Solid-State Sensors and Actuators* (1993)
10. S. Kumar, T. Thorsen, S.K. Das, Thermal modeling for design optimization of a microfluidic device for continuous flow polymerase chain reaction (PCR), in *ASME 2008 Heat Transfer Summer Conference collocated with the Fluids Engineering, Energy Sustainability, and 3rd Energy Nanotechnology Conferences* (American Society of Mechanical Engineers, 2008)
11. Y. Zhang, P. Ozdemir, Microfluidic DNA amplification—a review. Anal. Chim. Acta. **638**(2), 115–125 (2009)
12. M.U. Kopp, A.J. De Mello, A. Manz, Chemical amplification: continuous-flow PCR on a chip. Science **280**(5366), 1046–1048 (1998)
13. N. Crews, C. Wittwer, B. Gale, Continuous-flow thermal gradient PCR. Biomed. Microdevice **10**(2), 187–195 (2008)
14. D. Moschou et al., All-plastic, low-power, disposable, continuous-flow PCR chip with integrated microheaters for rapid DNA amplification. Sens. Actuators B: Chem. **199**, 470–478 (2014)
15. I. Schneegaß, R. Bräutigam, J.M. Köhler, Miniaturized flow-through PCR with different template types in a silicon chip thermocycler. Lab Chip **1**(1), 42–49 (2001)
16. M. Hashimoto et al., Rapid PCR in a continuous flow device. Lab Chip **4**(6), 638–645 (2004)
17. Q. Zhang et al., Temperature analysis of continuous-flow micro-PCR based on FEA. Sens. Actuators B: Chem. **82**(1), 75–81 (2002)
18. J.J. Chen, C.M. Shen, Y.W.J.B.M. Ko, Analytical study of a microfludic DNA amplification chip using water cooling effect **15**(2), 261–278 (2013)
19. K. Toh, X. Chen, J. Chai, Numerical computation of fluid flow and heat transfer in microchannels. Int. J. Heat Mass Transf. **45**(26), 5133–5141 (2002)
20. D.B. Tuckerman, *Heat-Transfer Microstructures for Integrated Circuits* (Lawrence Livermore National Lab Ca, 1984)

Modelling and Evaluation of the Thermohydraulic Performance of Finned-Tube Supercritical Carbon Dioxide Gas Coolers

Lei Chai, Konstantinos M. Tsamos, and Savvas A. Tassou

1 Introduction

Supercritical carbon dioxide (sCO$_2$) is becoming an important commercial and industrial fluid due to its environmental credentials and its advantageous characteristics, such as being nontoxic and non-flammable and having low viscosity and a large refrigeration capacity. Thermodynamic systems using sCO$_2$ as the working fluid have been widely studied in many engineering applications, such as refrigeration and heat pumps, and power generation and conversion systems. In these systems, the thermohydraulic performance of gas coolers has a significant influence on the system efficiency. The understanding of the thermohydraulic performance under different operating conditions is essential for design and optimization purposes. To achieve this objective, Yin et al. [1] and Asinari et al. [2], respectively, developed a first principle-based model and a fully three-dimensional simulation model for microchannel gas coolers. Ge et al. [3], Gupta et al. [4], Marcinichen et al. [5], and Singh et al. [6] proposed detailed mathematical models utilizing distributed methods for finned-tube gas coolers. Li et al. [7, 8] developed a simplified mathematical model for an air-cooled coil gas cooler. In this paper, a detailed mathematical model employing the distributed modelling approach and the ε-NTU method is discussed, and the thermohydraulic performance of two finned-tube sCO$_2$ gas coolers [9] is investigated.

L. Chai (✉) · K. M. Tsamos · S. A. Tassou
RCUK Centre for Sustainable Energy Use in Food Chain (CSEF), Institute of Energy Futures, Brunel University London, Uxbridge, Middlesex UB8 3PH, UK
e-mail: Lei.Chai@brunel.ac.uk

© Springer Nature Singapore Pte Ltd. 2021
C. Wen and Y. Yan (eds.), *Advances in Heat Transfer and Thermal Engineering* ,
https://doi.org/10.1007/978-981-33-4765-6_71

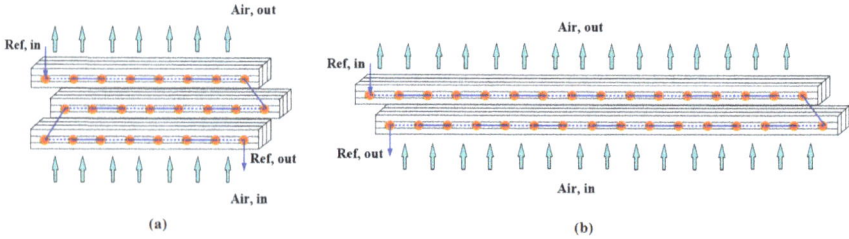

Fig. 1 Model of finned-tube sCO$_2$ gas cooler: **a** gas cooler A, and **b** gas cooler B

2 Methdology

Figure 1 demonstrates the distributed method (segment by segment) design for the two finned-tube sCO$_2$ gas coolers. Gas cooler A has three tube rows with eight tubes per row, and gas cooler B, two tube rows, and 16 tubes per row. The finned copper tubes have 8 mm outer diameter, 0.68 mm wall thickness, and 1.6 m length. The thickness of the wavy aluminum fins is 0.16 mm and the fin spacing 2.12 mm. For the model, each tube is divided into 20 equal segments. For each segment, the ε-NTU method [10] is employed for the heat transfer and pressure drop calculations. The thermophysical properties of the sCO$_2$ and the cooling air in each segment using the NIST REFPROP v9.1 software.

The thermohydraulic performance of the fluid flowing over or through the tubes is calculated using empirical correlations for Nusselt number and friction factor: Krasnoshchekov and Protopopov [11] correlation for sCO$_2$ and Wang et al. [12] correlation for cooling air. Since the density of sCO$_2$ undergoes a significant change with temperature in the near-critical region, the pressure drop of sCO$_2$ takes into consideration the influence of flow deceleration.

3 Results

The models were verified against the experimental data from Santosa et al. [9]. The comparison between the modelling predictions and the experimental results is illustrated in Table 1. Considering the uncertainty of the K-type thermocouples and pressure transducers used in the test facility, the present modelling can predict the thermohydraulic performance of the gas coolers with good accuracy. For numbers 1–6, the sCO$_2$ pressure drop from the modelling $\Delta P_{ref,mol}$ is 12.5, 11.4, 11.8, 52.3, 66.2 and 70.7 kPa. Gas cooler B shows much larger pressure drop but similar heat transfer rate for per unit mass flow rate of sCO$_2$ compared to gas cooler A with the other similar operating conditions, due to the different engineering design.

Following model verification, the effect of sCO$_2$ mass flow rate m_{ref} on the thermo-hydraulic performance was investigated. For gas cooler A and test No. 3, increasing the m_{ref} by 100% increased the sCO$_2$ outlet temperature $T_{ref,out,mol}$ from 34.3 °C to

Table 1 Comparison of modelling results with experiment data

Test	Operating condition					Results					
	$v_{air,in}$ (m/s)	$T_{air,in}$ (°C)	$P_{ref,in}$ (bar)	$T_{ref,in}$ (°C)	m_{ref} (kg/s)	$\Delta P_{air,exp}$ (Pa)	$\Delta P_{air,mol}$ (Pa)	$T_{ref,out,exp}$ (°C)	$T_{ref,out,mol}$ (°C)	Q_{exp} (kW)	Q_{mol} (kW)
	Gas cooler A					Gas cooler A					
No. 1	1.7	32.8	85.1	105.5	0.0105	26.6	26.9	33.2	32.83	2.4	2.46
No. 2	2.0	32.8	84.2	99.2	0.01	34.2	34.8	32.8	32.82	2.23	2.31
No. 3	2.4	34.3	86.6	116.8	0.0103	41.4	46.6	34.9	34.31	2.4	2.43
	Gas cooler B					Gas cooler B					
No. 4	1.7	35.1	86.3	100.8	0.019	13.9	17.6	35.3	35.38	4.1	4.12
No. 5	2.0	35.2	86.5	104.6	0.0215	25.6	22.7	35.0	35.54	4.6	4.75
No. 6	2.4	33.0	83.9	101.3	0.021	27.6	30.4	33.0	33.15	4.65	4.94

35.0 °C, and the heat transfer rate Q_{mol} from 2.43 kW to 4.9 kW. $\Delta P_{ref,mol}$ increases by 4 times from 11.8 kPa to 47.1 kPa. For gas cooler B and test No. 6, doubling m_{ref} increased $T_{ref,out,mol}$ from 33.2 °C to 35.3 °C, and Q_{mol} from 4.9 kW to 8.7 kW. The pressure drop $\Delta P_{ref,mol}$ quadruples from 70.7 kPa to 282.4 kPa.

The higher pressure drop in gas cooler B is due to the higher mass flow rate and total length of tubes in each row of the gas cooler. Even though the heat transfer for per unit mass flow rate is similar to gas cooler A, the higher pressure drop will lead to a higher compressor power consumption which will impact on the overall efficiency of the CO_2 refrigeration system.

4 Conclusions

To investigate the thermohydraulic performance of finned-tube sCO_2 gas coolers, a detailed mathematical model employing the distributed modelling approach and the ε-NTU method was developed and presented in this study. The model uses empirical correlations for the Nusselt number and friction factor for the heat transfer and pressure drop calculations on the sCO_2 and cooling air sides. The model was validated against experimental data and then employed to investigate the influence of design and operating parameters on overall gas cooler performance. The pressure drop in the gas cooler is an important parameter in the optimization of the performance of CO_2 refrigeration and heat pump systems particularly in the drive to reduce the footprint of components and the system as a whole. The model will be used as a tool in the investigation of the potential of 5.0 mm tube and microchannel heat exchangers as CO_2 gas coolers.

Acknowledgements The work presented in this paper is supported by a number of funders as follows: (i) The Engineering and Physical Sciences Research Council (EPSRC) of the UK under research grants EP/P004636/1 'Optimising Energy Management in Industry—OPTEMIN', and EP/K011820/1 (Centre for Sustainable Energy Use in Food Chains) and (ii) the European Union's Horizon 2020 research and innovation programme under grant agreement No. 680599, -Industrial Thermal Energy Recovery and Management—I-TERM'. The authors would like to acknowledge the financial support received from the funders and industry partners.

References

1. J.M. Yin, C.W. Bullard, P.S. Hrnjak, R-744 gas cooler model development and validation. Int. J. Refrig **24**, 692–701 (2001)
2. P. Asinari, L. Cecchinato, E. Fornasieri, Effects of thermal conduction in microchannel gas coolers for carbon dioxide. Int. J. Refrig **27**, 577–586 (2004)
3. Y.T. Ge, R.T. Cropper, Simulation and performance evaluation of finned-tube CO_2 gas coolers for refrigeration systems. Appl. Thermal Eng. **29**, 957–965 (2009)
4. D.K. Gupta, M.S. Dasgupta, Simulation and performance optimization of finned tube gas cooler for trans-critical CO_2 refrigeration system in Indian context. Int. J. Refrig **38**, 153–167 (2014)

5. J.B. Marcinichen, J.R. Thome, R.H. Pereira, Working fluid charge reduction. Part II: Super-critical CO_2 gas cooler designed for light commercial appliances. Int. J. Refrig. **65**, 273–286 (2016)
6. V. Singh, V. Aute, R. Radermacher, Investigation of effect of cut fins on carbon dioxide gas cooler performance. HVAC&R Research **16**, 513–527 (2010)
7. W. Li, S. Xuan, J. Sun, Entropy generation analysis of fan-supplied gas cooler within the framework of two-stage CO_2 transcritical refrigeration cycle. Energy Convers. Manage. **62**, 93–101 (2012)
8. W. Li, Optimal analysis of gas cooler and intercooler for two-stage CO_2 trans-critical refrigeration system. Energ. Convers. Manage. **71**, 1–11 (2013)
9. I.M.C. Santosa, B.L. Gowreesunker, S.A. Tassou et al., Investigations into air and refrigerant side heat transfer coefficients of finned-tube CO_2 gas coolers. Int. J. Heat Mass Transf. **107**, 168–180 (2017)
10. A.L. London, R.A. Seban, A generalization of the methods of heat exchanger analysis. Int. J. Heat Mass Transf. **23**, 5–16 (1980)
11. E.A. Krasnoshchekov, V.S. Protopopov, Experimental study of heat exchange in carbon dioxide in the supercritical range at high temperature drops. High Temp. **4**, 375–382 (1966)
12. C.C. Wang, W.S. Lee, W.J. Sheu, A comparative study of compact enhanced fin-and-tube heat exchangers. Int. J. Heat Mass Transf. **44**, 3565–3573 (2001)

Modeling Fouling Process on Tubes with Lattice Boltzmann Method and Immersed Boundary Method

Zi-Xiang Tong, Dong Li, Ya-Ling He, and Wen-Quan Tao

1 Introduction

Increasing the energy utilization efficiency is essential for alleviating energy shortage and environmental problems. In industrial productions, large amounts of energy are wasted in the form of low-to-medium temperature liquid and gas flows. The recovery of the waste heat is important for improving the efficiency of energy utilization. The heat exchangers are typical equipment for the waste heat recovery systems. However, a common feature of the heat exchangers is that the liquid and gas flows are not clean and fouling on the surface of heat exchangers will occur. For example, the ash particles in the flue gas of the boilers can deposit on the tube bundles of the heat exchangers and form granular fouling layers. The fouling layer will significantly reduce the efficiency of the heat exchangers as well as the whole system. Also, the cleaning of the fouling will lead to shutdown of the system and cause further costs. Therefore, it is important to understand the fouling processes and invent new heat exchanger designs for fouling reduction.

One of the difficulties in numerical investigation of the fouling problems is the changing of the shapes of the fouling layer, which will influence the flow fields and particle deposition. In the previous numerical studies, the dynamic meshes were widely used to deal with the growing boundary [1–5]. The dynamic meshes will become unstable during shape-changing processes because the negative volume cells and cracks will be generated. In some other researches, the regular mesh is used, and

Z.-X. Tong
School of Human Settlements and Civil Engineering, Xi'an Jiaotong University, No. 28 Xianning West Road, Xi'an 710049, Shaanxi, People's Republic of China

D. Li · Y.-L. He (✉) · W.-Q. Tao
Key Laboratory of Thermo-Fluid Science and Engineering of Ministry of Education, School of Energy and Power Engineering, Xi'an Jiaotong University, No. 28 Xianning West Road, Xi'an 710049, Shaanxi, People's Republic of China
e-mail: yalinghe@mail.xjtu.edu.cn

© Springer Nature Singapore Pte Ltd. 2021
C. Wen and Y. Yan (eds.), *Advances in Heat Transfer and Thermal Engineering* ,
https://doi.org/10.1007/978-981-33-4765-6_72

the evolution of fouling layer is implemented by changing the flow grid points into solid grid points [6]. The boundary of the regular mesh is not smooth, which can have large influences on the flow and particle collision near the fouling layer. Therefore, in this work a new numerical scheme is proposed for the simulation of fouling processes, in which the immersed boundary (IB) method is employed to deal with the evolution of fouling layers. With the proposed method, the morphology of fouling layer is obtained, and the details of the fouling processes on tubes are discussed.

2   Numerical Method

The proposed model contains a multiple-relaxation-time (MRT) lattice Boltzmann (LB) model for gas flow. The multiblock grids are used in which the finest grids are used in the region near the tubes to save the computational time. A large eddy simulation (LES) based on MRT-LB scheme is also employed to deal with the turbulence. The strain rates can be obtained from the velocity distribution functions of LB model, and the Smagorinsky model is employed to calculate the turbulent viscosity. Then, the MRT-LB model is coupled with the IB method to deal with the fouling layer. In IBM, the shape of the fouling layer is represented by a chain of points. The external forces are exerted to the flow field at the position of IB points to insurance the non-slip boundary condition on the IB. Then, the motion of particle is simulated by Lagrangian particle tracking method. In order to take into consideration of the effects of sub-grid scale fluctuations, a stochastic force is added to the drag force in the motion equation of particles.

The most important part of the model is the collision and deposition mechanisms of particles and the evolution of fouling layer. When the particles collide with the fouling layer, the force and momentum analyses are firstly employed to determine whether the kinetic energy of the particle is large enough to cause removal of the deposited particles. Then, the energy analyses are employed to determine whether the particle will deposit to form new fouling layer or bounce back from the layer. The velocity after bouncing back can be therefore determined by calculating the restitution coefficient [6]. Finally, the immersed boundary is moved according to the number of deposited particles. In the simulations, the flow, particle motion and deposition are simulated for a time interval which includes several vortex shedding periods. The numbers of deposited particles are counted for each position, and the points on the IB are moved according to the numbers of deposited particles.

The computational domain is given in Fig. 1. The left boundary is the inlet boundary of flow and particles. The right boundary is the outlet boundary. The top and bottom boundary is periodic boundary. Near the tube cylinder two domains with different mesh fineness are used.

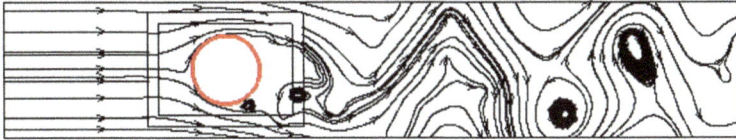

Fig. 1 Computational domain of the simulation

3   Results

In Fig. 2a, the streamlines and the particle distributions for 4 m/s flow velocity are shown. It can be seen that the flow field carries the particles to collide with the interface. Because of the centrifugal forces, there are less particles concentrated in the center of the vortex. On the windward side of the tube, the particles collide on the tube with the fluid flow. On the leeward side of the tube, as shown in Fig. 2a, several vortexes are generated, which can carry the particles back to the surface and cause deposition. Therefore, Fig. 2b and c show that the particles deposited on both windward side and leeward side, but less deposition happens on the lateral side with the angle around 90° or 270°. In Fig. 3, the distribution of particle deposition for different flow velocities after the same period of simulation is compared. It can be

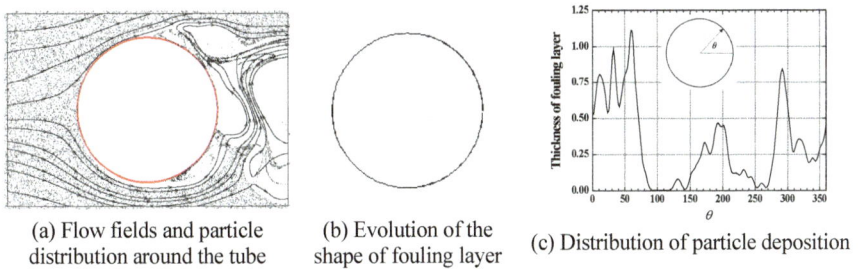

| (a) Flow fields and particle distribution around the tube | (b) Evolution of the shape of fouling layer | (c) Distribution of particle deposition |

Fig. 2 Simulation results of the numerical methods for 4 m/s flow velocity

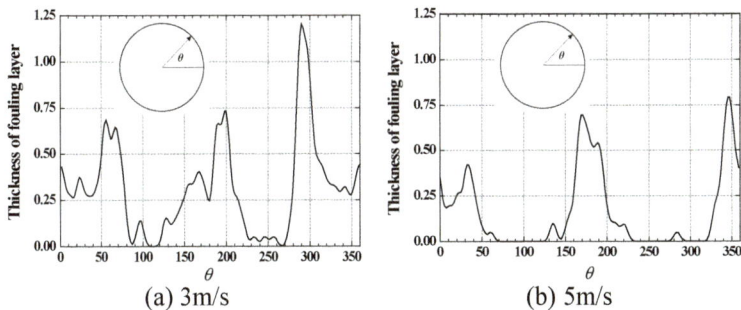

(a) 3m/s   (b) 5m/s

Fig. 3 Distribution of particle deposition for different flow velocities

seen that on the windward side, the deposition rate increases with the flow velocity because the boundary layer becomes thinner and more particles collide with the tube. On the contrary, the deposition on the leeward side decreases with the increase of flow velocity. Also, the region that the deposition happens shrinks with the increase of velocity. When the flow velocity is 5 m/s, it can be found that the deposition only happens on some narrow angular regions.

4   Conclusions

In this work, a numerical method has been proposed for the simulation of the fouling processes on tubes, in which the IBM is coupled with LBM and the shape of the boundary is evolved according to the amount of deposited particles. Based on the proposed numerical model, the evolution of the shape of fouling layer is obtained. The effects of air velocity on the growth of fouling layer are also studied. The results show that the deposition happens on both the windward side and leeward side of the tube. As the flow velocity increases from 3 m/s to 5 m/s, the deposition on the windward side experiences a slight growth while the deposition on the leeward side decreases. The angular range that the deposition occurs also shrinks. The proposed numerical model can be further used to predict the fouling processes on heat exchangers and facilitate the design of heat exchangers.

Acknowledgements This work was supported by the Basic Science Center Program for Ordered Energy Conversion of the National Natural Science Foundation of China (No.51888103).

The authors would also like to thank the Foundation for Innovative Research Groups of the National Natural Science Foundation of China (No.51721004).

References

1. M.G. Pérez, E. Vakkilainen, T. Hyppänen, 2D dynamic mesh model for deposit shape prediction in boiler banks of recovery boilers with different tube spacing arrangements. Fuel **158**, 139–151 (2015)
2. M.G. Pérez, E. Vakkilainen, T. Hyppänen, Fouling growth modeling of kraft recovery boiler fume ash deposits with dynamic meshes and a mechanistic sticking approach. Fuel **185**, 872–885 (2016)
3. M.G. Pérez, E. Vakkilainen, T. Hyppänen, Unsteady CFD analysis of kraft recovery boiler fly-ash trajectories, sticking efficiencies and deposition rates with a mechanistic particle rebound-stick model. Fuel **181**, 408–420 (2016)
4. S.Z. Tang, M.J. Li, F.L. Wang, Z.B. Liu, Fouling and thermal-hydraulic characteristics of aligned elliptical tube and honeycomb circular tube in flue gas heat exchangers. Fuel **251**, 316–327 (2019)
5. M.J. Li, S.Z. Tang, F.L. Wang, Zhao Q.X., W.Q. Tao, Gas-side fouling, erosion and corrosion of heat exchangers for middle/low temperature waste heat utilization: a review on simulation and experiment. Appl. Therm. Eng. **126**, 737–761 (2017)
6. Z.X. Tong, M.J. Li, Y.L. He, H.Z. Tan, Simulation of real time particle deposition and removal processes on tubes by coupled numerical method. Appl. Energ. **185**, 2181–2193 (2017)

Numerical Study of the Impacts of Forced Vibration on Thermocapillary Bubble Migration

Mohammad Alhendal and Yousuf Alhendal

1 Introduction

In a non-uniform temperature gradient fluid medium, the surface tension varies according to the local temperature conditions. The surface tension in the cold fluid region is greater than that in the hot region. Therefore, in a zero-gravity stagnant fluid of a non-uniform temperature field, this difference in surface tension creates a net force acting on the fluid particles, which leads to a general fluid motion from the hot region to the cold region. This phenomenon is known as Marangoni or thermocapillary migration phenomenon. Vibrations on ground always compose with the earth gravity field. It is a matter of fact that the phenomenology of vibrations effects when gravity is absent is almost lacking, and nobody knows exactly what will be the consequences of vibrations on the thermocapillary flow of bubble/droplet migration and agglomeration or brakeage on board. In space, vibrations can have marked effects in fluids flow and migration. These effects result from small-amplitude vibrations such as g-jitter aboard space platforms can have significant effects on various fluids and interface motions. We are interested in understanding the effects of vibrations on Marangoni bubble motion and identifying the parameters that could be controlled to minimize the vibration effects in future fluid experiments conducted in space. On the other hand, vibrations might play a critical role on thermocapillary bubble flow and coalescence on board space. We are interested in conducting further space simulation to obtain additional data on bubble translation and shape deformation to validate the theoretical model covering different vibration frequencies. Better understanding of the vibration effects will help us better understand the onset of oscillation

M. Alhendal (✉)
Australian College of Kuwait (ACK), Kuwait city, Kuwait
e-mail: moh_hendal@hotmail.com

Y. Alhendal
College of Technological Studies (CTS), Public Authority for Applied Education and Training (PAAET), Kuwait city, Kuwait

© Springer Nature Singapore Pte Ltd. 2021   427
C. Wen and Y. Yan (eds.), *Advances in Heat Transfer and Thermal Engineering* ,
https://doi.org/10.1007/978-981-33-4765-6_73

mechanisms and ways to design g-jitter-resistant multiphase flow in the future. In this research work, we are interested in conducting new zero-gravity simulations to discover further data and thermocapillary flow pattern on bubble behavior under different vibration.

2 Methodology

The thermocapillary motion of a bubble was first examined experimentally by Young, Block and Goldstein [1], when Reynolds number (Re) and Marangoni number (Ma) are small, which means that both convective momentum and energy transport are negligible, who also found an analytical expression for its terminal velocity in the creeping flow:

$$V_{YGB} = \frac{2|d\sigma/dT|r_b \lambda dT/dx}{(2\mu + 3\mu')(2\lambda + \lambda')} \tag{1}$$

commonly called the YGB model, which is suitable for small Reynolds and Marangoni numbers:

$$Re_T = r_b V_T / \nu \tag{2}$$

$$Ma_T = r_b V_T / \alpha = Re_T . Pr \tag{3}$$

where Prandtl number is the ratio of kinematic viscosity to thermal diffusivity:

$$Pr = \nu / \alpha \tag{4}$$

(4).

and ν is the kinematic viscosity in m²/s: $\nu = \mu / \rho$. The velocity V_T derived from the tangential stress balance at the free surface is used for scaling the migration velocity (m/s) in Eqs. (2) and (3):

$$V_T = \frac{(d\sigma/dT).(dT/dx).r_b}{\mu} \tag{5}$$

where μ and μ', λ and λ' are the dynamic viscosity and thermal conductivity of continuous phase and bubble, respectively. ρ is the density, and r_d is the radius of the bubble. The constant $d\sigma/dT$ or σ_T is the rate of change of interfacial tension, and dT/dx is the temperature gradient imposed in the continuous phase fluid.

3 VOF Model and Computational Procedure

The governing continuum conservation equations for two-phase flow were solved using the Ansys Fluent commercial software package [2], and the volume of fluid (VOF) method was used to track the liquid–gas interface. This method deals with completely separated phases with no diffusion. The geometric reconstruction scheme, based on the piece-wise linear interface calculation (PLIC) method of Young's [3] in Ansys Fluent, was chosen for the current investigation. Geo-reconstruction is an added module to the already existing VOF scheme that allows for a more accurate definition of the free surface [4].The movement of the gas–liquid interface is tracked based on the distribution of the volume fraction of the gas, i.e.,α_G, in a computational cell, where the value of α_G is 0 for the liquid phase and 1 for the gas phase. Therefore, the gas–liquid interface exists in the cell where α_G lies between 0 and 1. A single momentum equation, which is solved throughout the domain and shared by all the phases, is given by

$$\frac{\partial}{\partial t}(\rho \vec{v}) + \nabla.(\rho \vec{v}\vec{v}) = -\nabla p + \nabla.[\mu(\nabla \vec{v} + \nabla \vec{v}^T)] + \vec{F} \tag{6}$$

In Eq. 6, $\vec{F}$ represents volumetric forces at the interface, resulting from the surface tension force per unit volume. The continuum surface force (CSF) model proposed by Brackbill et al. [5] is used to compute the surface tension force for the cells containing the gas–liquid interface:

$$\vec{F} = \sigma\left(\rho k \vec{n} \bigg/ \frac{1}{2}(\rho_L + \rho_G)\right) \tag{7}$$

where σ is the coefficient of surface tension,

$$\sigma = \sigma_0 + \sigma_T(T_0 - T). \tag{8}$$

and σ_0 is the surface tension at a reference temperature T_0, and T is the liquid temperature.

4 Results and Discussion

Several simulations were analyzed to study the effect of container acceleration frequency on the flow field and time required for the thermocapillary bubble flow and behavior in a vibrated cylinder. Figures 1, 2 and 3 show the results of the numerical simulations for constant amplitude at $A_P = 0.005$ m/s² and different frequency. In the case where vibration frequency (f) is 0.025 Hz, the bubbles rose, with slight diversion at the top of the cylinder toward the wall of the container. In the case where vibration

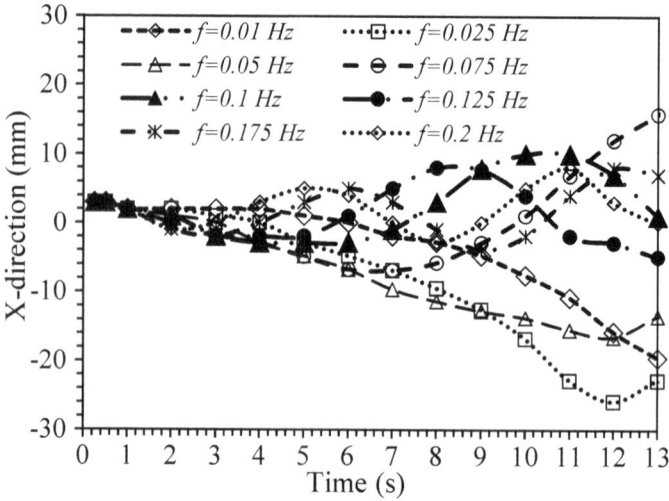

Fig. 1 Bubble migarion and oscilation around the Y-axis. (under the effect of both Marangoni and vibration forces)

Fig. 2 Migraton of gas bubble toward the hotter side for different frequency

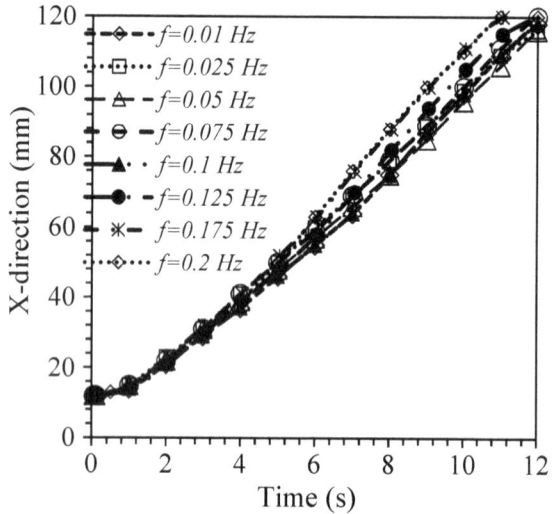

frequency (f) is 0.05 Hz, the bubble moves toward the center before reaching the top of the cylinder. These figures also illustrate that at the vibration frequency (f) of 0.075 Hz and beyond, the bubble moves from the wall to the center again before reaching the top of the cylinder. A major bubble osculation can be observed by increasing (f) to 0.1 Hz, which resulted in a large divergence of the bubbles toward the axis of cylinder with a small shift in the direction of the warmer region. The figures show the variation in the thermocapillary bubble speed as it moves toward

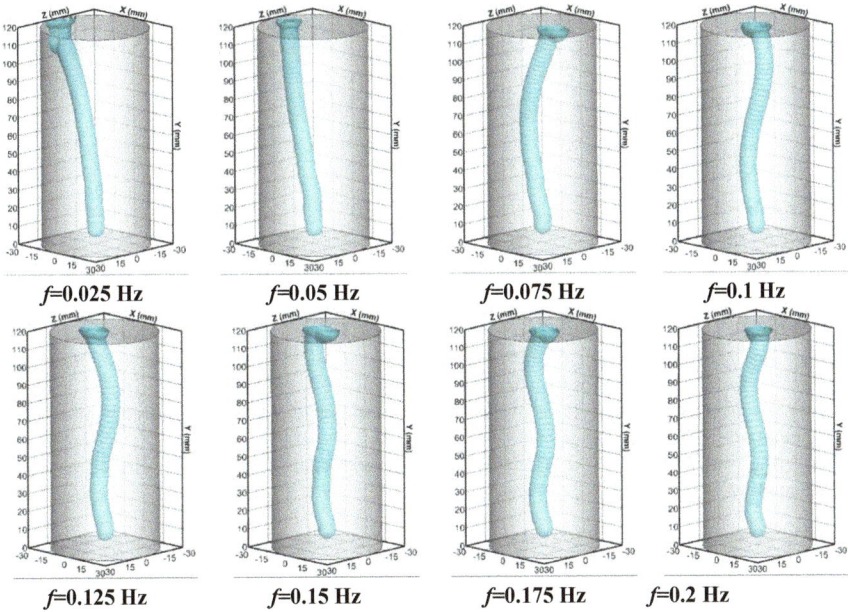

Fig.3 Flow of gas bubble toward the hotter side for different freguancy

the hotter side due to vibration frequencies. As frequency gets larger while keeping the container amplitude constant, bubble goes from small oscillations to deformation.

5 Conclusion

The thermocapillary flow of isolated bubble in a vibrating container was studied. The amplitude of bubble translation and nonlinear shape oscillation characteristics were determined for single bubble migrating with different frequency. Small-amplitude vibrations aboard space platforms can have significant effects on bubble movement. Conducting future experiments on this topic aboard the space station will be of interest to us since better understanding of the vibration effects will help us better understand the onset of oscillation mechanisms and ways to design g-jitter-resistant thermocapillary bubble migration in the future. Since zero gravity is difficult to achieve in a laboratory setting, one can demonstrate the relevant phenomena using numerical simulations. Simulating these phenomena also allowed one to study the effects of altering the sensitivities of different parameters. It therefore may be concluded that computer simulations proving their worth as a valuable tool to study the complex problems in zero-gravity conditions and one can observe the credentials of numerical modeling to simulate realistic 3D Marangoni cases. This has important implications on the Marangoni flow characteristics in space.

Acknowledgements The principal investigator would like to express his sincere gratitude to Kuwait Foundation for the Advancement of Sciences (KFAS), Kuwait, for supporting and funding this research work.

References

1. N.O. Young, J.S. Goldstein, M.J. Block, The motion of bubbles In a vertical temperature gradient. J. Fluid Mech. **6**, 350–356 (1959)
2. ANSYS-FLUENT 2011. Users Guide
3. D.L. Youngs, *Time-dependent Multi-material Flow with Large Fluid Distortion. Numerical Methods for Fluid Dynamics* (Academic Press, 1982) 273–285
4. C.W. Hirt, B.D. Nichols, Volume of fluid (Vof) method for the dynamics of free boundaries. J. Comput. Phys. **39**, 201–225 (1981)
5. J.U. Brackbill, D.B. Kothe, C. Zemach, A continuum method for modeling surface tension. J. Comput. Phys. **100**, 335–354 (1992)

CFD Modelling and Experimental Calibration of Concentrated Windings in a Direct Oil-Cooled Segmented Stator

Robert Camilleri

1 Introduction

This paper presents a new approach in performing CFD modelling for electrical machine stator windings. Thermal modelling of the stator windings is important as the winding insulation is rated to a maximum operating temperature which limits the machine life or the torque ratings. While CFD analysis has been applied to investigate the flow path [1, 2], its application to electrical machine windings has been limited as this is complex and challenging. Electrical machine stator windings are a composite of materials, typically consisting of multiple copper conductors wound around a stator iron bobbin. Each conductor is coated with a two-part insulation of approximately 2-micron thickness. The windings may be sometimes infiltrated with epoxy resin or a process of vacuum pressure impregnation (VPI) to increase mechanical rigidity. Representing the individual insulated conductors in FEA or CFD models requires an inefficient meshing process due to the difference in scale between the main conductor and its insulation. This adds significant complexity and leads to large computational demands and long solution times. To simplify the problem, researchers have often treated the windings as a homogeneous material with equivalent thermal properties [3]. However, this approach lacks detailed information and an accurate physical representation.

This paper addresses this problem by simplifying concentrated windings into multi-layered structures on which CFD analysis is performed. A thermal resistance is applied between the winding layers, to represent the thermal contact between the windings. The author believes that this provides a better physical representation of the conductive process in machine windings. The technique enables to achieve a detailed temperature map of the machine windings, thereby providing more information to the thermal designer. The paper uses a direct liquid-cooled axial flux yokeless and

R. Camilleri (✉)
Institute of Aerospace Technologies, University of Malta, Msida MS2050, Malta
e-mail: robert.c.camilleri@um.edu.mt

© Springer Nature Singapore Pte Ltd. 2021
C. Wen and Y. Yan (eds.), *Advances in Heat Transfer and Thermal Engineering* ,
https://doi.org/10.1007/978-981-33-4765-6_74

433

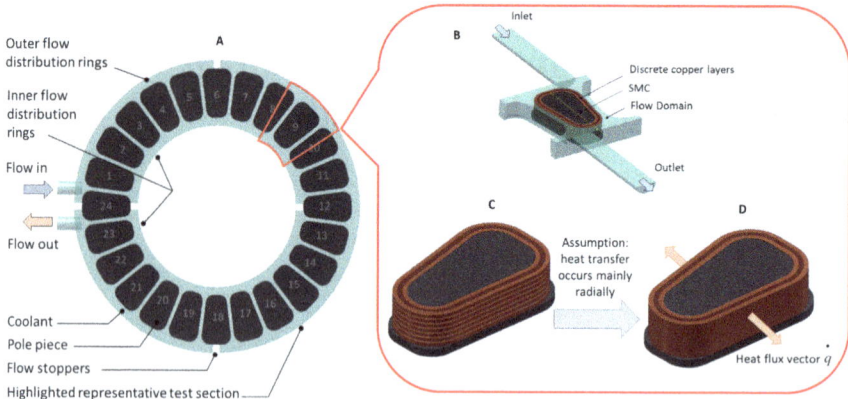

Fig. 1 a Schematic of the direct oil-cooled stator with segmented pole pieces. **b** Representative model of the direct liquid-cooled pole piece. **c** CAD model of a typical concentrated winding. **d** A simplification of the windings that was used for CFD analysis

segmented armature (YASA) permanent magnet machine as a test case study. Axial flux permanent magnet machines enable higher torque densities and higher efficiencies making them suitable for applications such as road transportation or wind energy generation [4]. This architecture has recently received increasing attention, and several variants can be found in the literature. In the YASA machine, the stator is enclosed in a glass fibre casing allowing liquid coolant to be injected into the stator, in direct contact with the windings as shown in Fig. 1a.

2 Methodology

2.1 CFD Prediction

Segmented stators provide a unique advantage in thermal analysis of machines as a segment of the windings can be used to represent a portion of the entire machine stator. This approach was also taken in this paper as shown in Fig. 1b. In direct liquid-cooled machines, heat generated in the windings is transferred radially across the winding layers and into the coolant. By neglecting axial heat transfer, complex concentrated windings can hence be simplified into a radial multi-layered structure as shown in Fig. 1c and d. Adjacent winding layers are thermally interfaced to each other through a thermal contact resistance. This technique requires an initial calibration process, described in the following experimental section.

A 3D CAD model of the flow path and windings was developed. The flow domain was set with OptiCool properties while the pole pieces components were set as SMC material and copper, respectively. The properties of OptiCool coolant, SMC iron and copper windings can be found in existing literature [5]. The flow rate and fluid temperature were set as inlet boundary conditions. The outlet was set with a gauge static pressure of 0 Pa, to simulate the flow out of the component. This is used as reference by the software to compute the pressure upstream. A heat loss was applied to the winding layers and iron. A thermal contact resistance achieved during calibration was also set between the winding layers.

The 3D mesh was formed from tetrahedral elements with a minimum refinement length set to 0.5 and the maximum size set to 1.8 at the surfaces. The boundary layer mesh was made of 15 layers with a factor of 0.75 and gradation of 1.15. A grid sensitivity analysis was performed with the final mesh resulting in a total of 4.3 M elements, to be most accurate with a conjugate, laminar model using ADV 4 and temperature-dependent coolant properties. Details of the advection schemes can be found in [6]. The simulation was repeated with tighter mesh settings until the temperature variation of the pole piece was <2%. A solution of the 3D model is solved in approximately 12 h. The simulations were run for a number of inlet flow rates, oil inlet temperature and pole piece heat loss.

2.2 Experimental Calibration and Validation

A concentrated wound pole piece was manufactured using a winding machine that is made of two main components:

1. A lathe machine with controllable speed and modified jaws to grab pole pieces and.
2. A copper reel clamp coupled to fixed speed electric machine via a clutch mechanism with controllable slip.

During the winding process, the copper is subjected to a tension by having an electric motor holding the winding reel and the lathe winding the bobbin rotating in opposite directions. The tension is adjusted by adjusting the slippage on the clutch.

Thermocouples were mounted on the pole piece during the winding process so that two are mounted on the iron, two between the iron and the first copper layer, two between the first and second copper layers and two between the second and third copper layers. A representative flow path was set up, recreating the flow channels between the pole pieces. The coolant was re-circulated from a reservoir into the test rig and then pumped through a heat exchanger, expelling heat to ambient. The reservoir was also fitted with a heating element to control the inlet temperature of the fluid into the test section. The power input to the coolant heater was also regulated through a transformer. The flow rate was regulated using a globe valve and measured using a variable area flow metre.

The heat losses in the copper windings and the SMC iron bobbin were simulated by Joule heating. Hence, these were each connected to a DC power supply and energized separately. The iron losses were kept at 20% of the total pole piece losses as per manufacturer recommendations. Current metres and a differential voltage

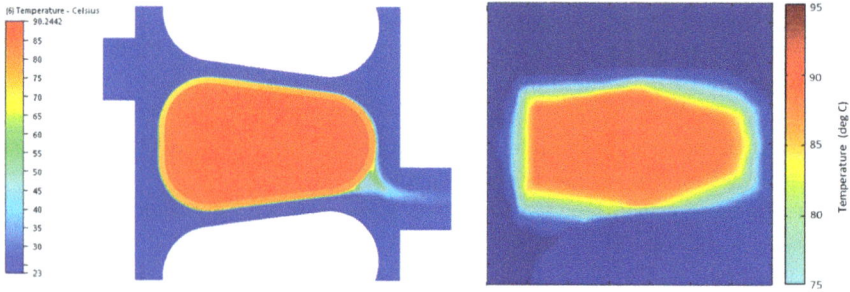

Fig. 2 (left) CFD results showing the temperature profile of the windings, (right) an experimental measurement of the windings' temperature map when subjected to total losses of 61 W and coolant flow rate of 2.5 l pm

probe were used to measure the current and voltage supplied to the windings using a PicoScope 3000 series oscilloscope recording at 1 Hz. The experiment was run in transient mode, thereby calibrating the inter-winding resistances. This was followed through a series of steady-state experiments which were used to create a temperature map of the windings and compared it to the CFD model.

3   Results

Figure 2 shows a typical example of the temperature map of the windings, when subjected to a winding losses of 50 W, and SMC iron losses of 11 W, with a coolant flow rate of 2.5 lpm and an inlet temperature of 24 °C.

The results show that the CFD model is within ± 2 °C of measured temperatures. It can be seen that concentrated windings are subjected to a temperature gradient across its windings. Heat from each winding layer is transferred across the outer layers prior to being dissipated into the coolant. The inner windings therefore tend to be the hottest as they are subjected to a longer thermal path.

4   Discussion and Conclusion

This paper presents a simplified layered CFD model for simulating complex concentrated windings for electrical machine stators. The model was applied to a direct liquid-cooled machine with segmented stator. A calibration technique was required to establish the inter-layer thermal resistances, which then acted as an input to the CFD model. The technique demonstrated in this paper extends the use of CFD to the analysis of electrical machine windings, thereby enabling to achieve a temperature map of the stator windings. While the model achieves a good accuracy when compared with experimental measurements, it provides a significant reduction in computational effort and time. The new CFD model also provides a significant insight into the effects of the inter-winding thermal resistances on the temperature map of the windings. Thus, it allows the thermal designer to consider solutions that would reduce the effect of the inter-winding resistances and the temperature of the inner windings. This would allow further current input and an improvement in electrical machine current density.

References

1. P.H. Connor, S.J. Pickering, C. Gerada, C.N. Eastwick, C. Micallef, CFD modelling of an entire synchronous generator for improved thermal management, in *6th IET International Conference on Power Electronics, Machines and Drives (PEMD 2012)* 27–29 March 2012, Bristol, UK
2. K. Bersch, P.H. Connor, C.N. Eastwick, M. Galea, R. Rolston, CFD optimisation of the thermal design for a vented electrical machine, in *2017 IEEE Workshop on Electrical Machines Design, Control and Diagnosis (WEMDCD2017)*, 20–21 April 2017, Nottingham, UK
3. S. Ayat, R. Wrobel, J. Goss, D. Drury, Estimation of equivalent thermal conductivity for impregnated electrical windings formed from profiled rectangular conductors, in *8th IET Int. Conf. on Power Electron. Machines and Drives (PEMD2016)*, 19–21 April, 2016, Glasgow, UK
4. T.J. Woolmer, M.D. McCulloch, Analysis of the yokeless and segmented armature machine, in *IEEE International Electric Machines & Drives Conference, IEMDC'07*, 3–5 May 2007, Antalya, Turkey, pp. 704–708
5. R. Camilleri, D.A. Howey, M.D. McCulloch, Predicting the temperature and flow distribution in a direct oil-cooled electrical machine with segmented stator. IEEE Trans. on Ind. Electron. **63**(1), 82–91 (2016)
6. Autodesk CFD Simulation Help, https://knowledge.autodesk.com/support/cfd. Accessed on 11 Nov 2015

Numerical and Experimental Investigation on Single-Point Thermal Contact Resistance

Anliang Wang, Zhenyu Liu, Hongwei Wu, Andrew Lewis, and Tahar Loulou

1 Introduction

In the engineering applications, the thermal contact resistance has an important effect of heat transfer design and operation of systems and devices. For general contact heat transfer, as long as the geometry, mechanics and boundary conditions are known, the steady thermal contact resistance of the interface will be "unique," which is independent of theoretical prediction and experimental measurement methods [1]. In other words, for the same thermal contact resistance problem, the experimental value and calculation value should be completely consistent. The logical starting point of macrothermal contact resistance is single-point thermal contact resistance, which can be simplified as series connection of thermal constriction resistance, heat conduction and thermal spreading resistance [2]. Although there were many experimental methods for measuring thermal contact resistance, there are few experimental data on thermal spreading (or constriction) resistance [3]. Most of the models of thermal contact resistance have made a comprehensive analysis of the main physical mechanism, such as surface topography and deformation; however, the predicted results and experimental values are still quite different, and the predicted models and numerical methods are still difficult to be directly applied to engineering problems. Based on the single-point cylinder–cylinder contact model, this paper provides a standard method for verifying the accuracy of numerical simulation results and experimental apparatus.

A. Wang (✉) · Z. Liu
School of Astronautics, Beihang University, Beijing 100191, China
e-mail: wanganliang@buaa.edu.cn

A. Wang · H. Wu · A. Lewis
School of Engineering and Computer Science, University of Hertfordshire, Hatfield AL10 9AB, UK

T. Loulou
UMR CNRS 6027, Université Bretagne Sud IRDL, Lorient, France

© Springer Nature Singapore Pte Ltd. 2021
C. Wen and Y. Yan (eds.), *Advances in Heat Transfer and Thermal Engineering* ,
https://doi.org/10.1007/978-981-33-4765-6_75

2 Experimental Apparatus and Numerical Methodology

We have designed and set up an apparatus for measuring thermal contact resistance, which has four subsystems: heater and cooler, loading measurement, vacuum pump subsystem, data acquirement and process subsystem, which is shown in Fig. 1 [4]. The single-point contact specimens are stainless steel 304, and the radius ratios ε are 0.2, 0.3, 0.4, 0.5 and 0.6.

A single-point geometric and physical model of a cylinder–cylinder contact is established to study the thermal spreading (or constriction) resistance. The total thermal contact resistance R of the single point is the sum of thermal constriction resistance R_c, intermediate cylinder body conduct resistance R_d and thermal spreading resistance R_s that is

$$R = R_c + R_d + R_s \tag{1}$$

Because of the symmetry of the geometric model, the heat flow distribution of the upper and lower part is also symmetrical, which means the constriction resistance can be regarded as equal to the spreading resistance:

$$R_c = R_s \tag{2}$$

For the isoflux tube model, the dimensionless thermal spreading resistance can be calculated as follows [5]:

$$\psi_{th} = k_s \xi R_s = \frac{0.27\xi}{a} F(a/b) \tag{3}$$

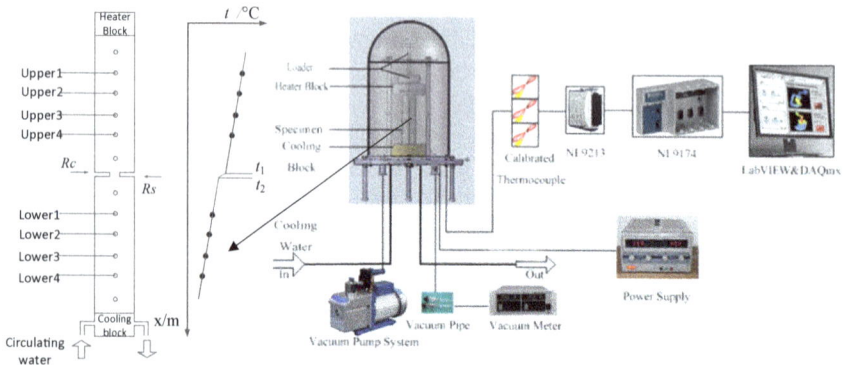

Fig. 1 Experimental model and apparatus

The calculation method of experimental values of thermal spreading resistance is as follows:

$$\psi_e = 0.5\,k\sqrt{A_c}(R - R_{d\text{-}v}) \tag{4}$$

For the steady-state heat conduction of a simple cylinder–cylinder body, the 2D model and the 3D model are simulated at FLUNT platforms. We have verified the numerical solution methods and the grid independence.

3 Results

A comparison of theoretical, simulated and experimental values of dimensionless thermal spreading resistance under different area ratios is shown in Fig. 2. With the increase of area ratio, the dimensionless thermal spreading resistance decreases gradually, and the overall trend of experimental and simulation values is consistent. The simulation results and theoretical values have little difference in the same area ratio. The differences between simulation values and theoretical values show the characteristics of complexity under different area ratio ε. There are two major reasons for the deviation between the experimental and theoretical values of dimensionless thermal spreading resistance: (A) deviations between the experimental sample and the theoretical model and (B) measurement error.

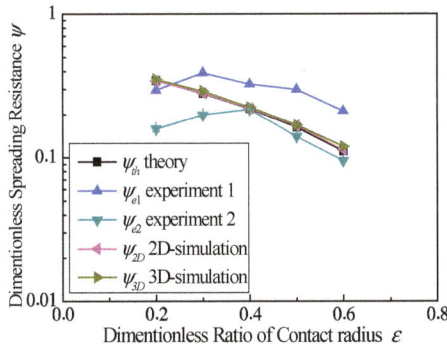

Fig. 2 Comparisons of theoretical values with numerical and experimental results

4 Conclusions

(1) The single-point contact thermal resistance method can be used to "calibrate" the precision of the test rig. Once the precision of measuring thermal contact resistance is determined, the effects of heat flux, temperature distribution, surface roughness, interface contact pressure, material properties and other factors can be studied quantitatively.

(2) Experiments and numerical simulations show that the dimensionless thermal spreading resistances decrease ψ with the increase of the contact area ratio ε, which are consistent with the theoretical values.

(3) The measurement values of thermal contact resistance are influenced by both macroscopic geometric model and experimental measurement error. Accurate calibration of thermocouples and reduction of heat leak can improve the precision of experimental measurement, but the deviation between numerical solution, experimental value and theoretical value cannot be completely eliminated. Air or other interface medium between the contact surfaces has a direct impact on the thermal contact resistance, and in some cases, there can be an order of magnitude difference.

Acknowledgements A. Wang thanks the grant of China Scholarship Council for this visiting at the UK. We also thank Mr. Fan Peisheng, Liu Liangtang and Gaotian for their hard work in building the test rig and measurement data.

References

1. A. Wang, J. Zhao, Review of prediction for thermal contact resistance. Sci. China Tch. Sci. **53**, 1798–1808 (2010)
2. M. Razavi, Y. Muzychka, S. Kocabiyik, Review of advances in thermal spreading resistance problems. J. Thermophys. Heat Transfer **30**, 863–879 (2016)
3. Y. Xian, P. Zhang, S. Zhai et al., Experimental characterization methods for thermal contact resistance: a review. Appl. Therm. Eng. **130**, 1530–1548 (2018)
4. P. Fan, *Experimental Research of Thermal Contact Resistance in Vacuum Environment (Master's Degree Thesis in Chinese)* (Beihang University, Beijing, 2013).
5. C. Madhusudana, *Thermal Contact Conductance* (Springer-Verlag, New York, 1996).

Optimization of Conformal Cooling Channels for Rapid Prototyped Mould Inserts

Tongyan Zeng, James Jewkes, and Essam Abo-Serie

1 Introduction

High-pressure die casting (HPDC) is a well-established manufacturing field that has received renewed attention because of the design opportunities that have arisen as a consequence of the development of additive manufacturing methods such as selective laser sintering (SLS). These novel processes are uninhibited by the conventional limitations of computer numerical control (CNC) machining methods and allow the cooling channel layout of the mould cavity to be created freely. HPDC bears many similarities to injection moulding; however, metal forming engenders more challenging operating conditions than those associated with polymers. Therefore, the development of an optimal cooling channel layout for a rapid prototyped mould insert is of considerable interest to industry. Previous work has shown that the cooling of corners can be challenging, as typical cooling channels with circular cross sections fail to provide sufficient heat flux in these regions. This is a consequence of the small surface area of the sections of the cooling pipes exposed to the corners and their distance from the cast.

Recent studies have explored various shape optimisation approaches [1, 2] that have demonstrated improvements over conventional cooling channels. The adjoint optimisation approach developed by Lions and Pironneau [3, 4] is a powerful numerical method that can be used to calculate a pre-defined mesh sensitivity based on an objective function that can then be used to deform the shape of the geometry. Whilst adjoint optimisation has been successfully applied in aeronautical applications [5], its potential application in heat transfer has received limited attention.

This extended abstract presents preliminary adjoint optimisation results as a 'proof of concept' for the development of a cooling channel optimisation strategy for HPDC mould inserts. We are interested in achieving uniform heat transfer/cooling,

T. Zeng · J. Jewkes (✉) · E. Abo-Serie
Fluids and Complex Systems Research Centre, Coventry University, Coventry CV1 5F, UK
e-mail: ac3541@coventry.ac.uk

© Springer Nature Singapore Pte Ltd. 2021
C. Wen and Y. Yan (eds.), *Advances in Heat Transfer and Thermal Engineering* ,
https://doi.org/10.1007/978-981-33-4765-6_76

minimising pressure drop between inlet and outlet flow and reducing thermal fatigue to extend tooling life.

2 Methodology

This preliminary investigation was focused upon applying the adjoint method to a simplified cooling geometry, a single L-shaped pipe with constraints and boundary conditions chosen to reflect those encountered in aluminium HPDC. The commercial finite volume computational fluid dynamics (CFD) software STAR-CCM+® was used to compare the pressure drop between the flow inlet and outlet, before and after the application of mesh deformation using adjoint optimisation. The geometry is shown in Fig. 1 and consists of a single 6 mm pipe with one inlet and outlet that passes through a solid part representing the mould. The highlighted areas on the solid region indicate the interface between the aluminium cast and the mould, where the temperature was assumed to be a constant 450 °C. Adiabatic boundary conditions were applied to the remaining external faces of the solid region. The properties of the mould material and coolant were set as H13 steel and water, respectively. A constant mass flow rate of 0.075 kg/s was applied with a uniform velocity profile at the inlet and a no-slip boundary condition applied at the pipe interface. A constant average pressure boundary condition was applied at the outlet.

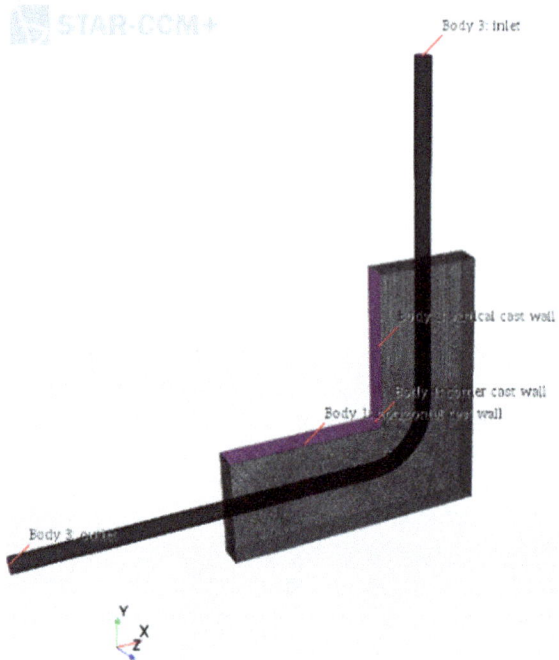

Fig. 1 Meshed schematic diagram of the single pipe flow through the solid body

The geometry was discretised using polyhedral cells, and mesh convergence study was carried out to ensure that results were independent of resolution. The base sizes of the mesh in the solid and fluid regions were set as 0.0005 m and 0.001 m, respectively, giving a mesh quality of 99.6% according to STAR-CCM+®'s diagnostics. The simulations were steady and constant density, a conjugate heat transfer (CHT) model was run, with the Reynolds averaged Navier–Stokes (RANS) equations used to solve the fluid region, with a K-Epsilon turbulence closure. The use of a coupled flow model enabled the adjoint flow solver to be used for both solid and fluid regions. The coupled implicit solver solves the governing equations implicitly in local pseudo-time steps ΔT, which generate a much more stable and accurate solution. The simulation was initialised with a potential flow solver. A well-converged primal result was required before running the adjoint solver. A point set was defined around the pipe surface, and an objective function relating cumulative displacement to pressure drop was defined, to control the mesh deformation before running each subsequent simulation.

3 Results

After solving the baseline simulation of a straight pipe, the adjoint solver was run and the required deformations to the point set around the pipe were calculated. The vector scene in Fig. 2 indicates the resulting sensitivity of the pressure field to pipe deformations, based upon the defined pressure drop objective function. Vector colour and size indicate the relative magnitude of displacement (scale of vectors increased for clarity) that should occur at a given point. Once this deformation was applied, then the simulation was rerun and subsequent iterations of adjoint optimisation were performed. Table 1 shows the overall reduction in pressure drop between pipe inlet/outlet for

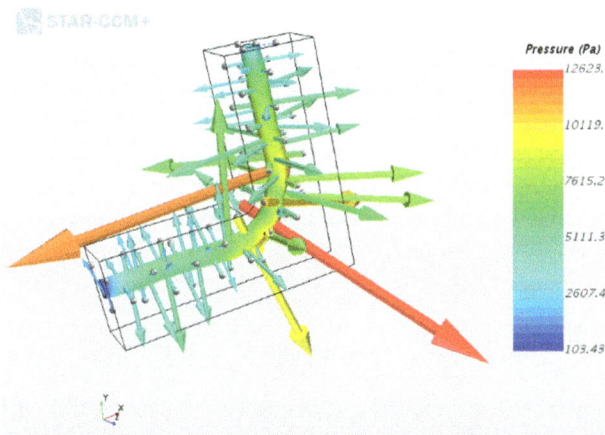

Fig. 2 Mesh sensitivity vector view for the adjoint solver to minimise inlet/outlet pressure drop

446   T. Zeng et al.

Table 1 Iteration results of
pressure drop

Run	Pressure drop (Pa)	Overall Reduction (%)
Baseline	12,728.5	N/A
Deformation 1	12,271.6	3.59
Deformation 2	10,848.3	14.8
Deformation 3	10,190.6	19.9

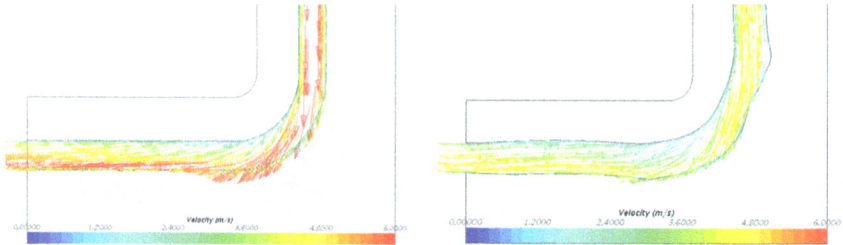

Fig. 3 Baseline vs deformation 3, velocity vector field (sub-sampled for clarity) and pipe cross section resulting from changes to the pipe geometry

consecutive adjoint optimisation runs. The initial deformation resulted in a relatively small reduction in pressure drop of 3.59%. Subsequent optimisation and deformation runs resulted in more significant decreases to 14.8% and 19.9% below the baseline value. The results show a significant overall reduction in pressure drop by the end of the optimisation process.

Figure 3 shows the spatial distribution of the velocity vectors for the optimised case (Deformation 3) versus the baseline simulation; it also indicates some of the changes that have occurred to the pipe cross section. Clearly, the centreline velocity of the flow within the pipe has been reduced because of the increased cross-sectional area of the pipe. This has resulted in a lower velocity gradient across the pipe cross section, yielding lower skin friction and cross-stream turbulent velocity causing a reduction in the pressure loss. Achieving a lower velocity gradient will lead to more uniform velocity profile, and therefore the fluid particles retention times will be similar which can eliminate the sub-cooled local boiling which is one of the major challenges when dealing with water as a coolant for aluminium casting. However, this also may lead to lower turbulence and therefore less heat transfer.

4 Conclusions

Adjoint optimisation has been applied to a simple 'proof of concept' L-shape conjugate heat transfer pipe simulation with pressure drop as the objective function. It has yielded a 19.9% reduction in pressure drop between the pipe inlet and outlet by increasing the cross-sectional area of the pipe in appropriate locations. The next step for this work will be to validate this approach against an experimental geometry and then to modify the objective function to have a blended function to target boundary heat flux as well as pressure drop for application to more realistic industrial geometries.

Acknowledgements This work is supported by Innovate UK Grant (TSB TS/R011249/1): '3D-printed conformal cooling channels for extended tooling life', in conjunction with CastAlum Ltd., and Renishaw Plc.

References

1. C. Othmer, A continuous adjoint formulation for the computation of topological and surface sensitivities of ducted flows. Int. J. Numer. Meth. Fluids **58**, 861–877 (2008)
2. S. Kitayama, H. Miyakawa, Multi-objective optimisation of injection moulding process parameters for short cycle time and warpage reduction using conformal cooling channel. Int. J. Adv. Manuf. Technol. (2017)
3. J.L. Lions, *Optimal Control of Systems Governed by Partial Differential Equations (Grundlehren Der Mathematischen Wissenschaften)*, vol. 170 (Springer, Berlin, 1971).
4. O. Pironneau, On optimum design in fluid mechanics. J. Fluid Mech. **64**(1), 97–110 (1974)
5. S. Xu, W. Jahn, J.D. Müller, CAD-based shape optimisation with CFD using a discrete adjoint. Int. J. Numer. Meth. Fluids **74**(3), 153–168 (2014)

Numerical Analysis of a Segmented Annular Thermoelectric Generator

Samson Shittu, Guiqiang Li, Xudong Zhao, and Xiaoli Ma

1   Introduction

The shape of the heat source usually influences the choice for the optimal geometry of a thermoelectric generator. When round-shaped heat sources are used, the conventional flat-shaped thermoelectric geometry becomes unsuitable, as it will lead to geometry mismatch with high thermal contact resistance and reduced thermoelectric generator (TEG) performance. Therefore, this study presents a detailed numerical investigation of a segmented annular thermoelectric generator (SATEG) using three-dimensional finite element analysis and COMSOL 5.4 Multiphysics software. Temperature-dependent thermoelectric material properties are considered and a comparison is made with non-segmented annular thermoelectric generator (ATEG). The influence of the ratio of length of hot segment to the total length of the leg on the power output of the SATEG is studied and the optimized geometry is compared with the ATEG. Bismuth telluride and Skutterudite are used as cold segment and hot segment thermoelectric materials and results show that the power output of the optimized SATEG is 54.55% and 13.94% greater than that of the Bismuth telluride ATEG and Skutterudite ATEG, respectively, at a temperature difference of 300 °C. In addition, results show that the optimum ratio of length of hot segment to the total length of leg (θ) is 0.35 for maximum power output.

A thermoelectric generator is a solid-state device that can convert waste heat into electricity via the Seebeck effect. The advantages of a TEG includes small size, lightweight, low maintenance, high reliability, and silent operation [1]. Therefore, thermoelectric generators have been used for several applications such as waste heat recovery, wearable sensors, and micro-power generation. However, the main limitation of the TEG is its low conversion efficiency and power output. Thermoelectric geometry optimization is an effective method to improve the conversion

S. Shittu · G. Li (✉) · X. Zhao · X. Ma
School of Engineering, University of Hull, Hull HU6 7RX, UK
e-mail: Guiqiang.Li@hull.ac.uk

© Springer Nature Singapore Pte Ltd. 2021
C. Wen and Y. Yan (eds.), *Advances in Heat Transfer and Thermal Engineering* ,
https://doi.org/10.1007/978-981-33-4765-6_77

efficiency and power output of a TEG [2]. In addition, segmentation of different thermoelectric materials can enable the combination of highly efficient thermoelectric materials based on their optimum operating temperature range. Consequently, segmented thermoelectric generators can provide enhanced performance compared to non-segmented thermoelectric generators [3]. The conventional flat-shaped TEG configuration is only suitable for applications in which heat flow is perpendicular to the ceramic plates; however, this configuration is unsuitable for systems in which the heat flow is in a radial direction or for cylindrical heat sources. Therefore, research on annular thermoelectric generators has increased recently due to the unique advantages they offer. Shen et al. [4] performed a theoretical analysis to study the performance of annular thermoelectric couples under a constant heat flux. In their study, they used finite element method to investigate the temperature dependency of thermoelectric materials and Thomson effect. Their results indicated that the level of the input heat flux must be controlled to protect the device from damage. Asaadi et al. [5] studied the thermal and electrical performance of an ATEG under pulsed heat flux using finite element method. They concluded that transient pulsed heating enhanced the output power and efficiency of the ATEG compared to constant steady-state heating.

2 Numerical Model

The governing equation of the thermoelectric analysis is solved using COMSOL 5.4 Multiphysics software which is based on finite element method. The thermoelectric analysis employed takes into account the Peltier effect, Fourier effect, Joule effect, and Thomson effect.

The heat flow equation in the thermoelectric analysis can be expressed as:

$$\rho_d C_p \frac{\partial T}{\partial t} + \nabla \cdot \vec{q} = \dot{q} \tag{1}$$

where ρ_d is the density, C_p is specific heat capacity, T is temperature, $\vec{q}$ is heat flux vector, and $\dot{q}$ is the heat generation rate per unit volume.

The electric charge continuity equations can be expressed as

$$\nabla \cdot \left(\vec{J} + \frac{\partial \vec{D}}{\partial t} \right) = 0 \tag{2}$$

where $\vec{J}$ is the electric current density vector and $\vec{D}$ is the electric flux density vector.

Equations (1) and (2) are coupled using the following thermoelectric constitutive equations,

$$\vec{q} = T[\alpha] \cdot \vec{J} - [\kappa] \cdot \nabla T \tag{3}$$

$$\vec{J} = [\sigma] \cdot \left(\vec{E} - [\alpha] \cdot \nabla T \right) \tag{4}$$

where $[\alpha]$ is the Seebeck coefficient matrix, $[\kappa]$ represents the thermal conductivity matrix, and $[\sigma]$ is the electrical conductivity matrix.

$$\vec{E} = -\nabla \varphi \tag{5}$$

where $\vec{E}$ is the electric field intensity vector and φ is the electric scalar potential.

The coupled thermoelectric equations can be obtained by combining the above equations as,

$$\rho_d C_p \frac{\partial T}{\partial t} + \nabla \cdot \left(T [\alpha] \cdot \vec{J} \right) - \nabla \cdot ([\alpha] \cdot \nabla T) = \dot{q} \tag{6}$$

$$\nabla \cdot \left([\varepsilon] \cdot \nabla \frac{\partial \varphi}{\partial t} \right) + \nabla \cdot ([\sigma] \cdot [\alpha] \cdot \nabla T) + \nabla \cdot ([\sigma] \cdot \nabla \varphi) = 0 \tag{7}$$

where $[\varepsilon]$ represents the dielectric permittivity matrix.

Finally, the coupled thermoelectric governing equations can rewritten as,

$$\nabla \cdot \left(T \alpha \vec{J} \right) - \nabla \cdot (\lambda \nabla T) = \dot{q} \tag{8}$$

$$\nabla \cdot (\sigma \alpha \nabla T) + \nabla \cdot (\sigma \nabla \varphi) = 0 \tag{9}$$

The electrical performance of the thermoelectric generator is determined from its output power and efficiency, which are defined as,

$$P_{out} = I^2 R_L \tag{10}$$

$$\eta = \frac{P_{out}}{Q_{in}} \tag{11}$$

where P_{out} is the output power and I is the circuit electric current.

The total thermoelectric leg length is given as,

$$L = L_1 + L_2 = 2 \, \text{mm} \tag{12}$$

The ratio of length of hot segment to the total length of leg is given as,

$$\theta = \frac{L_1}{L} \tag{13}$$

The thermoelectric generator is made of ceramic layers, copper layers, solder layers, and semiconductor materials (p-type and n-type). The inner radius of the

(a) Hot surface   (b)

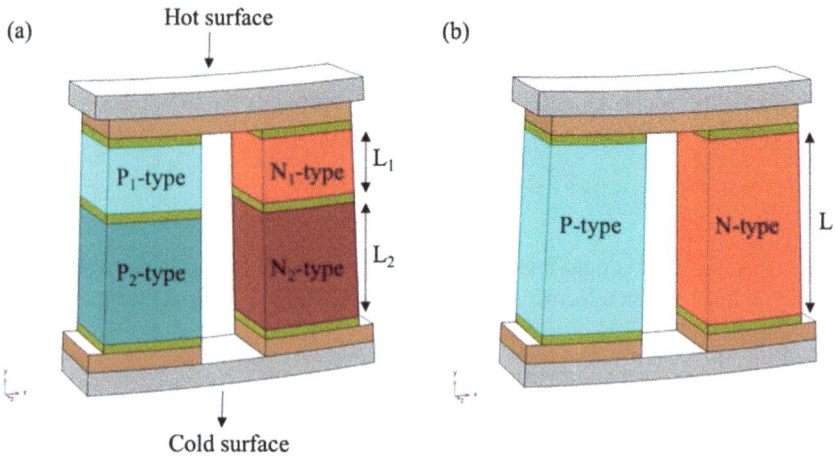

Cold surface

Fig. 1 Schematic diagram of **a** SATEG and **b** ATEG

ceramic is 30 mm, the thickness of the ceramic, copper and solder are 0.25 mm, 0.2 mm, and 0.1 mm, respectively. The schematic diagram of the segmented thermo-electric generator and annular thermoelectric generator is shown in Fig. 1. The hot segment is made up of Skutterudite material (P_1-type and N_1-type) while the cold segment is made up of Bismuth telluride material (P_2-type and N_2-type). The bottom copper on the P-type leg is grounded while that on the N-type leg is connected to an external load resistance, which is varied to obtain maximum power output.

3 Results

In this study, constant temperature boundary conditions are applied at the hot surface and cold surface of the thermoelectric generator. The temperature of the cold surface is kept constant at 20 °C while that of the hot surface is varied according to the study. To optimize the length of the hot segment and cold segment of the SATEG, the ratio of length of hot segment to the total length of leg (θ) is varied from 0.05– 0.45 for different hot side temperature values and the result is shown in Fig. 2a. It can be seen clearly that the maximum power output of the SATEG increases as the hot side temperature increases. This is because the power output of a thermoelectric generator is directly related to the temperature difference across its hot and cold surfaces. Therefore, since the cold surface temperature is constant, increasing the hot side temperature leads to an increase in the temperature difference across the TEG and consequently an increase in power output. In addition, it is obvious from Fig. 2a that the optimum θ ratio for the SATEG is 0.35 for obtaining maximum SATEG power output. Therefore, this optimum ratio is used to compare the SATEG with the ATEG and the result is shown in Fig. 2b. It can be seen that the optimized

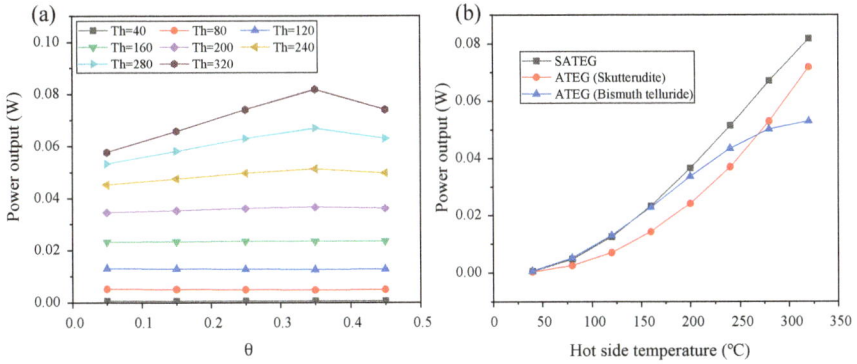

Fig. 2 Variation of maximum power output with **a** hot side temperature and θ **b** hot side temperature of optimized SATEG and different ATEG

SATEG provides the best performance compared to the Bismuth telluride ATEG and Skutterudite ATEG.

4   Conclusions

This study presented a detailed numerical investigation of a segmented annular thermoelectric generator (SATEG) using three-dimensional finite element analysis and COMSOL 5.4 Multiphysics software. Bismuth telluride and Skutterudite thermoelectric materials were used as the cold segment and hot segment materials. Temperature-dependent thermoelectric materials were considered, and the external load resistance was varied to obtain the maximum power output. The geometry of the SATEG was optimized and the power output of the optimized SATEG was compared with that of the Bismuth telluride ATEG and Skutterudite ATEG. Results showed that the power output of the optimized SATEG is 54.55 and 13.94% greater that of the Bismuth telluride ATEG and Skutterudite ATEG, respectively, at a temperature difference of 300 °C. In addition, results showed that the optimum ratio of length of hot segment to the total length of leg (θ) is 0.35 for maximum power output.

Acknowledgements The first author would like to acknowledge the PhD studentship received from the University of Hull.

References

1. S. Shittu, G. Li, Y.G. Akhlaghi, X. Ma, X. Zhao, E. Ayodele, Advancements in thermoelectric generators for enhanced hybrid photovoltaic system performance. Renew. Sustain. Energy Rev.

109, 24–54 (2019). https://doi.org/10.1016/j.rser.2019.04.023
2. S. Shittu, G. Li, X. Zhao, X. Ma, Y.G. Akhlaghi, E. Ayodele, Optimized high performance thermoelectric generator with combined segmented and asymmetrical legs under pulsed heat input power. J. Power Sources **428**, 53–66 (2019). https://doi.org/10.1016/j.jpowsour.2019.04.099
3. S. Shittu, G. Li, X. Zhao, X. Ma, Y.G. Akhlaghi, E. Ayodele, High performance and thermal stress analysis of a segmented annular thermoelectric generator. Energy Convers. Manag. **184**, 180–193 (2019). https://doi.org/10.1016/j.enconman.2019.01.064
4. Z.-G. Shen, X. Liu, S. Chen, S.-Y. Wu, L. Xiao, Z.-X. Chen, Theoretical analysis on a segmented annular thermoelectric generator. Energy **157**, 297–313 (2018). https://doi.org/10.1016/j.energy.2018.05.163
5. S. Asaadi, S. Khalilarya, S. Jafarmadar, Numerical study on the thermal and electrical performance of an annular thermoelectric generator under pulsed heat power with different types of input functions. Energy Convers. Manag. **167**, 102–112 (2018). https://doi.org/10.1016/j.enconman.2018.04.085

Heat Transfer Modelling of APAA Transmit-Receive Modules

V. I. Zhuravliov, N. M. Naumovich, and V. S. Kolbun

1 Introduction

Synthetic aperture radars occupy a special place in space systems for radar monitoring of the earth's surface, along with optical systems. They allow conducting round-the-clock observation in bad weather conditions. Most modern satellite-borne systems are based on phased antenna arrays, which make it possible to enhance the abilities by the use of electronic beam scanning [1].

Currently, the most important task is to increase the radiated microwave power level of active phased antenna array (APAA) to extend the range and improve accuracy of target detection. This problem places demand on the overall dimensions of transmit/receive modules (TRM) – the main elements of APAA. Solving the general module, layout task must necessarily take into account the values of operating temperatures of using components and the system capabilities for thermal conditions in a space environment. TRMs in the APAA design are the components that generate the most heat. The sources of excess heat in TRM are crystals of output power amplifiers. To decrease the active element maximum temperature, highly heat-conducting bases are used for such microcircuits contributes in contrast to methods for increasing the heat transfer coefficient by convection and radiation. Since convection heat transfer in space conditions is practically absent and local thermal radiation is difficult and

V. I. Zhuravliov (✉)
Fundamental Electrical Engineering, Belarusian State University of Informatics and Radioelectronics, Minsk, Belarus
e-mail: vadzh@bsuir.by

N. M. Naumovich
Scientific Center 1.6, Belarusian State University of Informatics and Radioelectronics, Minsk, Belarus

V. S. Kolbun
Department of Information and SCAD, Belarusian State University of Informatics and Radioelectronics, Minsk, Belarus

© Springer Nature Singapore Pte Ltd. 2021
C. Wen and Y. Yan (eds.), *Advances in Heat Transfer and Thermal Engineering* ,
https://doi.org/10.1007/978-981-33-4765-6_78

leads to heating of structural components, obviously, the main way to remove the excess power from the TRM is the conductive path.

2 Thermal Model of VHF Distributed Heat Sources

The thermal model under consideration includes the unpackaged microcircuits of the VHF printed circuit board of the TRM, which are located on a separate duralumin base, and their pins are connected to contact pads of the PCB.

We consider the power amplifier chip's small thickness in comparison with the base and TRM body when calculating the thermal field of the VHF board. Therefore, it is possible to neglect the heat transfer from the chip side surfaces. Thus, these surfaces are considered to be thermally insulated, i.e., there are boundary conditions of the second kind. The total thermal power released in a single chip can be determined as follows:

$$P = \frac{P_{\text{input}} - P_{\text{emit}}}{N} \tag{1}$$

where P_{input} is the input power; P_{emit} is emitted microwave power; N is the number of amplifier chips in the TRM.

Equation (1) allows us to determine the specific density of active element heat sources:

$$q_v = \frac{P}{V_0} \tag{2}$$

where V_0 is the crystal volume.

Since the emitted power is irregular in time, the chip case is represented as a pulsed power source P_0 in the rectangular parallelepiped form with side lengths a, b, c, and the heat equation particular solution by the Green's function for the temperature instantaneous value at time temperature looks like this [2]:

$$T(t) = T_0 + \frac{P_0}{\rho \, C_p \, K} \int_0^t \text{erf}\left(\frac{a}{4\sqrt{\chi\tau}}\right) \cdot \text{erf}\left(\frac{b}{4\sqrt{\chi\tau}}\right) \cdot \text{erf}\left(\frac{c}{4\sqrt{\chi\tau}}\right) dt \tag{3}$$

where K, ρ, C_p, χ are respectively the thermal conductivity coefficient, the material density, the semi-finite body heat capacity, and heat transfer coefficient; T_0 is the ambient temperature.

3 Results

By using of numerical methods to solve an Eq. (3), it possible to more accurately estimates the heating of both the source and the PCB area adjacent to the amplifiers (Fig. 1a).

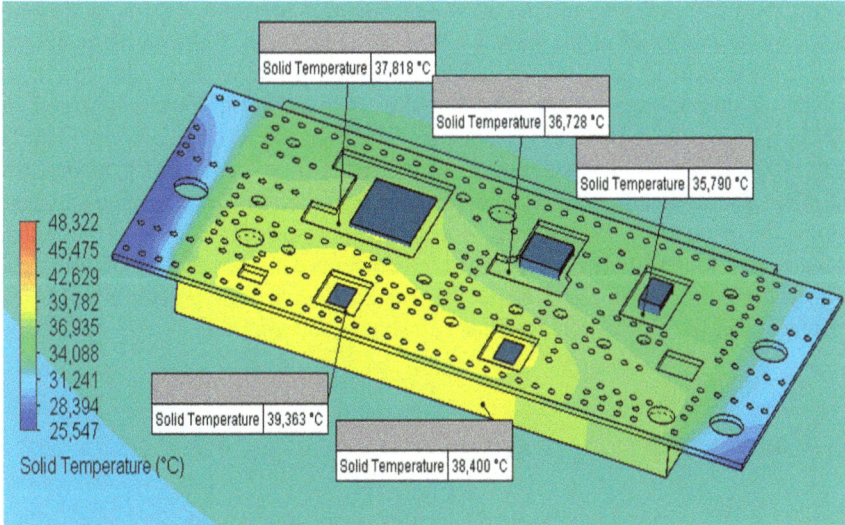

a

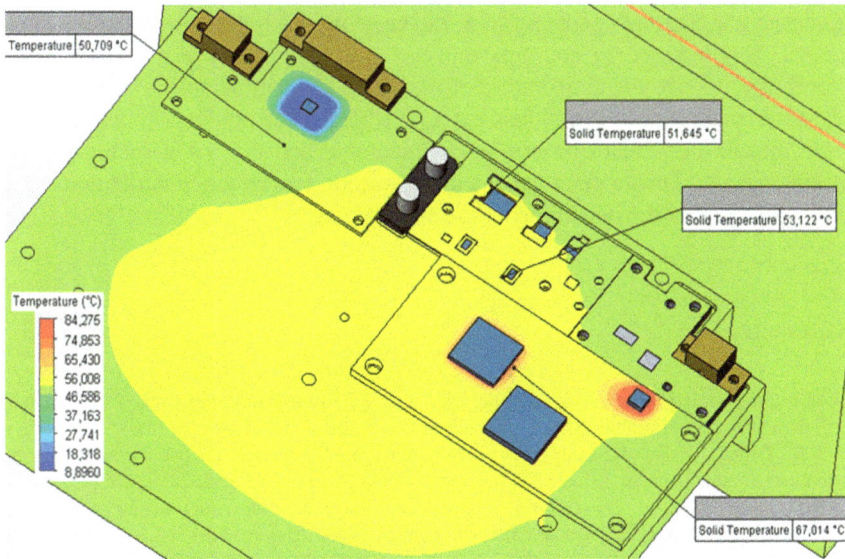

b

Fig. 1 Temperature distribution on the VHF board (**a**) and in the TRM (**b**)

As can be seen, the use of a metal divider in addition to reducing the peak temperature makes it possible to avoid large temperature gradients for the PCB.

Further, this area is represented as an independent heat source in the grid thermal model. It is possible to obtain the distribution of thermal fields for the entire non-uniform TRM structure containing several boards (Fig. 1b).

In the further thermal chain, a part of the excess power is diverted to the body of the TRM and radiated, and the other part (main) is transmitted through the body to a honeycomb panel of the APAA. Direct installation of TRM on the panel can lead to formation of gaps, which makes it difficult to remove heat and requires to reduce the thermal resistance of the "TRM body–panel" [3]. Towards this end, all TRMs are placed on a thermal spreader, made of high thermal conductivity material with increased mechanical strength, but sufficiently plastic to fill the gaps between the TRM bodies and the APAA panel [4]. Due to some of its own volume and heat-conducting properties, the spreader reduces the working temperature of the hottest components. It contributes to the uniform temperature distribution due to transferal of heat from hotter to cooler areas of the panel. Moreover, the probability of bends decreases and the mechanical resistance to shocks and vibration.

4   Conclusions

Thus, more effective way to reduce the heating temperature of satellite-borne APAA TRMs and to increase the maximum achievable radiated VHF power is to improve the process of heat removal from the power amplifier chips. This is achieved by increasing the thermal conductivity of the base on which the power amplifiers are located. This reduces the non-uniformity of the thermal field of the TRM, which allows us to reduce the temperature of the active element thanks to more efficient heat transfer through the module's external surface. The cooling process involves those module parts that previously remained cold and did not participate in the cooling process. On the other hand, a more uniform temperature distribution inside the module lowers the maximum temperature values.

References

1. W. Pitz, Miller the TerraSAR-X Satellite. IEEE Trans. Geosci. Remote Sens. **48**, 615–622 (2010)
2. V.M. Dwyer, A.J. Franklin, D.S. Campbell, Solid-St. Electr. **33**, 553–560 (1990)
3. A.M. Elhady Design and analysis of a LEO micro-satellite thermal control including thermal contact conductance, in *2010 IEEE Aerospace Conference*, 6–13 March 2010, (MT, USA)
4. N.M. Naumovich, V.I. Zhuravliov et al. Constructive scheme of the experimental base panel of APAA fragment in *VII Belarusian Space Congress* (Minsk, Belarus, 2017) T.1, 228–231

Numerical Optimisation of the Regenerator of a Multi-stage Travelling Wave Thermoacoustic Electricity Generator

Wigdan Kisha and David Hann

1 Introduction

In a travelling wave thermoacoustic engine, the regenerator is the heart of the engine [1], where the thermal interaction between the working gas and the solid material takes place. Unlike the stack in standing wave engine, the regenerator is more of a solid porous material which offers an advantage of better heat transfer and exhibits an enhanced performance. At the frequency of interest, excellent thermal contact can be achieved by having a regenerator of very narrow channels. However, this can increase the pressure drop, and therefore, less acoustic power will be produced. Within the regenerator, the amplification of the acoustic power is determined by the temperature gradient along its end, as well as the phase difference between the pressure and volumetric velocity of oscillation. The temperature gradient is maintained by having a high-temperature source in one end of the regenerator, and a heat sink in the other, all these parts form the so-called thermoacoustic core. On the other hand, the correct time phasing condition can be met by having excellent thermal contact between the gas parcels and the solid wall. Several attempts have been made over the past two decades to better understand the mechanisms of acoustic power amplification within the regenerator [2–4]. It has observed a strong relationship between design parameters of the regenerator and the thermoacoustic performance. However, the process of optimisation of the regenerator for a particular application is complex, particularly, when the cost of the system is the main driver. Additionally, the complexity increases when the thermoacoustic system of more than one regenerator. The current research focuses on the regenerators of a double core travelling wave thermoacoustic electricity generator designed by SCORE project [5, 6]. The effect of the regenerator design parameters on the power conversion will be discussed. The motivation behind this work is driven by the interest in improving the power output of the engine.

W. Kisha (✉) · D. Hann
Faculty of Engineering, University of Nottingham, Nottingham NG7 2RD, UK
e-mail: wigdan.kisha@nottingham.ac.uk

© Springer Nature Singapore Pte Ltd. 2021
C. Wen and Y. Yan (eds.), *Advances in Heat Transfer and Thermal Engineering* ,
https://doi.org/10.1007/978-981-33-4765-6_79

2 Methodology

The system is shown schematically in Fig. 1. It has two identical engines, two stub branch pipes, a loudspeaker and feedback pipes (loop). Each engine has a main ambient heat exchanger MAHE, a regenerator, a hot heat exchanger HHX, a thermal buffer tube TBT and a secondary ambient heat exchanger SAHE.

The loudspeaker is used to harvest the electricity, while the two tuning stubs are installed to reduce the acoustic reflections resulted from the existence of the power extractor. The change in the acoustic power, $dP_{Ac.}$, through a length dx of the regenerator channel, in x-direction can be expressed as [7]:

$$\frac{dP_{Ac.}}{dx} = \frac{1}{2}\mathrm{Re}[g\tilde{p}_1 U_1] - \frac{1}{2r_k}|p_1^2| - \frac{r_v}{2}|U_1^2| \qquad (1)$$

Where: the first, second and third terms in the right-hand-side of Eq. 1 represent the generated acoustic power, the thermal-relaxation loss, and the viscous dissipation, respectively. p_1 and U_1 represent the pressure and volumetric velocity amplitudes, respectively, ~ denotes a complex conjugate, Re[ ] indicates the real element of a complex number, r_k and r_v represent the thermal and viscous resistances, respectively. The system receives heat from two identical custom-made electrical heaters, and heat is removed by water circulation method. Details of the system components can be found elsewhere [6]. Atmospheric air is selected as the working gas, and the operational frequency is about 71 Hz. A total amount of heating power of 2.5 kW is added into the two engines unevenly; 1 kW in engine 1, and 1.5 kW in engine 2 because it shows the possibility of improving the amplification of the acoustic power [8]. The simulation of the thermoacoustic system is carried out using DeltaEC

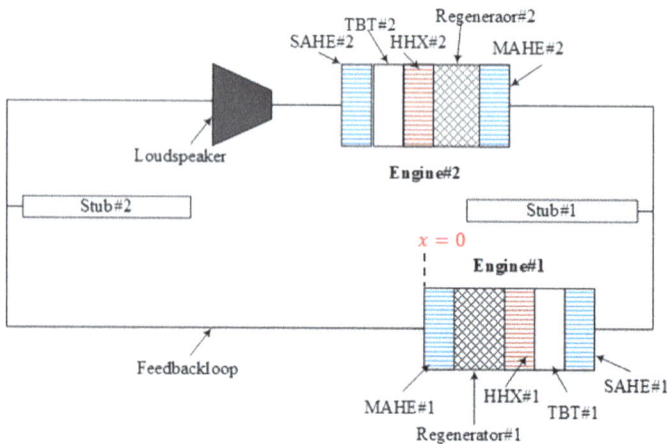

Fig. 1 Schematic diagram of the two-stage thermoacoustic electricity generator showing the main system components. The distance between the two engines is approximately quarter the wavelength

software [9]. The software enabled carrying out a systematic study for the geometrical parameters of the regenerator. The model starts at $x = 0$ at the first main ambient heat exchanger MAHE#1 and continuous anticlockwise. The pressure and volumetric velocity of oscillation, and their time-phasing adjusted to be the same at the starting and ending points using guess-target strategy [9]. The thermal and viscous losses were calculated together with the acoustic gain and their dependency on the tested design parameters.

3   Results

The net acoustic power running through the feedback loop is investigated theoretically to show the influence of the regenerator parameters on the performance. This net acoustic power flow is called loop acoustic power [10]. The loop acoustic power, $P_{Ac.}$ of the double core thermoacoustic system with various lengths and hydraulic radii of the regenerators is plotted in Fig. 2a. $P_{Ac.}$ starts at relatively low level of the loop acoustic power, but grows rapidly with the hydraulic radius and subsequently levels off and the sudden drop in the loop acoustic power corresponding to amplitude drop. The electrical power output as shown in Fig. 2b also has a similar trend.

As it presents in Fig. 2c and d, for each length of the regenerator, there is an optimal hydraulic radius that corresponds to the optimal loop acoustic power. The optimal loop acoustic power was found to be 131 W when each regenerator is 28 mm long and both have a hydraulic radius $r_h = 180$ mm. The corresponding electrical power output is 52.4 W. These findings reflect an increase of about 71.5% in the acoustic power output and 67% increase in electricity achieved before the optimisation. A summary of the numerical results of the optimised regenerators is provided in Table 1.

4   Conclusions

This research explores many of the facets of the design of the core of a two-stage thermoacoustic engine using numerical calculations. Bearing in mind, the type of the application (for developing world) and the affordability, attention has been drawn into the unpressurised air-filled resonator, with a simple design for the thermoacoustic system. One of the most critical design features in travelling wave thermoacoustic engines is the choice of the regenerator, due to its function as an acoustic amplifier. This work aims to improve the power output of the system. Consequently, an optimisation procedure is proposed for the regenerator's geometry, and it is shown to enhance the thermoacoustic performance of the two-stage thermoacoustic engine. By varying the length and the pore shape of the regenerator, the numerical data realised that it is possible to boost the acoustic and electrical power outputs up to 131 and 52.4 W, respectively. This reflects percentages of increase in the outputs of 72% and 67%, respectively, as a result of increasing the lengths from 12 to 28 mm,

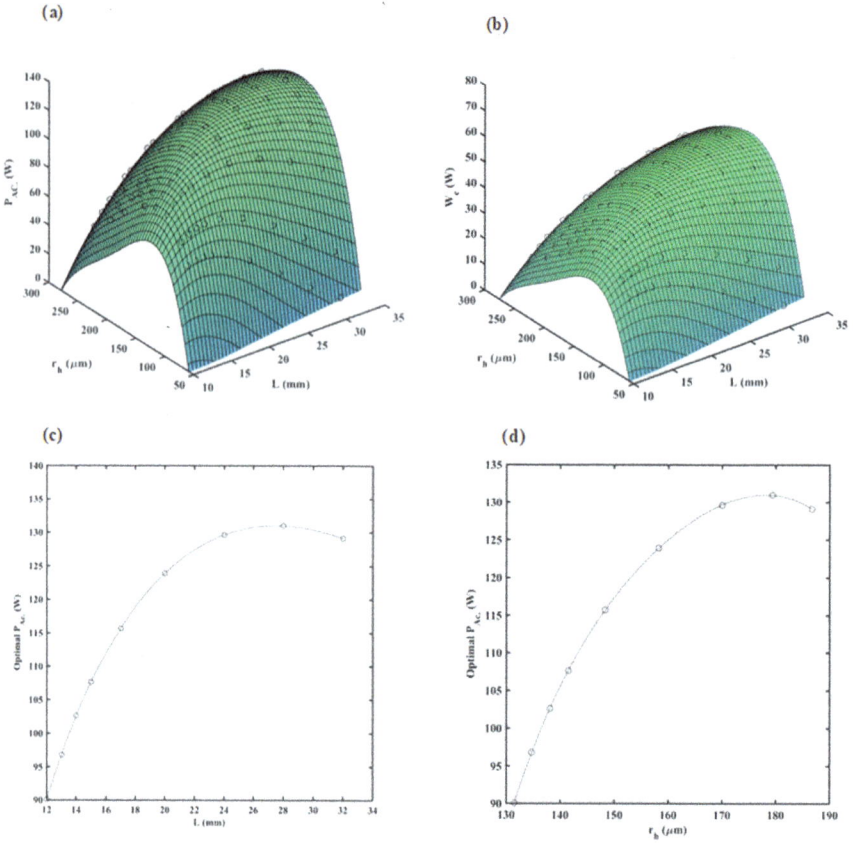

Fig. 2 Optimisation results of the double core thermoacoustic engine for lengths and hydraulic radii ranges of 12–32 mm and 86–260 µm, respectively; **a** the acoustic power flowing within the thermoacoustic loop, **b** the extracted electrical power, **c** the optimal length of the regenerator and **d** the optimal hydraulic radius of the regenerator

Table 1 Results of DeltaEC optimisation of the double core thermoacoustic engine

Parameters	Engine #1
Heat input power into engine #1	1
Heat input power into engine #2	1.5
Acoustic power production within reg. 1	33
Acoustic power production within reg. 2	59
Acoustic power of the loop	131
Electrical power output, W_e (W)	52.4
Thermal-to-acoustic efficiency, $\eta_{t\text{-}a}$ (%)	5.2
Thermal-to-electrical efficiency, $\eta_{t\text{-}e}$ (%)	2.1

and the radii from 97 to 180 µm. Even though the acoustic amplification could be increased by having extra mesh layers, but this would enhance the attenuation effect. Therefore, the length of the regenerator must be selected wisely by compromising between the amount of useful work and attenuation.

Acknowledgements The author would like to thank the University of Nottingham and Gordon Memorial College Trust Fund for supporting this research.

References

1. P.H. Ceperley, Gain and efficiency of a short traveling wave heat engine. J Acoust. Soc. Am. **77**, 1239–1244 (1985)
2. Z. Yu, A.J. Jaworski, Impact of acoustic impedance and flow resistance on the power output capacity of the regenerators in travelling-wave thermoacoustic engines. Energ. Convers. Manag. **51**, 350–359 (2010)
3. A. Kruse, A. Ruziewicz, M. Tajmar, Z. Gnutek, A numerical study of a looped-tube thermoacoustic engine with a single-stage for utilization of low-grade heat. Energy Convers. Manag. **149**, 206–218 (2017)
4. C. Lawn, Acoustic pressure losses in woven screen regenerators. Appl. Acoust. **77**, 42–48 (2014)
5. B. Chen, A.A. Yousif, P.H. Riley, D.B. Hann, Development and assessment of thermoacoustic generators operating by waste heat from cooking stove. Engineering **4**, 894–902 (2012)
6. P.H. Riley, *Designing a Low-Cost electricity-Generating Cooking stove for High-Volume Implementation* (University of Nottingham, 2014)
7. G. W. Swift, *Thermoacoustics: A Unifying Perspective for Some Engines and Refrigerators* (Springer, 2017)
8. W. Kisha, P.H. Riley, J. McKechnie, D. Hann, The influence of heat input ratio on electrical power output of a dual-core travelling-wave thermoacoustic engine, in *The Heat Powered Cycles Conference* (Bayreuth, 2018)
9. J.P. Clark, W.C. Ward, G.W. Swift, Design environment for low-amplitude thermoacoustic energy conversion (DeltaEC). J. Acoust. Soc. Am. **122**, 3014–3014 (2007)
10. K. De Blok, Low operating temperature integral thermo acoustic devices for solar cooling and waste heat recovery. J. Acoust. Soc. Am. **123**, 3541–3541 (2008)

Experimental and Numerical Investigation of the Effects of Different Heat Transfer Modes on the Sublimation of Dry Ice in an Insulation Box

Abhishek Purandare and Srinivas Vanapalli

1 Introduction

Pharmaceuticals and biopharmaceuticals requiring controlled temperature conditions must be transported in a way that does not affect their quality adversely. The cold chain has become an important part of the overall pharma supply chain as a result of regulatory requirements, temperature monitoring, risk assessment, and good storage and distribution practices. Over the last few decades, dry ice has been increasingly used for temperature control due to its high energy density and ability to maintain constant temperature during solid-vapor phase change process (sublimation). It is reported in [1] that in 2003, FedEx was shipping approximately 15,000 packages per day containing dry ice, out of which the maximum packages were related to biopharmaceutical industry. An extensive amount of literature [2] is available on the numerical and experimental investigation of the insulated boxes using phase change materials as the cold source as opposed to the use of dry ice as the cold source for similar application. Therefore, a thorough understanding about the sublimation of dry ice contained in an insulating package usually made of polyurethane and polystyrene foams is required in order to improve the effectiveness of the shipper. The sublimation rate of the dry ice in an insulation box is determined by different transient heat transport phenomena like heat diffusion, buoyancy-driven convection and radiation. Therefore, this study focuses on experimentally and numerically investigating the thermal performance of the shipper containing dry ice as a cold source.

A. Purandare (✉) · S. Vanapalli
Laboratory of Applied Thermal Sciences, Faculty of Science and Technology, University of Twente, 7500, AE Enschede, The Netherlands
e-mail: a.s.purandare@utwente.nl

© Springer Nature Singapore Pte Ltd. 2021

C. Wen and Y. Yan (eds.), *Advances in Heat Transfer and Thermal Engineering* ,
https://doi.org/10.1007/978-981-33-4765-6_80

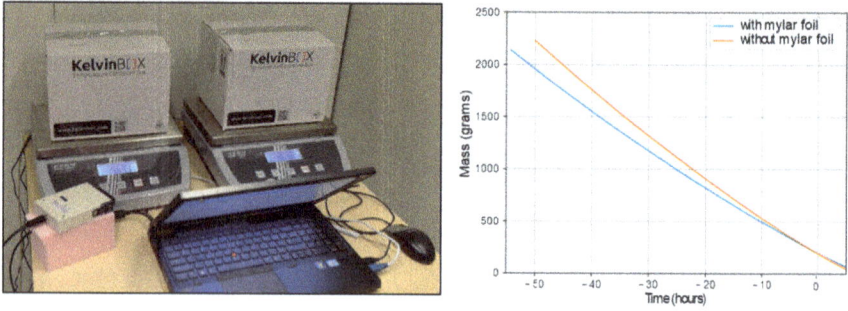

Fig. 1 Experimental setup (left), measured values of dry ice mass as a function of time (right)

2 Methodology

The investigation performed in this work is based on an extensive comparison with experiments. For this purpose, an extended polystyrene (EPS) box with external dimension of $0.265 \times 0.215 \times 0.190$ m and wall thickness of 0.04 m is used which can hold up to 2.2 kg of dry ice pellets. Preliminary experiments were performed and thermal images were captured to see that the cold CO_2 gas was escaping from the gap between the lid and the box. The temperatures at different locations on the EPS box and the mass of dry ice were measured as a function of time. The measurements are carried out on two different EPS boxes, one of which has its interior walls covered with thin layer of aluminized mylar foil for investigating the effects of radiation. Before starting the measurements, each of the two EPS boxes is sealed in a cardboard box as shown in Fig. 1 (left).

3 Results

Figure 1 (right) shows one set of measured values of the mass of the two EPS boxes as a function of time. Since the initial mass of the two boxes filled with dry ice pellets could be different from each other, we note the time at which the mass of the two boxes reduces to 0.2 kg. This time is set to time $t = 0$ and all the previous values of time are considered to be negative, hence, the negative scale on the x-axis of Fig. 1 (right). After 50 h (from 0 to -50 on x-axis), the amount of dry ice that sublimates in the EPS box with mylar foil is approximately 14% less as compared to the EPS box that does not contain the foil layer.

Figure 2 shows the measured values of temperature at seven different locations on the EPS boxes. Thermocouple no. 8 depicts ambient temperature in both cases as shown in Fig. 2. The temperature at the bottom center (thermocouple 7) stays nearly constant till the end of the sublimation process while the temperatures measured by thermocouple number 3, 4, 5, and 6 start to increase at an early stage. This signifies

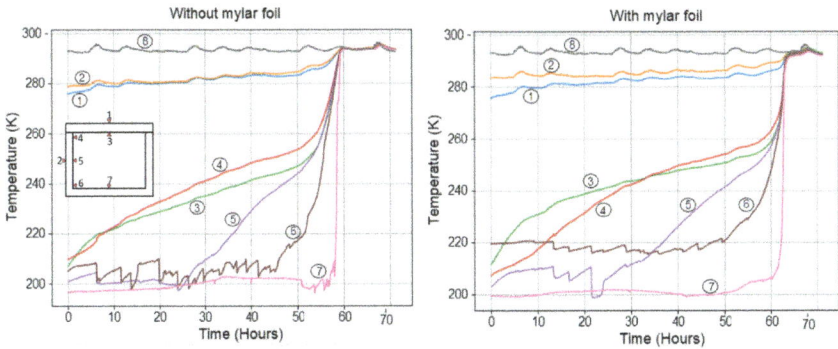

Fig. 2 Measured temperature values at different locations on the EPS boxes

that dry ice, while sublimating inside the box forms a dome shape resulting in higher amount of dry ice in the center than near the walls. The dome shape can be due to stronger convection currents near the vertical walls of the EPS box indicating that the sublimation rate is higher for the pellets near the wall. The rapid rise in the temperature values of thermocouple number 3, 4, 5, 6, and 7, and their values reaching the ambient temperature indicates that the dry ice is completely sublimated. It can be seen from Fig. 2 that the dry ice present in the box without mylar foil sublimates early than the dry ice present in the box with the mylar foil (over 60 h). This implies that the radiation has a considerable effect on the sublimation rate of the dry ice pellets as the mylar foil layer reflects a portion of the radiation falling on it. A quasi-steady numerical model based on heat conduction and radiation is developed using a comercial software package COMSOL 5.4. Further details on modelling and experimental results can be found in journal article [3].

4 Conclusions

In this work, the problem of sublimation of dry ice in an insulation box is investigated with the aim of increasing the effectiveness of the shipper. Although the results are briefly described in this abstract, the numerical and the experimental results will be explained extensively during the presentation. Both numerical and experimental results reveal that radiation has a considerable effect on the sublimation rate. The simplified two-dimensional model shows that sublimation rate is not largely influenced by convection as the maximum velocity calculated from the model has a smaller magnitude of approximately 0.003 m/s. In practice, during sublimation of dry ice, the concentration of CO_2 gas in the voids of EPS foam will increase with time. As a result, the thermal conductivity of the foam will decrease because the CO_2 gas has lower thermal conductivity than air. For the more exact prediction of the experimental results, a further numerical investigation is needed to model the aforementioned phenomena and to study the effect of convection in three dimensions.

References

1. D. Caldwell, R. Lewis, R. Shaffstall, R. Johnson, Sublimation rate of dry ice packaged in commonly used quantities by the air cargo industry (Federal Aviation Administration, 2006)
2. Y. Kozak, M. Farid, G. Ziskind, Experimental and comprehensive theoretical study of cold storage packages containing PCM. Appl. Therm. Eng. **115**, 899–912 (2017)
3. A.S. Purandare, S.W. Lohuizen, R.M.A. Spijkers, S. Vanapalli, Experimental and numerical study of insulation packages containing dry ice pellets. Appl. Therm. Eng. 1359–4311 (2021), https://doi.org/10.1016/j.applthermaleng.2020.116486

Heat Transfer and Thermal Stress Analysis of PDC Cutter in Rock Breaking Process

Zengzeng Zhang, Yan Zhao, and Congshan Zhang

1 Introduction

With the continuous increase of human demand for oil and natural gas, the exploration and development of oil and gas wells have gradually moved from conventional formation to deep formation. PDC bits are widely used in oil and gas exploration and development because of its high rate of penetration (ROP) and drilling efficiency [1]. Nevertheless, PDC bits suffer thermal abrasive wear and impact damage while drilling hard and brittle formations, resulting in serious failure of cutters, which will greatly reduce the service life, drilling efficiency and increase drilling costs [2]. Cutting temperature is generated by high-speed drilling of PDC cutters during the rock breaking process. More than half of the energy supplied to the bits is converted into friction heat [3], which causes the temperature of the bits to rise sharply. During the drilling process, the bits are subjected to a large impact load, which easily causes the cutter to withstand overload in a short time, resulting in large-scale macroscopic fracture. Therefore, the temperature and stress of cutters are one of the important factors affecting the failure of drill bits. It is of great value to thoroughly study the temperature and stress field of PDC cutters during the rock breaking process in order to explore the failure mechanism of PDC bits comprehensively and improve the service life of PDC bits.

In recent years, experiment or numerical research was mainly focused on the two-dimensional linear cutting process of a single cutter, and there were few studies on full-sized PDC bits. In addition, most of the simulations assumed that the drill bit was a rigid body, ignoring the deformation and wear of the drill bit, which was quite different from the actual drilling process. Therefore, the numerical simulation

Z. Zhang · Y. Zhao (✉) · C. Zhang
College of Construction Engineering, Jilin University, Changchun 130026, China
e-mail: zhaoyan1983@jlu.edu.cn

Key Laboratory of Drilling and Exploration Technology in Complex Conditions of Ministry of Natural Resources, Changchun 130026, China

© Springer Nature Singapore Pte Ltd. 2021
C. Wen and Y. Yan (eds.), *Advances in Heat Transfer and Thermal Engineering* ,
https://doi.org/10.1007/978-981-33-4765-6_81

analysis and test of the temperature, stress field distribution and deformation of full-sized PDC cutters are seldom studied, and the specific process of rock breaking and temperature rise of the cutter is not described in detail.

2 Methodology

The finite element ABAQUS was used to simulate the thermal-structural coupling of rock breaking process (Fig. 1), and the distribution and the specific changes of the temperature and stress field of the cutter are analyzed during the drilling process. The fragmentation of rock can be considered as brittle failure, and the heat generated is limited to friction heat generation. According to Fourier's law, the partial differential governing equation of heat conduction on the working interface is as follows:

$$\rho c \frac{\partial T}{\partial t} = \left[\lambda_x \frac{\partial^2 T}{\partial x^2} + \lambda_y \frac{\partial^2 T}{\partial y^2} + \lambda_z \frac{\partial^2 T}{\partial z^2} \right] + \dot{q}_f \tag{1}$$

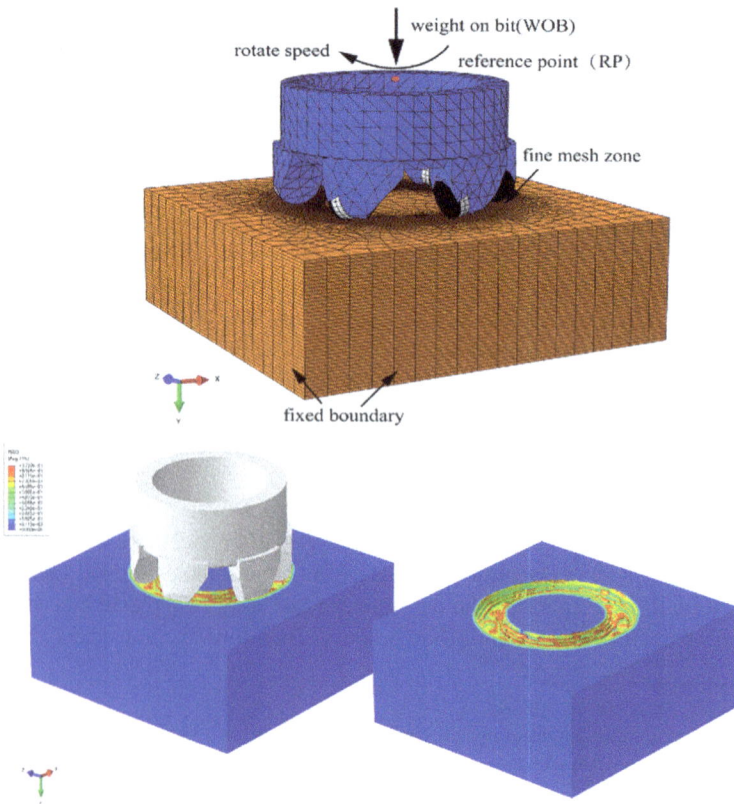

Fig. 1 Bit–rock interaction model

When the PDC cutters slide across rock surface at a constant speed, assume that the cutters are subjected to uniform and stable frictional heat flow, and all mechanical friction is converted into heat at the working interface. The total heat generated on the unit wear area in a per unit time can be described as follows [4]:

$$\dot{q}_f = \frac{\mu F_n v}{A_w} \tag{2}$$

The part of the heat entering the cutter can be expressed as:

$$q_1 = \frac{\alpha \dot{q}_f}{A_w} \tag{3}$$

The thermal response function f of the bit is:

$$f = \frac{\bar{T} - T_0}{q_1} \tag{4}$$

Combining Eqs. (2)–(4) yields an equation for the mean cutter and rock temperature:

$$\bar{T} = T_0 + \frac{\alpha \mu F_n v f}{A_w} \tag{5}$$

where α is the distribution coefficient, T_0 is the initial rock temperature, μ is the coefficient of friction, F_n is the normal force, v is the cuter speed, A_w is the wear area.

3 Results

The simulation results showed that the high-temperature region shifted from the top of the cutting crown to the outside of the cutter during drilling, and the high-temperature region of the cutter distributed in a fan-shaped shape, mainly distributed in the contact with rock and the cutting speed was the highest. With the increase of rotational speed, the difference between the temperature at the outermost end of the cutter and the temperature at the top of the crown will be greater and greater. The temperature of cutting teeth decreased with the increase of cutting angle, as in shown in Fig. 2(a).

It can be seen clearly from Fig. 2(b), the stress distribution on the cutter of PDC bit fluctuated from the interaction between the cutter and rock. The maximum stress in the diamond layer occurred at the edge of the diamond layer. The high-stress concentration and fluctuating impact at the joint surface were the main reasons leading to the fracture failure of PDC cutter. The crown top of PDC bit cutter was most likely to fracture and delamination failure due to high stress.

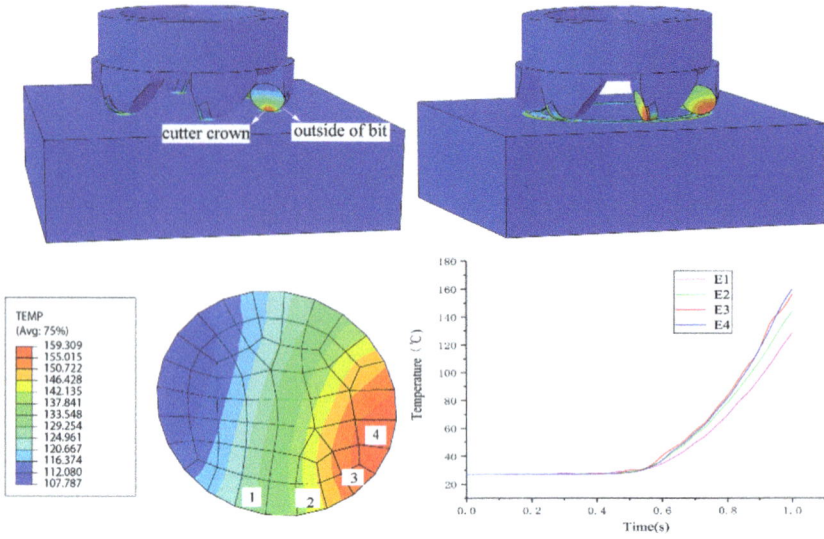

(a) Temperature nephogram of cutting teeth and Temperature curves of 4 units.

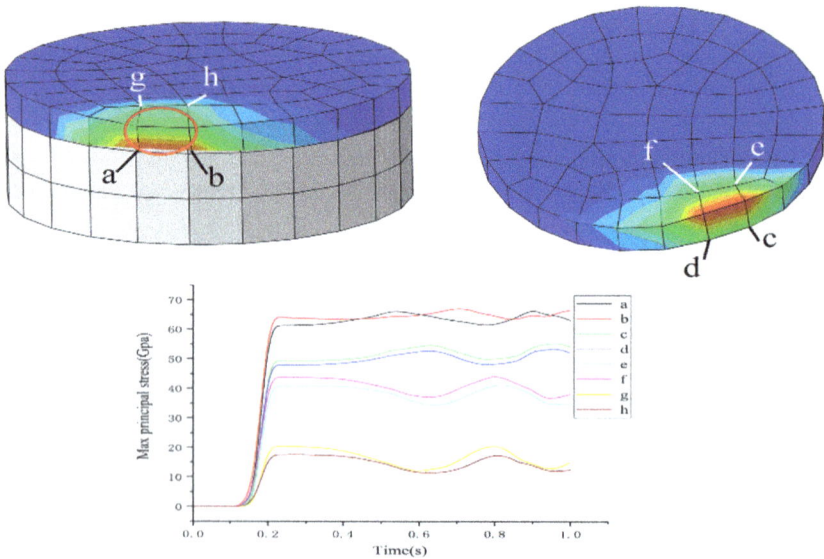

(b) Stress and deformation nephogram of cutting teeth

Fig. 2 Temperature and stress simulation results

4 Conclusions

The high-temperature region of the cutter was offset from the top of the crown to the outside of the cutter during drilling process, and the high-temperature region of the cutter was fan-shaped. The maximum stress in the diamond layer occurred at the edge of the diamond layer. The high-stress concentration and fluctuating impact at the joint surface were the main reasons leading to the fracture failure of PDC cutter.

The outside of the cutter and the top of the crown are the two places where the failure is most serious. Therefore, it was an effective way to prolong the service life of PDC bit to strengthen the material of wear-resistant layer on the surface of PDC cutter and to increase the thickness of the wear-resistant layer on the top of the cutting crown.

Acknowledgements This work was financially supported by National Natural Science Foundation of China (Grant No. 41602370).

References

1. Z. Cheng et al., Analytical modelling of rock cutting force and failure surface in linear cutting test by single PDC cutter. J. Petrol. Sci. Eng. **177**, 306–316 (2019)
2. M. Yahiaoui, et al., A study on PDC drill bits quality. Wear (2013)
3. K. Abbas, A review on the wear of oil drill bits (conventional and the state of the art approaches for wear reduction and quantification). Eng. Failure Anal. **90**, 554–584 (2018)
4. D.A. Glowka, The thermal response of rock to friction in the drag cutting process. J. Struct. Geol. **11**(7), 919–931 (1989)

Thermal Performance Evaluation Methodology of Grooved Heat Pipes with Rectangular, Trapezoidal and Wedge Cross Sections

Gökay Gökçe and Zafer Dursunkaya

1 Introduction

In this study, a mathematical model for a rectangular grooved heat pipe is developed using a commercial CFD program (Fluent®) along with Python®-based models developed to simulate phase change phenomena, which cannot be handled with a standard commercial CFD code.

Modeling of the heat pipes is a complex task, requiring a physically based mathematical model to address evaporation, condensation, and free surface flow phenomena. Moreover, some of the input parameters to the CFD code and the geometry of the working fluid inside the channel are not known a priori. Therefore, it is not possible to solve the entire problem in a single step by using any currently available commercial software. To address this, the problem is divided into incremental steps, namely the calculation of evaporation and condensation mass fluxes, solution of the free surface, resolving the in-groove flow and solving the heat transfer in both materials. Incremental steps are dependent on each other in such a way that numerous output parameters of a particular incremental step are inputs in consecutive steps. For this reason, an iterative process is followed, and some of these steps (free surface flow and heat transfer) are solved with the help of a commercial CFD program.

Various models of heat pipes are available in the literature [1–3]. Zhang [1] developed a 1-D model based on thermal resistances in order to optimize the performance of a heat pipe with Ω-shaped grooves. Effect of axial conduction and details of fluid flow are neglected in [1]. Ternet et al. [2] investigated the effect of microgroove shape on flat miniature heat pipe efficiency; their model predicts the heat transport limitations and maximum heat capacity of a grooved heat pipe. The effect of microgroove

G. Gökçe (✉) · Z. Dursunkaya
Mechanical Engineering Department, Middle East Technical University, 06800 Ankara, Turkey
e-mail: gokaygokce1@gmail.com

G. Gökçe
ASELSAN A.Ş., 06172 Yenimahalle, Ankara, Turkey

© Springer Nature Singapore Pte Ltd. 2021
C. Wen and Y. Yan (eds.), *Advances in Heat Transfer and Thermal Engineering* ,
https://doi.org/10.1007/978-981-33-4765-6_82

width and depth is also investigated in [2]. However, in these studies [1–3], some correlations are used in order to model micro-evaporation and skin friction.

Due to the excessive data to be loaded in every iteration, a manual implementation is impractical, rendering automatization of the iterative process crucial. The methodology is packaged in a single code which performs the consecutive iterative processes automatically. The sequence of events involves the generation of the CAD model, meshing the model, running the code, post-processing results to feed the next iterative step, and repeating the entire process until the defined convergence criteria are satisfied. The driver that runs the entire iterative loop and individual models is developed in Python environment.

2 Methodology

A schematic view of a rectangular grooved heat pipe and half of a channel is given in Fig. 1. Since the model is symmetric, half of one channel is modeled and investigated.

At the beginning of the analysis, geometry of the liquid–vapor interface, wall temperatures and vapor temperature are not known. Therefore, initial estimates are assigned to these operating parameters. The solution procedure begins with the calculation of evaporation and condensation mass fluxes along the channel. The computer code based on kinetic theory to calculate mass fluxes is written in Python environment. These fluxes are assigned as conditions on the boundaries of the CFD model which is generated using a commercial CFD program, Fluent. CFD model is solved in to find the pressure distribution at the vapor–liquid interface. With the pressure distribution, the shape of liquid–vapor interface can be obtained solving Young–Laplace equation along the channel. The computed geometry of the liquid–vapor interface may differ from the initial estimate; therefore, the solution procedure is continued with the new geometry until the shape of the liquid–vapor interface converges to a final geometry. After obtaining the shape of the interface, energy equation is similarly solved in Fluent to find the temperature variation along the channel. In this model, a constant vapor temperature is assumed and assigned to the fluid–vapor interface, the value of which is a variable in the iterative process. Convection boundary condition is given to surfaces where evaporation and condensation occur. As a result of

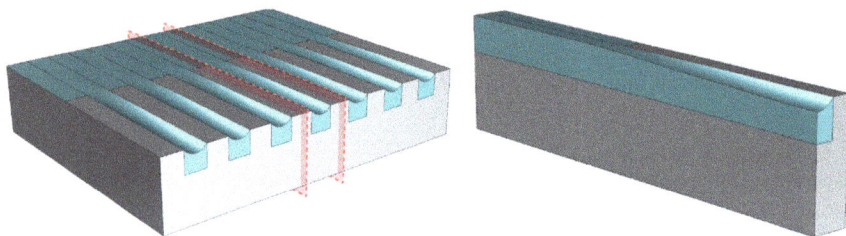

Fig. 1 Schematic view of the rectangular grooved heat pipe

this analysis, temperature distribution inside the channel is obtained. The process is repeated until the convergence of wall temperatures is reached.

With the calculated temperature distribution and shape of the liquid–vapor interface, one can calculate the new evaporation and condensation mass fluxes with the computer code, written in Python environment. Vapor temperature is iteratively altered until the difference between the evaporative and condensation mass fluxes is equal. The solution procedure is given in Fig. 2.

The results of the rectangular cross-sectional channel are in close agreement with studies in the literature [3, 4], and the same solution procedure is applied to channels with different cross sections. Liquid cross-sectional areas and the thicknesses of the fin top are kept identical for all the models to enable a comparison. Models of the channels simulated are given in Fig. 3.

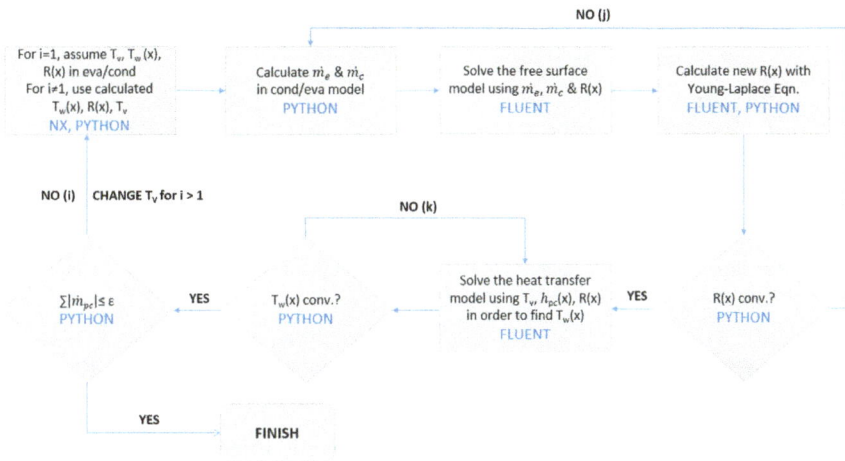

Fig. 2 Solution procedure

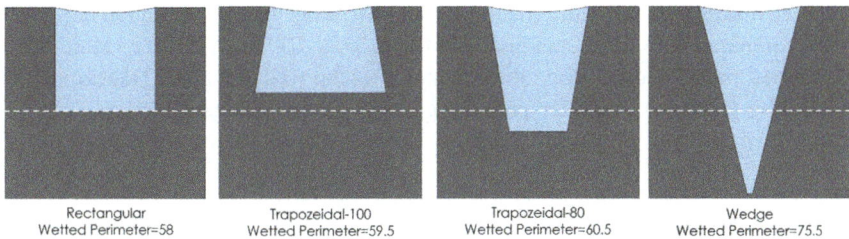

Fig. 3 Cross-sectional views of different groove geometries

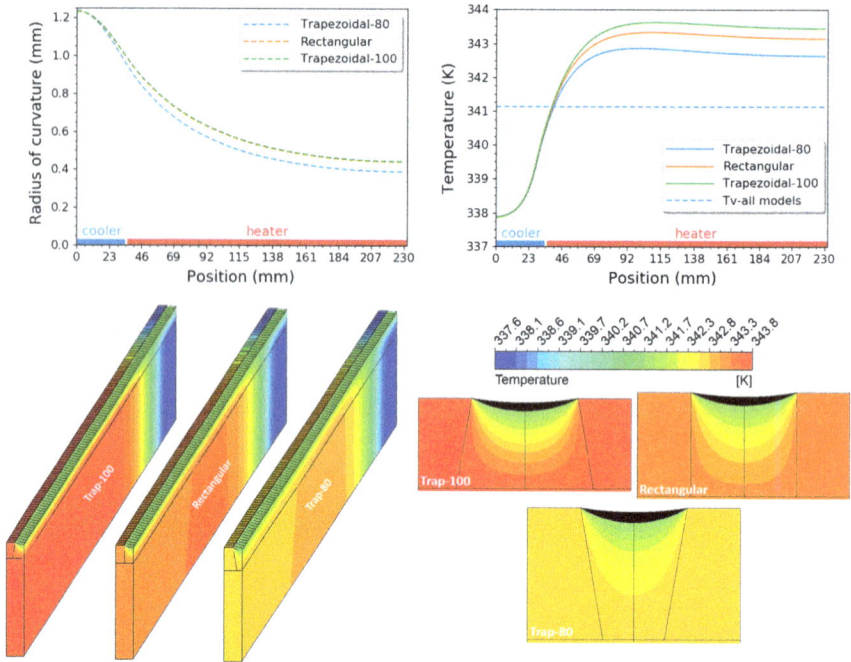

Fig. 4 Variation of wall temperatures, radii of curvature along the channel, and CFD temperature distributions

3 Results

Radius of curvature and wall temperature variations for different groove geometries is plotted in Fig. 4. Dry-out phenomenon occurs in wedge cross-sectional groove due to its long perimeter; therefore, these results are not plotted in the related figures. Evaporation rate increases with increasing end angle—the angle the liquid–vapor interface makes with the wall. The trapezoidal-80 model has the minimum end angle when compared with the rectangular and trapezoidal-100 models. As a result, it has the highest pressure drop along the channel after the wedge model. Trapezoidal-80 model also gives the lowest temperature gradient along the channel.

4 Conclusions

An iterative methodology using a commercial code and Python-based driver and physical models are developed to simulate the performance of flat grooved heat pipes. With the help of a commercial CFD program and a computer code, pressure drop along the channel and microevaporation are calculated inside the code without using any approximate correlations. The approach is validated using results available in the literature [3, 4] and applied to new geometries to assess the effect of channel profile on the performance of the proposed heat pipes.

References

1. C. Zhang et al., Optimization of heat pipe with axial 'Ω'-shaped micro grooves based on a niched pareto genetic algorithm (NPGA). Appl. Therm. Eng. **29**(16), 3340–3345 (2009)
2. F. Ternet, H. Louahlia-Gualous, S. Le Masson, Impact of microgroove shape on flat miniature heat pipe efficiency. Entropy **20**, 44 (2018)
3. F. Lefèvre, R. Rullière, G. Pandraud, M. Lallemand, Prediction of the temperature field in flat plate heat pipes with micro-grooves—experimental validation. Int. J. Heat Mass Transf. **51**(15–16), 4083–4094 (2008)
4. G. Odabaşı, Modeling of multidimensional heat transfer in a rectangular grooved heat pipe. Ph.D. thesis, 20 June 2014

Thermal Management

Development of High Heat Dissipation and Low Thermal Expansion Printed Wiring Boards

Yohei Ito and Sohei Samejima

1 Introduction

In recent years, demands for higher performance, smaller size, and lighter weight of electric and electronic equipment are getting higher and higher. To meet these demands, high power parts such as large size ceramic parts and IC packages are used, and high density mounting of parts is developed. As a result, the heat generation density on the printed wiring board (PWB) is increasing. An increase in the heat generation density leads to an increase in the component temperature. And, if enough exhaust heat cannot be performed, instability of a part and malfunction are caused. Therefore, it is necessary to efficiently exhaust the heat generated from the parts. Radiation, convection, and heat conduction are considered as methods for dissipating heat from parts mounted on the PWB. Among them, we pay attention to heat dissipation due to the heat conduction of the PWB.

In the glass epoxy PWB, the heat generated from the parts cannot be easily dissipated, since thermal conductivity of glass epoxy materials is low. A metal core PWB is known to have high heat dissipation ability. Aluminum alloy has generally been used as core material, since aluminum alloy has high thermal conductivity and the low specific gravity. However, since the coefficient of thermal expansion (CTE) of the aluminum alloy is high, the solder joint may be broken due to the difference in CTE between the mounted component and the aluminum.

It is difficult to achieve compatibility between component mounting reliability and heat dissipation by conventional PWB such as glass epoxy board and aluminum core PWB. So we developed a new PWB with a core, which can achieve both high heat dissipation and component mounting reliability.

Y. Ito (✉) · S. Samejima
Composite Material Group, Materials and Processing Technology Department, Advanced Technology R&D Center, Mitsubishi Electric Corporation, Amagasaki, Japan
e-mail: Ito.Yohei@eb.MitsubishiElectric.co.jp

© Springer Nature Singapore Pte Ltd. 2021
C. Wen and Y. Yan (eds.), *Advances in Heat Transfer and Thermal Engineering*,
https://doi.org/10.1007/978-981-33-4765-6_83

2 Developed PWB

SiC-dispersed Al-based composite material (Al-SiC) is used for core material of
the new developed PWB. Al-SiC is composite material in which silicon carbide
particles are dispersed in aluminum. In Al-SiC, physical properties can be adjusted
by changing SiC content. For the PWB developed this time, Al-SiC with a content
of 30% SiC manufactured by the star cast method was adopted. A thin plate shape
is required as core material of a PWB, but Al-SiC manufactured by the star casting
method has a block shape and it is difficult to directly obtain a thin plate shape.
Therefore, this time, a thin plate was produced by cutting from thick block-shaped
Al-SiC.

Figure 1 shows the layer structure of the PWB that was prototyped this time. Since
Al-SiC is conductive, it needs to be insulated from through-holes that connect upper
and lower wirings. So, alumina-containing high thermal conductivity insulating resin
having a thermal conductivity of 3 W/m K and a coefficient of thermal expansion of
30 ppm/K was adopted for the insulating part. Figure 2 shows a cross-sectional view
of the prototyped PWB. No problems such as cracks have been identified.

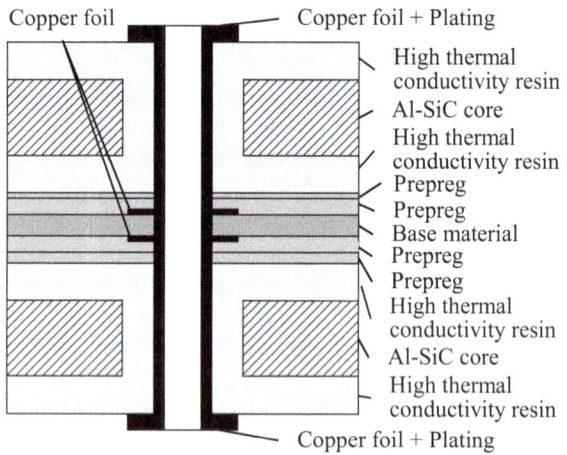

Fig. 1 Layer structure of the prototyped PWB

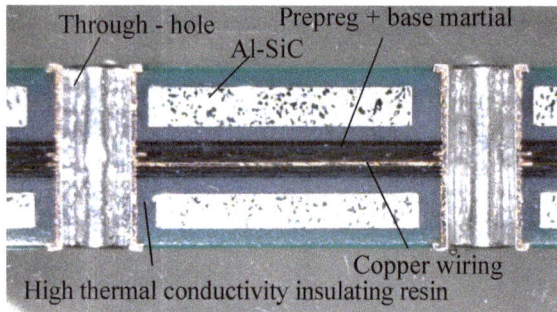

Fig. 2 Cross-sectional view of the prototyped PWB

3   Thermal Properties Evaluation

The PWB used for thermal conductivity measurement is two types, all-layer copper-covered Al-SiC core PWB in which all wiring layers are occupied by copper and all-layer copper-free Al-SiC core PWB in which no copper is present in the wiring layer as shown in Fig. 3. Assuming that all the input power from the power supply is supplied to the heater, the thermal conductivity is calculated from the temperature gradient and the input power. The temperature gradient is determined from three thermocouples put on the surface of PWBs.

The obtained thermal conductivity was 96.3 W/m K for all-layer copper-free Al-SiC core PWB and 130.1 W/m K for all-layer copper-covered Al-SiC core PWB. It can be seen that the thermal conductivity changes up to about 30% depending on the residual copper ratio. With the PWB developed this time, thermal conductivity of about 100 W/m K or more can be secured. The thermal conductivity of the developed PWB is about twenty times as high as conventional PWB which is same layer structure without core material whose thermal conductivity is about 5 W/m·K. Therefore, it is considered that a PWB with high heat dissipation properties can be obtained.

The CTE of the prototyped PWB was also measured with a thermomechanical analyzer (TMA).

The obtained in-plane CTE of Al-SiC core PWB was 14 ppm/K, and out-of-plane was 40 ppm/K. It can be said that low CTE of the PWB can be achieved by using Al-SiC as the core. Further, since the CTE of the Al-SiC core PWB is almost the same as the other material used for the PWB, and the reliability of the PWB is also improved.

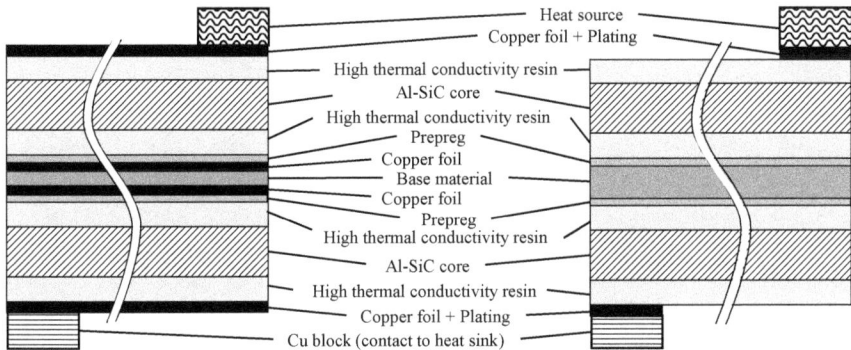

Fig. 3 Layer structure of PWBs for thermal conductivity measurement. Left: all-layer copper-covered Al-SiC core PWB, Right: all-layer copper-free Al-SiC core PWB

In order to confirm the reliability of the Al-SiC core PWB, the electrical resistance of the Al-SiC core PWB with the 100-hole daisy chain pattern was measured before and after the heat cycle test in the temperature range from −65 to 125 °C. It was confirmed that the change in the electric resistance value was 1% or less, and the electrical connection reliability was secured. Furthermore, after the heat cycle test, the cross section of the Al-SiC core PWB was observed and it was confirmed that there were no problems such as the occurrence of cracks.

4 Summary

We developed a printed wiring board using SiC-dispersed Al-based composite material that has high heat dissipation and component mounting reliability and examined thermal conductivity, coefficient of thermal expansion, and electrical connection reliability by heat cycle. The results are shown below.

(1) The in-plane thermal conductivity of the Al-SiC core PWB is 96–130 W/m K.
(2) The coefficient of thermal expansion (room temperature) of the Al-SiC core PWB is 14 ppm/K. in the in-plane direction, 40 ppm/K in the out-of-plane direction.
(3) In the heat cycle test in the temperature range of −65 to 125 °C, it was confirmed that there was no crack or the like and there was no problem with the electrical connection reliability.

Optimization of Thermal Management of Li-Ion Cells with Phase Change Materials

S. Landini, R. Waser, A. Stamatiou, J. Leworthy, J. Worlitschek, and T. S. O'Donovan

1 Introduction

Li-ion cells will promote the exploitation of renewable energy sources, playing an important role in helping societies to reach their emissions reduction targets. They do this in three ways, by balancing the mismatch between electricity production and demand, by permitting the use of distributed renewable energy systems, and by providing power backup. However, while Li-ion cells are characterized by high voltages, high energy/power density and moderate operating life, their performance is quite sensitive to temperature [1–3]. Adverse temperature operation leads to capacity fade, faster aging effect, and thermal runaway. Therefore, Li-ion cells should operate in the range of 25–40 °C. Also, a minimal temperature disuniformity within the cell [3, 4] can avoid localized cell deterioration and battery pack performance defects.

Four main thermal management systems (TMS) for Li-ion cells are reported in the literature: air-cooling, liquid-cooling, flow/pool boiling, and phase change materials (PCM). Among these, PCMs can be an effective TMS. When used as a passive cooling technique, they are characterized by moderate capital and O&M costs, compactness, high efficiency and low parasitic power consumption [5]. However, when Li-ion cells are operated continuously at high discharge rates, PCMs could recover only part of

S. Landini (✉) · T. S. O'Donovan
Institute of Mechanical, Process and Energy Engineering, School of Engineering and Physical Sciences, Heriot-Watt University, Edinburgh EH14 4AS, UK
e-mail: sl30@hw.ac.uk

R. Waser · A. Stamatiou · J. Worlitschek
Thermal Energy Storage Research Group, Lucerne University of Applied Sciences and Arts (HSLU), Lucerne 6048, Horw, Switzerland

J. Leworthy
Dukosi Ltd, Quantum Court Research Avenue South, Heriot-Watt University Research Park, Edinburgh EH14 4AP, UK

© Springer Nature Singapore Pte Ltd. 2021
C. Wen and Y. Yan (eds.), *Advances in Heat Transfer and Thermal Engineering* ,
https://doi.org/10.1007/978-981-33-4765-6_84

their latent heat potential by solidification and this can lead to thermal runaway after a certain number of cycles. Therefore, details on the electricity demand are essential.

This research project focuses on the optimization of Li-ion cells' TMS by PCM. Different designs of aluminum-sintered blocks filled with octadecane as PCM are tested to determine the TMS benefits on the Li-ion cells' electrical and thermal performance.

2 Methodology

Two kinds of tests were conducted: single cycles and stress sequences. Single cycles consisted in single charge-discharge cycles at constant charge rate (CR) of 1C and discharge rate (DR) of 1, 2, 3, 4, 5 C with resting periods after each charge and discharge to cool down the PCM TMS Li-ion cell system. These tests were designed to investigate the effect of DR on PCM effectiveness as TMS in the case of intermittent demands (e.g., power tools, electronics, stationary electricity storage). In the stress sequences, Li-ion cells were cycled 10 consecutive times at $CR = DR = 3C$ without resting periods, similar to the constant fast charge-discharge load present in automotive applications.

By keeping track of Li-ion cell voltage, current, surface temperatures (4 RTD PT100 sensors, 2 per side) and heat flux (one sensor on one side), the authors evaluated discharge capacity, discharge electrical energy, average/maximum/minimum Li-ion cell's surface temperatures, Li-ion cell's surface temperature disuniformity, volumetric heat generation (being the sum of the enthalpy variation and heat losses towards the surroundings) and electro-chemical efficiency. To compare the benefits of the PCM TMS proposed, tests were run without PCM (i.e., air natural convection) and with PCM as TMS. All devices were operated by LabVIEW code and data were sampled at 0.25 Hz.

The Li-ion pouch cells tested were AKKU300 characterized by nominal voltage of 3.7 V, capacity of 300 mAh, dimensions 32 mm × 22 mm × 4.8 mm, mass 7.02 gr, and specific heat capacity of 1479 ± 205 J/kg K. The batteries were cycled using an HM8143 Arbitrary Power Supply, which could be operated as power supply (cell charge) and electronic load (cell discharge). The PCM selected for these tests was a 99% pure octadecane, a paraffin characterized by melting temperature in the range of 26-29°C, specific latent heat of 250 kJ/kg, thermal conductivity of 0.144 W/m K, specific heat capacity of 1990 J/kg K, and solid density of 777 kg/m^3. The PCM was provided by Sigma Aldrich (O652), and its thermo-physical properties were measured at the HSLU Thermal Energy Storage Competence Center.

The passive PCM TMS were composed of aluminum blocks designed by the authors and sintered by direct metal laser sintering (DMLS) technique using a Concept Laser Mlab Cusing. The aluminum powder was a Concept Laser CL 31AL 86% Al pure of equivalent thermal conductivity around 120–180 W/m K. Two designs were proposed: simple and finned (Fig. 1). In the simple design, the PCM was located within the two lateral pockets while the Li-ion pouch cell was positioned in the center.

Fig. 1 Aluminum-sintered TMS blocks: **a** simple design dimensions in mm, **b** simple (left) and finned (right) designs, **c** finned design with octadecane located

Thermal paste provided by Electrolube was used to minimize the thermal contact resistance between Li-ion cell and aluminum block. The masses of Al and octadecane were, respectively, 18.3 (63%) and 10.55 (37%) gr. The finned design was an improved version of the simple one, characterized by fins surrounding the Li-ion cells and enclosing the PCM in separate pockets. This model should improve the equivalent thermal conductivity of the PCM TMS and foster the PCM crystallization by solidification while decreasing the energy density and increasing the total TMS weight. In this design, the masses of Al and octadecane were, respectively, 36.6 (74%) and 12.85 (26%) gr. Moreover, to evaluate the PCM TMS cooling effectiveness without the contribution of external air natural convection, the blocks were located at the center of a thermally-insulating box made by ethylene propylene rubber foam of external dimensions 70 mm × 75 mm × 100 mm and thermal conductivity at 25 °C of 0.037 W/m K.

3 Results

The results of the stress sequences are reported here as these tests show the PCM TMS effectiveness when working in the most extreme conditions, being continuous operation and no cooling periods. In Fig. 2, the temperature profiles for Li-ion cell and PCM are reported for air natural convection, the PCM TMS simple and finned designs. During these tests, the median temperature of the laboratory was 22.8 °C.

Compared to air natural convection, where the cell experienced maximum temperatures up to 42 °C, the TMS PCM simple design was capable to keep the cell in the range of 26–32 °C once the PCM started melting. This condition was reached in between the second and the third cycles. Also, the maximum surface temperature disuniformity was always kept lower than 1.33 °C compared to 2.09 °C of air natural convection. It must be pointed out that the lower temperatures of the PCM TMS cooled Li-ion cell led to a decrease of discharge capacity and energy compared to the air-cooled cell of 28.3% and 28.7%. However, this operating condition would

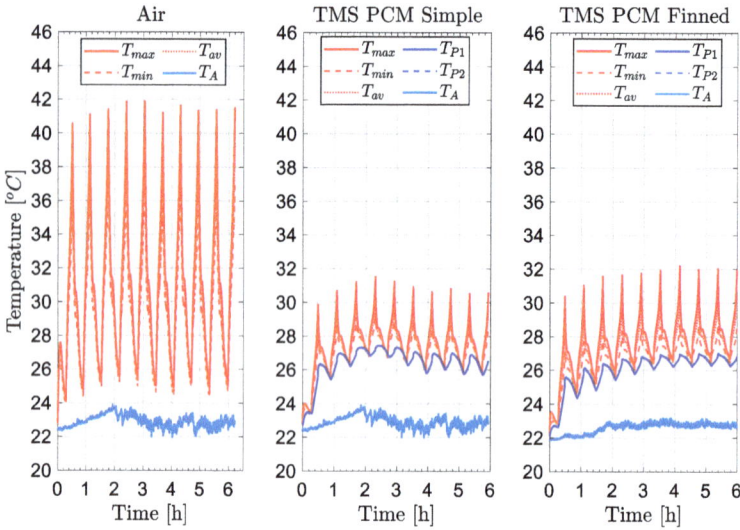

Fig. 2 Stress sequences: 10 consecutive charge-discharge cycles at CR = DR = 3C. Air natural convection (left), TMS PCM simple design (center, see Fig. 1, **a** and finned design (right, see Fig. 1, **b**. For each case, Li-ion pouch cell (red) minimum, average, maximum temperatures and PCM (blue) temperatures for both pockets are shown

lead to potentially 27% longer Li-ion cell operating life, as demonstrated in a study by Waldmann et al. [3]. The same test was reproduced using the PCM TMS finned design. The Li-ion cell was kept at 25–32 °C while the maximum surface temperature disuniformity was kept lower than 1.6 °C. As with the previous design, the lower temperatures of the PCM TMS cooled Li-ion cell led to a decrease of discharge capacity and energy compared to the air-cooled cell of, respectively, 15.4 and 16.2%.

4 Conclusions

Thermal management of Li-ion 300-mAh pouch cells by passive PCM cooling has been investigated. Results of stress sequences consisting of 10 consecutive charge CCCV-discharge CC cycles at CR = DR = 3C without resting periods for two designs of PCM-filled sintered aluminum TMS blocks have been proposed and compared to air natural convection. The simple and finned designs lead to average cell surface temperatures, respectively, in the range of 26–32 °C and 25–32 °C compared to peaks of 42 °C for air natural convection. Also, surface temperature disuniformities lower than 1.33 and 1.6 °C were measured for, respectively, simple and finned designs compared to 2.09 °C for air natural convection. Previous literature show that the specific thermal condition obtained by using PCM TMS would potentially lead to 27% longer Li-ion cell operating life.

Acknowledgements The authors would like to express their sincere gratitude to ETP PECRE and Swiss Competence Center for Energy Research Storage of Heat and Electricity (SCCER) who funded this research. We would also like to acknowledge the contribution of Ms. Rebecca Ravotti and Mr. Marcel Furrer who provided guidance into respectively the chemistry of PCM and Direct Metal Laser Sintering.

References

1. S. Panchal, I. Dincer, M. Agelin-Chaab, R. Fraser, M. Fowler, Experimental and simulated temperature variations in a LiFePO$_4$—20 Ah battery during discharge process. Appl. Energy **180**, 504–515 (2016)
2. S. Panchal, I. Dincer, M. Agelin-Chaab, M. Fowler, R. Fraser, Uneven temperature and voltage distributions due to rapid discharge rates and different boundary conditions for series-connected LiFePO$_4$ batteries. Int. Commun. Heat Mass Transf. **81**, 210–217 (2017)
3. T. Waldmann, M. Wilka, M. Kasper, M. Fleischhammer, M. Wohlfahrt-Mehrens, Temperature dependent ageing mechanisms in Lithium-ion batteries—a post-mortem study. J. Power Sources **262**, 129–135 (2014)
4. T.M. Bandhauer, S. Garimella, T.F. Fuller, Temperature-dependent electrochemical heat generation in a commercial lithium-ion battery. J. Power Sources **247**, 618–628 (2014)
5. A.R.M. Siddique, S. Mahmud, B. Van Heyst, A comprehensive review on a passive (phase change materials) and an active (thermoelectric cooler) battery thermal management system and their limitations. J. Power Sources **401**(August), 224–237 (2018)

The Simulation of the Motor Temperature Distribution with Spray Cooling

Xuehui Wang, Bo Li, and Yuying Yan

1 Introduction

With increasing concerns on the environment problems caused by using of traditional fossil energies, the electric vehicles are drawing a lot of attention from the scholars and companies in this field. The technology improvements make the power of the motor get higher and higher. During the operation of the motor, there are a lot Joule heat and eddy losses in the copper winding and the stators [1, 2]. Generally, the motor is a very complex thermal system, with different materials of different temperature limitations and distributed thermal sources. The heat must be dissipated efficiently. Otherwise, it will cause the deterioration of the motor efficiency, shorter life span of the motor. Meanwhile, there are some materials are used as the insulation materials, if they are overheated, they will burn out and caused the short-circuit problems [3]. The current cooling methods (water jacket cooling or forced convection air cooling) are not qualified for the high-speed e-motor.

To meet the change of the higher heat flux of the motor and decrease the temperature, a lumped parameter thermal network for the electric motor cooled by water jacket and spray hybrid method was proposed. The temperature distribution of the motor and the spray cooling characteristics were analyzed based on the thermal model.

X. Wang · B. Li · Y. Yan
Fluids and Thermal Engineering Research Group, Faculty of Engineering, University of Nottingham, Nottingham NG7 2RD, UK

Y. Yan (✉)
Research Centre for Fluids and Thermal Engineering, University of Nottingham Ningbo China, 315100 Ningbo, China
e-mail: yuying.yan@nottingham.ac.uk

© Springer Nature Singapore Pte Ltd. 2021
C. Wen and Y. Yan (eds.), *Advances in Heat Transfer and Thermal Engineering* ,
https://doi.org/10.1007/978-981-33-4765-6_85

2 Methodology

A lumped parameter thermal network was proposed in this paper to calculate the temperature distribution in an electric motor. The thermal model was a 3D thermal model, in which the transfer in axial direction, radial direction, and circumferential direction were all considered. The motor domain was divided into water jacket, housing, teeth, yoke, slot winding, and end winding unit. Each of the former four units had five nodes along the axial direction. The proposed thermal network was shown in Fig. 1.

In this analysis, the motor was cooled by a water jacket outside of the housing part and spray cooling at the end winding. The thermal resistances between different nodes were calculated by the thermal conduction or the convection/spray heat transfer laws, while the radiation heat transfer was neglected. For all the nodes considered here, the governing equations were given by

$$\sum_{j=1}^{n_i} \frac{T_j - T_i}{R_{ij}} + \dot{q}_{vi} = 0 \tag{1}$$

By solve all the governing equations of the thermal nodes, all the temperature information and the temperature response with the time could be obtained. The length of the motor was 93 mm and with 30 slots. The heat losses of the yoke, teeth, slot winding, and end winding was obtained from electromagnetic analysis and changed with the input current density, and three levels of the current density of 15A/mm^2, 20A/mm^2, and 26A/mm^2 were considered in this analysis. The thermal conductivities of the materials used in the motor were dealt as constants in the whole temperature range.

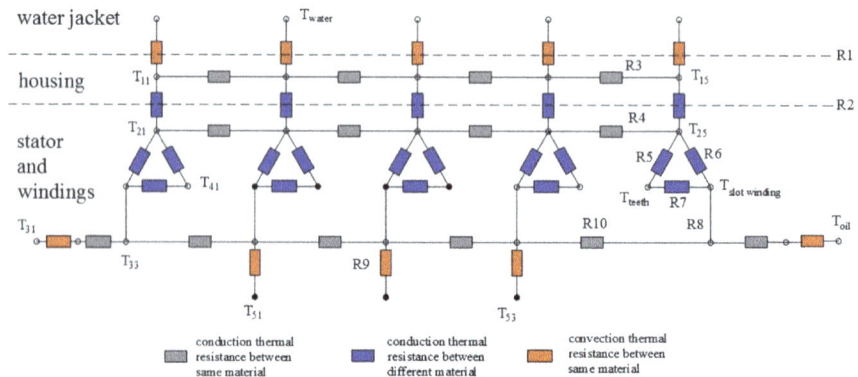

Fig. 1 Lumped parameter thermal network of the motor

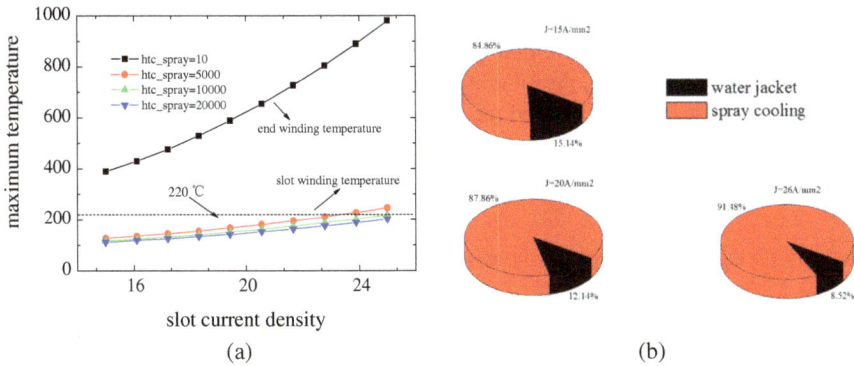

Fig. 2 Cooling characteristics of the spray cooling, **a** maximum temperature with different spray cooling heat transfer coefficient; **b** the ratio of heat dissipated by spray cooling

3 Results

In this paper, the temperature distribution of the motor with different cooling capability of the spray cooling was simulated based on the proposed thermal network. The results were presented in Fig. 2.

To evaluate the effect of the spray cooling, the natural convection used for the end winding was selected as the baseline case. When the end winding was cooled by the natural convection, the highest temperature in the motor is the end winding, the temperature could be as high as 1000 °C when the current density over 25 A/mm². However, when it comes to the spray cooling, the situation was totally different. Take the spray cooling with the heat transfer coefficient of 10000 W/(m² K) as an example, the highest temperature of the motor was in the middle of the slot winding, rather than the end winding. To be more important, the temperature of it was lower than 220 °C when the current density up to 24 A/mm².

Further analysis showed that the majority of heat generated in the motor was dissipated by the spray cooling, and the ratio could be higher than 84.86%. Meanwhile, the ratio increased with the current density, which meant the cooling capability of the spray cooling could be made better use when the current density was higher.

4 Conclusions

A lumped parameter thermal model for a typical electric motor was proposed to analyze the temperature distribution cooled by the water jacket and spray cooling hybrid method. The temperature distribution and cooling characteristics of the spray cooling were analyzed in different conditions. The main conclusions of this paper were:

(1) When the motor was cooled by spray cooling at the end winding, the highest temperature was in the middle of the slot winding instead of the end winding, and the temperature of the winding could be decreased a lot by the spray cooling.

(2) The spray cooling dissipated the majority heat generated in the motor, and the ratio increased with the input current density. It means we can make better use of the spray cooling when the current density was higher.

Acknowledgements This work was financially sponsored by the following research grants: H2020-MSCA-RISE-778104—ThermaSMART, Innovate UK (ACeDrive No. 113167), and Ningbo Science and Technology Bureau Technology Innovation Team (No. 2016B10010).

References

1. C. Yung, Cool facts about cooling electric motors: improvements in applications that fall outside the normal operating conditions. IEEE Ind. Appl. Mag. **21**, 47–56 (2015)
2. S. Mezani, N. Takorabet, B. Laporte, A combined electromagnetic and thermal analysis of induction motors. IEEE Trans. Magn. **41**, 1572–1575 (2005)
3. R. Saidur, A review on electrical motors energy use and energy savings. Renew. Sustain. Energy Rev. **14**, 877–898 (2010)

Influence of Nanofluids Inlet and Outlet Positions on CPU Cooling Performance

Cong Qi, Jinghua Tang, Tiantian Chen, Maoni Liu, Guiqing Wang, and Yuying Yan

1 Introduction

In order to promote the development of CPU, the intensity of heat emission in limited heat transfer surface becomes larger and larger. Air, water, and ethylene glycol as the common cooling mediums cannot meet the requirement of CPU. Due to the distinguished thermal performance of nanofluids [1, 2], nanofluids have been used to cool the CPU instead of common fluid [3–5].

However, the influence of nanofluids inlet and outlet positions on CPU cooling performance is less investigated. This paper aims at numerically and experimentally studying the influence of nanofluids inlet and outlet positions (top and two sides) on CPU cooling performance. The results show that the numerical simulation results are in good agreement with the experimental results, and the max error between them is only 3.9%. In addition, average and local maximum temperature distributions of water and nanofluids at different inlet and outlet locations (top and sides) are analyzed and their cooling performance is compared.

C. Qi · J. Tang · T. Chen · M. Liu · G. Wang
School of Electrical and Power Engineering, China University of Mining and Technology, Xuzhou 221116, China

C. Qi (✉) · Y. Yan (✉)
Faculty of Engineering, University of Nottingham, Nottingham NG7 2RD, UK
e-mail: qicong@cumt.edu.cn

Y. Yan
e-mail: yuying.yan@nottingham.ac.uk

© Springer Nature Singapore Pte Ltd. 2021
C. Wen and Y. Yan (eds.), *Advances in Heat Transfer and Thermal Engineering* ,
https://doi.org/10.1007/978-981-33-4765-6_86

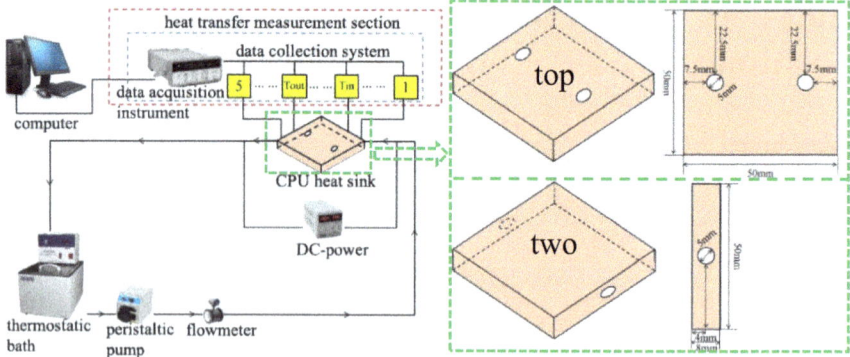

Fig. 1 Schematic diagram of CPU cooling system

2 Methodology

2.1 CPU Cooling System

Figure 1 presents the schematic diagram of CPU cooling system. TiO$_2$-water nanofluids with volume fraction 9% are used as the cooling medium, and the inlet and outlet locate at the top and two sides of CPU heat sink severally. The influence of inlet and outlet positions on CPU cooling performance is numerically and experimentally studied.

2.2 Numerical Simulation Model

A two-phase lattice Boltzmann model is adopted to study the CPU cooling performance, and the main equations are given in this paper.

The equations for flow fields are as follows:

$$f_\alpha^\sigma (r + e_\alpha \delta_t, t + \delta_t) - f_\alpha^\sigma (r, t)$$

$$= -\frac{1}{\tau_f^\sigma} \left[f_\alpha^\sigma (r, t) - f_\alpha^{\sigma eq} (r, t) \right] + \frac{2\tau_f^\sigma - 1}{2\tau_f^\sigma} \cdot \frac{F_\alpha^\sigma \delta_t e_\alpha}{B_\alpha c^2} + \delta_t F_\alpha^{\sigma'} \qquad (1)$$

$$f_\alpha^{\sigma eq} = \rho^\sigma w_\alpha \left[1 + \frac{e_\alpha \cdot u^\sigma}{c_s^2} + \frac{(e_\alpha \cdot u^\sigma)^2}{2c_s^4} - \frac{u^{\sigma 2}}{2c_s^2} \right] \qquad (2)$$

The equations for temperature fields are as follows:

$$T_\alpha^\sigma (r + e_\alpha \delta_t, t + \delta_t) - T_\alpha^\sigma (r, t) = -\frac{1}{\tau_T^\sigma} \left[T_\alpha^\sigma (r, t) - T_\alpha^{\sigma eq} (r, t) \right] \qquad (3)$$

$$T_\alpha^{\sigma eq} = w_a T^\sigma \left[1 + 3\frac{e_a \cdot u^\sigma}{c^2} + 4.5\frac{(e_a \cdot u^\sigma)^2}{2c^4} - 1.5\frac{u^{\sigma 2}}{2c^2} \right] \tag{4}$$

2.3 Experimental Data Processing

Equivalent diameter of CPU heat sink is:

$$D = \frac{4A}{L} \tag{5}$$

Density of nanofluids and specific heat capacity of nanofluids are respectively:

$$\rho_{nf} = (1 - \varphi)\rho_{bf} + \varphi\rho_{np} \tag{6}$$

$$c_{pnf} = (1 - \varphi)c_{pbf} + \varphi c_{pnp} \tag{7}$$

Based on Eqs. (6) and (7), the density and specific heat of TiO_2-water can be easily calculated. The thermal conductivity and viscosity are measured by thermal conductivity apparatus (TC3000L, measurement error: 2%) and Kinexus rotational rheometer (Kinexus Pro), respectively.

Heat dissipating capacity of CPU is:

$$Q_{nf} = c_{pnf}q_{nf}(T_{out} - T_{in}) \tag{8}$$

The temperatures of outside and inside surfaces of CPU bottom are respectively:

$$T_{wls} = \frac{(T_1 + T_2 + \cdots + T_9)}{9} \tag{9}$$

$$T_{wus} = T_{wls} - \frac{Q_{nf}\delta}{A\lambda_w} \tag{10}$$

Average temperature of nanofluids is:

$$T_{nf} = \frac{(T_{out} + T_{in})}{2} \tag{11}$$

Convective heat transfer coefficient is:

$$h = \frac{Q_{nf}}{A(T_{wus} - T_{nf})} \tag{12}$$

Nusselt number can be obtained as follows:

$$Nu = \frac{h \cdot D}{\lambda_{nf}} \tag{13}$$

3 Results

In this paper, a grid independence verification is conducted. Table 1 shows the grid numbers (No. 1, No. 2, No. 3, No. 4, No. 5, No. 6) adopted in this paper. Results indicate that the results between No. 5 and No. 6 are close to each other; in order to save time, No. 5 is chosen as the grid number in the simulation. Also, an experimental verification is done to ensure the reliability of this model. The max error between the numerical simulation and the published literature [3] is 0.15%, which explains that this model has a high accuracy.

Figure 2a–d gives the overall temperature distributions and surface temperature distributions, respectively. It can be found from Figs. 2a–d that CPU surface temperature with inlet and outlet at top is more homogeneous and lower than that at both sides. CPU surface with inlet and outlet at both sides appears high temperature at some places. High temperature and uneven temperature distribution are all disadvantageous to CPU cooling.

Temperature distribution can intuitively show the cooling effect, in order to quantificationally show it further; Fig. 2e presents the average temperatures of CPU surface under different conditions. The CPU is cooled by TiO_2-water nanofluids and water under different inlet and outlet locations (top and two sides). It is found that the average temperature can be decreased significantly by 4.54 °C at most cooled by nanofluids compared with water under the same condition.

It is also found that CPU with inlet and outlet at top can reduce the average temperature by 8.49 °C at most compared with that at both two sides. Figure 2f also shows the variation of local maximum temperature on CPU surface under different Reynolds numbers. Results show that the local maximum temperature can be decreased by 7.95 °C at most cooled by nanofluids compared with water. It is also found that inlet and outlet at top can reduce the local maximum temperature by 11.37 °C at most compared with inlet and outlet at two sides. Nanofluids show excellent cooling performance due to not only the high thermal conductivity but also the large brown force of nanoparticles.

Table 1 Grid numbers

Location of inlet and outlet	No. 1	No. 2	No. 3	No. 4	No. 5	No. 6
Top	785,424	108,452	156,424	180,458	215,280	254,124
Two sides	847,521	109,870	134,856	158,438	187,960	201,254

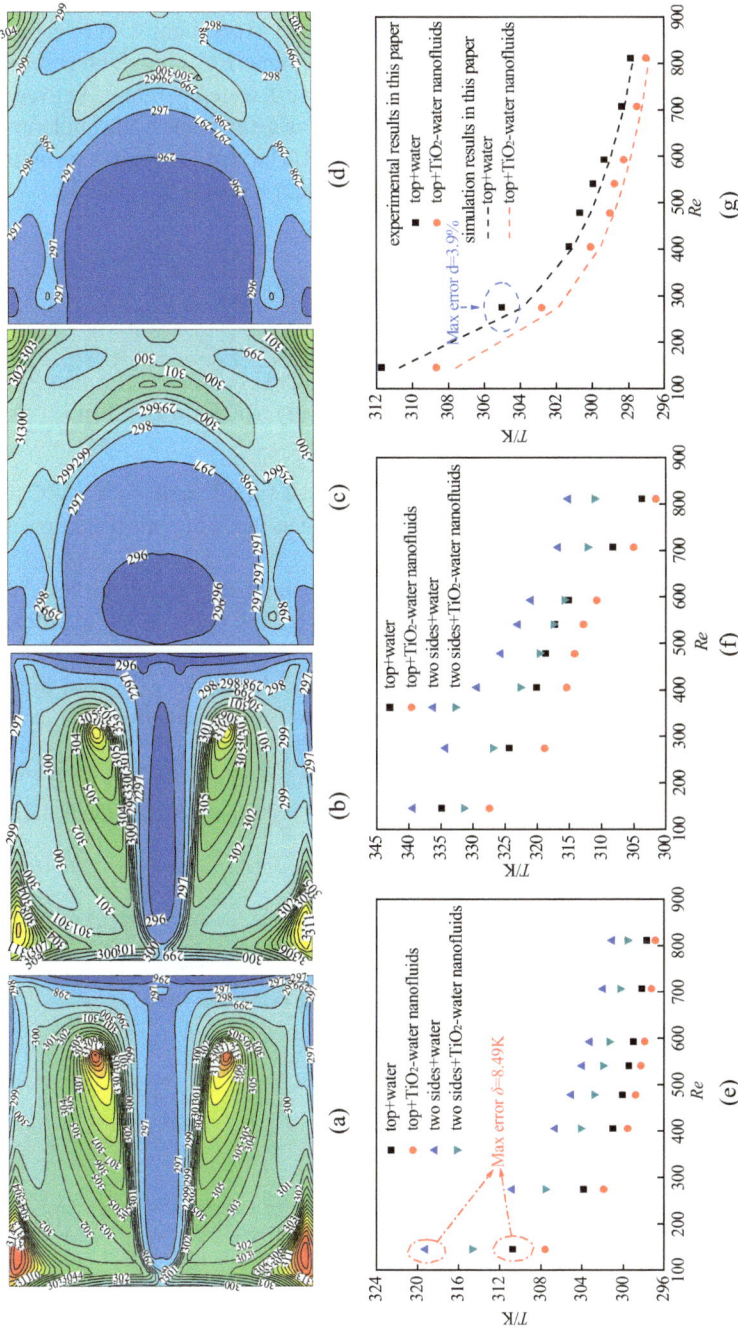

Fig. 2 CPU surface temperature distributions at $Re = 700$ with different locations of inlet and outlet [both sides (**a**, **b**) and top (**c**, **d**)], **a** deionized water, **b** nanofluids, **c** deionized water, **d** nanofluids, **e** average temperatures, **f** local maximum temperatures, **g** comparison between experimental and simulation results

The specific reasons for the enhancement of heat transfer are that nanofluids scour the bottom of CPU, the laminar boundary layer is destroyed when the inlet and outlet locate at top. In addition to the numerical simulation, an experiment with inlet and outlet at top is also conducted and compared with the numerical simulation results. Figure 2g shows the comparison of the average CPU surface temperatures between experimental and simulation results. Through the analysis, the numerical simulation results are well consistent with the experimental results. The max error between them is only 3.9%, which explains that the numerical simulation results are reliable.

4 Conclusions

Influence of nanofluids inlet and outlet positions on CPU cooling performance is numerically and experimentally researched, respectively. Relevant conclusions are got as follows:

(1) The average temperature of CPU can be decreased significantly by 4.54 °C at most cooled by nanofluids than that of water when the same pre-condition is adopted.
(2) CPU with inlet and outlet at top can reduce the average temperature by 8.49 °C at most compared with that at both two sides. Inlet and outlet at top are more suitable for CPU cooling.
(3) Nanofluids can reduce the local maximum temperature by 7.95 °C at most compared with water. CPU temperature distribution is more uniform when using nanofluids.
(4) Inlet and outlet at top can reduce the local maximum temperature by 11.37 °C at most compared with those at two sides. CPU temperature distribution is more uniform when inlet and outlet locate at top of CPU.

Acknowledgements This work is financially supported by "National Natural Science Foundation of China" (Grant No. 51606214), "Natural Science Foundation of Jiangsu Province, China" (Grant No. BK20181359) and "EU ThermaSMART project, H2020-MSCA-RISE (778104)-Smart thermal management of high power microprocessors using phase-change (ThermaSMART)."

References

1. L. Shi, Y. He, Y. Hu, X. Wang, Thermophysical properties of Fe_3O_4@CNT nanofluid and controllable heat transfer performance under magnetic field. Energy Convers. Manage. **177**, 249–257 (2018)
2. H. Li, L. Wang, Y. H, Y. Hu, J. Zhu, B. Jiang, Experimental investigation of thermal conductivity and viscosity of ethylene glycol based ZnO nanofluids. Appl. Therm. Eng. **88**, 363–368
3. B. Sun, H. Liu, Flow and heat transfer characteristics of nanofluids in a liquid-cooled CPU heat radiator. Appl. Therm. Eng. **115**, 435–443 (2017)

4. C. Qi, J. Hu, M. Liu, L. Guo, Z. Rao, Experimental study on thermo-hydraulic performances of CPU cooled by nanofluids. Energy Convers. Manage. **153**, 557–565 (2017)
5. C. Qi, N. Zhao, X. Cui, T. Chen, J. Hu, Effects of half spherical bulges on heat transfer characteristics of CPU cooled by TiO_2-water nanofluids. Int. J. Heat Mass Transf. **123**, 320–330 (2018)

The Temperature Uniformity Analysis and Optimization on the Lithium-Ion Battery with Liquid Cooling

Guohua Wang, Yuying Yan, and Qing Gao

1 Introduction

The electric vehicle is as one of the potential alternatives. And as one of the electric vehicle power sources, the lithium-ion battery has a high energy density, high power density, long life, and environmental friendliness. So it is widely applied for hybrid electric vehicle (HEV) and electric vehicle (EV). But the battery's performance can be impacted by temperature [1]. In order to ensure the battery operating temperature in the optimum range, battery thermal management strategies are proposed to make the battery pack work effectively [2]. So the battery liquid cooling is used as the battery thermal management mean to deal with the battery temperature uniformity. In this paper, a water jacket with multi-channels is used for cooling the prismatic lithium-ion battery. The liquid cold plate is divided into four parallel mini-channels and set on the bottom of the battery module for cooling the battery uniformly. The simulation model of the battery module is built for simulating the different working conditions. The working conditions include flow rates, liquid temperatures, and discharge rates. The temperature difference is utilized to judge the battery temperature uniformity.

G. Wang (✉) · Q. Gao
State Key Laboratory of Automotive Simulation and Control, Jilin University, Changchun 130025, China
e-mail: Guohua.Wang@nottingham.ac.uk

Y. Yan (✉)
Faculty of Engineering, University of Nottingham, Nottingham NG7 2RD, UK
e-mail: Yuying.Yan@nottingham.ac.uk

© Springer Nature Singapore Pte Ltd. 2021
C. Wen and Y. Yan (eds.), *Advances in Heat Transfer and Thermal Engineering* ,
https://doi.org/10.1007/978-981-33-4765-6_87

2 Methodology

In order to meet the power requirements of the battery pack, the circuit connection method of the battery module is set as 14S5P. And the battery module will have 70 cells. Each mini-channel of the liquid cold plate can be used for cooling 14 battery cells. Battery property parameters are described by the equivalent parameter because there are several materials in the battery. The density is 2055.2 kg/m^3, the specific heat is 1399 J/(kg K), and the thermal conductivity are 18.3 W/(m k), 18.3 W/(m k), and 1.1 W/(m k). The 1.1 W/(m k) is the thermal conductivity of battery thickness direction. The material of the liquid cold plate is aluminum alloy. The density is 2720 kg/m^3, the specific heat is 893 J/(kg K), and the thermal conductivity is 163 W/(m k). For the battery heat generation, the experimental data are supplied with the discharge rate changes (Table 1).

For reducing the load of calculation, the structural grid and the interior surface is utilized for meshing. The meshes of the cold plate and battery cells are built separately. The process of meshing is shown in Fig. 1. The cold plate and battery cells are meshed separately. The sample points are positioned on the battery surface near the cold plate and farthest point in the central axis.

Table 1 Battery heat generation data

Discharge rate	Heat generation W/m^3
1C	8916.7
2C	24966.9
3C	52291.558
4C	104236.9576

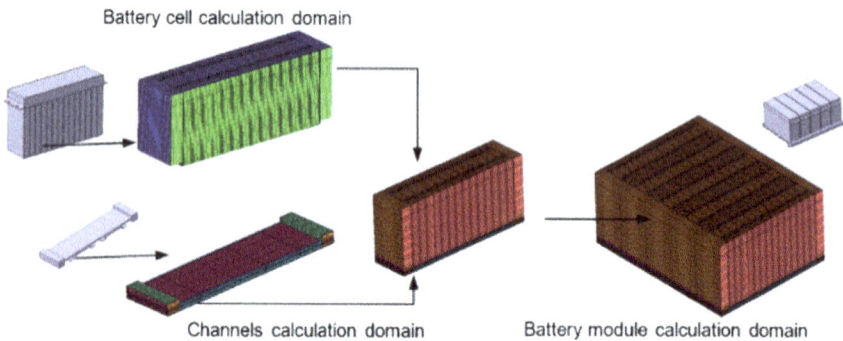

Battery cell calculation domain

Channels calculation domain   Battery module calculation domain

Fig. 1 Schematic of the lithium-ion battery module calculation mesh

3 Results

The analysis condition change includes fluid flow rates, temperatures, and discharge rates. The battery surface points which are near the cold plate and farthest point are selected to compare the temperature difference. The fluid flow rate change conditions are analyzed firstly. The values are 10, 20, 30, and 40 L/min. The liquid temperature is 20 °C, and the battery initial temperature is 45 °C as shown in Fig. 2a. Because of the big difference between the battery initial temperature and the liquid temperature, the temperature difference between surface points is quite big. The temperature difference is reduced with time. On the same time, the temperature difference is very small between the different flow rates. The reason is that the thermal conductivity of the battery is smaller compared with the convective heat transfer coefficient on the cold plate surface. So the Biot number has no significant change, the temperature curve also has been changed obviously. In the thermal management designing, the liquid flow rate cannot be selected in high value for the battery pack cooling.

For analyzing the influence of the liquid temperatures, the different liquid temperatures are selected on the same liquid flow rate and discharge rate. They are 15, 20, 25, 30, 35, and 40 °C as shown in Fig. 2b. The temperature difference between the surface points is quite big at the initial cooling stage. And the temperature difference can be reduced with the liquid temperature increasing. The reason is that the temperature difference between the battery cell and the liquid makes a big difference. The temperature difference between the battery cell and the liquid is greater, and the temperature difference on the battery surface points is the greater. So in the thermal management designing, the low-temperature difference between the battery and the liquid should be selected. The battery will be discharged at different rates. So the temperature difference should be analyzed in the different discharge rates as shown in Fig. 2c. The liquid flow rate is 30 L/min, and the liquid temperature is 20 °C. In the different discharge rates, the temperature differences are the difference with the times. High discharge rate makes the temperature difference bigger. And the existing structure cannot reduce the temperature and the temperature difference. The heat transfer enhancement measures should be adopted for enhancing the thermal conduction from the battery to the cold plate. For improving the temperature uniformity, the aluminum sheet and graphite sheet are arranged between the batteries

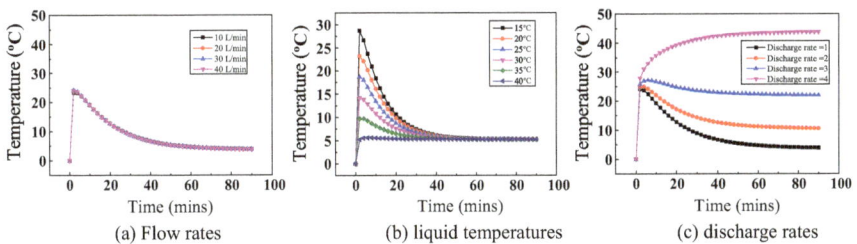

(a) Flow rates   (b) liquid temperatures   (c) discharge rates

Fig. 2 Temperature curves in different conditions

surface. The results show that the graphite sheet is better than the aluminum alloy fin with the same thickness. The quality of the graphite sheet is only one-third of the aluminum alloy. Those can reduce the gravimetric power density of the whole battery pack.

4 Conclusions

In the battery thermal management, the liquid flow rate has a little influence on the temperature difference between battery surface points. The more temperature difference between the liquid and the battery can make the temperature difference between battery surface points bigger on the initial cooling stage. With the discharge rate increases, the temperature difference between battery surface points will be big. So, the heat transfer enhancement measures should be adopted for enhancing the thermal conduction from the battery to the cold plate.

Acknowledgements The work is supported by Double Ten "Science and Technology Innovation Project of Jilin Province of China" No. 17SS022. The work is also supported by China Scholarship Council (CSC) for the first author's scholarship.

References

1. A.A. Pesaran, Battery thermal models for hybrid vehicle simulations. J. Power Sources **110**, 377–382 (2002)
2. Y. Deng, C. Feng, E. Jiaqiang, H. Zhu, J. Chen, M. Wen, H. Yin, Effects of different coolants and cooling strategies on the cooling performance of the power lithium-ion battery system: a review. Appl. Therm. Eng. **142**, 10–29 (2018)

Heat Spreading Performance of Integrated IGBT Module with Bonded Vapour Chamber for Electric Vehicle

Bo Li, Yiyi Chen, Yuying Yan, Xuehui Wang, Yong Li, and Yangang Wang

1 Introduction

Single-phase cooling technologies included air, water glycol, etc., are most-used in thermal management of IGBT modules. With increasing heat dissipation, however, it is predicted that conventional air cooling solution is difficult to meet cooling demand sufficiently, especially when the power dissipation exceeds up to 1500 W [1]. Micro- and mini-channel liquid cooling solutions was focused over the past decade. It has a major advantage of simplicity of heat sink design due to compact and light in weight. It can also offer a higher cooling capacity typically 120 W/cm^2 in comparison with that of air cooling (typically 50 W/cm^2) [2]. In order to improve temperature uniformity, array impingement cooling was paid more attention, The jet impingement cooling reduces junction to coolant thermal resistance by 2.8 and 1.7 times compared to cooling plate and microchannel cooler [3, 4]. In general, heat flux levels of around 250 and 1000 W/cm^2 for water have been reported for single-phase and phase change impingement, respectively [5, 6]. Although jet impingement can achieve a very heat transfer coefficient, the applicability of this strategy is limited by problems of high cost, complex cooling flow redistribution, cooling loop leakage, and channel blockage. Therefore, economical but more reliable thermal management device is needed for power module cooling purpose. Phase change cooling strategy becomes more attractive in higher thermal demands. Vapour chamber/heat pipe, as

B. Li · Y. Chen · Y. Yan (✉) · X. Wang
Faculty of Engineering, University of Nottingham, Nottingham NG7 2RD, UK
e-mail: yuying.yan@nottingham.ac.uk

Y. Li
School of Mechanical and Automotive Engineering, South China University of Technology, Guangzhou, China

Y. Chen · Y. Wang
Dynex Semiconductor Limited, Doddington Road, Lincoln, UK

© Springer Nature Singapore Pte Ltd. 2021
C. Wen and Y. Yan (eds.), *Advances in Heat Transfer and Thermal Engineering* ,
https://doi.org/10.1007/978-981-33-4765-6_88

phase change cooling components, is a super heat conductive device, and it now represents an appealing option for efficient cooling electric components. Vapour chamber is a vacuum metallic container with wick structure lining the internal walls saturated with working fluid. It utilizes cyclic phase change heat transfer process which can absorb much more heat than single-phase heat transfer process. Generally, a vapour chamber effective thermal conductivity is in range of 1000 to 50,000 W/cm^2

2 Methodology

In a semiconductor device such as an IGBT or a diode, there are three categories of power losses as shown in Eqs. (1)–(5). The first category is the conduction loss which consists of the heat dissipation from IGBT P_{SS} and diode P_{DC}. The second category is switching losses P_{SW} and includes the energy lost when the device is switching ON or OFF. The third category is named recovery loss of diodes P_{rec} and regards the losses when the device is fully OFF.

$$P = P_{SS} + P_{SW} + P_{DC} + P_{rr} \tag{1}$$

where conduction losses from IGBT PSS and diode P_{DC} are defined, respectively, as

$$P_{SS} = I_{CP} \times V_{CE(SAT)} \times \left(\frac{1}{8} + \frac{D}{3\pi} \cos\varphi\right) \tag{2}$$

and

$$P_{DC} = I_{CP} \times V_f \times \left(\frac{1}{8} - \frac{D}{3\pi} \cos\varphi\right) \tag{3}$$

where I_{CP}, $V_{CE(SAT)}$, D, $\cos\varphi$,V_f are peak value of output current, forward voltage drop in IGBT, duty ratio, power factor, and forward voltage drop in the diode.

Switching losses P_{SW} of each IGBT typically contribute a large amount to the total system losses. It can be described as

$$P_{SW} = \left(E_{SW(ON)} + E_{SW(OFF)}\right) \times \frac{f_s}{\pi} \tag{4}$$

where $E_{SW(ON)}$, $E_{SW(OFF)}$ is the switching on loss of IGBT and switching off loss of IGBT. f_s is the switching frequency.

During switching from the conduction to block state, a diode stored charge which must be discharged. The power loss from this stage is called recovery loss of diodes P_{rec} given by Eq. (5).

$$P_{rec} = E_{rec} \times \frac{f_s}{\pi} \tag{5}$$

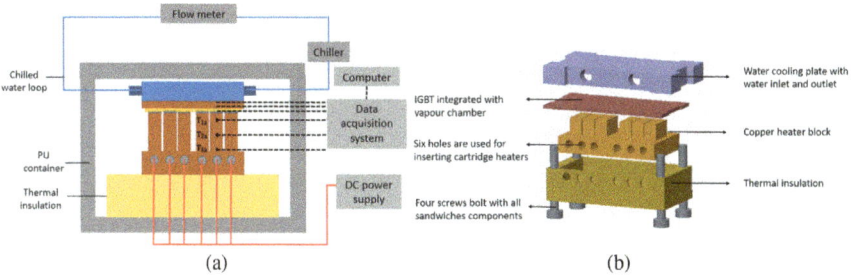

Fig. 1 **a** Schematic diagram of experiment apparatus. **b** Schematic diagram of IGBT power model with heat sink and heat source

where E_{rec} is the recovery loss of diode.

Figure 1 displays entire schematic diagram of experiment apparatus and structure of IGBT semiconductor module with vapour chamber thermal management system. The vapour chamber is directly soldered on DCB substrate by using Sn-Pb solder material, and it is bolted with a water cooled plate. Constant heat load is directly transferred to each diode and IGBT chip by six cartridge heaters through the copper heating blocks. The effective heating areas are the same as that of IGBT chips and diode. The IGBT chips and diodes are directly attached to the copper heating block, and the whole module is sandwiched between heating block and copper water cooling plate by screw installations.

3 Results

A. *Junction temperature*

Junction temperature is the crucial factor which affects the reliability of IGBT semiconductor significantly. As shown in Fig. 2, temperature distributions of IGBT integrated with copper and VC are obtained by simulation results. The thermal results are the same at the stage of top and bottom switch power loss, and therefore, only the results at the stage of top switch power loss are presented here. It can be seen that the hot spots mainly distribute on IGBT and diodes, and the maximum temperature is much higher at IGBT power loss stage than at diodes stage. Compared with IGBT semiconductor with copper baseplate, vapour chamber can improve the junction temperature and temperature uniformity. The junction temperature reduces from 85.6 to 54.6 °C and from 41.1 to 31.8 °C at stage of IGBT power loss and diodes power loss, respectively.

Figure 3 depicts temperature at junction, DCB layer, evaporation surface of vapour chamber/copper, and condensation surface of vapour chamber/copper under different heat flux. In this experiment, the heat power input of copper and VC case varied from 30 to 240 W in a step of 30 W, respectively. It can been seen that DCB temperature T_{DCB} is close to junction temperature T_j for both cases, since thermal resistance

(a)   (b)

(c)   (d)

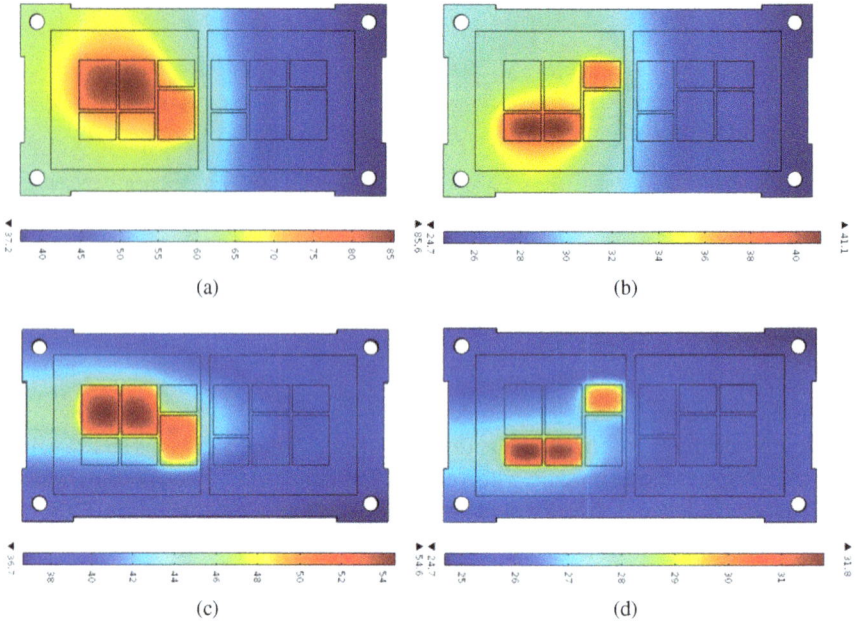

Fig. 2 Temperature distribution **a** IGBT semiconductor with copper baseplate at stage of top switch IGBT power loss. **b** IGBT semiconductor with copper baseplate at stage of top switch diodes power loss. **c** IGBT semiconductor with VC baseplate during top switch IGBT at stage of top switch IGBT power loss. **d** IGBT semiconductor with VC baseplate during top switch diodes at stage of top switch diodes power loss

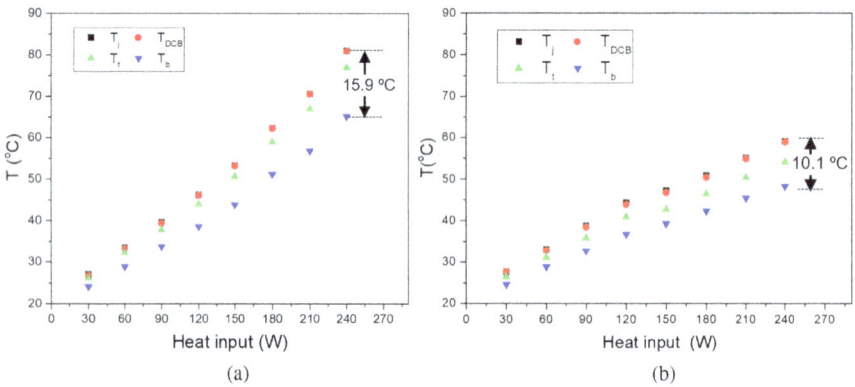

(a)   (b)

Fig. 3 Temperatures measured on IGBT module under different heat input power. **a** copper baseplate. **b** VC baseplate

of chip and chip solder layer have low thermal resistances. Temperature difference is closely related to thermal resistance. As heat input increased, the temperature difference between junction and vapour chamber condensation surface increases more slowly than the temperature difference between junction and cold surface of copper baseplate. The highest junction temperature of copper case is as high as 81.9 °C compared with 53.5 °C of vapour chamber case at the same applied power of 240 W. To compare with simulation results, the junction temperature of copper and VC case obtained by finite element method are 85.6 and 54.6 °C, respectively. It can be found that the simulation model provides a good agreement with the experimental data in term of junction temperature. Furthermore, vapour chamber module has a smaller temperature difference between junction and condenser surface than copper baseplate module. These results demonstrate that VC is more effective at cooling than copper baseplate.

B. Thermal resistance and heat transfer coefficient

The one-dimensional thermal resistance of VC R_{VC}, thermal resistance between chip and baseplate $R_{th(j-c)}$, heat sink and cooling media $R_{th(c-a)}$ and the total thermal resistance between chip and cooling media $R_{th(j-a)}$ are plotted as a function of heat power input in Fig. 4. The thermal resistance of copper baseplate reduced from 0.068 to 0.055 K/W with heat input power ranged between 30 and 60 W, and then, it is kept between 0.045 and 0.049 K/W with heat input increase from 90 to 240 W. The thermal resistance of VC continuously decreases from 0.053 to 0.032 K/W when heat input ranges between 30 and 240 W, which is attributed to phase change heat transfer. At the low heat input, the water film in VC is thick, and boiling is at incipient stage, which leads to high evaporation and condensation thermal resistance. With increase in heat input, water film recession and more water turns into vapour by absorbing heat from evaporator wall of vapour chamber, and the boiling process is enhanced. Therefore, the water film becomes thinner, and the evaporation and condensation thermal resistance reduces [3, 5]. Regarding $R_{th(b-a)}$, it mostly depends

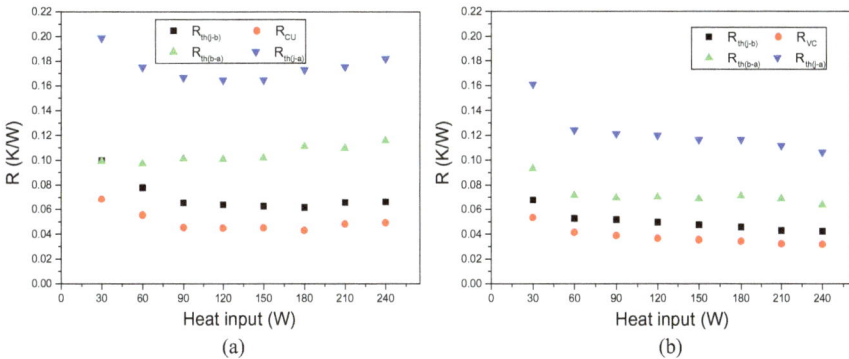

Fig. 4 Thermal resistance of IGBT power module under different heat loads. **a** Copper baseplate **b** VC baseplate

Table 1 Baseplate heat sink transfer coefficient for copper and vapour chamber baseplate module

Heat input	30 W	90 W	150 W	210 W	240 W
Copper	1333	1305	1296	1206	1140
Vapour chamber	1416	1900	1920	1923	2069

on the thermal performance of baseplate since the water flow rate, water temperature, and other boundary conditions are kept the same between IGBT with VC and copper cases. Therefore, IGBT with VC baseplate has a lower thermal resistance of water cold plate than IGBT with copper baseplate, and it is found that $R_{th(b-a)}$ is reduced with R_{VC}. In general, the total resistance $R_{th(j-a)}$, $R_{th(j-b)}$, and $R_{th(b-a)}$ also decrease with increase in heat input in vapour chamber module. In general, VC module can reduce 34.9, 35.8, 44.9, and 41.6% of thermal resistance of baseplate, $R_{th(j-b)}$, $R_{th(b-a)}$, and total thermal resistance $R_{th(j-a)}$ compared with copper module, respectively.

To evaluate the baseplate heat sink thermal performance, heat transfer coefficients of vapour chamber and copper baseplate heat sinks are calculated as given in Table 1. The results demonstrate that the phase change heat transfer coefficient for vapour chamber baseplate heat sink is much higher than single phase heat transfer coefficient for copper baseplate heat sink, especially for high heat flux. Heat transfer coefficient of vapour chamber heat sink increases with heat flux since more nucleation sites become active and bubble formation increases.

4 Conclusions

Experiments are conducted to investigate thermal performance of IGBT integrated with vapour chamber and original copper baseplate. Furthermore, a computational heat transfer model is built to predict the junction temperature from the experimentally measured from heat transfer coefficient data. Based on the results, the new IGBT power module with VC provided a higher heat dissipation ability and better temperature uniformity compared with traditional IGBT power module with copper baseplate. At stationary heat load of 240 W, the junction temperature of IGBT with copper baseplate is up to 81.0 °C compared with 53.6 °C of vapour chamber case. High temperature uniformity is also important requirement for power module since large temperature difference will increase thermal stress and lead to many thermal reliability problems. The experimental results shows temperature difference on the hotter surface of copper and VC baseplate are 5.5 and 1.3 °C. This indicates that vapour chamber has better temperature uniformity than copper baseplate. Besides, lower thermal resistance of vapour chamber could also increase the heat removal rates of cooling water. The results show that VC module can reduce 34.9, 35.8, 44.9, and 41.6% of thermal resistance of baseplate, $R_{th(j-b)}$, $R_{th(b-a)}$, and total thermal resistance $R_{th(j-a)}$, respectively, compared with copper module.

Acknowledgements This project has been supported in the frame of the ECPE Joint Research Programme and Innovate UK project (102287).

References

1. S.G. Kandlikar, C.N. Hayner, Liquid cooled cold plates for industrial high-power electronic devices—thermal design and manufacturing considerations. Heat Transf. Eng. **30**(12), 918–930 (2009)
2. P. Wang, P. McCluskey, A. Bar-Cohen, Two-phase liquid cooling for thermal management of IGBT power electronic module. J. Electron. Packag. **135**(2), 021001 (2013)
3. K. Oliphant, B. Webb, M. McQuay, An experimental comparison of liquid jet array and spray impingement cooling in the non-boiling regime. Exp. Thermal Fluid Sci. **18**(1), 1–10 (1998)
4. M. Fabbri, S. Jiang, V.K. Dhir, Experimental investigation of single-phase micro jets impingement cooling for electronic applications. In: Proceedings of the HT2003 ASME summer heat transfer conference, July 2003
5. A. Bhunia, Q. Cai, C.-L. Chen, Liquid impingement and phase change for high power density electronic cooling. In: Proceeding of the 41st AIAA aerospace sciences meeting and exhibit, Reno, NV, January 2003
6. J. Ditri et al., Embedded cooling of high heat flux electronics utilizing distributed microfluidic impingement jets. **56901**, V003T10A014 (2015)

Predicting the Temperature Distribution in a Lithium-Ion Battery Cell for Different Cooling Strategies

Mahmoud Sawani and Robert Camilleri

1 Introduction

This paper presents the temperature distribution in lithium-ion battery cells under various cooling strategies. Lithium-ion batteries have been presented as a major enabler in energy storage, particularly for electrical transportation. The manufacture of safe and reliable battery cells, with repeatable performance, quick charge and discharge capability, and high energy density is necessary [1]. When dis/charged at high rates, heat generated within the electrodes causes a rapid increase in core temperature. This may accelerate degradation and may lead to thermal runaway and battery failure [2]. The overall battery performance is significantly affected by temperature with the battery comfort temperature being between 15 and 35 °C. Excessively low or high temperatures can reduce battery life and may lead to thermal runaway, posing a threat to the safety of the battery and/or surroundings. To address this problem, several ideas for a battery thermal management system have been developed. While every cooling strategy has benefits and shortfalls [3, 4], they also affect the temperature distributions within the battery pack and within each cell. While the temperature distribution within the battery pack has been investigated, to the author's knowledge, the effect of the temperature distribution for the different cooling strategies has not been thoroughly explored. This is therefore the subject of this paper which investigates this through CFD simulations.

M. Sawani (✉) · R. Camilleri
Institute of Aerospace Technologies, University of Malta, Msida MSD2050, Malta
e-mail: Mahmoud.sawani@um.edu.mt

© Springer Nature Singapore Pte Ltd. 2021
C. Wen and Y. Yan (eds.), *Advances in Heat Transfer and Thermal Engineering* ,
https://doi.org/10.1007/978-981-33-4765-6_89

2 Methodology

2.1 CFD Prediction

This paper performs CFD simulations on battery cells operated under four cooling systems:

- Natural convection,
- Air cooling (Forced convection),
- Indirect liquid cooling, and
- Direct oil cooling.

The 3D CAD geometry of a single lithium-ion 18650 cell was modelled. A heat loss ranging from 1 to 5 W was applied to the cell, thereby representing the losses from battery cells during various charging and discharging rates.

Following an initial validation process, the battery cell material properties were set as reported in the literature [5]. In each cooling mode, the adequate flow geometry was designed as shown in Fig. 1, and the respective flow properties were set. For example, in natural convection, the flow domain was set to be air with temperature dependent properties to model the buoyancy effects. Conversely in the forced air cooling, the flow was set to act perpendicular to the battery axial dimensions. The flow rate and fluid temperature of 20 °C were set as inlet boundary conditions. The outlet was set with a gauge static pressure of 0 Pa, to simulate the flow out of the domain. Conversely indirect liquid cooling required a cooling block which was maintained at a steady temperature of 20 °C. Finally, for direct oil coolant, the flow domain was set to flow parallel to the battery axial direction. The coolant was set with Opticool properties, whose properties where made to vary with temperature. The properties of Opticool coolant were reported in [6]. The inlet flow rate and fluid temperature, and the outlet static pressure of 0 Pa were set.

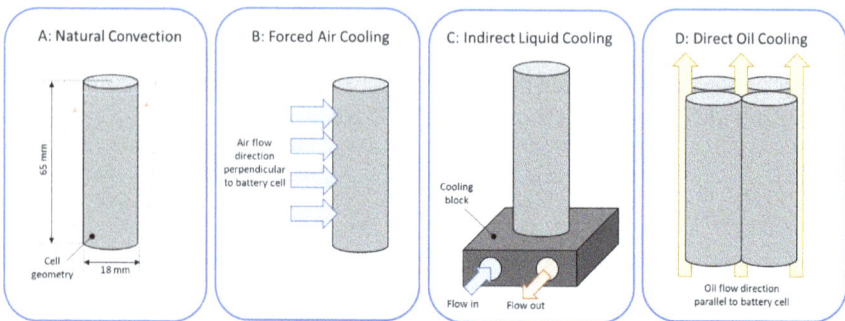

Fig. 1 Schematic showing the various cooling techniques for lithium-ion batteries including; **a** Natural convection, **b** Forced air cooling, **c** Indirect liquid cooling, **d**: Direct liquid cooling

2.2 Experimental Validation

The simulations were validated experimentally validated using a dummy Li-ion 18,650 cell. The dummy cell was manufactured using an aluminium rod of the same geometry with an internal heating rod which simulates the internal heat generation. During these tests, a DC power supply was used to power the heating rod. The voltage and current were monitored and recorded, from which the heat loss was calculated. The dummy cell was mounted with six surface mount T-type thermocouple sensors whose values were also logged. The experiment was repeated for the various cooling solutions, and found to be within 4% of simulations.

3 Results

The temperature profile of the battery cells under the various cooling modes are shown in Fig. 2. The results show that natural cooling is very limited and requires careful control of the heat load to avoid overheating. As the cooling mode shifts to forced air cooling, indirect liquid cooling, and direct oil cooling, heat transfer improves, and the battery cell surface temperature decreases.

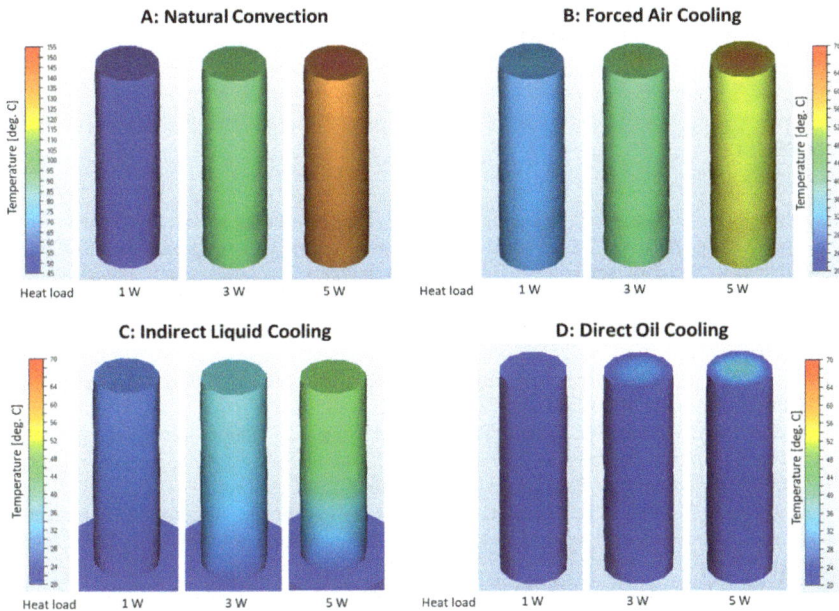

Fig. 2 CFD results comparing the temperature profile for the various cooling modes, at different battery cell heat generation

However, it can be seen that switching from natural cooling to forced air cooling causes a radial temperature gradient across the cell. Conversely, switching to indirect liquid cooling causes a steep temperature gradient along the axial direction. Direct oil cooling is seen to be able to offer the minimum low battery surface temperatures, independent of battery heat generation. However, at high heat loads (beyond 3 W), a temperature gradient in the radial direction starts to emerge.

The reduction in battery temperature allows for higher charging and discharging rates. This may have interesting implications. For example, while the early generation battery electric vehicles such as the Nissan LEAF made use of natural convection, the customer focus has now shifted and demands quick charging mechanisms. This increases the heat losses in the battery cells, and therefore, electric vehicle manufacturers have to consider more efficient cooling modes. The cooling circuit should be designed to remain powered even while the vehicle is turned off and charging. However, the paper also demonstrates that the different cooling mechanisms provides a characteristic battery cell temperature distribution, along the radial or axial direction (dependent on cooling modes). In modelling battery systems, it is often assumed that every cell is at a single temperature. Large temperature gradients across battery cells may affect their electrochemical performance and/or life [7]. The extent of this effect is still relatively unknown. The authors therefore argue for caution when designing battery cell cooling systems and for further studies in this field.

Acknowledgements This research is a result of project NEVAC, financed by the Malta Council for Science and Technology through FUSION: The R&I Technology Development Programme 2017.

References

1. D. Howell, B. Cunningham, T. Duong, P. Faguy, US Department of Energy, Vehicle Technologies Office. 2016. pp. 24. Available: https://www.energy.gov/sites/prod/files/2016/06/f32/es000_howell_2016_o_web.pdf
2. R. Kantharaj, A.M. Marconnet, Heat generation and thermal transport in lithium-ion batteries: a scale-bridging perspective. Nanoscale Microscale Thermophys. Eng. **23**(2), 128–156 (2019). https://doi.org/10.1080/15567265.2019.1572679
3. J. Li, Z. Zhu, *Battery Thermal Management Systems in Electric Vehicles*, Masters Thesis (Chalmers University of Technology, Göteborg, Sweden, 2014)
4. L.H. Saw, A.A.O. Tay, L.W. Zhang, Thermal management of lithium-ion battery pack with liquid cooling, in *31st Conference: Thermal Measurement, Modeling & Management Symposium (SEMI-THERM)*, San Jose, CA., USA, 15–19 March 2015
5. S.J. Drake, D.A. Wetz, J.K. Ostanek, S.P. Miller, J.M. Heinzel, A. Jain, Measurement of anisotropic thermophysical properties of cylindrical Li-ion cells. J. Power Sources **525**, 298–304 (2014)
6. R. Camilleri, D.A. Howey, M.D., McCulloch, Predicting the temperature and flow distribution in a direct oil-cooled electrical machine with segmented stator. IEEE Trans. Ind. Electron **63**(1), 82–91 2016
7. I.A. Hunt, Y. Zhao, Y. Patel, G.J. Offer, Surface cooling causes accelerated degradation compared to tab cooling for lithium-ion pouch cells. J. Electrochem. Soc. **163**(9), A1846–A1852

Electro-osmotic Non-isothermal Flow in Rectangular Channels with Smoothed Corners

Marco Lorenzini

1 Introduction

Microchannel heat sinks are able to provide high cooling capabilities in terms of heat flux rates. This makes them particularly interesting for the thermal management of electronic devices, owing to the latter's ever-increasing compactness and, consequently, power density. Even for very low flowrates, pressure drops across such devices may be significant, especially when liquids are employed as coolants, with the decrease in hydraulic diameter of the ducts quickly leading to viscous heating of the fluid circulated and unviable pumping costs. When the coolant is a polar fluid and the channel walls have a net electrical charge, an uneven charge distribution develops in the fluid, which may be used to move the liquid by applying an electric field at the two ends of the channel. The flow which originates is called electro-osmotic (EOF). EOF represents an alternative method to circulate the fluid, which is applicable just when channel dimensions drop. EOFs do not require moving parts, do not produce noise nor vibrations, and do not need lubrication. Also, they have tiny volumes, making them ideal for direct connection to the chips [1–3]. EOF have been reported to yield experimental Nusselt numbers about 10% larger than pressure-driven flow (PDF) for the same geometry, although Joule heating may become significant when applied voltage between the electrodes increases beyond a certain threshold [4]. Further enhancement can be obtained by introducing nanoparticles in the fluid [5]. Channel geometries employed are often rectangular or, in alternative, trapezoidal, but the current development of micro-fabrication allows other forms to be produced, which may offer better performance in terms of heat transfer, pressure drop, or entropy production. Taking this as its starting point, this work investigates numerically the influence of smoothing the corners of the cross section at fixed hydraulic diameter

M. Lorenzini (✉)
Department of Industrial Engineering, University of Bologna, Via Fontanelle 40, Forlì 47121, Italy
e-mail: marco.lorenzini@unibo.it

© Springer Nature Singapore Pte Ltd. 2021
C. Wen and Y. Yan (eds.), *Advances in Heat Transfer and Thermal Engineering* ,
https://doi.org/10.1007/978-981-33-4765-6_90

and several aspect ratios on the values of the Poiseuille, Nusselt, and entropy generation numbers for the laminar, steady, and fully developed electro-osmotic flow in a
rectangular channel subject to uniform heat flux and Joule heating.

2 Methodology

The geometry investigated is that of a microchannel of rectangular cross section,
whose corners are progressively rounded: several aspect ratios (shorter-to-larger-
side ratio), β, namely 1, 0.5, 0.25, and 0.10 and radii of curvature, R_c, (from 0 to β)
are investigated. The momentum, energy, and entropy transport equations, Eqs. (1)–
(3), are written in non-dimensional form, superscript '*'. Owing to the presence of
an electric field and of a polar fluid, prior to solving them, the charge and potential
distribution, ψ, in the fluid must be computed [3]. Knowledge of the non-dimensional
velocity, u^*, and temperature, T^*, fields, and entropy production per unit length of
the duct, $\dot{S}^*_{gen}$, allow computation of the Poiseuille, Nusselt, and entropy generation
numbers [6], as highlighted in [7].

$$\nabla^{2^*} u^* = M_z E_z^* \sinh(\psi^*) \tag{1}$$

$$\frac{u^*}{u_m} U = \nabla^{2^*} T^* + M_z \tag{2}$$

$$d\dot{S}^*_{gen} = ds^* - \frac{q' dz}{\dot{m} T_w s_i} \tag{3}$$

In Eq. (1) M_z represents the ratio of heat generation due to Joule heating within
the fluid to convective heat transfer and E_z is the electric field acting on the fluid,
whereas U and u_m in Eq. (2) represent two reference velocities used to make the
energy equation non-dimensional. Finally, T_w in Eq. (3) is the local wall temperature
(dependent on the axial coordinate, z), $\dot{m}$ is the mass flow rate, s_i is the inlet specific
entropy, and q' is the perimeter heat flux. To solve Eqs. (1)–(3), the no-slip boundary
condition is implemented for velocity at the walls, and H1 (i.e., uniform heat flux
over the duct walls and uniform temperature along the heated perimeter of any cross
section) boundary conditions are given for the energy equation. The Poiseuille and
Nusselt numbers are expressed in polynomial form as

$$Po = \sum_{i=0}^{3} c_i R_c^{*^i}, \quad Nu = \sum_{i=0}^{3} d_i R_c^{*^i} \tag{4}$$

where c_i, d_i, are suitable coefficients.

3 Results

Velocity and temperature fields are computed for two different dimensional hydraulic diameters ($D_h = 24\,\mu\text{m}$ and $D_h = 3\,\mu\text{m}$, respectively) for several values of M_z (from 0.001 to 1), of β (0.1, 0.25, 0.5 and 1), and of the non-dimensional radius of curvature (from 0.001 to 1). It can be appreciated from the results that the velocity profile is much flatter in the case of the higher hydraulic diameter, owing to the constancy of the Debye–Hückel parameter, i.e., of the thickness of the electric-double layer, which occupies a smaller portion of the cross section for the larger D_h. This dependence on the hydraulic diameter carries over to the non-dimensional friction factors and heat transfer coefficients, i.e., the Poiseuille and Nusselt numbers. They also depend in the same qualitative way by M_z, β, and R_c^*. Smoothing of the corners modifies the velocity and temperature fields, as shown in Fig. 1.

The change of the aspect ratio and of the radius of curvature has little effect on the variation of the Poiseuille number (see Fig. 2, left) which is almost independent of it, whilst it is strongly affected by D_h, ($f\,Re = 164.65$ for $D_h = 3\,\mu\text{m}$ and $f\,Re = 956.30$ for $D_h = 24\,\mu\text{m}$). The Nusselt number (Fig. 2, right) increases with the hydraulic diameter at fixed β and M_z and decreases with the aspect ratio for given D_h and M_z. Smoothing the corners enhances Nu invariably, although the benefit is more sensitive at lower aspect ratios (e.g. at $\beta = 1$ and $M_z = 0.001$ from $Nu = 5.52$ for sharp corners, and $Nu = 6.27$ at $R_c^* = 1$), whereas it is less significant for lower β (e.g., at $\beta = 0.1$ and $M_z = 0.001$ from $Nu = 8.60$ for sharp corners, and $Nu = 8.91$ at $R_c^* = 1$). The Nusselt number is also proven to be decreasing linearly with M_z.

The non-dimensional entropy is obtained through integration over the whole channel of the contributions due to heat transfer and pressure drop. In this case,

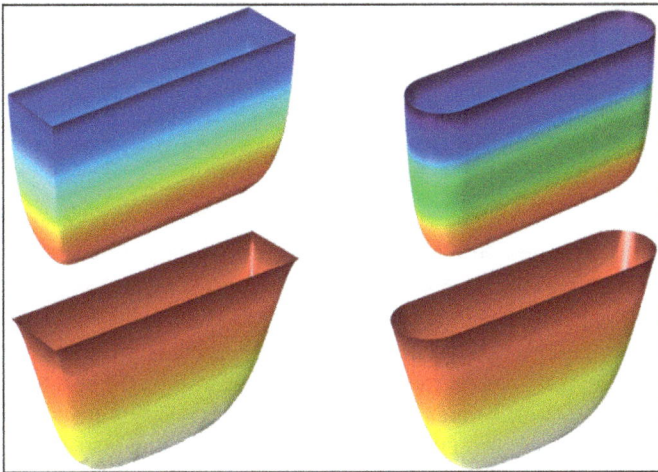

Fig. 1 Velocity (above) and temperature (below) profiles for a channel of aspect ratio $\beta = 0.25$, $M_z^* = 10^{-3}$ with sharp (left) and rounded (right, $R_c^* = 0.25$) corners

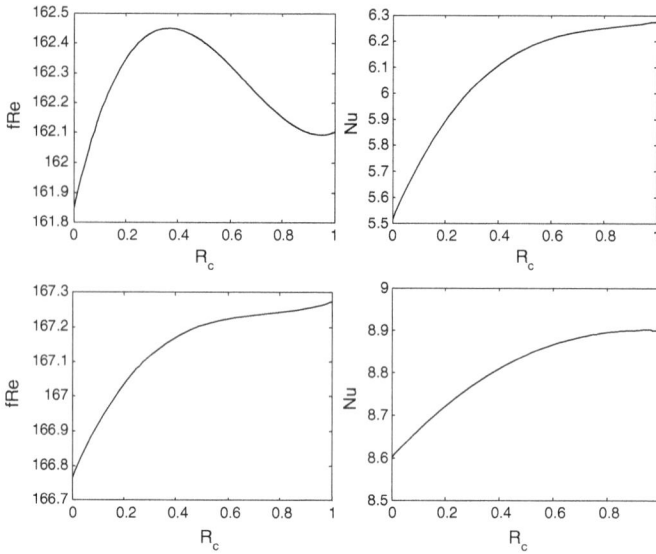

Fig. 2 Poiseuille (left) and Nusselt (right) numbers as a function of the non-dimensional radius of curvature for $\beta = 1$ (above) and $\beta = 0.1$ (below) and $M_z = 10^{-3}$

an analytical expression is obtained which is strongly dependent on M_z, which also enters through the Nusselt number. For optimization purposes, it is expedient to express the entropy generation for a modified configuration (i.e., with smoothed corners) as a non-dimensional quantity, obtained by dividing it by the entropy generated at the reference condition (i.e., with sharp corners), thus computing the so-called entropy generation number, [6], which can be used in conjunction with the results obtained by the first-law analysis in terms of Poiseuille and Nusselt numbers, as in [8].

4 Conclusions

The electro-osmotic flow in microchannels with smoothed corners was studied in order to highlight the dependence of the Poiseuille, Nusselt, and entropy generation number on the radius of curvature and other quantities typical of electro-osmotic flow with Joule heating and convective heat transfer at the walls. The results may be applied to performance-evaluation criteria in order to maximize, e.g., the heat transfer rate at fixed temperature difference and pumping power or flowrate or minimize temperature difference at fixed heat transfer rate and pumping power or flowrate without changing the total area for heat transfer (so-called FG criteria, [9]) nor the bulk dimension of the heat exchanger.

References

1. K. Pramod, A.K. Sen, Flow and heat transfer analysis of an electro-osmotic flow micropump for chip cooling. J. Electr. Packag. Trans. ASME **136**, 031012 (2014)
2. M. Geri, M. Lorenzini, G.L. Morini, Effects of the channel geometry and of the fluid composition on the performances of DC electro-osmotic pumps. Int. J. Therm. Sci. **55**, 114–121 (2012)
3. G.L. Morini, M. Lorenzini, S. Salvigni, M. Spiga, Thermal performance of silicon micro heat-sinks with electro-kinetically-driven flows. Int. J. Therm. Sci. **45**, 955–961 (2006)
4. M.F. Al-Rjoub, A.K. Roy, S. Ganguli, R.K. Banerjee, Assessment of an active-cooling micro-channel heat sink device, using electro-osmotic flow. Int. J. Heat Mass Tr. **55**, 4560–4569 (2011)
5. M.F. Al-Rjoub, A.K. Roy, S. Ganguli, R.K. Banerjee, Enhanced heat transfer in a micro-scale heat exchanger using nano-particle laden electro-osmotic flow. Int. Comm. Heat Mass Tr. **68**, 228–235 (2015)
6. A. Bejan, *Entropy generation through heat and fluid flow* (Wiley, New York, 1982)
7. M. Lorenzini, N. Suzzi, The influence of geometry on the thermal performance of microchannels in Laminar flow with viscous dissipation. Heat Transf. Eng. **37**, 1096–1104 (2016)
8. M. Lorenzini, Electro-osmotic flow in rectangular microchannels: geometry optimization. J. Phys.: Conf. Ser. **923**, 012002 (2017)
9. R.L. Webb, N.H. Kim, *Principles of enhanced heat transfer* (CRC Press, Boca Raton, 2005)

Transient Simulation of Finned Heat Sinks Embedded with PCM for Electronics Cooling

Adeel Arshad, Pouyan Talebizadehsardari, Muhammad Anser Bashir, Muhammad Ikhlaq, Mark Jabbal, Kuo Huang, and Yuying Yan

1 Introduction

A shear usage of electronic mobile devices during gaming, Web browsing and high-power consuming applications, the portable devices get considerably hot enough to be damaged. Previously, active cooling methods adopted to cool the electronic devices had some limitations. When the size increases, the active cooling methods stay ineffective and that is the case with higher power dissipation level too. In such instances, a novel technology is needed which could ensure thermal management (TM) in best way possible. In recent years, the invention of passive TM of electronics devices using phase change materials (PCMs) and various thermal conductivity enhancers (TCEs) has introduced a new direction in cooling technologies. The significant outcomes of cooling through PCMs in electronics equipment have been achieved as they emanate heat during heavy usage due to their high latent of fusion. To facilitate the reliable TM of high-speed electronic devices, latent heat storage unit (LHSU) containing high thermal conductivity finned heat sinks filled with PCMs are to be used to operate the devices with ease for maximum possible time. Till yet, several studies have been reported which introduced the finned heat sinks as a TCEs, filled with PCMs which have high specific and latent heat of fusion with small change in volume. A numerical study is proposed by Nayak et al. [1] for TM of electronics using porous matrix and finned (plate-type and rod-type) as a TCEs for PCM-filled heat sinks. TCEs 5,

A. Arshad (✉) · P. Talebizadehsardari · M. Jabbal · K. Huang · Y. Yan
Fluids & Thermal Engineering (FLUTE) Research Group, Faculty of Engineering, University of Nottingham, Nottingham NG7 2RD, UK
e-mail: adeel.arshad@nottingham.ac.uk

M. A. Bashir
Department of Mechanical Engineering, Mirpur University of Science & Technology (MUST), Mirpur 10250, AJK, Pakistan

M. Ikhlaq
School of Engineering, Edith Cowan University, Joondalup, WA 6027, Australia

© Springer Nature Singapore Pte Ltd. 2021
C. Wen and Y. Yan (eds.), *Advances in Heat Transfer and Thermal Engineering* ,
https://doi.org/10.1007/978-981-33-4765-6_91

10 and 15% volume fractions of TCE were selected in case of finned heat sinks. Authors marked two findings; firstly, 10% had the better thermal performance, and secondly rod-type TCE was more significant outcomes in convection of melt front of PCM to transfer heat which in turn lead to lower chip temperature. Shatikan et al. [2] conducted the transient numerical simulation of a PCM-based heat sink at constant heat flux and performed the dimensional analysis. The results were presented with good agreement with melt fraction in terms of Nusselt, Fourier, modified Stefan and Rayleigh numbers. Further, Saha and Dutta [3] proposed the heat transfer correlation of a plate-fin PCM-based heat sink of various enclosures and heat fluxes. The authors suggested that a single correlation of Nu and Ra numbers is not enough to relate the performance of heat sink with melt fraction. The most recent parametric experimental studies are based on the square and circular profile pin-fin heat sinks reported by Arshad et al. [4–6]. The authors selected the 9% volume fraction of TCEs and concluded that a 2 and 3-mm fin thickness of square and circular fins, respectively, had the maximum thermal performance comparing with no fin, 1, 3 and 4-mm fin thickness PCM-filled heat sinks.

This research activity presents the numerical solution of combined 2 and 3 mm square and circular, respectively, configurations of finned heat sinks, acting as TCEs, having the constant volume fraction of 9% resulting in different numbers of fins. Both heat sinks are filled with PCM, namely n-eicosane, and three different input power levels are applied at the base of heat sink to explore the thermal enhancement performance of LHSU. This will eventually provide the better picture to select the optimum configuration finned heat for solution of efficient TM of electronic gadgets as passive cooling technology.

2 Methodology

The transient simulations have been performed considering 2D geometry of the system using commercial CFD code Ansys Fluent 19.1. The flow of the molten PCM in heat sink assumed to be laminar and unsteady. The conjugate heat transfer has been solved using continuity, momentum and energy equations. For the phase transition problem, "melting and solidification" model has been applied. The conservation equations for this problem can be written as

$$\frac{\partial \rho}{\partial t} + \frac{\partial}{\partial x_i}(\rho u_i) = 0 \tag{1}$$

$$\frac{\partial}{\partial t}(\rho u_i) + \frac{\partial}{\partial x_j}(\rho u_i u_j) = -\frac{\partial p}{\partial x_i} + \frac{\partial}{\partial x_{ij}}(\tau_{ij}) + \rho g_i + F_i \tag{2}$$

$$\frac{\partial(\rho H)}{\partial t} + \frac{\partial(\rho u_i H)}{\partial x_i} = \frac{\partial}{\partial x_i}\left(k\frac{\partial T}{\partial x_i}\right) + S_h \tag{3}$$

where τ is the stress tensor, ρg is the gravitational body force and F represents the external body forces. In the energy Eq. (3), k, H and S represent the thermal conductivity, PCM enthalpy and source term, respectively. The enthalpy H (Eq. 3) is the sum of sensible enthalpy and latent heat. The solidification/melting model is based on the enthalpy-porosity method. The liquid volume of each cell is represented as liquid fraction (β), where $\beta = 0$ for solid and $\beta = 1$ for liquid.

The computational domain consists of two domains: one liquid domain containing PCM and second solid domain for heat sink. The outer wall has been made insulated, and constant heat flux has been provided at the base. The appropriate initial and boundary conditions have been defined before the start of the time-depended simulations. For the pressure-velocity coupling, SIMPLE algorithm has been used. For the pressure correction equation, Presto scheme is used while QUICK scheme is employed for the momentum and energy equations. A second-order upwind differencing scheme was employed for the discretization of convective terms in momentum and energy equations. Gravity effects were also considered. The convergence criteria are set to 10^{-4}, 10^{-4} and 10^{-6} for continuity, momentum and energy equations, respectively.

3 Results

The results are presented by comparing the temperature–time and melt fraction history of 2 and 3-mm fin thickness heat sinks at three different input power levels of 4, 5 and 6 W. Figure 1 presents the temperature–time curves of 2 and 3-mm fin thick-

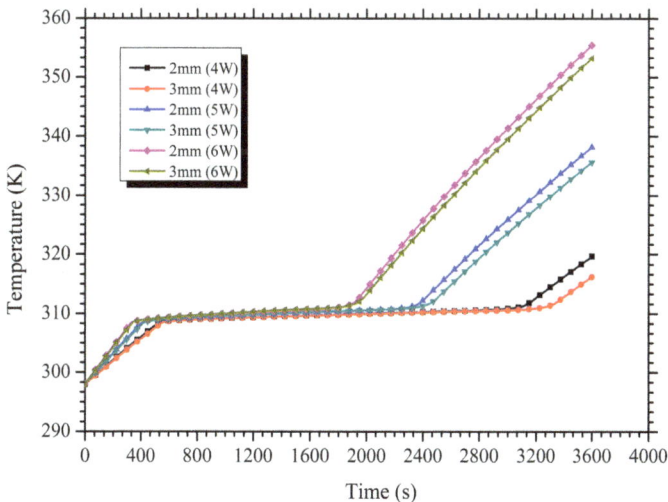

Fig. 1 Temperature-time history of 2 and 3 mm fin thickness PCM filled heat sinks at different power levels

ness heat sinks at various input power levels. From lower to higher value of power level, the time to complete the latent heat of fusion reduces as expectedly. Hence, the selection of PCM-based heat sink cooling media is based on the operating power of the electronic device. At higher input power, the one should need more amount of the PCM or a PCM with higher latent heat of fusion which can prolong the operating duration at comfortable temperature level. Further, it can be seen from Fig. 1 that at each value of power level, the 3-mm fin thickness heat sink takes more time to complete the latent heat of fusion compared to 2-mm fin thickness heat sink at fully filled with PCM. This ultimately leads to lower the base temperature of the heat sink, and thus 3-mm fin thickness heat sink has the best thermal performance to enhance the passive cooling capability of electronic devices filled with PCMs. The melt fraction curves of 2 and 3-mm fin thickness heat sinks at input power levels of 4, 5 and 6 W are shown in Fig. 2. At each power level, the melt fraction of both fin thickness heat sinks finishes the melting phase uniformly which shows the natural convection heat transfer within the melt region. In addition, the 3-mm fin thickness heat sink takes the higher time to complete the melting phase compared to the heat sink having fin thickness of 2 mm. For all input power levels, the melt fraction curves remain clearly separated for both fin thickness heat sinks, as expectedly. Furthermore, 3-mm fin thickness heat sink shows the more uniform melting during natural convection heat transfer due to the uniform and optimum fin arrangement.

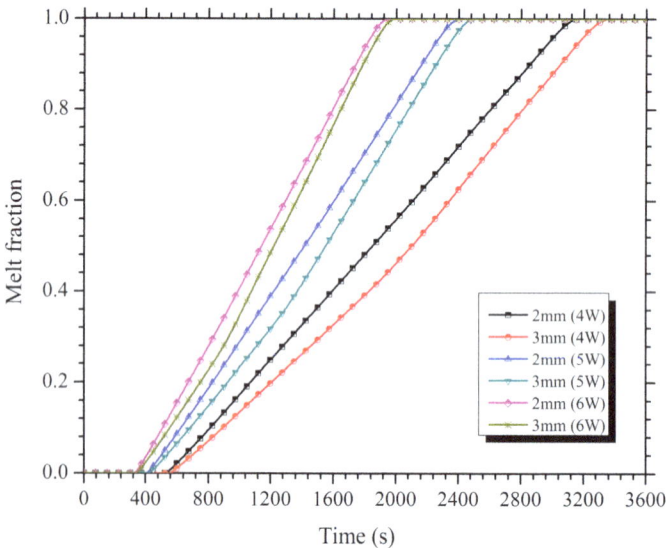

Fig. 2 Melt fraction curves of 2 and 3 mm fin thickness PCM filled heat sinks at different power levels

4 Conclusions

The transient two-dimensional numerical simulation of PCM-filled finned heat sinks is carried out to investigate the heat transfer and flow-field characteristics having the fin thickness of 2 and 3 mm. Both fin thicknesses of finned heat sinks have been optimally explored from the previous experimental investigations of Arshad et al. [4–6]. Three different input power levels of 4, 5 and 6 W are provided at base of each heat sink. The numerical results show that aluminum-made heat sink filled with PCM has the capability of uniform heat distribution inside the LHSU containing finned heat sinks as TCEs. Further, the results indicated that 3-mm fin thickness heat sink has the better heat transfer performance to reduce the base temperature, to enhance the better cooling capability and to lower the maximum operating temperature as compared to 2-mm fin thickness heat sink.

References

1. K.C. Nayak et al., A numerical model for heat sinks with phase change materials and thermal conductivity enhancers. Int. J. Heat Mass Transf. **49**(11–12), 1833–1844 (2006)
2. V. Shatikian, G. Ziskind, R. Letan, Numerical investigation of a PCM-based heat sink with internal fins: Constant heat flux. Int. J. Heat Mass Transf. **51**(5–6), 1488–1493 (2008)
3. S.K. Saha, P. Dutta, Heat transfer correlations for PCM-based heat sinks with plate fins. Appl. Therm. Eng. **30**(16), 2485–2491 (2010)
4. A. Arshad et al., An experimental study of enhanced heat sinks for thermal management using n-eicosane as phase change material. Appl. Therm. Eng. **132**, 52–66 (2018)
5. A. Arshad et al., Experimental investigation of PCM based round pin-fin heat sinks for thermal management of electronics: effect of pin-fin diameter. Int. J. Heat Mass Transf. **117**, 861–872 (2018)
6. M.J. Ashraf et al., Experimental passive electronics cooling: parametric investigation of pin-fin geometries and efficient phase change materials. Int. J. Heat Mass Transf. **115**(Part B), 251–263 (2017)

Experimental Investigation of Mini Pin-Fin Heat Sink Filled with PCM for Thermal Management

Adeel Arshad, Luke Jackman, Mark Jabbal, and Yuying Yan

1 Introduction

The effective and efficient cooling of current faster and smarter electronic devices has become a key challenge for thermal engineers. The revolutionary development in electronic technologies related to integrated circuits has aggravated the inherent heat generation level. As a result of high heat augmented inside electronic devices, the overall temperature increases at the worst and uncomfortable level which ultimately reduces the operating performance of electronics. Despite the internal heat from the components, there is an extensive need of efficient and reliable cooling system to keep the devices at lower and comfortable level. From the last decade, the extensive research on latent heat thermal management system (LHTMS) by employing the phase change materials (PCMs) has proven very good traits because of its sustainability and thermal performance as a passive cooling technology. Heat sinks with extended surfaces, called fins, generally used thermal conductivity enhancers (TCEs) in LHTMS for passive thermal management (TM) of electronics under national convection heat transfer. Until now, many configurations and fin arrangement have been introduced to enhance the thermal performance of LHTMS. Ali et al. [1, 2] firstly investigated the square and round pin-fin heat sinks filled with paraffin wax and n-eicosane. The results revealed the optimum configurations of 2-mm square and 3-mm round pin-fin heat sink. Later, the authors found the optimum configuration of triangular pin-fin heat sink of constant volume fraction of 9%. Six various PCMs of different melting temperatures were filled in staggered and inline fin arrangement of square and round pin-fin heat sinks by Ashraf et al. [3, 4]. The findings explored the better heat transfer performance on inline fin arrangement for both round and square pin-fin heat sinks. Haghighi et al. [5] investigated the plate-fin and pin-fin

A. Arshad (✉) · L. Jackman · M. Jabbal · Y. Yan
Fluids & Thermal Engineering (FLUTE) Research Group, Faculty of Engineering, University of Nottingham, Nottingham NG7 2RD, UK
e-mail: adeel.arshad@nottingham.ac.uk

© Springer Nature Singapore Pte Ltd. 2021
C. Wen and Y. Yan (eds.), *Advances in Heat Transfer and Thermal Engineering* ,
https://doi.org/10.1007/978-981-33-4765-6_92

heat sinks under natural convection regime. The results showed the plate cubic pin-fin heat sink had the lower thermal resistance and higher heat transfer rate. The maximum heat transfer rate 10–41.6% is compared to normal pin-fin heat sink. In recent experimental studies, Arshad et al. [6, 7] performed the parametric investigation of square and circular profile pin-fin heat sinks of various fin thicknesses and found the maximum heat capacity and thermal conductance of 3.1 kJ/K and 5.7 × 10^{-1} W/K for round pin-fin heat sink.

The present study explores the natural convection heat transfer of solid and hollow round mini pin–fin heat sink having constant volume fraction of 9% filled with PCM, namely RT-28HC at three input powers of 2, 2.5 and 3 W to explore the thermal enhancement performance of LHTMS. This will eventually provide the better picture to select the suitable configuration of mini pin-fin heat sink for solution of efficient TM of electronic gadgets as passive cooling technology.

2   Methodology

In this experimental setup, LHTMS contains the two different configurations of circular pin-fin (solid and hollow) heat sinks, plate-type silicon rubber heater (OMEGALUX), K-type thermocouples (OMEGA), rubber mat for insulation, perspex sheet to visualize the physical melt fractions of PCMs. The RT-28HC is used as a PCM from RUBITHERM® having melting temperature of 27–29 °C, heat storage capacity of 250 kJ/kg and thermal conductivity of 0.2 W/m K. The heat sink with dimensions of 55 × 55 × 5 mm³ was made from copper using computer numerical control (CNC) machine. A constant volume fraction of 9% for both solid and hollow round pin-fin heat sinks having fin diameter of 3 mm and 8 mm, respectively, act as TCEs, as shown in Fig. 1. A DC power supply of 35 V/5A is connected to the heater to supply the constant of power levels 2, 2.5 and 3 W at each heat sink.

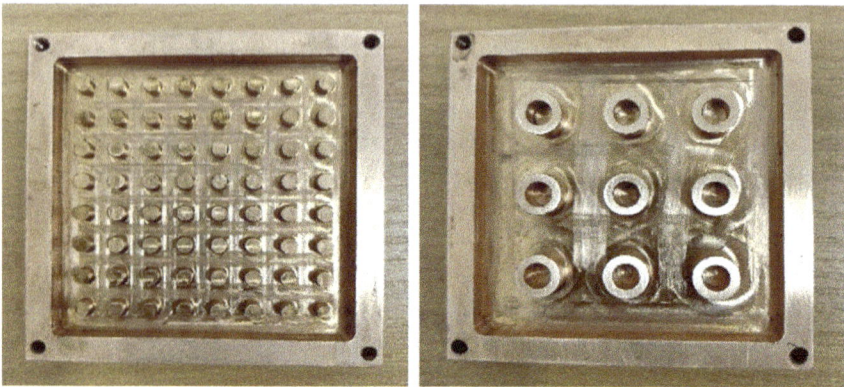

Fig. 1   Photographic images of **a** soild and **b** hollow pin-fin heat sink

To measure the temperature variations at different locations of heat sinks, seventeen thermocouples are fixed using Araldite™ and positions of all thermocouples. Thermocouples are connected to a PC-based digital data acquisition system (Agilent 34972A) to record the temperature variation in LHTMS at various locations. Calibration of all thermocouples of temperature range of 0–100 °C as performed, and maximum variation was found about ±0.1 °C. The rate of heat transfer across each heat sink is calculated by measuring the temperature at the base and top surface of the heat sink through the thermocouples that connect with the data logger. To prevent excess heat escaping to the atmosphere, the heat sink is insulated in rubber insulation and placed on a rubber mat under heating and discharging. Each heat sink was tested for 0% and 100% volumetric fraction of PCM at an input power of 2, 2.5 and 3 W.

3   Results

The results are shown in Fig. 2 for two different heat sinks tested in current study for input power levels of 2 and 3 W. The base temperature curves of each heat sink are compared at volumetric fraction of 0% and 100% of PCM at each input power. Figure 2 presents a very clear difference of heat transfer performance between the solid and hollow pin-fin heat sinks at input power levels of 2 and 3 W without and with filling the PCM. The heating curves are obtained at constant input powers while recording temperature at base of each heat sink which is directly attached with the heat-generating source of the electronic devices. The clear evidence is achieved

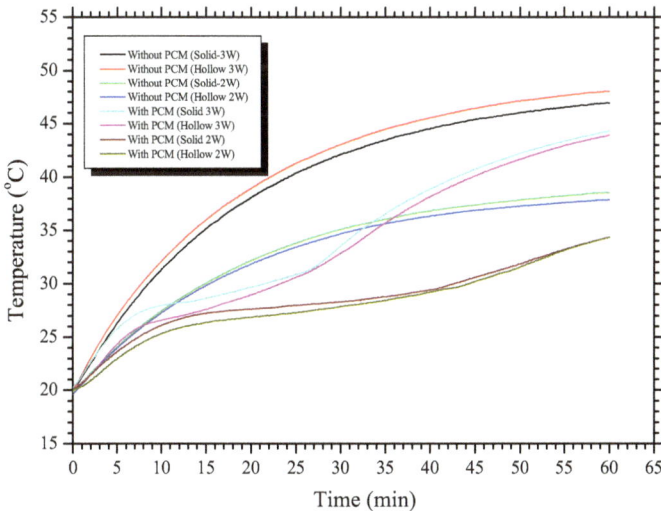

Fig. 2 Temperature–time history of solid and hollow pin-fin heat sinks at different input power levels

against each input power level filling either with or without PCM that hollow pin-fin heat sink has the better heat transfer characteristics. The base temperature of heat sink reaches at higher level compared with the PCM-filled case for both input power levels. Another observation reveals that the inclusion of PCM reduces the heat sink base temperature at comfortable level which ensures the cooling performance of LHTMS. Furthermore, PCM-based heat sinks have the tendency to increase the operating duration of electronic devices which increases the reliability and working performance of electronic device. In addition, the pin-fins enhance the latent heating phase of the PCM which is due to the natural convection heat transfer through the PCM.

4 Conclusions

The current study explores the cooling performance of pin-fin heat sink having solid and hollow configurations at three different input power levels of 2, 2.5 and 3 W. Both heat sinks are filled at 0% and 100% volume of the PCM, and melting heat transfer phenomenon is observed. The result evidences that filling either with or without PCM, the hollow pin-fin heat sink has the better cooling performance. In addition, uniform melting phenomenon is observed for both configurations of pin-fin heat sink.

References

1. H.M. Ali et al., Thermal management of electronics devices with PCMs filled pin-fin heat sinks: a comparison. Int. J. Heat Mass Transf. **117**, 1199–1204 (2018)
2. H.M. Ali et al., Thermal management of electronics: An experimental analysis of triangular, rectangular and circular pin-fin heat sinks for various PCMs. Int. J. Heat Mass Transf. **123**, 272–284 (2018)
3. M.J. Ashraf et al., Experimental passive electronics cooling: parametric investigation of pin-fin geometries and efficient phase change materials. Int. J. Heat Mass Transf. **115**, Part B, pp. 251–263 (2017)
4. H. Usman et al., An experimental study of PCM based finned and un-finned heat sinks for passive cooling of electronics. Heat Mass Transf. pp. 1–12 (2018)
5. S.S. Haghighi, H.R. Goshayeshi, M.R. Safaei, Natural convection heat transfer enhancement in new designs of plate-fin based heat sinks. Int. J. Heat Mass Transf. **125**, 640–647 (2018)
6. A. Arshad et al., An experimental study of enhanced heat sinks for thermal management using n-eicosane as phase change material. Appl. Therm. Eng. **132**, 52–66 (2018)
7. A. Arshad et al., Experimental investigation of PCM based round pin-fin heat sinks for thermal management of electronics: effect of pin-fin diameter. Int. J. Heat Mass Transf. **117**, 861–872 (2018)

Experimental and Numerical Heat Transfer Investigation of Impingement Jet Nozzle Position in Concave Double-Wall Cooling Structures

Edward Wright, Abdallah Ahmed, Yuying Yan, John Maltson, and Lynda Arisso Lopez

1 Introduction

This research is one aspect of a larger investigation into the optimisation of cooling air usage in a double-wall structure, with the aims being to maximise the internal heat transfer enhancement, whilst considering the required cooling air. Fundamentally, this enhancement of heat transfer can generally be achieved in one of three ways: decreasing the boundary layer thickness, increasing the boundary layer turbulence and decreasing the boundary layer viscosity. The viscosity is limited by the working fluid's properties and temperature, but the boundary layer thickness and turbulence are contributed to by several flow phenomena, and they are the focus of this study.

Jet impingement is a technique, whereby a jet of air is focussed through a nozzle, and onto a target surface whereby the jet then performs heat or mass transfer against that surface. This method allows for a very high coefficient of heat or mass transfer that can be adapted to modify the resultant boundary layer properties, therefore specifically achieving desired levels of transfer. Of particular interest in this paper, are the effects of a concave-type target surface on the effectiveness of a row of impinging jets, specifically in applications when cooling hot end turbine aerofoil components. Within a hollow turbine blade or nozzle, the impingement geometry can be produced using additive manufacturing (AM), or traditional casting and machining methods, possibly using an insert pushed into the aerofoil base to implement the impingement jet configuration. With the insert, modern AM techniques, of other methods, the impingement layout is easily modified; therefore, it is beneficial to decide upon the most effective design. On a flat plate, the orientation of the jet is somewhat limited

E. Wright (✉) · A. Ahmed · Y. Yan · L. A. Lopez
Faculty of Engineering, University of Nottingham, Nottingham NG7 2RD, UK
e-mail: edward.wright1@nottingham.ac.uk

J. Maltson
Siemens Industrial Turbomachinery Ltd, Firth Road, Lincoln LN6 7AA, UK

© Springer Nature Singapore Pte Ltd. 2021   537
C. Wen and Y. Yan (eds.), *Advances in Heat Transfer and Thermal Engineering* ,
https://doi.org/10.1007/978-981-33-4765-6_93

to achieve optimum cooling, but with any sort of irregular concave surface; more options are presented.

Early investigation on heat transfer due to an impingement on a concave surface was conducted to gather Nusselt number data dependant on the jet Reynolds number and various geometry parameters [1]. With more recent investigation also further looking at the effect of the jet's location on the heat transfer in a symmetrical concave target section, this research was conducted to a maximum Re of 20,000 [2]. Whilst these papers give valuable insight into the heat transfer and flow phenomena, and develop increasing relevant models to those required by industry, they do not fully model the conditions seen within an advanced turbine aerofoil leading edge cooling system.

In this study, experimental and numerical techniques are utilised to investigate jet impingement cooling on a realistic leading edge region of a gas turbine blade geometry at a higher, more typical jet Reynolds number of up to 50,000. Experimental work relies primarily on data from a transient thermochromic liquid crystal (TLC) measurement technique at the target surface. To further investigate and understand the fluid flow and heat transfer within the geometry, and optimise cooling, a numerical study is conducted. The numerical study is conducted with the same geometry using commercial computational fluid dynamics (CFD) software. The study is carried out with a row of five circular jets varied between two target positions on the leading edge. A standard flat target plate impingement test is also experimentally conducted and compared against existing literature for method validation.

2 Methodology

A perspex model of an advanced, typical leading edge geometry has been developed in partnership with Siemens Industrial Turbomachinery Ltd. TLCs are painted on the inside of the jet target surface and then coated in a black backing to give good contrast. Heated air enters the test section when a fast action valve is activated. As in the real situation, when the air flows into the impingement jets, it turns a 90° angle towards the coated target plate. The impingement jet configuration used in this test is an in-line array of 5 nozzles with radiused profile as shown in Fig. 1. The jet temperatures are measured using calibrated thermocouples, whilst a high-resolution and high-speed camera is used to monitor the colour response from the TLCs over

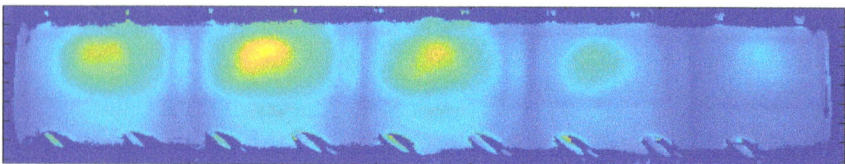

Fig. 1 Target Nusselt number distribution for jet location A

time. Total uncertainty for the Nusselt number based on the resolution of the camera and thermocouple error is calculated at just over ±8%. Pressure measurements are taken at the inlet to the test section, before the jet nozzles and after the jet nozzles.

As the primary aim of this study is to mimic typical industry idealised leading edge geometries, the dimensions cannot be expressed perfectly but the nondimensional parameters of the geometry are in the region of distance from nozzle to target $1.8 < H/D < 4.0$, pitch between jets $2.2 < P/D < 5.0$ and concave target diameter $0.0 < C/D < 8.0$. The jet Reynolds number during testing ranged from 10,000 to 70,000.

A transient TLC method is used to analyse the heat transfer coefficient of each test, with narrow band crystals from LCR Hallcrest being applied (R36C01W). This method is based on the assumption of a 1D conduction of the thermal pulse through semi-infinite plane of the 12-mm perspex target surface. These LCRs are calibrated through perspex to negate any effects of distortion or refraction during testing. The processing of the data is performed using a MATLAB script, which also uses a subtraction of the first reference image, from every subsequent image during testing, and this removes any 'zero' imaging errors. The heat transfer coefficients are measured by timing the hue reaction of the TLCs against the time of this reaction, to further reduce error, and a set hue value must be achieved for two consecutive frames in an effort to reduce any error caused by camera noise, or any other external influences such as fast-moving dust particles.

The experimental set-up for this study will be conducted with two variations of jet location (A and B), these are chosen because they impact the target surface in different ways and represent two possible primary variations in the aerofoil design, and the relative locations of A and B can be seen in Fig. 3.

A CFD analysis using Ansys CFX with SST turbulence model was conducted and validated against the experimental data for target surface Nu. These results are used to further analyse the flow and eddies within the leading edge geometry based on the jet's target location.

3 Results

The results obtained are validated against both Chupp et al. [1] and Cheng et al. [3] showing good agreement of Nu(s)/Nu(s = 0) for a row of jets impinging on a concave target surface.

To assess the effectiveness of jet location on the heat transfer coefficient, Nusselt number distributions across the surface were generated across the parameter ranges of the test. An example of these results is shown for two jet location variations (A and B) at a jet Reynolds number of 22,000.

The experimental Nusselt number distributions for the two variations of jet location are shown in Figs. 1 and 2. Some regions close to the sidewall appearing with a very low Nusselt number (dark blue colour) are caused by a lack of TLC as they do not reach the transition temperature within the time limits of the experiment. A higher jet temperature could improve the colour response in this region, but this would cause

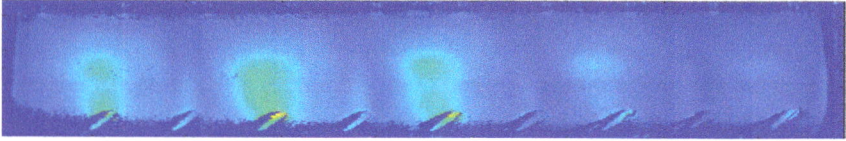

Fig. 2 Target Nusselt number distribution for jet location B

Fig. 3 Average jet centreline
velocity distribution with
indicated jet locations

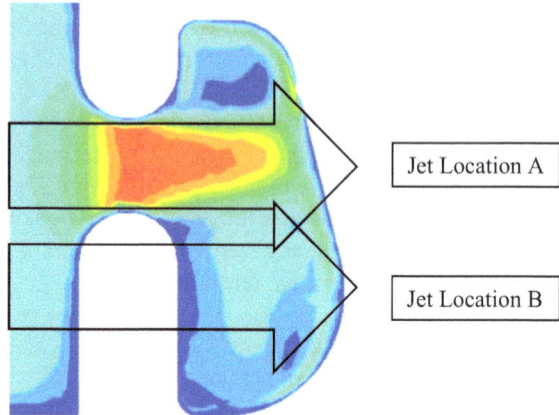

a faster response in the stagnation area and therefore lead to increased error in this region. As the stagnation region has by far the highest heat transfer coefficient and is therefore of particular interest to this study, it is chosen to concentrate here to the detriment of data at the sidewall region.

Both jet locations show similar Nusselt number distributions but with a larger magnitude of Nu number seen especially in the stagnation zones of jet location A. The higher Nu values seen in jet location A are contributed to by the slightly reduced H/D value compared to when the jet is in location B, but the contribution from each parameter be judged by varying the H/D and P/D values. To that end, correlations containing the specific target curvature effect on Nu can be produced. Based on the Nu distribution results, further analysis and variations can easily be validated and tested with numerical modelling.

With validation from the Nu0/NuAve experimental data and results from the literature, the numerical model is then built so that the flow phenomena related to each jet location can be observed.

From the CFD analysis, a side view of the average jet velocity can be seen in Fig. 3, and this also notes the two primary jet location variations. As expected, the highest velocity is seen after the jet nozzle, which then stagnates on the target surface, and the flow then turns 90 degrees and forms a large eddy throughout the concave cavity.

4   Conclusions

The current experimental and numerical investigation is conducted into the flow and heat transfer characteristics of a typical turbine aerofoil geometry. TLC is used to gather a distribution of Nusselt number over the entire target surface, and this is then compared against numerical data using commercial CFD techniques, allowing for further analysis of the internal flows. The primary focus of this testing is to analyse the effect of the jet's target location relative to the concave leading edge wall of a cooled gas turbine aerofoil. The following conclusions are made:

1. At these conditions, the location of highest Nusselt number is seen in the stagnation point regardless of its location relative to the target geometry. Nusselt number distributions show that the nozzle position does however modify the distribution in the wall jet regions.
2. Cross-flow effects between the nozzles and target diminish the heat transfer effectiveness of each subsequent jet. Some effect also experienced the inlet of the first jet nozzle, where there is an increased cross-flow, causing a pressure reduction in the flow direction in the inlet manifold.
3. Results show that the Nusselt number distribution is similar between the two jet locations with no notable differences in the distribution other than with a higher stagnation Nusselt number is seen in jet location A, primarily driven by the optimised H/D.

Acknowledgements Thank you to Siemens Industrial Turbomachinery Ltd for continued support and guidance in this project.

References

1. R.E. Chupp, H.E. Helms, P.W. McFadden, T.R. Brown, Evaluation of internal heat-transfer coefficients for impingement-cooled turbine airfoils. J. Aircr. **6**(3), 202–208 (1969)
2. L. Liu, X. Zhu, H. Liu, Z. Du, Effect of tangential jet impingement on blade leading edge impingement heat transfer. Appl. Therm. Eng. **130**, 1380–1390 (2018)
3. J.-R. Cheng, B.-G. Wang, Experimental investigation of heat transfer by a single and a triple-row round jets impinging on semi-cylindrical concave surfaces, in *Heat Transfer Conference* (Munich, 1982)

Critical Review and Ranking of Novel Solutions for Thermal Management in Electric Vehicles

Marco Bernagozzi, Anastasios Georgoulas, Nicolas Miché, Cedric Rouaud, and Marco Marengo

1 Introduction

We are at the dawn of a new era for passenger transportation, where internal combustion engine (ICE)-powered cars are bound to be replaced by electric vehicles (EVs). Latest awakening by different governments worldwide regarding global warming has encouraged the development and adoption of novel solutions to reduce CO_2 emissions. In particular, the British government set the end of the sale of conventional ICE cars and vans by 2040 [1].

However, although progress has been made in recent years, EVs still represent less than 1% of the global fleet for passenger cars. The most prominent barriers preventing the customers to purchase an EV are, in order of importance: limited all-electric range (confirmed by the range anxiety phenomenon); elevated cost; limited number of charging infrastructure; and prohibitive charging time [2].

At the moment, research on novel battery technologies is still at early stages; hence, the performance of Li-ion batteries needs to be improved. A critical factor influencing the performance, operation and safety of this type of battery is temperature. Due to limitation in the Li-ion intercalation process at low temperatures and the dissolving of the solid electrolyte interphase at high temperature, a critical decrease in capacity, number of cycles and power output have been observed. In fact, at $-20\,°C$ the available energy is 60% of the room-temperature value [3], whereas capacity losses of 36% were reported at 45 °C [4]. Moreover, a potentially destructive situation known as thermal runaway may arise as response to different fault scenarios, such as mechanical abuse (tearing of the separator), electrical abuse (overcharging)

M. Bernagozzi (✉) · A. Georgoulas · N. Miché · M. Marengo
School of Computing, Engineering and Mathematics, Advanced Engineering Centre, University of Brighton, Lewes Road, Brighton BN2 4GJ, UK
e-mail: MBernagozzi2@brighton.ac.uk

C. Rouaud
Ricardo Innovations, Shoreham-by-Sea BN43 5FG, UK

© Springer Nature Singapore Pte Ltd. 2021
C. Wen and Y. Yan (eds.), *Advances in Heat Transfer and Thermal Engineering* ,
https://doi.org/10.1007/978-981-33-4765-6_94

or thermal abuse (insufficient cooling), leading to a self-sustained reaction that can escalate into smoke and fire and eventually into an explosion.

As a consequence, there is a need to develop a smart thermal management system (TMS) and to push it towards a solution that is: efficient, light, consuming the least parasitic power possible (to increase the range of the vehicle); cheap, durable and simple (to reduce the cost of the vehicle); and able to work at high charging rates (to allow for shorter charging time).

The TMS of the battery needs to work on two levels, namely at cell level and module/pack level. The cell temperature must be maintained between 25 and 40 °C for best operative life and performance; at module/pack level, the temperature difference between cells must be maintained below 5 °C, as otherwise it creates an electrochemical imbalance, deteriorating the performance of the whole pack.

2   Thermal Management Solutions

Air cooling is the simplest and cheaper of the EV TMS solutions. It consists in a series of fans and intakes directing air from the surroundings and/or the cabin to the battery pack [5]. It can come in different design layouts such as series, where in the first case the same airstream encounters a single line of cells, and parallel, where the coolant firstly goes into a wedged distributor, gets split in equal branches and then exits in a collector. Albeit this solution outstands for its simplicity, its main limitations are threefold: firstly, air-scarce thermal properties and temperature rise along the flow path fail to maintain the two-level requirements of an EV TMS; secondly, the need of a minimum spacing between the cells, to allow for sufficient air mass flow rate, reduces significantly the energy and power density, leading to a bulkier battery pack; finally, thanks to the fans, it consumes the most parasitic power.

Liquid is the most powerful solution in terms of maximum temperature reduction and uniformity. There are two types of liquid TMS: direct and indirect. In the first one, the coolant is in actual contact with the battery pack, where it flows in a plenum and is then distributed in channels sandwiched between cells. This strategy requires dielectric working fluids that are usually expensive and have poor thermal properties. The different types of vector for indirect cooling are: jacket, which is a coolant-filled casing surrounding the module; tube, where the coolant is controlled in a bent tube travelling through the cells; cold plate, which is a thin-walled metal pressing with inbuilt channels; minichannels, where several small diameter channels are dug into metallic plates that are sandwiched in between the cells; and refrigerant cooling, where a refrigeration cycle is used [6]. The drawbacks of this technology are the added complexity given by additional moving parts such as expansion tanks, valves, pumps, compressors and so on. This reflects in a cost and weight increase, hindering the car price competeveness and performance.

Amongst the technologies not yet employed in EVs, solid-to-liquid phase change material (PCM) is the most investigated choice. Here, heat is removed due to sensible heat in the solid; then due to latent heat during the phase transition from solid to liquid, then the heat stored in the phase change process is released when the cell temperature decreases, improving hence temperature homogeneity. The most investigated PCM is paraffin wax due to high latent heat (179 kJ/kg) and low melting temperature (26–28 °C), as this should be chosen in the desired battery pack operative range. However, paraffin flammability represents an issue, but there are viable alternatives with comparable physical properties [7]. Thickness is also a very important parameter, as increasing the mass of the PCM increases the amount of stored heat but at the same time it increases weight and cost, hence reducing the overall EV performance. Moreover, the volume change associated with the phase transition needs to be considered, as it leads to poor mechanical properties. It is a passive technology, where no moving parts or additional electrical power is required to remove heat from the cell surface. However, it is not a fully independent solution, as when full liquefaction happens, an additional source of heat removal is needed to prevent an uncontrolled temperature increase. PCMs are also limited by their low values of thermal conductivity (<1 W/m K), which slows the response to high C-rate applications. Nevertheless, improvement methods such as the addition of conductive nanoparticles, metal fins or the use of porous material can potentially increase the thermal conductivity by two orders of magnitude.

Boiling immersion is a form of two-phase direct TMS. As in the case of PCM, it does not need moving parts or additional electrical power to remove heat from the cell surface. The most used working fluid is NOVEC7000©, with a boiling temperature of 34 °C and environmentally friendly, dielectric and non-flammable characteristic. The most important design parameters are working fluid selection in terms of latent heat, boiling point and physical properties, but also immersion percentage, which will influence the boiling performance but also the weight and the cost of the system [8]. The system pressure is also a crucial parameter, as the pressure will change the boiling point of the fluid hence allowing to operate within the nucleate boiling regime, which is the most favourable operating condition where heat is removed not only by the bubble formation (latent heat) but also due to the circulation motion of the liquid filling the void created by the detaching bubbles. However, this technology is generally difficult to control and purchasing large volume of fluid is critically expensive.

Heat pipes (HPs) have been investigated but not yet employed for EV TMS, despite their popular use for electronics cooling (laptops and smartphones). They are a passive technology where the fluid motion is ensured by capillarity and/or gravity and heat is removed due to liquid-to-vapour phase change. While the term HP is generally referring to an entire family of devices, there exist various types of HPs like standard sintered heat pipes, loop heat pipes (LHPs) and pulsating heat pipes (PHPs). In sintered HPs, a porous structure (wick) runs along the whole length of the device, being responsible for the return of the condensate from the condenser to the evaporator [9]. LHPs have the wicked structure in the evaporator only, allowing for heat transportation over longer distances at reduced costs (as the wick is the most expensive part) [10]. PHPs are a partially filled meandered tube characterized by a lower than critical inner diameter; hence, the fluid is arranged in an alternation of liquid slugs and vapour plugs, the latter being responsible of the typical pulsating motion due to their expansion and contraction at the evaporator and condenser, respectively [11]. The most important design parameters are the working fluid selection, the solid material selection and the evaporator/condenser configuration. Despite being passive devices, they cannot be applied in a totally independent manner for TMS in EVs and, in the case of large dimensions sintered HPs, the cost is critical.

3 Ranking and Decision Matrix

In order to identify the optimum solution for an efficient TMS, the critical factors were identified: design simplicity, accounting for the ease of integration and the adaptability to different configurations; independency, considering if an additional cold source may be needed as well as the dependency from environmental conditions; possibility to be used for heating; heat transfer efficiency; maximum temperature reduction capability; degree of achieved temperature homogeneity; operational life; cost; parasitic power, which will influence all-electric range; and safety. The grades used went from 3 being the most beneficial to 1 being the least beneficial.

Albeit two-phase passive devices seem to be the most promising, none of the technologies comply 100% with the requirements. Hence, it could be inferred that a combination of these solutions might lead to a better performing TMS. In Fig. 1, examples on how combining different TMS solution, it is possible to improve the competitiveness of the selected design. In fact, the outer ring symbolizes the maximum fulfilment of the requirements, where the inner ring means the inability to fulfil the requirement. Hence, in all the six proposed combinations, the outer ring is more populated than in case of the single solution.

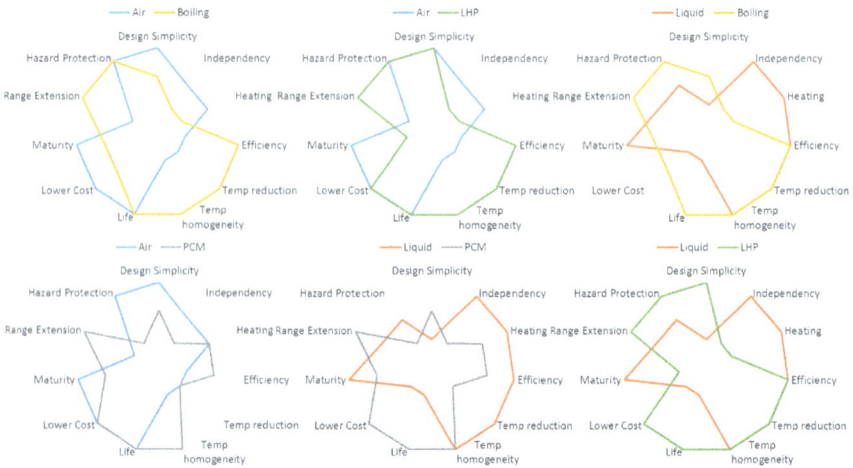

Fig. 1 Examples of how the decision matrix has been used to rate the different hybrid combinations of TMS

From Fig. 1, the most promising solution seems to be the bottom right one, where liquid cooling is combined with LHPs. In this case, having LHPs as heat vector from the module/pack to a liquid-cooled cold plate could allow to reach the maximum grade in all the considered factors (the outer ring represents the best condition). The LHP can be exploited as heat removal vector that takes the heat from the battery pack, and thanks to its outstanding long-length transportation properties, it could be connected to a remote condenser. This condenser is envisaged to be embedded in a heat exchanger with liquid cooling on the other side. This solution allows to reduce the amount of liquid coolant needed, as it will be confined only to cool down the condenser only, not the entire battery pack. This leads to choosing the working fluid of the liquid loop disregarding its dielectric properties, hence with a much better thermally performing refrigerant.

Summarizing, there is a strong need for an efficient and cost-effective TMS system to improve the spreading of EVs globally. None of the technologies so far envisaged can alone meet all the requirements needed to improve significantly the EV competitiveness; hence, a combination might be the best strategy to follow, i.e. a hybrid TMS.

References

1. UK Department for Transport, Air quality plan for nitrogen dioxide (NO2) in UK, (2017). doi:https://doi.org/10.1056/NEJMc1503870.
2. L. Noel, G. Zarazua de Rubens, B.K. Sovacool, J. Kester, Fear and loathing of electric vehicles: The reactionary rhetoric of range anxiety. Energy Res. Soc. Sci. **48**, 96–107 (2019). https://doi.org/10.1016/j.erss.2018.10.001
3. R. Bugga, M. Smart, J. Whitacre, W. West, Lithium ion batteries for space applications, IEEE Aerosp. Conf. Proc. (2007) 1–7. doi:https://doi.org/10.1109/AERO.2007.352728.
4. P. Ramadass, B. Haran, R. White, B.N. Popov, Capacity fade of Sony 18650 cells cycled at elevated temperatures. J. Power Sourc. **112**, 614–620 (2003). https://doi.org/10.1016/s0378-7753(02)00473-1
5. R.D. Jilte, R. Kumar, L. Ma, Thermal performance of a novel confined flow Li-ion battery module. Appl. Therm. Eng. **146**, 1–11 (2019). https://doi.org/10.1016/j.applthermaleng.2018.09.099
6. W. Wu, S. Wang, W. Wu, K. Chen, S. Hong, Y. Lai, A critical review of battery thermal performance and liquid based battery thermal management. Energy Convers. Manag. **182**, 262–281 (2019). https://doi.org/10.1016/j.enconman.2018.12.051
7. L. Ianniciello, P.H. Biwolé, P. Achard, Electric vehicles batteries thermal management systems employing phase change materials. J. Power Sourc. **378**, 383–403 (2018). https://doi.org/10.1016/j.jpowsour.2017.12.071
8. R.W. Van Gils, D. Danilov, P.H.L. Notten, M.F.M. Speetjens, H. Nijmeijer, Battery thermal management by boiling heat-transfer. Energy Convers. Manag. **79**, 9–17 (2014). https://doi.org/10.1016/j.enconman.2013.12.006
9. J. Smith, R. Singh, M. Hinterberger, M. Mochizuki, Battery thermal management system for electric vehicle using heat pipes. Int. J. Therm. Sci. **134**, 517–529 (2018). https://doi.org/10.1016/j.ijthermalsci.2018.08.022
10. M. Bernagozzi, S. Charmer, A. Georgoulas, I. Malavasi, N. Michè, M. Marengo, Lumped parameter network simulation of a Loop Heat Pipe for energy management systems in full

electric vehicles. Appl. Therm. Eng. **141**, 617–629 (2018). https://doi.org/10.1016/j.appltherm aleng.2018.06.013

11. J. Qu, C. Wang, X. Li, H. Wang, Heat transfer performance of flexible oscillating heat pipes for electric/hybrid-electric vehicle battery thermal management. Appl. Therm. Eng. **135**, 1–9 (2018)

System Simulation on the Refrigerant-Based Lithium-Ion Battery Thermal Management Technology

Qing Gao, Ming Shen, and Yan Wang

1 Introduction

With the rapid development of electric vehicles and the increasing performance requirements of lithium-ion power batteries, an efficient battery thermal management system is urgently presented. Refrigerant-based battery thermal management system, in which a battery evaporator is connected in parallel with the air-conditioning evaporator to introduce refrigerant directly into the battery pack for heat exchange, has been widely developed and applied to actual products. Xian et al. [1] and Tan et al. [2] elaborated the structure of refrigerant-based battery thermal management system. Park et al. [3] not only analyzed the effects of refrigerant temperature and mass flow on the cooling performance of refrigerant-based thermal management system, but also compared the cooling characteristics of refrigerant-based and phase change material thermal management under single discharge and cyclic operation conditions. The result showed that refrigerant-based thermal management system had better cooling performance. Heckenberg [4] studied the integration of the Mercedes-Benz S400 Blue Hybrid battery cooling system and air-conditioning system. Neudecker [5] simulated and confirmed the better temperature uniformity of the cold plate of BMW i3 battery.

From open patents to journal articles, most open literatures are conducted around the battery branch, and the cell or battery module is separately extracted. The thermal behavior of the battery under stable conditions are analyzed by the method of definite boundary conditions, and the systematic dynamic performance analysis is lacking.

Q. Gao (✉) · M. Shen · Y. Wang
State Key Laboratory of Automotive Simulation and Control, Jilin University, Changchun 130022, China
e-mail: gaoqingjlu@163.com; gqing@jlu.edu.cn

College of Automotive Engineering, Jilin University, Changchun 130022, China

© Springer Nature Singapore Pte Ltd. 2021
C. Wen and Y. Yan (eds.), *Advances in Heat Transfer and Thermal Engineering*,
https://doi.org/10.1007/978-981-33-4765-6_95

549

2 Methodology

Based on the analysis of the structure principle of electric vehicles, a vehicle system simulation platform was built by using LMS Imagine. Lab Amesim, which mainly involved drive control library, battery thermal management library, air-conditioning library, controller, battery pack and passenger cabin. In terms of temperature control, the PID controller was used to maintain the power of the compressor DC motor at the proper operating point. The controller read the temperature signal of the battery, determined the torque and speed of the compressor motor referring the ambient temperature and the vehicle speed information collected by the sensor, and then regulated the flow rate of the refrigerant, thereby reducing the deviation between the actual temperature and the predetermined value of the battery. The same was true for the control mode of the cabin. In the thermal management system, r-134a refrigerant was used as the cooling medium, and the simulation analysis was carried out with a cooling capacity of 6 kW. In the aspect of performance analysis, the default set temperatures of cabin and battery pack were 25 °C and 30 °C, respectively.

3 Results

The NEDC dynamic working condition was selected to carry out the simulation calculation. Considering that the battery pack is placed in the enclosed space of the chassis and close to the ground surface, the initial temperature of the battery pack is about 10 °C higher than the ambient temperature. The effect of battery thermal management system on cabin air-conditioning system was discussed. And the optimized control strategy was proposed to the coupling between the two systems.

Figure 1a shows the average temperature change curve of the battery pack and cabin at an ambient temperature of 35 °C under the NEDC conditions. Since the cabin and the battery are connected in parallel, the two will inevitably influence each other. The larger thermal mass of the battery cause its good temperature stability and

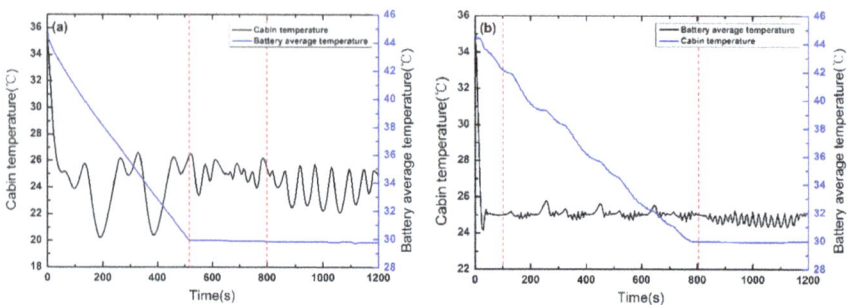

Fig. 1 Battery average temperature and cabin average temperature: **a** routine control and **b** optimal control

control robustness. Instead, the smaller thermal mass of the cabin bring about a large temperature fluctuation according to the load-based simultaneous thermal management control strategy. In order to avoid the interaction between two branches in parallel loop at the same time, a thermal control strategy with a certain priority was proposed. Specifically, it judged the load requirements of the battery pack and cabin, and the cabin branch was taken as a priority for thermal management. Meanwhile, the other branch valve was kept running at the minimum opening degree. And the judgment was made at a certain time interval and perform alternate cooling. The system would stop running until there is no thermal management demand in both branches. There was no bypass flow due to the change of control mode. In Fig. 1b, the cabin was first stabilized at 25 °C around 100 s, and then the battery temperature was gradually reduced to 30 °C. The later period of running range had a small temperature fluctuation due to the large change of the vehicle condition. But in general, the temperature of each component was well controlled. Inevitably, because of the priority control of the cabin, the battery temperature response time lagged, reaching the preset value by about 750 s.

With regard to the effect of the change in control strategy on the energy consumption of the system, since the same system components operated under the same initial conditions and driving conditions for 1200 s, the heat generation of the battery module was the same. The energy consumption of the system was judged from the viewpoint of the compressor (the power of the fan or the like was ignored). For the long-term dynamic condition, the energy consumption of the system was compared by integrating the power consumption, as shown in Eq. 1.

$$e = \int Pdt \qquad (1)$$

At the end of NEDC condition, the system's energy consumption increased by 7.83% under the proposed control strategy. This is due to the use of an optimized control strategy to make the number of branch cycles more frequent, resulting in an increase in the number of refrigerant migrations in the system and a slight increase in system energy consumption.

4 Conclusions

When both the battery evaporator and the cabin evaporator require thermal management, they inevitably interact with each other. Since the cabin has a smaller thermal mass than the battery pack, the controlled stability of the cabin temperature is poor when the external parameters change. The thermal management strategy based on cabin priority improves the temperature fluctuations of the cabin to a certain degree. But it also causes an increase in system energy consumption. This system simulation platform is the first step in the research of refrigerant-based battery thermal management system. Based on this model, the problem of energy consumption increase caused by optimizing system stability can be further explored, such as the dynamic sampling interval method. And the motor liquid cooling can be added, and a more comprehensive vehicle thermal management system is established.

Acknowledgements Thanks to the National Natural Science Foundation of China (51376080 and U1864213), Jilin Province's High-tech Industry Development Project (2016C022), Changchun Science and Technology Plan's Major Science and Technology Innovation Project (17SS022) and Jilin Province's Youth Talents Fund (No. 20180520066JH) for their research funding and support.

Reference

1. M. Xian, J Xie, H Liu, Cooling device for power lithium battery and cooling plate, China patent: 107482279 (2017)
2. H.S. Tan, H.J. Ye, X.F. Chen, Electric vehicle Thermal Management system and Electric vehicle, China patent: 205930310 (2017)
3. Seonggi, J. Dong soo, L. Dongchan et al., Simulation on cooling performance characteristics of a refrigerant-cooled active thermal management system for lithium ion batteries. Int. J. Heat Mass Transf. **135**, 131–141 (2019)
4. T. Heckenberg, *Li-ion Battery Cooling: More Than Just Another Cooling Task* (Technical Press Day, 2009)
5. B. Gohla-Neudecker, V.S. Maiyappan, S. Juraschek et al., Battery 2nd life: Presenting a benchmark stationary storage system as enabler for the global energy transition, in *International Conference on Clean Electrical Power* (IEEE, 2017)

Investigation on Battery Thermal Management Based on Phase Change Energy Storage Technology

Hongtao Gao, Yutong Liu, Jiaju Hong, Yuchao Song, and Yuying Yan

1 Introduction

Under the dual pressure of environmental pollution and energy crisis, electric vehicles have a tendency to replace fuel vehicles. As the only power source of electric vehicles, power batteries directly affect the performance of the whole vehicle, and the temperature is an important factor affecting the performance of the battery. Each battery type works better in a particular temperature range, for example, lead/acid, nickel metal hydride (NiMH) and lithium-ion (Li-ion) batteries operate best at temperatures between 25 and 40 °C, and at these temperatures, they achieve a good balance between performance and life [1]. The lifespan of the Li-ion cell, for example, is reduced by about two months for every degree of temperature rise in an operating temperature range of 30–40 °C [2]. The battery thermal management (BTM) is of significant importance to power battery.

Air-cooled heat dissipation is one of the most common and most traditional battery cooling methods. It includes natural convection and forced convection. Mahamud and Park [2] simplified the calculation of the influence of the reciprocating air flow on the temperature characteristics of the power battery array arranged in a two-dimensional model. It was found that the reciprocating air flow arrangement can reduce the internal temperature difference of the battery pack by 4 °C. Wang et al. [3] discussed factors that influence the cooling capability of forced air cooling by 3D CFD method.

H. Gao (✉) · Y. Liu · J. Hong · Y. Song
Institute of Refrigeration & Cryogenics Engineering, Dalian Maritime University, Dalian 116026, China
e-mail: gaohongtao@dlmu.edu.cn

Y. Yan
Fluids & Thermal Engineering Research Group, Faculty of Engineering, University of Nottingham, University Park, Nottingham NG7 2RD, UK

© Springer Nature Singapore Pte Ltd. 2021
C. Wen and Y. Yan (eds.), *Advances in Heat Transfer and Thermal Engineering* ,
https://doi.org/10.1007/978-981-33-4765-6_96

Liquid cooling has outstanding advantages in cooling capacity and temperature uniformity. However, as a non-contact indirect heat exchange system, liquid cooling system requires higher sealing and insulation of the system, which increases economic costs. Jarrett and Kim [4] using a numerical optimization algorithm incorporating a CFD analysis optimized the serpentine plate liquid cooling channel, changed the runner position and optimized its width. It was found that the optimum design for lowest average temperature is almost identical to that for lowest pressure drop. The design for temperature uniformity has a narrow inlet channel widening toward the outlet. Xu and Zhao [5] conducted an experimental study on the battery cooling system using liquid-cooled plates and analyzed the effects of flow rate, water-cooled plate flow path and external ambient temperature on the performance of the liquid-cooled system.

Many researches focused on the performance of BTM's heat dissipation at the excessive ambient temperature, since the thermal runaway will occur under extreme condition. However, when the ambient temperature is low, the energy density and power densities of battery decrease sharply [6]. Consequently, the demand of temperature maintaining in low-temperature environment must be reached for BTM. The thermal performance of BTM in low-temperature environment is divided into two parts, the heat preservation and heating. The phase change material (PCM) has received more and more attention in temperature maintaining [7]. Khateeb et al. [8] tried to design and simulate a lithium-ion battery with a PCM thermal management system for an electric scooter, and their results demonstrated the feasibility of a well-designed PCM system as a simple, cost-effective thermal management solution for Li-ion batteries in all applications including hybrid electric vehicle. Kizilel et al. [9] investigated small format cells surrounded by a wax composite material of expanded graphite and wax, showing that the latent heat of the wax helped prevent a propagating failure, where passive air cooling was unable to prevent propagation. Li et al. [10] conducted experimental investigation of a passive thermal management system for high-powered lithium-ion batteries using porous metal foam saturated with PCM. They found that the use of pure PCM can dramatically reduce the surface temperature and maintain the temperature within an allowable range due to the latent heat absorption and the natural convection of the melted PCM during the melting process. The foam-paraffin composite further reduced the battery's surface temperature and improved the uniformity of the temperature distribution caused by the improvement of the effective thermal conductivity.

In this paper, PCM thermal management for one cell and one pack consisting of five cells are constructed, respectively, to fully illustrate the effect of PCM on the power battery. Capric acid is chosen as the PCM. For the single cell, PCM is arranged around the cell. For the cell pack, sandwiched PCM and cell system is built. The condition of high discharge rate is selected. The change of thermal parameter before and after phase change process is also considered. The impact of natural convection is considered. Battery cells are constructed as substitute volumetric heat source. The effects of PCM on battery thermal behavior including effective thermal management time and temperature uniformity were investigated quantitatively. The melting process of PCM was analyzed meanwhile. The difference of whether considering the change of thermal parameters of PCM or not was illustrated quantitatively.

2 Methodology

Commercial CFD software STAR-CCM+ is used for simulation. Surface wrapper is used for the generation of surface mesh. Trimmed mesh is used for the generation of volume mesh. Interfaces are created between different regions. The Euler multiphase flow model is applied for simulating the existence of multiple phases, gravity model for loading different gravity directions, VOF model for simulating the existence and movement of interface between two phases, laminar flow model for simulating the flow state of liquid PCM, separate multiphase temperature model for simulating heat transfer in a phase change process. Enthalpy method is used to simulate the phase change process. The model assumes the following:

1. Liquid PCM is Newton incompressible fluid and laminar flow.
2. The thermal resistance between battery cells and PCM is negligible.
3. The effect of PCM expansion rate is negligible.
4. Convective heat transfer and radiation inside the battery are ignored.
5. The battery is a uniform volume heat source, generating the same amount of heat in all directions.
6. The density, specific heat capacity and thermal conductivity of the battery do not change with temperature.

The governing equations in the phase change process are as follows:

$$\frac{\partial}{\partial t}(\rho H) = \nabla \cdot (k \nabla T) \tag{1}$$

$$H = h + \Delta H \tag{2}$$

$$h = H_{\text{ref}} + \int_{T_{\text{ref}}}^{T} C_{\text{p}} \mathrm{d}T \tag{3}$$

$$\Delta H = \beta L \tag{4}$$

where β refers to the liquid volume fraction of PCM, H refers to enthalpy of PCM, k refers to the thermal conductivity of PCM, ρ refers to the density of PCM, L refers to latent heat of PCM.

3 Results

3.1 Single Battery Cell

For the single battery cell, due to the existence of phase change materials, the temperature rise curve of cell is slower than that of natural convection after melting begins, and the time of maximum temperature reaching 50 °C is 18.67% (Fig. 1) longer than the natural convection with maximum temperature difference of 4.8 °C at the time of 1000 s (Fig. 2).

The slope of curves of minimum temperature changes obviously at the time of about 200, 700, 1000 s. At about 200 s, a small amount of liquid PCM appears in the vicinity of interface between PCM and cell, the interface of solid–liquid phase moves

Fig. 1 Comparison of maximum temperature

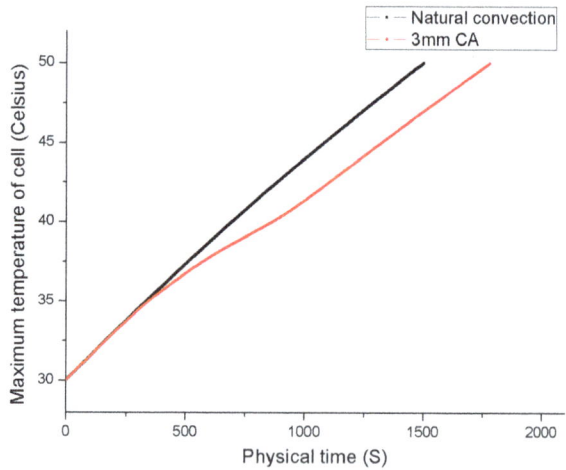

Fig. 2 Maximum temperature difference with PCM

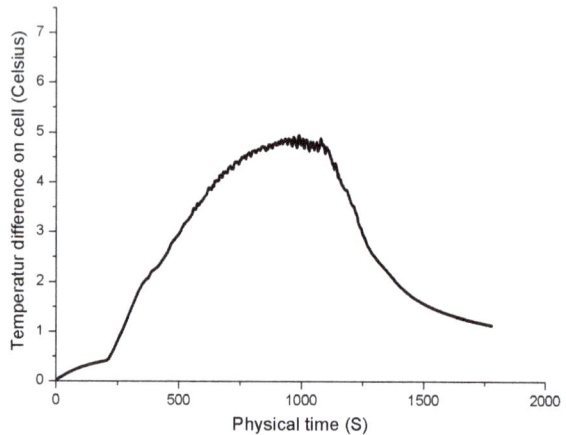

(a) Temperature distribution at t=207.5s

(b) Temperature distribution at t=720s

(c) Temperature distribution at t=1035s

(d) Temperature distribution at t=1220s

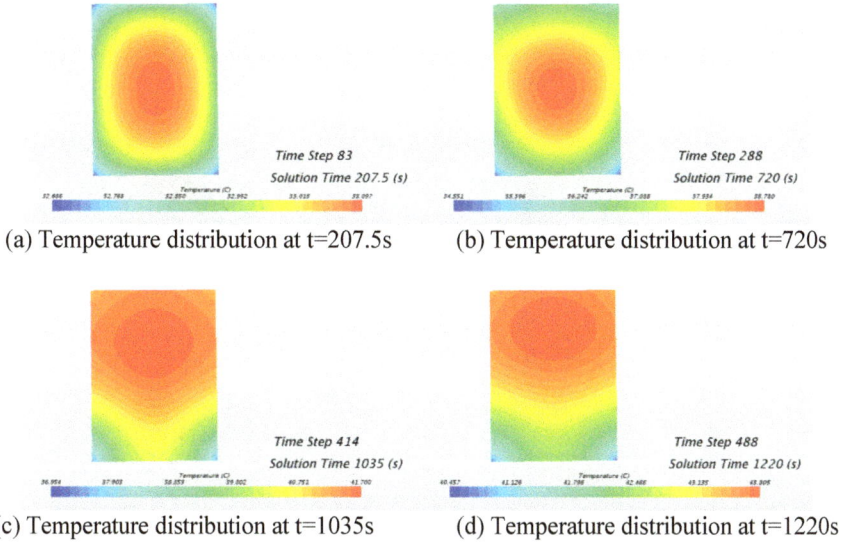

Fig. 3 Temperature change of single cell while PCM is melting

along the heat flux direction at the beginning than moving toward the outer lower direction because of the temperature difference and density difference between solid phase PCM and liquid phase PCM. At about 700 s, the top region is almost completely filled with liquid phase change material, and the solid–liquid phase interface moves outward in the downward direction. At about 1000 s, the PCM in the entire area is almost completely melted, and all become liquid. The temperature change of single cell while PCM is melting is shown in Fig. 3.

One of the future development directions of the automobile is lightweight. Considering the volume and economy of the battery pack system, for the battery thermal management system of the electric vehicle, the rational use of the phase change material is critical for the compactness of the battery thermal management system. So, the effect of different thickness tantalum domains of 3, 5 and 7 mm on the temperature distribution of the battery is checked. Figure 4 depicts the battery heat output with a heat output of 63,970 W/m^3 as a function of time at different thicknesses of tantalum. When the thickness of the PCM is 3 mm, the effective thermal management time of the battery cell is 1780s, which is 18.67% longer than the natural convection cooling time; when the PCM thickness is 5 mm, the effective thermal management time is 1890s, increased by 26%; when the PCM thickness is 7 mm, the effective thermal management time is 2020s, 34.67% longer than the natural convection cooling time. Figure 5 shows the effect of PCM with different thickness on the temperature difference of the battery. As shown, PCMs of different thickness have a significant effect on the temperature difference of the battery cells. This is because during the melting process, when the PCM at the proximal end of the heated surface gradually becomes liquid, under the effect of the temperature difference and the difference in density

Fig. 4 Effect of thickness
on maximum temperature

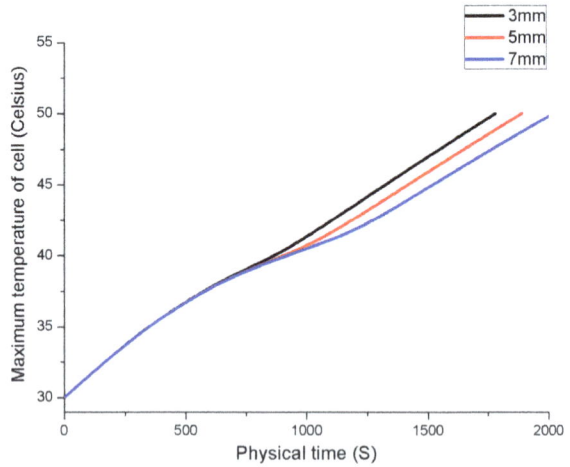

Fig. 5 Effect of thickness
on temperature difference

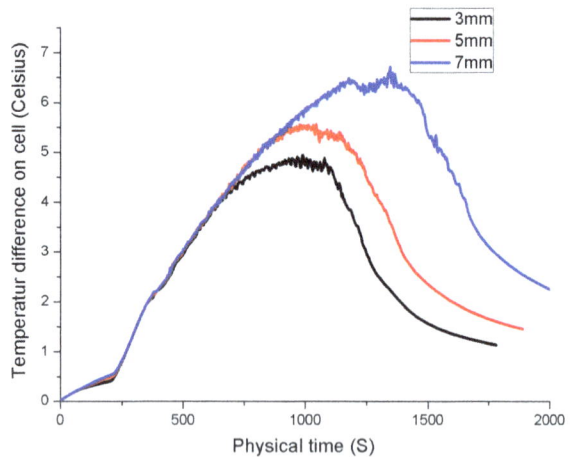

caused by the change of the physical properties of PCM before and after phase change, the upper part is gradually filled with liquid phase. The more the thickness of the PCM, the slower the two-phase interface moves down.

3.2 Battery Pack

The battery modules without PCM are modeled in STAR-CCM + with a spacing of 2 mm between each cell, and the geometry model is shown in Fig. 6. The temperature distribution of the battery pack is not a simple superposition of the temperature distribution of the single cell. The temperature distribution is generally characterized

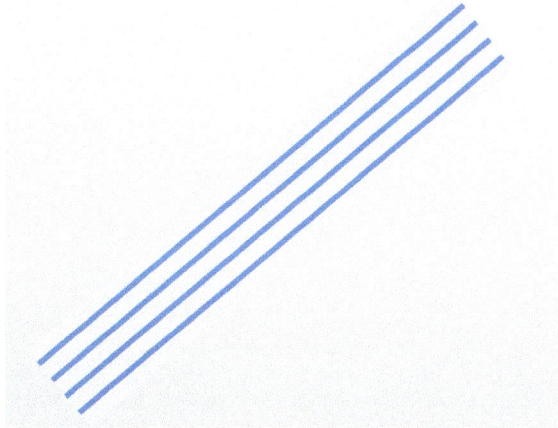

Fig. 6 Battery module without PCM

by a higher central temperature and a lower ambient temperature, while the temperatures of upper and lower surfaces are slightly lower due to the large convective heat transfer area, and the temperature gradient of the whole pack is about 2 °C (Fig. 7).

Figure 8 shows the thermal management system geometry model with PCM, in which white is the battery domain, blue is the PCM domain, and the whole is composed of five battery domains and six PCM domains. Wherein, the thickness of the four PCM domains in the middle part is 2 mm, and the thickness of the outer two PCM domains is 1 mm.

Figure 9 shows the maximum and minimum temperature rise curves in the battery pack. For the battery pack, compared to natural convection, PCM thermal management system can lengthen the time of maximum temperature reaching 50 °C twice. Before 300 s, the battery temperature rises, and the heat generated by cell during the discharge process is absorbed by the sensible heat of the PCM. From 300 to 2200 s,

Time Step 301

Solution Time 1505 (s)

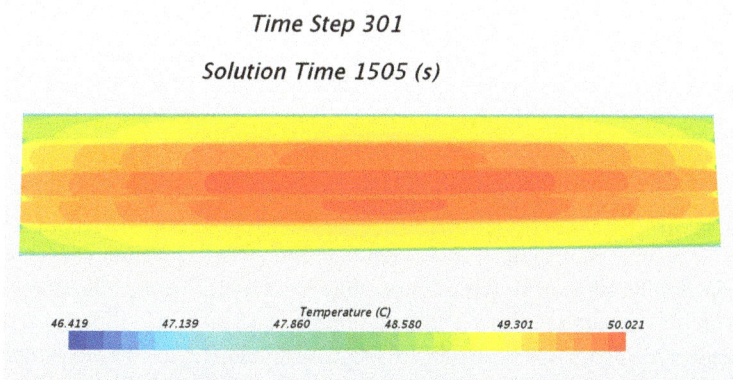

Temperature (C)

| 46.419 | 47.139 | 47.860 | 48.580 | 49.301 | 50.021 |

Fig. 7 Overall temperature distribution of YOZ section

Fig. 8 Battery module with
PCM

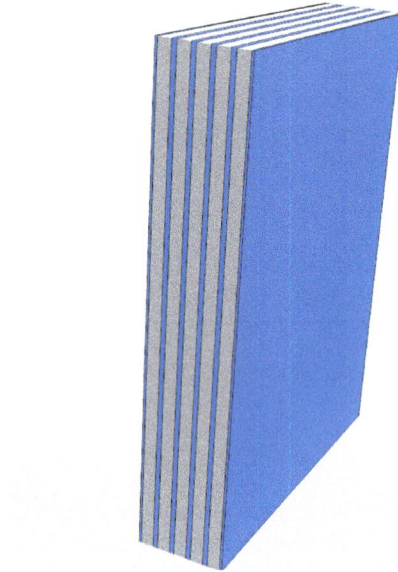

Fig. 9 Maximum and
minimum temperature profile

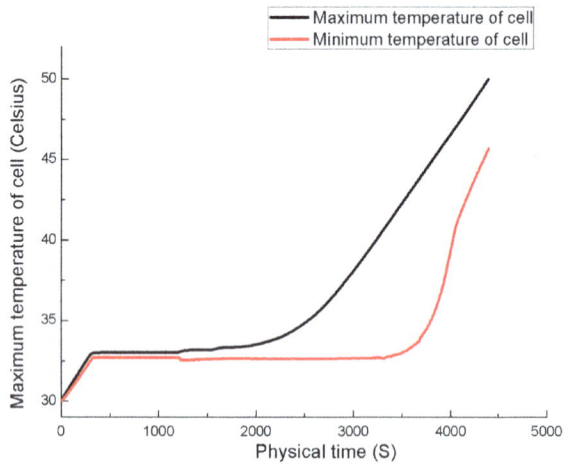

the heat generated by the battery is absorbed by the latent heat of the PCM, and
the maximum temperature of the battery pack and the minimum temperature remain
basically constant. At around 2200 s, the maximum battery temperature suddenly
increased rapidly, but the minimum temperature remained the same. At around 3700 s,
the minimum temperature curve suddenly jumped and increased rapidly. When the
temperature difference curve reaches its critical point, the maximum temperature
difference is about 10 °C. During time interval from 2950 to 4240 s, it is in the
thermal management failure status, accounting for about 43% of the total physical
time.

Fig. 10 Effect of different gravity directions on battery temperature difference

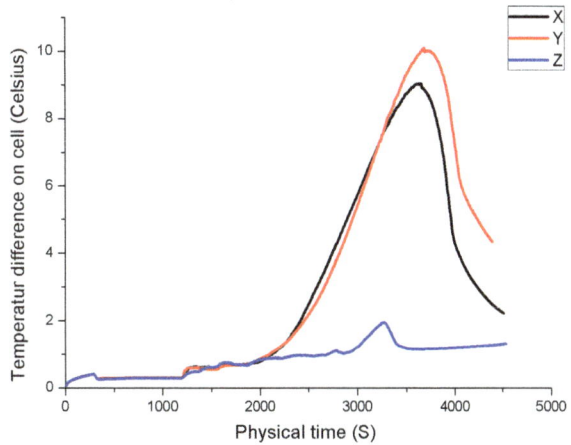

Fig. 10 Effect of different gravity directions on battery temperature difference

The direction of gravity, the way of placement of system has great influence on the cell temperature uniformity as shown in Fig. 10. When the gravity direction loads on X negative half-axis direction, the biggest temperature difference is about 10 °C. When loading on Y negative half-axis direction, the biggest temperature difference is about 9 °C. The thermal management failure time accounts for half of the total time. When loading on Z negative half-axis direction, the biggest temperature difference is about 2 °C.

4 Conclusions

The PCM thermal management system models have been established for single cell and cell pack, respectively. Based on enthalpy method, the melting process of PCM has been simulated, and its impacts on the thermal behavior of cells have been quantitatively analyzed.

1. Natural convection is the main heat transfer mode inside the PCM after melting begins. The density difference and temperature difference are the main factors affecting the solid–liquid interface distribution in the domain and hence determine the temperature distribution of cell.
2. For a single cell with PCM, the temperature distribution of the cell will vary greatly due to the state of the PCM. As the thickness of PCM increases, the effective thermal management time increases linearly.
3. For a cell pack with PCM, the gravity load along the direction of shortest geometric length direction will bring the best thermal behavior for the cell pack. And PCM system can strongly lower the battery temperature rise rate.

Acknowledgements This work was financially supported by Research funds of the Maritime Safety Administration of the People's Republic of China (2012_27) and the Fundamental Research Funds for the Central Universities (3132019305).

References

1. A.A. Pesaran, Battery thermal models for hybrid vehicle simulations. J. Power Sourc. **110**, 377–382 (2002)
2. R. Mahamud, C. Park, Reciprocating air flow for Li-ion battery thermal management to improve temperature uniformity. J. Power Sourc. **196**, 5685–5696 (2011)
3. T. Wang, K.J. Tseng, J. Zhao, Z. Wei, Thermal investigation of lithium-ion battery module with different cell arrangement structures and forced air-cooling strategies. Appl. Energy **134**, 229–238 (2014)
4. A. Jarrett, I.Y. Kim, Design optimization of electric vehicle battery cooling plates for thermal performance. J. Power Sourc. **196**, 10359–10368 (2011)
5. X. Xu, Y. Zhao. An experimental study on the heat dissipation performance of cooling-plate liquid cooling system for electric vehicle in different conditions. Automot. Eng. **36**, 1057–1062,1092 (2014)
6. S.S. Zhang, K. Xu, T.R. Jow, The low temperature performance of Li-ion batteries. J. Power Sourc. **115**, 137–140 (2003)
7. Y. Huo, Z. Rao, Investigation of phase change material based battery thermal management at cold temperature using lattice Boltzmann method. Energy Convers. Manage. **133**, 204–215 (2017)
8. S.A. Khateeb, M.M. Farid, J.R. Selman, S. Al-Hallaj, Design and simulation of a lithium-ion battery with a phase change material thermal management system for an electric scooter. J. Power Sourc. **128**, 292–307 (2004)
9. R. Kizilel, R. Sabbah, J.R. Selman, S. Al-Hallaj, An alternative cooling system to enhance the safety of Li-ion battery packs. J. Power Sourc. **194**, 1105–1112 (2009)
10. W.Q. Li, Z.G. Qu, Y.L. He, Y.B. Tao, Experimental study of a passive thermal management system for high-powered lithium ion batteries using porous metal foam saturated with phase change materials. J. Power Sourc. **255**, 9–15 (2014)

Design and Simulation of the Thermal Management System for 5G Mobile Phones

Zhaoshu Chen, Yong Li, Wenjie Zhou, and Yuying Yan

1 Introduction

With the rapid development of microelectronics and communication technology, smartphone has changed from simple communication tools to multi-functional portable devices integrating office, entertainment, payment, and health. The increasing power consumption of the system makes the heat dissipation problem become the bottleneck restricting the development of smartphone. Cengel [1] proposed a rule of thumb that the failure rate of electronic components will increases by 50% for every 10 °C rise in their operating temperature.

Ultra-thin heat pipes (UTHP) are widely concerned in the heat dissipation field of smartphone due to their advantages such as high thermal conductivity, light weight, thin thickness, and no need for additional energy drive. Chen et al. [2] designed a vapor–liquid channel-separated UTHP (0.4-mm-thick) fabricated via etching and diffusion bonding. The maximum heat transfer capacity and effective thermal conductivity of UTHP were 4.50 W and 12,000 W/(m K). Ahamed et al. [3] designed a sintered wick structure called "center fiber wick" and investigated the effects of the width, flattened thickness, and effective length of a UTHP on its thermal performance. The maximum heat transport capacity of a 0.4-mm UTHP was found to be only 1.5 W. Zhou et al. [4] designed a UTFP with a spiral woven mesh wick and tested its thermal performance, and the thickness and the maximum heat

Z. Chen · Y. Li (✉) · W. Zhou
School of Mechanical and Automotive Engineering, South China University of Technology, Guangzhou, China
e-mail: meliyong@scut.edu.cn

Y. Yan (✉)
Fluids & Thermal Engineering Research Group, Faculty of Engineering, University of Nottingham, Nottingham, UK
e-mail: yuying.yan@nottingham.ac.uk

Centre for Fluids & Thermal Engineering Research, University of Nottingham, Ningbo, China

© Springer Nature Singapore Pte Ltd. 2021
C. Wen and Y. Yan (eds.), *Advances in Heat Transfer and Thermal Engineering* ,
https://doi.org/10.1007/978-981-33-4765-6_97

transport capacity of the UTHP were 0.4 mm and 3.6 W, respectively. Singh et al. [5] designed a UTHP cooling module with a piezo fan and tested the thermal performances of UTHPs with thicknesses of 0.6–2 mm. As the UTHP thickness increased, the maximum heat transfer capacity of the UTHP also increased.

In summary, many researchers have studied the structure and thermal conductivity of UTHPs, but few have studied the heat dissipation effect of UTHPs in smart phone. In the present work, thermal simulation technology is adopted to analyze the heating performance of a 5G smartphones under different conditions, such as play game by 5G data, browser by 5G data, and 5G data transfer with max transmitted power. The cooling system of 5G smartphones with UTHP is analyzed, and the proposed cooling solution is promising for heat dissipation problems in 5G smartphones.

2 Methodology

ANSYS icepak 19.0 is used for thermal simulation; the smartphone model and heat source location are shown in Fig. 1a. In order to improve the grid precision and reduce the computational load, some features and elements that have less impact on heat dissipation in the model are removed. The simplified model is shown in Fig. 1b. The materials and thermal conductivity of the components of the smartphones are given in Table 1. The ambient temperature is 25 °C, and radiative heat transfer is considered. The effective thermal conductivity of UTVC is taken from the data in literature [3].

The geometry model was meshed with a hexahedral mesh using icepak mesh tool. The mesh type was Mesher-HD and allowing multi-level meshing. To assess the accuracy of computations, the grid independence test was conducted by adopting different grid distributions of $2 \times 2 \times 2$ mm^3, $2 \times 2 \times 1$ mm^3, $2 \times 2 \times 0.5$ mm^3 for the model. The grid independence test indicated that a grid system of $2 \times 2 \times 1$ can ensure a satisfactory solution for the smartphone analysis, and error of the maximum temperature in the smartphone was not greater than 1%.

3 Results

Figure 1c shows the temperature distribution of the back cover of the smartphone with and without UTHP when playing game by 5G data with charging. In this case, the heating power of the smartphone is the highest. It can be seen that when there is no UTHP, the temperature difference of the back cover of the smartphone is relatively large. This is because the back cover of the smartphone is glass with low thermal conductivity, which cannot effectively distribute heat evenly. When UTHP is used, the temperature of the heat source can be effectively reduced to prevent overheating. Figure 1d shows the maximum temperature of the smartphone under different conditions and the maximum temperature of the back cover of the

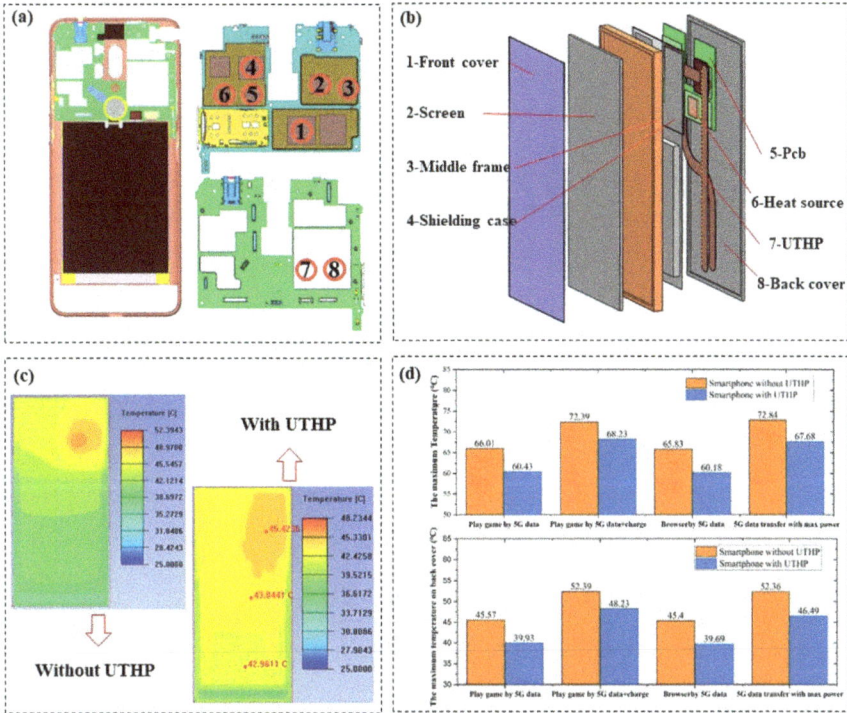

Fig. 1 **a** Original model of smartphone and heat source distribution; **b** explosion picture of simplified model of smartphone; **c** temperature distribution of the back cover in play game by 5G data with charge; **d** maximum temperature of the smartphone and the back cover under different conditions

smartphone. Under different conditions of use, the smartphone with UTHP can reduce the chip temperature by 4–6 °C and the back cover temperature by 5 °C.

4 Conclusions

In this paper, a cooling module based on UTHP is designed for a 5G smartphone, and the heating condition of the smartphone with or without UTHP is simulated in different conditions. With the increasing power consumption of smartphones, using phase change heat transfer components with high thermal conductivity can effectively reduce chip temperature and improve the uniformity of temperature.

Table 1 Parameters of the thermal simulation

Name	Material	Thermal conductivity W/(m·K)
1-Front cover	Glass	0.12
2-Screen	Liquid crystal display	0.12
3-Middle frame	Al 6063	201
4-Shielding case	Steel	36
5-Pcb		15
6-Heat source	Si	150
7-UTHP	0.4 mm-thick	10,000
8-Back cover	Glass	0.12

Situation (mW)	1	2	3	4	5	6	7	8	Total
Play game by 5G data	565	141	230	506	166	160	1370	228	3366
Play game by 5G data + charge	565	141	2030	506	166	160	1370	228	5166
Browser by 5G data	551	170	170	506	166	160	1370	228	3321
5G data transfer with max power	122	70	0	1818	1049	160	1370	228	4817

Acknowledgements This work was supported by the National Natural Science Foundation of China (grant number 51675185), the Natural Science Foundation of Guangdong Province (grant number 2018B030311043), the Project of the Guangzhou Science and Technology Plan (grant numbers 201807010074 and 201707010071), the Project of Tianhe District Science and Technology Plan (grant number 201705YX263), and the Fundamental Research Funds for the Central Universities, SCUT.

References

1. Y.A. Cengel, *Heat Transfer: A Practical Approach* (McGraw-Hill, New York, 2003).
2. Z. Chen, Y. Li, W. Zhou et al., Design, fabrication and thermal performance of a novel ultra-thin vapour chamber for cooling electronic devices. Energ Convers. Manage **187**, 221–231 (2019)
3. M.S. Ahamed, Y. Saito, K. Mashiko et al., Characterization of a high performance ultra-thin heat pipe cooling module for mobile hand held electronic devices. Heat Mass Transf. **53**, 3241–3247 (2017)
4. W. Zhou, Y. Li, Z. Chen et al., A novel ultra-thin flattened heat pipe with biporous spiral woven mesh wick for cooling electronic devices. Energ Convers. Manage **180**, 769–783 (2019)
5. R. Singh et al. Low profile cooling solutions for advanced packaging based on ultra-thin heat pipe and piezo fan, in *Cpmt Symposium Japan IEEE* (2014)

Investigation of Critical Overheating Behavior in Batteries for Thermal Safety Management

Mengdi Zhao, Qing Gao, Tianshi Zhang, and Yubin Liu

1   Introduction

The demands of dynamic performance, fast charge, lightweight, compact, energy efficiency for electric vehicles and the application of large capacity and high specific energy batteries make the safety of power batteries receive more attention. The critical identification and diagnosis become the key technology for battery thermal safety control. When exploring analytical methods for overheating diagnostic prediction, the temperature rise and venting events are mainly considered. By analyzing the identification of thermal safety control and the ability of thermal management system, the criticality of thermal safety prevention and control diagnosis is defined. Through analyzing the thermal and gas generation characteristics, the overheating controllable criticality is subdivided into three stages, including the initial, medium and late stage.

Currently, a lot of experimental research and theoretical analysis on the exothermic side reaction and gas production mechanism during power battery overheating have been conducted by domestic and foreign scholars. Roth et al. [1] conducted thermal test for 18,650 lithium-ion battery using accelerating rate calorimeter, the results showed that the battery thermal runaway starts from SEI decomposition, and the profiles can be characterized in three stages. Abraham et al. [2] explored the gas production performance on high-power 18,650 lithium-ion cells and found that the self-heating of the cell begin at 84 °C, and the gases generated in the cell included CO_2 and CO and smaller quantities of H_2, C_2H_4, CH_4 and C_2H_6.

The research work focuses on the characteristics of heat and gas production during overheating process as well as analyzes the characteristics of the critical control. Through the external heating method, the monitoring experiments of gas production

M. Zhao · Q. Gao (✉) · T. Zhang · Y. Liu
State Key Laboratory of Automotive Simulation and Control, Jilin University, Changchun 130022, China
e-mail: gqing@jlu.edu.cn

College of Automotive Engineering, Jilin University, Changchun 130022, China

© Springer Nature Singapore Pte Ltd. 2021
C. Wen and Y. Yan (eds.), *Advances in Heat Transfer and Thermal Engineering* ,
https://doi.org/10.1007/978-981-33-4765-6_98

and temperature change are carried out. Meanwhile, a thermal model is built to describe the heat production performance, which includes the main chemical reaction during the battery overheating process. By seeking powerful diagnosis, accurate prediction and sensor identification of overheating, the thermal safety prevention and control system will be more accurate, which can lay the theoretical foundation for the future thermal safety control and overheating prevention technology.

2 Methodology

Lithium-ion battery is a complex energy system, the heat generation and temperature changes of the battery follow energy balance equation [3], which is given by Eq. (1).

$$\frac{\partial\left(\rho c_{\mathrm{p}}T\right)}{\partial t} = \nabla \cdot \lambda \nabla T + q \tag{1}$$

where the expression on the left side of the equation represents the increment of thermodynamic energy of battery unit, the first item on the right side represents the energy increase of the battery unit due to heat conduction per unit time, the last item on the right side indicates battery heat source, which include the heat generated from Joule heating, electrochemical reactions that take place during the normal operating conditions and other chemical reactions during thermal runaway. The energy balance equation can be used to analyze the overheating process of power battery. Meanwhile, a specific commercial 18,650 lithium-ion battery is selected to conduct the heating experiments under three working conditions, respectively.

3 Results

Through analyzing the thermal and gas generation characteristics, the overheating controllable criticality is subdivided into three stages, including the initial, medium and late stage [4]. During the research of heat generation characteristics, the species of side reactions, the thermal reaction capacity and the temperature range during overheating are analyzed. The staged upward trend of heat generation in battery is found. Research shows that by taking SEI decomposition as the effective boundary, battery thermal safety management can be achieved. After separator melts, the strong cooling capacity is needed to suppress the temperature rise of battery. When the battery enters the third stage, emergency cooling measures are required to avoid battery thermal safety accidents. Generally, it is necessary to identify the typical characteristics in the early stage of overheating. Combined with the cooling capacity of thermal management system, a dynamic control boundary can be formed to prevent battery overheating. At the same time, the dynamic thermal control threshold is subdivided to construct the thermal basis for judgment.

Fig. 1 Temperature and exhaust signal in low heating process

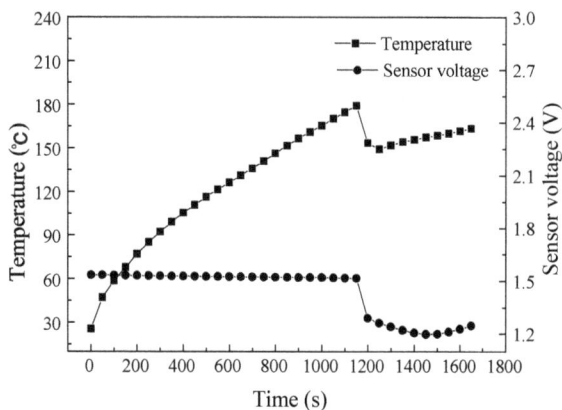

In the process of researching gas generation characteristics, by analyzing the mechanism, the variation of gas species and content, the staged increasing trend of gases is pointed out. Research shows that there will be two significant venting events during the overheating. The first venting event is due to safety vent opening, and the second venting event owing to the reaction between cathode and electrolyte. According to the venting events and the variation of gas generation, the overheating process can be subdivided. In addition, the temperature and exhaust signals in experiment are shown in Fig. 1. Through the battery overheating monitoring experiment, the basic rules of gas generation, venting events and temperature variation during the overheating process are verified. After safety vent opening, battery will exhaust a large amount of CO_2 and other gases, and the obvious signal can be detected. Gas monitoring will not only assist in overheating diagnosis for thermal runaway suppression and fire safety protection, but also provides the basis for implementing emergency cooling. The research work provides assistance for analyzing the common character of battery overheating, which also offers theoretical basis for feature recognition to protect battery from overheating, even prevent and suppress thermal runaway.

4 Conclusions

The research work analyzes the critical overheating behavior in batteries, a thermal model is built to describe the heat production performance, and the experiment for gas and temperature monitoring is conducted. The research indicates that the heat generation capacity of the side reactions is gradually increased during overheating, and it is helpful to understand the process of battery overheating and design the enhanced cooling system. Furthermore, as the heating load rises, the exhaust temperature and the temperature rise rate of the battery increase gradually, the voltage signals of the sensors under three conditions show a gradual downward trend, then rise after the sudden drop, and the decreasing amplitude of the exhaust signal slowly reduces.

Acknowledgements This work is supported by the National Natural Science Foundation of China (51376080 and U1864213), the "Double Ten" Science and Technology Innovation Project of Jilin Province of China (17SS022), Science Foundation for Excellent Youth Scholars of Jilin Province of China under the research grant of 20180520066JH and Jilin Development and Reform Committee under the research grant of 2016C022.

References

1. E.P. Roth, C.C. Crafts, D.H. Doughty, Advanced technology development program for lithium-ion batteries: thermal abuse performance of 18650 Li-ion cells, Office of Scientific & Technical Information Technical Reports (2004)
2. D.P. Abraham, E.P. Roth, R. Kostecki et al., Diagnostic examination of thermally abused high-power lithium-ion cells. J. Power Sourc. **161**(1), 648–657 (2006)
3. S. Santhanagopalan, P. Ramadass, J. Zhang, Analysis of internal short-circuit in a lithium ion cell. J. Power Sourc. **194**(1), 550–557 (2009)
4. B. Ravdel, K.M. Abraham, R. Gitzendanner et al., Thermal stability of lithium-ion battery electrolytes. J. Power Sourc. **119–121**, 805–810 (2003)

Experimental Study on Direct Expansion Cooling Battery Thermal Management System

Yuan Meng, Gao Qing, Zhang Tianshi, and Wang Guohua

1 Introduction

With the continuous development of electric vehicles and the increasing requirements of power performance, large specific energy and safety of power batteries, electric vehicles urgently need an efficient thermal management system. According to the classification of heat transfer media, power battery thermal management system can be roughly divided into air cooling, liquid cooling, phase-change cooling and direct expansion cooling [1]. At present, liquid-cooled battery thermal management system is mainly used in common electric vehicles on the market. The new type of direct expansion cooling thermal management system directly uses the evaporating plate to cool the battery system instead of the complex liquid pipeline in the traditional liquid cooling system. Compared with the traditional liquid cooling system, the system has lower quality and higher energy density and avoids the potential leakage of thermal liquid, thus greatly improving the safety of the battery system.

In this paper, a new type of direct expansion cooling battery thermal management system is established and relevant experiments are carried out.

2 Methodology

It is found that the performance of direct expansion cooling thermal management system is mainly related to the compressor and condenser. Based on the experimental bench of direct expansion cooling thermal management system, the influence characteristics of compressor speed and condenser fan speed in refrigeration cycle

Y. Meng · G. Qing (✉) · Z. Tianshi · W. Guohua
State Key Laboratory of Automotive Simulation and Control, Jilin University, Changchun, China
e-mail: gqing@jlu.edu.cn

College of Automotive Engineering, Jilin University, Changchun, China

© Springer Nature Singapore Pte Ltd. 2021
C. Wen and Y. Yan (eds.), *Advances in Heat Transfer and Thermal Engineering* ,
https://doi.org/10.1007/978-981-33-4765-6_99

were investigated. By adjusting the main control parameters of the direct expansion cooling system, the cooling performance of the direct expansion cooling system and the cooling characteristics of the power battery system in the direct expansion cooling process were studied from the perspectives of the compressor speed and the condenser fan speed.

3 Results

In the experiment, the compressor speed was increased from 1500 to 2500 r/min, while the condenser fan speed was increased from 50% to full speed (Fig. 1).

It is found in the experiment that the speed of the compressor will affect the temperature difference of the battery module. The higher the speed of the compressor, the greater the temperature difference of the battery module, especially at the bottom of the module. In the process of direct expansion cooling start-up, the surface temperature difference of the direct expansion cooling plate will fluctuate greatly under the influence of the starting characteristics of the refrigeration system [2] (Table 1).

The cooling capacity of the refrigeration system can be effectively increased by the speed increase of the compressor and the condenser fan, but the input power of the compressor in the refrigeration system will also increase correspondingly, thus affecting the COP of the refrigeration system. It was found that COP of refrigeration system decreased with the increase of compressor speed. The compressor speed increased from 1500 to 2500 r/min, and the COP decreased from 4.10 to 2.80. When the speed of condenser fan increases from 50 to 100%, COP increases from 2.49 to 2.73. At the same time, the energy efficiency ratio of the system decreases greatly with the increase of the input power of the condensing fan.

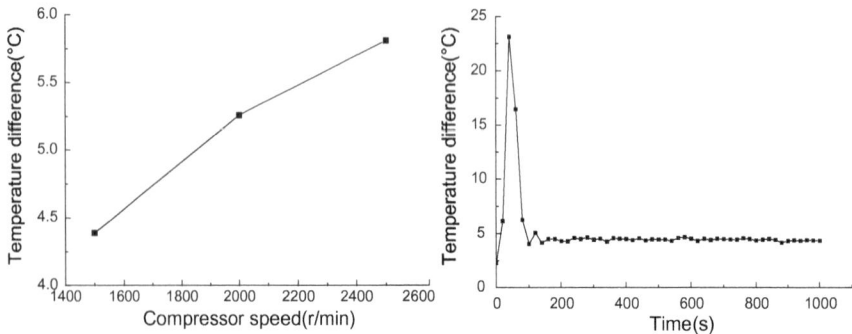

Fig. 1 Temperature difference of the battery module

Table 1 Performance parameters of the direct expansion cooling thermal management system

Working conditions		Performance parameters		
Compressor speed (r/min)	Condensing fan speed (% of full speed)	Refrigerating capacity (w)	COP	EER
1500	100	861	4.10	–
2000	100	1024	3.24	–
2500	100	1179	2.80	–
2500	50	1047	2.49	1.89
2500	66.7	1088	2.59	1.82
2500	83.3	1108	2.64	1.70
2500	100	1149	2.73	1.61

4 Conclusions

The refrigeration performance of the direct expansion cooling battery thermal management system is related to the compressor speed and the condenser fan speed. The higher the compressor speed, the more obvious the refrigeration effect; at the same time, improving the condenser fan speed is also beneficial to improve the cooling capacity of the system. In addition, changes in the compressor speed and the condenser fan speed will also cause changes in the COP of the system refrigeration. The COP of direct expansion cooling system decreases with the increase of compressor speed and increases with the increase of condensing fan speed.

The cooling characteristics of the power battery system are also affected by the compressor speed and the condensing fan speed. The higher the compressor speed, the faster the battery module cooling, and slightly less consistent. When the speed of the condensing fan is increased, the cooling effect of the battery module is slightly increased, and the overall temperature difference does not change much.

Acknowledgements Thanks to the National Natural Science Foundation of China (51376080 and U1864213), Jilin Province's High-tech Industry Development Project (2016C022), Changchun Science and Technology Plan's Major Science and Technology Innovation Project (17SS022), Jilin Province's Youth Talents Fund (No. 20180520066JH) for their research funding and support.

References

1. Z. Guobin, W. Yuping, W. Chao, Current status of thermal management technology for large capacity Li-ion battery energy storage systems. Int. Energy Storage Sci. Technol. **7**(02), 203–210 (2018)
2. L. Yongqiang, Z. Zhenya, Huangdong, Dynamic characteristics of refrigerant migration during start-up and shutdown of horizontal refrigerators. Int. J. Harbin Univ. Technol. **49**(01), 155–159 (2017)

Numerical Analysis of Thermal Systems of a Bus Engine

Konstantinos Karamanos, Yasser Mahmoudi Larimi, Sung In Kim, and Robert Best

1 Introduction

Transportation is one of the main sectors contributing to climate change via the production of greenhouse gas emissions. In particular, 27% of UK emissions are attributed to transport [1]. European Union, in order to tackle this problem, has introduced the Euro VI regulations for heavy-duty vehicles [2]. Vehicle thermal management systems have been a critical part of improving vehicle fuel economy, as more than 50% of fuel energy is going to heat. Thermal systems are very complex in terms of thermo-fluid analysis due to the different shape and size of the components, which make experimental testing challenging [3]. Wang et al. [4] developed an integrated (1D/3D) model consisting of three sub-models, an engine cooling system including the oil circuit, a charge-air cooling system and a HVAC system. With this method, both the heat transfer and transient characteristics of the thermal systems can be simulated and analysed. Similarly, Bayraktar [5] used the integrated modelling method to analyse the powertrain and HVAC systems. The results showed that low-order modelling provides a positive first representation of a proposed system, however, 3D simulations are required to analyse the system's behaviour at transient state conditions. Bolehovsky et al. [6] compared the 1D and 1D/3D simulations for an engine cooling system. The results showed that the 1D model overestimates the heat transfer rate at radiator by 10%.

K. Karamanos (✉) · Y. M. Larimi · S. I. Kim
School of Mechanical and Aerospace Engineering, Queen's University Belfast, Belfast BT9 5AH, United Kingdom
e-mail: kkaramanos01@qub.ac.uk

R. Best
Wrights Group Ltd., Ballymena, Co., Antrim BT42 1PY, United Kingdom

© Springer Nature Singapore Pte Ltd. 2021
C. Wen and Y. Yan (eds.), *Advances in Heat Transfer and Thermal Engineering* ,
https://doi.org/10.1007/978-981-33-4765-6_100

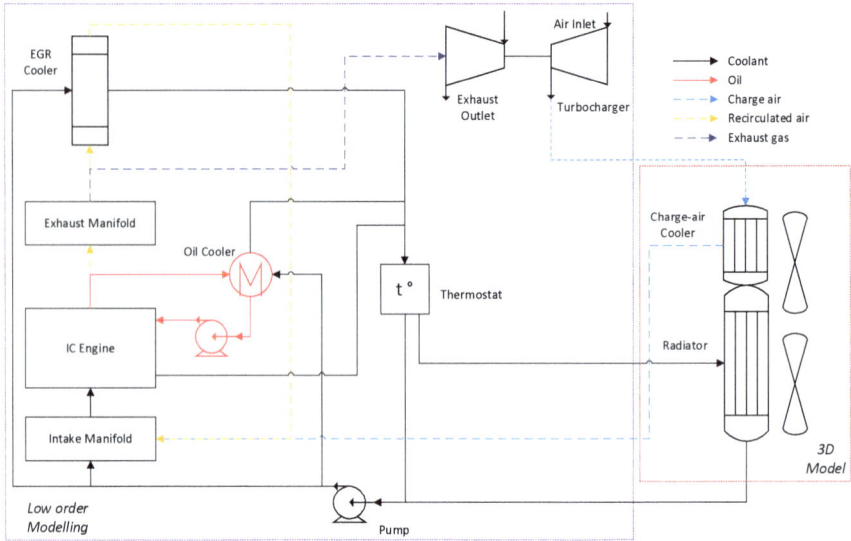

Fig. 1 Schematic diagram of the integrated model

2 Methodology

An integrated 0D/3D modelling approach is adopted in this research. A MATLAB-based 0D model is developed which consists of six sub-models: combustion chamber model, cooling system, charge-air cooling system, lubricant system, EGR cooler and turbocharger (Fig. 1). All the components placed in the under-hood are modelled using laws of thermodynamics and heat transfer equations. In a bus, the engine bay is at the back in order to maximise the interior space. Radiator and charge-air cooler are at the side of the vehicle and they are exposed to highly transient and turbulent airflow. Therefore, low-order modelling approaches cannot resolve property the flow and thermal characteristics of these two elements. Thus, 3D computation is required to model the effect of the external air flow as well as the internal water's cooling flow on the thermal feature of the system. In this regard, to minimise the computational time, two 3D models have been developed. One that simulates the airflow around the bus and one that analyses the heat and fluid flow of radiator and charge-air cooler in the engine bay. The first model establishes the boundary conditions for the engine bay model, which feeds data to the 0D model and complete the calculation.

3 Results

A closed fluid domain to replicate the wind tunnel was created to generate the airflow features in the engine bay. Figure 2 shows that the airflow captured at the grilles is

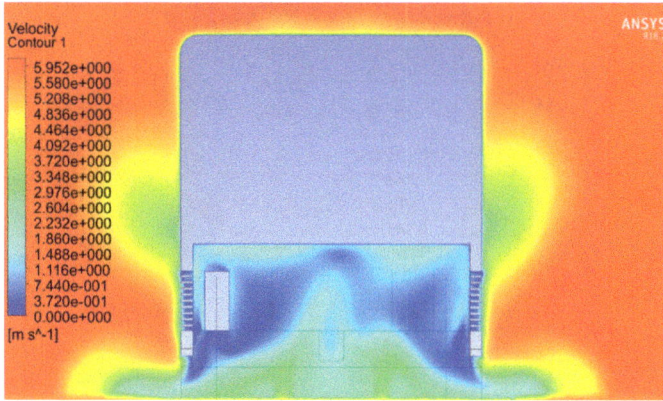

Fig. 2 Air velocity contours at the engine bay

significantly reduced. This results in lower heat rejection at radiator and charge-air cooler for buses comparing to cars. Thus, the main source of airflow is the fans which cause a transient and non-uniform flow at the downstream of the fans which subsequently go through the fins of the radiator and charge-air cooler (Fig. 3).

A full scale 0D and a coupled 0D/3D model were tested under the same operating condition based on the MLTB drive cycle. The bus speed was 55 km/h and the engine speed and torque were 1500 rpm and 850 Nm, respectively. As it is illustrated in Fig. 4, the heat transfer rate at radiator and charge-air cooler are overestimated. This occurs because the 0D model cannot capture accurately the airflow features at the rear of the bus, specifically at the grilles where radiator and charge-air cooler are placed. Therefore, the heat transfer coefficient between the external air and the

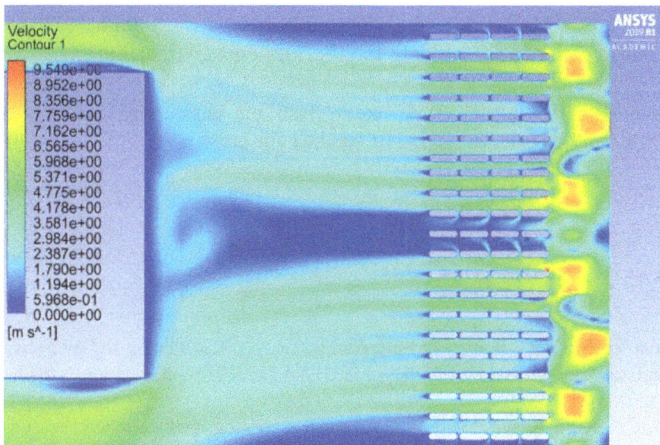

Fig. 3 Air velocity contours between the tubes

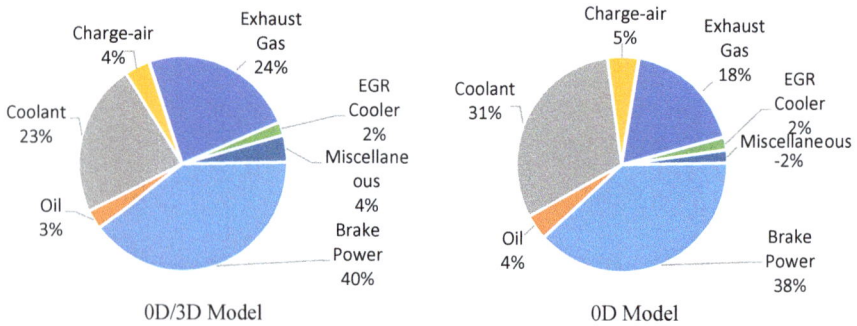

Fig. 4 Thermal energy balance

surface of the components is not precisely estimated. Moreover, the results show that more than 50% of the heat loss can be recovered.

4   Conclusions

The integrated simulation method is implemented in vehicle thermal systems of a bus engine. A MATLAB-based low-order (0D) model and 3D CFD models using FLUENT have been developed. The 0D model considers heat transfer aspects and thermodynamic laws of all the under-hood thermal systems. Through the CFD simulations, the heat transfer coefficient of radiator and charge-air cooler is calculated and fed to the low-order model.

As the engine is placed at the rear of the bus, the heat and fluid flow features are difficult to be estimated using full scale 0D model. Therefore, more accurate results can be generated using the proposed approach. Moreover, the parametric study can be carried out in order to re-evaluate the thermal energy distribution.

Acknowledgements The authors would like to thank the Engineering and Physical Research Council (EPSRC) and Wrights Group Ltd for funding this project.

References

1. Department for Business, Energy & Industrial Strategy, *UK Greenhouse Gas Emissions* (2017)
2. EU Commission, Commission Regulation (EU) No 582/2011—heavy duty. Off. J. Eur. Union, **L 167**, 1 (2011)
3. H.J. Kim, C.J. Kim, A numerical analysis for the cooling module related to automobile air-conditioning system. Appl. Therm. Eng. **28**(14–15), 1896–1905 (2008)
4. G. Wang, Q. Gao, H. Zang, Y. Wang, A simulation approach of under-hood thermal management. Adv. Eng. Softw. **100**, 43–52 (2016)

5. I. Bayraktar, Computational simulation methods for vehicle thermal management. Appl. Therm. Eng. **36**(1), 325–329 (2012)
6. O. Bolehovsky, A.N. Novotny, Influence of underhood flow on engine cooling using 1D and 3D approach (2015)

Robust Optimisation of Serpentine Fluidic Heat Sinks for High-Density Electronics Cooling

Muyassar E. Ismaeel, N. Kapur, Z. Khatir, and H. M. Thompson

1 Introduction

The goal of fluidic heat sinks for electronics cooling is to provide effective and energy-efficient cooling which ensures that the processors are below critical temperatures with minimal power input [1]. Although air-cooled heat sinks currently dominate the market, a number of recent studies have concluded that increasing densities of integrated circuits (up to 10 kW/cm^2 by 2020 [2]) will require effective liquid-cooling heat sink technologies. These trends have stimulated much recent interest in single-phase flows in fluidic channel devices for cooling high heat flux electronics encountered in, e.g. aircraft and in RF and microwave applications [3]. Single-phase flow in serpentine channel heat sinks is particularly well-suited to providing uniform processor temperatures for high-density electronics cooling applications. Al-Neama and his co-researchers [4] demonstrated recently that combining serpentine channels with fin structures can provide very good temperature uniformity while reducing significantly the pressure drop associated with serpentine systems. Several research groups are working to optimise the performance of microchannel heat sinks; however, most of these optimisation studies are deterministic and do not take into account the uncertainties associated with manufacturing processes and operating conditions [1]. Uncertainty could be classified as aleatory uncertainty and epistemic uncertainty. The former refers to inherent randomness in system behaviour, like randomness in operating conditions and geometric parameters, which is irreducible unless it is taken

M. E. Ismaeel (✉) · N. Kapur · H. M. Thompson
School of Mechanical Engineering, University of Leeds, Leeds LS2 9JT, UK
e-mail: muyassar.alhasso@uomosul.edu.iq; pmmei@leeds.ac.uk

M. E. Ismaeel
Department of Mechanical Engineering, Faculty Engineering, University of Mosul, Mosul, Iraq

Z. Khatir
School of Engineering and the Built Environment, Birmingham City University, Birmingham B4 7XG, UK

© Springer Nature Singapore Pte Ltd. 2021
C. Wen and Y. Yan (eds.), *Advances in Heat Transfer and Thermal Engineering* ,
https://doi.org/10.1007/978-981-33-4765-6_101

583

into consideration through the design stage, while the latter is a result of limited data and information about the system, such as those due to lack of knowledge about a model structure, which could be reduced by gathering more information about the system [5]. This study is the first to consider the effect of robustness considerations on the design and optimisation of the liquid-cooled serpentine heat sinks with chevron fin structures. It provides a contrast between the results of deterministic and robust optimisation of the liquid-cooled serpentine fluidic heat sinks with chevron fin structures introduced by Al-Neama et al. [4].

2 Problem Description and Methodology

A 3D geometrical model of the serpentine microchannel heat sink (SMCHS), investigated by Al-Neama et al. [4], is shown in Fig. 1. It consists of 12 main channels with 10 secondary channels dividing the walls between the main channels. These secondary channels have a chevron shape. All the dimensions are depicted in the figure below. Two heaters are attached at the base of this heat sink to mimic the heat generated by the chip processors of the electronic systems. The substrate of the heat sink is manufactured from copper, and water is used as a coolant. COMSOL V5.3a multiphysics software has been used to solve the fluid flow heat transfer governing Eqs. (1–4), using finite element methods, and obtained the numerical solution for this conjugate heat transfer problem.

$$\nabla.u = 0 \tag{1}$$

$$\rho\left(\frac{\partial u}{\partial t} + u.\nabla u\right) = -\nabla p + \nabla.\left(\mu\left(\nabla u + (\nabla u)^{\mathrm{T}}\right) - \frac{2}{3}\mu(\nabla.u)I\right) + F \tag{2}$$

$$\rho C_p\left(\frac{\partial T}{\partial t} + u.\nabla T\right) = \nabla.(k\nabla T) + Q \tag{3}$$

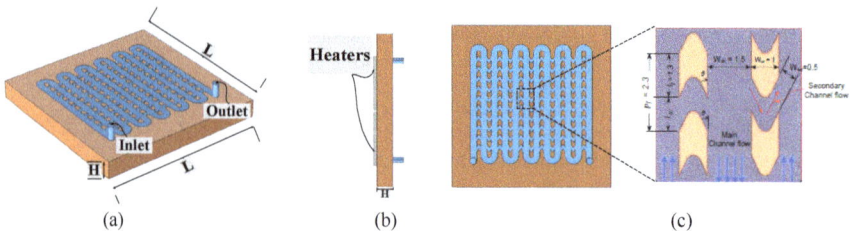

Fig. 1 Geometrical model: **a** 3D geometry, **b** side view and **c** top view with enlarged details, Al-Neama [4]

$$-k_s \left.\frac{\partial T_s}{\partial n}\right|_{\text{interface}} = -k_f \left.\frac{\partial T_f}{\partial n}\right|_{\text{interface}} \qquad (4)$$

Here, $\mathbf{u}$, ρ, p, μ, I, C_p and T are velocity vector [m/s], density [kg/m^3], pressure [Pa], dynamic viscosity [Pa s], turbulent intensity [1], specific heat [J/(kg K)]and absolute temperature of the fluid [K], respectively. t, F and Q are the time [sec], external force applied on the fluid [N] and the heat flux [W/m^3], respectively.

The boundary conditions associated with this problem are as follows: the inlet temperature was taken as 20 °C and the outlet pressure has been set to be ambient pressure. A heat flux is applied at a part of the bottom surface of the heat sink could take values in the range 25–100 W/cm^2. A no-slip boundary condition was applied at all walls in contact with the fluid. Further, a number of assumptions have been made for simplifying the CFD simulation which are: (1) the flow is steady and single-phase, (2) the fluid is incompressible, (3) viscous dissipation is neglected, (4) radiative heat transfer is neglected, (6) all the outer surfaces exposed to the surroundings are insulated and (6) the channel surfaces are smooth. The results are validated against a previous study, and a good agreement between the present work and those in the literature has been produced.

Deterministic and probabilistic optimisation for the performance of the SMCHS with chevron is carried out. Surrogate-based optimisation is used, i.e. the strategy of optimisation is based on replacing the costly CFD evaluation with a cheaper mathematical model to implement the optimisation process on the microchannel heat sink. This requires creating a sample of the design of experiment points (DoE) in the design space, run the CFD model at these points to generate the build points that are used to construct the surrogate model and finally performing the optimisation on the objective functions, thermal resistance and pressure drop of the SMCHS of the present investigation. Three design variables, namely the width of the main channel ($1.0\ mm \leq W_{\text{ch}} \leq 2.0\ mm$), the length of the secondary channel ($0.75\ mm \leq l_{\text{sc}} \leq 1.25\ mm$) and the oblique angle of the fins ($20° \leq \theta \leq 45°$), have been considered.

The open-source DAKOTA toolkit has been utilised to implement this study. It is a powerful toolkit as it has algorithms for optimisation, uncertainty quantification, parameter estimation and sensitivity analysis. Latin hypercube sampling (LHS) is used, and Gaussian processes are employed to build the surrogate meta-models. A multi-objective function deterministic optimisation, using Eqs. (5) and (9–11), is conducted first, to obtain a Pareto front of non-dominated solutions. After that, a probabilistic optimisation, using Eqs. (6–11), is performed to explore the effect of the uncertainties in the input variables of the serpentine microchannel heat sink with chevron fin on its performance criteria. This will be done by propagating the uncertainty in the design variables, which is set to be ($\pm 0.025°$) for θ angle and ($\pm 0.025\ mm$) for W_{ch} and l_{sc}, into the quantities of interest, i.e. the thermal resistance and pressure drop, and this can be accomplished using Monte Carlo simulation and normalising the mean and standard deviation of each objective function with respect to its maximum value to produce Eq. (6). Further, a weighting factor (wi) will be introduced to examine the relative influence of the mean and the standard deviation on the performance.

Deterministic Optimisation :
$$\textbf{Minimise}\{R_{\text{th}}(W_{\text{ch}}, l_{\text{sc}}, \theta) \text{ and } \Delta P(W_{\text{ch}}, l_{\text{sc}}, \theta)\} \tag{5}$$

Normalised(R_{th}) :
$$\textbf{OF\_Rth} = wi \times N\mu_{R_{\text{th}}} + (1 - wi) \times N\sigma_{R_{\text{th}}} \textbf{Normalised}(\Delta P):$$
$$\textbf{OF\_}\Delta\textbf{P} = wi \times N\mu_{\Delta P} + (1 - wi) \times N\sigma_{\Delta P} \tag{6}$$

Probabilistic Single Objective Optimisation for the Normalised(R_{th}) :

$$\textbf{minimise}\{\textbf{OF\_Rth}(W_{\text{ch}}, l_{\text{sc}}, \theta)\} \tag{7}$$

Probabilistic Single Objective Optimisation for the Normalised(ΔP) :

$$\textbf{minimise}\{\textbf{OF\_}\Delta\textbf{P}(W_{\text{ch}}, l_{\text{sc}}, \theta)\} \tag{8}$$

Subjected to:
$$1.0 \ mm \leq W_{\text{ch}} \leq 2.0 \ mm \tag{9}$$

$$0.75 \ mm \leq l_{\text{sc}} \leq 1.25 \ mm \tag{10}$$

$$20° \leq \theta \leq 45° \tag{11}$$

3   Results

The results of the multi-objective function deterministic optimisation along with CFD validation points are illustrated in Fig. 2. The validation points are in good agreement with the selected points with a maximum difference between the values does not exceed 4%. Figure 3 shows the geometries for two cases chosen for a comparison between an optimal design obtained by multi-objective optimisation of thermal resistance and pressure drop, represented by point A in Fig. 2, and a non-optimal design from set of DoE points, represented by point B in Fig. 2, working with the same boundary conditions. It can be noticed the remarkable difference in the shape and size of the microchannel passages between the two designs which affects the performance of the microchannel heat sink positively where the optimum design

Fig. 2 Deterministic optimisation of the serpentine MCHS with chevron fins

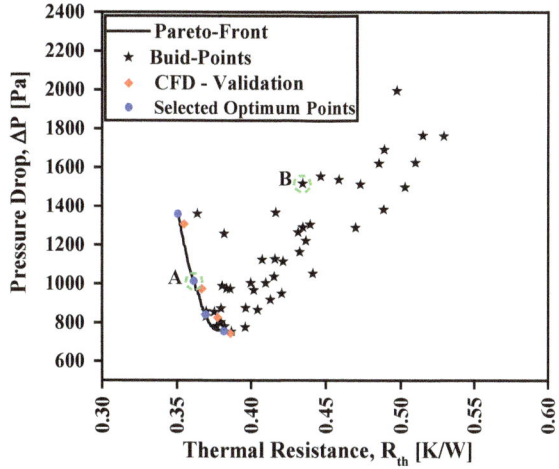

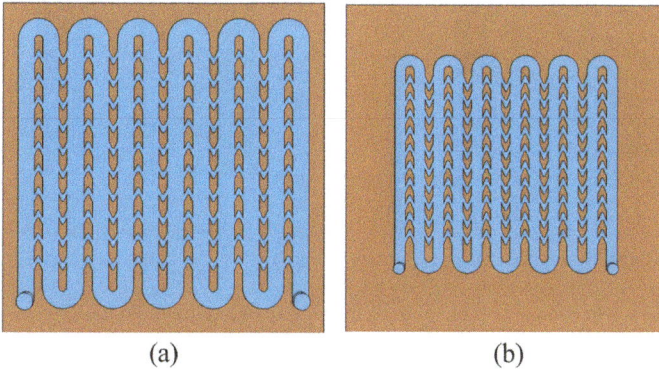

Fig. 3 Geometry for: **a** optimum design, point A in Fig. 2 and **b** non-optimum design, point B in Fig. 2

has reduced the thermal resistance, heat sink temperature and the pressure drop of the fluid in comparison with the non-optimum design. These differences have been illustrated in Fig. 4 and Table 1. Therefore, it could be concluded that the hydrothermal characteristics for optimum designs are all better than those of the non-optimum design. This could be attributed to the fact that the optimum design has a larger microchannel passage volume, wider channel width and a good aerodynamic fin shape in comparison with the non-optimum design. All of these factors enhanced heat transfer and reduced the pressure drop.

Figure 5 depicts 3D contours for the thermal resistance as a function of the main channel width (W_{ch}), the secondary channel length (l_{sc}) and the fin oblique angle (θ). Preliminary results for the probabilistic optimisation of single objective optimisation for the normalised thermal resistance (OF_Rth) with a weighting factor of $wi = 0.5$

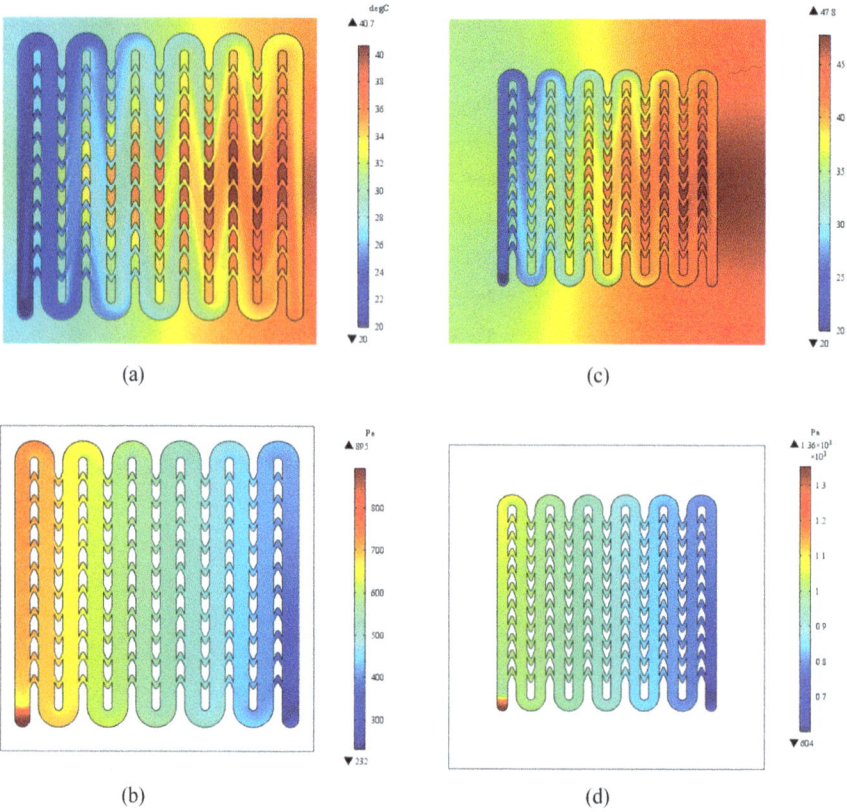

(a)

(c)

(b)

(d)

Fig. 4 a and **b** temperature and pressure disterbuitions for optimum design [Point A] and **c** and **d** for non-optimum design [Point B], respectively

Table 1 Comparison between optimum and non-optimum design

Case	Optimum design variables			Performance criteria		Max. temperature, (°C)
	l_{sc}	W_{ch}	θ	R_{th}(K/W)	ΔP (Pa)	Heat sink Base
Optimum design, point A	0.844	1.955	25.733	0.361	1014.12	40.6
Non-optimum design, point B	0.865	1.283	21.99	0.434	1516.7	47.1

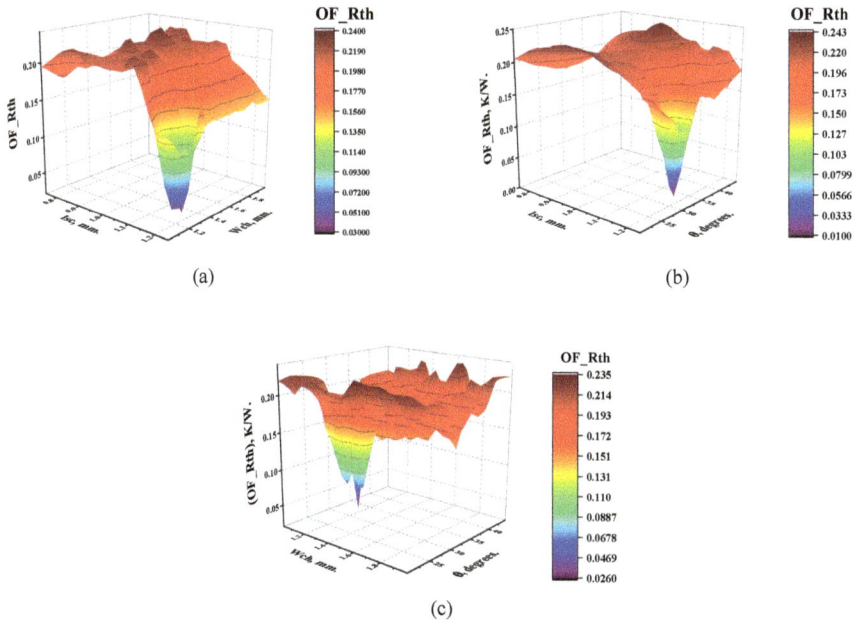

Fig. 5. 3D presentation of thermal resistance as function of the main channel width, secondary channel length and fin angle for $w1 = 0.5$

has been presented here. These results have revealed that the optimum R_{th} (0.375 K/W) occurred with l_{sc}, W_{ch} and θ of respectively 0.852 mm, 1.939 mm and 39.21°. In contrast, the deterministic optimisation results showed that the optimum R_{th} was 0.347, and the corresponding design variables were $l_{sc} = 0.751$ mm, $W_{ch} = 1.987$ mm and $\theta = 20.04°$. The R_{th} obtained from the probabilistic optimisation was higher than that of the deterministic optimisation by 8%. Similarly, the results for the probablistic single objective optimisation of the normalised pressure drop have shown that the optimum ΔP was 777.2 Pa which is higher than that obtained using the deterministic optimisation (741.63 Pa) by 4.79%.

4   Conclusions

Deterministic and probabilistic optimisation for the liquid-cooled serpentine heat sinks with chevron fin shape have been conducted by considering the main channel width (1.0 mm $\leq$ W_{ch} $\leq$ 2.0 mm), the secondary channel length (0.75 mm $\leq$ l_{sc} $\leq$ 1.25 mm) and fins oblique angle (20° $\leq \theta \leq$ 45°) as the design variables for the optimisation problem. The results show that these design variables have a vital effect on the performance of the heat sink under investigation. Furthermore, the optimum designs produced using probabilistic optimisation have thermal resistance and pressure drop penalty higher than that obtained using the deterministic optimisation process by 8% and 4.8%, respectively.

Acknowledgements The first author would like to acknowledge the sponsorship of the Ministry of Higher Education and Sceintific Research of Iraq and the University of Mosul for the financial support.

Reference

1. H. Ahmed, B. Salman, A. Kherbeet, M. Ahmed, Optimization of thermal design of heat sinks: A review. Int. J. Heat Mass Transf. **118**, 129–153 (2018)
2. Int. Technol. Roadmap or Semiconductors 2.0 (2015)
3. G. Agarwal, T. Kazior, T. Kenny, D. Weinstein, Modeling and analysis for thermal management in Gallium Nitride HEMTs using microfluidic cooling. J. Electron. Pack. **139**(1), 011001 (2017)
4. A. Al-Neama, Z. Khatir, N. Kapur, J. Summers, H. Thompson, An experimental and numerical investigation of chevron fin structures in serpentine minichannel heat sinks. Int. J. Heat Mass Transf. **120** 1213–1228 (2018)
5. L.P. Swiler, A.A. Giunta, Aleatory and epistemic uncertainty quantification for engineering applications (No. SAND2007–4655C),*Sandia National Lab.(SNL-NM), Albuquerque, NM (United States)* (Sandia National Laboratories,, Washington, DC, 2007)

Numerical Study on the Fluid Flow and Heat Transfer Performance of Flat Miniature Heat Pipe for Electronic Devices Cooling

Fei Xin, Ting Ma, Yuying Yan, and Qiuwang Wang

1 Introduction

With the miniaturization of electronic devices, high heat flux has been a great threat to the devices. Electronic devices require to have high heat transfer capacity and temperature uniformity to avoid hot spot. It was reported that about 55% electronic devices failed from thermal effect [1]. It is necessary to develope an effective way to cool the microelectronic devices with small space. Flat miniature heat pipe with compact structure, no moving parts and high heat conduction ability is very suitable for the effective cooling. Grooved wick of heat pipe integrated with wall can be easy to be produced. It can decrease the thermal and flow resistances of heat pipe.

Experimental study on mini-grooves flat heat pipe heat transfer performance has been published by some researchers. Lips et al. [2] experimentally study the thermal behavior of a flat heat pipe with grooves, focusing on the effect of boiling in the grooves. However, as flat miniature heat pipe is a sealed container with small space, the fluid flow and heat transfer inside the heat pipe can be hardly to be measured by experiment. In order to clearly know about the fluid flow and heat transfer characteristics inside the flat miniature heat pipe, a three-dimensional numerical model is established to study the performance of the flat copper heat pipe with mini-grooves under transient and steady states. The effects input power on the heat pipe were studied.

F. Xin · T. Ma · Q. Wang (✉)
Key Laboratory of Thermo-Fluid Science and Engineering, MOE, Xi'an Jiaotong University, Xi'an 710049, Shaanxi, P.R. China
e-mail: wangqw@mail.xjtu.edu.cn

Y. Yan (✉)
Faculty of Engineering, University of Nottingham, Nottingham NG7 2RD, UK
e-mail: Yuying.Yan@nottingham.ac.uk

© Springer Nature Singapore Pte Ltd. 2021
C. Wen and Y. Yan (eds.), *Advances in Heat Transfer and Thermal Engineering* ,
https://doi.org/10.1007/978-981-33-4765-6_102

2 3d Numerical Model Description

A heat pipe consists of airtight container, wick structure and working fluid, which is separated into evaporation, adiabatic and condensation sections, as shown in Fig. 1. This paper refers to the equations in Reference [3] and employs software Fluent through SIMPLE algorithm to simulate the fluid flow and heat transfer inside the heat pipe. Constant heat flux is added to the evaporation zone, and heat convection is conducted at the condensation zone. Some assumptions are made in this model:

1. The working fluid is incompressible and laminar flow.
2. Boling phenomenon is ignored.
3. Wick is filled with liquid and the liquid–vapor interface keeps unchanged during the simulation.
4. Working fluid velocity at the liquid–vapor interface is normal to the interface.
5. Temperature at the liquid–vapor interface is calculated through saturation pressure at vapor chamber side.
6. Using porous media model to calculate the wick structure.

Based on the above assumptions, equations are built to study the fluid flow and heat transfer inside the heat pipe, and Eqs. (1)–(4) are the mass, momentum conservation equations at the vapor and wick sides.

$$\varepsilon \frac{\partial p}{\partial t} + \nabla \cdot \left(\rho \vec{V}\right) = 0 \tag{1}$$

$$\frac{\partial \rho u}{\partial t} + \nabla \cdot \left(\rho \vec{V} u\right) = \frac{\partial \varepsilon p}{\partial x} + \nabla \cdot (\mu \nabla u) - \frac{\mu \varepsilon}{K} u - \frac{C_E \varepsilon}{K^{\frac{1}{2}}} \rho \left|\vec{V}\right| u \tag{2}$$

$$\frac{\partial \rho v}{\partial t} + \nabla \cdot \left(\rho \vec{V} v\right) = \frac{\partial \varepsilon p}{\partial y} + \nabla \cdot (\mu \nabla v) - \frac{\mu \varepsilon}{K} v - \frac{C_E \varepsilon}{K^{\frac{1}{2}}} \rho \left|\vec{V}\right| v \tag{3}$$

$$\frac{\partial \rho w}{\partial t} + \nabla \cdot (\rho \vec{V} w) = -\frac{\partial \varepsilon p}{\partial z} + \nabla \cdot (\mu \nabla w) - \frac{\mu \varepsilon}{K} w - \frac{C_E \varepsilon}{K^{\frac{1}{2}}} \rho \left|\vec{V}\right| w \tag{4}$$

where ε and K are porosity and permeability at every side, respectively. $K = \infty$ and $\varepsilon = 1$ are at vapor chamber.

Equation (5) is the energy conservation equation at the wall, vapor and wick sides:

Fig. 1 Schematic of mini-flat heat pipe

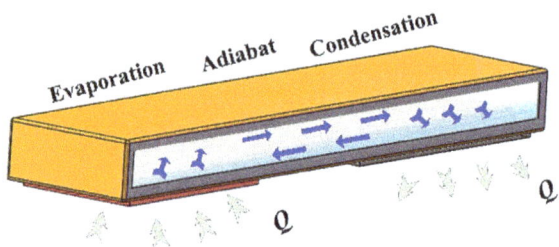

$$\frac{\partial(\rho C)_m T}{\partial t} + \nabla \cdot [(\rho C)_l \vec{V} T_l] = \nabla \cdot (k_{\text{eff}} \nabla T) \tag{5}$$

The interface temperature T_i at liquid–vapor interface is obtained from an energy balance at the interface:

$$-k_{\text{wick}} A_i \frac{\partial T}{\partial y} + m_i C_l T_i = -k_v A_i \frac{\partial T}{\partial y} + m_i C_v T_i + m_i h_{fg} \tag{6}$$

where $m_i < 0$ denotes evaporation and $m_i > 0$ denotes condensation.
The interface pressure P_i is obtained from the Clausius–Clapeyron equation [4]:

$$\frac{R}{h_{fg}} \ln\left(\frac{p_i}{p_0}\right) = \frac{1}{T_0} - \frac{1}{T_i} \tag{7}$$

where p_0 and T_0 are the reference temperature and pressure of vapor, respectively. In this paper, they are given as heat pipe initial saturated vapor temperature and pressure, respectively.

3 Results and Discussion

Deionized water is employed as working fluid to study the fluid flow and heat transfer performance of mini-grooves flat heat pipe. The input heating power Q at evaporation section is varied from 4 to 40 W, and the heat source area is 20*20 mm^2. The cold source is convection heat transfer with the convective heat transfer coefficient 1000 W m^{-2} K^{-1} and the convective temperature 293.15 K. And the cold source area is 40 * 20 mm^2. The initial temperature is 298.15 K.

Firstly, as porous media model is used in this simulation, the effective thermal coefficient of heat pipe is unknown. The heat transfer performance of wick in the heat pipe is simulated at first. Obtained from simulation, the effective thermal coefficients of wick filled with deionized water is 12.92 W m^{-1} K^{-1}.

Then, the transient variations of the operation pressure, the average velocity at vapor chamber center $y = 2.5$, and the heat transfer rates at heat and cold source areas are shown in Fig. 2. The simulation time is 400 s. It can be seen that the heat pipe reaches steady state within 200 s. It can be seen that the operation pressure increases with the input heating power. And the vapor average velocity first increases with the input heating power but then decreases due to the change of vapor density.

The temperature distribution and mass flow rate at liquid–vapor interface in the wick side of mini-grooves flat heat pipe under input heating power 40 W after 400 s are shown in Fig. 3. It can be seen that the temperature changes a little in the vapor chamber under input heating power 40 W. It clearly displays that when the evaporation section receives external heat, the working fluid in the wick begins to evaporate

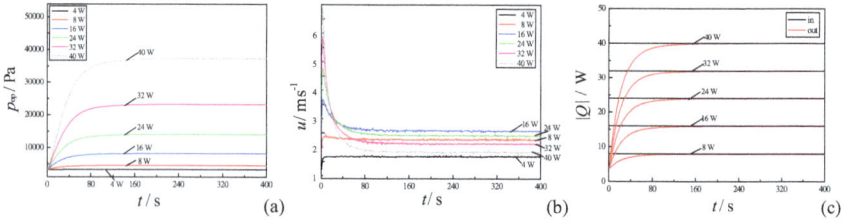

Fig. 2 Parameters transient variations versus time: **a** operation pressure; **b** average velocity; **c** heat transfer rates at heat and cold source areas

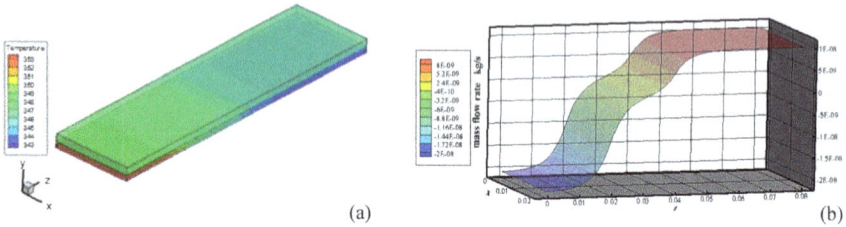

Fig. 3 Parameters distribution of mini-grooves flat heat pipe under input heating power 40 W: **a** temperature; **b** mass flow rate at liquid–vapor interface

and the vapor flows into the vapor chamber; in the condensation section, vapor condensates into liquid and releases heat.

The thermal resistance, maximum temperature at evaporation section and temperature uniformity of mini-grooves flat heat pipe under different input heating power with pure copper and deionized water as working fluid after 400 s are shown in Fig. 4. As shown in Fig. 4, the thermal resistance, maximum temperature at evaporation section and temperature uniformity of mini-grooves flat heat pipe are obviously lower than those of pure copper, and the effective thermal conductivity of the heat pipe is greatly higher than that of pure copper. It manifests that heat pipe has better heat transfer ability than copper plate to cool the electronic devices, removing the hot spots and improving the temperature distribution uniformity. And with the increase

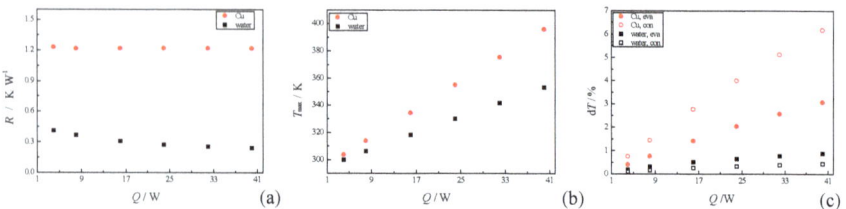

Fig. 4 Evaluation parameters of heat pipe heat transfer performance under different input heating powers: **a** thermal resistance; **b** maximum temperature at evaporation section; **c** temperature uniformity

of input heating power, though the maximum temperature at evaporation section of the heat pipe increases a lot, the thermal resistance decreases and temperature uniformity increases a bit.

4 Conclusions

A three-dimensional numerical model is established to study the performance of the flat copper heat pipe with mini-grooves under transient and steady states. The fluid flow and heat transfer characteristics inside the flat miniature heat pipe are clearly displayed in the simulation. It is found that the heat pipe filled with deionized water has better heat transfer performance than copper plate, which can realize the electronic devices cooling rapidly and effectively.

Acknowledgements This work is financially supported by China Scholarship Council (No.201806280100) during a visit of Fei XIN to University of Nottingham.

References

1. D. Yu, Fatigue life assessment and prediction of ball grid array package under environmental conditions.Dissertations & Theses—Gradworks (2012)
2. S. Lips, F. Lefèvre, J. Bonjour, Nucleate boiling in a flat grooved heat pipe. Int. J. Therm. Sci. **48**, 1273–1278 (2009)
3. U. Vadakkan, S.V. Garimella, J.Y. Murthy, Transport in flat heat pipes at high fluxes from multiple discrete sources. ASME J. Heat Transf. **126**, 347–354 (2004)
4. Y.M. Hung, Q. Seng, Effects of geometric design on thermal performance of star-groove micro-heat pipes. Int. J. Heat Mass Transf. **54**, 1198–1209 (2011)

Thermal Performance on a Vapor Chamber-Based Battery Thermal Management System

Dan Dan, Hongkui Lian, Yangjun Zhang, Yuying Yan, and Chengning Yao

1 Introduction

The performance of LIB is subject to its operating temperature. High temperature and low temperature may both lead to a capacity loss and affect the life expectancy of a battery. Temperature difference within a battery pack also has a negative effect on the charging and discharging efficiency of a battery. Therefore, battery thermal management system (BTMS) is essential to ensure the battery pack operate within a temperature range of 25–40 °C, and to limit the temperature difference within 5 °C [1, 2].

Heat pipe, operating on the principle of gas–liquid phase change principle, is a high-efficiency heat transfer element. It has received extensive attention in the field of battery thermal management in recent years [3]. The existing research shows that BTMS using heat pipes has a higher heat dissipation efficiency at high temperature and a faster heating rate at low temperature [4, 5]. In the case of thermal runaway, the heat pipe-based BTMS also shows an advantage [6]. Wang et al. [7] placed an L-shaped flat heat pipe between each two adjacent cells, with the cooling end soaking in the water tank, which dissipated the absorbed heat by direct heat exchange with water. Liang et al. [8] designed a similar structure and further studied the effects of ambient temperature, cooling power, and control strategy on cooling efficiency. In order to increase the contact area between heat pipes and batteries, some researchers used flat

D. Dan · Y. Zhang (✉) · C. Yao
State Key Laboratory of Automotive Safety and Energy, School of Vehicle and Mobility, Tsinghua University, Beijing 100084, China
e-mail: yjzhang@tsinghua.edu.cn

H. Lian
Beijing Thermatech Co. Ltd., Beijing 100084, China

Y. Yan
Fluids and Thermal Engineering Research Group, Faculty of Engineering, University of Nottingham, Nottingham NG7 2RD, UK

© Springer Nature Singapore Pte Ltd. 2021
C. Wen and Y. Yan (eds.), *Advances in Heat Transfer and Thermal Engineering*,
https://doi.org/10.1007/978-981-33-4765-6_103

micro-heat pipe for battery thermal management [9, 10]. Xu et al. [11] designed a flat micro-heat pipe-water cooling BTMS, by arranging several heat pipes in parallel at the bottom of a battery module. The heat was then removed by water cooling on both sides of the heat pipe. However, the micro-heat pipe array is composed of a series of microchannels in parallel, which means the thermal conductivity along its flow direction is much higher than that along the width direction. Therefore, it is difficult to ensure the overall temperature difference when applied to the battery pack.

Compared with the one-dimensional heat pipe, vapor chamber (VC) is a 2D flat-plate heat pipe, which can flatten the temperature distribution and reduce the spreading thermal resistance accordingly. Gou and Liu [12] used a copper water vapor chamber in a battery module and studied the effects of heat generation rate and the inclination of the system. The results proved the excellent thermal performance of the vapor chamber. This paper adopts an aluminum vapor chamber for BTMS in order to provide a weight reduction and a higher energy density for the system. Electric heating plate is adopted to simulate the heat generation of the battery. The thermal resistance of the vapor chamber under different input power is theoretically calculated.

2 Methodology

The heat transfer performance of the vapor chamber is experimentally studied. Figure 1 illustrates our experimental setup. Five heating plates, the resistances of which are 21.6 Ω, are placed on the surface of the VC (226 * 170 * 7) in parallel. Different input powers are applied on the heater by adjusting voltages. Aerogel material is used on the evaporation surface in order to ensure thermal insulation. The temperature on the surface of the vapor chamber is measured by Ω-type thermocouples, and the data are collected by the data acquisition (Agilent 34972A) and synchronized to a computer. With the aid of an electric fan, air coolant is supplied to the condensation surface with different velocities and the ambient temperature (25 °C).

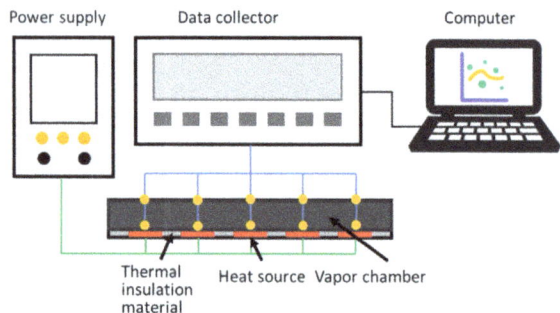

Fig. 1 Experimental setup

The thermal performance of the VC is evaluated by the maximum temperature, temperature difference of the heaters and the thermal resistance of the vapor chamber.

The maximum temperature and temperature difference of the VC are directly obtained through experimental data; the thermal resistances are further calculated.

$$R_t = \frac{T_{max} - T_o}{Q} \tag{1}$$

$$R_o = \frac{T_a - T_o}{Q} \tag{2}$$

$$R_s = R_t - R_o = \frac{T_{max} - T_a}{Q} \tag{3}$$

where R_t, R_o and R_s are, respectively, the total, external and diffusion thermal resistance of the vapor chamber.

3 Results

Figure 2 shows the temperature of the VC under different cooling air velocities. The T_{max} and T_{min} are reduced with the increase of cooling air velocity. T_{max} is reduced to 40 °C when the air velocity is increased to 2 m/s. However, further increase of the air velocity slightly affects T_{max}. It worth noting that higher air velocity leads to a greater temperature difference in the meanwhile. It is important to consider the trade-off between the temperature and $\triangle T$ when improving the heat transfer capacity. Normalization processing is carried out for the maximum temperature: The initial temperature (ambient temperature T_o) is defined as 0, and the temperature after stabilization is defined as 1. The temperature rise curve under different cooling

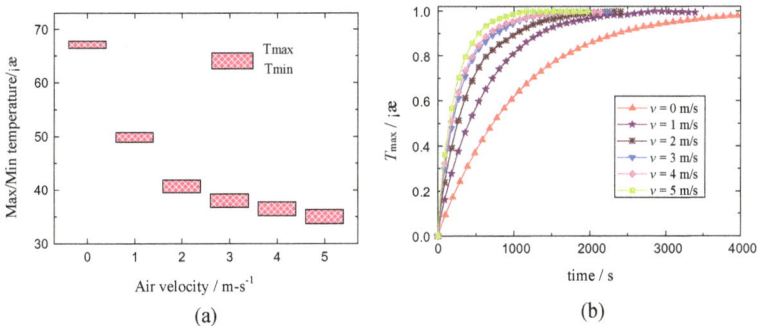

Fig. 2 Temperature of the VC under different air velocities **a** T_{max} and T_{min} of the five heaters; **b** T_{max} after normalization

Fig. 3 Thermal resistance of
the vapor chamber

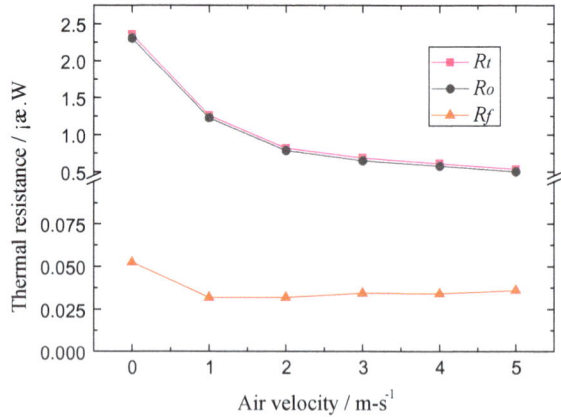

condition is shown in Fig. 2b. The higher the cooling velocity is, the faster it takes
to reach stability.

Figure 3 shows the total thermal resistance (R_t), the external thermal resistance
(R_o) and the diffusion thermal resistance (R_f) of the vapor chamber under different
cooling conditions. With the increase of air velocity, R_t and R_o decrease significantly.
However, the diffusion thermal resistance remains unchanged when the velocity is
higher than 1 m/s.

Cooling strategy has a significance on the thermal performance of vapor chamber.
It needs to be taken into consideration when the vapor chamber is applied to a battery
thermal management system that the heat transfer enhancement could lead to a lower
temperature and a greater temperature difference.

4 Conclusions

Efficient thermal management technology is the key to control the temperature of the
battery and ensure the temperature difference. A vapor chamber is a high-efficiency
heat transfer component that is based on two-dimensional heat transfer, which helps
to improve the temperature difference of the battery pack and simplify the system.
An aluminum VC was experimentally studied, and its thermal resistances were theoretically calculated. The results showed that VC has a good thermal performance in
the field of battery thermal management.

Acknowledgements The authors would like to thank the supports from National Natural Science
Foundation of China (Grant number U1864212).

References

1. A. Pesaran, Battery thermal models for hybrid vehicle simulations. J. Power Sources **110**, 377–382 (2002)
2. A. Mills, S. Al-Hallaj, Simulation of passive thermal management system for lithium-ion battery packs. J. Power Sources **141**, 307–315 (2005)
3. D. Dan, C. Yao, Y. Zhang, Research progress and future prospects of battery thermal management system based on heat pipe technology. Sci. Bull. **64**, 682–693 (2019)
4. A. Greco, D. Cao, X. Jiang, A theoretical and computational study of lithium-ion battery thermal management for electric vehicles using heat pipes. J. Power Sources **257**, 344–355 (2014)
5. X. Ye, Y. Zhao, Z. Quan, Thermal management system of lithium-ion battery module based on micro heat pipe array. Int. J. Energy Res. **42**, 648–655 (2018)
6. H. Hata, S. Wada, T. Yamada, Performance evaluation of a battery-cooling system using phase-change materials and heat pipes for electric vehicles under the short-circuited battery condition. J. Therm. Sci. Technol. **13**, JTST0024 (2018)
7. Q. Wang, B. Jiang, Q. Xue, Experimental investigation on EV battery cooling and heating by heat pipes. Appl. Therm. Eng. **88**, 54–60 (2015)
8. J. Liang, Y. Gan, Y. Li, Investigation on the thermal performance of a battery thermal management system using heat pipe under different ambient temperatures. Energy Convers. Manage. **155**, 1–9 (2018)
9. R. Zhao, J. Gu, J. Liu, An experimental study of heat pipe thermal management system with wet cooling method for lithium ion batteries. J. Power Sources **273**, 1089–1097 (2015)
10. X. Ye, Y. Zhao, Z. Quan, Experimental study on heat dissipation for lithium-ion battery based on micro heat pipe array (MHPA). Appl. Therm. Eng. **130**, 74–82 (2018)
11. X. Xu, W. Tang, J. Fu et al., Plate flat heat pipe and liquid-cooled coupled multistage heat dissipation system of Li-ion battery. Int. J. Energy Res. **43**, 1133–1141 (2019)
12. J. Gou, W. Liu, Feasibility study on a novel 3D vapor chamber used for Li-ion battery thermal management system of electric vehicle. Appl. Therm. Eng. **152**, 362–369 (2019)

.

Conformal Cooling of Aluminium Flat Fins Using a 3D Printed Water-Cooled Mould

Y. Liang, R. Sharma, E. Abo-Serie, and J. Jewkes

1 Introduction

Additive manufacturing (AM) provides significant geometric design freedom in comparison with traditional manufacturing methods for the cooling of high-pressure die casting (HPDC) tools. Designing cooling channels that can achieve a uniform temperature throughout the tool–cast interface during the cooling and the demoulding phases are a desirable aim for researchers and engineers in this field. Problems such as part warping and sink marks, internal part stresses, and reduced tool life due to thermal stresses are related to uneven surface temperature. Nevertheless, the cycle time can be significantly reduced. To design a conformal cooling system, the heat flux from the casted part should be evenly transferred to the tool and finally to the coolant. Various design parameters can affect the heat transfer from the cast to the cooling channels. The depth of the channel from the interface, the distance between the channels (pitch), and the channel cross-sectional shape and size are among these design parameters [1]. Various studies have been carried out for plastic tools. However, the design techniques that are used cannot always be applied to aluminium casting as the coolant channel has to be short to avoid boiling. Limited publications are available for HPDC conformal cooling for aluminium casting. Having a uniform interface temperature entails varying heat flux that can be related to the casted part thickness and geometry. Despite having flexibility in the design of cooling channels using AM, conformal cooling remains challenging. For example, molten cast in corners is constrained by limited available surface area to allow heat to pass through to the cooling channels [2]. In addition, fluid flowing through bends can result in flow separation with a recirculation zone that can become hotter; this may result in subcooled local boiling [3]. Another challenge is the rapid increase in water temperature that results from the high temperature of the molten aluminium in comparison

Y. Liang · R. Sharma · E. Abo-Serie (✉) · J. Jewkes
Fluid and Complex Systems Research Centre, Coventry University, Coventry CV1 2JH, UK
e-mail: aa3426@coventry.ac.uk

© Springer Nature Singapore Pte Ltd. 2021
C. Wen and Y. Yan (eds.), *Advances in Heat Transfer and Thermal Engineering* ,
https://doi.org/10.1007/978-981-33-4765-6_104

with plastic injection moulding. The large amount of heat transfer required to solidify the aluminium part limits the length of cooling pipes, and therefore, a greater number of separate water supply branches have to be considered.

To understand the heat flux from the molten aluminium source to the coolant, a complex transient conjugate heat transfer model that includes phase change and three regions is required. To achieve reasonable results, considerable mesh refinement is necessary close to the wall, in addition to a small time step. This model is not only complex but also computationally demanding, especially if various design parameters are to be examined. In this work, a simplified model has been proposed based on experimental observations and applied to cooling a fin on a flat surface for various geometries. The simple geometry represents an idealized subset of a more complex geometry. The challenges to achieve a uniform interface temperature have been highlighted, and heat flux distribution at the corners has been evaluated relative to a reference flat surface. The effect of the cooling channel depth, pitch, and Reynolds number on heat flux distribution around flat fins has been evaluated. Finally, an allocation table has been generated to identify the design parameters to match various fin geometries.

2 Methodology

For simplification, the cast model has been decoupled from the model of the tool and its cooling channels. The previous work shows that the injection phase for molten aluminium casting is very short when compared to the cooling time, and therefore, the cooling during this phase can be neglected. Experimental observation has shown that the temperature of the interface does not change much during the cooling stage and its value can be 400–500 °C [3]. A value of 450 °C is used as an interface temperature and kept constant during the cooling stage; this value can be estimated if the Riemann interface temperature is calculated assuming the mould temperature initially at 200 °C [4]. The heat flux distribution from the cast was first evaluated using the volume of fluid method (VOF) assuming no convection and only conduction. Similar results can be obtained for the heat flux through a steady simulation with uniform heat generated per unit volume, which can be calculated from an assumed demoulding temperature and target cooling time. To show the distribution and effect of curvature, the interface average heat flux at different sections around the fin is calculated.

For the tool with impeded cooling pipes, a steady and 3D conjugate heat transfer (CHT) model was run, with the Reynolds-averaged Navier–Stokes (RANS) equations used to solve the water coolant region, with a low Reynolds number k-ε turbulence closure. The simulation was initially run for a reference case to identify the heat flux over a flat surface with identical coolant design parameters as that for the flat fins. The properties were set to H13 steel for the solid region and water for the fluid region. These reference values were used to compare the effect of heat flux relative to the reference cases. Two types of cooling channels were simulated, a U-shaped cooling pipe running between the two fins and a straight pipe between the fins. The effect of depth and pitch in addition to Re on heat flux distribution has been examined.

3   Results

Reference Cases—Flat Surface
The cooling a flat sheet of aluminium is first considered as a reference case. Cooling pipes with various depth, pitch, and Reynolds number were simulated and validated experimentally to evaluate average heat flux on the interface surface between the cast and mould. The relative heat flux is the ratio between heat flux and the reference value.

U-shaped Pipe
Figure 1a shows the layout of the cast, mould, and cooling pipes and highlights the fact that the external curvature has larger area for the heat to be dissipated, while opposite scenario exists for an internal curvature. This is reflected in the results of heat flux distribution based upon a simulation of the molten aluminium alone, as shown in Fig. 1b. To create a uniform interface temperature, the cooling channels should be designed to achieve a matching heat flux distribution on the tool side. However, the heat flux on the mould surface due to the U-shaped cooling channels reveals a significant difference in heat flux distribution. The internal curvatures require more cooling, and the external curvature requires less cooling as Fig. 1b indicates. Unfortunately, the U-tube cooling layout provides the opposite trend for the heat flux. The low heat flux at the internal curvature is attributed to having large ratio of mould surface area to cooling pipe surface area compared to flat surfaces. Moreover, there is a high thermal resistance due to a greater distance from the curved mould surface to the pipe surface. Low water flow velocities in the pipe corners and flow separation are also additional factors that can lead to a reduction in heat flux as shown in Fig. 1e. Also, associated with higher water temperatures as shown in Fig. 1d. Flow separation near the external curvature can be an advantage if boiling can be avoided. All these factors lead to a higher temperature at the internal curvature which affect cast quality and also the mould life [5].

Straight Pipe Between the Fins
The simulation for a single pipe passing between the fins is a reasonable approach when the distance between the fins is small. Internal curvatures encounter a similar problem; limited heat flux is provided from the cooling pipe. To identify the layout of cooling pipes in the internal curvature for achieving conformal cooling, the heat flux for various cast thicknesses and internal curvature radii has been evaluated and presented in Fig. 2 a. To match the heat flux from the cast, the depth of the cooling pipes in the corners has to be less than 1.6 mm from the cast/mould interface surface as shown in Fig. 2b, c. Clearly, this cannot practically be achieved due to the high thermal fatigue and the requirement for increased tool life. Locating the pipes further

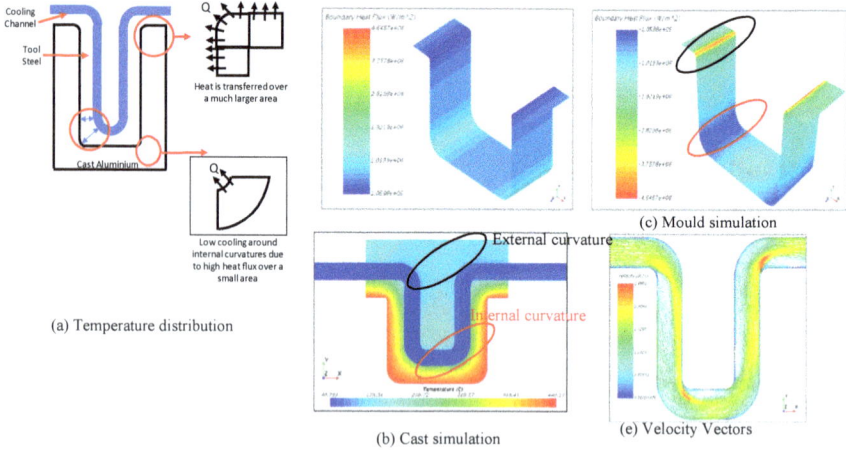

(a) Temperature distribution

(c) Mould simulation

(b) Cast simulation

(e) Velocity Vectors

Fig. 1 Simulation results for the U-shaped pipes

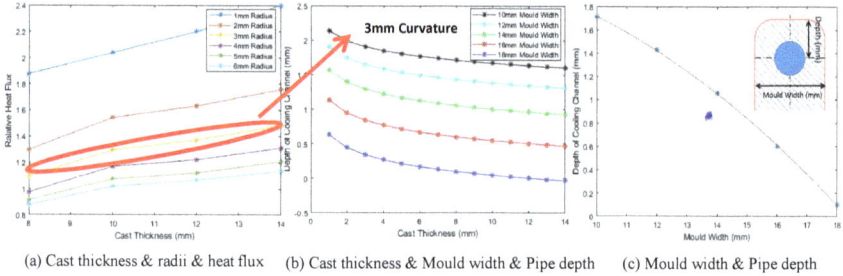

(a) Cast thickness & radii & heat flux   (b) Cast thickness & Mould width & Pipe depth   (c) Mould width & Pipe depth

Fig. 2 Heat flux and cooling channel geometries to achieve conformal cooling at the internal corners

down the surface will cause the corners to be hotter than other parts of the cast, and consequently, the cooling uniformity index will drop. Further work needs to be carried out to propose different ways to enhance corner cooling while keeping the cooling of other sections low.

4 Conclusions

By decoupling the cast simulation model from the mould model, it is possible to examine a number of design variables without having to solve a transient multiphase problem. This assumption can be made for HPDC during the cooling stage, based on experimental observation. Based upon this methodology, the effect of various design variables can be identified for a specific cast geometry. The results show that internal corners need enhanced cooling that cannot be achieved by the changing the common parameters used to achieve conformal cooling. Further work needs to be carried out to examine various physics and designs for enhancing the cooling, by considering the 3D printing flexibility of cooling channels.

Acknowledgements This work is supported by Innovate UK Grant (TSB TS/R011249/1): "3D-printed conformal cooling channels for extended tooling life", in conjunction with CastAlum Ltd., and Renishaw PLC.

References

1. M. Mazur, M. Leary, M. McMillan, J. Elambasseril, M. Brandt, SLM additive manufacture of H13 tool steel with conformal cooling and structural lattices. Rapid Prototyp. J. **22**(3), 504–518 (2016)
2. J. Kovács, B. Sikló, Investigation of cooling effect at corners in injection molding. Int. Commun. Heat Mass Transfer **38**(10), 1330–1334 (2011)
3. A.E. Bergles, J.H. Lienhard V, G.E. Kendall, P. Griffith, Boiling and evaporation in small diameter channels. Heat Transf. Eng. **24**(1), 18-40 (2003)
4. G. Dour, M. Dargusch, C. Davidson, A. Nef, Development of a non-intrusive heat transfer coefficient gauge and its application to high pressure die casting: effect of the process parameters. J. Mater. Process. Technol. **169**(2), 223–233 (2005)
5. A. Long, D. Thornhill, C. Armstrong, D. Watson, Predicting die life from die temperature for high pressure dies casting aluminium alloy. Appl. Therm. Eng. **44**, 100–107 (2012)

A Hybrid Microchannel and Slot Jet Array Heat Sink for Cooling High-Power Laser Diode Arrays

Zeng Deng, Jun Shen, Wei Dai, Ke Li, and Xueqiang Dong

1 Introduction

High-power laser diode arrays (HPLDAs) have been widely used in pumps for solid-state lasers, medicine, optical data storage, and free-space optical communication due to their high efficiency, compact package, and long-term reliability. Similar to other high-power electronic chips, cooling is of great significance in maintaining the superior performance of HPLDAs including electro-optic conversion efficiency, beam quality, reliability, and long life span. Recently, Sung and Issam [1] proposed that micro/mini-channel and jet impingement cooling can be combined in pursuit of a much higher heat removal ability. They found the vorticity effect when the coolant impinges to the microchannel, and this vorticity effect enhances the heat transfer.

In this paper, the hybrid microchannel and slot jet heat sink is designed and fabricated for the HPLDAs. A commercial laser bar is packaged on the heat sink. Experiments are performed to assess the cooling capability of the new heat sink. In the experiments, the forward voltage method (FVM) is used to obtain transient temperature behavior of the chip, and the structure function method is applied to analyze the thermal resistance distribution from the chip to the coolant inlet. The pressure drop of the heat sink and optical power of the laser chip are also observed. The designed heat sink is compared to published solutions for cooling the HPLDAs.

Z. Deng · J. Shen (✉) · W. Dai (✉) · K. Li · X. Dong
Key Laboratory of Cryogenics, Technical Institute of Physics and Chemistry, Chinese Academy of Sciences, Beijing 100190, China
e-mail: jshen@mail.ipc.ac.cn

W. Dai
e-mail: cryodw@mail.ipc.ac.cn

Z. Deng · J. Shen · W. Dai · X. Dong
University of Chinese Academy of Sciences, Beijing 100049, China

© Springer Nature Singapore Pte Ltd. 2021
C. Wen and Y. Yan (eds.), *Advances in Heat Transfer and Thermal Engineering* ,
https://doi.org/10.1007/978-981-33-4765-6_105

2 Methodology (Length and Layout)

The thermal resistance defined in Eqs. (1) and (2) is used to evaluate the cooling performance of the heat sink. It should be noted that the thermal resistance of the soldering layer may be different for each laser because voids are always produced during the soldering process. For better evaluation of the cooling performance, this data reduction is dedicated to calculate the thermal resistance of the heat sink itself.

$$R(i)_{tot} = \frac{T_{i\,max} - T_0}{Q} \tag{1}$$

$$R_i = R(i)_{tot} - R(i-1)_{tot} \tag{2}$$

where T_{imax} is the maximum temperature of i layer; T_0 is the water inlet temperature; Q is the heating power; R_i is the thermal resistance of each layer; $R(i)_{total}$ is the total thermal resistance from water inlet to the i layer.

The structure function method [2] is used to calculate the thermal resistance of every layer. For a laser system, the temperature train set behavior (temperature drop curve) can be expressed as Eq. (3) after removing the heating load.

$$\Delta T_j(t) = Q \sum_i R_i(1 - \exp(-t/\tau_i)) \tag{3}$$

where τ_i is the thermal time constant of each layer; ΔT_j is junction temperature rise of the chip.

Using the method proposed by Székely [2], then,

$$R_i = \frac{1}{Q} \int_{z_i - \Delta z}^{z_i + \Delta z} \left(1 - \frac{|\zeta - z_i|}{\Delta z}\right) R(\zeta) d\zeta \tag{4}$$

$$C_i = \exp(z_i)/R_i \tag{5}$$

where $z = \ln t$; ζ is the valuable between $z_i - \Delta z$ and $z_i + \Delta z$.

Then, the following parameter K is used to distinguish the layer boundaries. Due to K changes dramatically at the interface, it can be used to distinguish the thermal resistance of each layer.

$$K = dC/dR = \frac{cA dx}{dx/(A\lambda)} = c\lambda A^2 \tag{6}$$

where c is the volume specific heat; λ is heat conductivity; A is the heat transfer area.

3 Results

Figure 1 shows the schematic diagram of the complished HPLDAs. We processed heat sinks of two different slot jet lengths ($L = 0.5$ mm, 1.0 mm), and we set up an experimental system to measure temperature transient behavior of the laser chip after removing the heat load using our published method (forward voltage method) [3].Then, it was transferred to the thermal resistance distribution as shown in Fig. 2. The peak in Fig. 3 represents the resistance boundary between the chip and the heat sink. Then, the thermal resistance of the heat sink (R_h) can be obtained by

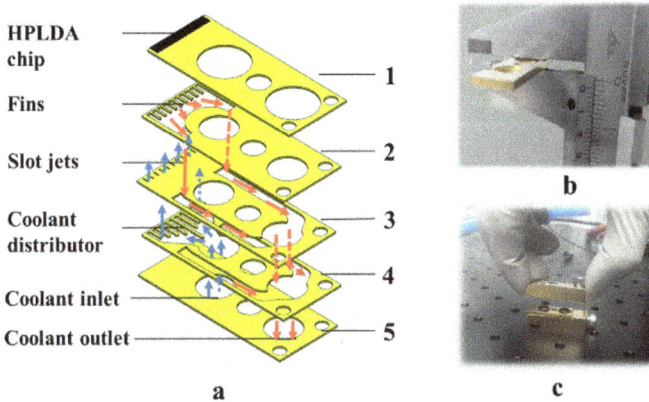

Fig. 1 Structure of heat sink. **a** Schematic diagram of the heat sink. **b** The soldered heat sink. **c** The test module

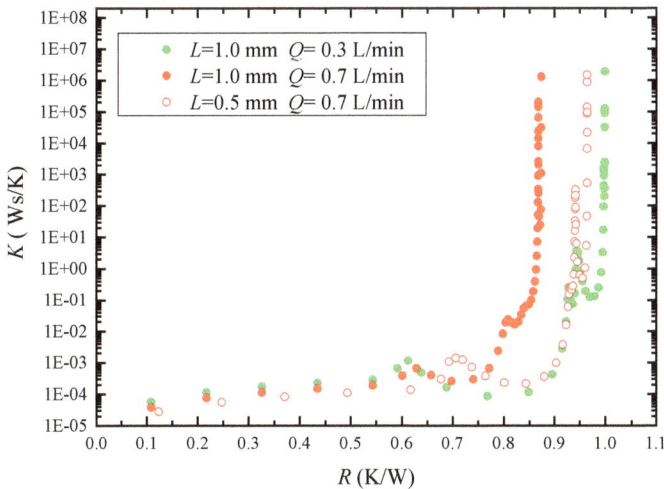

Fig. 2 Thermal resistance versus K parameter

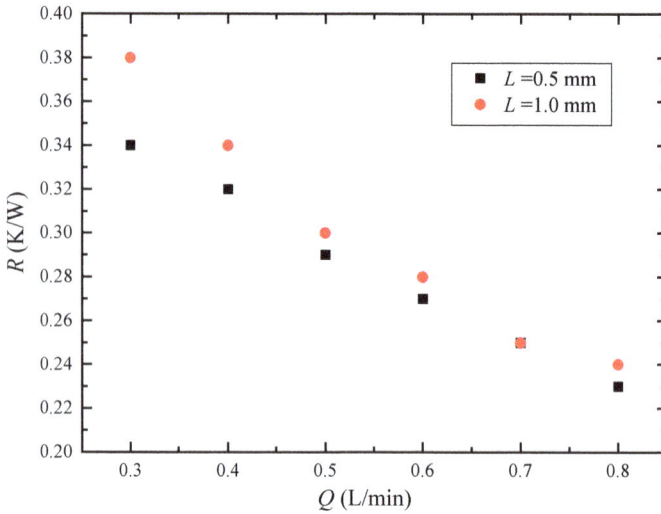

Fig. 3 Thermal resistance of the heat sink versus flowrate

subtracting the thermal resistance of the chip and the soldering layer from the total thermal resistance (R_t). R_h decreases with the increasing flowrate as shown in Fig. 3. The minumum thermal resistance of the heat sink is 0.23 K/W. Figures 4 and 5 present the variations of pressure drop and optical power versus the flowrate corresponding to a pressure drop of 396 kPa. The maximum optical power is up to 85.9 W with related to an electro-optic conversion efficiency of up to 60.6%

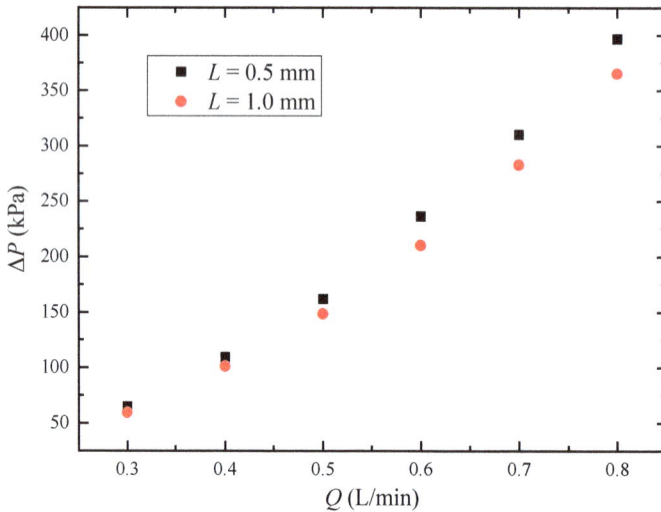

Fig. 4 Pressure drop versus flowrate

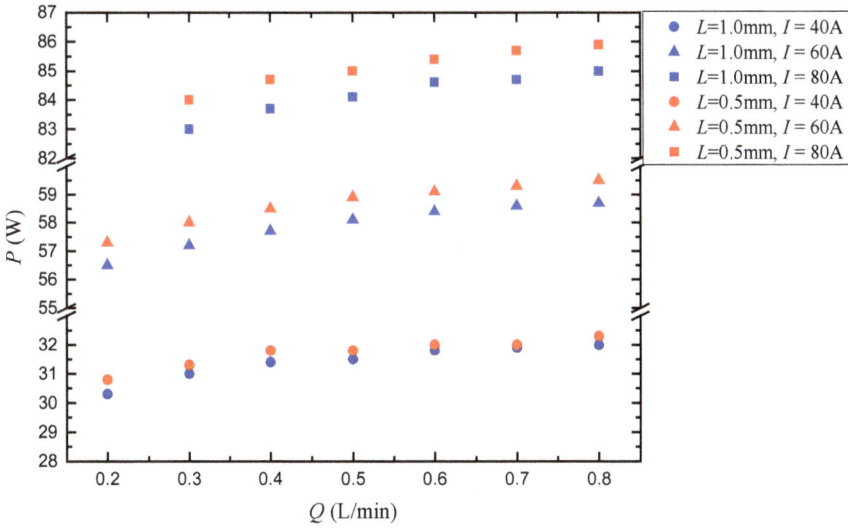

Fig. 5 Optical power versus flowrate

The present work is compared with the published cooling solutions which are concerned with the same type of HPLDAs, as presented in Table 1. Our previous work [3] gave thermal resistance distribution of the popular and long-term reliable two-layer heat sink. The thermal resistance of heat sink is up to 0.71 K/W which is much larger than the present heat sink. Compared with microchannel cooler designed by Kozłowska et al. [4], the total thermal resistance decreases about 15%. But their

Table 1 Illustrative comparison of the proposed design with other existing solutions

Authors	Brief description	ΔP (kPa)	Q (L/min)	R_h (K/W)	R_t (K/W)
Present study	$L = 0.5$ mm	397	0.8	0.23	0.94
	$L = 1.0$ mm	365	0.8	0.24	0.87
Deng et al. (experimental)	Two-layer macrochannel heat sink	–	1.7	0.71	1.50
Kozłowska et al. (experimental)	Microchannel heat sink with micro-pipes inside	–	0.09	–	1.02
Dix and Jokar [5] (numerical)	Zig-zag microchannel cooler	98.4	0.2	0.44	–
	Straight microchannel cooler	38	0.2	0.44	–
Farsad et al. [6] (numerical)	Micro-channel cooler filled with metal foam medium	1400	0.2	0.33	–
Beni et al. [7] (numerical)	Sinusoidal micro-channel	17.6	0.2	0.28	–

flowrate is only 0.09 L/min. This is because their cooler uses micropipe to input the coolant and the gap between the pipe and channel to output the coolant, and the pressure drop increases greatly with the increasing flowrate. A large flowrate with the corresponding big pressure drop challenges the sealing performance of the heat sink. Compared with some numerical studies, the hybrid heat sink shows a decrease in thermal resistance and an increase in the flowrate. The thermal resistance of heat sink decreases 48% compared with the straight microchannel cooler. When the microchannel is filled with metal foam medium, the thermal resistance decreases, but it is still 33% larger than the present study. Moreover, the pressure drop increases greatly up to 1400 kPa which is not acceptable for the current processing technology. Sinusoidal micro-channel study may be an option for cooling of the HPLDAs. It has a small thermal resistance and pressure drop, but the numerical conclusion should be further validated by experimental results.

4   Conclusions

The present study gives the design, fabrication, package and test of a new heat sink. The heat sink combines the microchannel and slot jet, and it is successfully applied in the HPLDAs. The forward voltage method was introduced to measure the chip temperature transient behavior after removing the heating load. Using the structure function method, the thermal resistance can be easily obtained. Under the condition of the maximum flowrate (0.8 L/min), the thermal resistance of the heat sink ($L = 0.5$ mm) is only 0.23 K/W with a pressure drop of 397 kPa. The thermal resistance decreases greatly compared with published solutions for cooling the HPLDAs, indicating that the hybrid heat sink will be a promising solution for improving cooling the high-power electronic devices.

Acknowledgements This work was supported by the National Key Research and Development Program of China (No. 2016YFB0402102), the Scientific and Equipment Research Program, CAS (No. YZ201532) and the Critically Arranged Program, CAS (No. KGZD-SW-T01-1).

References

1. M. Sung, M. Issam, Single-phase and two-phase cooling using hybrid micro-channel/slot-jet module. Int. J. Heat Mass Transfer **51**, 3825–3839 (2008)
2. V. Székely, Identification of RC networks by deconvolution: chances and limits. IEEE Trans. Circ. Syst. I Fundam. Theory Appl. **45**, 244–258 (1998)
3. Z. Deng, J. Shen, W. Gong, W. Dai, M. Gong, Temperature distribution and thermal resistance analysis of high-power laser diode arrays. Int. J. Heat Mass Transfer **134**, 41–50 (2019)
4. A. Kozłowska, P. Łapka, M. Seredyński, M. Teodorczyk, E. Dąbrowska-Tumańska, Experimental study and numerical modeling of micro-channel cooler with micro-pipes for high-power diode laser arrays. Appl. Therm. Eng. **91**, 021011-1–7 (2015)

5. J. Dix, A. Jokar, Fluid and thermal analysis of a microchannel electronics cooler using computational fluid dynamics. Appl. Therm. Eng. **30**, 948–961 (2010)
6. E. Farsad, S.P. Abbasi, M.S. Zabihi, J. Sabbaghzadeh, Fluid flow and heat transfer in a novel microchannel heat sink partially filled with metal foam medium. J. Therm. Sci. Eng. Appl. **6**, 479–490 (2014)
7. S.B. Beni, A. Bahrami, M.R. Salimpour, Design of novel geometries for microchannel heat sinks used for cooling diode lasers. Int. J. Heat Mass Transfer **112**, 689–698 (2017)

◆

Thermal Design for the Passive Cooling System of Radio Base Station with High Power Density

Kai-Wen Duan, Ji-Wei Shi, and Wen-Quan Tao

1 Introduction

As communication systems are gradually transferred to 5G, communication base station (CBS) is developing toward large capacity, high power density, and high integration. The system's heat dissipation is getting larger while its size is turning to be smaller. In this case, thermal reliability has become the bottleneck of reliability design. According to statistics, 55% of failure rate of electronic equipment is caused by thermal failure. Therefore, the thermal design problem of the device has become an important issue in the design of the outdoor base station system.

The studied case is a radio base station (RBS) of high power density. Operating in outdoor scenarios, RBS requires unattended duty, maintenance-free, and long life-time. Compared with active heat dissipation, passive cooling scheme is the optimal choice for reducing temperature of RBS. The purpose of thermal design is to achieve the lowest average temperature of key components and heat sink (HS), the optimal size and weight of whole system, under the assigned power density in the limited space. This article will discuss various improvement methods via numerical simulation, including installation method, metal conduction rod [1], emissivity, geometric optimization of HS, graphene film, vapor chamber, metal foam, and wire-plate ionic generator.

K.-W. Duan · W.-Q. Tao (✉)
School of Energy and Power Engineering, Xi'an Jiaotong University, Xi'an 710049, China
e-mail: wqtao@mail.xjtu.edu.cn

J.-W. Shi
Key Laboratory of Thermo-Fluid Science and Engineering, Ministry of Education, Xi'an 710049, China

© Springer Nature Singapore Pte Ltd. 2021
C. Wen and Y. Yan (eds.), *Advances in Heat Transfer and Thermal Engineering* ,
https://doi.org/10.1007/978-981-33-4765-6_106

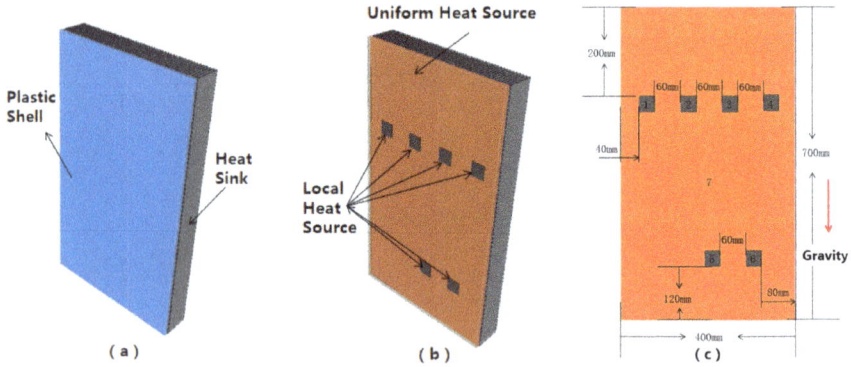

Fig. 1 RBS with plastic shell **a**, RBS without plastic shell **b**, Layout of chips **c**

2 Methodology

The studied RBS consists of PCB, six chips, plastic shell, and HS as shown in Fig. 1. PCB produces heat dissipation of 560 W which is simplified to an uniform heat source, while 60 W heat dissipation for each chip is simplified to local heat source. The layout and size of chips on PCB are shown in Fig. 1C. The PCB and chips on it are enclosed in the plastic shell with size of 700 mm × 400 mm × 10 mm and thickness of 2 mm, while the other side of PCB is attached to the substrate of HS for cooling. The overall height of initial aluminum extruded HS is 60 mm, while the thickness of substrate is 15 mm. This HS has 20 fins with 2 mm thickness. The optimization of HS will be described step by step below.

The ambient temperature is 45 °C while the limiting high temperature of key components is 85 °C.

The CFD software Icepak 14.0 was used to solve the 3D steady-state heat transfer and fluid flow equations based on several assumptions. The turbulent flow was modeled with standard $k-\varepsilon$ model, and the Boussinesq hypothesis was introduced to deal with the buoyancy effect caused by temperature difference. All heat sources work as surface sources. The surface-to-surface radiation model in Icepak [2] is used here with 0.85 emissivity for all solid surface.

3 Results

The grid independence study is performed by three different grid systems within 2% difference in results. The thermal design process is divided into nine steps shown in Table 1 with its result. The evaluation indices K_1 and K_2 are calculated by Eq. (1) to evaluate the thermal performance of system in terms of mass and size.

$$K_1 = (85 - T_{\text{global\_max}})/ M_{\text{HS}} \quad K_2 = (85 - T_{\text{global\_max}})/ H_{\text{HS}} \qquad (1)$$

Table 1 Thermal design results and evaluation indices

Case	Method	T_{global_max}	T_{ave_local}	M_{HS}	H_{HS}	K_1	K_2
1	Initial	103.5 (°C)	100.8 (°C)	15.3 (kg)	60 (mm)	0.10	0.02
2	Downside	102.7	99.7	15.3	60	0.15	0.04
3	Conduction rod	102.4	99.6	15.8	60	0.17	0.04
4	Graphene film	97.1	95.3	16.4	60	0.48	0.13
5	Vapor chamber	92.1	91.4	11.3	60	1.14	0.22
6	radiation	91.0	90.5	11.3	60	1.24	0.23
7	Fin thickness	92.7	92.1	9.1	60	1.35	0.21
8	Fin number	92	91.3	9.4	60	1.39	0.22
9	Fin height	81.2	80.5	11.5	110	2.07	0.22

Case 1 shows the thermal performance of the initial model. It can be concluded that the global temperature highly exceeds 85 °C in this stage with maximum global temperature of 103.5 °C. It is observed that the temperature of local heat source at the upper part of HS is significantly higher than that at bottom.

Due to the buoyancy lift, the high temperature air will rise. Therefore, it is recommended in Case 2 to install PCB with high-power dissipation chips at bottom. Without additional input, 0.8 °C decrease of T_{global_max} is attained in Case 2.

Observing that the temperature distribution of plastic shell is uniform and its temperature is relatively lower, it is considered to enhance heat transfer in Case 3 by adding copper conduction rods between chips and shell to rapidly conduct part of heat dissipation from chips to plastic shell. 0.3 °C global temperature decrease is attained in Case 3 which seems not to be significant. However, this actually helps to implement the next step in Case 4.

It is observed in Case 3 that local hot spots appear on the outside of shell which should be spread uniformly with the application of graphene film between inner surface of shell and rods in Case 4. High planar thermal conductivity of graphene film helps to decrease diffusion thermal resistance and consequently reduce global temperature by 5.3 °C.

After uniforming the surface temperature of plastic shell, it is observed that hot spots on HS side need to be handled as well. Application of vapor chamber could spread heat in the plane rapidly and provide bigger heat capacity. The relative simulation was performed in Case 5 which shows the best thermal optimization till now with maximum temperature decrease of 5 °C. Simultaneously, the mass of whole system is reduced to 11.3 kg with the same size.

Enhancement of radiation heat transfer by improving emissivity from 0.85 to 1 is simulated in Case 6. Even though it attains quite slight decrease in temperature, high surface treatment cost rejects this optimization method.

The following steps are in order to optimize the HS.

In Case 7 and Case 8, we figured out separately the optimal fin thickness of 1 mm and the optimal fin number of 22 with the T_{global_max} of 92 °C.

As for the height of fin, the results show that evaluation indices increase stably within the overall height from 60 to 110 mm. Eventually, we choose Case 9 with 5 mm thickness of vapor chamber substrate and 105 mm fin height as optimal case with best indices. In this stage, the T_{global_max} has been optimized to 81.2 °C which is already below the limiting high temperature of 85 °C.

As power density of device increases, more advanced cooling technologies should be applied to thermal design. Attached to double sides of fins of HS, metal foam could greatly increase the surface area of heat transfer and promote air disturbance, while the corresponding pressure drop leads to significant decrease of air velocity which decreases heat transfer coefficient. Wire-plate ionic generator could enhance air disturbance and generate ionic wind as well to augment air velocity. Therefore, the combination of metal foam and wire-plate ionic generator [3] and its application on HS could significantly improve the thermal performance of HS. According to relative studies, its surface heat transfer coefficient can be estimated to increase by 3 times at least under the same conditions. This kind of combination can be regarded as the advanced cooling technology in future for RBS.

4 Conclusions

Several thermal design methods are studied in this article to enhance passive cooling, including installation method, metal conduction rod, geometric optimization of HS, and application of graphene film, vapor chamber. With these methods, the T_{global_max} drops from 103.5 to 81.2 °C, and the mass of system drops from 15.3 to 11.5 kg, while the overall height of HS increases from 60 to 110 mm. This passive cooling design is the optimal one in terms of temperature, mass and size. As power density of device increases gradually, combination of metal foam and wire-plate ionic generator can be considered as a new technology to improve thermal performance of system.

Acknowledgements This work was supported by Xi'an Science and Technology Bureau (2019218714SYS002CG024).

References

1. C. Wei, Z.J. Liu, Z.Y. Li, Z.G. Qu, W.Q. Tao, Numerical study on some improvements in the passive cooling system of a radio base station. Numer. Heat Transfer Part A **62**, 319–335 (2012)
2. Fluent Inc., Icepak 14.0 documentation (2014)
3. A. Ramadhan, N. Kapur, J.L. Summers, H.M. Thompson, Numerical development of EHD cooling systems for laptop applications. Appl. Therm. Eng. **139**, 144–156 (2012)

Design and Experimental Study on a New Heat Dissipation Method for Watch-Phones

Wenjie Zhou, Yong Li, Zhaoshu Chen, Yuying Yan, and Hanyin Chen

1 Introduction

With the development of communication and electronic technology, the functions of watch-phones have become more and more powerful, such as photography, video call, positioning and navigation which leads to high power consumption and heat dissipation issue of the chips. The traditional way of using thin copper sheet to dissipate the heat could not effectively solve the issue of the current watch-phones. Ultra-thin flattened heat pipe (UTHP) has excellent temperature uniformity and heat transfer performance [1], so it has been widely used in the heat dissipation of thin and light portable electronic devices, such as smartphones, ultra-thin computers and tablets [2]. Some researchers have conducted useful work to improve the heat dissipation performance of electronic devices by using UTHP. In 2016, Li et al. [3] proposed a type of UTHP with the spiral woven mesh (SWM) wick structure for cooling mobile phones and disclosed a stable and economical manufacturing process for the UTHP. Ahamed et al. [4] developed an UTHP with a center fiber wick and fabricated a module made by the UTHP and a metal sheet for cooling electronics. However, how to apply UTHP to the thermal management of watch-phone has not been reported. The main difficulty lies in the compact structure and small heat dissipation space

W. Zhou · Y. Li (✉) · Z. Chen
School of Mechanical and Automotive Engineering, South China University of Technology, Guangzhou 510640, China
e-mail: meliyong@scut.edu.cn

Y. Yan (✉)
Fluids and Thermal Engineering Research Group, Faculty of Engineering, University of Nottingham, Nottingham NG72RD, UK
e-mail: yuying.yan@nottingham.ac.uk

H. Chen
Guangdong Newidea Technology Co., Ltd, Guangzhou 510610, China

© Springer Nature Singapore Pte Ltd. 2021
C. Wen and Y. Yan (eds.), *Advances in Heat Transfer and Thermal Engineering* ,
https://doi.org/10.1007/978-981-33-4765-6_107

inside the watch-phone. Therefore, designing an appropriate UTHP cooling module for watch-phone is an important issue to improve its heat dissipation performance.

In the present work, a new heat dissipation method based on UTHP is developed for cooling watch-phones. A UTHP cooling module was designed and fabricated by soldering a UTHP and a copper plate. The thickness of the UTHP was 0.4 mm, which was fabricated by flattening a cylindrical heat pipe with an outer diameter of 2 mm, in order to adapt to the small cooling space inside watch-phone. The heat dissipation performance of the UTHP cooling module was experimentally investigated.

2   Experiment

The UTHP samples were fabricated by flattening the cylindrical heat pipes with an outer diameter of 2 mm and thickness of 0.08 mm. The tube material was C1020. The length of UTHP was 86 mm. The wick of the UTHP was SWM, which was woven using 96 copper wires with an outer diameter of 0.04 mm. The 96 copper wires were divided into 24 strands, and every strand had six copper wires. The strands were spirally woven into a hollow ring shape to form the SWM structure. The main manufacturing process for the UTHPs was same with the reference [3, 5]. As shown in Fig. 1a, the UTHP was designed to be a three-dimensional shape according to the watchcase shape to improve the heat dissipation performance of the cooling module. The head and tail ends of the UTHP were bent inward by 90° and perpendicular to the bottom. These two ends were bonded to the inner wall of the watchcase through a thermal conductive adhesive. The thickness of the copper plate was 0.15 mm, and it was soldered on the bottom of the UTHP to increase the contact area with heating chip and the assemblability of the UTHP cooling module (HPM). A copper sheet with the same dimensions was used as a comparison to study the heat

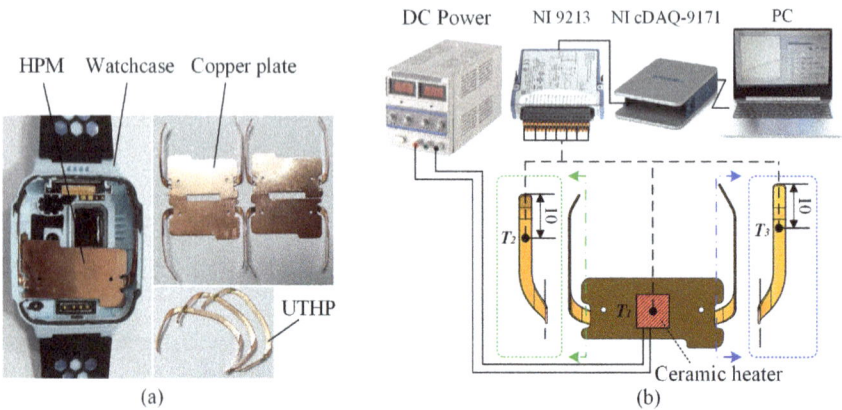

Fig. 1 **a** Cooling module and **b** heat transfer performance testing system (unit: mm)

transfer performance of the UTHP. Analogously, the heat dissipation performance of the copper sheet cooling module (CSM) with the same dimensions was made for comparison with the UTHP cooling module. The masses of the HPM and CSM were 2.93 and 3.65 g, respectively.

As shown in Fig. 1b, the heat transfer performance testing system mainly consists of a heating module and a temperature data acquisition module. The heating module is made of a ceramic heater (10 mm × 10 mm × 1 mm) and a DC power supply, which regulated the heat input power to the heater. The temperature data acquisition module consists of a PC, a temperature input module and three testing points (T-type thermocouples with a diameter of 0.125 mm). The heater was fixed in the middle of the copper plate, and T_1 was tightly attached to the heater surface. T_2 and T_3 were tightly attached to the surface of the UTHP two ends. The system used natural convection to dissipate the heat.

The cooling modules were placed horizontally with the copper plate and the ceramic heater attached tightly, and thermal grease with a thermal conductivity of 3.5 W/(m K) was used to reduced their contact thermal resistance. The heat input power started from 0.2 W and increased by increment of 0.1 W. The temperature data of the testing points were recorded until the UTHPs reached an equilibrium state, at which point the temperature change was less than 0.5 °C over 30 s. The maximum heat dissipation power is defined as the maximum heat input power at which the cooling module still operates in at a normal state. The surface temperature of the heater T_h, namely T_1, was set to not exceed 65 °C, which was consistent with the actual application.

$$\Delta T = T_h - \max(T_2, T_3) \tag{1}$$

where ΔT is the maximum temperature difference of cooling module.

3 Results

The curve CH in Fig. 2 indicates the temperature change of the ceramic heater without the HPM and CPM cooling modules. Figure 2a shows the changes of T_h with time. T_h raised rapidly at the beginning of each test and then finally reached a stable state. As the heating power increases, T_h of CH rises fastest. When the heating power was greater than 0.4 W, T_h of CH had exceeded 65 °C. At the lower heating powers (0.2 and 0.3 W), the T_h changes of CPM and HPM were similar, such as start-up time (from start-up to stable states) and equilibrium temperature. When the heating power was greater than 0.3 W, HPM had a shorter start-up time and lower equilibrium temperature compared with CPM. This phenomenon was more obvious when the heating power was greater than 0.5 W. Figure 2b presents T_h of different cooling systems under various heating powers. As the heating power increases, the temperature of the heater rises rapidly. The temperature that did not use cooling module rises the fastest, while the temperature rise with HPM is the

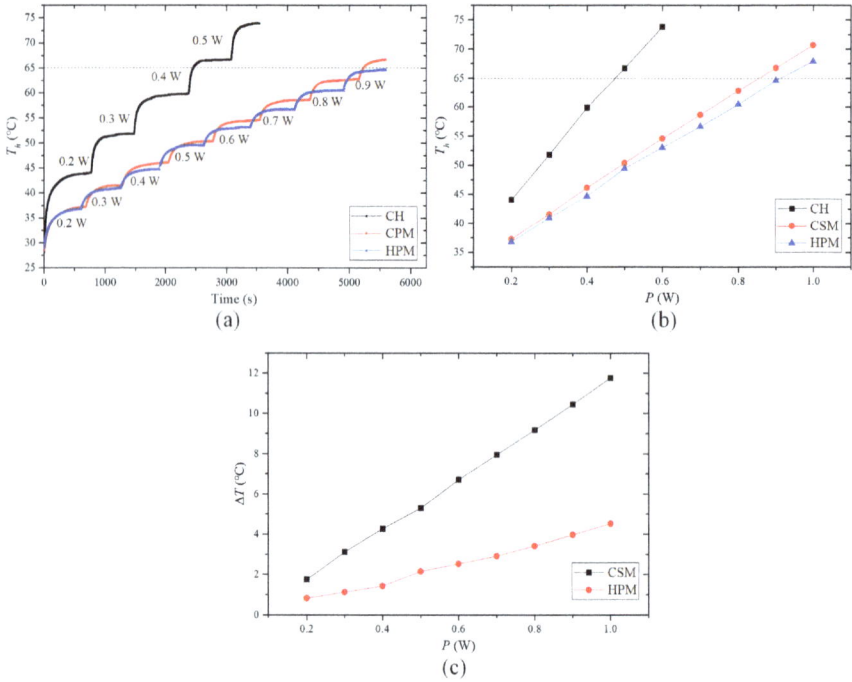

Fig. 2 Temperature data of different cooling systems: **a** T_h varies with time, **b** T_h under various heating powers and **c** ΔT of CSM and HPM under various heating powers

slowest. The maximum heat dissipation power of the CH, CSM and HPM were 0.4, 0.8 and 0.9 W, respectively, at T_h not exceeding 65 °C. Compared with the system without cooling module, the maximum heat dissipation power of HPM had increased by 125%; compared with CSM, the maximum heat dissipation power of HPM had increased by 12.5%, but the mass had been reduced by 19.7%. Figure 2c shows ΔT of CSM and HPM under various heating powers. As the heating power increases, ΔT of CSM rises rapidly, while ΔT of HPM rises slowly. When the heat input powers of CSM and HPM were their respective maximum heat dissipation powers, ΔT of them are 9.18 and 3.98 °C, respectively. The results show that HPM has excellent temperature uniformity, which can rapidly spread the heat/temperature of the heater with a small temperature difference and increase the average temperature of the cooling module, which increases the temperature difference between the cooling module and the environment, thereby improving the heat dissipation efficiency of the cooling module.

4 Conclusions

A new heat dissipation method based on UTHP is developed for cooling watch-phones. A UTHP cooling module was designed and fabricated by soldering a UTHP and a copper plate. The heat dissipation performance of the UTHP cooling module was experimentally investigated. The main conclusions are as follows:

Compared with the system without cooling module, the maximum heat dissipation power of HPM had increased by 125%; compared with CSM, the maximum heat dissipation power of HPM had increased by 12.5%, but the mass has been reduced by 19.7%. The results show that the UTHP cooling module has a better heat dissipation performance, which can quickly reduce the chip temperature, eliminate the chip hotspot and solve the heat dissipation issue of the watch-phone.

Acknowledgements This work was supported by the National Natural Science Foundation of China (grant number 51675185), the Natural Science Foundation of Guangdong Province (grant number 2018B030311043), the Project of the Guangzhou Science and Technology Plan (grant numbers 201807010074 and 201707010071), the Project of Tianhe District Science and Technology Plan (grant number 201705YX263) and the Fundamental Research Funds for the Central Universities, SCUT.

References

1. X. Yang, Y.Y. Yan, D. Mullen, Recent developments of lightweight, high performance heat pipes. Appl. Therm. Eng. **33–34**(1), 1–14 (2012)
2. H. Tang, Y. Tang, Z.P. Wan et al., Tang, Review of applications and developments of ultra-thin micro heat pipes for electronic cooling. Appl. Energy **223**, 383–400 (2018)
3. Y. Li, W.J. Zhou, G.W. Huang et al., An ultra-thin heat pipe for cooling mobile phone. China patent (2016)
4. M.S. Ahamed, Y. Saito, K. Mashiko et al., Characterization of a high performance ultra-thin heat pipe cooling module for mobile hand held electronic devices. Heat Mass Transf. **53**(11), 3241–3247 (2017)
5. W.J. Zhou, Y. Li, Z.S. Chen et al., Ultra-thin flattened heat pipe with spiral woven mesh wick for smartphone cooling, in: *Proceedings of the 19th International Heat Pipe Conference and 13th International Heat Pipe Symposium*. Pisa, Italy, 10–14 June 2018

Heat Exchanger

Experimental and Numerical Investigation on Fouling and Heat Transfer Performance of a Novel H-type Finned Heat Exchanger

Song-Zhen Tang, Zhan-Bin Liu, Ming-Jia Li, and Wen-Quan Tao

1 Introduction

Gas-side medium of the heat exchanger is usually unclean, and it is easy to form fouling on the heat transfer surface, which seriously reduces the utilization efficiency of the waste heat resources [1]. Therefore, it is necessary to study the fouling mechanism and the effect of fouling on the heat transfer performance under dusty conditions.

Experimental studies on the fouling characteristics of heat exchangers include laboratory experiments and field experiments. In laboratory research, Nuntaphan and Kiatsiriroa [2] experimentally studied the heat transfer performance of spiral finned tube heat exchangers under dusty conditions and obtained the correlation of the fouling thermal resistance. However, simulated flue gas in the laboratory is different from the actual flue gas composition at the site. Therefore, some scholars have carried out online experimental research on the fouling characteristics of flue gas heat exchangers. Shi et al. [3] also studied the effect of fouling on the heat transfer performance of the spiral finned tube heat exchanger in CFB boiler and obtained the fouling thermal resistance correlation of the different arrangements. However, the experimental section is embedded inside the boiler and has certain limitations. To this end, Wang and He et al. [4] built an online experimental system for fouling characteristics of heat transfer elements outside the heating boiler and studied the fouling and heat transfer characteristic of five different H-type finned tube heat exchangers. It can be seen from the above review that it is very necessary to propose a novel heat transfer surface suitable for dusty environment based on the numerical prediction of fouling characteristics, carry out online experimental

S.-Z. Tang · Z.-B. Liu · M.-J. Li (✉) · W.-Q. Tao
Key Laboratory of Thermo-Fluid Science and Engineering of Ministry of Education, School of Energy and Power Engineering, Xi'an Jiaotong University, Xi'an 710049, Shaanxi, China
e-mail: mjli1990@mail.xjtu.edu.cn

© Springer Nature Singapore Pte Ltd. 2021
C. Wen and Y. Yan (eds.), *Advances in Heat Transfer and Thermal Engineering* ,
https://doi.org/10.1007/978-981-33-4765-6_108

Fig. 1 Online experiment system of fouling characteristics

research on the fouling characteristics in the actual environment and perform the comparative verification with the numerical prediction method.

2 Methodology

The chain boiler of the thermal power company's urban heating station is selected for on-site experiment, and the online experimental system for studying the fouling and heat transfer performance of heat transfer components in industrial processes is built. The experimental system follows the connection section of the gravity settling chamber and the bag filter in the heating boiler. The experimental system (as shown in Fig.1) includes a flue gas circulation loop, a cooling water circulation loop and an experimental data acquisition system.

3 Results

The experimental elements studied in this paper are divided into circular tube, elliptical tube, H-type finned tube and H-type finned tube with longitudinal vortex generator (LVG), as shown in Fig. 2, which constitute six different tube bundle arrangements: aligned circular tube bundle (ACTB), aligned elliptical tube bundles (AETB), staggered circular tube bundles (SCTB), aligned H-type finned circular tube bundles (AHCTB), aligned H-type finned elliptical tube bundles with LVGs (AHETB) and

(a) circular tube   (b) elliptical tube

(c) H-type finned circular tube   (d) H-type finned elliptical tube

（d）H-type finned elliptical tube with LVGs

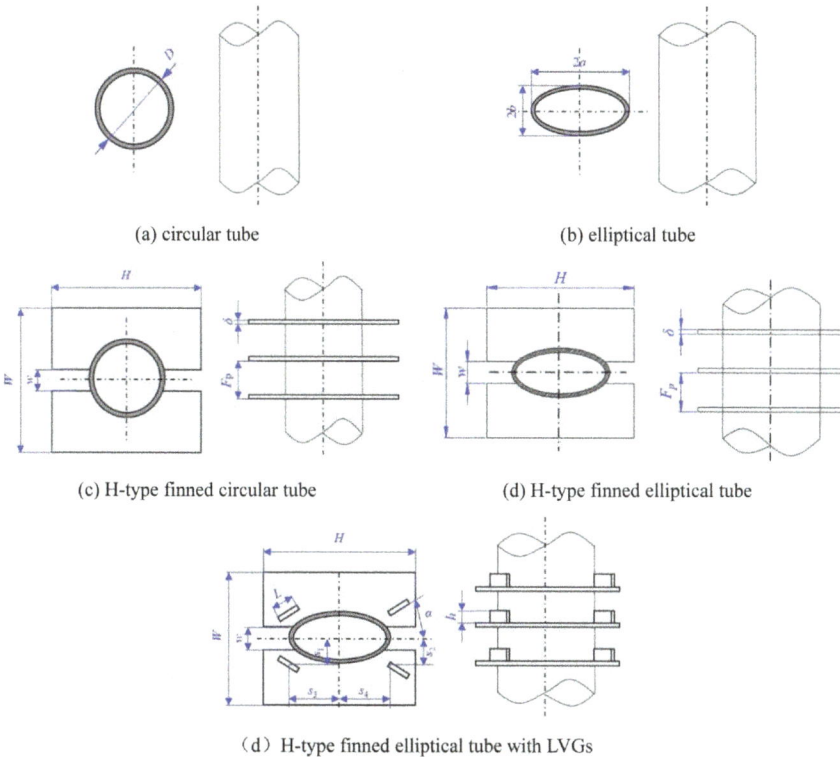

Fig. 2 Online experiment system of fouling characteristics

staggered *H*-type finned elliptical tube bundles with LVGs (SHETB). The relative lateral and longitudinal pitches of all tube bundles are consistent.

Through the online experimental study on the fouling characteristics of different tube bundles, the heat transfer coefficients in the dusty environment are obtained with time. Figure 3 shows the heat transfer coefficients of the different tube bundles in the clean state and the weaken degree of the heat transfer coefficient when the fouling is stable. It can be seen that in the clean state, the order of heat transfer coefficients is SETB > AETB > ACTB > SHETB > AHETB > AHCTB, that is, the staggered elliptical tube bundle arrangement is the best, followed by aligned elliptical tube bundle, and aligned circular tube bundle is the worst. In addition, it can be seen from Fig. 3 that when the fouling reaches a substantially stable state, the weaken degrees of ACTB, AETB, SCTB, AHCTB, AHETB and SHETB are reduced by 21.5%, 11.1%, 10.5%, 9.0%, 6.8% and 3.6%, respectively. This means that the elliptical tube arrangement can effectively reduce the effect of fouling on heat transfer performance. Overall, SHETB has the best heat transfer and anti-fouling performance, followed by AHETB. The reasons can be explained as follows: (1) Slit structure of *H*-type finned tube can not only improve the velocity near the stagnation point on the windward side and enhance the axial scavenging shunt, but also increase the fluid disturbance

Fig. 3 Comparison results
of heat transfer coefficient
and weaken degrees

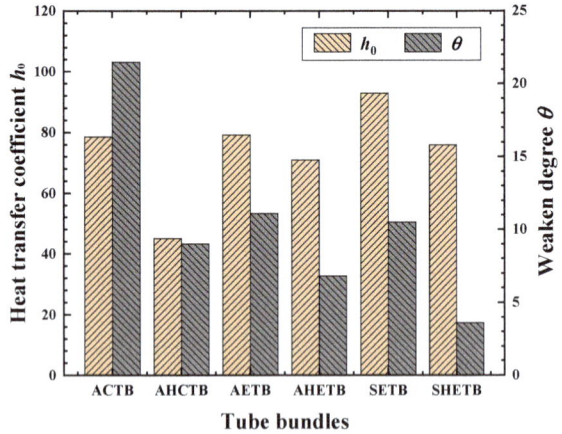

on the leeward side; (2) Elliptical tube has a small windward area, which can reduce the vortex area behind the tube and reduce the probability of particle deposition; (3) longitudinal vortex generator generates strong three-dimensional spiral flow, which increases the probability of particle removal.

On this basis, in order to further analyze the causes and distribution of the fouling, the fouling prediction method established by our team is used to numerically predict the fouling characteristics of the ACTB, AETB and SETB. The fouling morphologies of three tube bundle arrangements are obtained. The main reason for the difference in the effect of fouling on heat transfer performance is revealed.

4   Conclusions

In this study, experimental and numerical investigations on the fouling and heat transfer characteristics in cross flow heat exchangers were carried out. The on-site online experimental system was established, and the heat transfer coefficients for different heat transfer surfaces at the fouling state are monitored in real time. Results showed that compared with the aligned circular tube bundle, using elliptical tube bundle can improve the anti-fouling performance, especially for the staggered H-type finned elliptical tube bundles with LVGs, which has the best heat transfer and anti-fouling performance and suitable for waste heat utilization of dusty flue gas.

Acknowledgements The present work is supported by the National Key R&D Program of China (2018YFB0605901).

References

1. M.J. Li, S.Z. Tang, F.L. Wang et al., Gas-side fouling, erosion and corrosion of heat exchangers for middle/low temperature waste heat utilization: a review on simulation and experiment. Appl. Therm. Eng. **126**, 737–761 (2017)
2. A. Nuntaphan, T. Kiatsiriroat, Thermal behavior of spiral fin-and-tube heat exchanger having fly ash deposit. Exp. Therm. Fluid Sci. **31**, 1103–1109 (2007)
3. Y.T. Shi, M. Gao, G.H. Tang, F.Z. Sun, W.Q. Tao, Experimental research of CFB ash deposition on helical finned tubes. Appl. Therm. Eng. **37**, 420–429 (2012)
4. F.L. Wang, S.Z. Tang, Y.L. He, F.A. Kulacki, Y. Yu, Heat transfer and fouling performance of finned tube heat exchangers: experimentation via on line monitoring. Fuel **236**, 949–959 (2019)

.

Improvement of Multi-objective Optimization Tool for Shell-and-Tube Heat Exchanger Design

Thomas McCaughtry and Sung in Kim

1 Introduction

Heat exchangers (HE) have a wide range of industrial applications, and a thorough understanding of this equipment is one of key engineering knowledge. "Heat XChange Designer" is a multi-objective optimization computer program for shell-and-tube heat exchanger (STHE) design, successfully developed and validated for teaching purposes [1]. The log mean temperature difference (LMTD) method and the effectiveness–number of transfer units (ε-NTU) method were used for the sizing and rating of a STHE, respectively. A teaching–learning-based optimization (TLBO) algorithm was implemented to enable effective multi-objective (effectiveness, cost) optimization procedure.

"Heat XChange Designer" used the Kern method to calculate pressure drop in a shell-side heat exchanger. However, inaccuracies in this model are widely acknowledged. From the previous results, the pressure drop was overestimated by 25% in shell-side and underestimated by 10% in tube-side against a reference case [1]. The equations for calculating shell-side heat transfer coefficient and pressure drop using the Kern method are given in Eqs. (1) and (2), respectively [2].

$$h_s = 0.36(\text{Re}_s)^{0.55} \frac{k}{D_e} \left(\frac{c_p \mu}{k} \right)^{1/3} \left(\frac{\mu_s}{\mu_w} \right)^{0.14} \tag{1}$$

$$\Delta p_s = \frac{f G_s^2 (N_b + 1) D_s}{2 \rho D_e \phi_s} \tag{2}$$

T. McCaughtry · S. Kim (✉)
School of Mechanical and Aerospace Engineering, Queen's University Belfast, Belfast BT7 1NN, UK
e-mail: s.kim@qub.ac.uk

© Springer Nature Singapore Pte Ltd. 2021
C. Wen and Y. Yan (eds.), *Advances in Heat Transfer and Thermal Engineering* ,
https://doi.org/10.1007/978-981-33-4765-6_109

The Kern method reduces calculation time and requires relatively fewer user inputs, however, it does not take into account the effect of the leakage and bypass streams of the shell-side flows, which are complex, combining crossflow and baffle window flow [2]. The Bell–Delaware method (BDM) is an alternative, more reliable method for shell-side analysis. BDM accounts for the various shell-side flow streams including tube baffle leakage, crossflow, crossflow bypass, shell baffle leakage and tube pass partition bypass [2]. This was implemented in for both "Design Mode" of the sizing problem and "Performance Mode" of the rating problem for the improvement of the program accuracy. Fouling (deposits on the tube wall, reducing conductivity) and tube wall resistivity were considered in the term of R_{dw} in Eq. (3) for the overall heat transfer coefficient.

$$U_f = \frac{1}{\frac{A_o}{A_i h_i} + R_{dw} + \frac{1}{h_s}} \tag{3}$$

A compact solution for shell-side heat transfer coefficient and pressure drop by Serna and Jiménez was more readily applied to "Design Mode" than the standard BDM, but has the same range of application [3]. It can calculate STHE geometry given the required heat duty and could also be modified for application to "Performance Mode".

2 Methodology

The relationships between shell-side and tube-side heat transfer coefficients and pressure drops in the compact solution are shown in Eqs. (4) and (5), respectively [3]. In "Design Mode", there is an inner loop that determines tube-side heat transfer coefficient and an outer loop which calculates shell-side heat transfer coefficient and the corresponding geometry [3]. The equation of the inner loop is given by Eq. (6) which is solved using the Newton–Raphson method.

$$\Delta p_s = K_s A_o (h_s)^m \tag{4}$$

$$\Delta p_t = K_t A_o (h_t)^n \tag{5}$$

$$h_t - \left[\frac{(\Delta p_t F_t \Delta T_{lm}/K_t Q)}{\left(K_s \Delta p_t / K_t \Delta p_s (h_t)^n\right)^{1/m} + R_{dw} + D_t / D_{ti} h_t} \right]^{1/n} = 0 \tag{6}$$

The compact solution results in a non-integer number of baffles; therefore, baffle cut was modified depending on the difference to the nearest integer, causing the number of baffles to converge to an integer value as the loop iterates [3]. Thermal calculations using the BDM are conducted at mean fluid temperature, but in "Performance Mode", the outlet temperatures are unknown. The updated program features a loop that recalculates fluid properties (temperature-dependent) once the first iteration has completed and the outlet temperatures are known. For all loops used in the program, there are upper limits for the number of iterations above which the analysis is stopped and the user alerted. Otherwise, the analysis stops when convergence criteria are met. For "Design Mode", these values are the "tear variables", and for "Performance Mode", these values are the outlet temperatures and heat transfer coefficients [3].

An accurate knowledge of the maximum number of tubes for a given shell size is critical for the BDM, especially for the optimization when the program checks whether geometry is feasible. In the existing program, the calculation for maximum number of tubes was unsuitable because it determined how many tubes could fit in a tube row along a chord, then moving to the next tube row and repeating these steps. However, because there is no relation between tube rows, the desired tube layout (e.g. in-line, triangular) becomes skewed, and the total is overestimated. An approach similar to [4] was taken where the program checks if the centre of each tube lies inside the maximum radius (moving along every tube row and then through each tube row) and counting the number of acceptable tubes accounting for multiple tube passes, tube bundle supports (fixed tubesheet, U-tube) and geometry (tube bundle clearances, pass partition lane width). For STHE with more than four tube passes, it was assumed that there must always be the same number of tubes in each pass and one tubesheet configuration was available for each number of tube passes.

The tube-side heat transfer coefficient was previously calculated using the Dittus–Boelter equation, while both the Colburn equation and Petukhov equation (Eq. (7)) were used in [3]. The Petukhov equation is the most accurate so this was used in the program although the friction factor calculation was modified to use the Blasius equation below a Reynolds number of 30×10^4 and McAdams equation above a Reynolds number of 30×10^4.

$$Nu = \frac{(f/2)\text{Re}\,\text{Pr}}{1.07 + 12.7(f/2)^{0.5}(\text{Pr}^{2/3} - 1)} \tag{7}$$

Static fouling factors allow the user to model both clean and fouled STHE which increases the program's utility. The effects of fouling on the STHE design optimization were investigated. The GUI (shown in Fig. 1) was updated to ensure it remained user friendly with additional inputs and outputs, along with the original data, displayed on a single form rather than multiple forms in the previous program. This simplified the program, and the accessibility of separate functions is ensured through various buttons and menu options.

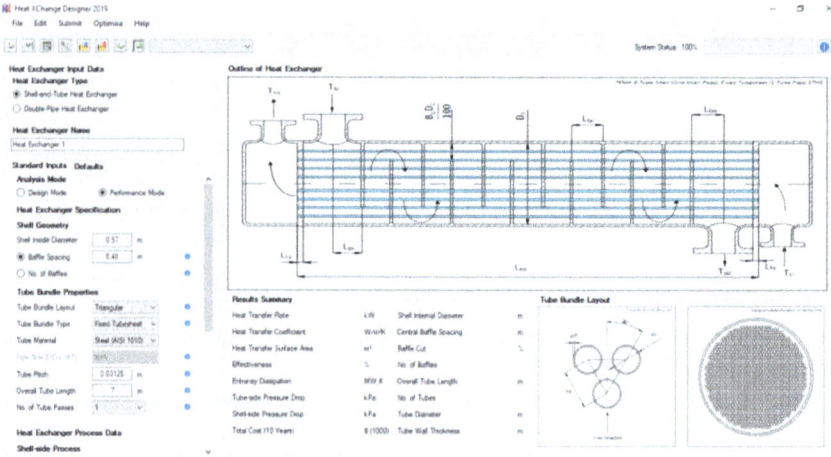

Fig. 1 Updated program GUI (Graphical User Interface)

3 Results

The method for determining the maximum number of tubes in a shell was validated against tube count tables, the previous program and the scale drawings. The algorithm corresponded exactly with the scale drawings (0.0% error), therefore, the program was functioning as intended and particularly for STHE with fewer tube passes, good accuracy was achieved against the tables (1-Tube Pass: 0.6%, 2-Tube Passes: 3.6%). There was significant improvement over the previous program where accuracy was poor (average deviation from the tables was 44.2%) due to the inferior methodology causing skewness in tube layout.

The updated program with BDM was validated against a real heat exchanger, a TEMA E-Type, single tube pass STHE [5]. The improvement in accuracy over the previous program was also shown in Table 1. The results from the new program were very close to the reference, especially with regard to thermal analysis, and much more accurate than previous one. Shell-side pressure drop (+3.81%) was slightly overestimated, although it showed higher accuracy than the previous program (+40.34%), which may be because some of the inputs required for the BDM (e.g. shell baffle, shell-bundle clearances) were calculated using default values as these were not provided in the reference.

STHE design optimization study was carried out to find an optimum design, which has best effectiveness within the given cost (build and operating cost). The Pareto-optimal front (the effectiveness vs the total cost) was shown in Fig. 2. The effect of fouling on design optimization was also compared in Fig. 2: one clean and the other with considering the fouling factor. The cost to achieve the maximum effectiveness (48.36%) in a clean STHE is about $2.747 m. Trends are generally similar with the majority of STHE clustered along a straight line before a sharp increase in cost relative to effectiveness. To achieve the same effectiveness, the cost in a fouled STHE

Table 1 Program validation

Parameter	Units	Reference	Previous program		Current program	
			Value	Accuracy (%)	Value	Accuracy (%)
Shell-side outlet temp	°C	80	86.0	+7.50	79.8	−0.25
Tube-side outlet temp	°C	148.5	144.9	−2.42	148.7	+0.13
Heat transfer rate	kW	5258.1	6918.3	+31.57	5195.6	−1.19
Shell-side pressure drop	kPa	31.5	52.8	+40.34	32.7	+3.81
Tube-side pressure drop	kPa	38.3	19.2	−49.87	38.6	+0.78
Heat transfer coefficient	W/m^{2}K	500.9	695.8	+38.91	496.7	−0.84

Shell/tube-side fluid properties from program fluid database/reference (with/without viscosity correction factor), respectively

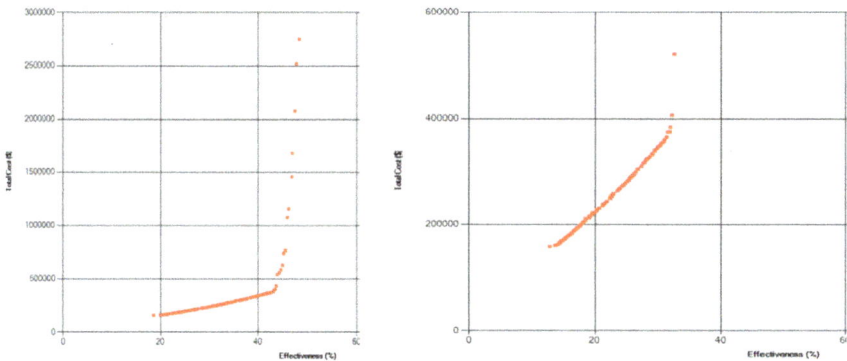

Fig. 2 Effect of fouling on TLBO pareto-front (left: clean, right: fouled)

is greater. The effectiveness of a fouled STHE is limited to 32.64%. Eventually, the fouled STHE has a narrower optimization range compared to the clean STHE.

4 Conclusions

The BDM greatly improves the accuracy and is an effective prediction model, requiring limited user inputs for STHE design. The program was successfully validated against tube count tables and a real STHE with high levels of accuracy. In the design optimization study, the effect of fouling was significant: resulting in a limited performance range compared to a clean STHE.

References

1. N. Sweeney-Ortiz, S. Brockmann, S. Kim, Multi-objective optimization tool for shell-and-tube heat exchanger design, in *15th UK Heat Transfer Conference* (2017)
2. S. Kakaç, H. Liu, A. Pramuanjaroenkij, *Heat Exchangers: Selection, Rating, and Thermal Design*, 3rd edn. (Taylor & Francis Group, Boca Raton, 2012).
3. M. Serna, A. Jimenez, A compact formulation of the bell-delaware method for heat exchanger design and optimisation. Chem. Eng. Res. Des. **83**, 539–550 (2005)
4. S. Murali, Y. Bhaskar Rao, A simple tubesheet layout program for heat exchangers. Am. J. Eng. Appl. Sci., 131–135 (2008)
5. T. Labbe, *E22-2 Reactor Primary Heat Exchanger—F1 Case* (M. Labbe, 2013). Available at https://www.seine-et-marneinvest.com/sites/default/files/labbe_design.pdf. Accessed 30 Apr 2019

Performance Analysis on Compact Heat Exchangers for Activated Carbon-Ethanol Adsorption Heat Pump/Thermal Storage

Takahiko Miyazaki, Nami Takeda, Yuta Aki, Kyaw Thu, Nobuo Takata, Shinnosuke Maeda, and Tomohiro Maruyama

1 Introduction

Adsorption heat pumps can be driven by low-grade thermal energy, such as jacket cooling water of engines. Even though electrification of automobiles will reduce waste heat from engines, internal heat recovery within electric or hybrid vehicles is considered important to improve the mileage because considerable amount of energy is consumed for cabin cooling and heating. In fact, electric devices of electric or hybrid vehicles, such as batteries, inverters/converters, motors, require temperature control so that the devices work efficiently. For this purpose, adsorption heat pumps can provide cooling, heating and thermal storage within the thermal management system of the vehicle.

Although vast amounts of studies have been done in this research field, experimental verification of adsorption heat pumps for automobile is limited [1]. Therefore, the study developed a prototype of an adsorption heat pump/thermal storage system for automobile to demonstrate the performance of a compact adsorption heat exchanger. The experiment was carried out at nominal operating conditions for both cooling operation and thermal storage operation, and the effects of heat exchanger size and of the porous characteristics of activated carbons on the performance indices were clarified.

T. Miyazaki (✉) · N. Takeda · Y. Aki · K. Thu · N. Takata
Interdisciplinary Graduate School of Engineering Sciences, Kyushu University, 6-1 Kasuga koen, Kasuga-shi, Fukuoka 816-8580, Japan
e-mail: tmiyazak@kyudai.jp

T. Miyazaki · K. Thu
International Institute for Carbon-Neutral Energy Research, Kyushu University, 744 Motooka, Nishi-ku, Fukuoka 819-0395, Japan

S. Maeda · T. Maruyama
Global Technology Division, Calsonic Kansei Corporation, 8 Sakae-cho, Sano-shi, Tochigi 327-0816, Japan

© Springer Nature Singapore Pte Ltd. 2021
C. Wen and Y. Yan (eds.), *Advances in Heat Transfer and Thermal Engineering* ,
https://doi.org/10.1007/978-981-33-4765-6_110

(a) Experimental prototype   (b) Heat exchanger

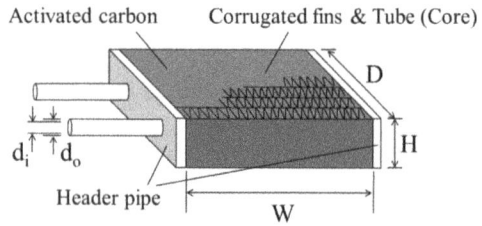

Fig. 1 Experimental prototype and the heat exchanger

2 Adsorption Heat Pump/Thermal Storage Prototype

A photo of the experimental prototype and an illustration of the heat exchanger are shown in Fig. 1. The activated carbon was confined between corrugated fins of a multiport mini-channel flat tube type heat exchanger. Two different sizes of heat exchangers, volumes $(D \times W \times H)$ of which were approximately 534 and 1166 cm^3, were used for comparison of system performances. In addition, two different types of activated carbons, a commercialized activated carbon with particle diameter of 75–125 μm and the average pore width of 1.1 nm, and a laboratory-based activated carbon with particle diameter of 10–30 μm and the average pore width of 1.6 nm, were also tested.

The experimental prototype consisted of one adsorber, in which the adsorption heat exchanger was confined, and one evaporator/condenser. Ethanol was used as refrigerant, and ethylene glycol water solution was used as a heat transfer medium to the heat exchangers. Temperatures at the inlet and outlet of each heat exchanger and flow rate of the heat transfer medium were measured to calculate heat input and output.

The adsorber and the evaporator/condenser were heated or cooled by heat transfer medium from constant temperature water circulators. The operation mode, either cooling, heating or thermal storage, could be defined by the heat transfer medium temperature.

3 Experiment

For the cooling operation experiment, firstly the temperatures of the adsorber and the evaporator/condensers were initialized at a desorption process condition, where temperature at the inlet to the adsorber was 80 °C and that to the evaporator/condenser

Table 1 Maximum and average cooling rate

Adsorption heat exchanger type		I	II	III
Adsorbent type		Commercialized AC	Laboratory-based AC	Commercialized AC
Adsorbent mass (g)		44.7	47.1	113.3
Volume (cm^3)		534	534	1166
Maximum cooling rate	(W/L)	846.0	788.8	1039.7
	(W/g)	10.1	8.9	10.7
Average cooling rate (180 s)	(W/L)	218.6	443.2	422.3
	(W/g)	2.6	5.0	4.3

was 25 °C. The adsorbent became at adsorption equilibrium of the given temperature conditions. Then, the temperatures at the inlets to the adsorber and evaporator/condenser were changed to 30 °C and 25 °C, respectively, for evaporation-adsorption process. The evaporation-adsorption process was tested until it reached the equilibrium. Cyclic performance of the prototype was estimated from temperature profiles.

The experimental results are summarized in Table 1. The maximum cooling rate and the average cooling rate during 180 s were compared among three different types of adsorption heat exchangers. The adsorption heat exchanger type I and type II used the same type of the heat exchangers, but the activated carbon types were different. It was shown that the type II showed slightly less maximum cooling rate, but the average cooling rate for 180 s was almost doubled. It was due to a large adsorption capacity of the laboratory-based activated carbon.

The adsorption heat exchanger type I and III used the same type of activated carbon, but the heat exchanger size was different, especially, the thickness of the heat exchanger was 25 mm in type I and that was 29 mm in type III. As a result, type III was about 2.2 times larger in volume and contained 2.5 times more activated carbon in weight compared with type I. The experimental results revealed that type III was superior to type I in average cooling rate per volume as well as that per adsorbent mass. The results implied that the adsorption capacity of the activated carbon was more effectively used with type III.

4 Conclusions

A prototype for testing an adsorption heat pump and thermal storage system was developed, and the performance was evaluated in cooling operation mode. Three different types of adsorption heat exchangers were compared in terms of cooling rate. The main findings of the study were summarized as follows.

- The commercialized activated carbon and the laboratory-based activated carbon resulted in a similar maximum cooling rate. The adsorption capacity of the laboratory-based activated carbon was larger, however, which caused a better average cooling rate by type II heat exchanger.
- Type III heat exchanger, which used thicker adsorption heat exchanger, improved the average cooling rate.
- Among three types of adsorption heat exchanger, type II achieved the highest average cooling rate. It was expected that further improvement would be achieved by the laboratory-based activated carbon with the thicker heat exchanger used by type III.

Acknowledgements The study was carried out under collaborative research project with Thermal Management Materials and Technology Research Association (TherMAT), NEDO, Japan.

Reference

1. S. Maeda et al., Critical review on the developments and future aspects of adsorption heat pumps for automobile air conditioning. Appl. Sci. **8**, 2061 (2018)

Experimental Validation of a Numerical Model of a Corrugated Pipe-Phase Change Material (PCM)-Based Heat Exchanger to Harness Greywater Heat

Abdur Rehman Mazhar, Shuli Liu, and Ashish Shukla

1 Introduction (Details for Submitting Paper)

Annually about 83 TWh worth of energy ends up in the sewage system of the UK, contained in wastewater from non-industrial buildings [1]. All water except toilet water is categorized as greywater (GW), and it is estimated that a mere 10–20% of the original thermal energy is lost before water is converted into GW in buildings. The harnessing of heat from this GW has tremendous potential from the exergetic point of view to enhance overall building efficiency [1]. Commercial heat exchangers and heat pumps are the only technologies at the moment to harness this heat. However, the storage of this harnessed heat into phase change materials (PCMs) would decouple demand and supply while integrating both heat recovery and storage in a single unit, unlike in commercial technologies. At the same time, GW heat harnessing would allow non-industrial buildings to be a potential source of decentralized heat generation especially useful for the concept of future low-temperature fourth generation district heating/cooling grids [2].

Conceptually a counter-flow heat exchanger of GW and cold water flow embedded in an appropriate PCM within the drain lines of conventional buildings would recover this otherwise lost low-grade heat. However, due to the limited thermal conductivity of PCMs, passive heat transfer enhancement techniques have to be investigated in this application both internal and external to the heat exchanger pipes, as the first step of an analysis. The use of corrugated pipes compared to simple pipes is quite effective to serve this purpose. These pipes enhance internal turbulence within the pipe, the surface area of contact and helically guide the flowing fluid along the circumferential length of the pipe to enhance the overall heat transfer coefficient with a minimum pressure loss penalty [3]. At the same time, the probability of biofilm deposits and

---

A. R. Mazhar (✉) · S. Liu · A. Shukla
Centre for Research in the Built and Natural Environment, Coventry University, Coventry CV1 2HF, UK
e-mail: mazhara4@uni.coventry.ac.uk

© Springer Nature Singapore Pte Ltd. 2021
C. Wen and Y. Yan (eds.), *Advances in Heat Transfer and Thermal Engineering* ,
https://doi.org/10.1007/978-981-33-4765-6_111

fouling is the least when using corrugated pipes, as normally there is a high prospect in GW flow. Additionally, these pipes also have a doubly enhancing effect, in which heat transfer is also improved external to the pipe walls that is to the PCM.

To investigate this concept, an experimental model is used to validate a 3D numerical simulation of this corrugated pipe GW-PCM configuration.

2 Methodology (Length and Layout)

An experimental test bench is developed using typical boundary conditions of GW flow [1]. A tank is used to store and heat laboratory synthesized GW using ohmic heating rods to 325 K. In a closed circuit arrangement, a centrifugal pump delivers 0.1 kg/s of GW through the test section consisting of the corrugated copper pipe embedded into the PCM as illustrated in Fig. 1. The commercially available corrugated pipe has a rib height of 0.7 mm, a pitch of 5 mm and a thickness of 1 mm. The organic PCM used for this application is RT-25, manufactured by Rubitherm having a phase change temperature between 295 and 299 K and a latent heat capacity of 170 kJ/kg [4]. A steel container with 10 cm of Styrofoam insulation encloses this test section. Using K-type thermocouples calibrated as per ISO 71025:2005, the temperature readings at three different positions in the test bench are recorded including the (a) inlet and outlet of the GW flow (b) a specific point in the PCM and (c) at three discrete points on the copper pipe containing the GW.

Similarly, a 3D layout of this set-up is also replicated in the CFD software Star-CCM v11.04. The geometric layout consists of three distinct regions namely the GW fluid domain, the corrugated copper pipe and the PCM, with the dimensions outlined in Fig. 1. An unstructured conformal polyhedral mesh is used to discretize the three domains. To capture the details of the corrugations, a surface mesh is used prior to a volume mesh followed by a 1.1 mm first cell height of a prism layer. After a grid independent study, a mesh size of about 20 million cells is used for all three regions of the geometry. The Eulerian multiphase model based on the enthalpy

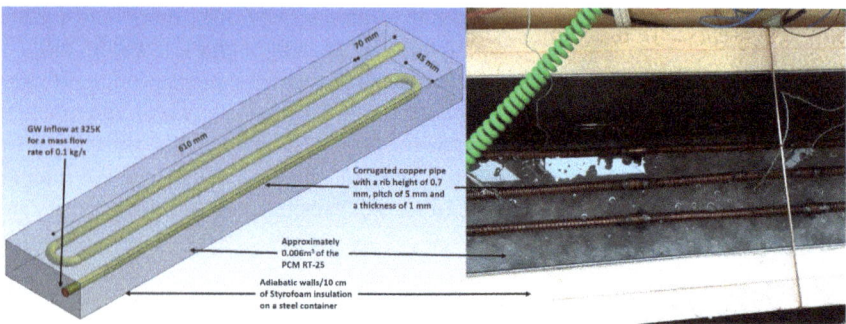

Fig. 1 Numerical layout of the corrugated test section replicated in the experimental model

formulation is used to predict the behaviour of the PCM. Replicating the experimental measurements, the temperature at the same positions is computed in this numerical simulation, for a comparison.

Both the experiment and simulation have a duration of 900 s, after which the corresponding values are compared for validation. The average percentage error between both sets of data is computed as follows:

$$\text{Average percentage error} = \sum \left(\frac{T_{\text{exp}} - T_{\text{sim}}}{T_{\text{exp}}} \right)_t \times 100 \tag{1}$$

3 Results

Based on the similar boundary conditions and geometric dimensions, the differences in values of both sources of data are presented in Table 1.

The error in the difference between inlet and outlet temperatures of the GW flow is trivial. Most deviation in this outcome along with the average copper pipe temperature is in the initial 200 s when the flow is in the transient zone as depicted in Fig. 2.

The main reasons for the deviation, which are amplified during this transient zone, are as follows:

- The minor variations in the environmental conditions of the experiment coupled with the systematic errors in instrument readings. The cumulative uncertainty in the thermocouple readings is about 0.4% while 3.15% for the flow metres, coupled with the data loggers.
- The tolerable fluctuations in the boundary conditions of the experiment whilst being consistent in the simulation and the errors due to the assumptions undertaken in the simulation.

Table 1 Comparison of the experimental and simulation results

	Description	Average percentage error (%)	Mean difference between values (K)
1	Difference between inlet and outlet temperatures	8.5	0.40
2	Average copper pipe temperature	0.85	2.70
3	Temperature at a sample point in the PCM: 100 mm from the inlet and 2 mm from the edge of the pipe	0.93	2.80

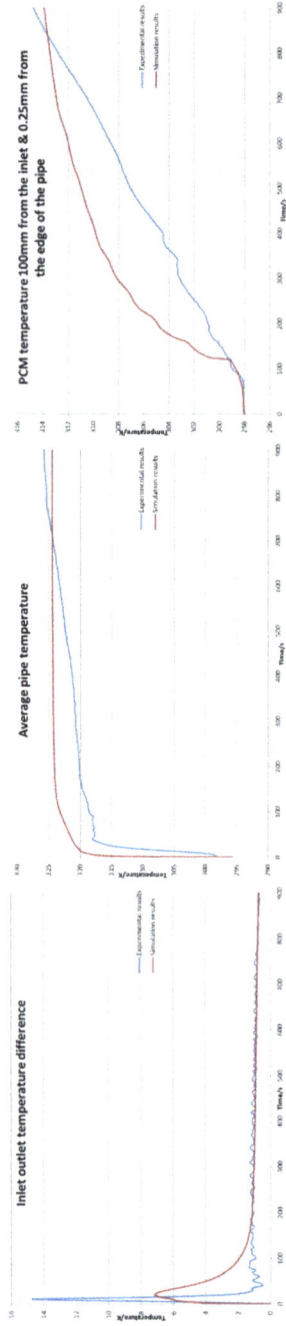

Fig. 2 Results for the temperature readings of both sets of data for the 900 s duration

Similarly, the results of the sample point in the PCM have a same trend and end values, with the errors attributed to the following reasons:

- Due to the temperature differential, the K-type thermocouples have reduced effectiveness as a solid low-conductive layer of PCM is developed over them.
- Additionally the PCM has a naturally amorphous structure with non-consistent thermo-physical over its domain which are modelled as constant values in the simulation.

Nevertheless the variations in the results are inevitable systematic errors. In spite of which the results of the numerical simulation are in acceptable agreement with the experimental data, especially during steady state conditions.

4   Conclusion

Compared to a plane pipe, the usage of the corrugated pipe results in a mere increment in the heat transfer rate by about 2% and the melted PCM by about 4%. This clearly advocates in favour of the fact that a custom-based solution of a corrugated pipe for this specific application is to be designed. This passive enhancement technique cannot be predicted and must be tested for any specific application either through CFD simulations or experimentally [3]. However, since commercially available corrugated pipes are limited for experimental testing, this numerical simulation would serve as the basis of a future sensitivity analysis to discover the dimensional parameters of a corrugated pipe delivering optimum results within this GW harnessing application.

References

1. A.R. Mazhar, S. Liu, A. Shukla, A key review of non-industrial greywater heat harnessing. Energies **11**(386) (2018)
2. H. Lund et al., 4th generation district heating (4GDH). Integrating smart thermal grids into future sustainable energy systems. Energy **68**, 1–11 (2014)
3. W.T. Ji, A.M. Jacobi, Y.L. He, W.Q. Tao, Summary and evaluation on single-phase heat transfer enhancement techniques of liquid laminar and turbulent pipe flow. Int. J. Heat Mass Transf. **88**, 735–754 (2015)
4. M. Iten, S. Liu, A. Shukla, Experimental validation of an air-PCM storage unit comparing the effective heat capacity and enthalpy methods through CFD simulations. Energy **155**, 495–503 (2018)

.

Fluid Flow

A Heat Transfer Analysis of Separated and Reattached Flow Around a Heated Blunt Plate

Christopher D. Ellis, Hao Xia, and Gary J. Page

1 Introduction

Heat transfer within the turbulent flow of separated and reattached flow regimes is of practical interest to the engineering community. Modelling such complex flow physics is difficult with the cheap, commonly used Reynolds-averaged Navier–Stokes (RANS) modelling of turbulence. High fidelity, eddy-resolving methods, such as large eddy simulations (LES), can capture the rich physics of separation and reattachment providing valuable heat transfer knowledge. However, such tools are computationally expensive, and their use within industry design loops is impractical for the foreseeable future. Within research, efforts are not wasted, and the use of LES data can complement the design process and further the understanding of unsteady heat transfer mechanisms and turbulence anisotropy behaviour.

Separated and reattached flow is present within blunt plate, backwards and forwards facing steps, abrupt expansion and contraction geometries. Relevance of such geometries presents themselves within gas turbine combustor liners, heat exchangers, cooling channels and nuclear reactors. Investigations into separated and reattached flow and their respective heat transfer have been present over the last half century. The present work compares results to the experimental studies of Ota and Kon [1] investigating the Nusselt number distribution over a blunt flat plate for a range of turbulent Reynolds numbers. Yanaoka et al. [2] conduct a numerical study into the time-averaged and root-mean-square heat transfer about the blunt plate geometry at a Reynolds number of 5000.

The following work aims to extend upon those before it. Comparisons are made to the earlier work of Ota and Kon [1] for validation of the LES approach. Extensions to the work of Yanaoka et al. [2] are made by studying an increased Reynolds

C. D. Ellis (✉) · H. Xia · G. J. Page
Department of Aeronautical and Automotive Engineering, Loughborough University,
Loughborough L11 3TU, UK
e-mail: c.d.ellis@lboro.ac.uk

© Springer Nature Singapore Pte Ltd. 2021 653
C. Wen and Y. Yan (eds.), *Advances in Heat Transfer and Thermal Engineering* ,
https://doi.org/10.1007/978-981-33-4765-6_112

number of 21,600. Investigations into the coupled behaviour of near-wall turbulent eddies and unsteady heat transfer are established. Finally, an assessment of turbulence anisotropy and the turbulent heat flux are used to improve upon the knowledge of model discrepancies within the use of RANS modelling.

2 Methodology

LES modelling using the *rhoPimpleFoam* solver in the open-source CFD solver OpenFOAM has been performed for this case. Conservation equations for continuity, momentum and energy are solved in their Favre-filtered forms. Wall-adapting local eddy-viscosity (WALE) modelling [3] of the sub-grid scale stresses is used and provides correct near-wall scaling of eddy viscosity with y^3 without the requirement of a dynamic formulation. Time stepping is discretised using a first-order implicit Euler scheme with a time step providing a maximum CFL of 0.8 achieving time accurate and numerically stable results. Convective fluxes are discretised using an upwind-biased central scheme achieving the second-order spatial accuracy. The discretisation of diffusive fluxes is achieved using a fully central scheme.

Sub-grid scale heat fluxes are obtained with a sub-grid scale Prandtl number of 0.4 as recommended by Grötzbach and Wörner [4]. Although other simulations have been known to use greater values, the effects on surface heat transfer showed that results were insensitive for sub-grid scale Prandtl numbers between 0.4 and 1.0 for this fine mesh.

Boundary conditions are used to replicate the experimental conditions of Ota and Kon [1]. No-slip conditions are applied to the top, bottom and leading edge surfaces of the plate with a wall-normal pressure gradient of zero. A heat flux of 668.28 W/m^2 is applied to the top and bottom plate surfaces. A velocity inlet is used with $U_\infty = 27$ m/s and $T_\infty = 300$ K with zero pressure gradient. The pressure outlet boundary condition defines a pressure of 101,325 Pa. Periodic conditions are applied to the spanwise constraining boundaries simulating an infinite span permitting vortical structures to pass between. Slip walls are applied to the top and bottom boundaries representing experimental blockage conditions.

Tetrahedral octree meshing is used to satisfy shear layer and near-wall refinements without high aspect ratio, skewed cells and unnecessary far-field refinements associated with structured meshing techniques. The near-wall cell height was selected to achieve a Δy^+ of less than 1, while streamwise and spanwise spacing were selected to achieve Δx^+ and Δz^+ of less than 25. Volumetric refinements were used in the recirculation region, the shear layer and near-wall regions to appropriately resolve the three-dimensional vortical structures and the turbulent behaviour of the shear layer. The resulting mesh contains 16 million cells.

3 Results

Time-averaged reattachment length, defining the length of the recirculation region, is found at a value of 9.2 H in good agreement with experimental correlations. Velocity profiles reflect those of the Ota and Kon [1] experiment showing the recirculation region and the redeveloping boundary layer behaviour. Non-dimensional temperature profiles show elevated values within the recirculation region reflecting the influence of the recirculating flow field. Small differences to the experiment are present in the recirculation region profiles reflecting temperature sensitivity to the differing Reynolds number shown for the experimental profiles. Time-averaged Nusselt number (Fig. 1) is in good agreement to the results of Ota and Kon [1] for consistent Reynolds number. Peak Nusselt number reflects the results of experimental correlations at a location just upstream of the reattachment. Root-mean-square values of Nusselt number hold similar trends to the mean with peak values occurring at coherent streamwise positions and the decaying magnitude in the redeveloping region seen in Fig. 1.

Figure 2a shows a snapshot of the temperature field and the unsteady behaviour influenced by vortical structures. The turbulent behaviour of the flow influences the surface heat transfer. Instantaneous Nusselt number (Fig. 2b) shows large differences in the maxima and minima Nusselt number encountered at spanwise locations. At reattachment, large variations in the heat transfer coefficient show instantaneous Nusselt numbers from 30 to above 100. In the redeveloping region, the progressing decline in Nusselt number is present, with values between 20 and 50. Instantaneous surface heat transfer is strongly correlated to local near-wall vortical structures. Within the recirculation region, peaks and troughs of Nusselt number number reflect the corresponding peaks and troughs in near-wall wall-normal velocity. These near-wall velocities reflect the small near-wall turbulent eddies and their strong influence upon convection of heat from the surface. In the redeveloping boundary layer, the near-wall structures are influenced by the strong shear forces. Turbulent hair pin structures are stretched and elongated through the boundary layer providing streamwise velocity streaks downstream. The streamwise velocity is correlated to the Nusselt

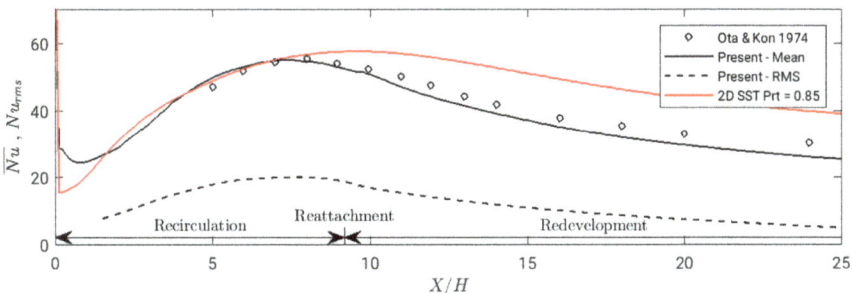

Fig. 1 Time-averaged and root-mean-square Nusselt number distributions across the blunt plate with comparisons to a RANS modelling approach

(a)

(b)

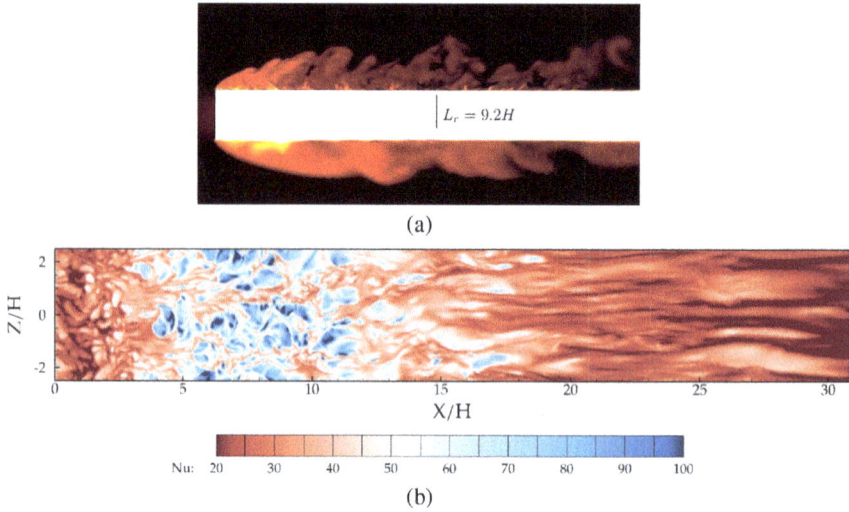

Fig. 2 a Instantaneous temperature field and **b** instantaneous Nusselt number distribution across the plate

number in this region. Local increases in streamwise velocity in the near-wall region reflect localised increased Nusselt number.

Turbulent heat flux presents the turbulent transport of heat represented by the correlations of velocity and temperature fluctuations. The present LES data showed regions of increased transport within the shear layer where turbulent fluctuations greatly enhance mixing. Streamwise and wall-normal components of the turbulent heat flux vector are most prominent with negligible spanwise component. Streamwise contributions are of a greater value within the turbulent shear layer and the redeveloping boundary layer. As the boundary layer redevelops, the magnitude of the turbulent heat flux vector reduces and the streamwise component increases adjacent to the wall. The turbulent heat flux vector in RANS simulations is commonly modelled with alignment to the direction of the thermal gradient vector. The present LES results for this heated plate flow showed that this is not the case, and a significant contribution from the streamwise component is present. Near the wall, the heated plate influences the thermal gradient and presents further deviations from common RANS assumptions.

4 Conclusions

LES with WALE sub-grid scale modelling is shown to be in good agreement with a range of experimental studies. Such a study justifies its use for further work within the realm of heat transfer and separated and reattached flow regimes. The present work advances the computational results for such case studies by extending the Reynolds number achieved by other studies.

The presence of near-wall structures has a strong contribution to the instantaneous Nusselt number to the wall and their influence on the time-averaged state. Root-mean-square Nusselt number distribution follows the same trends as the time-averaged distribution, with peak values close to the reattachment location. Calculation of instantaneous Nusselt number and root-mean-square value allows for advanced knowledge of the heat transfer conditions in similar geometric configurations.

Turbulent heat flux results show the contribution of shear layer mixing on the turbulent transport of heat with a strong streamwise component. Decay in the vector magnitude is seen as the flow redevelops from reattachment. The present LES results show that the turbulent heat flux vector is not aligned to the thermal gradient for the shear layer, recirculation region and redeveloping boundary layer within this flow regime. Future work investigates the modelling knowledge captured to improve upon the RANS modelling of convective heat transfer where common assumptions deviate from the physics found in separated and reattached flow.

Acknowledgements The authors would like to acknowledge the use of HPC Midlands Plus (https://www.hpc-midlands-plus.ac.uk) for the computational time used in this work (EP/P020232/1). This research was supported by EPSRC (Engineering and Physical Sciences Research Council) Centre for Doctoral Training in Gas Turbine Aerodynamics (EP:L015943/1).

References

1. T. Ota, N. Kon, Heat transfer in the separated and reattached flow on a blunt flat plate. J. Heat Transf. 459–462 (1974)
2. H. Yanaoka, H. Yoshikawa, T. Ota, Direct numerical simulation of turbulent separated flow and heat transfer over a blunt flat plate. J. Heat Transfer **125**, 779–787 (2003). https://doi.org/10.1115/1.1597623
3. F. Nicoud, F. Ducros, Subgrid-scale stress modelling based on the square of the velocity gradient tensor. Flow Turb. Combust. **62**, 183–200 (1999). https://doi.org/10.1016/j.jcp.2004.10.018
4. G. Grötzbach, M. Wörner, Direct numerical and large eddy simulations in nuclear applications. Int. J. Heat Fluid Flow **20**(3), 222–240 (1999). https://doi.org/10.1016/S0142-727X(99)00012-0

Flow and Heat Transfer Characteristics of Piezoelectric-Driven Synthetic Jet Actuator with Respect to Their Stroke Length

Muhammad Ikhlaq, Mehmet Arik, Adeel Arshad, and Mark Jabbal

1 Introduction

The computational power of microelectronic devices is increasing, which creates a need to dissipate more heat per unit area compared to the last decade [1]. The recent development in electronics and the device sizes is shrinking, while circuit density is increasing. Furthermore, a cheap, reliable with high performance and compact cooling solution is needed in order to cater the future needs of the industry. The synthetic jet actuator (SJA) is a cheaper, reliable and simple cooling solution available for highly functional and small electronic equipment, while many researches have been conducted for the last two decades [2].

The alternate inhalation and exhalation of air from an orifice (circular/slot) form a synthetic jet. Synthetic jets work on zero-mass-net-flux principle along with the positive momentum transfer in the system. Usually, synthetic jets utilize ambient air which make them suitable for many applications, since they do not need complex plumbing system. The current experimental study is aimed to characterize the flow and heat transfer behavior of low- and high-frequency central orifice synthetic jet driven by the piezoelectric actuators with respect to their deflection. As the operating frequency of SJA increases, the size decreases. This study will discuss the flow, heat transfer and mechanical response of high- and low-frequency synthetic jet with respect to jet formation criterion (stroke length (L) and Strouhal number (St)). For

M. Ikhlaq (✉)
School of Engineering, Edith Cowan University, 270 Joondalup Dr, Joondalup, Perth, Australia
e-mail: mikhlaq@our.ecu.edu.au

M. Arik
EVATEG Center, Ozyeğin University, TurkeyIstanbul, Nişantepe Mahallesi, Orman Sk., 34794 Çekmeköy/İstanbul, Turkey

A. Arshad · M. Jabbal
Fluids and Thermal Engineering (FLUTE) Research Group, Faculty of Engineering, University of Nottingham, Nottingham NG7 2RD, UK

© Springer Nature Singapore Pte Ltd. 2021
C. Wen and Y. Yan (eds.), *Advances in Heat Transfer and Thermal Engineering* ,
https://doi.org/10.1007/978-981-33-4765-6_113

each individual jet, the synthetic jet formation criterion ($L > 2$) is calculated with respect to their heat transfer for the range of operating frequency at plate-to-orifice distance (H/D) equals to 10 [3].

2 Methodology

Heat transfer and velocity measurements are conducted for the current study. An automated experimental setup is utilized to measure heat transfer data for SJA assembly [4]. A custom-developed LabVIEW program is developed to collect and control the experimental setup. A waveform generator, a data acquisition system, a power supply and a 3D robot are controlled by the LabVIEW-based program. An indigenously manufactured and calibrated heater is used to determine the heat transfer coefficient and Nusselt number (Nu) of the synthetic jet. Jet exit velocity measurements were taken using a single TSI IFA300 constant-temperature hot-wire anemometer system with a TSI 1210 a single-axis hot-wire probe. The synthetic jet is excited using an Agilent function generator along with an amplifier. The heater was powered by an Agilent DC power supply. Figure 1a is showing experimental setup that is used to get data for the characterization of the synthetic jets. Figure 1b is showing characteristic curve for the heater.

$$\text{Re} = \frac{\rho U d}{\pi \mu} \tag{1}$$

$$\text{Stk} = \sqrt{\frac{2\pi f \rho d^2}{\mu}} \tag{2}$$

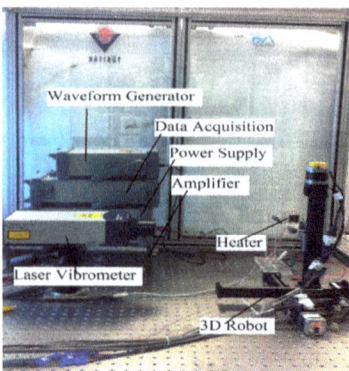

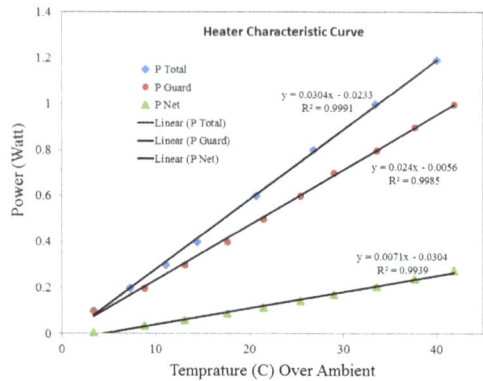

(a)   (b)

Fig. 1 a Experimental setup **b** heater characteristic curve

$$\frac{1}{St} = \frac{Re}{Stk^2} \tag{3}$$

$$S.L = \frac{\pi}{St} \tag{4}$$

Reynolds number (Re), Stokes number (Stk), inverse Strouhal number (1/St) and stroke length (S.L) are calculated using Eq. (1–4), respectively. The ρ and μ are the density and dynamic viscosity of air, respectively. d is the synthetic jet orifice diameter, and f is the operating frequency of SJA. The uncertainty in the heat transfer experiments is quantified as 5% [4].

3   Results

Figure 2a shows the variation of Nusselt number of low-frequency synthetic jet along with Stokes number and stroke length with respect to operating frequency, while Fig. 2b illustrates results for high-frequency synthetic jet. All heat transfer measurements are taken plate-to-orifice distance of 10. The SJA deflection with respect to frequency is measured in order to find out the natural structural frequency using laser vibrometer [4]. The driving voltage for both SJAs is kept at 20 V_{pp} (volt peak-to-peak).

It can be seen that Stk increases as the operating frequency increase, Stk range varies for low-frequency jet from 10 to 40, while for high-frequency jet, it varies approximately between 96 and 105. In contrast to the Nu and S.L, they showed increase and decrease behavior with respect to different operating frequencies. The maximum Nu is measured at 700 Hz and 24.8 kHz for low- and high-frequency synthetic jet, respectively. S.L is maximum at 200 Hz for low frequency, and it continues to decrease as the operating frequency increases. The peak S.L almost matches with Nu peak for high-frequency SJA and also follows the same trend compared to the Nusselt number. For low-frequency SJA, S.L is more than 5, while for high-frequency SJA, the maximum S.L is less than 0.8.

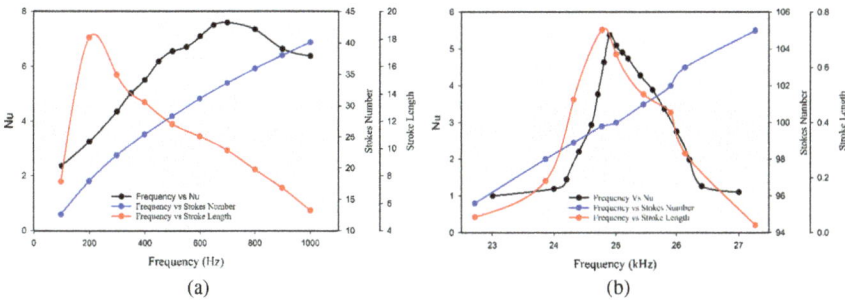

Fig. 2 Nusselt number, Stokes number and stroke length for **a** low-frequency SJA **b** high-frequency SJA

It is observed that at high-frequencies, synthetic jets act similar to continuous impinging jet ($L < 0.82$) since vortex ring does not have enough time to grow and the second vortex ring comes and coalesces with the previous one. The heat transfer enhancement due to disturbance of the boundary layer created by the vortex ring apart from the main jet stream cannot be benefitted in the case high-frequency synthetic jet. For low-frequency synthetic jet, the calculated stroke length is more than 5 for all operating frequencies, and hence, it exhibits strong synthetic jet over a target surface lie in the regime which is the best for heat transfer. According to the Pawel [5], the variation in Nusselt number with respect to different frequencies because of high actuator deflections and velocity outputs near the resonance frequency of SJA.

4 Conclusions

In this experimental study, low- and high-frequency SJAs are examined with respect to its heat transfer and synthetic jet formation criterion.

- For low-frequency SJA, all operating frequencies lie in the best regime for heat transfer, since system is benefited from the direct impingement as well as the disturbance of boundary layer from vortex ring.
- The high-frequency jet acts like a continuous jet, as the operating frequencies are too high and system generated vortices do not have enough time to generate and propagate properly.

Acknowledgements Authors thank to EVATEG Center at Ozyegin University for the testing and computational facilities.

References

1. M. Ikhlaq, B. Dogruoz, O. Ghaffari, M. Arik, A computational study on the momentum and heat trasnfer distribution of a low frequency round impinging synthetic jet, in: *InterPACK2015* July 6–9, 2015, San Fr. California, USA, ASME (2015), p. V003T10A016. https://doi.org/10.1115/IPACK2015-48094
2. M. Arik, Local heat transfer coefficients of a high-frequency synthetic jet during impingement cooling over flat surfaces. Heat Transf. Eng. **29**, 763–773 (2008). https://doi.org/10.1080/01457630802053769
3. R. Holman, Y. Utturkar, R. Mittal, B.L. Smith, L. Cattafesta, Formation criterion for synthetic jets. AIAA J. **43**, 2110–2116 (2005). https://doi.org/10.2514/1.12033
4. M. Ikhlaq, O. Ghaffari, M. Arik, Predicting heat transfer for low- and high-frequency central-orifice synthetic jets. IEEE Trans. Comp. Packag. Manuf. Technol. **6**, 586–595 (2016). https://doi.org/10.1109/TCPMT.2016.2523809
5. P. Gil, Morphology of synthetic jet, in *Zesz. Nauk. Politech. Rzesz. Mech.* (2017), pp. 43–51.

RANS Model Validation of Natural Circulation in Differentially Heated Cavities

Constantinos Katsamis, Tim Craft, Hector Iacovides, and Juan Uribe

1 Introduction

The phenomenon of natural circulation in which fluid motion is driven solely by density variations due to temperature difference within an internal space occurs in numerous applications, including domestic heating and passive cooling systems of nuclear reactors. The first case considered in this study is that of a rectangular cavity with high aspect ratio, which has attracted a lot of research attention, with the experiments of Betts and Bockhari [1] being widely used as a benchmark case for both studying the flow circulation, the turbulence generation due to buoyancy, and the near wall heat transfer. From the experiments, temperature and velocity measurements are obtained using thermocouples and Laser Doppler Anemometry which show the flow to be two-dimensional and anti-symmetric about the diagonal across the cavity at a Rayleigh number of 0.86×10^6. The large aspect ratio of the cavity leads to relatively high turbulence levels in the core region, which promotes mixing of the fluid and hence results in a fairly uniform temperature over much of the cavity core. An exception is the near top and bottom wall regions where the local thermal stratification causes the decay of turbulence levels. The near wall regions are the main challenge to model with Omranian et al. [2] carrying an investigation on the performance of the standard wall function approach and the more recently developed Analytical Wall Function (AWF) [3] on both rectangular and square cavities. Both Eddy Viscosity Models (EVMs) and Reynolds Stress Transport models (RSMs) are compared, where the latter showed a strong dependence on the heat flux model used.

C. Katsamis (✉) · T. Craft · H. Iacovides
School of Mechanical, Aerospace and Civil Engineering, University of Manchester, Manchester M13 9PL, UK
e-mail: constantinos.katsamis@manchester.ac.uk

J. Uribe
EDF Energy R&D UK Centre, Low Carbon Generation, Manchester, UK

© Springer Nature Singapore Pte Ltd. 2021
C. Wen and Y. Yan (eds.), *Advances in Heat Transfer and Thermal Engineering* ,
https://doi.org/10.1007/978-981-33-4765-6_114

The second geometry considered here, that of a square cavity, is found by [2] to be rather more challenging in terms of turbulence model performance. The study also exhibits the sensitivity of the Low-Re EVM in laminarising the flow and the significant improvements that the AWF has to offer in the prediction of the non-dimensional Nusselt number and the turbulent kinetic energy levels with lower grid requirements. Experimental data for such flows at high Rayleigh numbers are rather limited, however, recent DNS data have been reported in [4] for an infinitely deep square cavity at Rayleigh numbers from 10^8 to 10^{11}. In this case, strong gradients exist at the boundaries and the flow exhibits a largely motionless and stratified laminar core region. Thin and unsteady boundary layers develop along the walls, which undergo transition from laminar to turbulent state. The current work assesses the ability of RANS strategies to predict the mean and turbulent flow field of the above statistically steady two-dimensional cases, with particular emphasis given to the validation of the square cavity case, which poses the more severe modelling challenges. The high-Reynolds-number (High-Re) $k - \varepsilon$ scheme, and RSMs, the latter based either on the Launder Reece and Rodi redistribution model [5] or the Speziale, Sharkar and Gatski SSG [6] are assessed with standard log-law based wall function as well as some low-Reynolds-number (Low-Re) alternatives, namely the $k - \omega$ SST, a $v^2 f$ version, $k - \varepsilon - \overline{v}^2/k$ (BL $- \overline{v}^2/k$), and a more complex form of a RSM with elliptic blending, EBRSM, with suitably fine grids for the resolution of the near-wall layers.

2 Methodology

2.1 Mathematical Formulation

The enthalpy and mean momentum transport equations characterise the flow field, where the latter incorporates a buoyancy-force term which is approximated using the Boussinesq formulation:

$$F_i^b = \rho g_i = \rho_{\text{ref}} g_i (1 - \beta(T - T_0)) \tag{1}$$

This approximation assumes a linear variation of density with temperature and can generally be made when $\beta(T - T_0) \ll 1$. Constant fluid properties are assumed to match a Prandtl number of 0.71 and for the ideal gas case, the thermal expansion coefficient (β) is assumed to be the inverse of the absolute fluid temperature ($1/T$).

The EVMs relate the Reynolds stresses with the mean strains through the turbulent viscosity, μ_t, and the turbulent heat fluxes for the turbulence generation rate term due to buoyancy with the effective diffusivity approach (SGDH):

$$\overline{u_j t'} = -\frac{\mu_t}{\text{Pr}_t} \frac{\partial \overline{T}}{\partial x_j} \tag{2}$$

In the case of the RSMs, the more complex generalized gradient diffusion hypothesis is adopted (GGDH):

$$\overline{u_i t'} = -c_\theta \frac{k}{\varepsilon} \overline{u_i u_j} \frac{\partial \overline{T}}{\partial x_j} \tag{3}$$

2.2 Numerical Details

The RANS transport equations are discretised and solved iteratively using EDF's unstructured finite-volume CFD code, *Code_Saturne v5.0.8*. Two dimensional Cartesian grids are designed, for the rectangular cavity (2.18 m × 0.076 m) only low-Re models are applied, and an 80 × 120 grids is used, whilst for the square one (1 m x 1 m) a coarse grid with 100 × 100 and a near wall refined one with 250 × 250 nodes for the High-Re and the Low-Re schemes, respectively. The non-active walls are assumed adiabatic for the rectangular cavity and with a monotonic variation of temperature ($T = T_{\text{hot}} - (\Delta T \cdot x)$) for the square one, to match the experimental and DNS set-ups, respectively. The temperature difference between the two active walls was chosen to give a Ra ~ 0.86×10^6 and Ra ~ 1×10^{11} for the two cases, respectively, where the Rayleigh number is Ra $= g\beta \Delta T h^3 / \alpha\nu$. The steady RANS approach is adopted for the low Ra case, whereas for the square case, the flow remains unsteady in time and thus the URANS approach was adopted, with a suitably-chosen small timestep, and time-averaged quantities are compared against the DNS statistics. Non-dimensional quantities presented in the results below include:

$$k^+ = \frac{kh^2}{\alpha^2 \text{Ra}}, \, x^+ = \frac{x}{h}, \, \overline{u_i t}^+ = \frac{uh(T - T_{\text{avg}})}{\alpha \text{Ra}^{1/2} \Delta T}, \, \text{Nu} = \frac{q_w l}{\Delta T \lambda} \tag{4}$$

where the thermal diffusivity is denoted by $\alpha (= \lambda / \rho c_p)$ and Nu, the Nusselt number.

3 Results

3.1 Long Rectangular Cavity

All the Low-Reynolds number models tested here predict the expected fully turbulent core and thermally stratified near wall top and bottom regions. The maximum convective heat transfer, corresponding to the peak of the Nusselt number, occurs at the lower corner of the hot wall, and Nu then reduces as the fluid accelerates toward the top of the cavity (see Fig. 1a). The two EVMs capture the peak though when Nu becomes constant between $y/h = 0.3$ and 0.7, an underprediction is observed. In constrast, the

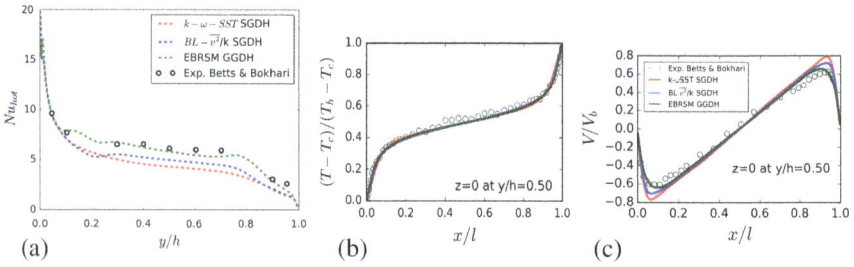

Fig. 1 (**a**) Nusselt number along the hot wall from $k - \omega - $ SST, $k - \varepsilon - \overline{v}^2/k$, and EBRSM at Ra $\sim 0.86 \times 10^6$. (**b**) Normalized temperature $((T - T_c/T_h - T_c))$ and (**c**) Vertical velocity where $V_b = \left(\sqrt{\beta g \Delta T H}\right)$ at $y/h = 0.5$

EBRSM computes a profile very close to the experimental one, exhibiting greater accuracy. From the non-dimensional temperature field, $(T - T_c/T_h - T_c)$ at the same location, it can be seen that the steep near-wall gradients agree closely with the experimental data of [1] (see Fig. 1b), whilst a linear variation of temperature across the central portion of the cavity suggests high levels of turbulent mixing and largely isothermal conditions. The normalised vertical velocity profile is captured fairly reasonably by the models at half the height of the cavity, $y/h = 0.5$, showing anti-symmetry across the cavity (see Fig. 1c), although close to the active walls the EVMs do slightly overestimate the velocity peak values, indicating that the convective transport in the wall parallel direction is slightly overpredicted.

3.2 Infinitely Wide Square Cavity

From the URANS computations, the time-averaged Nusselt number profiles from the different models show rather significant variations in the square cavity case (see Fig. 2a). The predictions by the High-Re schemes with the log-law-based wall function predict too low Nu levels compared to the DNS data whereas the Low-Re EVMs follow a similar trend to the data, although still underpredicting it. The

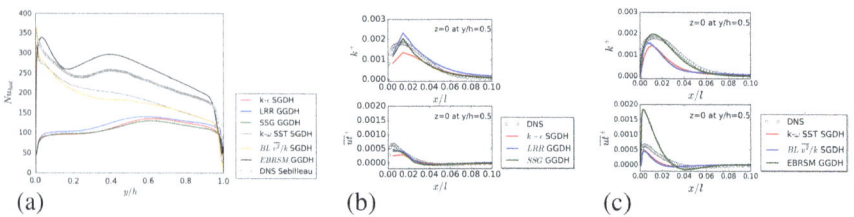

Fig. 2 (**a**) Nusselt number distribution along the hot wall at Ra $\sim 1 \times 10^{11}$. Normalized turbulent kinetic energy and wall normal turbulence heat flux near the hot wall at half the height of the cavity predicted by (**b**) high-Re models and (**c**) low-Re models

EBRSM results shown exhibit a slight overprediction, although it should be noted that in this case, the convective discretization scheme had to be locally switched to first order; using purely second order, as for the other models resulted in a completely laminar flow for this model. Figure 2b, c show profiles close to the hot wall, at half the cavity height. The low-Re EVMs in Fig. 2c under-estimate the turbulence levels within the boundary layers, failing to capture the right time that the transition from laminar to turbulent occurs as the fluid moves up the wall. Although not shown here, this predictive deficiency of the low-Re EVM models is even more pronounced at the lower Ra case of 1.58×10^9, where, in agreement with the findings of Omranian et al. [2], the predicted flow is entirely laminar. In contrast, as can be seen in Fig. 2b, all High-Re models tested return near-wall turbulence levels which follow the DNS profiles, with the two RSM profiles being the closest. The near-wall profiles of the wall-normal heat flux vector computed by the High-Re and the Low-Re EVMS have much lower peaks compared to the DNS whereas the profile predicted by the EBRSM exhibits the opposite behavior. The underestimation of the wall normal turbulent heat flux is explained by the underprediction of the buoyancy generation term which contributes to the production in the boundary layer region. The outcome is an underestimation of the turbulent kinetic energy and a delay of the location of the turbulence enhancement and the maxima of the Nusselt number. This suggests that the choice, and accuracy, of the turbulent heat flux model may be quite crucial in capturing these flows, together with the challenges in modelling the transition region. Further work to be reported will also examine more advanced wall function strategies, such as that proposed by Craft et al. [3], which include some more direct physics such as the convective and the pressure gradient effects.

4 Conclusions

In the present study, two geometrically simple but challenging cases of turbulent buoyant flows have been simulated using a range of RANS models. The 2D steady RANS of the rectangular cavity of [1] predict the isothermal central circulation region and the behavior of the near-wall boundary layers as expected. Most of the models tested reproduce adequately the thermal and velocity fields despite their inherent limitations. Regarding the second case, the square cavity one, the complexity of the flow physics demonstrates the greater challenge in modelling flows with significant unsteadiness and transition. The full paper presentation will contain a more detailed analysis of the models' performance, including results from the more advanced wall function of [3].

Acknowledgements The funding by the MACE BEACON scholarship and support by EDF Energy are gratefully acknowledged.

References

1. P.L. Betts, I.H. Bokhari, Experiments on turbulent natural convection in an enclosed tall cavity. Int. J. Heat Fluid Flow **21**, 675–683 (2000)
2. A. Omranian, T.J. Craft, H. Iacovides, The computation of buoyant flows in differentially heated inclined cavities. Int. J. Heat Mass Transf. **77**, 1–16 (2014)
3. T.J. Craft, A.V. Gerasimov, H. Iacovides, B.E. Launder, Progress in the generalization of wall-function treatments. Int. J. Heat Fluid Flow **23**(2), 148–160 (2002)
4. F. Sebilleau, R. Issa, S. Lardeau, S.P. Walker, International journal of heat and mass transfer direct numerical simulation of an air-filled differentially heated square cavity with Rayleigh numbers up to 10 11. Int. J. Heat Mass Transf. **123**, 297–319 (2018)
5. B. Launder, G. Reece, W. Rodi, Progress in the development of a reynolds stress turbulence closure. J. Fluid Mech. **68**, 537–566 (1975)
6. C.G. Speziale, S. Sarkar, T.B. Gatski, Modelling the pressure-strain correlation of turbulence: an invariant dynamical systems approach. J. Fluid Mech. **227**, 245–272 (1991)

Experimental and Numerical Investigation of the Heat Transfer Characteristics of Laminar Flow in a Vertical Circular Tube at Low Reynolds Numbers

Suvanjan Bhattacharyya, Marilize Everts, Abubakar I. Bashir, and Josua P. Meyer

1 Introduction

Laminar flow is very common in practical situations and vertical tubes are used for many industrial applications ranging from cooling of thermal systems such as compact heat exchangers, solar energy collectors, boilers, and nuclear reactor. For forced convection heat transfer in circular tubes, the literature [1] stated that Nusselt number is constant at 4.36 in the fully developed laminar flow regime for a constant heat flux boundary condition. By assuming constant fluid properties, the Nusselt number of 4.36 was derived analytically. However, for mixed convection conditions, the value of Nusselt numbers is effortlessly 160–500% higher than 4.36 [2–4].

Little computational and experimental studies have been done to investigate the laminar heat transfer characteristics of fully developed flow at low Reynolds numbers in a vertical smooth circular tube. The objective of this study was therefore to experimentally investigate the heat transfer characteristics of fully developed forced and mixed convection laminar flow in a tube. Experiments were conducted for vertically upward ($+90°$) and downward ($-90°$) flow directions between Reynolds numbers of 200–2000 at a constant heat flux of 2 kW/m². Water was used as a working fluid and the Prandtl numbers varied between 3 and 7. The inner tube diameter was 5.1 mm and the heated length-to-diameter ratio was 886. It has been found that it is challenging to conduct forced convection experiments at Reynolds number lower than 600 because of the buoyancy effects. The thermal energy transport temperature difference between the channel wall and working fluid rised with the increase of Reynolds number in the laminar flow regime, and decreased slowly as the flow approached the transitional flow regime. It was found that although the flow was

S. Bhattacharyya · M. Everts · A. I. Bashir · J. P. Meyer (✉)
Clean Energy Research Group, Department of Mechanical and Aeronautical Engineering,
University of Pretoria, Hatfield, South Africa
e-mail: Josua.Meyer@up.ac.za

© Springer Nature Singapore Pte Ltd. 2021
C. Wen and Y. Yan (eds.), *Advances in Heat Transfer and Thermal Engineering* ,
https://doi.org/10.1007/978-981-33-4765-6_115

fully developed, the Nusselt number was a function of Reynolds number. A sudden decrease in the heat transfer coefficients was also found for Reynolds numbers less than 500 and 250 for downward and upward flows, respectively. This was caused by the opposing and assisting buoyancy effects. Furthermore, it was found that thermal unsteadiness increased by changing orientation of the channel.

2 Experimental Setup and Procedure

The schematic of the experimental setup is described in Fig. 1. The detail on the experimental setup is already described by Bashir and Meyer [5] and thus only important parts are briefly described in this paper. The experimental setup includes of a closed-loop fluid flow system, calming section, test section, and the test bench. The structure of the test bench helps to change the orientation of the test section (+90° and −90°), and it has been presented by Bashir and Meyer [5] and schematically in Fig. 1

Schematic representation of the test section indicating the pressure taps, thermo-couple stations, the flow directions and a cross section of the test section tube that shows the thermocouple positions per station are shown in Fig. 2. The test tube was made from copper with outer and inner diameters of 6.3 mm and 5.1 mm, respectively. The length of the test section was 4.6 m. The wall temperatures were measured at 21 thermocouple stations. The thermocouple stations were positioned very close to each other near the entrance of the test section and at the end of the test section to capture enough data point in the developing and developed part. Copper constantan

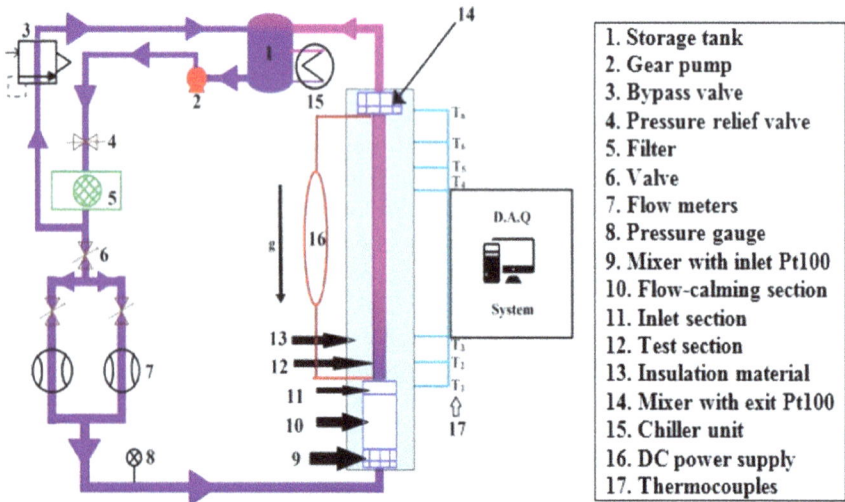

1. Storage tank
2. Gear pump
3. Bypass valve
4. Pressure relief valve
5. Filter
6. Valve
7. Flow meters
8. Pressure gauge
9. Mixer with inlet Pt100
10. Flow-calming section
11. Inlet section
12. Test section
13. Insulation material
14. Mixer with exit Pt100
15. Chiller unit
16. DC power supply
17. Thermocouples

Fig. 1 Schematic diagram of the experimental setup

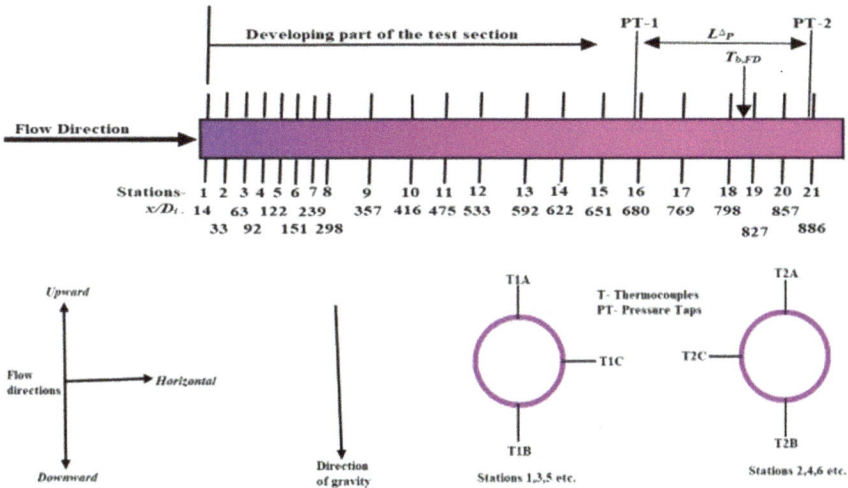

Fig. 2 Schematic representation of the test section indicating the pressure taps and thermocouple stations, the flow directions and a cross section of the test section tube that shows the thermocouple positions per station

T-type thermocouples were used with a diameter of 0.25 mm. Three thermocouples were used at each station as shown in Fig. 2. The maximum Nusselt number uncertainties were 3.6% in the laminar flow regimes at the maximum heat flux of 2 kW/m^2.The detailed experimental setup, procedure, and data reduction can be found in the previous research works of the same group [5].

3 Mathematical and Numerical Modelling

Steady-state, incompressible flow of a Newtonian fluid with temperature-dependent fluid properties is investigated. Radiative heat transfer is neglected. The simulation was performed in the laminar regime for Prandtl number 7–3 (water) and Reynolds number ranging from 200 to 2000. The numerical simulation was performed by solving continuity, momentum and energy equations for an incompressible fluid. The usual no-slip boundary conditions are applied at the wall. For the computational analysis, the finite volume method-based general purpose computational fluid dynamics software ANSYS Fluent 19.2 is employed. The convergence criteria for continuity, momentum and energy equations were set at 10^{-4}, 10^{-5}, and 10^{-7}, respectively, following Bhattacharyya et al. [6]. At the inlet a uni-directional uniform velocity profile is prescribed as boundary condition, along with a constant heat flux. The detail numerical technique and meshing is well described in the author's previous work [6, 7].

4   Results

Figure 3 shows the result of fully developed average heat transfer in terms of Nusselt number as a function of Reynolds numbers in the vertical upward and downward orientations of the test section under the same flow conditions. The experiment runs from Reynolds number of 200–2000 first with the test section in vertical upward position and then vertical downward position at a heat flux of 2 kW/m^2.

The Nusselt number is increased with an increase in Reynolds number (Re ≤ 400 for upward flow, and Re ≤ 600 for downward flow). For upward flow, Nusselt number is almost 4.36 when Reynolds number ranged from 400 to 1000, with a maximum deviation $\pm 3.1\%$ and for downward flow, the Nusselt number is close 4.36 when the Reynolds number varied between 600 and 1000, with a maximum deviation $\pm 2.9\%$.

It is also interesting to note from Fig. 3 that the Nusselt number decreased when the Reynolds number decreased (after Re 500 for vertically downward flow). This is due to buoyance force which is acting opposite of the main flow direction and forced convection is getting weaker and weaker and it is approaching the lower limit of conduction (Nu $= 1$). Similarly, for upward flow, heat transfer also decreases with the decrease of Reynolds number (after Re 500 for vertically upward flow). This is due to very low flow rate and the buoyance force dominates the flow and it is also approaching conduction.

After the short distance of the tube, instability begins in the buoyancy secondary flow, which prevents the continuance decline heat transfer. This is due to the buoyancy strength, which becomes sufficient to instability the boundary layer. It can be observed from Fig. 3 that the Nusselt number has the highest value at high Reynolds

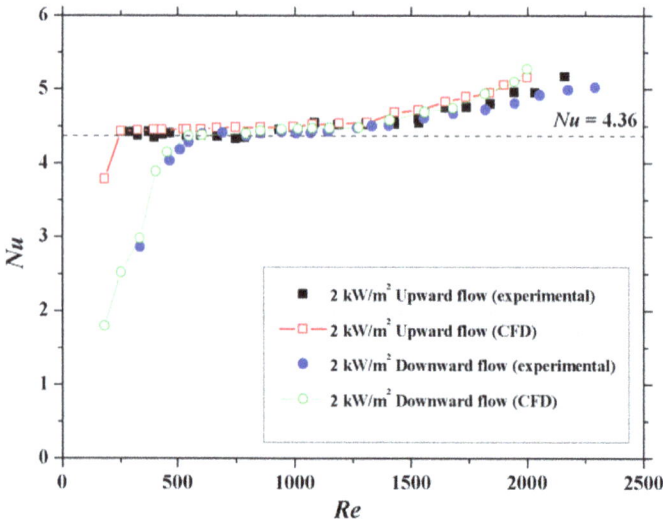

Fig. 3 Nusselt number as a function of Re for upward and downward flow at heat flux of 2 kW/m^2

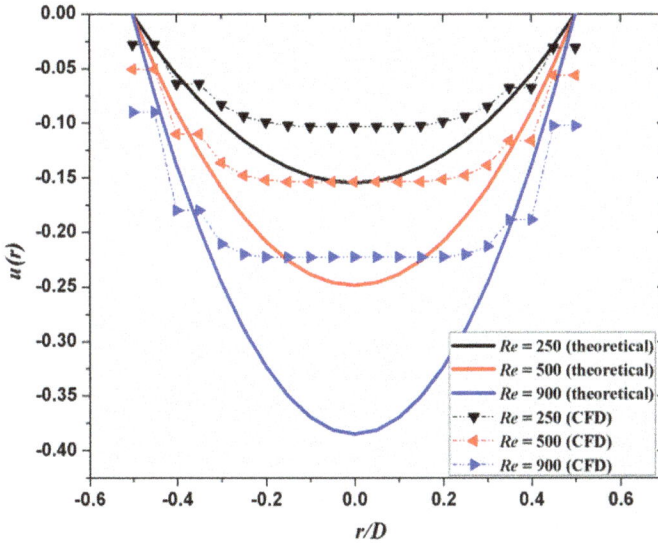

Fig. 4 Axial velocity as a function of dimensionless radial distance (r/D) for downward flow at different Re at constant heat flux 2 kW/m^2 at $x/D = 883$

number (above Re 600 for downward flow and above Re 300 for upward flow) due to a thin boundary layer and supremacy-forced convection. Then, the Nusselt number is gradually decreased because there is a balance between the effect of all convection and buoyancy. Thus, the buoyancy effect controls over the forced convection effect at the end part of the test section, leading to a decrease in the heat transfer.

The cross-sectional flow distribution inside the tube is shown as Fig. 4 with three different Reynolds number (Re = 250, 500, and 900) and axial position at $x/D = 833$. The high velocity zone is located in the center of the tube when theoretical equation is applied. But, it is interesting to see for the figure that the maximum velocity no longer appears in the center of the tube in the simulation cases which shifts toward the wall. The high velocity zone move toward the wall of the tubes because of the longitudinal swirl developed due to buoyance forces. As the height of cubes increases, the mean velocity near the rough wall decreases and the tube leads to a higher component velocity in Y–Z plan.

5 Conclusions

It was observed that fluid flow direction had a significant effect on Nusselt numbers for Reynolds numbers lower than 500. Also, it was found that Nusselt number was not constant at 4.36 at fully developed laminar flow forced convection at a constant wall heat flux boundary condition and it is also dependent of Grashof number. Nevertheless, the Reynolds numbers increased with increasing heat flux and Reynolds number above 600 due to increasing temperature with decreasing working fluid viscosity.

Acknowledgements The work was conducted by the first author as a post-doctoral research fellow as the experimental set-up of the third author. It was supervised by the second author (post-doctoral fellow) and last author (Professor).

References

1. Y.A. Çengel, A.J. Ghajar, *Heat and Mass Transfer: Fundamentals and Applications*, 5th edn. (McGraw Hill, New York, 2015).
2. M. Everts, J.P. Meyer, Heat transfer of developing and fully developed flow in smooth horizontal tubes in the transitional flow regime. Int. J. Heat Mass Transf. **117**, 1331–1351 (2018)
3. J.P. Meyer, M. Everts, Single-phase mixed convection of developing and fully developed flow in smooth horizontal circular tubes in the laminar and transitional flow regimes. Int. J. Heat Mass Transf. **117**, 1251–1273 (2018)
4. J.P. Meyer, M. Everts, A.T.C. Hall, F.A. Mulock-Houwer, M. Joubert, L.M.J. Pallent, E.S. Vause, Inlet tube spacing and protrusion inlet effects on multiple circular tubes in the laminar, transitional and turbulent flow regimes. Int. J. Heat Mass Transf. **118**, 257–274 (2018)
5. A.I. Bashir, J.P. Meyer, Heat transfer in the laminar and transitional flow regimes of smooth vertical tube for upward direction, in *13th International Conference on Heat Transfer, Fluid Mechanics and Thermodynamics (HEFAT2017)*, 17th–19th July 2017, Portoroz, Slovenia
6. S. Bhattacharyya, H. Chattopadhyay, A. Guin, A.C. Benim, Investigation of inclined turbulators for heat transfer enhancement in a solar air heater. Heat Transf. Eng. Online Published (2018). https://doi.org/10.1080/01457632.2018.1474593
7. S. Bhattacharyya, H. Chattopadhyay, A.C. Benim, Numerical Investigation on Heat Transfer in Circular tube with inclined ribs, progress in computational fluid dynamics. Int. J. **17**(6), 390–396 (2017)

The Role of Turbulence Models in Simulating Urban Microclimate

Azin Hosseinzadeh, Nima shokri, and Amir Keshmiri

1 Introduction

Rapid increase in construction of high-rise buildings brings the need of wind microclimate assessment. Basically, wind microclimate assessment is done during the design stage to indicate the results of the wind impact on the design which is followed by the measures that can be taken to mitigate the wind in the areas of high velocities to be in the acceptable range. Assessment of wind conditions around buildings has been carried out using observational techniques and computational fluid dynamics (CFD) method. Observational techniques are conducted through measurement (e.g. wind tunnel testing) and are widely used for validation of simulations conducted by CFD techniques. However, not all the cases can be validated. Wind tunnel test for real cases with complex geometry is problematic due to the difficulty related to similarity of real wind condition with wind tunnel chamber [1]. Nowadays, with the help of computational aid there has been growing interest in the use of computational fluid dynamics (CFD) methods to predict the wind around buildings, and numerous works have been conducted in this area for generic or real case [2–4]. Among various works in this area, the work conducted for urban environment by Yoshihide [5] demonstrates the use of five different turbulence models including standard $k\text{-}\varepsilon$, RNG $k\text{-}\varepsilon$, realizable $k\text{-}\varepsilon$, standard $k\text{-}\omega$ and $k\text{-}\omega$ SST under steady and unsteady-state RANS method. Although this work can be used as a well comparative study, but it is only validated for single high-rise building and if the difference of each method for complex geometry is negligible or not is still not clear. Among CFD analysis over urban areas, none

A. Hosseinzadeh (✉) · A. Keshmiri
School of Mechanical, Aerospace and Civil Engineering, The University of Manchester, Manchester, UK
e-mail: Azin.hosseinzadeh@manchester.ac.uk

N. shokri
School of Chemical Engineering and Analytical Science, The University of Manchester, Manchester, UK

© Springer Nature Singapore Pte Ltd. 2021
C. Wen and Y. Yan (eds.), *Advances in Heat Transfer and Thermal Engineering* ,
https://doi.org/10.1007/978-981-33-4765-6_116

of them compared the effect of various turbulence methods on the same configuration, same wind direction and same grid size for complex geometry. Different results are obtained using different models, but the order of this difference is not clearly mentioned in any previous works. The main aim of this work is to show the impact of various factors that can affect numerical simulations around urban environment. To do so, wind microclimate assessment for east Village of London Olympic Park consists of 67 blocks simulated using CFD techniques. The accuracy of each CFD simulation is determined by its validation. In order to validate this case, initially the wind simulation for two buildings is conducted. The results of the simulation vary that depends on the input parameters including: Grid size, type of grid (e.g. structured or unstructured), the ratio of prism layer thickness to the cell size far from walls, type of turbulence model, methods of solving the near wall velocity (e.g. low Re number or wall function), solver setting, shape and size of computational domain. The results of the CFD simulation for two buildings are validated by the work that was done by Blocken [6]. After validation, the same setting for solver and input parameters is used for the case of east village.

2   Numerical Simulations: Wind Speed Between Parallel Buildings and East Village

This section starts with the CFD analysis of wind speed for genetic case consists of two parallel buildings. The input parameter which gives the best results according to the experimental data is used for the case of east village.

2.1   Computational Domain and Mesh

Based on CFD guideline for CFD simulation of pedestrian comfort, the outflow boundaries must be $15H_{max}$, where H_{max} is the height of the tallest building. The top and inlet boundaries must be at least $5H_{max}$ from the target area [7, 8]. The building blocks are surrounded by circular subdomain to distinguish between the mesh size in the vicinity of area of interest and far from it as well as roughness height. The grid resolution must meet the criteria based on CFD guideline. There should be a reasonable growth rate between outer prism layer and first core cell to avoid a big jump between two areas. To evaluate this, prism layer thickness close to the wall is defined as 10, 20 and 30% of the core cell size of for each section. The number of prism layer varied between 2 and 5.

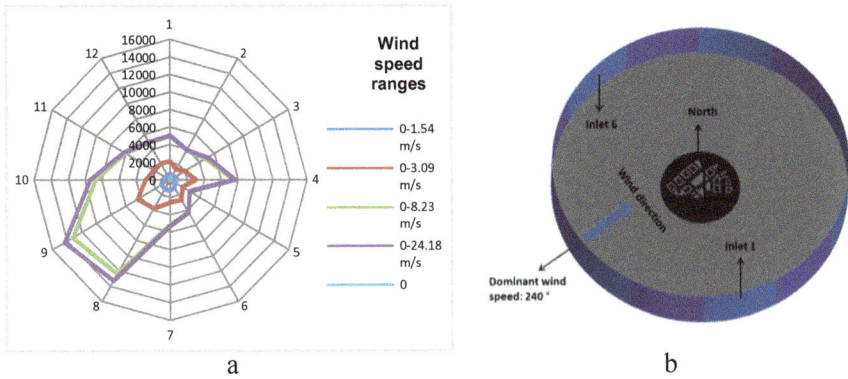

Fig. 1 **a** Wind data analysis, **b** computational domain and boundary conditions for the case of east village

2.2 Governing Equations and Boundary Conditions

Using RANS methods, the equations that need to be coupled to resolve are: continuity and Navier–Stokes. Additional terms are required for closure to calculate eddy viscosity model which varies depend on the type of turbulence model. To solve all these equations, boundary conditions need to be defined at each surface of computational domains, on building walls, roof and ground. The computational domain is divided into two sections of inlet and outlet. The boundary condition that is imposed on the flow for inlet is velocity, while for the outlet zero static pressure is imposed. [9]. The inlet boundary conditions need to show the velocity dependency to the height. [10]. Wall function for rough surfaces is imposed on the ground, and for building walls and roof, aerodynamic roughness length is set as zero. The wind speed is given at different direction at 10 m above the ground level using wind rose which is shown in Fig. 1. It can be clearly observed that velocity is dominant in south-west (SW) direction, and at angle of 240° it reaches the maximum. The inlet boundary is then defined in the range of SW-90 °C < Inlet < SW + 90 °C.

3 Results

To compare the results of CFD simulation between two buildings passage quantitatively, 5 points between building passages are taken between buildings which is shown in Fig. 2. As it can be observed from this figure, realizable and standard k-ε model matches the experimental data more with less than 10% difference with the wind tunnel data. The difference of turbulence models is also compared for the case of London Olympic Park. Figure 3 demonstrates the velocity contour at pedestrian level of 2 m using realizable k-ε model. The other turbulence models are not shown

Fig. 2 Comparison of amplification factors [U/U_0]: U_0: 5.9 [m/s]

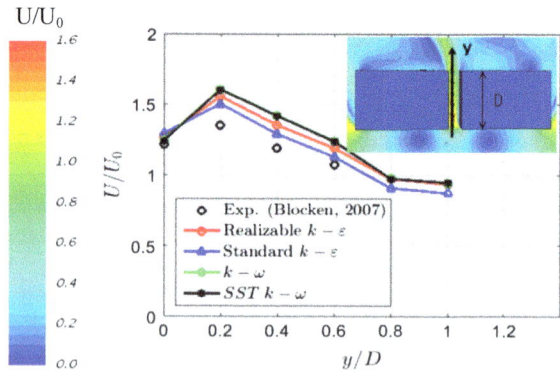

Fig. 3 Velocity contour at pedestrian level of 2 m using realizable k-ε model, wind direction: 240°

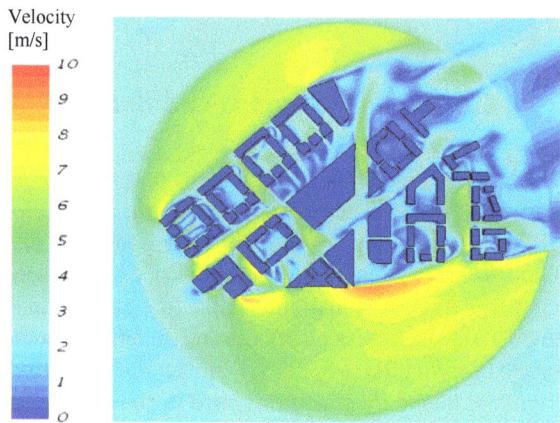

in this figure due to difficulty of comparison through contour. Thus, the variation of velocity contours is monitored at different height in Figs. 3 and 4.

4 Conclusions

With analysis of the results of the predicted wind at pedestrian level using various turbulence models, no significant difference can be found. However by comparing the wind speed variation with height at three different locations on the proposed model, the difference of each model is more observable. Below is some conclusion from this study:

- Unstructured polyhedral mesh type results in more accurate simulation than tetrahedral
- By using wall function to predict the velocity around buildings, there should be a reasonable growth rate between outer prism layer and first core cell. More robust

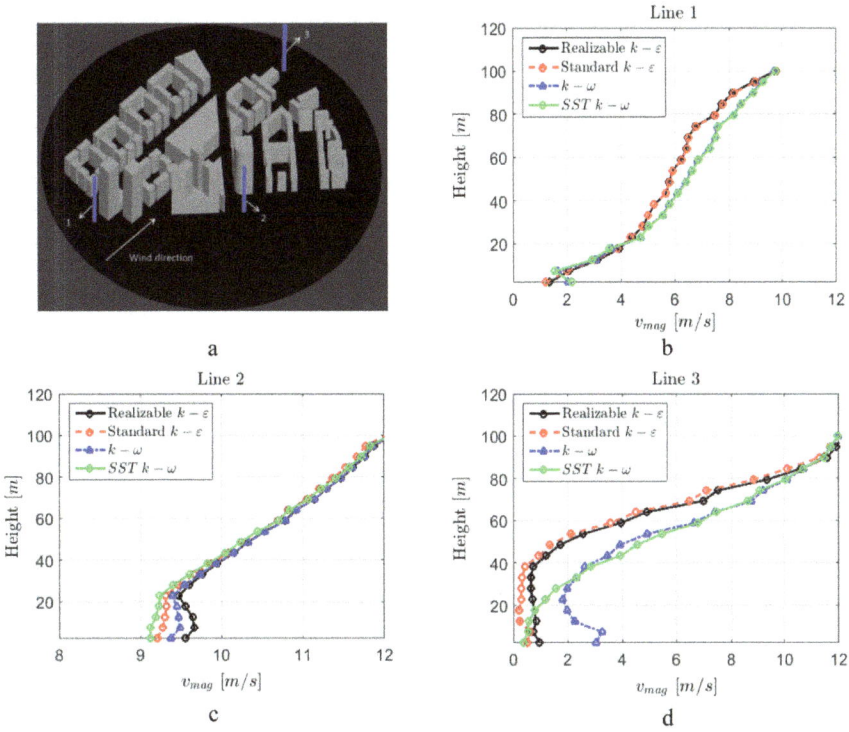

Fig. 4 Comparison of various turbulent models at different height for 3 parts of the geometry, **b** line1, **c** line2, **d** line3

results are obtained from the simulation of two buildings, if the prism layer total thickness is 20% of the core cell size.

- Grid dependency test shows 2.7 million polyhedral cells are fine enough for CFD simulation of east village.
- Realizable and standard k-ε show more accurate results as it fits experimental data better
- Realizable and standard k-ε models have similar trend at the entrance and in the middle of buildings model. However, results from k-ω SST model at the entrance of buildings show velocity with lower order of magnitude in comparison to the other models. This fact has been proven by the work done by Yoshihide [5] which was evaluating the accuracy of this model around one high-rise building. It has been confirmed that k-ω SST underestimates the turbulent kinetic energy around buildings, and as a result, flow separation is exceeded around the building corners.

By comparing the different turbulence model regarding the accuracy and computational cost, easier decision can be made for industrial application. Among the evaluated methods, k-ω and k-ω SST show the poorest performance, while realizable and standard k-ε method shows a better trend. The work that has been done in this paper will be followed by implementing mitigation strategies including canopies, trees and edges to slow down the wind in the areas of high velocities which is as a result of downwash, corner and funnelling effect. In addition,

heat transfer will be added to the model for computational modelling of urban heat island effects for the case study of London Olympic Park.

References

1. B. Blocken, T. Stathopoulos, J.P.A.J. van Beeck, Pedestrian-level wind conditions around buildings: review of wind-tunnel and CFD techniques and their accuracy for wind comfort assessment. Build. Environ. **100**, 50–81 (2016)
2. Y. Toparlar, B. Blocken, B. Maiheu, G.J.F. van Heijst, A review on the CFD analysis of urban microclimate. Renew. Sustain. Energy Rev. **80**, 1613–1640 (2017)
3. B. Blocken, T. Stathopoulos, CFD simulation of pedestrian-level wind conditions around buildings: past achievements and prospects. J. Wind Eng. Ind. Aerodyn. **121**, 138–145 (2013)
4. A. Mochida, I.Y.F. Lun, Prediction of wind environment and thermal comfort at pedestrian level in urban area. J. Wind Eng. Ind. Aerodyn. **96**(10–11), 1498–1527 (2008)
5. Y. Tominaga, Flow around a high-rise building using steady and unsteady RANS CFD: effect of large-scale fluctuations on the velocity statistics. J. Wind Eng. Ind. Aerodyn. **142**, 93–103 (2015)
6. B. Blocken, J. Carmeliet, T. Stathopoulos, CFD evaluation of wind speed conditions in passages between parallel buildings-effect of wall-function roughness modifications for the atmospheric boundary layer flow. J. Wind Eng. Ind. Aerodyn. **95**(9–11), 941–962 (2007)
7. B. Blocken, Computational fluid dynamics for urban physics : importance, scales, possibilities, limitations and ten tips and tricks towards accurate and reliable simulations computational fluid dynamics for urban physics: importance, scales, possibilities, lim. Build. Environ. **91**, 219–245 (2015)
8. Y. Tominaga et al., AIJ guidelines for practical applications of CFD to pedestrian wind environment around buildings. J. Wind Eng. Ind. Aerodyn. **96**(10–11), 1749–1761 (2008)
9. A. Parente, C. Benocci, On the RANS simulation of neutral ABL flows. Cwe **2010**, 1–9 (2010)
10. R.P.H.P.J. Richards, *Appropriate Boundary Conditions for Computational Wind Engineering Models Using the k-ε Turbulence Model no. 1993* (1993)

Temperature Effect on Falling Behaviour of Liquid Gallium Droplet

M. Sofwan Mohamad, C. M. Mackenzie Dover, and K. Sefiane

1 Introduction

The motion and deformation of liquid droplets are crucial to various applications and consequently have drawn the attention of many researchers. In industrial engineering, knowing the characteristics of liquid droplets is highly relevant to numerous applications, for instance liquid sprays injected in combustion engine, ink-jet printers, microfluidics and cooling systems. Among many droplet elements, gallium and its alloys receive massive interest in recent times due to the relevant physical and chemical characteristics such as the low-melting points (liquid state at or near room temperature) and high thermal conductivities. These qualities promote huge potentials for establishing advanced technologies in newly growing fields which include electronic cooling, 3D printing and printed electronics, flexible device, soft actuator in robotic application and adhesion.

A large amount of literature is available about liquid droplet behaviour covering in-depth evaluation of theory, experimental data and also applicable approximations describing the behaviour of individual droplet in fluid systems [1, 2]. Nevertheless, reported data on the characteristics of deformable droplet are very little. Up to now, to the knowledge of author, there was little to no systematic research on the influence of liquid gallium droplet shape on its motion and velocity. In this article, the characteristics of a free-falling droplet of gallium in water are studied along with the effect of temperature on the droplet shape, and velocity is investigated. The comprehension

M. Sofwan Mohamad (✉) · C. M. Mackenzie Dover · K. Sefiane
Institute for Multiscale Thermofluids, School of Engineering, University of Edinburgh, Faraday Building, King's Buildings, Edinburgh E9 3DW, UK
e-mail: Muhammad.Mohamad@ed.ac.uk

M. Sofwan Mohamad
School of Mechatronic Engineering, Universiti Malaysia Perlis (UniMAP), Kampus Pauh Putra, 02600 Arau, Perlis, Malaysia

© Springer Nature Singapore Pte Ltd. 2021
C. Wen and Y. Yan (eds.), *Advances in Heat Transfer and Thermal Engineering*,
https://doi.org/10.1007/978-981-33-4765-6_117

of this knowledge might stretch to an improved manipulation in the usage of this
material and even stimulate different possible utilizations.

2 Methodology

An experimental set-up was designed to capture the behaviour of a free-falling droplet
in water (Fig. 1). The set-up comprises a 500 mm-height square cross section, straight-
walled column with inner side width of 70.45 mm made of clear Perspex. These
dimensions were selected such that the falling droplet with diameter of less than 7 mm
could reach terminal velocity with negligible wall effects and end effects [3, 4]. The
motion of the falling droplet was recorded by a high-speed camera which is mounted
on vertical and horizontal sliders for flexibility in positioning the camera. A custom-
made high-power LED light was used to illuminate the test section. The bottom
part of the column is attached to a liquid gallium retrieval mechanism which allows
the liquid gallium to be removed from the column easily. The temperature of the
water is controlled by an electric heater and a proportional–integral–derivative (PID)
controller connected to two thermocouple probes. The liquid gallium is preheated to
the desired temperature and then dispersed into the column just below the water free

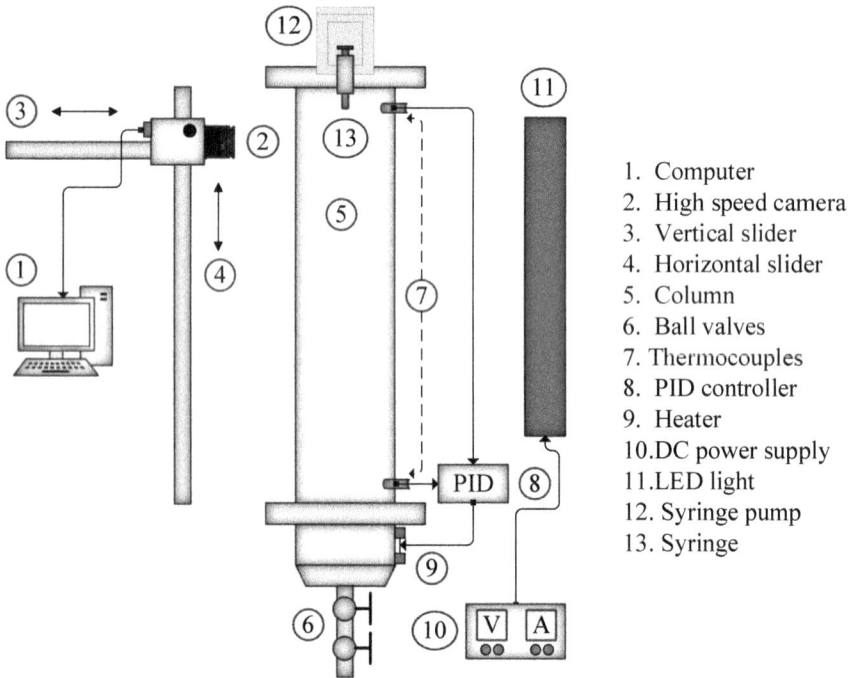

Fig. 1 Schematic diagram of the experimental set-up

surface by a syringe connected to a programmable syringe pump. Different sizes of needle can be attached to the syringe to produce various sizes of droplet. The experiment is conducted under isothermal condition in the temperature range from 30 to 70 °C.

The captured video is processed using an in-house ImageJ macro to obtain the relative position, velocity and dimensions of the droplet. Since the droplet is constantly deformed during the free-fall, the droplet size is determined by the sphere volume equivalent diameter, $d_{eq} = \sqrt[3]{3V/4\pi}$ using the pre-determined liquid volume, V set to the syringe pump. The shape of the droplet is categorized by the particle aspect ratio, $E = d_h/d_v$ calculated from the largest horizontal dimension, d_h to the largest vertical dimension, d_v detected from the image-processing procedure. The droplet with aspect ratio more or less than 1.1 is classified as oblate-spheroid or prolate-spheroid, respectively. The effect of temperature is discussed in terms of viscosity ratio, $\lambda = \mu_{Ga}/\mu_w$ between the viscosity of liquid gallium, μ_{Ga} and water, μ_w.

3   Results

The experiments were conducted for four different droplet sizes. However, we will only present the result of 4.86 mm droplet, which exhibit most of the characteristic intended to be discussed. Figure 2a–e show the typical falling behaviour of the droplet in water at different temperature. The droplet images in the same figure are the same droplet, and the time interval between two consecutive images is 0.01 s. It can be seen in all the cases that the droplets experience a significant deformation as soon as they

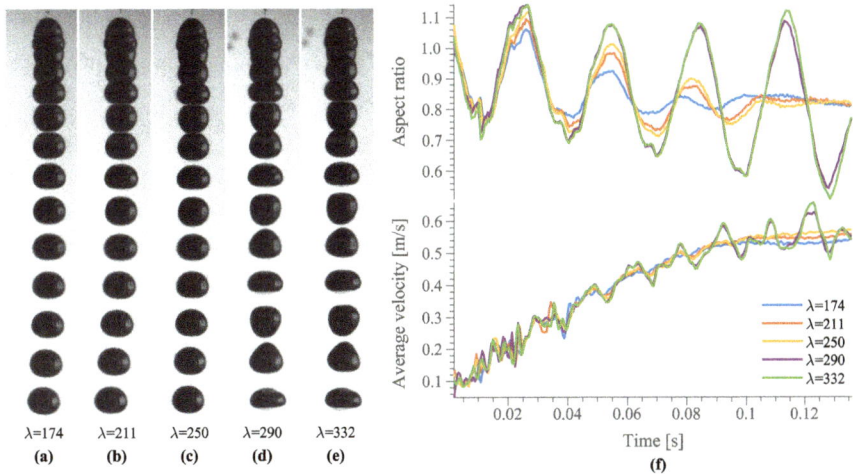

Fig. 2 Typical trajectory of free-falling liquid gallium droplets with equivalent diameters of 4.86 mm at different temperatures: **a** 30 °C, **b** 40 °C, **c** 50 °C, **d** 60 °C, **e** 70 °C and **f** the corresponding time evolution of the aspect ratio (top) and velocity (bottom) of the droplets

are released from the syringe. The droplet continues to deform (mainly the vertical dimension) while falling in rectilinear fashion. In order to get a clear comparison of the shape of the droplets, the time evolution of the droplet aspect ratio while falling in the column is plotted in the top graph of Fig. 2f. The initial shape of the droplets after detachment was found to be prolate-spheroid. Prolate-oblate oscillation began directly after the detachment of the droplet prior to dampening of the liquid–metal into final shape of an oblate-spheroid. It can be clearly seen that the lower viscosity ratios yield a smaller oscillation amplitude than the higher viscosity ratios. The lowest three viscosity ratios exhibit similar behaviour, in which the aspect ratio oscillation amplitude is slowly dampened to steady-state condition at about 0.12 s. In contrast, for the remaining two higher viscosity ratio cases, the oscillation keeps on amplifying. It is likely that the temperature between 50 and 60 °C or viscosity ratio between 250 and 290 contain a critical point, at which a 4.86 mm droplet will significantly change it deformation behaviour before reaching an equilibrium shape. Unfortunately, the droplets went out of the camera view before reached steady-state condition. However, as discussed in [5] the final shape of the droplets with viscosity ratio of 290 and 322 is also in the shape of oblate-spheroid.

The vertical velocity of the droplets was calculated from the time derivative of their centroid coordinate. The bottom graph of Fig. 2f depicts the transient vertical velocity of the droplets derived from trajectories in Fig. 2a–e. It can be seen that the vertical velocity oscillate in about double frequency of the aspect ratio oscillation. The velocity of the droplets oscillates in a similar fashion during the acceleration period. Approaching terminal velocity at about 0.09 s, the oscillation in velocity for lower viscosity ratios droplets starts to dampened, although the aspect ratio still have small oscillation. On the other hand, the higher viscosity ratio droplet velocity continues to oscillate at an asymptote terminal velocity value. Both graphs in Fig. 2f are plotted from the ensemble average of 5 trials for each case. All experiments are highly reproducible with the highest average standard deviation of 0.02 and 0.05 for aspect ratio and vertical velocity, respectively. The error bars are not shown in the figure for the sake of clarity.

4 Conclusions

An experimental study has been conducted on the effect of temperature on the behaviour of free-falling liquid gallium droplet in water. The temperature of both liquids was varied in the range from 30 to 70 °C to get viscosity ratio in the range between 174 and 332. For droplets with equivalent diameter of 4.86 mm, initial shape of the droplets after detachment was found to be prolate-spheroid at all viscosity ratio. In transient state, the droplets' shape was oscillated between prolate and oblate-spheroid shapes towards a terminal shape of oblate-spheroid. The amplitude of aspect ratio oscillation is found to increase with viscosity ratio. The trajectories of the droplet are found to be rectilinear in all cases. The falling vertical velocity accelerated and decelerated as the droplet aspect ratio oscillates. In conclusion, the variation in liquids properties caused by increasing the liquid temperature has significant influence on the falling behaviour of liquid gallium droplet.

Acknowledgements The author acknowledges the financial support from Kementerian Pendidikan Malaysia and Universiti Malaysia Perlis for the sponsorship which enables me to pursue this study.

References

1. M.E. Clift, R. Grace, J.R. Weber, *Bubbles, Drops and Particles* (Academic Press Inc. (London) Ltd., New York, 1978)
2. E. Loth, Quasi-steady shape and drag of deformable bubbles and drops. Int. J. Multiph. Flow **34**(6), 523–546 (2008)
3. P. Uhlherr, R. Chhabra, Wall Effect for the fall of spheres in cylindrical tubes at high Reynlolds number. Can. J. Chem. Eng. **73**, 918–923 (1995)
4. P.P. Brown, D.F. Lawler, Sphere drag and settling velocity revisited. J. Environ. Eng. **129**(3), 222–231 (2003)
5. M. Sofwan Mohamad, C.M. Mackenzie Dover, K. Sefiane, Experimental investigation of drag coefficient of free-falling deformable liquid gallium droplet. Eur. Phys. J. Appl. Phys. (2018)

Needle-Based Formation of Double Emulsion Encapsulating Multiple Cores in Parallel Mode

Zheng Lian, Yong Ren, Kai Seng Koh, Jun He, George Z. Chen, Yuying Yan, and Jing Wang

1 Introduction

Multiple-cored double-emulsion microdroplet generation has drawn great attention and interest recent years especially for the formation of stable and tunable inner cores in the core–shell structured double emulsions [1–3]. In this work, needle-based microfluidic device was applied as multiphase system to form microdroplets possessing core–shell structure and encapsulating multiple cores within shell phase. By tuning the flow rates of three different phases and using flow diverters Y-shaped

Z. Lian · Y. Ren (✉)
International Doctoral Innovation Centre, University of Nottingham Ningbo China, Ningbo, China
e-mail: yong.ren@nottingham.edu.cn

Department of Mechanical, Materials and Manufacturing Engineering, University of Nottingham Ningbo China, Ningbo, China

Y. Ren · Y. Yan
Research Group for Fluids and Thermal Engineering, University of Nottingham Ningbo China, Ningbo, China

K. S. Koh
School of Engineering and Physical Sciences, Heriot-Watt University Malaysia, No. 1 Jalan Venna P5/2, Precinct 5, 62200 Putrajaya, Malaysia

J. He (✉) · G. Z. Chen
Department of Chemical and Environmental Engineering, University of Nottingham Ningbo China, Ningbo, China
e-mail: jun.he@nottingham.edu.cn

Y. Yan
Research Group for Fluids and Thermal Engineering, University of Nottingham, Nottingham NG7 2RD, UK

J. Wang
Department of Electrical and Electronic Engineering, University of Nottingham Ningbo China, Ningbo, China

© Springer Nature Singapore Pte Ltd. 2021
C. Wen and Y. Yan (eds.), *Advances in Heat Transfer and Thermal Engineering* ,
https://doi.org/10.1007/978-981-33-4765-6_118

connector, multiple cores could be encapsulated inside the polydimethylsiloxane (PDMS) shell in parallel connection. The size of microdroplets could be manipulated either by flow rates or applying various needle combinations with different sizes. The needle-based microfluidic device is versatile, reliable and cost-effective. The platform exhibits exquisite control over the structure, overall droplet size and the size of cores, leading to formation of microdroplets with high monodispersity. The PDMS microdroplets were thermally cured off-site to form functional microcapsules and stored for application such as encapsulation of various activities in the same microcapsule which can be applied in drug delivery, allowing for encapsulation and release of multiple drugs with different dosages. Furthermore, the developed microcapsules can be used in energy and environmental applications such as water treatment and carbon capture.

2 Methodology

The commercially available Y-shaped connecter was feasible to be linked with syringes to divert flows into two identical streams attributed to its special design that one such unit was equipped with inlet entrance and two equal-sized outlet exits shown in Fig. 1a. Hence, this unit allowed an extra connection of one needle-based device to form a parallel connection and making flexible connection possible upon different designs. The needle-based microdroplet generator could produce micro-droplets either using single- or double-emulsion templates [4]. The single-emulsion microdroplet generator was composed of two different types of plate-tipped needles defined as inlet needle and outlet needle, respectively. While the double-emulsion droplet generator was based on the structure of single-emulsion microdroplet gener-ator with inserting another needle, defining these three as inner, middle and outer needles, respectively. Those needles were linked by simple accessories such as polypropylene (PP) T-shaped link, cross-link, silicone tubes and so on. In this study, DI water with PVA surfactant was employed as inner and outer phases, and PDMS precursor with curing agent was used as middle phase. The generated droplets were collected in a beaker that contained a mixture of 100 mL DI water with 5% PVA. The droplets were captured by an N-800F optical microscope (Ningbo Novel Optics, China), and sizes of 50 drops were measured by ImageJ software. CV could be calcu-lated based on the values of standard deviation and average diameter of droplets, as below:

$$\tag{1}$$

The collected droplets were placed on a heating oven under 70 °C for 2 h to ensure the polymer to be cross-linked from liquid to solid state. Flow rates of each phase could be manipulated to assist the investigation on its impacts on droplet morphologies [3].

Fig. 1 **a** Y-shaped connecter
linked with two soft tubes for
diverting fluids, **b** 25G-21G
needle-based microfluidic
device for production of
single emulsions,
c 34G-21G-20G
needle-based microfluidic
device for production of
double emulsions

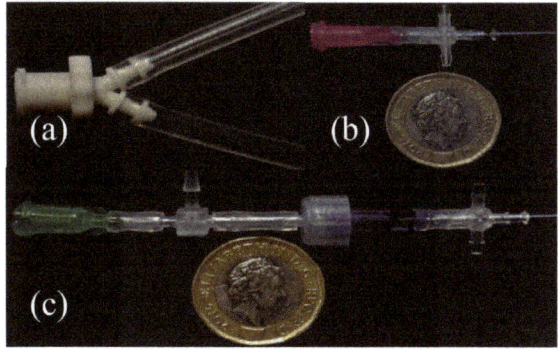

Fig. 1 **a** Y-shaped connecter linked with two soft tubes for diverting fluids, **b** 25G-21G needle-based microfluidic device for production of single emulsions, **c** 34G-21G-20G needle-based microfluidic device for production of double emulsions

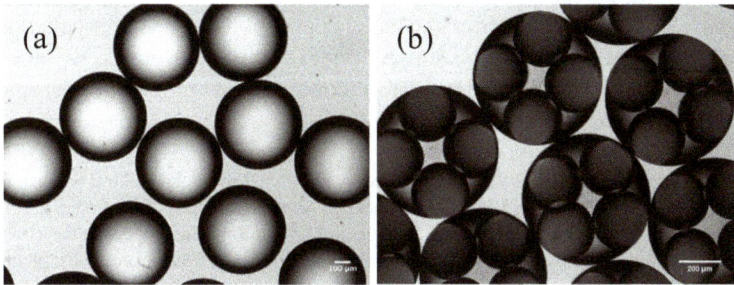

Fig. 2 **a** Single-emulsion PDMS microdroplets by parallel connection of two 25G-21G needle-based devices, under flow rates of 30 μL/min for dispersed phase and 200 μL/min for continuous phase. The average sizes were 540 μm with CV as 1.3%, **b** double-emulsion water-in-PDMS microdroplets with four cores generated by parallelly connecting two needle-based microfluidic devices with needle combinations of 34G-21G-20G. The flow rates for inner, middle and outer phases were 12, 20 and 200 μL/min, respectively. The average diameter was measured as 605 μm with CV as 1.8%

3 Results

Monodispersed PDMS droplets using single-emulsion template were generated by two parallel connected 25G-21G (25G inner diameter 250 μm, 21G inner diameter 510 μm, the assembled device is shown in Fig. 1, leading to formation of single emulsion (see Fig. 2a), the average diameter was 540 μm with CV values calculated as 1.3% by Eq. 1, indicating that the droplets were highly monodispersed in this parallel mode. Flow rates imposed to the dispersed phase and continuous phase were 30 and 200 μL/min, respectively. The production rates under this parallel connection were almost doubled compared with a single device.

Meanwhile, double-emulsion microdroplets with stable four inner cores were fabricated based on flow rate adjustment as shown in Fig. 2b. A 34G-21G-20G (34G, inner diameter 60 μm, 20G, inner diameter 600 μm) needle device was applied and operated under flow rates of 12, 20 and 200 μL/min for inner, middle and outer phases, respectively. The generated droplets contained four cores, and the overall average diameter of droplets was 605 μm with CV as 1.8%. The number of inner cores was increased when increasing the flow rate of middle phase, while inner and outer phases were fixed. In addition, the parallel connected devices could extend the droplet production rate twofold by shunting two identical devices and given the same flow conditions.

4 Conclusions

The Y-shaped connecter enabled two microfluidic devices to be connected in a shunt manner. Monodispersed PDMS single and double emulsions were generated by needle-based microfluidic devices in parallel connection mode. The production rates could be enhanced double than a single device, while the overall CV values were remained low (<3%). The droplets were thermally cured into solid microparticles which could be used in other applications.

Acknowledgements This research was supported by Zhejiang Provincial Natural Science Foundation of China under Grant No. LY19E060001, National Natural Science Foundation of China under Grant No. NSFC51506103/E0605, Ningbo Science and Technology Bureau, under the project code 2019F1030, as well as Research Seed and Supplementary Support Fund from Nottingham China Health Institute.

References

S.H. Kim, H. Hwang, C.H. Lim, J.W. Shim, S.M. Yang, Packing of emulsion droplets: structural and functional motifs for multi-cored microcapsules. Adv. Funct. Mater. **21**, 1608–1615 (2011)

L. Chu, A.S. Utada, R.K. Shah, J. Kim, D.A. Weitz, Controllable monodisperse multiple emulsions. Angew. Chem. Int. Ed. Engl. **46**, 8970–8974 (2007)

S.A. Nabavi, G.T. Vladisavljevic, V. Manovic, Mechanisms and control of single-step microfluidic generation of multi-core double emulsion droplets. Chem. Eng. J. **322**, 140–148 (2017)

T. Li, L. Zhao, W. Liu, J. Xu, J. Wang, Simple and reusable off-the-shelf microfluidic devices for the versatile generation of droplets. Lab Chip, 4718–4724 (2016)

Flow-Visualization Experimental Research on the Plume Pattern Above a Horizontal Heated Cylinder

Guopeng Yu, Haiteng Ma, Weihong Li, Li He, and Zhibin Yu

1 Introduction

Natural convection heat transfer of horizontal heated cylinders has been studied for nearly a century. However, the flow physics of the evolving plume above heat horizontal cylinders has not been taken as a research focus and still needs investigation for clearer insights [1]. In recent years, more advanced numerical and experimental methods have been applied to further study the fundamental subject from different perspectives [2].

In this work, flow-visualization experiments have been conducted to investigate the flow plume pattern above a heated horizontal cylinder. Firstly, the experimental flow pattern is compared to numerical results with the same scale of Rayleigh number; secondly, plume pattern variation against cylinder temperature is investigated.

2 Methodology

Experiments were carried out on a self-designed smoke visualization test rig. Smoke is generated from burning materials. Hot smoke is directed through a cooling channel to reach/close to ambient temperature. Smoke of ambient-similar temperature is then lead to a collecting chamber to buffer the uneven speed and density caused by the burning process. At last, an even and stable smoke is discharged through a converging round nozzle. The formed smoke line flows normal to a horizontal cylinder for tests. A cartridge heater is used as the heating cylinder. Adhesive thermal couples are attached

G. Yu · W. Li · Z. Yu (✉)
School of Engineering, University of Glasgow, Glasgow G12 8QQ, UK
e-mail: Zhibin.Yu@glasgow.ac.uk

H. Ma · L. He
Dept of Engineering Science, University of Oxford, Oxford OX2 0ES, UK

© Springer Nature Singapore Pte Ltd. 2021
C. Wen and Y. Yan (eds.), *Advances in Heat Transfer and Thermal Engineering* ,
https://doi.org/10.1007/978-981-33-4765-6_119

on the surface for temperature monitoring and control. A voltage transformer and PID controller are applied to regulate the surface temperature of the cartridge heater. High speed camera and LED illumination are equipped for flow images capture.

3   Results

Flow plume pattern tested is firstly compared to previous numerical work. Numerical calculations of Ma et al. [3] provide the reference of velocity contour at Rayleigh number of 10e4, as shown on the left of Fig. 1. The isothermal cylinder has a surface temperature (T_w) of 120 °C. The current test rig in this work has similar setup. The average cylinder surface temperature (T_w) and the ambient temperature (T_{air}) are 120.0 °C and 15.5 °C respectively, as shown on the right of Fig. 1. The Ra number is also at the same order of magnitude (10e4).

The comparison indicates that both plumes are laminar at first, and then sway and transit to turbulence due to flow instability. According to criteria by Noto et al. [4], the transition to turbulence occurs at 9D for Ra = 10e4. However, the transition starts a bit earlier in the test than it does in the numerical results. One possible reason is that the inflow smoke below the cylinder has a speed that cannot easily be eliminated and should not be ignored (Fig. 2).

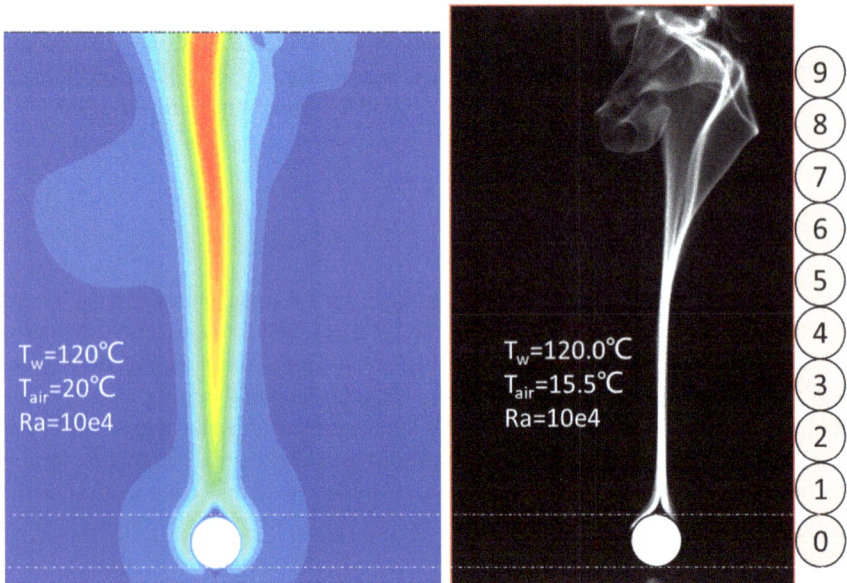

Fig. 1 CFD calculated contour of velocity magnitude at Ra = 10e4 from Ref. [3] (left) versus experimental tested smoke flow pattern at Ra = 10e4 (right)

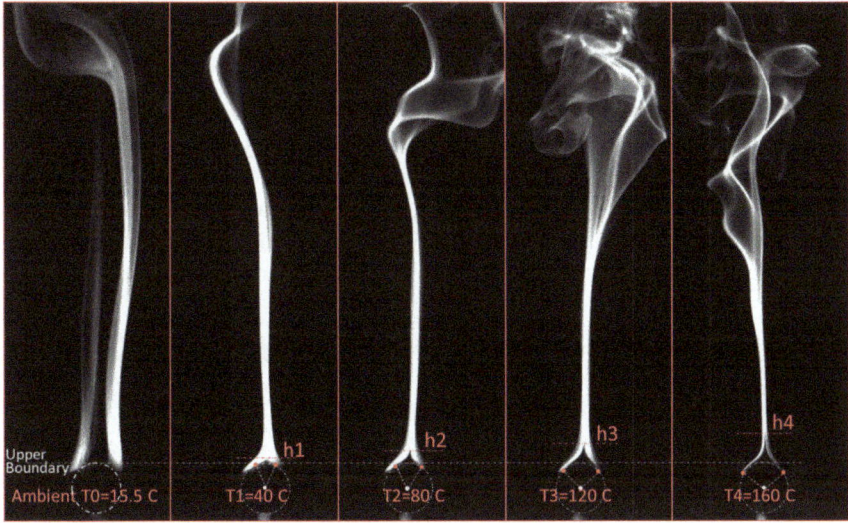

Fig. 2 Comparison of flow plume patterns with increasing cylinder surface temperature

The plume developments under different cylinder surface temperature (average) were then studied by increasing the temperature from ambient state to 160 °C. The smoke is separated by the cylinder when it flows across at the ambient state. However, by inputting heat to the cylinder, the air above the cylinder is heated and lower air density is caused. Density difference leads to pressure difference between the middle region and side flows, and two separated smoke lines are thus drawn together and become a single plume.

Bigger buoyancy force is generated by increasing the cylinder temperature (from T_1 to T_4), leading to increasingly higher smoke velocity and greater flow instability. Consequently, the transition onset from laminar to turbulence occurs earlier along the stream of the smoke. At the same time, higher flow velocity leads to earlier separation of the smoke from the cylinder, as the increasing sector angles indicate. As a consequence of the earlier separation, the converging point of two side flows is getting further from the cylinder. It gradually increases from h1 to h4, as demonstrated in the figure.

4 Conclusions

Flow plume pattern and its variation above a heated horizontal cylinder are studied experimentally for better insights into its flow mechanism. Main conclusions are drawn as following:

(1) Flow plume pattern of experiments is similar to numerical results from previous
     researchers. The thermal plume is laminar at first, then sways and transits to
     turbulence due to instability.
(2) The separating flows under ambient condition are drawn together and a single
     smoke plume is formed because of heat input to the cylinder.
(3) Higher surface temperature of the cylinder induces greater buoyance force and
     bigger smoke velocity, leading to earlier transition onset from laminar to turbu-
     lence. Meanwhile, earlier separation of the smoke from the cylinder and further
     converging point of two-side flows from the cylinder are caused.

Acknowledgements This research is funded by EPSRC (EP/P028829/1) in United Kingdom.

References

1. S. Grafsrønningen, A. Jensen et al., PIV investigation of buoyant plume from natural convection
   heat transfer above a horizontal heated cylinder. Int. J. Heat Mass Transf. **54**, 4975–4987 (2011)
2. S. Grafsrønningen, A. Jensen, Large eddy simulations of a buoyant plume above a heated
   horizontal cylinder at intermediate Rayleigh numbers. Int. J. Therm. Sci. **112**, 104–117 (2017)
3. H. Ma, L. He, S. Rane, Heat transfer-fluid flow interaction in natural convection around heated
   cylinder and its thermal chimney effect. in *IAPE 2019* (Oxford, United Kingdom, 2019). ISBN:
   978-1-912532-05-6
4. N. Katsuhisa, K. Teramoto, T. Nakajima, Spectra and critical Grashof numbers for turbulent
   transition in a thermal plume. J. Thermophys. Heat Transf. **13**, 82–90 (1999)

Experimental Study of Thermal Field of a Pebble Bed Channel with Internal Heat Generation

Meysam Nazari and Yasser Mahmoudi

1 Introduction

Currently, nuclear energy is an important part of the energy mix, which generates roughly 10% of the world's electricity, making up around one-third of the world's low-carbon electricity supply. The basic design of pebble bed reactors features spherical uranium spheres as fuel (pebbles), which create a reactor core and are cooled by a gas such as helium that does not react chemically with the fuel elements. The core of the pebble bed reactor can be considered as a packed bed. Packed bed reactor design is based upon the mechanisms of heat and mass transfer and pressure drop of the fluid through the bed of solids. The capability of removing the heat produced in the core is a key point of safety and reliability of the pebble bed reactor. Present work focuses on the core of a packed bed reactor by performing a fundamental experimental heat transfer analysis of a pebble bed cylindrical channel with internal heat generation. The main objective of the present work is to analyze the effect of the system pertinent parameters, such as flow Reynolds number, diameter of the spheres in the packed bed and generate heat on the fluid and solid temperature filed in the system.

M. Nazari
Department of Forest Biomaterials and Technology, Swedish University of Agricultural Sciences, Vallvägen 9c, Uppsala, Sweden

Y. Mahmoudi (✉)
Department of Mechanical, Aerospace and Civil Engineering, The University of Manchester, Manchester M13 9PL, UK
e-mail: yasser.mahmoudi@manchester.ac.uk

© Springer Nature Singapore Pte Ltd. 2021
C. Wen and Y. Yan (eds.), *Advances in Heat Transfer and Thermal Engineering* ,
https://doi.org/10.1007/978-981-33-4765-6_120

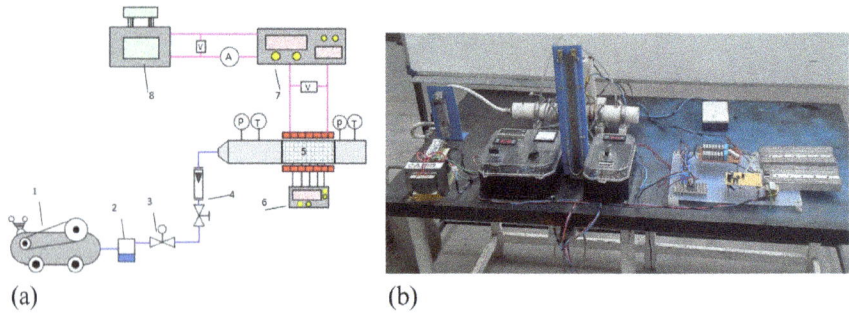

(a)   (b)

Fig. 1 Experimental setup; **a** schematic diagram 1: compressor, 2: water filter, 3: pressure regulator, 4: volumetric flow meter, 5: test section, 6: digital thermometer, 7: heating system controller, and 8: power supply. **b** Real picture of the setup

2 Methodology

To study temperature distribution in the porous medium with internal heat generation, an experimental work is conducted. As shown in Fig. 1, a cylindrical channel (with diameter is 27 mm with a length of 133 mm) filled with metallic spheres is used as the pebble bed media.

The spheres are heated using an electromagnetic heating method [1, 2]. Dry air is used as the working fluid to cool the heated spheres. The tests are performed for different spheres diameters (5.5, 6.5, and 7.5 mm), flow Reynolds number (based on the channel hydraulic diameter) in the range of [4000–10000] and generated heat of [55–82] W. After reaching a steady condition, the temperatures along the channel for both solid particles and fluid flow are measured directly using LM 35 thermosensors. To measure the local fluid temperatures, five thermos sensors were inserted in the air flow near the wall of the test section. Furthermore, five sensors were glued to the surface of spheres embedded inside the test section to measure the local temperatures of the spheres surfaces. Temperatures of the fluid at the inlet and outlet of the channel were also measured. By making use of constant temperature oil bath, all of the thermos sensors were calibrated before installing. The measured temperatures were monitored by digital temperature indicators (SAMWON). The overall accuracy was within ±0.1 °C. The inlet air flow rate was measured using a rotameter type flowmeter. Schematic diagram and real picture of the experimental setup are illustrated in Fig. 1a and 1b, respectively.

3 Results

Figure 2a shows the mean temperature of solid and fluid phases for three different sphere diameters and for generated heat of 55 W. The figure shows that both solid and fluid temperatures experience a decrease by increasing the flow Re number and

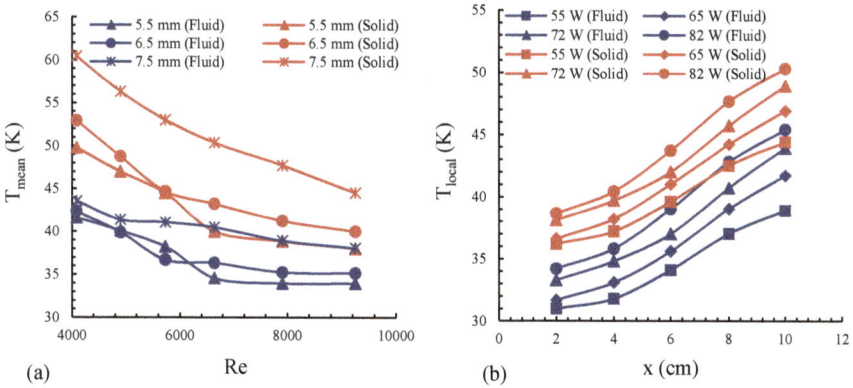

Fig. 2 Temperature behavior of two phases; **a** mean temperature of solid and fluid for different spheres diameter and for generated heat of 55 W versus Re number based on channel diameter. **b** Fluid and solid local temperatures along the channel length for different generated heat and sphere diameter of 5.5 mm and Re number based on channel diameter of 6600

decreasing the spheres diameter. Increasing the Re number is achieved by increasing the cool air mass flow rate. Hence, under steady condition and fixed generated heat, it is expected that the air temperature decreases with the increase of the mass flow rate. At the same time, increasing the Re number (i.e., increasing the air velocity) enhances the rate of heat transfer from the solid to the coolant air and hence reduces the solid temperature. In addition, for constant flow rate, increasing the spheres diameters, increases the pore sizes between particles and reduces the local velocity, hence, decreases the mean temperature in the system. It is further seen that by increasing Re number, the temperature difference between the solid and fluid phases gradually decreases. Because by increasing the fluid velocity, the rate of heat transfer between two phases increases [3]. Figure 2b illustrates the fluid and solid temperatures along the channel length for different heat generated with sphere diameter of 5.5 mm and Re number of 6600. The figure shows that by increasing the generated heat, the fluid and solid temperatures increase.

4 Conclusions

In this work, an experimental study was conducted to investigate the thermal characteristics of a pebble bed porous channel with internal heat generation. Results show that increasing the fluid velocity decreases the temperature of the solid and fluid phases in the packed bed. The temperature difference between the two phases decreases by increasing the velocity.

Reference

1. M. Nazari, D. Jalali Vahid, R. Khoshbakhti Saray, Y. Mahmoudi, Experimental investigation of heat transfer and second law analysis in a pebble bed channel with internal heat generation. Int. J. Heat Mass Transf. **114**, 688–702 (2017)
2. X. Meng, Z. Sun, G. Xu, Single-phase convection heat transfer characteristics of pebble-bed channels with internal heat generation. Nucl. Eng. Des. **252**, 121–127 (2012)
3. J.H. Du, B.X. Wang, Forced convective heat transfer for fluid flowing through a porous medium with internal heat generation. Heat Transf. Asian Res.**30**(3) (2001)

Numerical Simulation for a Single Bubble on the Vertical Flat Surface by an Immiscible Two-Phase LBM

Tomohiko Yamaguchi and Satoru Momoki

1 Introduction

The motion of a bubble on a vertical flat surface depends on the thermophysical and transport properties of fluid, the gravitational acceleration, the liquid flow, and the wettability on the solid surface. Though the authors have studied about a lattice Boltzmann model (LBM) for immiscible two-phase flow with large density difference, we have not controlled the wettability on the solid surface [1]. In this paper, the contact angle on the solid surface, bounce back condition with extrapolation boundary scheme and modified interpolation function for the viscosity coefficient in transition region between liquid and gas phases have been introduced into the lattice Boltzmann athermal model for immiscible two-phase flow. The wetting potential theory published by Yan and Zu [2] is embedded into our method in order to take into account the wettability on the solid surface, because their model employs a three-dimensional 15-velocity (D3Q15) model which is the same model of ours. While many studies about LBM simulations for droplets or bubbles on the solid surface are published, there are a few literatures about the bubble(s) with large density difference. We chose the simplest case that is a single bubble on the vertical wall surface as a condition of simulation to confirm the validity of this numerical method. The bubble motions with two contact angles, 30° and 150°, on the vertical solid flat surface were simulated by the improved LBM. The advancing and receding contact angles were reproduced in the results of simulation.

T. Yamaguchi (✉) · S. Momoki
Graduate School of Engineering, Nagasaki University, 1-14 Bunkyo-machi, Nagasaki 852-8521, Japan
e-mail: tomo@nagasaki-u.ac.jp

2 Numerical Method

We have chosen the lattice Boltzmann method as the numerical method from these points. Since the lattice Boltzmann method is based on the lattice gas cellar automaton, it is suitable for the complex boundary and parallel computing. Swift et al. proposed their free energy model in 1996 [3]. They simulated isothermal multi-component two-phase flows in their papers. Inamuro et al. proposed LBM with pressure projection, and they succeeded to simulate isothermal two-phase flow with large density difference [4]. Yan et al. [2, 5, 6] had enhanced the ability of this Inamuro's model to include the contact angle and the wettability of liquid on the functional surfaces. Takada et al. [7] had solved Cahn–Hilliard and Allen–Cahn advection equations of two-phase flow with large density difference by means of this model effectively and had discussed about the area conservation errors. In this study, a single bubble on the vertical flat surface was simulated by the Lattice Boltzmann athermal model for immiscible two-phase fluids proposed by Inamuro et al. [4]. Non-dimensional variables defined in Ref. [4] are also used. The 3D15V model is adopted in order to simulate three-dimensional two-phase flows. Please refer to Ref. [4] for detailed information about governing equations.

There are two modifications from the Inamuro's model [4]. μ^L and μ^G are given as the viscosities of liquid and gas phase. Zu and He [8] claimed that the interpolation method for the viscosity in the phase interface μ should be the following function to ensure the continuity of viscosity flux across the interface.

$$\mu = \frac{\mu^L \mu^G (\phi^L - \phi^G)}{(\phi - \phi^G)\mu^G + (\phi^L - \phi)\mu^L} \tag{1}$$

where ϕ is the order parameter to use for the phase separation. If the lattice space is extended by one layer to the outside of the boundary layer as proposed by Chen et al. [9], the following derivatives on the boundary layer,

$$\frac{\partial \lambda}{\partial x_\alpha} \approx \frac{1}{10\Delta x} \sum_{i=2}^{15} c_{i\alpha} \lambda(\boldsymbol{x} + \boldsymbol{c}_i \Delta x), \quad \frac{\partial^2 \lambda}{\partial x_\alpha^2} \approx \frac{1}{5\Delta x} \left[\sum_{i=2}^{15} \lambda(\boldsymbol{x} + \boldsymbol{c}_i \Delta x) - 14\lambda(\boldsymbol{x}) \right] \tag{2}$$

can be calculated by the simple programming codes. In order to adapt the programming codes to complex 3D structures easily, the following extrapolation scheme is introduced to variables on boundary layer.

$$\xi^0 = 2\xi^1 - \xi^2 \tag{3}$$

The liquid contact angle on the solid surface θ is implemented into the partial derivative for z (the coordinate normal to the solid surface) direction as follows [2].

$$\Omega = 2\mathrm{sgn}\left(\frac{\pi}{2} - \theta\right)\left\{\cos\left(\frac{\gamma}{3}\right)\left[1 - \cos\left(\frac{\gamma}{3}\right)\right]\right\}^{1/2}, \quad \lambda = \frac{1}{4}\Omega\left(\phi^L - \phi^G\right)^2\sqrt{2\kappa_f\beta},$$

$$(4)$$

$$\left.\frac{\partial\phi}{\partial z}\right|_{z=0} = \frac{\lambda}{\kappa_f}, \frac{\partial^2\phi}{\partial z^2} \approx \frac{1}{2}\left(-3\left.\frac{\partial\phi}{\partial z}\right|_{z=0} + 4\left.\frac{\partial\phi}{\partial z}\right|_{z=1} - \left.\frac{\partial\phi}{\partial z}\right|_{z=2}\right),$$

$$(5)$$

$$\left.\frac{\partial\phi}{\partial z}\right|_{z=2} \approx \frac{1}{2}\left(3\phi|_{z=0} - 4\phi|_{z=1} + \phi|_{z=2}\right)$$

$$(6)$$

where $\gamma = \arccos\left(\sin^2\theta\right)$, k_f is a constant parameter for determining the width of interface and the strength of surface tension, β is the constant relating to interfacial thickness.

3   Results

Conditions of simulation are shown in Table 1. Superscripts "G", "L" indicate "gas phase" and "liquid phase", and subscripts "f" and "g" indicate "the lattice Boltzmann equation for order parameter" and "velocity", respectively. The density ratio equals to 50. This ratio is as same as the density ratio between saturation liquid and vapor of water at approximately 240 °C. The 3D simulation space is divided by $48 \times 48 \times 48$ lattices. The numerical values of the cut-off values of the order parameter, the viscosity coefficients, the coefficients of equation of state, surface tension between liquid and vapor, the constant parameters for the width of interface, and the relaxation times are fixed as same as the values using in Ref. [4].

Figure 1 shows the results of simulation. A hemisphere single bubble located on the vertical wall with different surface wettabilities whose contact angles are $\theta = 30°$ and $\theta = 150°$, respectively. The figure shows the fluid density on the vertical cross section across the center of the bubble. The density altitude is expressed by colors from dark violet for the lowest density of 1 to orange for the highest density of 50. The both of bubbles are moving upward by the buoyancy force. The different

Table 1 Conditions of simulation (non-dimensional parameters)

Parameters	Values
Density ρ^L, ρ^G	50, 1
Cut-off order parameter ϕ^L, ϕ^G	$9.2 \times 10^{-2}, 1.5 \times 10^{-2}$
Viscosity coefficient μ^L, μ^G	$8.0 \times 10^{-3}, 1.6 \times 10^{-4}$
Coefficents and temperature in EOS a, b, T	$1, 6.7, 3.5 \times 10^{-2}$
Parameters for the width of interface κ_f, κ_g	$0.5\Delta x^2, 2.5 \times 10^{-4}\Delta x^2$
Relaxation time τ_f, τ_g	1, 1

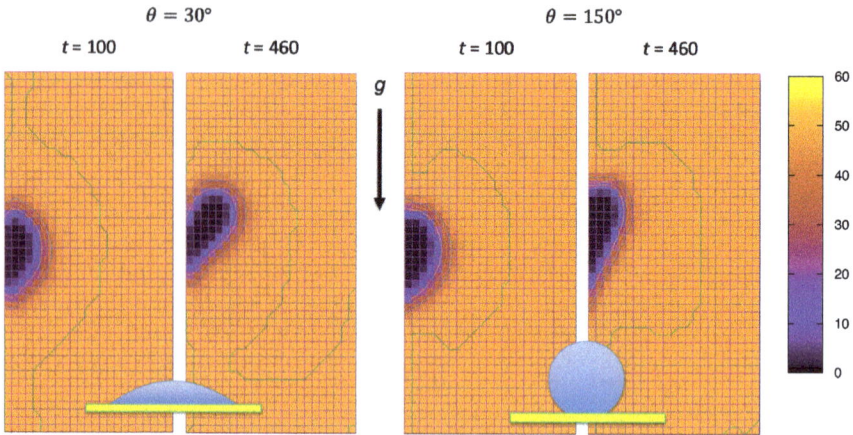

Fig. 1 Single bubble behaviors on the vertical flat surface at the contact angles $\theta = 30°$ and $\theta = 150°$. Dark color lattices indicate lower density region as gas phase in a bubble and orange lattices indicate higher density region as liquid phase

advancing and receding contact angles are observed in each surface characteristics. In this study, since the moving phase is gas phase, the advancing contact angles are smaller than receding contact angles at both surface wettabilities. The advancing and receding contact angles on the surface of $\theta = 30°$ are smaller than each contact angle on surface of $\theta = 150°$ at both time steps $t = 100$ and $t = 460$.

4 Conclusions

The contact angle on the solid surface and bounce back condition with extrapolation boundary scheme and modified interpolation function for the viscosity coefficient are introduced into the lattice Boltzmann athermal model for immiscible two-phase flow. The behaviors of a single bubble on the solid walls with different surface wettabilities were simulated, and the physical consistency was observed between the results of simulation and actual phenomena. It is confirmed that LBM is one of the promising methods to simulate the two-phase flow that has the large density difference in the 3D structure.

Acknowledgements This study is supported by JSPS KAKENHI Grant Number 23560238 and 19K04241, JSPS "Strategic Young Researcher Overseas Visits Program for Accelerating Brain Circulation" and the parallel computing system of the Graduate School of Engineering, Nagasaki University.

References

1. T. Yamaguchi, Q. Wan, Y. Yan, J. Hong, Numerical simulation of liquid-gas two-phase flow with large density difference in multi-layered sintered wick by the lattice Boltzmann method, in *Proceedings of the 15th International Heat Transfer Conference IHTC-15*, Kyoto, Japan (2014), IHTC15–9543
2. Y.Y. Yan, Y.Q. Zu, A lattice Boltzmann method for incompressible two-phase flows on partial wetting surface with large density ratio. J. Comput. Phys. **227**, 763–775 (2007)
3. M.R. Swift, E. Orlandini, W.R. Osborn, J.M. Yeomans, Lattice Boltzmann simulations of liquid-gas and binary fluid systems. Phys. Rev. E **54**(5), 5041–5052 (1996)
4. T. Inamuro, T. Ogata, S. Tajima, N. Konishi, A lattice Boltzmann method for incompressible two-phase flows with large density differences. J. Comput. Phys. **198**(2), 628–644 (2004)
5. Y.Y. Yan, Y.Q. Zu, B. Dong, LBM, a useful tool for mesoscale modelling of single-phase and multiphase flow. Appl. Therm. Eng. **31**, 649–655 (2011)
6. Y.Q. Zu, Y.Y. Yan, J.Q. Li, Z.W. Han, Wetting behaviours of a single droplet on biomimetic micro structured surfaces. J. Bionic Eng. **7**, 191–198 (2010)
7. N. Takada, J. Matsumoto, S. Matsumoto, Phase-field model-based simulation of motions of a two-phase fluid on solid surface. J. Comput. Sci. Technol. **7**, 322–336 (2013)
8. Y.Q. Zu, S. He, Phase-field-based lattice Boltzmann model for incompressible binary fluid systems with density and viscosity contrasts. Phys. Rev. E **87**, 043301 (2013)
9. S. Chen, D. Martínez, R. Mei, On boundary conditions in lattice Boltzmann methods. Phys. Fluids **8**, 2527–2536 (1996)

Experimental Study on Fuel Spray Impingement Process Against Metal Surface with Different Wettability

Liang Guo, Jianyi Wei, Wanchen Sun, Xuejiang Hao, Deigang Li, and Kai Fang

1 Introduction

Combustion of internal combustion engines is a complex turbulent process in finite space. The phenomenon of wall wetting is almost inevitable; it is one of the main reasons that affects the combustion boundary controllability and combustion stability of homogeneous and premixed combustion, resulting in high HC and CO emissions. So far, some researchers used blend fuels with high volatility to reduce wall wetting [1–3]; and other trying to reduce the possibilities of wall wetting or utilizing wall impingement to improve combustion quality by improving combustion chamber structures design [4]. In order to further improve the controllability and stability of the new combustion modes, active control need to be developed to exert influence to the follow-up process after wall wetting. In this study, the surface wettability, a parameter characterizing the tendency of one fluid to spread on, or adhere to, a solid surface in the presence of other immiscible fluids, and has been improved utilizing biomimetic methods. The development process of fuel spray after wall impingement to the surfaces with different wettability and temperatures was investigated experimentally under different back pressures with. The results of the present study will lay experimental foundation for improving controllability of wall wetting phenomenon and fuel evaporation process through surface property control.

L. Guo · J. Wei · W. Sun (✉) · X. Hao · D. Li · K. Fang
Jilin University State Key Laboratory of Automotive Simulation and Control, Changchun 13022, China
e-mail: sunwc@jlu.edu.cn

© Springer Nature Singapore Pte Ltd. 2021
C. Wen and Y. Yan (eds.), *Advances in Heat Transfer and Thermal Engineering* ,
https://doi.org/10.1007/978-981-33-4765-6_122

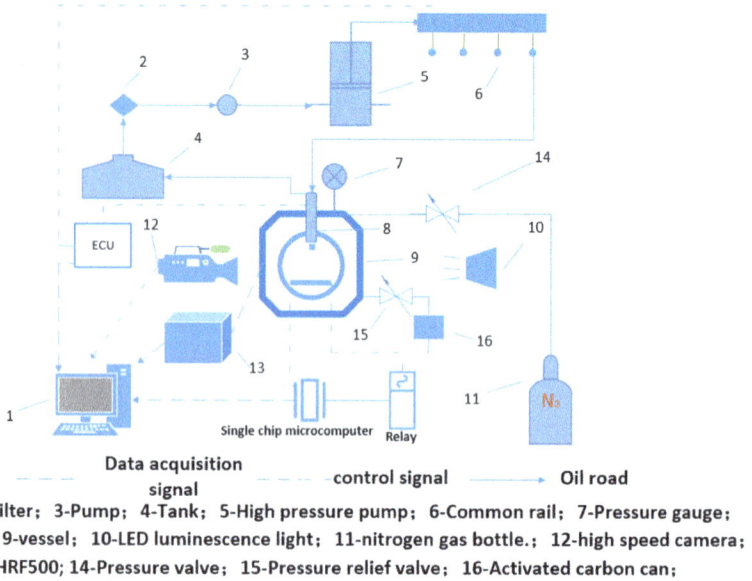

Fig. 1 Schematic diagram of the experimental device structure

1-PC; 2-Filter; 3-Pump; 4-Tank; 5-High pressure pump; 6-Common rail; 7-Pressure gauge; 8-Injector; 9-vessel; 10-LED luminescence light; 11-nitrogen gas bottle.; 12-high speed camera; 13-HRF500; 14-Pressure valve; 15-Pressure relief valve; 16-Activated carbon can;

2 Methodology

The experiment is carried out based on a self-developed fuel spray constant volume system. The sub-systems include a fuel supply system, a pressure and temperature control system, a mixed gas concentration measurement system, and a high speed camera system. The schematic of the experimental device is shown in Fig. 1.

For the fuel supply system, it is mainly composed of a DENSO high-pressure common rail supplied with a HP3 pump, a controller, a drag motor, a fuel tank, and a filter. T fuel injection is achieved with a single-hole injector. A direct injection driving system made by national instrument is selected as the lower controller of the injector. The pressure control for the constant volume vessel is realized through a high-pressure nitrogen gas inflation system with pressure feedback, and the wall temperature is adjusted by using a cartridge heater of 200 W, a K-type thermocouple, a relay, and a programmable controller.

3 Results

The mental surface material selected in this study is aluminum alloy. In total, four different surfaces were prepared and tested in this study. One is a chemical etched surface and the other three are the laser etched surfaces with different roughness of 1.5, 2.8, and 3.7 μm.

Table 1 Microscopic features of the tested surfaces

Surface treatment method	Magnification 500×	Magnification 1000×
Laser etched surface		
Chemical etched surface		

In order to avoid the influence of the original roughness of the material on the experiment, all the tested surfaces were pre-prepared to ensure that the roughness of them are constant value of 0.5 μm before mechanical/chemical processing. And then, the surfaces were further etched with laser and chemical reagents. As a result, the static contact angles of all tested fuels upon the laser etched surfaces are less than 2°. And the contact angles of diesel and n-butanol upon the chemically etched surface are 85° and 65°, respectively. The microscopic features of the different surfaces taken by the scanning electron microscope are shown in Table 1.

For convenience of analysis, different experiments were numbered in a format of XY according to the fuel and surface tested. The first half of the number, X, indicates the tested fuel. In this study, D represents diesel, and B represents butanol. The second half of the number, Y, indicates the surface conditions. If Y is a number, it represents the roughness value of the experimental surface. Example 1.5 indicates that the roughness of the tested surface is 1.5 um. If Y is the subscript, ch, it indicates that the test surface is a chemically treated oleophobic surface.

Figure 2 shows the development of the spreading radius of different fuel sprays after impinged onto the walls with different wettabilities. It can be seen from Fig. 2 that the highest spreading radius of both fuels is obtained from the chemically treated surface. And upon the laser etched surfaces with different roughness, the spreading radius of both fuel sprays increased with the decrease of the surface roughness. This is mainly because for the lipophilic walls, like mental surface, it normally has an adhesive effect to the tiny fuel droplets attached on it, and according to the Wenzel model, the surface roughness will increase the actual contact area of the liquid and therefore amplify the wettability of the solid surface. As a result, a higher surface roughness, smaller spreading radius of the fuel sprays. Additionally, for the surface with larger roughness, the small droplets of the fuel spray may then infiltrate into the surface texture more easily, and this may be another reason for the smaller spreading

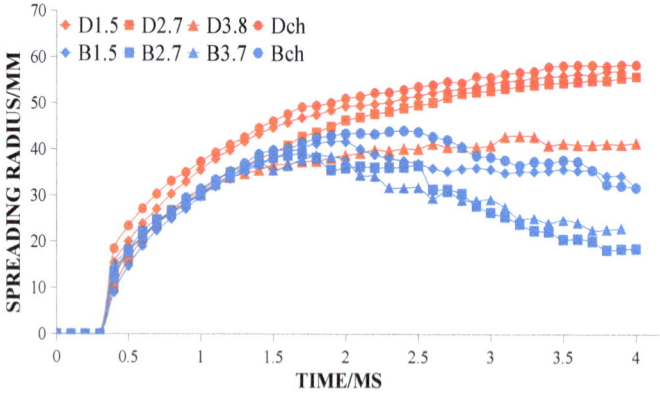

Fig. 2 Effect of wall different wettability on spreading radius of the both fuels

radius presented by the surface with a larger roughness. The oleophobicity of both fuels upon the chemical etched surface turns higher, comparing to the laser etched surfaces. As a result, the spreading radius of the fuel spray upon the chemical etched surface is also higher.

4 Conclusions

(1) The spreading radius and the entrainment height upon the chemically treated surface are highest, among all the surfaces used in the experiments. And for the laser etched surfaces, the spreading radius and the entrainment height of the both fuels decrease with the surface roughness.
(2) After impingement, the concentration of both fuel mixtures in the near-wall region is found increased as the surface wettability increases.

Acknowledgements The authors gratefully acknowledge financial support from the National Natural Science Foundation of China (51676084, 51776086); Industrial Innovation Special Funding Project of Jilin Province (2019C058-3); and Natural Science Foundation of Jilin Province, China, (20180101059JC).

References

1. H. Wu, K. Nithyanandan et al., Spray and combustion characteristics of neat acetone-butanol-ethanol, n-butanol, and diesel in a constant volume chamber. Energy Fuels **28**(10), 6380–6391 (2014)
2. H. Liu, C. Lee, H. Ming et al., Comparison of ethanol and butanol as additives in soybean biodiesel using a constant volume combustion chamber. Energy Fuels **25**(4), 1837–1846 (2011)

3. S. Lee, S. Jeong, O. Lim, An investigation on the spray characteristics of DME with variation of ambient pressure using the common rail fuel injection system. J. Mech. Sci. Technol. **26**(10), 3323–3330 (2012)
4. S. Kato, S. Onishi, New mixture formation technology of direct fuel injection stratified charge SI engine (OSKA)—test result with gasoline fuel. SAE Paper 871689 (1987)

Effect of Turbulence Models on Steam Condensation in Transonic Flows

Chuang Wen, Nikolas Karvounis, Jens Honore Walther, Hongbing Ding, Yan Yang, Xiaowei Zhu, and Yuying Yan

1 Introduction

Steam condensation in transonic flows is a common issue in various industries, such as Laval nozzles, ejectors, turbines, thermo-compressors and supersonic separators [1–5]. Various turbulence models have been used in wet steam flow simulations, from $k - \varepsilon$ models to SST $k - \omega$ model. Simpson and White [6] numerically calculated the steam nucleation and condensation in a converging–diverging nozzle using the standard $k - \varepsilon$ model. Ariafar et al. [7] employed the realizable $k - \varepsilon$ model to perform the numerical simulation on steam condensations in a primary nozzle used for an ejector. Wang et al. [8] adopted the RNG $k - \varepsilon$ model to predict the homogeneous condensation in a primary nozzle for the steam ejector. Mazzelli et al. [9] numerically studied the non-equilibrium condensation in a steam ejector by employing the $k - \omega$ shear stress transport (SST) model. In the present study, an assessment of the turnulence model on the steam condensation in transonic flows is performed using computational fluid dynamics (CFD) modelling.

C. Wen · N. Karvounis · J. H. Walther · Y. Yang · X. Zhu
Department of Mechanical Engineering, Technical University of Denmark, Nils Koppels Allé, 2800 Kgs., Lyngby, Denmark

C. Wen · Y. Yan (✉)
Faculty of Engineering, University of Nottingham, Nottingham NG7 2RD, UK
e-mail: yuying.yan@nottingham.ac.uk

J. H. Walther
Computational Science and Engineering Laboratory, ETH Zürich, Clausiusstrasse 33, C8092 Zürich, Switzerland

H. Ding
School of Electrical and Information Engineering, Tianjin University, Tianjin 300072, China

© Springer Nature Singapore Pte Ltd. 2021
C. Wen and Y. Yan (eds.), *Advances in Heat Transfer and Thermal Engineering* ,
https://doi.org/10.1007/978-981-33-4765-6_123

2 Numerical Model

The fundamental equations governing the non-equilibrium condensation under supersonic conditions are the compressible Navier–Stokes equations. Two transport equations are utilized to describe the phase change process during the steam condensation, including the liquid fraction (Y) and droplet number per volume (N):

$$\frac{\partial(\rho Y)}{\partial t} + \frac{\partial}{\partial x_j}(\rho Y u_j) = \Gamma \tag{1}$$

$$\frac{\partial(\rho N)}{\partial t} + \frac{\partial}{\partial x_j}(\rho N u_j) = \rho J \tag{2}$$

where J is the nucleation rate [10], N is the number of droplets per volume. Γ is the condensation mass due to phase changes:

$$\Gamma = \frac{4\pi r_c^3}{3}\rho_l J + 4\pi r^2 \rho_l N \frac{dr}{dt} \tag{3}$$

$$J = \frac{q_c}{1+\phi} \frac{\rho_v^2}{\rho_l} \sqrt{\frac{2\sigma}{\pi m_v^3}} \exp\left(-\frac{4\pi\sigma}{3k_B T_v}r_c^2\right) \tag{4}$$

The growth rate of droplets due to evaporation and condensation, dr/dt, is calculated by Young's model [11].

$$\frac{dr}{dt} = \frac{\lambda_v(T_s - T_v)}{\rho_l h_{lv} r} \frac{\left(1 - r_c/r\right)}{\left(\frac{1}{1+2\beta Kn} + 3.78(1-v)\frac{Kn}{Pr}\right)} \tag{5}$$

For the numerical implementation, the mass, momentum and energy conservation equations are directly solved by the ANSYS FLUENT 18 [12], while the User-Defined-Scalar (UDS) and User-Defined-Function (UDF) interfaces are used to solve two scalar equations and source terms. The pressure inlet and pressure outlet conditions are assigned for the entrance and exit of the supersonic nozzle. The computational simulation employs a structured mesh of 22,800 cells based on a mesh convergence test using 7920, 22,800 and 40,000 cells [13].

3 Results

Four turbulence models ($k - \varepsilon$ Standard, $k - \varepsilon$ RNG, $k - \varepsilon$ Realizable and $k - \omega$ SST) are evaluated against experimental data presented by Binnie and Green in a Laval nozzle [14] to evaluate the influence of the turbulence modelling both considering the steam condensation and shock waves in supersonic flows. Figure 1 displays the static pressure profile along the flow direction of the Laval nozzle clearly showing the occurrence of shock waves in supersonic flows. By comparing the numerical result and experimental data, it is shown that all of four turbulence models capture almost the same onset of the steam condensation and agree well with the experiments. However, the shock wave position differs significantly among four turbulence models. The shock wave appears in the downstream end of the Laval nozzle, around $x = 0.19$ m for the $k - \omega$ SST model. The $k - \varepsilon$ turbulence models predict the position of the shock waves further downstream, around $x = 0.21$ m for the standard $k - \varepsilon$ model and $x = 0.20$ m for the RNG and realizable $k - \varepsilon$ models. For this case, both the standard $k - \varepsilon$ model and the $k - \omega$ SST model capture the experimental static pressures.

To further investigate the $k - \varepsilon$ standard and $k - \omega$ SST turbulence models capabilities to predict the shock structure in supersonic flows, we consider experimental data in [15] which focused on the flow separation in a convergent-divergent nozzle without considering the nucleation behaviour. Figure 2 describes the comparisons

Fig. 1 Pressure profiles at the central line of the Laval nozzle [14]

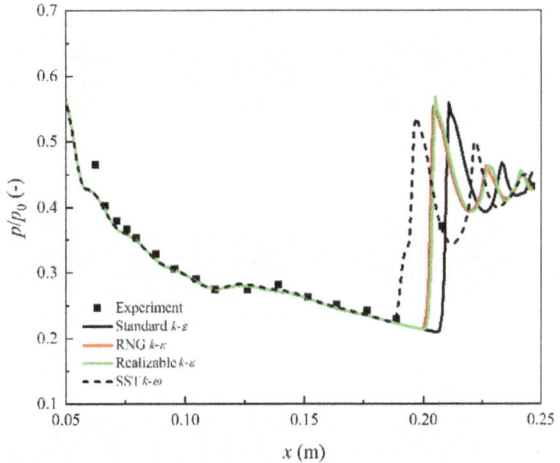

Fig. 2 Static pressure between computation and experiment

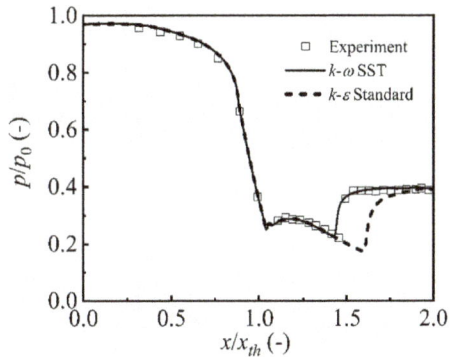

between the computed and experimental pressure profiles along the nozzle. The results show that the $k - \omega$ SST turbulence model accurately captures the shock position inside the convergent-divergent nozzle, while the standard $k - \varepsilon$ model predicts a later shock position. Hence, the standard $k - \varepsilon$ model is slow to respond to the flow separation in supersonic flows.

In general, by comparing the numerical results with experimental data from Binnie and Green Laval nozzle [14] and Hunter convergent-divergent nozzle [15], the $k - \omega$ SST turbulence model shows better performance on the prediction of the supersonic flow with non-equilibrium condensation and shock waves.

4 Conclusions

A computational fluid dynamics model is developed to evaluate the effect of turbulence models on steam condensation behaviour in transonic flows considering shock waves. The $k - \omega$ SST turbulence model shows good agreement with experimental measurements, which is recommended for the numerical simulation of the wet steam flows concerning both steam condensations and shock waves.

References

1. Y. Yang, J.H. Walther, Y. Yan, C. Wen, CFD modeling of condensation process of water vapor in supersonic flows. Appl. Therm. Eng. **115**, 1357–1362 (2017)
2. Y. Yang, X. Zhu, Y. Yan, H. Ding, C. Wen, Performance of supersonic steam ejectors considering the nonequilibrium condensation phenomenon for efficient energy utilisation. Appl. Energy **242**, 157–167 (2019)
3. S. Dykas, M. Majkut, K. Smołka, M. Strozik, Study of the wet steam flow in the blade tip rotor linear blade cascade. Int. J. Heat Mass Transfer **120**, 9–17 (2018)
4. S.N.R. Abadi, R. Kouhikamali, K. Atashkari, Non-equilibrium condensation of wet steam flow within high-pressure thermo-compressor. Appl. Therm. Eng. **81**, 74–82 (2015)
5. Y. Yang, C. Wen, CFD modeling of particle behavior in supersonic flows with strong swirls for gas separation. Sep. Purif. Technol. **174**, 22–28 (2017)
6. D. Simpson, A. White, Viscous and unsteady flow calculations of condensing steam in nozzles. Int. J. Heat Fluid Flow **26**, 71–79 (2005)
7. K. Ariafar, D. Buttsworth, N. Sharifi, R. Malpress, Ejector primary nozzle steam condensation: Area ratio effects and mixing layer development. Appl. Therm. Eng. **71**, 519–527 (2014)
8. C. Wang, L. Wang, T. Zou, H. Zhang, Influences of area ratio and surface roughness on homogeneous condensation in ejector primary nozzle. Energy Convers. Manage. **149**, 168–174 (2017)
9. F. Mazzelli, F. Giacomelli, A. Milazzo, CFD modeling of condensing steam ejectors: Comparison with an experimental test-case. Int. J. Therm. Sci. **127**, 7–18 (2018)
10. A. Kantrowitz, Nucleation in very rapid vapor expansions. J. Chem. Phys. **19**, 1097–1100 (1951)
11. J. Young, The spontaneous condensation of steam in supersonic nozzle. Physico Chemical Hydrodynamics **3**, 57–82 (1982)
12. *ANSYS Fluent Theory Guide* (ANSYS Inc., USA, 2017)

13. C. Wen, N. Karvounis, J.H. Walther, Y. Yan, Y. Feng, Y. Yang, An efficient approach to separate CO_2 using supersonic flows for carbon capture and storage. Appl. Energy **238**, 311–319 (2019)
14. M.A.M Binnie, J. Green, D. Be, An electrical detector of condensation in high-velocity steam. Proc. R. Soc. Lond. A **181**, 134–154 (1942)
15. C.A. Hunter, Experimental investigation of separated nozzle flows. J. Propul. Power **20**, 527–532 (2004)

An Experimental Study on Vent Locations in Road Racing Bicycle Helmet to Optimize Thermal Comfort and Aerodynamic Efficiency

Harun Chowdhury, Firoz Alam, Terence Woo, and Akshoy Ranjan Paul

1 Introduction

Aerodynamic design as well as thermal comfort of helmet plays an important role in the sport of cycling. At 30 km/h drag contributes to around 70–80% of the total resistance and of this 2–8% depends on the shape of the helmet [1]. Specifically, when comparing time trial (TT) helmets have significantly less aerodynamic drag (40–50% less) then more common road helmets [2]. One of the reasons for such significant aerodynamic differences between these helmet designs is due to the TT helmet designs sacrificing thermal considerations in the form of ventilation holes. The reduction in thermal considerations is possible with TT designs due to shorter race conditions which have shown to not significantly increase head and gastrointestinal temperatures over a race [3]. This difference in helmet design highlights the impact of ventilation on drag in helmet design as shown by Alam et al. [4] who identified that venting position needs to be selected based on aerodynamic and heat dissipation characteristics as a specific modified helmet with selectively covered ventilation holes clearly indicates position of the vent can increase aerodynamic efficiency while keeping thermal comfort intact. Therefore, the purpose of this study is to identify the key ventilation locations that are necessary in helmet design to identify the most optimal balance between thermal efficiency and reduction in aerodynamic drag in road cycling helmets.

H. Chowdhury (✉) · F. Alam · T. Woo · A. R. Paul
School of Engineering, RMIT University, Melbourne, VC 3000, Australia
e-mail: harun.chowdhury@rmit.edu.au

Department of Applied Mechanics, Motilal Nehru National Institute of Technology, Allahabad, India

© Springer Nature Singapore Pte Ltd. 2021
C. Wen and Y. Yan (eds.), *Advances in Heat Transfer and Thermal Engineering* ,
https://doi.org/10.1007/978-981-33-4765-6_124

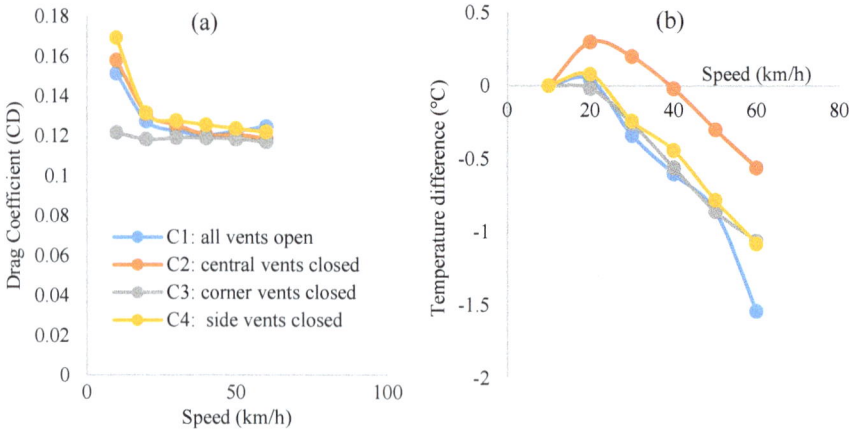

Fig. 1 a All vents open, **b** central vents closed, **c** corner vents closed and **d** side vents closed; **e** experimental setup

2 Methodology

In this study, commercially manufactured a road racing 'Giro Atmos' helmet was used. RMIT Industrial Wind tunnel was used to evaluate the aerodynamic and thermal performance of the helmet in four different ventilation configurations: (1) all vents open, (2) central vents closed, (3) corner vents closed and (4) side vents closed as shown in Fig. 1. Smooth duct tape was used to close the vents on the helmet.

A purpose made mannequin head was used for both the aerodynamic drag and thermal testing of the helmet as shown in Fig. 1e. The helmet was mounted to the mannequin head fitted with heating pad. The mannequin head was connected to the 6-component load sensor (type JR-3) with the maximum capacity 200 N. The drag forces at different test configurations of the helmet were recorded via a computer-controlled data accusation system. A temperature controller was used to maintain the constant temperature on the heat pad where 5 thermocouples were fitted on the heat pad and connected to a temperature measure data logger. The data (drag forces and temperature differences) were taken over a range of wind speeds (0 to 60 km/h with the increment of 10 km/h). Details about the specific tunnel physical properties including turbulence intensity and physical dimensions can be found in Alam et al. [5, 6].

3 Results

The aerodynamic drag coefficient as a function of wind speed of all 4 configurations is shown in Fig. 2a. The configuration-3 (corner vents closed) shows the lowest drag coefficient compared to all other configurations, particularly at speeds less than

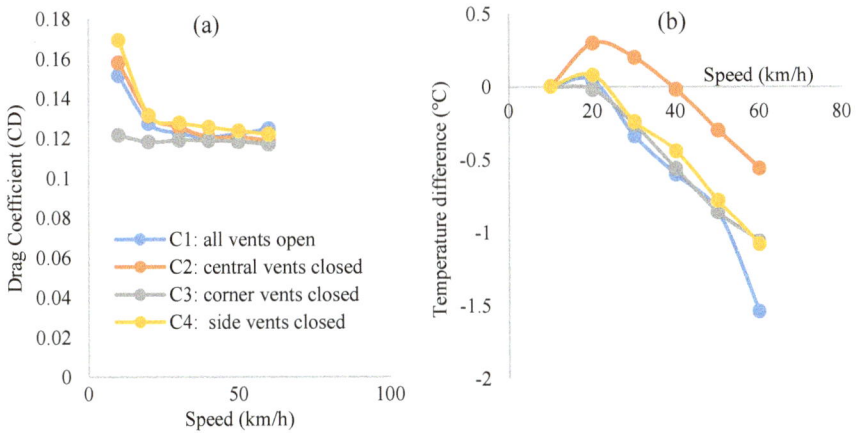

Fig. 2 **a** Drag coefficient as a function of speed; **b** temperature difference as a function of speed

30 km/h. At higher speeds, the data is more comparable yet overall the Intermediate C_D value remains on average 6.76% less than the standard helmet (all vents open). The stock and center configurations seem to change over at 40 km/h, where at the lower speeds the stock has a lower C_D, and above 40 km/h the center is lower. The configuration-4 (side vents closed) appears to be the worst performing configuration, which conceptually is counterintuitive to the notion that any reduction of vents on the frontal area would reduce the C_D value when compared to the fully vented helmet.

The data in Fig. 2b show that out of all the configurations, the configurations—1, 2 and 3 helmet conditions remained similar across the testing. The standard helmet (configuration-1) only demonstrates greater ability between speeds of 50–60 km/h. Conversely, the central modification performed the poorest of all the tests for temperature difference reduction. This was attributed to the significantly sized central vents that appear to play the largest role in thermoregulation design about the Giro Atmos Helmet.

4 Conclusions

Drag differences between the modified and standard conditions were marginal. Using these results alone are not enough to warrant a change in manufactured design. The central ventilation area provides the greatest difference in temperature regulation for Giro Atmos. More testing into whether optimal configuration of helmet design can be discovered, especially the intermediate area (corner vents) of the helmet which may have potential for significant aerodynamic performance gains without impedance to thermos-regulation.

References

1. I. Mustary, H. Chowdhury, B. Loganathan, M. Alharthi, F. Alam, Aerodynamic efficiency and thermal comfort of road racing bicycle helmets, in *AFMS 2014* (Australasian Fluid Mechanics Society (AFMS)), pp. 1–4
2. F. Alam, H. Chowdhury, H.Z. Wei, I. Mustary, G. Zimmer, Aerodynamics of ribbed bicycle racing helmets. Procedia Eng. **72**, 691–696 (2014)
3. F. Alam, S. Watkins, A. Akbarzadeh, A. Subic, A study of thermal comfort of a series of bicycle helmets, in *3rd International Conference on Thermal Engineering* (Bangladesh, 2006)
4. F. Alam, H. Chowdhury, Z. Elmir, A. Sayogo, J. Love, A. Subic, An experimental study of thermal comfort and aerodynamic efficiency of recreational and racing bicycle helmets. Procedia Eng. **2**(2), 2413–2418 (2010)
5. F. Alam, A. Subic, S. Watkins, J. Naser, M. Rasul, An experimental and computational study of aerodynamic properties of rugby balls. WSEAS Trans. Fluid Mech. **3**(3), 279–286 (2008).
6. F. Alam, A. Subic, S. Watkins, A.J. Smits. Aerodynamics of an Australian rules foot ball and rugby ball, in *Computational. Fluid Dynamics for Sport Simulation*, vol. 1, ed. by M. Peters. pp. 103–127

A Novel Horizontal Liquid–Liquid Flow Pattern Map Using Dimensionless Number Groups

Olusegun Samson Osundare, Liyun Lao, and Gioia Falcone

1 Introduction

In a horizontal pipeline, when two immiscible liquids flow together, various flow patterns are experienced. The factors responsible for these flow pattern configurations include density ratio, viscosity ratio, input phase ratio, mixture flow rate, wetting properties, surface tension and pipe geometry [3]. Various investigators presented their flow pattern maps in various coordinates, making the comparison and interpretation of the flow patterns difficult, due to different names or terminologies used by different investigators to describe each of the flow patterns. Flow patterns from different investigators but with common features are grouped and classified into slug, stratified, stratified mixed, dispersed oil-in-water, dispersed water-in-oil and annular flows. The flow patterns map by Trallero et al. [11] is one of the most adopted flow pattern maps among researchers, but it fails to report slug and annular flow patterns. However, the annular flow pattern tends to occur easily when oil viscosity is high, with a critical value of 35 mPa s [4]. Unlike the gas–liquid flow patterns, which are sensitive to pipe diameters, the oil–water flow patterns are more sensitive to the properties of the liquids such as density, viscosity and interfacial tension but less sensitive to pipe diameter [4].

Plotting the superficial water velocity ($v_{sw} = Q_w/A$) versus the superficial oil velocity ($v_{so} = Q_o/A$) is a common way of constructing the flow patterns maps [1], where Q_o and Q_w are volumetric flow rate of oil and water, respectively, and A is the cross-sectional area of the pipe. These types of flow pattern maps are usually

O. S. Osundare (✉) · L. Lao
Centre for Thermal Energy Systems and Materials, Cranfield University, Bedford MK43 0AL, UK
e-mail: o.osundare.1@research.gla.ac.uk

G. Falcone
School of Engineering, University of Glasgow, Glasgow G12 8QQ, UK

O. S. Osundare
Department of Industrial Chemistry, Ekiti State University, Ado-Ekiti, Nigeria

© Springer Nature Singapore Pte Ltd. 2021
C. Wen and Y. Yan (eds.), *Advances in Heat Transfer and Thermal Engineering* ,
https://doi.org/10.1007/978-981-33-4765-6_125

valid only for the particular pipe sizes, fluids and operating conditions, employed to construct the flow map [10]. Construction of flow pattern maps based on meaningful dimensionless parameter groups would enable the development of more generalised flow pattern maps, which are certainly beneficial for various applications involving horizontal liquid–liquid pipe flows.

2 Methodology

The dimensional analysis was used to determine the dimensionless parameter groups that control a liquid–liquid flow in horizontal pipelines, by applying the Buckingham π theorem to the independent system variables and fundamental dimensions of the liquid–liquid flow [8]. For a fully developed, steady-state, isothermal oil–water flow in a horizontal pipeline, there are eleven relevant independent variables. These include superficial oil and water velocities v_{so} and v_{sw}, oil and water densities ρ_o, and ρ_w, pipe diameter d, oil and water viscosities μ_o and μ_w, interfacial tension σ, pipe roughness ε, the acceleration due to gravity g, and wettability α (assumption: for horizontal orientation neglect inclination angle Θ), while the three fundamental dimensions are mass M, length L, and time T. Based on Buckingham π theorem, 8 dimensionless groups are expected from oil–water flow system, which [8] gave as (1) wettability, (2) relative roughness, (3) density ratio, (4) viscosity ratio, (5) water fraction, (6) mixture Reynolds number, (7) Eötvös number and (8) mixture Froude number. Some dimensionless groups governing the liquid–liquid flow are the ratio of the forces [inertia, viscous, gravity (buoyancy) and interfacial tension (capillary)] influencing the liquid–liquid flow in horizontal pipelines. Therefore, the effect of liquid properties, pipe diameter and velocity are all included in the dimensionless parameters.

3 Findings/Results

Data from some selected oil–water flow experimental studies [2, 5–7, 11] which are relevant to this study were extracted and used to construct flow pattern maps based on some dimensionless groups. In this project, the homogeneous flow model is employed, which means treating a liquid–liquid flow as a pseudo-single phase liquid with average liquid properties and average velocity. The mixture velocity v_m, mixture density ρ_m, and mixture viscosity μ_m can be expressed in terms of volume fraction or mass fraction assuming no-slip conditions [8].

$$v_m = \frac{Q_w + Q_o}{A} = v_{sw} + v_{so} \qquad \rho_m = \rho_w f_w + \rho_o(1 - f_w),$$

$$\mu_m = \mu_w f_w + \mu_o(1 - f_w) \qquad f_w = \frac{Q_w}{Q_{w+}Q_o} = \frac{v_{sw}}{v_{sw} + v_{so}}$$

where f_w means water fraction.

The flow map of the ratio of mixture Reynolds number to Eötvös number Re_m/Eo versus water fraction f_w

Ibarra et al. [8] characterised the co-current flow of immiscible liquids with a dimensionless flow pattern map constructed from the ratio of mixture Reynolds number to Eötvös number as a function of water fraction. The mixture Reynolds number Re_m can be expressed as:

$$Re_m = \frac{\rho_m v_m D}{\mu_m}$$

The lower the oil viscosity used in the experimental studies, the higher the Re_m/Eo obtained in the analysis. Based on the data analysed, two separate trends of flow pattern maps were formed on the graph; hence, none of the flow regimes can be predicted by using the combination of these dimensionless parameters. A critical analysis of the graph showed that oil viscosity was crucial to the trend of the flow maps formed; hence, the oil viscosity of the data should be within a close range to obtain a single trend flow pattern map.

The flow pattern maps for the ratio of Weber number to Eötvös number We/Eo and mixture Froude number Fr_m as functions of water fraction f_w.

The We/Eo and Fr_m can be expressed as (Fig. 1):

$$\frac{We}{Eo} = \frac{\rho_m v_m^2 D}{\Delta \rho g D^2}, \quad Fr_m = \frac{v_m}{\sqrt{gD}} \sqrt{\frac{\rho_m}{(\rho_w - \rho_o)}}$$

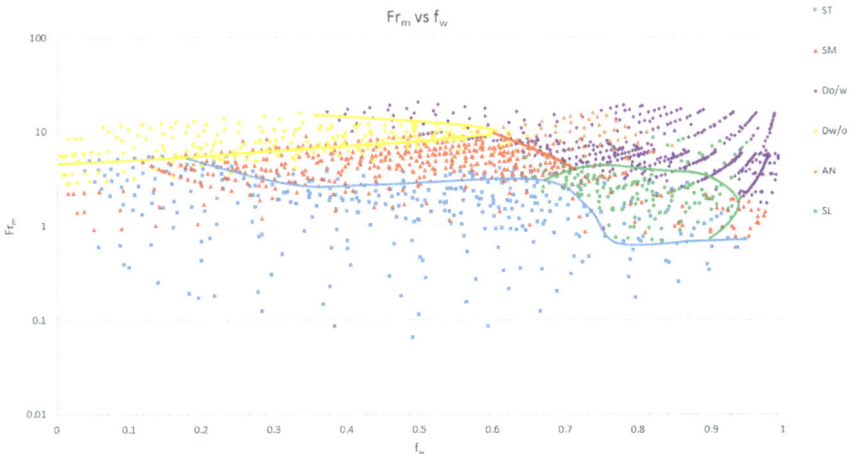

Fig. 1 Graph of mixture Froude number Fr_m versus water fraction f_w with boundary lines between the flow regimes. *Data source* [2, 5–7, 11]

$$\text{Therefore,} \sqrt{\frac{We}{Eo}} = \frac{v_m}{\sqrt{gD}} * \sqrt{\frac{\rho_m}{\Delta\rho}},$$

$$\text{hence} \, Fr_m = \sqrt{\frac{We}{Eo}} = \frac{v_m}{\sqrt{gD}} * \sqrt{\frac{\rho_m}{\Delta\rho}}$$

Therefore, the graphs of We/Eo and Fr_m as functions of f_w have the same pattern. A high We/Eo and Fr_m depict dominance of inertia forces, where the flow will be turbulent, and there will be instabilities at the interface between the two liquids. The dispersed flows are the most likely flow patterns in this region. At low We/Eo and Fr_m, the gravity forces dominate; hence, the stratified flow is the most likely flow pattern. Based on the data analysed, the flow pattern maps constructed with We/Eo and Fr_m as a function f_w predicted flow regions for stratified flow ST, stratified mixed flow SM, dispersed water-in-oil flow Dw/o and the slug flow SL. However, the flow region covered by annular flow was also mapped out, but with high interference of Do/w data points.

The flow pattern map of the ratio of Gravity force to Viscous force G/V versus water fraction f_w.

Shi and Yeung [9] showed the absence of viscous force in the characterisation of liquid–liquid flow in horizontal pipes done by Brauner [3] using the Eötvös number. Therefore, they proposed the ratio of the gravitational force to viscous force, G/V, which includes the effect of viscosity in the configuration of the liquid–liquid flows. The gravity force-to-viscous force ratio can be expressed as:

$$\frac{G}{V} = \frac{\Delta\rho g D^2}{\mu_m v_m}$$

The G/V replicates the completive role of gravity and viscous forces on the oil–water flows phase configuration. At high G/V, the gravity force dominates, while the viscous force is comparatively low. Hence, the fluid can easily be broken up when the kinetic energy is high, and the flow structures that can be developed are the stratified flow and the fine dispersions of one of the liquids in the continuum of the other. At low G/V, the stratified flow is less likely to form because the effect of the gravitational force is comparatively small. The probable flow configurations are oil core flowing inside annular water continuously to form annular (core annular flow) or discontinuously to form intermittent flows (slug, plug) in which shear stress is minimal. With high turbulence kinetic energy flow, the higher viscous (oil) phase is too viscous to be broken by the kinetic energy into fine drops but lumps of various sizes and shapes. Based on the flow pattern map constructed from the selected experimental studies, no flow regime was well mapped out, because 3 trends of maps were formed on the graph.

4   Conclusions

Based on the data analysed, four flow regimes' regions (stratified flow ST, stratified mixed flow SM, dispersed water-in-oil flow Dw/o and the slug flow SL) were well mapped out in the flow pattern maps constructed by plotting the We/Eo and Fr_m as function of f_w. While the flow pattern maps constructed by plotting the Re_m/Eo and G/V, functions f_w could not predict accurately any of the flow regimes.

References

1. S.A. Ahmed, B. John, Liquid-Liquid horizontal pipe flow—A review. J. Petrol. Sci. Eng. **168**(November 2017), 426–447 (2018). https://doi.org/10.1016/j.petrol.2018.04.012
2. T. Al-Wahaibi, Y. Al-Wahaibi, A. Al-Ajmi, R. Al-Hajri, N. Yusuf, A.S. Olawale, I.A. Mohammed, Experimental investigation on flow patterns and pressure gradient through two pipe diameters in horizontal oil-water flows. J. Petrol. Sci. Eng. **122**, 266–273 (2014). https://doi.org/10.1016/j.petrol.2014.07.019 (Elsevier)
3. N. Brauner, Liquid-liquid two-phase flow systems, in *Modelling and Control of Two-Phase Phenomena*, ed. by V. Bertola (Springer , Vienna, 2003)
4. J. Cai, C. Li, X. Tang, F. Ayello, S. R.S. Nesic, Experimental study of water wetting in oil-water two phase flow-Horizontal flow of model oil. Chem. Eng. Sci. **73**, 334–344 (2012). https://doi.org/10.1016/j.ces.2012.01.014 (Elsevier)
5. D.P. Chakrabarti, G. Das, S. Ray, Pressure drop in liquid-liquid two phase horizontal flow: experiment and prediction. Chem. Eng. Technol. **28**(9), 1003–1009 (2005). https://doi.org/10.1002/ceat.200500143
6. A. Dasari, A.B. Desamala, A.K. Dasmahapatra, T.K. Mandal, Experimental studies and probabilistic neural network prediction on flow pattern of viscous oil-water flow through a circular horizontal pipe. Ind. Eng. Chem. Res. **52**(23), 7975–7985 (2013). https://doi.org/10.1021/ie301430m
7. B. Grassi, D. Strazza, P. Poesio, Experimental validation of theoretical models in two-phase high-viscosity ratio liquid-liquid flows in horizontal and slightly inclined pipes. Int. J. Multiphase Flow **34**(10), 950–965 (2008). https://doi.org/10.1016/j.ijmultiphaseflow.2008.03.006
8. R. Ibarra, I. Zadrazil, C.N. Markides, O.K. Matar, Towards a universal dimensionless map of flow regime, in *11th International Conference on Heat Transfer, Fluid Mechanics and Thermodynamis*(2015)
9. J. Shi, H. Yeung, Characterization of liquid-liquid flows in horizontal pipes. AIChE J. **63**(3), 1132–1143 (2017). https://doi.org/10.1002/aic.15452
10. C.F. Torres, R.S. Mohan, L.E. Gomez, O. Shoham, Oil-water flow pattern transition prediction in horizontal pipes. J. Energy Res. Technol. **138**(022904–1), 1–11 (2016). https://doi.org/10.1115/1.4031608
11. J.L. Trallero, C. Sarica, J.P. Brill, A study of oil/water flow patterns in horizontal pipes, in *SPE Annual Technical Conference and Exhibition* (1997), pp. 165–172

Coupling Thermal–Fluid–Solid Modeling of Drilling Fluid Seepage into Coal Seam Borehole for Co$_2$ Sequestration

Shu-Qing Hao

1 Introduction

It has the potential risk and/or disadvantages when CO_2 sequestration in coal by physical adsorption principle, such as the environmental risk [1–4]. Hydrates-forming method shows the huge storage capacity for CO_2 sequestration in coal owing to the porous structure of the coal and solidifying sequestration principle of hydrates formation method which can be used to overcome the aforementioned problems. At present, several experimental investigations of CO_2 storage in coal by hydrates-forming method were carried out and the valuable results were obtained which testified its effectiveness [5]. However, there is still lots of obstacles for large-scale engineering application since lacking of the physical and chemical parameters of the coal seam which can be used to judge the targeted formation for CO_2 storage and sequestration.

To solve this problem, a direct and effective method, drilling and coring technique was considered to be used to collect the samples to analyze the key parameters which closely related to the CO_2 storage situation. However, a significant aporia existing in this process is the deep drilling borehole stability. As the complexity of the coal formation used for CO_2 storage and the high temperature, which results in the hardness to keep the borehole stability all the time when exploring for getting the cores. Moreover, one significant reason is the principle coupling the thermal –fluid–solid when drilling in the deep coal formation remains unclear.

S.-Q. Hao (✉)
Key Laboratory of CBM Resources and Reservoir Formation Process Ministry of Education of China, School of Resource and Geosciences, China University of Mining and Technology, Xuzhou 221116, China
e-mail: haoshuqing@cumt.edu.cn; jdhsq@163.com

State Key Laboratory for GeoMechanics and Deep Underground Engineering (SKLGDUEK1413), University of Mining and Technology, Xuzhou 221116, China

© Springer Nature Singapore Pte Ltd. 2021
C. Wen and Y. Yan (eds.), *Advances in Heat Transfer and Thermal Engineering* ,
https://doi.org/10.1007/978-981-33-4765-6_126

The objective of this study aimed to establish the coupling models to explain and calculate the thermal–fluid–solid coupling principle when drilling in the coal formation for CO_2 gas storage and sequestration using hydrates formation method especially considering the drilling fluid seepage into the borehole.

2 Methodology

The deformation constitutive equations of the coal samples and seepage equations were established to calculate the deformation values quantificationally. Combined this, the mathematical models coupling the high temperature, high pressure, drilling fluid seepage dynamic pressure, drilling fluid chemical substances, and time effects were established.

1. **The porous thermal mechanics of elasticity constitutive equations:**

$$\sigma_{ij} = 2G\varepsilon_{ij} + \frac{2Gv}{1-2v} - \alpha p\delta_{ij} - \frac{2G(1+v)\alpha_m^T}{3(1-2v)}T\delta_{ij} \tag{1}$$

$$\zeta = \frac{\alpha(1-2v)}{2G(1+v)}\left(\sigma_{kk} + \frac{3}{B}p\right) + (\alpha_m^T - \alpha_f^T)\phi \tag{2}$$

where σ_{ij} is the effective stress; G is the shear modulus; ε is the volumetric strain; v is the Poisson ratio; p is the pore pressure; δ_{ij} is Kronecker's function; α is the porous elastic constant; α_r^T is the porous elastic constant in radical direction; α_a^T is the porous elastic constant in axial direction; T is the temperature; ζ is the volumetric change of the fluid in the pores; ϕ is the porosity.

2. **Darcy seepage equation considering drilling fluid seepage into the coal seam**

$$n_f \cdot V_{fs} = -K\left[\nabla\left(p - \prod\right) + RT(-FC_{fc}\nabla\zeta)\right] \tag{3}$$

where K is the permeability of the coal; C_c is the chemical constituent concentration; P is the total pressure on the porous structural coal; $\prod$ is the swelling pressure; R is the gas constant; T is the absolute pressure; c_i is the ionic concentration.

The equation above implies the total pressure on the pores (including the pore pressure and the effective stress), drilling fluid concentration, and the thermal effective leads to the fluid flows and then results in the pressure upward.

3. **The chemical effect equation when drilling fluid seepage into the coal seam**

According to the Darcy law and the pressure diffusion equation, the instantaneous pore pressure of the coal is:

$$\frac{\partial p}{\partial t} = \frac{k}{\eta\varphi\beta}\left(\frac{\partial^2 p}{\partial r^2} + \frac{1}{r}\frac{\partial p}{\partial t}\right) - \frac{1}{\beta\varphi}\frac{\partial \varphi}{\partial t} \tag{4}$$

The pore fluid concentration is:

$$\frac{1}{\varphi}\frac{\partial C_s}{\partial r} + V\frac{\partial C_s}{\partial r} + \frac{C_s}{r}\left(V + r\frac{\partial V}{\partial r}\right) = D_{\text{eff}}\left(\frac{\partial^2 C_s}{\partial r^2} + \frac{1}{r}\frac{\partial C_s}{\partial r}\right) \tag{5}$$

The swelling pressure of the coal is:

$$\begin{cases} p_\pi = \dfrac{8\pi\, \text{CEC}^2}{\varepsilon s^2}\exp(-2k_n h) \times 10^{-10} \\[2mm] \left.\dfrac{\partial \ln p_s}{\partial t}\right|_{\text{V}} = -\dfrac{k_n h}{C_s}\dfrac{\partial C_s}{\partial t} \end{cases} \tag{6}$$

where p is the pore pressure, MPa; K is the permeability of the gas in the coal, D; p_s is the swelling pressure, MPa; η is the viscosity; φ is the porosity; β is the compressibility ratio of the fluid; r is the radial distance, cm; t is time, s; C_s is the average concentration of the constitute in the pore fluid (w/w); V is the origin volume of the coal, m³; CEC is the cation exchange capacity; ε is the dielectric constant; k_n is the layer thickness of electric double layer, Å^{-1}; h is half of the distance between the distance of the clay layer, Å.

3 Results

The CFD technology, including Ansys-Fluent existing codes, was employed to calculate the dynamic velocity and temperature evolution process when drilling fluid seepage into the borehole of coal seam which is considered to be the storage and sequestration medium for CO_2 by hydrate-forming method using coal samples collected from the depth of 1200 m from Qinshui Basin, China. Meanwhile, the experimental tests were carried out in the laboratory condition. The permeability testing instrument with the self-design accessorial instruments, including the pressure, temperature sensors as well as the data collector was employed to carry out the experimental tests under the boundary pressure is 10 MPa and the temperature is 50 °C. The velocity and the pore pressure when drilling fluid seepage into the coal seam are in good agreement with the results of the seepage experimental tests.

Figure (a) and figure (b) in Fig. 1 shows the temperature and seepage velocity varying situations with the observe points position to the center line of the hole, respectively. It implies the coupling models established is effective.

The results also show that, the numerical models established coupling the thermal–fluid–solid can be used to evaluate the borehole stability and the effectiveness of the optimal drilling fluid when considering the formation as the CO_2 storage and sequestration formation and is expected to write the human–computer interaction interface software to realize the intellectual application further. It should be noted that, in this investigation, only one case was compared about the simulation results with

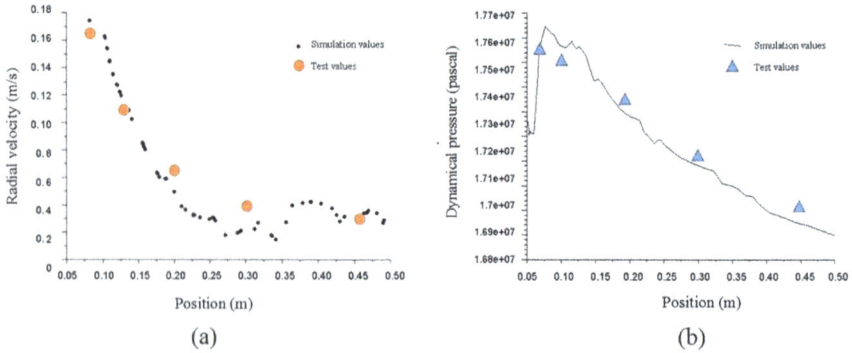

Fig. 1 **a** is the seepage radial velocity of the simulation and experimental tests results; **b** is the seepage dynamical pressure of the simulation and experimental tests results

experiments restricted by the test conditions, the increasing number of observation points are expected to be selected to testify the effectiveness of the numerical method.

4 Conclusions

The results show that, the numerical models established about coupling the thermal–fluid–solid can be used to evaluate the borehole stability and the effectiveness of the optimal drilling fluid when considering the coal seam as the CO_2 storage and sequestration formation and is expected to write the human–computer interaction interface software to realize the intellectual application further then encourages an engineering scale application immediately.

Acknowledgements This study is supported by the Fundamental Research Funds for the Central Universities (2012QNA64), National Natural Science Foundation of China (51674240), and Priority Academic Program Development of Jiangsu Higher Education Institutions (PAPD).

References

1. P. Brewer, E. Peltzer, G. Friderich et al., Experiment on the ocean sequestration of fossil fuel CO_2—pH measurement and hydrate formation. Mar Chem. **72**, 83–93 (2000)
2. H. Tajima, A. Yamasakij, F. Kiyono, Energy consumption estimation for greenhouse gas separation processes by clathrate hydrate formation. Energy **28**, 1713–1729 (2004)
3. N.H. Duc, F. Chauvy, J.M. Herri, CO_2 capture by hydrate crystallization- a potential solution for as emission of steel making industry. Energy Convers Manage. **4**, 1313–1322 (2007)
4. S.Q. Hao, Development of portable drilling fluid and seepage simulation for deep saline aquifers. J Petrol Sci Eng. **175**, 560–572 (2019)

5. S.Q. Hao, S. Kim, Y. Qin, X. Fu, Enhanced CO_2 gas storage in coal. Ind Eng Chem Res. **12**, 18492–18497 (2013)

.

Simulation and Experimental Research on Flow Field and Temperature Field of Diamond Impregnated Drill Bit

Baochang Liu, Shujing Wang, Shengli Ji, Zhe Han, Xinzhe Zhao, and Siqi Li

1 Introduction

Impregnated diamond bits (IDB) are essential tools used in geologic exploration, oil and gas exploitation, mining, construction, etc. [1–3]. Reasonable design of IDB structure is conductive to improve its service life and drilling efficiency. Number of water ways is an important parameter of IDB structure, which has a significant influence on cooling effect and cuttings discharge capacity [4–6]. In this paper, the bottom flow field and temperature field of IDB were investigated by combining numerical simulation with experimental study. The aim is to optimize the number of bit water ways to guide the design of IDB structure.

2 Methodology

A geometric model in accordance with the actual size of IDB was established, and finite element mesh was generated as shown in Fig. 1. The outer diameter of the IDB was 59.5 mm, and the inner diameter was 41.5 mm. The IDB had four water ways with a height of 4 mm and a width of 5 mm. The drilling fluid was set as water, the temperature was 20 °C, the pump volume was 60 L/min, and the rock was granite. The workflow of the flushing fluid water is as follows: water enters the drill pipe, flows through the annular gap between the inner surface of the drill bit and the core

B. Liu (✉) · S. Wang · S. Ji · Z. Han · X. Zhao · S. Li
College of Construction Engineering, Jilin University, Changchun, China
e-mail: liubc@jlu.edu.cn

Key Laboratory of Drilling and Exploitation Technology in Complex Conditions of Ministry of Natural Resources, Changchun, China

B. Liu
State Key Laboratory of Superhard Materials, Jilin University, Changchun, China

© Springer Nature Singapore Pte Ltd. 2021
C. Wen and Y. Yan (eds.), *Advances in Heat Transfer and Thermal Engineering* ,
https://doi.org/10.1007/978-981-33-4765-6_127

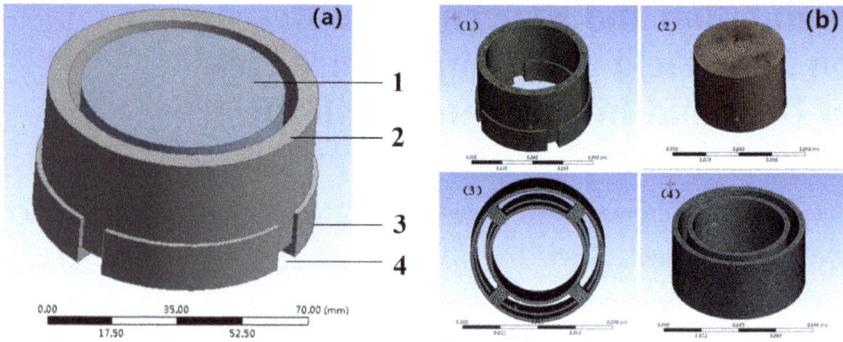

Fig. 1 **a** 3D model of bit: 1-rock core, 2-steel body, 3-matrix, 4-water way; **b** finite element mesh: 1-finite element mesh of bit, 2-finite element mesh of rock core, 3 and 4-finite element mesh of fluid channel

surface, supplies the flushing fluid to the working face to cool the drill bit, and carries rock cuttings, enters the annular gap between the drill pipe and the hole wall through the water ways and returns up.

The IDBs structure used in drilling experiment was the same as that used in numerical simulation. Three IDBs with four, six, and eight water ways were used for drilling experiment. Water was used as flushing fluid, and the flow rate was 60 L/min. The temperature measurement method is shown in Fig. 2. A number of K-type thermocouples were embedded in granite block to measure the temperature of contact surface between IDB and rock. With the increase of drill footage, the distance between the test point and the rock contact surface became closer and closer. When the IDB contacted the thermocouple, the paperless recorder can read the temperature of the contact surface.

Fig. 2 Arrangement of thermocouples in rock sample 1-granite block; 2-the borehole; 3-thermocouples

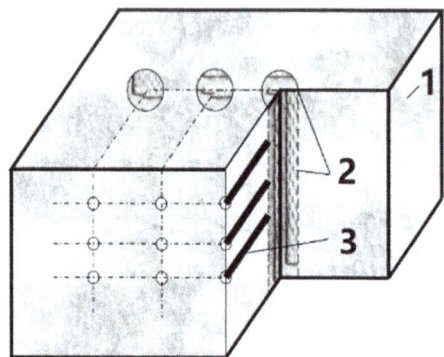

3 Results

The flow field through the water ways is asymmetrical due to the rotation of the IDB. As shown in Fig. 3, on the side of the water way with the same rotating direction of the bit, the velocity is faster, and the maximum velocity can reach 23.65 m/s. While on the other side of the water way, the velocity is relatively lower. When the rotation reaches steady state, the flushing pressure in the annular between the outer surface of core and the inner surface of bit reaches the maximum value, and the pressure at the water way decreases and the flow speed becomes faster. So severe convective velocity field will occur at the water way and lead to local high-speed low-pressure zone or vacuum zone. In this case, the cavitation is easy to be produced, which will increase the wear of the IDB.

The temperature distribution of the bit was simulated and the results indicated that the temperature of the bit is basically the same in the radial direction, and changes in the axial direction (Fig. 4). When the bit rotates, the turbulent energy of the flow field at the water ways is the largest. The temperature changes along the axis of drill bit. At the contact surface between the bit and rock, the temperature reaches the highest level of 210 °C (Fig. 5a). The temperature decreases obviously along the axial direction of the bit until it reaches the temperature of the flushing fluid.

The relationship of maximum contact temperature and the number of water ways is shown in Fig. 5b, and the relationship between drilling speed, wear height of bit matrix and the number of water ways is shown in Fig. 6. With the increase of the

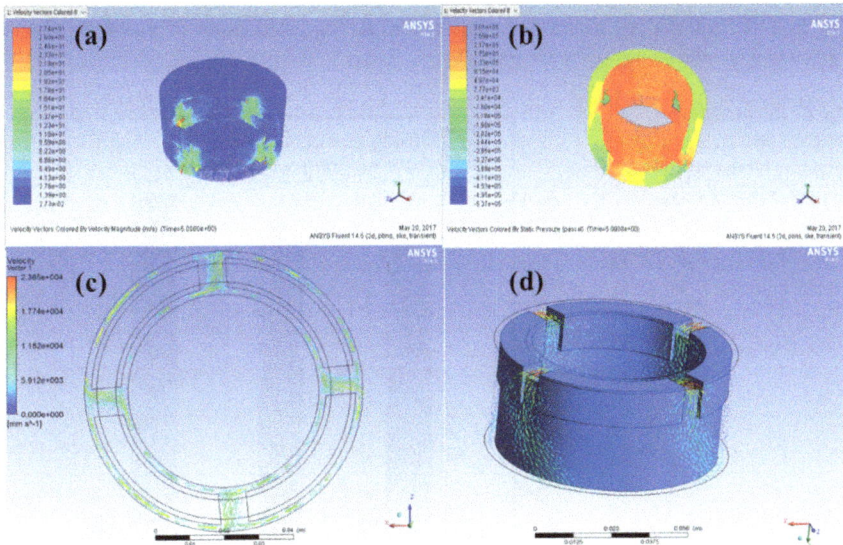

Fig. 3 Simulation result of fluid field. **a** The overall cloud graph of steady-state velocity vector; **b** the steady-state pressure field cloud graph; **c** the graph of steady-state velocity vector of 2 mm section from working surface; **d** The overall space graph of steady-state velocity vector

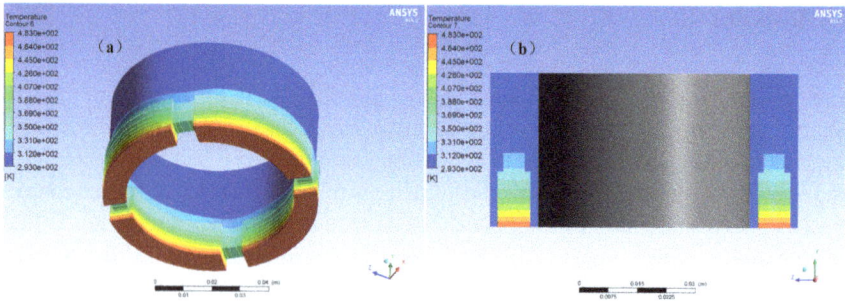

Fig. 4 Simulation result of temperature field. **a** Temperature distribution in bit. **b** Axial temperature field distribution in bit

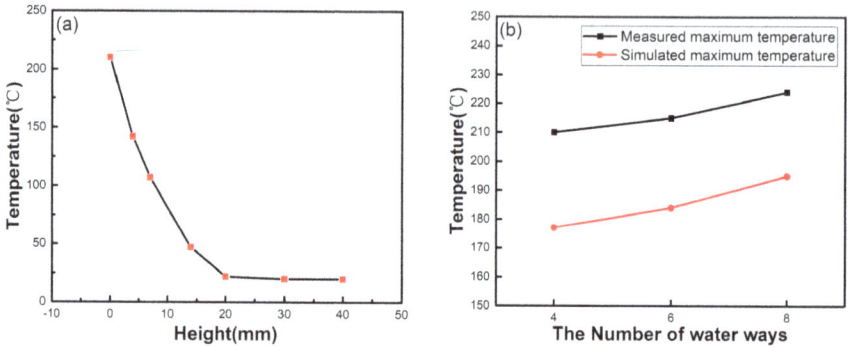

Fig. 5 a Relationship of height and temperature. **b** The relationship of maximum contact temperature and the number of water ways

Fig. 6 Relationships between drill speed, wear height of bit matrix and the number of water ways

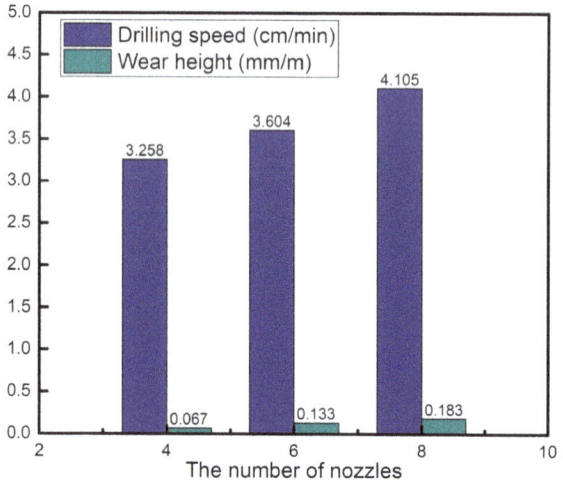

number of water ways, the temperature on the drill bit face rises and the drilling speed increases, but at the same time the wear of the matrix is more serious. The flow rate of drilling fluid in a single water way decreases due to the increase of the number of water ways, so the cooling effect decreases and the bottom temperature is higher. Good cooling effect can avoid thermal damage of diamond due to high temperature to ensure drilling efficiency. In this study, the IDB with four water ways has good cooling effect and a relative appropriate drilling speed while the wear height is minimal and the service life is largest. So the IDB with four water ways exhibited the best overall performance.

4 Conclusions

The bottom flow field and temperature field of IDB were investigated by combining numerical simulation with experimental study. The following conclusions can be drawn from the investigation.

1. For the IDB with 4 water ways, on the side of the water way with the same rotating direction of the bit, the velocity is faster, and the maximum velocity can reach 23.65 m/s. while on the other side of the water way, the velocity is lower.
2. For the IDB with 4 water ways, The temperature changes along the axis of drill bit. At the contact surface between the bit and rock, the temperature reaches the highest level of 210 °C, and drops rapidly from bottom to top until it is close to the temperature of the flushing fluid.
3. The temperature of the interface between 4 water ways bit and rock is lower than that of 6 and 8 water ways bit. The increase of the number of water ways leads to a small flow rate of drilling fluid at a single water way, which reduces the heat conduction efficiency at a single water way, resulting in poor cooling effect and serious bit wear. In the process of designing IDB structure, factors such as bottom hole temperature, drilling speed and bit wear should be considered comprehensively, and the number of water ways should be reasonably designed.
4. In this drilling experiment, the comprehensive performance of the IDB with 4 water ways bit is the best, which not only guarantees the cooling effect, but also has a good drilling speed. It can provide guidance for future IDB structure design.

Acknowledgements This work was supported by Research Foundation of Key Laboratory of Deep Geo Drilling Technology, Ministry of Natural Resources, China (No. KF201808)

References

1. S. Perez, M. Karakus, F. Pellet, Development of a tool condition monitoring system for impregnated diamond bits in rock drilling Applications. Rock Mech. Rock Eng. **50**(5), 1289–1301

(2017)
2. M. Mostofi, T. Richard, L. Franca et al., Wear response of impregnated diamond bits. Wear **04**, 010 (2018)
3. S. Tan, X. Fang, K. Yang et al., A new composite impregnated diamond bit for extra-hard, compact, and nonabrasive rock formation. Int. J. Refract. Metals Hard Mater. **43**, 186–192 (2014)
4. X. Cao, A. Kozhevnykov, A. Dreus, B.C. Liu, et al., Diamond core drilling process using intermittent flushing mode. Arab. J. Geosci. **12**(4) (2019)
5. A. Bhatnagar, M. Khandelwal, Laboratory investigations for the role of flushing media in diamond drilling of marble. Rock Mech. Rock Eng. **44**(3), 349–356 (2011)
6. T.C. Jen, G. Gutierrez, S. Eapen, et al., Investigation of heat pipe cooling in drilling applications.: part I: preliminary numerical analysis and verification. Int. J. Mach. Tools Manuf. **42**(5), 643–652 (2002)

Deposition from Waxy Crude Oils Flowing in Transportation Pipelines: A Numerical Study

Mirco Magnini and Omar K. Matar

1 Introduction

Crude oils are complex mixtures of hydrocarbons consisting of paraffins, aromatics, naphthenes, resins, and asphaltenes. Waxes are high-molecular weight paraffins, with carbon numbers ranging from 18 to 65 that have a low solubility in crude oil at reduced temperatures. When the oil is transported within pipelines across a cold environment, for example, within subsea pipelines where temperatures may drop to values around 5 °C, if the crude oil temperature falls below the wax appearance temperature (WAT) of the mixture, the wax may precipitate and deposit along the inner walls of the pipeline. The deposit layer consists of liquid and solid phases in a gel-like structure, whose composition and relative proportions vary across the deposit thickness because of the variations in temperature, concentration, and shear stress. The accumulation of the deposited material causes a significant flow-assurance problem, i.e., increased pumping power, decreased flow rate or even the total blockage of the pipeline. Under field conditions, the deposit layer formed on the pipe wall is usually removed mechanically by pigging the pipeline, with severe costs associated with pigging operations, and deferred production. Therefore, it is imperative to predict accurately the behavior of the deposited layer with time in order to schedule reliably the pigging operation, design effective remediation strategies, and mitigate the wax deposition problem effectively. Several zero- and one-dimensional models for predicting wax deposition have been proposed, with different levels of sophistication for what concerns the description of the deposition mechanism [1, 2]. However, the temporal growth of the deposit layer depends on the balance of deposition and

M. Magnini (✉)
Department of Mechanical, Materials and Manufacturing Engineering, University of Nottingham, Nottingham NG7 2RD, UK
e-mail: mirco.magnini@nottingham.ac.uk

O. K. Matar
Chemical Engineering Department, Imperial College London, London SW7 2AZ, UK

© Springer Nature Singapore Pte Ltd. 2021
C. Wen and Y. Yan (eds.), *Advances in Heat Transfer and Thermal Engineering* ,
https://doi.org/10.1007/978-981-33-4765-6_128

removal processes, and the available methods failed to properly account for the latter term. In fact, the removal term is still expressed following Ebert and Panchal [3], as a constant multiplying the wall shear stress. The actual deposit removal mechanism results from the complex interaction of fluid flow, two-phase dynamics, and heat transfer (due to temperature-dependent viscosities) and its understanding requires accurate resolution of the two-phase interface dynamics.

We have developed a numerical model to simulate wax deposition, based on the solution of the unsteady Navier–Stokes and energy equations for the oil–wax mixture fluid using a Volume-Of-Fluid method. Additional transport equations for the waxy and non-waxy components within each phase are simultaneously solved. The wax deposition process is described using a chemical equilibrium model based on the Gibbs free energy [4]. Density and viscosity of the fluid are dependent on the local temperature. Ageing effects are also included in order to capture the hardening of the deposit layer as time elapses. The numerical framework is validated by comparison with experimental wax deposition data from the literature. The model is utilized for a fundamental investigation of the wax deposition and removal processes and parametric analyses are carried out to study the effect of flow parameters such as crude oil flow rate, wall temperature, pipe size, and wax rheological parameters.

2   Methodology

We treat the crude oil and the deposit as two immiscible phases and we utilize a Volume-Of-Fluid (VOF) method to capture the dynamics of the interface separating the phases. Accordingly, the two-phase crude oil and deposit mixture is treated as a single mixture fluid. The single-fluid Navier–Stokes and energy equations are solved throughout the flow domain, where the oil is given liquid-like properties and the deposit is assumed to behave as a gel (i.e., much higher viscosity). Viscous forces are modeled via a Newtonian fluid model, except for the cases where deposit ageing is included. The capillary force (with constant surface tension coefficient) is included within the momentum equation and it acts at the oil–deposit interface. From a numerical perspective, it is unpractical to track the concentration of each component of the crude oil. Therefore, we adopt a simplified fluid model where both the crude oil and the deposit are assumed to be composed of only two lumped pseudo-species: the crude is composed of oil (non-wax) and dissolved wax; the deposit is composed of liquid oil (trapped in the wax crystals) and separated wax crystals. Hence, two additional transport equations are solved, one for the non-wax pseudo-species in the crude oil and another for the non-wax pseudo-species in the deposit.

Deposition occurs via two routes: from non-wax species in the crude to non-wax species in the deposit, and from wax species in the crude to wax species in the deposit. In order to evaluate the deposition rate locally at the interface, depending on the local temperature field and distribution of the pseudo-species, we adopt the model of Svendsen [4], which assumes that deposition is governed exclusively by molecular diffusion:

$$j_i = \Gamma \frac{\rho \eta_i}{T} \nabla T \cdot \mathbf{n} \tag{1}$$

where j_i indicates the deposition rate (units kg/m²s) via route i, Γ is a diffusion constant, ρ denotes the mixture fluid density, η_i represents the dimensionless variation of the mass concentration of species i at liquid state in the mixture per unit temperature [4], T is the local temperature at the crude–deposit interface, and $\mathbf{n}$ is the unit normal vector to the interface, or to the pipe wall if no deposit layer is present. Mass and energy source terms due to the phase separation are implemented within the flow equations based on Eq. (1).

We simulate the two-phase flow of crude oil and deposit within straight pipes; a snapshot of a simulation is shown in Fig. 1. We adopt 3D pipe geometries, meshed with block-structured grids which are refined near the wall to capture the thin-film dynamics. At the channel inlet, the crude enters in the domain with prescribed velocity, temperature, and mass fraction of each pseudo-component. No-slip conditions are set at the pipe wall, which is given a constant temperature, below the WAT. Zero-gradient conditions are imposed at the channel outlet. At $t = 0$, no deposit is present inside the pipeline. The initial velocity and temperature fields are obtained from a preliminary crude-only steady-state simulation run under the same flow conditions. Both laminar and weakly turbulent flow conditions are studied, with a $k - \omega$

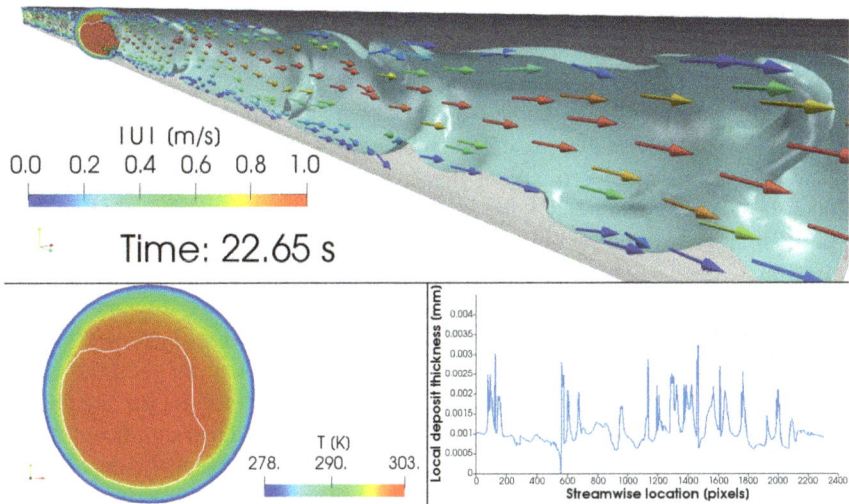

Fig. 1 (top) Snapshot of the flow (conditions from Svendsen [4]) in a $D = 9$ mm pipe. Half of the tube wall is clipped to allow visualization. Flow is from left to right. A layer of deposit, with the crude-deposit interface colored in light blue, covers the pipe wall. The vector field (not to scale) represents the velocity field on a vertical centreline plane. The upstream slice shows the temperature field on the cross-section, with a white line indicating the location of the oil-deposit interface. (bottom-left) Close-up image of the temperature field on the cross-sectional slice indicated on the top figure. (bottom-right) Instantaneous value of the thickness of the deposit upon the bottom wall, measured along a longitudinal plane passing through the pipe axis

turbulence model activated when the Reynolds number is above 2000. The flow equations are solved using ANSYS Fluent 18.1 and various user-defined functions are implemented to model deposition, temperature-dependent properties, and perform run-time data processing.

3 Results

We present the results of a representative numerical simulation to illustrate the capability of the numerical model. The flow conditions are taken from Svendsen [4]. The channel is horizontal ($g = 9.81 \, \text{m/s}^2$ acting downward) with diameter $D = 9$ mm and length $40D$. The crude oil enters the pipe at a temperature $T_c = 303$ K, while the wall is at $T_{\text{wall}} = 278$ K. The WAT is $T_{\text{WAT}} = 319$ K. The average oil velocity at the inlet is $U_c = 0.22 \, \text{m/s}$, corresponding to a Reynolds number of 193. No ageing effects are included in this case and therefore the deposit viscosity is independent of time (but dependent on temperature). At onset, a layer of deposit begins to grow at the pipe wall as $T_{\text{wall}} < T_{\text{WAT}}$. The deposit layer grows as time elapses, while being transported downstream due to the interfacial shear. The shear-removal rate increases with the deposit thickness, whereas the deposition rate decreases due to the increase of the interface temperature (and decrease of the temperature gradient) as the crude-deposit interface moves away from the wall. Eventually, a regime corresponding to a dynamic steady-state is reached with the rate of deposition being balanced by the rate of removal. The total deposit content within the pipe stops growing (on average), although the flow remains dynamic with waves being continuously created and travelling in the streamwise direction. The simulation is arrested at this stage. Figure 1 illustrates a representative image of the flow extracted during the steady regime. Under the conditions simulated, the average deposit layer thickness is $\delta/D = 0.09$, and the total pressure drop along the channel is about three times larger than the single-phase case.

We have analyzed the effect of the flow rate by increasing the Reynolds number up to 5000; although the deposit growth rate at the initial stages increases with Re, the steady-state deposit thickness decreases when increasing Re (non-monotonic behavior) due to the enhanced shear at the oil–deposit interface. Increasing the wall temperature leads to a reduction of the deposit thickness and to a significant change of its composition, i.e., an increase in the wax concentration of the deposit. The deposit thickness (made non-dimensional by the pipe radius) decreases when the pipe size increases, due to the increased interfacial shear and lower radial temperature gradients. The inclusion of ageing effects has a strong impact on the flow, as the continuous increase of the deposit viscosity as time elapses yields a continuous increase of the deposit thickness which does not necessarily reach a steady regime.

4 Conclusions

Wax deposition in crude oil flows is a complex process that involves fluid mechanics, heat transfer, thermodynamics, and phase separation. We have developed a numerical model that simulates the oil–deposit two-phase flow and heat transfer, and accounts for wax deposition at the pipe wall. The simulations show that the flow behaves like an annular flow, with a very viscous layer of deposit growing at the pipe wall and slowly moving downstream along the pipe, and a high-speed oil core flow traveling along the pipe axis. The effects of the oil flow rate, wall temperature, pipe size and wax ageing parameters can be accurately studied with this numerical model and provide a novel insight into the wax deposition and removal mechanisms.

Acknowledgements O. K. M. acknowledges funding from PETRONAS and the Royal Academy of Engineering for a Research Chair in Multiphase Fluid Dynamics.

References

1. O.C. Hernandez, H. Hensley, C. Sarica, J.P. Brill, M. Volk, E. Delle-Case, Improvements on single phase paraffin deposition modeling. SPE Prod. Facilities 237–243 (2004)
2. D. Eskin, J. Ratulowski, K. Akbarzadeh, A model of wax deposit layer formation. Chem. Eng. Sci. **97**, 311–319 (2013)
3. W. Ebert, C.B. Panchal, analysis of exxon crude-oil slip stream coking data, in *Fouling Mitigation of Industrial Heat Exchange Equipment*, ed, by C.B. Panchal, T.R. Bott, E.F.C. Somerscales, S. Toyama (Begell House, New York, NY, 1997), pp. 451–460
4. J.A. Svendsen, Mathematical modeling of wax deposition in oil pipeline systems. AIChE J. **39**, 1377–1388 (1993)

Simplified Layer Model for Solid Particle Clusters in Product Oil Pipelines

Dongze Li, Lei Chen, Qing Miao, Gang Liu, Shuyi Ren, and Zhiquan Wang

1 Introduction

Pipe corrosion caused by the pressure tests using water before starting the normal operation widely occurs in Chinese product oil pipelines because of remaining water [1, 2]. To explore the migration of the corrosion impurities in the product oil pipelines, this study started from the force balance principle and considered the entire particle cluster as the research object. This paper established a one-dimensional migration model and proposed the Fr equality criterion to calculate the particle cluster length in the equilibrium state. A loop was built to conduct the tests and obtain the migration velocities of the particle cluster from the non-equilibrium state to the equilibrium state in the pipeline. The proposed model was verified by experiments. Verification results demonstrate that the model can describe the development process from the non-equilibrium state to the equilibrium state of particle clusters after sudden external disturbance and accurately predict some important parameters, including the velocity of the particle cluster in the equilibrium state and the critical flow rate that leads to the transition from fixed bed flow to moving bed flow. The model provides the theoretical basis and calculation method to remove corrosion impurities from product oil pipelines.

D. Li · L. Chen (✉) · Q. Miao · G. Liu (✉) · S. Ren
Shandong Provincial Key Laboratory of Oil & Gas Storage and Transportation Safety, China
University of Petroleum (East China), Qingdao 266580, Shandong, China
e-mail: leo@upc.edu.cn

G. Liu
e-mail: liugang@upc.edu.cn

Q. Miao
CNPC Key Laboratory of Oil & Gas Storage and Transportation, Langfang 065000, Hebei, China

Z. Wang
Shandong Provincial Siwei Sefety Technology Center, Jinan 250000, China

© Springer Nature Singapore Pte Ltd. 2021
C. Wen and Y. Yan (eds.), *Advances in Heat Transfer and Thermal Engineering* ,
https://doi.org/10.1007/978-981-33-4765-6_129

2 Methodology (Length and Layout)

Based on the principle of force balance, the motion equation of particle cluster is established. The expressions of each force in the equation are derived based on a certain simplification [3]. The particle cluster is divided into two parts of equal length: the head and the tail, to describe the deformation characteristics after being impacted by the oil flow.

$$
\begin{cases}
m \dfrac{d^2 x_1}{dt^2} = C_D \cdot \dfrac{\rho \lambda Q^2 L R \cdot \left(2\pi - \theta + 2\sin\frac{\theta}{2}\right)}{8(1+\delta)\left(\pi R^2 - \frac{V_v}{L}\right)^2} \\
\qquad - (\mu_f \alpha \cos \varphi_0 + \sin \varphi_0) \cdot (mg - \rho g V_v) \\
\qquad - C_F \cdot \dfrac{6\eta\theta R L}{r} \dfrac{dx_1}{dt} + \dfrac{m\beta^2}{2}(L_e - L) \\
m \dfrac{d^2 x_2}{dt^2} = C_D \cdot \dfrac{\rho \lambda Q^2 L R \cdot \left(2\pi - \theta + 2\sin\frac{\theta}{2}\right)}{8(1+\delta)\left(\pi R^2 - \frac{V_v}{L}\right)^2} \\
\qquad - (\mu_f \alpha \cos \varphi_0 + \sin \varphi_0) \cdot (mg - \rho g V_v) \\
\qquad - C_F \cdot \dfrac{6\eta\theta R L}{r} \dfrac{dx_2}{dt} - \dfrac{m\beta^2}{2}(L_e - L)
\end{cases}
\tag{1}
$$

The expressions of each force in the equation are derived based on a certain simplification, and some important parameters including the length and velocity of particle cluster in the equilibrium state and the critical flow rate that leads to the transition from fixed bed flow to moving bed flow can be obtained:

$$
L_e = \frac{2V_v^2 \cdot g \cdot Fr_0}{Q(Q^2 + 4\pi R^2 \cdot V_v \cdot g \cdot Fr_0)^{0.5} + Q^2 + 2\pi R^2 \cdot V_v \cdot g \cdot Fr_0}
\tag{2}
$$

$$
Q_0 = \left(\frac{8(1+\delta)(\mu_f \alpha \cos \varphi_0 + \sin \varphi_0)(mg - \rho g V_v)\left(\pi R^2 - \frac{V_v}{L_e}\right)^2}{\rho \lambda C_D L_e R \cdot \left(2\pi - \theta + 2\sin\frac{\theta}{2}\right)} \right)^{0.5}
\tag{3}
$$

Experiments were carried out to obtain experimental data. Some experimental data were used to fit unknown parameters in the model and compared the remaining data with calculated values to verify the reliability of the model.

3 Results

It can be seen from Fig. 1 that the model can accurately describe the migration of the particle cluster after being suddenly disturbed by the oil flow.

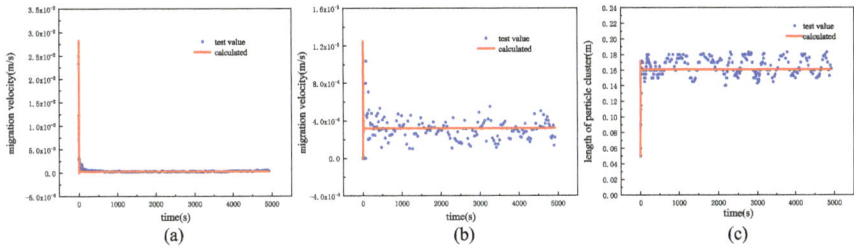

Fig. 1 **a** Forward velocity of particle cluster; **b** the tail velocity of particle cluster; **c** the length of particle cluster

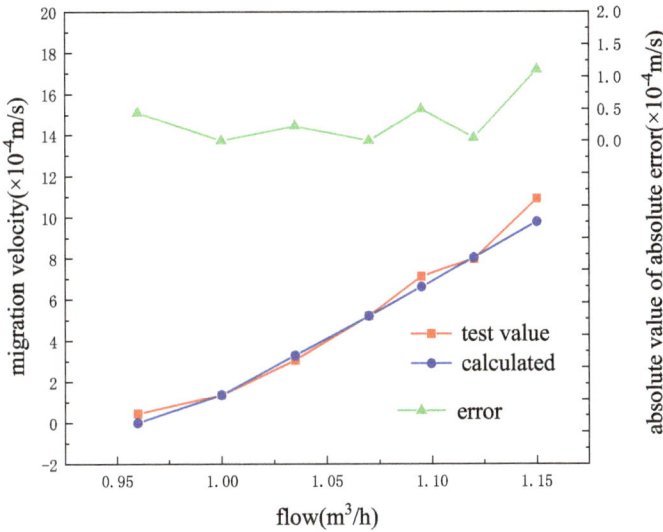

Fig. 2 Velocity comparison of test values and calculated values in the stable stage

It can be seen from Fig. 2 that the model can better predict the migration velocity of particle cluster under different flow rates.

4 Conclusions

1. The variation of particle length in equilibrium state follows the Froude number equality criterion.
2. The simplified layer model based on the force balance principle can better describe the transformation process from the non-equilibrium state to the equilibrium state and accurately predict the critical flow velocity and stable migration velocity of the particle cluster.

3. At the same flow rate, the larger the mass of the particle cluster or the inclination of the pipe is, the lower the migration velocity is.

Acknowledgements Support by "the Fundamental Research Funds for the Central Universities (18CX02172A)," "the National Natural Science Foundation of China (51774315)," "the National Natural Science Foundation of China (51704319)."

Reference

1. D.Z. Wang, W. L, Analysis of sediments in product oil pipelines. Oil Gas Storage Transport. **24**(2), 59–60
2. D.Z. Xu, Study on hydrodynamic characteristics of product oil pipelines (China University of Petroleum)
3. C.S. Campbell, Rapid granular flows. Annu. Rev. Fluid Mech. **22**(22), 57–90 (1990)

Flow and Heat Transfer Intensification in von Karmon Swirling Flow by Sucrose-Based Polymer Solution

Guice Yao, Jin zhao, and Dongsheng Wen

1 Introduction

Current cooling system is still struggling to meet the high demand of heat removal from highly integrated electronic devices in many industrial areas such as information computing technology, process intensification, and ultra-high heat flux encountered in aerospace field. A conventional way to enhance heat transfer performance is to use microchannel heat sink on the base of microelectronic devices [1]. However, the small channel size typically limits the flow in the laminar flow regime. Some complicated configurations of channels are designed in order to intensify the fluid motion, which, however, are always required a complex design of three-dimensional microchannel yet with increased pressure drop penalty [2]. One of the proposed approaches for flow intensification is to use viscoelastic fluids. Adding a small amount of high-molecular-weight polymer into a pure Newtonian solvent, even at very low Reynolds numbers, this viscoelastic fluid can exhibit turbulent-like features such as irregular flow patterns over a broad range of spatial and temporal scales and intensive velocity fluctuations. This turbulent-like phenomenon is called elastic turbulence [3].

The elastic turbulence is mainly attributed to polymer coil-stretch deformation and streamline curvature of the flow channel, and is always accompanied with a sharp growth of flow resistance. The occurrence of elastic turbulence has been observed in many geometries [4–7] and was proved to benefit the mixing performance [8] even between immiscible fluids [9]. Very recently, the chaotic flow motions caused by elastic instabilities were applied on enhancing heat transfer performance in both macro- and microscale due to the strong flow perturbations that weaken the insulating

G. Yao (✉) · J. zhao · D. Wen
Faculty of Engineering, University of Leeds, Leeds LS2 9JT, UK
e-mail: pmgy@leeds.ac.uk

D. Wen
Beihang University, Beijing 100191, People's Republic of China

boundary effect [10–12]. It was shown that at the same inlet velocity, the heat transfer efficiency was improved by four times compared to base fluid.

In fact, studies on heat transfer enhancement based on elastic turbulence are not well established. The limited previous studies were conducted in different geometry with different working fluids and even the working conditions and analysis methods were unique. It is hard to give a consistent insight of the relationship between elastic turbulence and heat transfer performance. What is more, due to the geometric limitation of curvilinear channel, the characterization of elastic turbulence is not sufficient, thereby whether the flow is in the elastic turbulence regime is not clearly validated. In particular, no Nussell number was determined in the experiments investigated in swirling flow, and only a heat transfer efficiency above conductive limits was performed with bulk fluids. The surface heat transfer coefficient was not considered. Addressing these limitations, this work aims to conduct a systematic research to reveal the surface heat transfer performance of elastic turbulence by conducting a macroscale swirling flow between two parallel plates.

2 Methdology

The experimental rig that was used to investigate the behavior of the flow and convectional heat transfer performance of viscoelastic fluids in swirling flow is shown in Fig. 1. It consists of an acrylic fluid container with inner diameter $D_{in} = 56$ mm and optically transparent walls. The thicknesses of side wall and bottom wall are 10 mm and 5 mm, respectively. The flow was driven by a rotating round disk with a radius $R_d = 25$ mm mounted on an electric motor whose controlling precision is up to 0.1 rpm. The distance between the top disk and the bottom of the fluid container was set at a constant value, $H = 40$ mm, for all experiments. A circulating fluid bath is attached to the bottom of the fluid container and the temperature within which was set to below the room temperature at a value of 5 °C to avoid thermal convection inside the bulk flow. The room temperature was maintained around 23 °C by an air conditioning system. The temperature profiles were conducted by thermocouples mounted inside the bulk.

Two types of working fluid were used in this study, Newtonian fluid and viscoelastic fluid. The Newtonian fluid was an aqueous solution of 65% sucrose and 1% sodium chloride (NaCl) in distilled water (referred to "sucrose solution" here after), which was regarded as a base fluid. The viscous fluids were consisted of 200 ppm high-molecular-weight hydrolyzed polyacrylamide (HPAM, Mw: 5–22 M g/mole), 65% sucrose, and NaCl solutions (referred to "HPAM solution" here after). Four salt concentrations of HPAM solutions were investigated from 0.01 to 1%. Hereinto, 65% sucrose was conducted as both the base fluid and the solvent of the HPAM solutions, due to the fact that it could maximize the relaxation time of the solution and minimize the Reynolds numbers, thereby the flow instabilities were attributed to elastic effect rather than inertial effect.

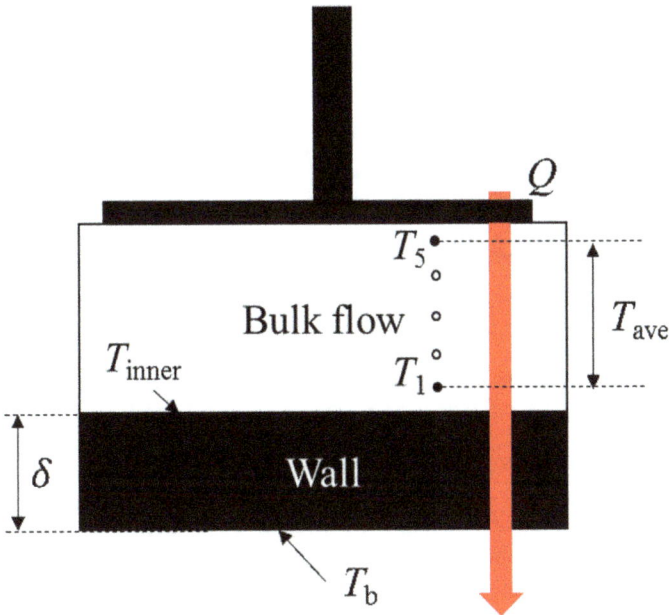

Fig. 1 Schematic view of the experimental setup

The longest or characteristic relaxation time was calculated as 3.1 s based on the Eq. (1):

$$\lambda = \lim_{\omega \to 0} \left\{ \frac{1}{\omega} \left[\frac{\eta_P''(\omega)}{\eta_P'(\omega)} \right] \right\} \tag{1}$$

3   Results

With seeding 1% lighting reflecting flakes into working fluids, the flow visualization was achieved. This particle-filled liquid, so called as Kalliroscope or rheoscopic liquid, is effective in capturing the flow patterns by reflecting differing intensities of light, making the movement of the streamline visible. Two representative snapshots of the sucrose solution at different rotating speed viewed from below are shown in Fig. 2. No obvious irregular flow pattern or vortex is observed even at largest applied rotating speed. The flow looks quite uniform and is completely laminar, which consistent perfectly with the temperature distribution profiles measured by thermocouples mentioned above. Therefore, it can be concluded that, for pure sucrose solution, the flow stays in laminar regime and the inertial effects could be neglected within the range of rotating angular speed applied during the experiments.

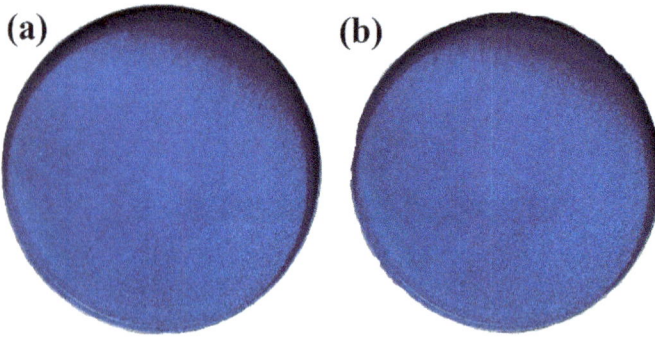

Fig. 2 Flow patterns observed from below at 1 rpm (**a**) and 10 rpm (**b**), respectively

A series of snapshots of flow patterns captured from the bottom of the fluid container is listed in Fig. 3. The patterns of the polymer solution at higher rotating speed look quite irregular and exhibit structures of different sizes. The evolution of these secondary flow patterns could be interpreted by the transition pathway to elastic turbulence in parallel-plate flow observed by Schiamberg et al. [13]. The flow sequentially develops as so-called base state, stationary ring mode, competing spirals mode, and multi-spiral chaotic mode, respectively, with increasing the driven shear forces. Compared with the final elastic turbulence mode, the spiral-like flow pattern at maximum applied rotating speed is less intensive, which consistent well with

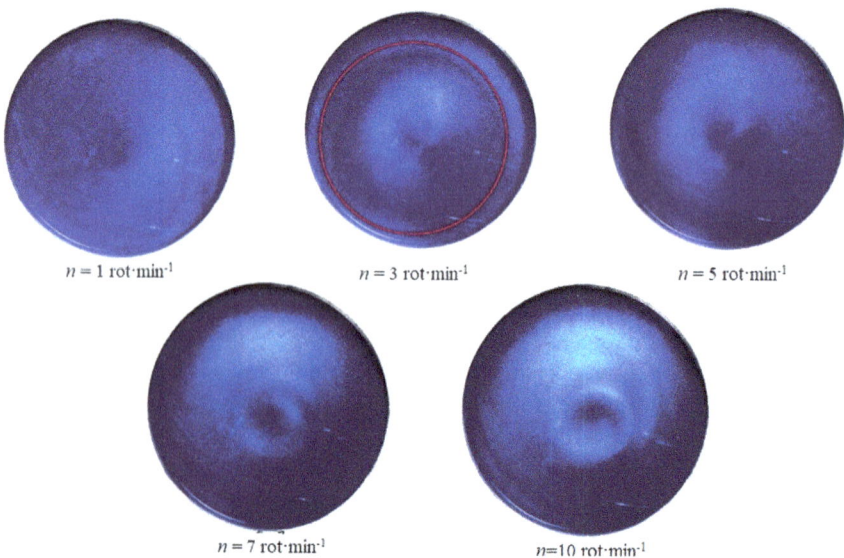

$n = 1$ rot·min^{-1} $n = 3$ rot·min^{-1} $n = 5$ rot·min^{-1}

$n = 7$ rot·min^{-1} $n = 10$ rot·min^{-1}

Fig. 3 Snapshots of flow patterns captured from bottom for HPAM solution at different rotating speed

statistical properties discussed in the later section, indicating the flow in still in the transition to elastic turbulence regime. These spiral-like forms are probably imposed by the average of azimuthal flow and circular symmetry of the setup. Furthermore, a peak point is observed in the middle at stationary ring mode, which corresponds to the center of a big persistent toroidal vortex and evolves to a spiral vortex latter. Direction of the bursting spiral motion is downward near the center and outward near the bottom, which is attributed to Weissenberg effect and is opposite to the motion in Newtonian fluids. The visual impression is consistent well with the temperature distribution and the existence of the vortexes recommend the elastic turbulence as a potential candidate to enhance heat transfer at least within swirling flow.

The temperature distribution profiles at maximum rotating speed for both sucrose solution and HPAM solution are shown in Fig. 4. The initial temperature of working fluids is homogenous and similar around 21 Celtics degree. For sucrose, as the heat removing by the cooled bottom, the temperature starts to decrease from bottom to top gradually and equilibrates at different values. Even at the maximum applied rotating speed, an obvious separation of temperature layer which is strongly dependent on the axial coordinates is still observed. In contrast, for HPAM solution, a fully homogenous distribution of the measured temperature profiles is obtained at largest degree of rotation. The temperature distribution along the vertical direction of the whole bulk entirely collapses into a single curve, which indirectly indicates the existence of elastic turbulence. These temperature profiles are consistent well with the flow dynamics observed above. The motion of chaotic vortex breaks the flow layers in the axial direction, moving fluids from one region to another, carrying energy between regions. However, for the sucrose solution, the laminar streamline makes the only way for heat to transfer from one layer to another that is through conduction at the shear layer, which requires more time and is insufficient.

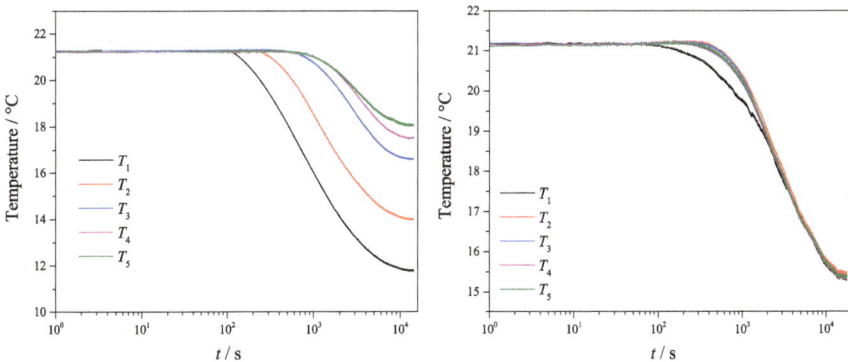

Fig. 4 Temperature distribution for sucrose solution and HPAM solution at maximum applied rotating speed

4 Conclusions

An experimental study of flow and heat transfer intensification by elastic turbulence in swirling flow is presented. At maximum applied rotating speed, the sucrose solution still exhibits laminar behavior while the HPAM solution was evolved to be turbulent-like. The corresponding temperature profiles are consistent well with the flow patterns. The temperature distributes in layers because the ordered flow streamline for sucrose but collapsed into a single curve for HPAM solution since the existence of elastic turbulence. This homogeneous temperature distribution indicates the elastic turbulence indeed has the capacity for enhancing heat transfer.

References

1. D.B. Tuckerman, R. Pease, High-performance heat sinking for VLSI. IEEE Electron Device Lett. **2**(5), 126–129 (1981)
2. S.W. Jones, O.M. Thomas, H. Aref, Chaotic advection by laminar flow in a twisted pipe. J. Fluid Mech. **209**, 335–357 (1989)
3. A. Groisman, V. Steinberg, Elastic turbulence in a polymer solution flow. Nature **405**(6782), 53–55 (2000)
4. A. Groisman, V. Steinberg, Elastic turbulence in curvilinear flows of polymer solutions. New J. Phys. **6**(1), 29 (2004)
5. H.Y. Gan, Y.C. Lam, N.-T. Nguyen, Polymer-based device for efficient mixing of viscoelastic fluids. Appl. Phys. Lett. **88**(22), 224103 (2006)
6. C. Scholz, F. Wirner, J.R. Gomez-Solano, C. Bechinger, Enhanced dispersion by elastic turbulence in porous media, EPL (Europhysics Lett.), **107**(5), 54003 (2014).
7. P.C. Sousa, F.T. Pinho, M.A. Alves, Purely-elastic flow instabilities and elastic turbulence in microfluidic cross-slot devices. Soft Matter **14**(8), 1344–1354 (2018)
8. T. Burghelea, E. Segre, I. Bar-Joseph, A. Groisman, V. Steinberg, Chaotic flow and efficient mixing in a microchannel with a polymer solution. Phys. Rev. E Stat. Nonlinear Soft. Matter Phys. **69**(6 Pt 2), 066305 (2004)
9. R.J. Poole, B. Budhiraja, A.R. Cain, P.A. Scott, Emulsification using elastic turbulence. J. Nonnewton. Fluid Mech. **177–178**, 15–18 (2012)
10. B. Traore, C. Castelain, T. Burghelea, Efficient heat transfer in a regime of elastic turbulence. J. Nonnewton. Fluid Mech. **223**, 62–76 (2015)
11. W.M. Abed, R.D. Whalley, D.J.C. Dennis, R.J. Poole, Experimental investigation of the impact of elastic turbulence on heat transfer in a serpentine channel. J. Nonnewton. Fluid Mech. **231**, 68–78 (2016)
12. D.-Y. Li, X.-B. Li, H.-N. Zhang, F.-C. Li, S. Qian, S.W. Joo, Efficient heat transfer enhancement by elastic turbulence with polymer solution in a curved microchannel. Microfluid. Nanofluid. **21**(1) (2017)
13. B.A. Schiamberg, L.T. Shereda, H.U.A. Hu, R.G. Larson, Transitional pathway to elastic turbulence in torsional, parallel-plate flow of a polymer solution. J. Fluid Mech. **554**(-1), 191 (2006)

Renewable Energy

Performance Evaluation of Liquid Air Energy Storage with Air Purification

Chen Wang, Xiaohui She, Ailian Luo, Shifang Huang, and Xiaosong Zhang

1 Introduction

The past decade has seen a significant growth in global power generation. As of 2016, modern renewables accounted for approximately 10.4% of total final energy consumption, a mild increase compared to 2015. The overall share of renewables in the total final energy consumption has increased slightly in recent years. A primary reason for this modest rise is the continued growth in overall energy demand, which offsets the strong forward momentum of renewable energy technologies. Although this is a positive development, it is slowing the growth of the total global share of renewables [1]. For accelerating the growth of renewables to combat the climate changes, energy storage can overcome the intermittent generation from renewables and contribute to resolve the mismatch between power demand and supply by shifting the peak-load [2]. There are many types of energy storage technologies for different applications at different scales. As a promising solution for grid-scale storage, liquid air energy storage (LAES) has attracted extensive attention over the years due to several advantages including high energy density, flexible adjustment, long life, no pollution and no geographical constraints [3, 4].

The principle of LAES was first proposed by Newcastle University in 1977 with an energy recovery efficiency of 72% [5]. Subsequently, Kishimoto et al. [6] proposed the first-generation LAES system, confirming the feasibility of the system by finding how to improve its performance. The world's first pilot plant (350 kW/2.5 MWh) was designed and established by the Highview Power Storage in collaboration with University of Leeds between 2009 and 2012, making a substantial progress in LAES

C. Wang · A. Luo · S. Huang · X. Zhang
School of Energy & Environment, Southeast University, Nanjing 210018, China

X. She (✉)
School of Mechanical Engineering, Shijiazhuang Tiedao University, Shijiazhuang 050043, China
e-mail: shexh19@hotmail.com

School of Chemical Engineering, University of Birmingham, Birmingham B15 2TT, UK

© Springer Nature Singapore Pte Ltd. 2021
C. Wen and Y. Yan (eds.), *Advances in Heat Transfer and Thermal Engineering* ,
https://doi.org/10.1007/978-981-33-4765-6_131

(a)   (b)

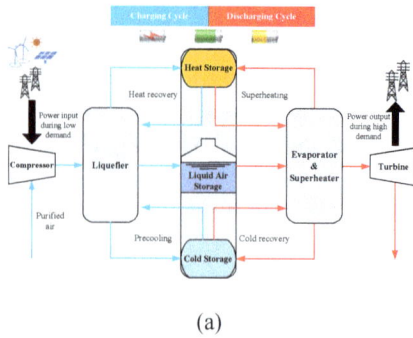

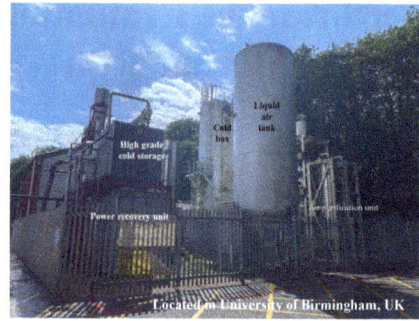

Fig. 1 The LAES: **a** basic concept; **b** the first pilot plant (350 kW/2.5 MWh)

research [7]. It was finally donated to the University of Birmingham in 2013, which was expected further technological improvements in the future [8]. As shown in Fig. 1a, the LAES plant operates in such a way that, at off-peak times, purified air is compressed and liquefied by the excess electricity through the charging cycle, with the heat of compression stored for the power enhancement during expansion in the discharging cycle; at peak times, liquid air is pumped and heated up to expand for power generation, with the cold energy of liquid air stored for recovery in the charging cycle (to be termed as baseline LAES in this paper).

There have been an increasing number of studies on the LAES particularly since 2010, including thermodynamics, process optimization, economic assessment, and integration with other systems [9–18]. Guizzi et al. [9] assessed the LAES performance through a thermodynamic analysis with the heat of compression stored during air liquefaction process and found that it can achieve a round trip efficiency of ~55%. Sciacovelli et al. [10] showed a dynamic study of a LAES plant containing the high-grade cold storage, with rated power of 100 MW and storage capacity of 300 MWh. It was suggested that dynamics of LAES should not be ignored. Xie et al. [11] analyzed the economic feasibility of adopting decoupled energy storage technologies. The model results indicated that introducing waste heat into the system or increasing system scale can improve the profitability of the LAES system. Li et al. [13] combined an LAES system with coal oxy-fuel combustion and carbon capture. The liquid nitrogen from the air separation unit was used for power generation instead of being discarded as a waste stream. Legrand et al. [14] conducted a 100 MW/300 MWh LAES plant applied to Spanish cases for electric grid load balancing, which can store photovoltaic energy in the daytime and release energy during the night time to maximize the use of storage plants. Al-Zareer et al. [15] showed a LAES system integrated with an adsorption cooling cycle based on a chemical solid–gas pair $SrCl_2 \cdot 8NH_3$ for cold production and heating, which has higher energy and exergy efficiencies than the standalone system. She et al. [16] proposed the integration of the LAES with the liquid natural gas regasification process to fully utilize the wasted high-grade cold energy of liquid natural gas, which shows an electrical round trip efficiency of $\sim70.6\%$. Kalavani et al. [17] conducted the optimal scheduling of the combination of wind power and LAES with considering the demand response program in the electricity market. The results also revealed that the total profit was increased by 33.65% and the total cost was decreased by 8.82% when integrating the LAES with an air separation unit. Yu et al. [18] combined a LAES system with the nuclear power plant for load shifting, which showed that the round trip efficiency was about 71%.

As aforementioned, the thermodynamic performance of LAES systems have been enhanced a lot during the past decade, especially the round trip efficiency which is normally over 50% at present. However, all the previous studies ignored the air purification process, leading to the overestimated performance of the LAES. As all we know, air purification unit is essential in the cryogenic process for the removal of H_2O, CO_2, hydrocarbons, and other impurities in air. It can prevent blocking the pipeline by impurities and guarantee the safe operation [19]. The industrial air purification methods basically utilize temperature swing adsorption (TSA) and pressure swing adsorption. TSA is the earliest and most technologically sophisticated air purification process, which is widely used in the world [20, 21].

This study focuses on the H_2O and CO_2 removal in the adsorber beds. The characteristics of H_2O adsorption on activated alumina can be obtained easily from the research of Shi et al. [22] and Li et al. [23], which shows activated alumina has good ability of H_2O adsorption in the air. Wang et al. [24] had investigated on post-combustion carbon dioxide capture using zeolite 13X-APG through both experiment and modeling, which revealed that the zeolite 13X-APG had excellent CO_2 adsorption, better than any other adsorbents. But it is difficult to find the characteristics of CO_2 adsorption on zeolite 13X-APG at specified state in the air purification process. The analysis of CO_2 adsorption is crucial to this work. Monte Carlo simulations of CO_2 adsorption are therefore carried out in order to verify feasibility and advantages of removing CO_2 by the zeolite 13X-APG. After conducting from molecular simulation to mathematical modeling, this paper takes into account the power consumption of air purification for regeneration of the adsorber to reassess the LAES performance.

2 Molecular Simulation of Co_2 Adsorption in the Air Purification

Molecular simulation enables the air composition range to be studied for adsorption equilibrium, which may be difficult to achieve experimentally. A series of grand canonical Monte Carlo (GCMC) simulations of the inlet air on zeolite 13X-APG was performed at diverse temperatures and pressures. Simulation box was constructed in the Materials Studio simulation package. In Monte Carlo algorithm, simulation of fluids, transitions between different states or configurations were achieved by: (1) generating a random trial configuration, (2) evaluating an acceptance criterion by calculating the change in energy and other properties, and (3) applying the acceptance

Table 1 Air composition of ambient air and inlet air on the zeolite

	N_2	O_2	Ar	H_2O	CO_2
Ambient air composition (Stdvol-fraction)	0.78	0.21	0.0092	0.0004	0.0004
Inlet air composition on the zeolite(Stdvol-fraction)	0.7804	0.21	0.0092	–	0.0004

Fig. 2 Simulated isotherms for CO_2 adsorption on zeolite 13X-APG

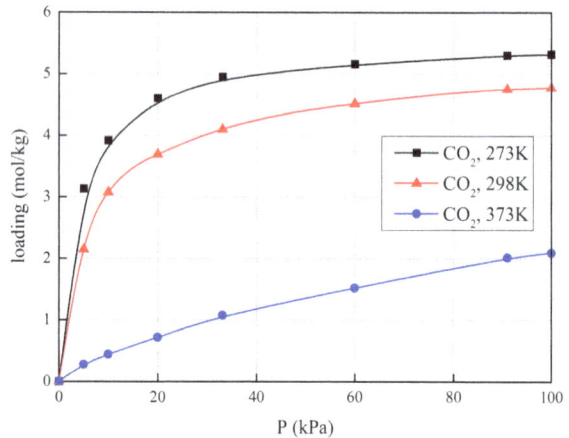

criterion for either accepting or rejecting the trial configuration [25, 26]. The influence of H_2O adsorption on zeolite 13X-APG is ignored, due to the good H_2O adsorption on activated alumina, even though zeolite 13X-APG may also have superb H_2O adsorption. The assumption of the inlet air on the zeolite is illustrated in Table 1, compared with ambient air.

Figure 2 shows that the zeolite 13X-APG increases its ability to remove CO_2, with increased CO_2 pressure from 0 to 100 kPa or decreased CO_2 temperature from 373 to 273 K. It means that when designing the adsorption step of the TSA process, a higher applied pressure and a lower applied temperature can make the purification of air better. Moreover, the required CO_2 adsorption data at the specified temperature and pressure can be obtained for establishing the mathematical model of adsorption and evaluating performance of LAES with air purification in the next sections.

3 Mathematical Modeling of the Air Purification

This paper establishes a mathematical model based on the molecular simulation, to describe intrinsic sorption mechanism and to verify the rationality of the TSA process. It is the theoretical basis of the air purification in the proposed LAES system.

The adsorption model of the adsorber beds is considered as cylindrical coordinate systems. In order to simplify the analysis, the following are assumed:

- A simplified one-dimensional model is used for the calculation, due to the axisymmetric beds.
- The inlet air follows an ideal gas law and flows along the radial direction.
- The thermal conductivity along the axial and circumference directions is ignored.
- The density and porosity of the activated alumina and zeolite 13X-APG are fixed.
- The concentration gradients along the axial and circumferential directions are neglected.

As the activated alumina has good ability of H_2O adsorption and the zeolite 13X-APG has obvious selectivity of CO_2 adsorption in the air based on the previous study, it can be assumed that the feed air is first removed H_2O totally on activated alumina at the bottom of the bed, and then removed CO_2 absolutely on the zeolite 13X-APG in the upper layer of the bed (presented in Fig. 3). Though the zeolite 13X-APG may also have good adsorption of H_2O, the adsorbed H_2O on the zeolite is ignored as almost all the H_2O has been adsorbed on the activated alumina, in order to simplify the mathematical model. For the same reason, the adsorbed CO_2 on the activated alumina is ignored.

The model includes the following equations, and the mass balance equation to a fixed bed is:

$$u\frac{\partial C_i}{\partial z} + \frac{\partial C_i}{\partial t} + \frac{1-\varepsilon}{\varepsilon}\frac{\partial q_i}{\partial t} = 0 \qquad (1)$$

Fig. 3 Sketch of adsorber bed

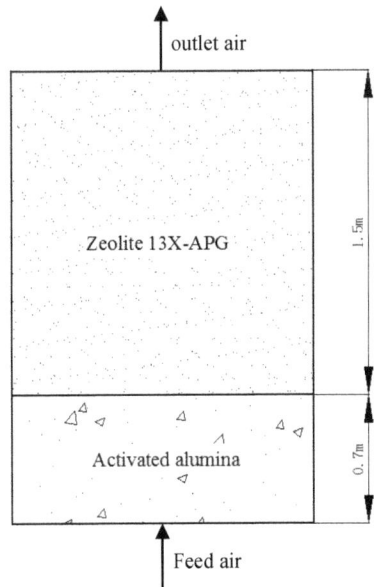

where C_i is the gas-phase concentration, u is the interstitial velocity (m/s), t is the time (s), ε is the column porosity, and q_i is the adsorbed concentration (g/g).

The LDF model is used for the predication rate as follows:

$$\rho_h \frac{\partial q_i}{\partial t} = K_v a_v \left(C_i - C_i^*\right) \tag{2}$$

where ρ_h is the packed density (kg/m^3), K_v is the mass transfer coefficient (kg/(m^2 s)), a_v is the surface area of particles per column volume (m^2/m^3) and C_i^* is the adsorption concentration in equilibrium.

The adsorption equilibrium equation is given by:

$$q_i = \frac{b q_{\text{max},i} C_i^*}{1 + b C_i^*} \tag{3}$$

where b is adsorption equilibrium constant, and $q_{\text{max},i}$ is the maximum amount adsorbed of component i.

The partial differential equations should be discretized before being solved in MATLAB R2014a using the finite volume method. Backward difference method is used to discretize the first-order partial derivative. The initial conditions and boundary conditions should be determined to solve the model. The initial feed air can be regarded as the boundary condition for the adsorption step.

Figure 4 shows the breakthrough curves of the adsorber beds for H_2O and CO_2. As time goes on, the outlet concentrations of H_2O and CO_2 from the adsorber bed increase slowly first, and then rise rapidly after 12,000 s, finally reach the same concentration as the intake. Since the requirement for the purification of H_2O and CO_2 is less than 1 ppm, the breakthrough points occur at about 14,200 s, which is consistent with the actual operation.

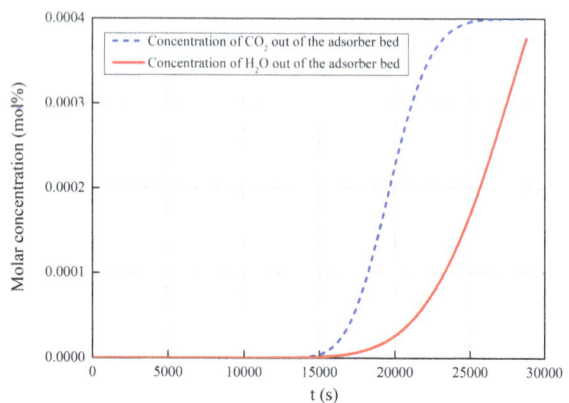

Fig. 4 Breakthrough curves of the adsorber bed for H_2O and CO_2

4 Simulation and Analysis of the Overall LAES system

After analyzing adsorption characteristics of adsorber beds, this study developed a simulation model of the LAES system with air purification. Figure 5 shows the flow diagram of the proposed LAES system, which consists of a charging cycle, a discharging cycle and a TSA process. At off-peak hours, the charging cycle is powered to produce liquid air by excess electricity; H_2O and CO_2 in the ambient air are removed in the adsorber beds. At peak hours, the discharging cycle works to generate electricity; the air exhausted from the discharging cycle goes through a heater to heat adsorber beds or a cooler to cool adsorber beds for regeneration through a TSA process.

In the charging cycle, ambient air (1) is first compressed to a pressure of 5.8 bar for H_2O and CO_2 adsorption, and then further compressed to the charging pressure of 90 bar, with the heat of compression harvested and stored in a heat storage tank using thermal oil. The compressed clean air (9) is gradually cooled down by methanol and propane in the cold storage tanks 2 and 1. Afterward, the liquid air (13) is produced through expansion in a cryo-turbine where gaseous air (14) is sent back via heat exchangers (HE#2 and HE#1) to mix with the ambient air. The liquid air is finally stored in the liquid air tank.

In the discharging cycle, the liquid air (37) is pumped to a pressure of 120 bar (38), and preheated by transferring the cold energy from air to propane and methanol. The cold energy is stored in the cold storage tanks 1 and 2 for the air liquefaction in the charging cycle. The pure air (40) is further heated up by streams of the heat of compression through heat exchanges with the thermal oil and then expands in the air turbine to generate power.

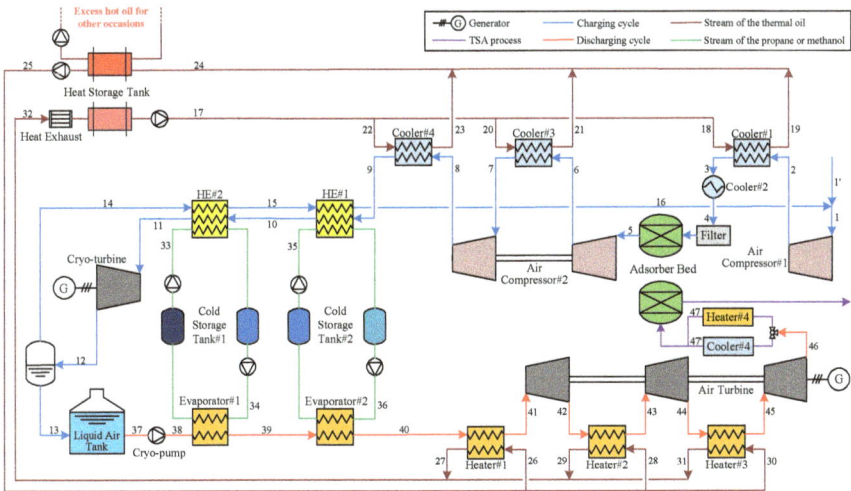

Fig. 5 Flow diagram of the LAES system with air purification

The TSA process consists of two steps: adsorption step and regeneration step. The adsorption step of the two adsorber beds (Bed#1 & Bed#2) operates at off-peak hours (assumed to be 8 h) and the regeneration step operates at peak hours (assumed to be 8 h). At off-peak hours, each bed adsorbs H_2O and CO_2 from the compressed air for 4 h. At peak hours, the air turbine generates electricity and meanwhile clean air (free of H_2O and CO_2) is exhausted. The regeneration step includes two processes: heating process and cooling process. In the heating process, clean air (46) exhausted by the discharging cycle is firstly heated up to the heating temperature (443 K) by an electric heater. Afterward, the hot air (47) flows through the adsorber for desorption. The heating process lasts for about 83 min. In the cooling process, the clean air (46) is cooled down to the cooling temperature (296 K) through heat exchange with the cooling water or ambient air, and the cool air (47′) then flows to the adsorber for cooling to near the adsorption temperature. The cooling step lasts for about 137 min.

4.1 Simulation Assumption

The model of the LAES system with air purification was developed in MATLAB software. Because of the complexity of the entire system, the model was divided into two parts, in order to improve the convergence. Some outputs of the charging cycle were as inputs of the discharging cycle, while some outputs of the discharging cycle were as inputs of the charging cycle. The following assumptions are made in the process simulation (further assumptions given in Table 2 for details).

- Dowtherm G as the thermal oil.
- Pure propane and methanol as the cold storage medium.
- H_2O and CO_2 can be completely adsorbed in the adsorption step, and adsorber beds can be regenerated completely to simplify the model for steady-state simulation.

4.2 Simulation Results

The simulation results at each point of the proposed LAES system are shown in Table 3. The calculations are based on a flowrate of 200 kg/s for inlet air (including 70.1 kg/s backflow air), with a flowrate of 0.16 kg/s H_2O and CO_2 captured in the adsorber beds.

4.3 Performance Evaluation

The round trip efficiency of the baseline LEAS system is an important indicator. It is defined as the ratio of the power output of the discharging cycle to the power input of the charging cycle:

Table 2 Simulation parameters of the proposed LAES system

Parameters	Value
Charging cycle	
Charging pressure P_c	90 bar
Initial thermal oil temperature T_{17}	293 K
Initial propane temperature T_{34}	214 K
Initial methanol temperature T_{36}	293 K
Liquid air temperature T_{13}	78.54 K
Liquid air pressure P_{13}	1 bar
Discharging cycle	
Discharging pressure P_d	120 bar
Air turbine inlet temperature P_{41}	458.5 K
Air turbine outlet pressure P_{46}	1 bar
Thermal swing adsorption process	
Adsorption temperature	298 K
Adsorption pressure	5.8 bar
Desorption (regeneration) pressure	1 bar
Heating temperature P_{47}	443 K
Cooling temperature $P_{47'}$	296 K
Overall system	
Isentropic efficiency of compressor	0.89
Isentropic efficiency of turbine	0.9
Isentropic efficiency of cryo-turbine	0.8
Pump efficiency	0.7

$$\eta_{RTE,baseline} = \frac{W_{dis,PO}}{W_{ch,PI}} \tag{4}$$

The net power input of the discharging cycle is given as:

$$W_{dis,PO} = W_{AT,PO} - W_{CP,PI} \tag{5}$$

where $W_{AT,PO}$ is the power output of air turbine (kW), and $W_{CP,PI}$ is the power input of cryo-pump (kW).

The net power input of the charging cycle is given as:

$$W_{ch,PI} = W_{AC,PI} - W_{CT,PO} \tag{6}$$

where $W_{AC,PI}$ is the power input of the compressors (kW), and $W_{CT,PO}$ is the power output of the Cryo-turbine (kW).

As the power consumption of the air purification process should be considered, the round trip efficiency of the overall system is defined as:

Table 3 Working fluid parameters in the proposed LAES system

No	Mass (kg/s)	Temperature (K)	Pressure (bar)	Medium
1	129.74	293.00	1.00	Air
2	200.00	506.79	5.80	Air
3	200.00	309.00	5.80	Air
4	200.00	298.00	5.80	Air
5	199.84	298.00	5.80	Air
6	199.84	446.19	20.85	Air
7	199.84	309.00	20.85	Air
8	199.84	490.20	90.00	Air
9	199.84	309.00	90.00	Air
10	199.84	221.38	90.00	Air
11	199.84	123.80	90.00	Air
12	199.84	78.74	1.00	Air
13	129.74	78.54	1.00	Air
14	70.10	79.24	1.00	Air
15	70.10	216.90	1.00	Air
16	70.10	293.00	1.00	Air
17	380.00	293.00	1.00	Thermal oil
18	129.20	293.00	1.00	Thermal oil
19	129.20	479.90	1.00	Thermal oil
20	125.40	293.00	1.00	Thermal oil
21	125.40	434.47	1.00	Thermal oil
22	125.40	293.00	1.00	Thermal oil
23	125.40	481.61	1.00	Thermal oil
24	380.00	465.85	1.00	Thermal oil
25	243.20	465.85	1.00	Thermal oil
26	86.82	465.85	1.00	Thermal oil
27	86.82	293.87	1.00	Thermal oil
28	79.04	465.85	1.00	Thermal oil
29	79.04	309.63	1.00	Thermal oil
30	77.34	465.85	1.00	Thermal oil
31	77.34	312.04	1.00	Thermal oil
32	243.20	304.86	1.00	Thermal oil
33	120.00	87.20	1.00	Propane
34	120.00	214.00	1.00	Propane
35	55.00	215.00	1.00	Methanol
36	55.00	293.00	1.00	Methanol

(continued)

Table 3 (continued)

No	Mass (kg/s)	Temperature (K)	Pressure (bar)	Medium
37	129.74	78.74	1.00	Air
38	129.74	86.11	120.00	Air
39	129.74	206.86	120.00	Air
40	129.74	290.42	120.00	Air
41	129.74	458.50	120.00	Air
42	129.74	304.51	24.33	Air
43	129.74	458.50	24.33	Air
44	129.74	306.77	4.93	Air
45	129.74	458.50	4.93	Air
46	129.74	307.53	1.00	Air
47	129.74	443	1.00	Air
47'	129.74	298	1.00	Air

$$\eta_{RTE,overall} = \frac{W_{dis,PO} \cdot t_{peak} - W_{ele} \cdot t_{heat}}{W_{ch,PI} \cdot t_{off\text{-}peak}} \tag{7}$$

where t_{heat}, t_{peak} and $t_{off\text{-}peak}$ are, respectively, the duration of heating process, peak hours and off-peak hours (s). W_{ele} is the power consumption of the electric heater (kW), which can be calculated by:

$$W_{ele} \approx k \cdot m_{air,dis} \cdot \int_{T_{46}}^{T_{47}} c_{air}|_T dT = k \cdot m_{air,dis} \cdot (h_{47} - h_{46}) \tag{8}$$

where k is the margin coefficient (set as 1.2 in this study), c_{air} is the heat capacity of the air (kJ/(kg K)), and $m_{air,dis}$ is the mass flowrate of the air in the discharging cycle (kg/s).

The exergy efficiency of the charging cycle is defined as:

$$\eta_{EE,ch} = \frac{m_{air,dis} \cdot e_{13} + m_{oil,ch} \cdot (e_{24} - e_{17})}{W_{ch,PI} + m_{propane} \cdot (e_{33} - e_{34}) + m_{methanol} \cdot (e_{35} - e_{36})} \tag{9}$$

where $m_{oil,ch}$, $m_{propane}$ and $m_{methanol}$ are, respectively, the mass flowrate of the thermal oil propane, and methanol (kg/s). By the way, the specific exergy can be calculated by:

$$e_i = (h_i - h_{a,i}) - T_a(s_i - s_{a,i}) \tag{10}$$

where h_i and s_i are the specific enthalpy (kJ/kg) and specific entropy (kJ/kg) at state i, respectively. T_a, $h_{a,i}$ and $s_{a,i}$ refer, respectively, to the ambient temperature (K), the specific enthalpy (kJ/kg), and specific entropy (kJ/kg) under ambient conditions.

The exergy efficiency of the discharging cycle is defined as:

$$\eta_{EE,dis} = \frac{W_{dis,PO} + m_{propane} \cdot (e_{33} - e_{34}) + m_{methanol} \cdot (e_{35} - e_{36})}{m_{air,dis} \cdot e_{13} + m_{oil,dis} \cdot (e_{24} - e_{17})} \tag{11}$$

where $m_{oil,dis}$ is the mass flowrate of the thermal oil in the discharging cycle (kg/s).

The liquid yield, which is a vital parameter for the LAES system, can be given as:

$$\text{Liquid yield} = \frac{m_{air,dis}}{m_{air,ch}} \tag{12}$$

where $m_{air,ch}$ is the mass flowrate of the air in the charging cycle (kg/s). A higher liquid yield means that the air is easier to be liquefied.

Moreover, exergy efficiency of the overall system is employed as another key performance for evaluation. The excess heat of compression can be used for other occasions, which is thought as an improvement of the thermodynamic performance.

$$\eta_{EE} = \frac{W_{dis,PO} \cdot t_{peak} + M_{excess\,oil} \cdot (e_{48} - e_{37}) - W_{ele} \cdot t_{heat}}{W_{ch,PI} \cdot t_{off-peak}} \tag{13}$$

The simulation results of the LAES system with air purification are illustrated in Table 4. It can be seen that the exergy efficiency of the charging cycle is 0.814 while that of the discharging cycle is 0.873. They are both over 80%, reflecting high-quality energy recovery. The liquid yield of this simulation is 0.649, which is significantly lower than 1. This leads to the excess heat of compression in the form of hot oil stored in the heat storage tank. Under the basic calculation, the ratio of mass flowrate of the thermal oil in the discharging cycle to that in the charging cycle is 0.64, with 36% heat of compression unused. The excess high-temperature hot oil can be utilized for

Table 4 Summary of simulation results

Parameters	Value
Compressor power consumption	109,243.864 kW
Cryo-turbine power output	4311.376 kW
Cryo-pump power consumption	2527.568 kW
Air turbine power output	59,068.701 kW
Liquid yield	0.649
Exergy efficiency of the charging cycle	0.814
Exergy efficiency of the discharging cycle	0.873
Round trip efficiency of the baseline LAES system	0.539
Round trip efficiency of the overall LAES system	0.485
Exergy efficiency of the overall system	0.571

other occasions (like driving other cycle for generation, regional heating) for more benefits to improve the exergy efficiency of the LAES system.

From Table 4, we can also see that the round trip efficiency of the baseline LAES system is 0.539, which is 5.4% higher than that of the proposed LAES system with air purification. It indicates that the electricity consumption of air purification can influence a lot on the round trip efficiency of the overall system, which should not be neglected by LAES simulations. The exergy efficiency of the overall system is 0.571 when considering that all the stored heat of compression can be utilized.

5 Conclusions

As recent research on the LAES generally neglects the energy consumption of air purification, a LAES system with air purification is proposed in this paper to reevaluate the performance. Monte Carlo simulations of CO_2 adsorption are carried out to show the zeolite 13X-APG is more selective to CO_2 in the feed air, when it is assumed that H_2O can be totally adsorbed on the activated alumina. A regeneration step at a high temperature of 443 K is a necessary and effective way to regenerate the adsorber. Based on the results of molecular simulation, the established mathematical model of the adsorber describes the air purification process well. With the above theoretical basis, the whole system is simulated for the thermodynamics analysis. The results show that the round trip efficiency of the proposed LAES system is 0.485 and the exergy efficiency is 0.571. Indicates the round trip efficiency of the LAES system is overrated by about 5.4% by the previous research.

Acknowledgements The research described in this paper is supported by the National Natural Science Foundation of China (No. 51520105009), the China National Key R&D Program (No. 2016YFC0700305) and Hundred Talents Program of Hebei Province (E2020050008).

References

1. F.S. Janet, L. Sawin Jr., *Renewables 2018 Global Status Report*. https://www.ren21.net/gsr-2018/chapters/chapter_01/chapter_01/2019. Accessed 1 Apr 2019 (2019)
2. E.M.G. Rodrigues, R. Godina, S.F. Santos, A.W. Bizuayehu, J. Contreras, J.P.S. Catalão, Energy storage systems supporting increased penetration of renewables in islanded systems. Energy **75**, 265–280 (2014). https://doi.org/10.1016/j.energy.2014.07.072
3. Y. Li, H. Chen, X. Zhang, C. Tan, Y. Ding, Renewable energy carriers: hydrogen or liquid air/nitrogen? Appl. Therm. Eng. **30**, 1985–1990 (2010). https://doi.org/10.1016/j.applthermaleng.2010.04.033
4. X. She, X. Peng, B. Nie, G. Leng, X. Zhang, L. Weng, L. Tong, L. Zheng, L. Wang, Y. Ding, Enhancement of round trip efficiency of liquid air energy storage through effective utilization of heat of compression. Appl. Energy **206**, 1632–1642 (2017). https://doi.org/10.1016/j.apenergy.2017.09.102

5. E.M. Smith, Storage of electrical energy using supercritical liquid air. Proc. Inst. Mech. En. **191**, 289–298 (2006). https://doi.org/10.1243/pime_proc_1977_191_035_02

6. K.H. Kenji Kishimoto, T. Asano, Development of generator of liquid air storage energy system. Mitsubishi JukoGiho**35**, 117–120 (1998)

7. (2013) Energy storage-revolution in the air. Mod. Power Syst. 32–33. https://viewer.zmags.com/publication/388070e3#/388070e3/32

8. R. Morgan, S. Nelmes, E. Gibson, G. Brett, Liquid air energy storage—Analysis and first results from a pilot scale demonstration plant. Appl. Energy **137**, 845–853 (2015). https://doi.org/10.1016/j.apenergy.2014.07.109

9. G.L. Guizzi, M. Manno, L.M. Tolomei, R.M. Vitali, Thermodynamic analysis of a liquid air energy storage system. Energy **93**, 1639–1647 (2015). https://doi.org/10.1016/j.energy.2015.10.030

10. A. Sciacovelli, A. Vecchi, Y. Ding, Liquid air energy storage (LAES) with packed bed cold thermal storage—From component to system level performance through dynamic modelling. Appl. Energy **190**, 84–98 (2017). https://doi.org/10.1016/j.apenergy.2016.12.118

11. C. Xie, Y. Hong, Y. Ding, Y. Li, J. Radcliffe, An economic feasibility assessment of decoupled energy storage in the UK: with liquid air energy storage as a case study. Appl. Energy **225**, 244–257 (2018). https://doi.org/10.1016/j.apenergy.2018.04.074

12. Y. Xie, X. Xue, Thermodynamic analysis on an integrated liquefied air energy storage and electricity generation system. Energies **11**, 2540 (2018). https://doi.org/10.3390/en11102540

13. Y. Li, Y. Jin, H. Chen, C. Tan, Y. Ding, An integrated system for thermal power generation. electrical energy storage and CO_2 capture. IJER**35**, 1158–1167 (2011). https://doi.org/10.1002/er.1753

14. M. Legrand, L.M. Rodríguez-Antón, C. Martinez-Arevalo, F. Gutiérrez-Martín, Integration of liquid air energy storage into the spanish power grid. Energy **187**, 115965 (2019). https://doi.org/10.1016/j.energy.2019.115965

15. M. Al-Zareer, I. Dincer, M.A. Rosen, Analysis and assessment of novel liquid air energy storage system with district heating and cooling capabilities, Energy **141**, 92–802 (2017). https://doi.org/10.1016/j.energy.2017.09.094

16. X. She, T. Zhang, L. Cong, X. Peng, C. Li, Y. Luo, Y. Ding, Flexible integration of liquid air energy storage with liquefied natural gas regasification for power generation enhancement. Appl. Energy **251**, 113355 (2019). https://doi.org/10.1016/j.apenergy.2019.113355

17. F. Kalavani, B. Mohammadi-Ivatloo, K. Zare, Optimal stochastic scheduling of cryogenic energy storage with wind power in the presence of a demand response program, Renewable Energy **130**, 268–280 (2019). https://doi.org/10.1016/j.renene.2018.06.070

18. Q. Yu, T. Zhang, X. Peng, L. Cong, L. Tong, L. Wang, X. She, X. Zhang, X. Zhang, Y. Li, H. Chen, Y. Ding, Cryogenic energy storage and its integration with nuclear power generation for load shift. Storage Hybrid. Nucl. Energy (2019). 249–273. https://doi.org/10.1016/b978-0-12-813975-2.00008-9

19. T. Hidano, M. Nakamura, A. Nakamura, M. Kawai, The downsizing of a TSA system for an air purification unit using a high flow rate method. Adsor**17**, 759–763 (2010). https://doi.org/10.1007/s10450-010-9291-5

20. L.G. Tong, P. Zhang, S.W. Yin, P.K. Zhang, C.P. Liu, N. Li, L. Wang, Waste heat recovery method for the air pre-purification system of an air separation unit, Appl. Therm. Eng. **143**, 123–129 (2018). https://doi.org/10.1016/j.applthermaleng.2018.07.072

21. P. Zhang, L. Wang, Numerical analysis on the performance of the three-bed temperature swing adsorption process for air prepurification. Ind. Eng. Chem. Res. **52**, 885–898 (2012). https://doi.org/10.1021/ie302166z

22. Y.F. Shi, X.J. Liu, Y. Guo, M.A. Kalbassi, Y.S. Liu, Desorption characteristics of H_2O and CO_2 from alumina F200 under different feed/purge pressure ratios and regeneration temperatures. Adsor**23**, 999–1011 (2017). https://doi.org/10.1007/s10450-017-9907-0

23. G. Li, P. Xiao, P. Webley, Binary adsorption equilibrium of carbon dioxide and water vapor on activated alumina. Langmuir **25**, 10666–10675 (2009). https://doi.org/10.1021/la901107s

24. L. Wang, Z. Liu, P. Li, J. Yu, A.E. Rodrigues, Experimental and modeling investigation on post-combustion carbon dioxide capture using zeolite 13X-APG by hybrid VTSA process. Chem. Eng. J. **197**, 151–161 (2012). https://doi.org/10.1016/j.cej.2012.05.017
25. A.J. Palace Carvalho, T. Ferreira, A.J. EstêvãoCandeias, J.P. Prates Ramalho, Molecular simulations of nitrogen adsorption in pure silica MCM-41 materials. J. Mol. Struct. Theochem**729**, 65–69 (2005). https://doi.org/10.1016/j.theochem.2005.01.057
26. A. Golchoobi, H. Pahlavanzadeh, Molecular simulation, experiments and modelling of single adsorption capacity of 4A molecular sieve for CO_2–CH_4 separation. SS&T**51**, 2318–2325 (2016). https://doi.org/10.1080/01496395.2016.1206571

Optimization of Optical Window of Solar Receiver by Genetic Algorithm Combined with Monte Carlo Ray Tracing

Xiao-Lei Li, Feng-Xian Sun, Jian Qiu, and Xin-Lin Xia

1 Introduction

Equipping with a fused silica window at the aperture of a solar receiver for concentrated solar application is a common way to reduce the re-radiation losses and create an enclosed cavity allowing high-pressure operational condition [1]. For that condition, the incoming radiation reflected away by the optical window occupies a major proportion of overall energy losses, which seriously restricts the efficiency of solar receivers. Therefore, the reduction of reflection losses is a dominating factor to improve the receiver performance in terms of the overall efficiency.

Concave window surface is one of the main ways to improve the optical efficiency for windowed receivers [1–4]. Hertel et al. [2] have carried out a comprehensive study on optical and mechanical properties of a dome window. Fernández and Miller [3] have presented a design optimization of a small particle receiver. The geometry of window is also included in the design space, and the 45° spherical-cap windows are demonstrated to be preferred over the ellipsoidal window in compromise between efficiency, mechanical behavior, manufacturing issues, and economic aspects.

These studies indicate that, to reduce reflection losses sufficiently, the window depth (the length of the concave window protruding into the receiver cavity) has to be large enough. However, increasing the window depth as much as possible is not always entirely desirable, because the deeper the window is, the closer it is to the high-temperature absorbers. Additionally, the flexibility of layout adjustment of

X.-L. Li · J. Qiu · X.-L. Xia (✉)
Key Laboratory of Aerospace Thermophysics of MIIT, School of Energy Science and Engineering, Harbin Institute of Technology, 92, West Dazhi Street, Harbin 150001, People's Republic of China
e-mail: xiaxl@hit.edu.cn

F.-X. Sun (✉)
School of Power and Energy Engineering, Harbin Engineering University, 145, Nantong Street, Harbin 150001, People's Republic of China
e-mail: fengxiansun@hrbeu.edu.cn

© Springer Nature Singapore Pte Ltd. 2021
C. Wen and Y. Yan (eds.), *Advances in Heat Transfer and Thermal Engineering* ,
https://doi.org/10.1007/978-981-33-4765-6_132

absorber will be largely limited if the window takes up too much space in receiver cavity. Therefore, it is necessary to explore the optimal shape of window with a given depth.

Since the transmittance of conventional concave window with quadric surface increases little, window depth is not so large [4]. Thus, an optimization approach based on the genetic algorithm (GA) and Monte Carlo ray tracing (MCRT) is proposed to find the optimal surface shape of window whose transmittance for concentrated solar irradiation is higher than that of windows in conventional shapes. The window is treated to be semitransparent medium with a pair of parallel boundary surfaces, which are represented by non-uniform rational B-spline (NURBS) surface with varying position and weight of control points. This approach is performed on the optimization of a window with a diameter of 52 mm and a thickness of 3 mm and located on the focal plane of a dish concentrator. The results show that multiple reflection improves the transmittance of a curve-surface window. The transmittance increases with the increasing window depth. The optimization approach established in this paper is demonstrated being effective in finding the best window shape with the highest transmittance at the same depth. The optimized window keeps reflection losses low while helping to avoid design limitations of a receiver due to excessive window depth.

2 Methdology

MCRT method implemented by using TracePro$^{®}$ is adopted to calculate window transmittance. The TracePro$^{®}$ is a powerful optical analysis software, which has been widely used in the field of solar energy. Absorption, reflection, and refraction of rays can be performed efficiently and accurately. The industry standard solid modeling engine, ACIS$^{®}$, is integrated at its core and gives it advanced graphic feature. Also, TracePro$^{®}$ and ACIS$^{®}$ engines provide a large number of commands that can be invoked by macros based on Scheme programming language so that their functions can be customized by users. The process of modeling and ray tracing of NURBS window is automatically executed by using macros. The genetic algorithm is carried out by using Matlab$^{®}$ optimal toolbox. Windows$^{®}$ Dynamic Data Exchange (DDE) is used to realize the interaction between these two applications.

NURBS is widely used in commercial CAD systems due to its ability to represent a large variety of complex surfaces and curves accurately and many other nice properties including affine invariance, convex hull, and local control properties [5]. The local control property allows that global shape of curve or surface will not be influenced by adjusting a specific control point. Thus, NURBS is especially adequate for shape optimization problems.

A NURBS curve C of degree p with $n + 1$ control points can be represented as a linear combination of a group of base function $\{R_{0,p}(u), R_{1,p}(u),..., R_{n,p}(u)\}$, as shown in Eq. (1), and the coefficients vector is a sequence of control points.

$$C(u) = [R_{i,p}][P_i]^T = \sum_{i=0}^{n} R_{i,p}(u) P_i \qquad (1)$$

$\{R_{i,p}(u)\}$ is called rational base function, which is piecewise rational function constructed on the base of another group of function $\{N_{i,p}(u)\}$ with a weights vector $\{w_i\}$.

With a weights vector $\{w_i\}$, the base function $\{R_{i,p}(u)\}$ in Eq. (2) is constructed on the base of another group of function $\{N_{i,p}(u)\}$, which is known as B-spline basis function. $N_{i,p}(u)$ is referred as the ith B-spline basis function, which is a piecewise polynomial function of degree p represented by Cox-de Boor recurrence formula as Eqs. (3) and (4).

$$R_{i,p}(u) = \frac{w_i N_{i,p}(u)}{\sum_{i=0}^{n} w_i N_{i,p}(u)} \qquad (2)$$

$$N_{i,0}(u) = \begin{cases} 1, & u_i \le u < u_{i+1} \\ 0, & \text{otherwise} \end{cases} \qquad (3)$$

$$N_{i,p}(u) = \frac{u - u_i}{u_{i+p} - u_i} N_{i,p-1}(u) \\ + \frac{u_{i+p+1} - u}{u_{i+p+1} - u_{i+1}} N_{i+1,p-1}(u) \qquad (4)$$

3 Results

In optimization cases, the incoming sunlight is redirected onto the focal plane by a parabolic dish concentrator with diameter of 2.4 m, rim angle of 45°, and surface slope error of 3.5 mrad. The window is treated as a transparent medium made of fused silica with absorbance coefficient $\kappa = 1.4 \text{ m}^{-1}$ and refraction index $n = 1.42$. The optimization is performed to find optimal shape of window with different depths $d = 10, 15, 20$, and 25 mm.

Window shape is represented with a surface of revolution whose generatrix C_0 is a NURBS curve of degree three with 12 control points ($p = 3$, $n = 11$). The knot vector is $U = \{0, 0, 0, 0, 1/9, 2/9, 3/9, 4/9, 5/9, 6/9, 7/9, 8/9, 1, 1, 1, 1\}$. The objective function is transmittance of window, which is calculated by using TracePro®. Control points vector $\{P_i\}$ and weight vector $\{w_i\}$ ($i = 0, 1, ..., 11$) need to be optimized to make the reflection losses as low as possible. In order to locate the window at focal plane and ensure smooth, four control points are prefixed as P_0 (0, 0, d), P_1 (1, 0, d), P_{10} (21, 0, 0), and P_{11} (26, 0, 0), and weights of these points are all unit.

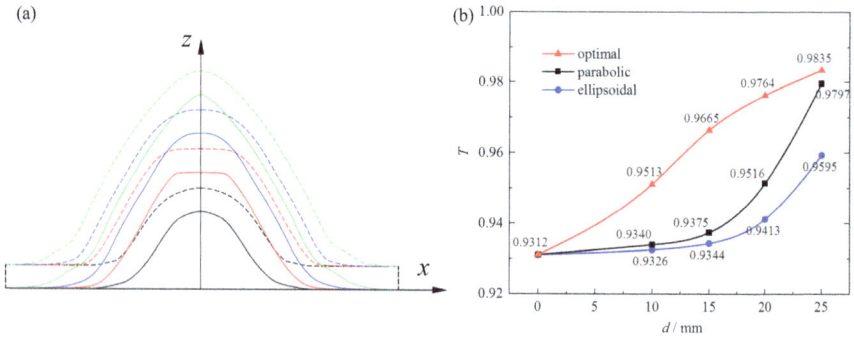

Fig. 1 **a** Shape of optimal windows and **b** transmittance as a function of depth for optimal, parabolic, and ellipsoidal windows

Transmittance of optimal window with different depths is shown in Fig. 1b, where that of parabolic and ellipsoidal window is also indicated. For comparison, the plane window is described as a special concave window with zero depth in the figure. The transmittance increases monotonically with the window depth due to the enhanced multiple reflection. Transmittance of the optimal NURBS window is higher than that of the other two types. The transmittance of optimal window with depth of 25 mm can be up to 0.9835, while that for ellipsoidal window is merely 0.9595. When d is equal to 15 mm, the optimal window has an efficiency improvement of more than 3% comparing to the ellipsoidal window, and the advantage of optimal window over conventional ones is the most obvious at this point.

As can be seen in Fig. 1b, higher transmittance can be gained by extending window depth. The optimal surface shape of window reduces the required depth to achieve the same transmittance. For instance, the transmittance of the 25-mm-depth ellipsoidal window is roughly equivalent to that of the 12-mm-depth optimal window, and cavity space occupied by window can be saved by 50% if replacing the ellipsoidal window with optimal window. Figure 1a shows the profiles of optimal windows schematically. The corresponding solutions to optimization of windows with different depths are indicated in Table 1, including location of control points and weight vectors.

Concave window changes the mean incidence angle of incoming solar rays and provides the reflected rays with chances to be redirected onto the window surface again. The latter is the main factor in increasing transmission. Although the incidence angle of NURBS windows is higher than that of conventional windows, the optimal windows reflect less solar flux. Thus, the high transmittance of the optimized window comes from efficiently redirecting the reflected light back to itself. This is difficult for traditionally shaped windows when the window depth is small. Taking windows of 10 mm depth as an example, the ellipsoid and parabolic windows re-project 2 and 9% of the reflected energy onto the window. For NURBS windows, this ratio is as high as 54%.

Table 1 Solutions to the transmittance optimization problem

d/mm		P_2	P_3	P_4	P_5	P_6	P_7	P_8	P_9
10	x	3.2222	5.4444	7.6667	9.8889	12.1111	14.3333	16.5556	18.7778
	z	9.222	5.983	2.386	0.865	0.280	0	0	0
	w	1.573	1.535	1.470	3.425	1.479	1.945	1.057	2.039
15	z	14.948	13.438	8.423	4.745	1.262	0.292	0.1149	0
	w	3.993	1.056	2.506	3.326	1.051	1.260	1.498	2.788
20	z	19.904	16.463	12.683	9.659	4.643	1.442	0.5182	0.1714
	w	0.394	0.584	2.600	0.496	1.265	0.783	0.525	1.289
25	z	22.571	20.044	16.869	13.081	9.244	3.529	2.9189	1.462
	w	3.492	2.626	1.676	3.892	3.051	1.040	1.365	1.454

4 Conclusions

In the present study, optimization of surface shape for optical window used in solar receiver is carried out by using the approach based on GA and MCRT. The window shape is represented by NURBS surface. Optical windows with depth of 10, 15, 20, and 25 mm are optimized to improve transmission performance as much as possible.

By comparing the transmittance of the conventional and optimal window, the optimization approach established in this paper is demonstrated being effective in finding the best window shape with the highest transmittance at the same depth. The transmittance of the optimal window with depth of 15 mm has a transmittance increment of more than 3% comparing to that of ellipsoidal window with the same depth. This work may provide a new way to reduce reflection losses without problems caused by large window depth.

Acknowledgements This work was supported by the project of National Natural Science Foundation of China (No. 51536001).

References

1. A.L. Ávila-Marín, Volumetric receivers in solar thermal power plants with central receiver system technology: a review. Sol Energy **85**, 891–910 (2011)
2. J. Hertel, R. Uhlig, M. Söhn, C. Schenk, G. Helsch, H. Bornhöft, Fused silica windows for solar receiver applications, in *AIP Conference Proceedings*, vol. 1734 (2016)
3. P. Fernández, F. Miller, M. McDowell, A. Hunt, Design space exploration of a 5 MWth small particle solar receiver. Energy Procedia **49**, 344–353 (2014)
4. P. Heller, Optimization of windows for closed receivers and receiver-reactors: enhancement of optical-performance. Sol. Energy Mater. **24**(1–4), 720–724 (1991)
5. L. Piegl, W. Tiller, *The NURBS Book* (Springer, Berlin, Heidelberg, 1995)

Application of the Superposition Technique in Conduction Heat Transfer for Analysing Arrays of Shallow Boreholes in Ground Source Heat Pump Systems

Carlos Naranjo-Mendoza, Muyiwa A. Oyinlola, Andrew J. Wright, and Richard M. Greenough

1 Introduction

The use of ground source heat pump (GSHP) systems for heating and cooling applications has increased notably in recent years. GSHP heating systems extract heat using ground heat exchangers (GHE), with long lengths of pipes in horizontal systems, or deep vertical boreholes (from 40 to 300 m deep) with a higher cost of installation. On the other hand, nowadays, with the application of policies towards low energy buildings, heating loads have significantly reduced. This, plus the development of new technologies like solar-assisted GSHP and dual source heat pumps, has helped to reduce the size of the GHE, enabling the use of arrays of shallow boreholes (up to 10 m) [1]. However, studying arrays of boreholes is challenging due to the complexity of modelling the interference of multiple boreholes in a defined volume, and this complexity increases when dealing with shallow boreholes as the ground cannot be treated anymore as an undisturbed medium. In fact, the ground close to the surface is highly affected by the ambient fluctuations in the short-term (hourly or sub-hourly basis) and the long-term (seasonal variations) [2]. Accurate analytical models have not been developed for this kind of systems. Alternatively, specialised heat transfer simulation packages based on finite volumes could be employed, but the complexity and computational cost are high. In order to deal with such complexity, a superposition technique for heat transfer can be a very helpful tool. In this technique, the thermal response of a system influenced by different phenomena in different

C. Naranjo-Mendoza (✉)
Departamento de Ingeniería Mecánica, Escuela Politécnica Nacional, Ladrón de Guevara E11-253, Quito, Ecuador
e-mail: carlos.naranjo@epn.edu.ec

C. Naranjo-Mendoza · M. A. Oyinlola · A. J. Wright · R. M. Greenough
Institute of Energy and Sustainable Development, De Montfort University, Leicester LE1 9BH, UK

© Springer Nature Singapore Pte Ltd. 2021
C. Wen and Y. Yan (eds.), *Advances in Heat Transfer and Thermal Engineering* ,
https://doi.org/10.1007/978-981-33-4765-6_133

directions is obtained by superimposing the thermal response of every phenomenon in a certain point of interest [3]. This technique has been used for the study of GSHP systems previously by combining multiple analytical models [4]. However, the application of this technique for the study of shallow boreholes in which the ambient thermal impact can be superimposed has not been previously studied. In this context, this paper aims to show the application of the superposition technique for the analysis of the conduction heat transfer in arrays of shallow boreholes that are highly influenced by the ambient conditions.

2 Methodology

The system of analysis consists of an array of 16 shallow boreholes (1.5 m depth) connected in series as shown in Fig. 1. The model is able to evaluate the temperature of the soil at the reference point where a temperature sensor is placed. This sensor is located at 2.25 m below the ground surface and 1 m in the radial direction from the

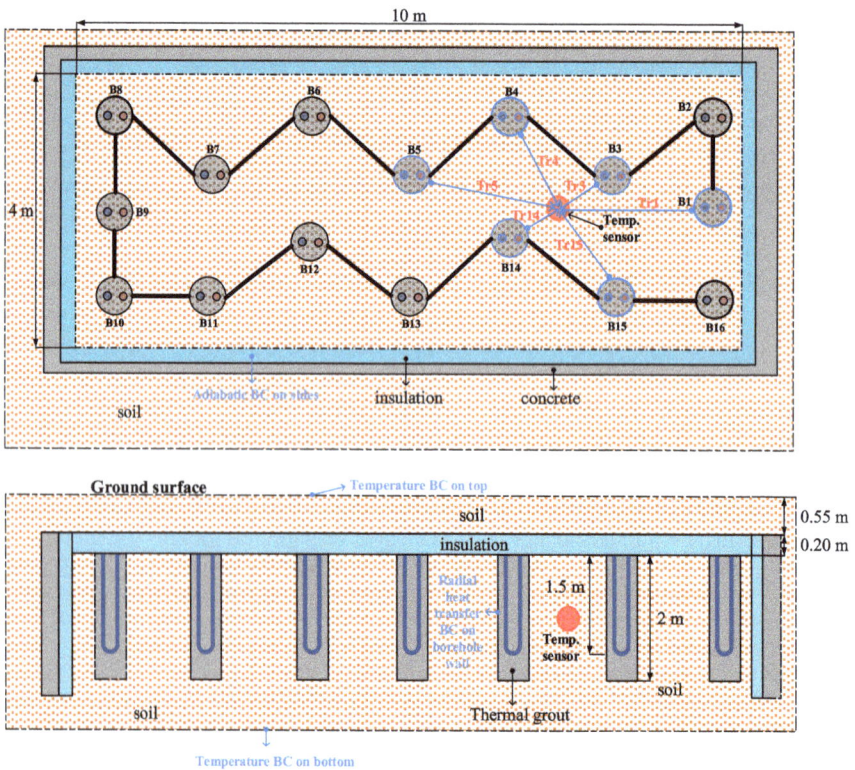

Fig. 1 Schematic of the system model and boundary conditions imposed

borehole wall of the bores B3 and B14, 2 m from the bores B4 and B15 and 3 m from the bores B1 and B5. In order to analyse the influence of the boreholes in the studied point, a numerical model using the finite difference method (FDM) in the radial direction was used. The boundary condition (BC) imposed was a time variable heat transfer rate from the borehole wall. As the boreholes are very shallow, a constant heat rate was imposed on each borehole and was calculated from monitored data of the flow rate and the inlet and outlet fluid temperature.

Two cases were analysed, Case A is the calculated temperature in the point of reference by superimposing only the heat transfer effects of the closest boreholes (B1, B3, B4, B5, B14 and B15). Hence, as seen in Fig. 1, the thermal responses (T_r) of the mentioned boreholes were included. Other boreholes were also superimposed in first calculations, but their effect was negligible, so they were removed for the final model. On the other hand, Case B is the calculated temperature in the point of reference by superimposing the heat transfer effects not only of the closest boreholes but also of the natural soil heat transfer effects by considering the top and bottom BC. The top BC is the ambient temperature while the bottom BC is the natural ground temperature at 3.75 m depth. Then, the thermal response of those BCs was determined by the FDM in the axial direction.

Equations 1 and 2 show in a simple way how the temperature of the reference point (T_{ref}) was calculated for Case A and B, respectively. In these equations, T_{r_n} represents the thermal response (temperature) at the reference point caused by the borehole (B_n). In Eq. 2, $T_{r_{nat}}$ represents the thermal response at the reference point caused by the influence of the BCs at the top and bottom of the system (relative to ambient), which represents the thermal response of the natural soil.

$$T_{ref_A} = T_{r_1} + T_{r_3} + T_{r_1} + T_{r_5} + T_{r_{14}} + T_{r_{15}} \tag{1}$$

$$T_{ref_B} = T_{r_1} + T_{r_3} + T_{r_4} + T_{r_5} + T_{r_{14}} + T_{r_{15}} + T_{r_{nat}} \tag{2}$$

3 Results

The modelling results of Case A and B were compared to experimental data from an RTD (Pt1000) located in the reference point. Temperature data were monitored from May 2016 to December 2017 on an hourly basis. In order to validate the accuracy of the models, the coefficient of determination (R^2), the root mean square error (RMSE), as well as the efficiency of the model (EF), which is the residual variance of the model predictions compared to the variance of the experimental data, were determined [5]. In this case, the closer to 1 is the value of EF the more accurate is the model. Figure 2 shows the results of the model predictions for Case A and B and the experimental data. From the results, Case A is less accurate with an R^2 of 0.942, an RMSE of 1.03 K and an EF of 0.7896, while Case B has much better accuracy

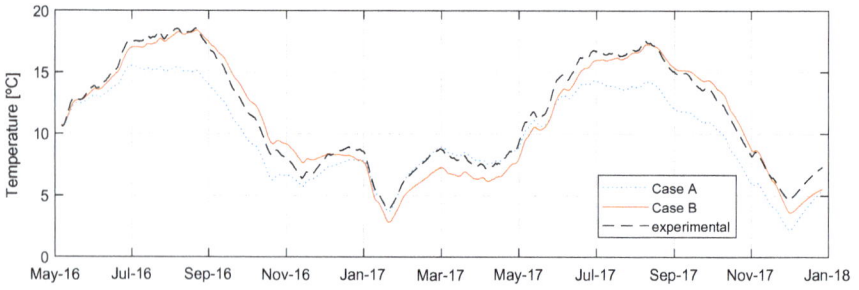

Fig. 2 Comparison of model predictions and experimental data

with an R^2 of 0.966, an RMSE of 0.79 K and an EF of 0.9563. It can be noticed, in Fig. 2, that the model in which the natural soil response is not superimposed lacks accuracy. This suggests that when modelling shallow geothermal systems, the natural soil temperature variation cannot be neglected, and the soil should not be treated as undisturbed. This is particularly important when designing shallow systems. It is worth highlighting that the currently available models like the infinite line source model (ILS) or the finite line source model (FLS) are the most common models to study boreholes. However, the direct application of these models should be avoided when dealing with shallow boreholes. Likewise, it is noted that the superposition technique in heat transfer is a simple and accurate approach that can be applied when dealing with multidimensional heat transfer problems.

4   Conclusions

This paper aimed to show the applicability of the superposition technique in heat transfer for analysing the thermal performance of arrays of shallow boreholes considering the influence of the ambient conditions. As seen in the results, the superposition technique is a simple and accurate approach that allows assessing multidimensional heat transfer systems. In the system of this study, when neglecting the influence of the ambient conditions, the efficiency of the model (EF) and the RMSE were found to be 0.7896 and 1.03 K, respectively. On the other hand, when considering the ambient conditions, the errors are reduced significantly to 0.9563 (EF) and 0.79 K (RMSE). The computational cost is low, and the accuracy is well enough to perform design and long-term study of shallow boreholes. Similarly, it was evidenced that the ambient conditions cannot be neglected when studying shallow geothermal systems as the consideration of an undisturbed ground leads to significant errors.

Acknowledgements The authors of this publication gratefully acknowledge De Montfort University, Caplin Homes and Vaillant UK for their support in this research.

References

1. M. Cimmino, P. Eslami-Nejad, A simulation model for solar assisted shallow ground heat exchangers in series arrangement. Energy Build. **157**, 227–246 (2017)
2. C. Naranjo-Mendoza, A.J. Wright, M.A. Oyinlola, R.M. Greenough, A comparison of analytical and numerical model predictions of shallow soil temperature variation with experimental measurements. Geothermics **76**, 38–49 (2018)
3. J. Taler, superposition method for multidimensional heat conduction problems, in *Encyclopedia of Thermal Stresses* (Springer, Dordrecht, Netherlands, 2014), pp. 4708–4718
4. M. Li, P. Li, V. Chan, A.C.K. Lai, Full-scale temperature response function (G-function) for heat transfer by borehole ground heat exchangers (GHEs) from sub-hour to decades. Appl. Energy **136**, 197–205 (2014)
5. M. Chalhoub, M. Bernier, Y. Coquet, M. Philippe, A simple heat and moisture transfer model to predict ground temperature for shallow ground heat exchangers. Renew. Energy **103**, 295–307 (2017)

Comparison of Direct Steam Generation and Indirect Steam Generation of Solar Rankine Cycles Under Libyan Climate Conditions

Amin Ehtiwesh, C Kutlu, Yuehong Su, and Jo Darkwa

Nomenclature

PTC	Parabolic Trough Collector
DSG	Direct Steam Generation system
HTF	Heat Transfer Fluid system
G	Solar beam irradiation W/m^2
W	Output power kW
ηth	Thermal efficiency of system
A	Aperture area m^2
m	Mass flow rate kg/s
Q_{in}	Inlet heat energy kW
Q_{out}	Outlet heat energy kW
T	Temperature °C
T_{am}	Ambient temperature °C
$1-8$	State points
h	Enthalpy kJ/kg
P	Pressure MPa
Q_{useful}	Useful heat energy of solar kW

1 Introduction

The great amount of proportion of the produced energy comes from environmentally unsustainable sources. Energy production through these sources increases the greenhouse gas emissions, which is directly responsible of environment degradation and

A. Ehtiwesh (✉) · C. Kutlu · Y. Su · J. Darkwa
Faculty of Engineering, University of Nottingham, Nottingham NG72RD, UK
e-mail: Amin.Ehtiwesh@nottingham.ac.uk

© Springer Nature Singapore Pte Ltd. 2021
C. Wen and Y. Yan (eds.), *Advances in Heat Transfer and Thermal Engineering* ,
https://doi.org/10.1007/978-981-33-4765-6_134

global warming. These environmental concerns have made pressure on governments and agencies to study for sustainable alternatives and actions that must be taken in order to change this issue. The solution basically includes political action, research, and development of sustainable energy production. All those reasons reinforce the relevance of clean and viable renewable energy sources, and among them, solar energy is one of the most promising options, due to its immense potential and lesser operation costs. Libya country is located at a high level of solar radiation. Location of Libya opens door to investment to using solar energy to produce electricity to overcome a weak electricity production with low cost. The one way to harness solar radiation is by using concentrating solar power plants (CSP), which makes useful of the thermal energy of solar radiation for electricity generation. This can be considered the promising technology to provide a considerable part of the future renewable energy demand [1]. The International Energy Agency has made two technologies roadmaps for CSP generation. The goal set in the 2010 roadmap was for CSP to reach about 11% of global electricity generation in 2050. In the revisited roadmap of 2014, that goal was maintained, which is a positive indicator toward CSP development [2]. Considering CSP and photovoltaic, solar power could generate 27% of the global electric demand by 2050. This roadmap aims to suggest actions in order to restrain the global warming limit to 2 °C in the long term [2]. The CSP have many design options. They can be used in both large- and small-scale installations, in solar only and hybrid configurations, using thermal oil or water/steam as HTF, etc. However, several questions still need to be answered and processes must be perfected to guarantee the success of this technology, especially in relation to the DSG operation because it is reported that DSG systems are less stable and complex compare to HTF. Thus, this study compares two configurations of CSP named as DSG and HTF according to their performances under Libyan weather conditions.

2 System Description

Figure 1 shows the DSG and HTF cycles. The cycle of DSG mainly consists of steam turbine, condenser, pump, and solar collectors. Pump increases the water pressure,

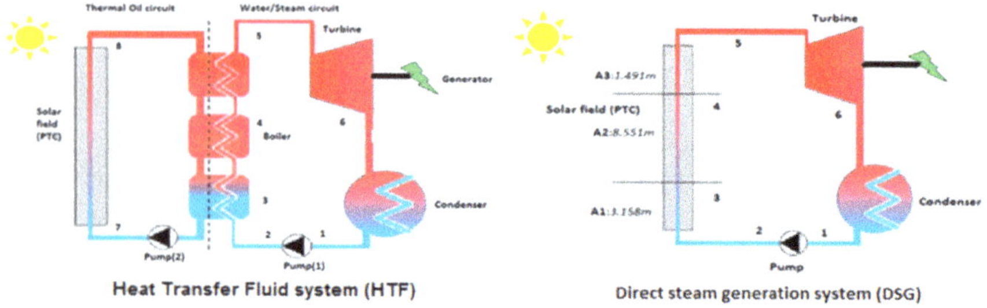

Fig. 1 Layout of direct and indirect steam generation system

Table 1 Comparison between DSG and HTF method of solar Rankine cycle

	HTF method	DSG method
Technology stage	Commercial	Demonstrative
Process stability	Stable	Less stable
Configuration	Simple	Complex
Scaling-up	Easier	With additional costs
Performance enhancement	Limited	Promising
Operating temperatures	Limited	Promising
Efficiency	Medium, limited	Higher, promising
Thermal storage	Less expensive, commercial	Expensive, demonstrative stage
Fluid toxicity	Yes	No
Initial investment	High	Low

and then high pressure water enters the PTC. The steam is generated in the PTCs and the steam goes into the turbine for electricity generation. Another type of CSP called indirect steam generation or the HTF technology, it has been commercially introduced in 1984 [3]. Water and Therminol-VP1 are the most used fluids. This type is same configuration with DSG but including heat exchanger between solar field and power block as shown in Fig. 1.

Table1 presents some advantages and disadvantages of using direct and indirect steam generation systems. It is very clear to see that DSG method is promising for high performance and efficiency with low cost as well as low environmental risks compared with HTF method according to [1].

3 Methdology

SolarGlobal software (https://globalsolaratlas.info/) has been used to determine the best location or site for solar radiation. A type of parabolic trough collectors (PTC) installed in the USA with up to 13.2 m^2 of aperture area is referenced [2] and it has been applied for both systems. The thermal efficiency of this solar collector is given in Eq. (1) [2].

$$\eta_{\text{PTC}}(T) = 0.762 - 0.2125 * \frac{T - T_a}{G} - 0.001672 * \frac{(T - T_a)2}{G} \tag{1}$$

The area of solar collector can be calculated by Eq. (2)

$$A_t = \int_{T_{in}}^{T_{out}} \frac{m_l.C_P(T)}{\eta_{PTC}(T).G_b} dT \tag{2}$$

The heat useful energy of solar can be calculated by Eq. (3)

$$Q_{useful} = G.A.\eta_{PTC} \tag{3}$$

The useful energy of PTC is given by Eq. (4)

$$Q_U = m_{col}c_p(T_{col, \ out} - T_{col, \ in}) \tag{4}$$

The thermal efficiency of PTC can be written by Eq. (5)

$$\eta_{th} = \frac{Q_u}{Q_{sol}} \tag{5}$$

The work required by pump is expressed by Eq. (6)

$$W_{pump1} = m(h_2 - h_1) \tag{6}$$

The heat inlet for system is expressed by Eq. (7)

$$Q_{in, useful} = m(h_5 - h_2) \tag{7}$$

The output heat energy of condenser is given by Eq. (8)

$$Q_{out} = m(h_6 - h_1) \tag{8}$$

The work generated by turbine is defined by Eq. (9)

$$W_{output} = m(h_5 - h_6) \tag{9}$$

The net output of system is defined by Eq. (10)

$$W_{net} = W_{output} - W_{pump} \tag{10}$$

The thermal efficiency of system is written by Eq. (11)

$$\eta_{thermal, cycle} = \frac{W_{net}}{Q_{in, useful}} \tag{11}$$

Fig. 2 Output of both systems with solar irradiation for 13.2 m^2 of PTC solar collector

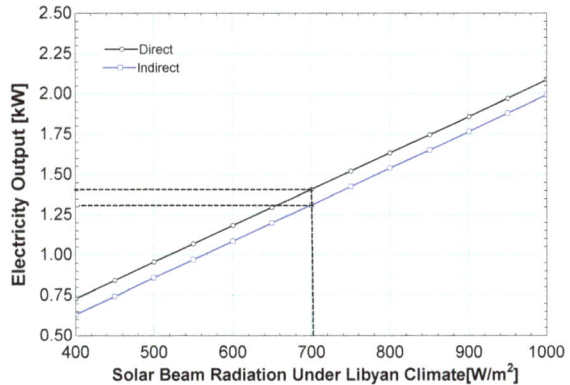

4 Results

According to SolarGlobal software, the solar irradiations for three cities in the south of Libya were given. Among them, ALKUFRAH city has highest annual solar irradiation with 2700 kWh/m^2, UBARE city has 2500 kWh/m^2, and MURSUG city has 2600 kWh/m^2. In this study, solar irradiation of ALKUFRAH city was taken as example. Some assumptions are made in the calculation as the turbine inlet temperature of 200 °C and inlet pressure of 1 MPa. These assumptions are considered for both DSG and HTF system with same total area 13.2 m^2 of PTC solar collector. The DSG system has a higher power output reaching 1.409 kW compared to the HTF system reach to 1.313 kW for the solar irradiation of 700 W/m^2. As shown in Fig. 2, the outputs for both systems are increasing with the increase of solar irradiation. The best type of method that was recommended through this study to install under Libyan climate conditions is the direct steam generation method in terms of low environmental risks and high efficiency.

5 Conclusions

A type of PTC with 13.2 m^2 of aperture area was chosen for direct steam generation and indirect steam generation of solar Rankine cycles under Libyan climate conditions of 700 W/m^2. DSG system has shown higher thermal performance with total power output 1.409 kW as compared to HTF system with total 1.313 kW. Water and thermal oil fluids are selected for both systems due to commonly used and their good thermal properties. Based on the higher efficiency of the DSG, it is better to be used under Libyan climates conditions to generate electricity. This difference will be considerably more important for large-scale productions and it is expected to reduce payback period of the system compared to other candidate HTF.

Acknowledgements This PhD study is sponsored by the Ministry of Education of Libya.

References

1. O. Behar, A. Khellaf, K. Mohammedi, S. Ait-Kaci, A review of integrated solar combined cycle system (ISCCS) with a parabolic trough technology. Renew. Sustain. Energy Rev. **39**, 223–250 (2014). https://doi.org/10.1016/j.rser.2014.07.066
2. J. Li, P.C. Li, G.T. Gao, G. Pei, Y.H. Su, J. Ji, Thermodynamic and economic investigation of a screw expander-based direct steam generation solar cascade Rankine cycle system using water as thermal storage fluid. Apply Energy **195**, 137–151 (2017). https://doi.org/10.1016/j.apenergy.2017.03.033
3. J. Li, P.C. Li, G. Pei, J.Z. Alvi, J. Ji, Analysis of a novel solar electricity generation system using cascade Rankine cycle and steam screw expander. Apply Energy **165**, 627–638 (2016). https://doi.org/10.1016/j.apenergy.2015.12.087
4. Global solar Atlas Source, https://globalsolaratlas.info/ 20 Mar 2019

Heat Transfer in Unconventional Geothermal Wells: A Double Numerical Modelling Approach

Theo Renaud, Patrick Verdin, and Gioia Falcone

1 Introduction

Geothermal energy aims at producing electricity or heat from underground resources. Worldwide geothermal energy extraction and use is still limited, despite its estimated high potential. To date, the efficiency and viability of enhanced geothermal system (EGS) and deep unconventional geothermal resources (e.g. superheated/supercritical systems) via conventional heat recovery techniques have led to limited success due to technology issues. Research on superheated/supercritical geothermal systems is highly active in Europe, notably triggered by the Iceland Deep Drilling Project (IDDP) [1]. Supercritical resources could deliver more energy than conventional resources thanks to the increase of enthalpy and the sharp decrease of density around the critical point of water [2]. The first well from IDDP was drilled at a depth of 2072 m after unintentionally drilling into magma between 2092 and 2104 m. The wellhead temperature reached 450 °C, with a superheated steam at a pressure of 140 bars [3].

As geothermal fluid extraction and injection from very hot environment is still a technical challenge, unconventional well designs (e.g. closed-loop wells) can offer more reliable engineering solutions by lowering the risks and the costs compared to the conventional geothermal fields [4, 5]. Deep borehole heat exchangers (DBHE) [6] aim at circulating a fluid into a closed well without contact with the geothermal fluids. Forced convection needs to be described to assess the feasibility of the innovative closed-wellbore concept in very hot geothermal conditions.

T. Renaud (✉) · P. Verdin
Energy and Power, Cranfield University, Cranfield MK43 0AL, UK
e-mail: theo.renaud@cranfield.ac.uk

G. Falcone
School of Engineering, University of Glasgow, Glasgow G12 8QQ, UK

© Springer Nature Singapore Pte Ltd. 2021
C. Wen and Y. Yan (eds.), *Advances in Heat Transfer and Thermal Engineering* ,
https://doi.org/10.1007/978-981-33-4765-6_135

2 Methodology

This study uses a double approach to understand the heat transfer taking place in the near-wellbore region of deep geothermal systems. While heat transfer processes in the well and in the reservoir have been investigated through several prior analytical and numerical approaches [5], this study makes a combined use of the 1D coupled wellbore–reservoir simulator T2Well [7] and a computational fluid dynamics (CFD) code to calculate the fluid flow and heat transfer in a closed-well. T2Well describes a 1D transient two-phase non-isothermal flow based on a drift-flux model [8]. For a single-component system such as water, it considers the following mass (1) and energy (2) conservation equations:

$$\frac{\partial}{\partial t}\left(\sum_\theta \rho_\theta S_\theta \, X_\theta\right) = q \; - \; \frac{1}{A}\frac{\partial}{\partial z}\left(A\sum_\theta \rho_\theta S_\theta \, u_\theta\right) \tag{1}$$

$$\frac{\partial}{\partial t}(\sum_\theta \rho_\theta S_\theta (U_\theta \, + \, \frac{1}{2}u_\theta^2)) \; = \; q \; - \; \lambda\frac{\partial T}{\partial t} - \frac{1}{A}\sum_\theta$$
$$\frac{\partial}{\partial z}\left(A\rho_\theta X_\theta S_\theta u_\theta\left(\, h_\theta + \frac{u_\theta^2}{2}\right)\right) - \sum_\theta (\rho_\theta S_\theta u_\theta g) \tag{2}$$

where the subscript θ refers to the liquid and gas phases of water, and z is the vertical axis [m]. The terms ρ [kg/m³] and X are the density and the local saturation of the phase considered, respectively. The latter is the ratio between the cross section occupied by the phase and the total cross section of the well A [m²]. In addition, q [W/m².K] represents the source or sink term, U [J] is the internal energy of the phase, u [m/s] is the velocity of the phase in the wellbore, λ [W/m.K] is the area-averaged thermal conductivity of the wellbore, h [kJ/kg] is the specific enthalpy, and g [m/s²] is the gravitational acceleration.

While the 1D flow formulation is based on an area-averaged value over the cross section of the pipe, CFD can consider the radial system of the well. The CFD code ANSYS FLUENT 17.1.0 is used to solve the mass conservation, momentum and energy equations for water: the realizable k-ε turbulence model is selected, which takes into account the shear stress at the wellbore walls [9]. Due to the nature of the flow and to reduce the overall simulation time, a 2D axisymmetric model is applied. The mass conservation equation for this system can therefore be written as:

$$\frac{\partial\rho}{\partial t} + \frac{\partial(\rho v_z)}{\partial r} + \frac{\partial(\rho v_z)}{\partial r} + \frac{\rho v_r}{r} = 0 \tag{3}$$

Further details on the radial r- and axis z-momentum equations for an axisymmetric model can be found in [9]. The energy equation is:

$$\frac{\partial(\rho E)}{\partial t} + \nabla.(\vec{v}(\rho E + p)) = \nabla.(k_{\text{eff}}\nabla T - \sum_j h_j \vec{J} + (\overline{\overline{\tau_{\text{eff}}}}.\vec{v})) \qquad (4)$$

where k_{eff} is the effective thermal conductivity [W/m.K] and T is the temperature [K]. The two terms on the right-hand side of Eq. 4 describe the diffusion and viscous dissipations. E is the energy calculated with the heat capacity [J/kg.K], detailed in [9].

T2well is based on the TOUGH2 formulation for porous media [10]. It calculates the water properties from the International Formulation Committee (1967). However, this code is currently limited to a maximum temperature of 350 °C, which is lower than the temperatures expected for supercritical/superheated geothermal resources. The CFD code does not show such temperature limitations, but simulations require a longer time to run. Similar to [11], constant water properties are applied in this work, and no phase change is considered in the CFD simulations. Both T2well and FLUENT are used to model a DBHE implemented in hot geothermal conditions, similar to the IDDP-1 system, with bottom boundary conditions of 350 °C at 2 km for T2well and 650 °C at 2100 m for FLUENT.

3 Results

Figure 1 (left) represents a DBHE where water is injected downwards at a temperature of 10 °C in an annular pipe and flowing upwards in a central pipe. Figure 1 (right) shows the CFD-based velocity distribution obtained in the well for an injected mass

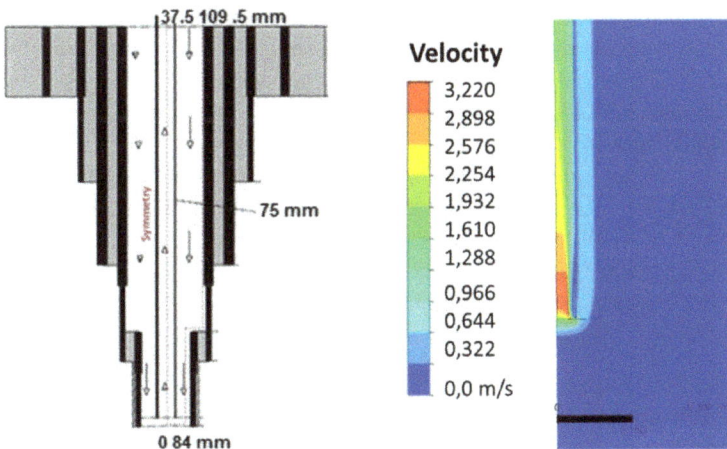

Fig. 1 (**Left**) DBHE geometry modelled for a depth of 1900 m with a 350 °C boundary condition at 2000 m (T2Well) and a depth of 2070 m with a 650 °C boundary condition at 2100 m (FLUENT). (**Right**) Velocity distribution in bottom hole region of DBHE for a mass flow rate injection of 6 kg/s

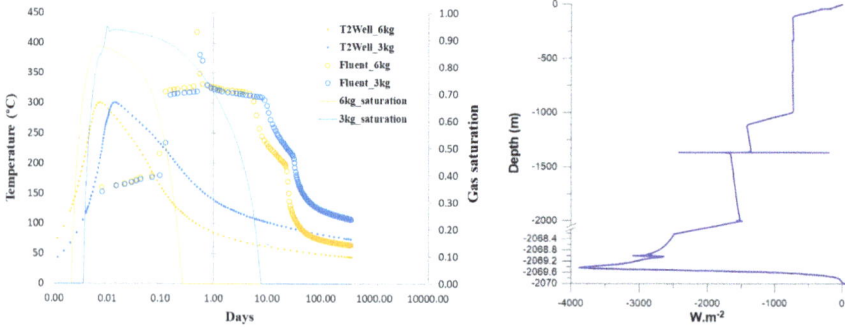

Fig. 2 (**Left**) One-year transient surface temperature of the DBHE for a 350 °C bottom temperature (T2WELL) and a 650 °C bottom temperature (FLUENT). The gas saturation at the outlet is presented on the right axis(**Right**) Horizontal heat flux on the external wellbore of the DBHE for a water injection mass flow rate of 3 kg/s

flow rate of 6 kg/s. As can be seen, water accelerates in the central pipe, from the bottom towards the surface, while no water circulation is present in the deeper region of the well system (navy blue zone).

Figure 2 (left) shows the transient temperature production at the surface for one year, with the DBHE modelled with both T2Well and FLUENT. Water mass flow rates of 3 kg/s and 6 kg/s are injected in the system. The gas saturation is presented on the right axis and is calculated with T2Well with an outlet pressure of 1 bar. Steam appears only during few days (10 days for 3 kg/s). Conventionally, DBHEs use water kept in liquid state, but their application in a very hot geothermal context can produce vapour at the outlet. The temperature recovery calculated with FLUENT is higher than using T2Well as it is not limited to 350°C. The FLUENT model describes a different behaviour than T2Well on the first days of simulation until showing a similar slow decrease regime after 100 days. Despite the assumptions made for water in the CFD calculations, the model shows the energy recovery increase, notably the high peak above 350 °C, by implementing a DBHE into superheated/supercritical geothermal context.

Figure 2 (right) highlights the horizontal heat transfer flux computed with FLUENT on the vertical external wall of the DBHE after 10 years. The heat flux reaches nearly 4 kW.m^{-2} at the bottom. The heat transfer is maximum at the bottom hole, mainly due to the temperature difference between the near-wellbore environment and the working fluid (around 100 °C).

4 Conclusions

Heat transfer and fluid flow in a closed-geothermal well in supercritical/superheated conditions have been discussed. T2Well provides a first assessment of the heat recovery from a hypothetical DBHE reaching a depth of 1900 m, when considering a constant temperature of 350 °C at a depth of 2000 m. To better model the conditions encountered in the IDDP first well, 2D axisymmetric-based CFD simulations have been performed to describe the estimated radial velocity and heat transfer (DBHE completed at a depth of 2070 m and with a temperature of 650 °C at a depth of 2100 m). These two modelling tools have provided a complementary overview of the deliverability enhancement of DBHEs in superheated/supercritical conditions. T2Well shows that vapour can be produced during the preliminary time of the system. CFD identifies a non-circulation zone at the bottom of the well, where the heat flux from the surroundings is maximum. While the surrounding geothermal medium is considered as either a single- or a two-phase flow system, better physical formulations are needed to accurately model the increase of enthalpy in supercritical/superheated settings and the effects on the heat transfer process for novel closed-loop well designs.

Acknowledgements The authors thank the UK Engineering and Physical Sciences Research Council (EPSRC) for its support [Project reference1878602].

References

1. G.O. Frideifsson, B. Palsson, A. L. Albertsson, B.Stefansson, E. Gunnlaugsson, J. Ketilsson, P. Gislason, IDDP-1 Drilled Into Magma - World's First Magma-EGS System Created, in *World Geothermal Congress 2015,* (2015)
2. M.-C. Suarez Arriaga, F. Samaniego, Deep Geothermal Reservoirs with Fluid at Supercritical Conditions, in *Twenty-seventh workshop on geothermal reservoir engineering, Stanford University* (2013)
3. B. Palsson, S. Holmgeirsson, A. Gudmundsson, H.A. Boasson, K. Ingason, H. Sverrisson, S. Thorhallsson, Drilling of the well IDDP-1. Geothermics **49**, 23–30 (2014)
4. G. Falcone, X. Lui, R.R. Okech, F. Seyidov, T. Catalin, Assessment of deep geothermal energy exploitation methods: the need for Novel single-well solutions. Energy **160**, 54–63 (2018)
5. C. Alimonti, E. Solo, D. Bocchetti, D. Berardi, The wellbore heat exchangers: A technical review. Renewable Energy **123**, 353–381 (2018)
6. T. Kohl, R. Brenni, W. Eugester, System performance of a deep borehole heat exchanger. Geothermics **31**, 687–708 (2002)
7. L. Pan, C.M. Oldenburg, T2Well-an integrated wellbore-reservoir simulator. Comput. Geosci. **65**, 46–55 (2014)
8. H. Shi, J. A. Holmes, L. J. Durlofsky, K. Aziz, L. Diaz, B. Alkaya, G. Oddie, Drift-Flux modeling of two-phase flow in Wellbores, *SPE-84228-PA* vol. 1 (2005)
9. ANSYS FLUENT Theory Guide, 17 (2016)
10. K. Pruess, C. Oldenburg, G. Moridis, TOUGH2 User's Guide Version 2.0 (1999)
11. Y. Noorollahi, M. Pourarshad, S. Jalilinasrabady, H. Yousefi, Numerical simulation of power production from abandoned oil wells in Ahwaz oil field in southern Iran. Geothermics **55**, 16–23 (2015)

An Experimental Investigation of a Thermochemical Reactor for Solar Heat Storage in Buildings

Cheng Zeng, Yang Liu, Xiaojing Han, Ming Song, Ashish Shukla, and Shuli Liu

1 Introduction

To improve the application of renewable energy such as solar energy in buildings [1], thermal energy storage techniques are being developed. Thermochemical energy storage is to store the renewable energy when it is available and supply it to users when demand is high. It involves reversible physical and chemical interactions between a thermochemical material and a working fluid such as water vapour. It features nearly zero energy loss and 2–10 times energy storage density compared to water [2]. However, the thermochemical reactor is a critical component when using the technique [3].

In the literature, experimental investigations have been conducting to understand and evaluate the performance of thermochemical reactors. However, issues such as airflow resistance, energy supply and extraction, and heat loss remain to be tackled to achieve optimal performance of thermochemical reactor in building's application. In 2018, van Alebeek et al. have reported an experimental study on a household scale thermochemical reactor [4]. A total capacity of 170 kg zeolite 13X has been separated and contained in 4 segments. The segmentation design is to enhance the operation feasibility of the reactors to allow separate charging or discharging process. According to the tests, the reactor achieves a maximum temperature lift of 24 °C and delivers maximum power at 4.4 kW. Stated by the authors, however, about 24% of the stored energy has been lost during a discharging process considering the conductive heat loss to the reactor wall and sensible heat taken up by the zeolite. In 2016, Tatsidjodoung et al. have developed and tested the thermal performance of a 40 kg zeolite 13X reactor [5]. The reactor has been divided into 2 segments for

C. Zeng (✉) · A. Shukla
School of Energy, Construction and Environment, Coventry University, Coventry CV1 2HF, UK
e-mail: zengc3@uni.coventry.ac.uk

Y. Liu · X. Han · M. Song · S. Liu
School of Mechanical Engineering, Beijing Institute of Technology, Beijing 100081, China

© Springer Nature Singapore Pte Ltd. 2021
C. Wen and Y. Yan (eds.), *Advances in Heat Transfer and Thermal Engineering* ,
https://doi.org/10.1007/978-981-33-4765-6_136

parallel and series operations. According to the authors, an average temperature lift of 38 °C has been achieved for 8 h of discharging. However, the air pressure drop through the reactor is relatively large, reaching to 230 Pa. This is due to the airflow resistance created by the packing of zeolite particles and the material supporting structure, the metallic mesh and grating sheet. In 2014, De Boer et al. have reported an experimental study of a thermochemical reactor for buildings. The authors separated 150 kg zeolite 13X into 2 segments and performed cycles of charging and discharging tests. However, due to energy loss such as air leakage and sensible heat loss, thermal storage efficiency of the reactor has been reported at 15%.

The recent studies in thermochemical reactors lead to the opportunity in improving the reactor structure to achieve better reactor performance. To tackle the issues, this paper illustrates experimental investigation of a three-phase reactor applied in buildings. The reactor design, experimental set-up, and the thermal performance of the reactor have been presented.

2 Reactor Experimental Set-Up

This section illustrates the experimental set-up of the thermochemical reactor including the involved facilities and the proposed reactor.

2.1 Illustration of the Experimental Set-Up

Figure 1 presents the schematic overview of the experimental set-up. The applied thermochemical material is zeolite 13X. In charging, an electric heater is used to

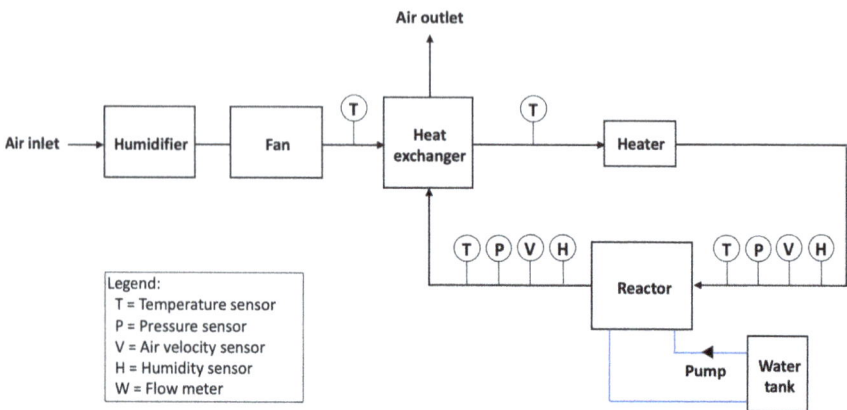

Fig. 1 Schematic overview of the experimental set-up

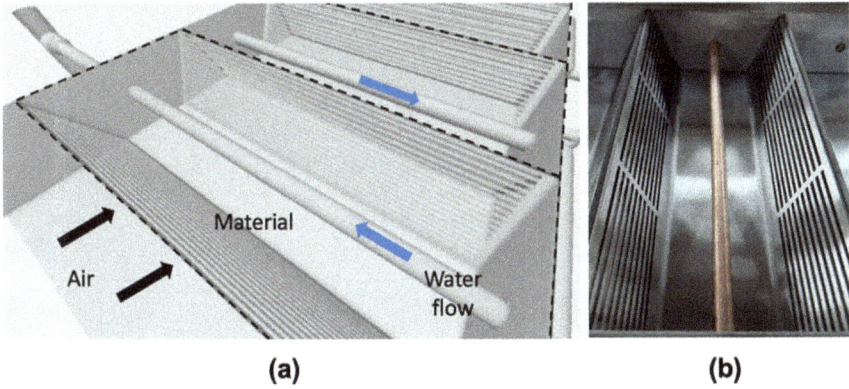

Fig. 2 Illustrations of the reactor: **a** design illustration and **b** reactor in the experiment

heat up the ambient air. In discharging, a humidifier is used to provide the moisture. Additionally, driven by a pump, water is used to extract energy from the reactor.

2.2 Reactor

Figure 2 presents the reactor container illustration (Fig. 2a) and the built reactor container (Fig. 2b). To promote airflow and reduce flow resistance, multiple side openings with the width of 4 mm have been created at the airflow entrance and exit path of the container. Additionally, a copper water flow pipe has been integrated in the reactor which is immersed in the thermochemical material. The water pipe connects the reactor, the pump, and the water tank.

There are four containers aligning in the reactor. To obtain temperature data of the reactor, each container has been installed with 8 thermocouples; inlet and outlet water temperature of a container has also been measured, as shown in Fig. 3. Additionally, the four containers are named from Container 1 to Container 4.

3 Experiment

3.1 Operating Conditions

The operation conditions applied in the charging and discharging process are presented in Table 1. Three cases have been conducted including charging with 120 °C air (Case 1), discharging with ambient air (Case 2), and discharging with heated air at 50 °C with water circulation (Case 3).

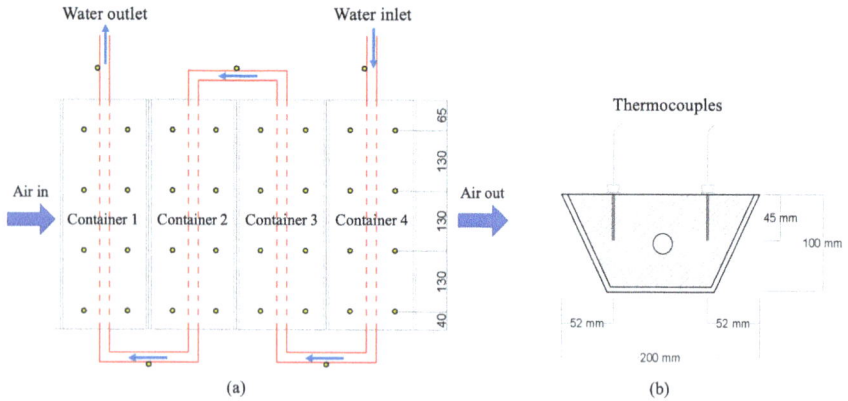

(a)   (b)

Fig. 3 Location of thermocouples in the reactor

Table 1 Operating conditions for charging and discharging processes

Parameter	Unit	Charging	Discharging	
		Case 1	Case 2	Case 3
Ambient temperature	°C	21.3	18.5	21.2
Airflow rate	kg/s	0.048	0.024	0.045
Charging inlet temperature	°C	120	–	–
Discharging inlet temperature		-	18.5	50
Water flow rate	L/min	-	–	0.5

3.2 Experiment Results

Figure 4 shows the average temperature of the zeolite for a 7-h charging test under Case 1. When the inlet air temperature increases to the set value at 120 °C, the reactor temperature increases. However, when comparing the temperature of the containers, Container 1 achieves the highest temperature and Container 4 is the lowest. The temperature difference between adjacent two containers becomes large from Container 1 to Container 4. This can be attributed to heat loss during the test including air leakage and conductive heat transfer from the zeolite to the reactor wall. According to the temperature difference, Container 1 reaches to a better charging state than that of Container 4.

When completing the charging test, the reactor cools to ambient temperature before the discharging test. Under the condition of Case 2, a stream of moist air with specific humidity at 12.27 g/kg is driven to the reactor. Figure 5 presents temperature of the containers. With the ambient temperature at around 18 °C, container temperatures increase while Container 1 reaches to the lowest figure among the containers, peaking at around 38 °C in 0.2 h. However, Container 4 achieves the highest reactor temperature at around 45 °C in 1.2 h. In terms of the discharging time, Container 1

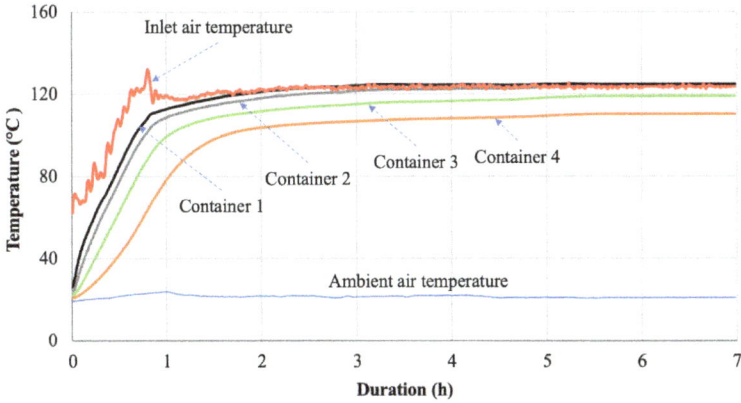

Fig. 4 Zeolite, inlet and outlet temperature during the charging process (Case 1)

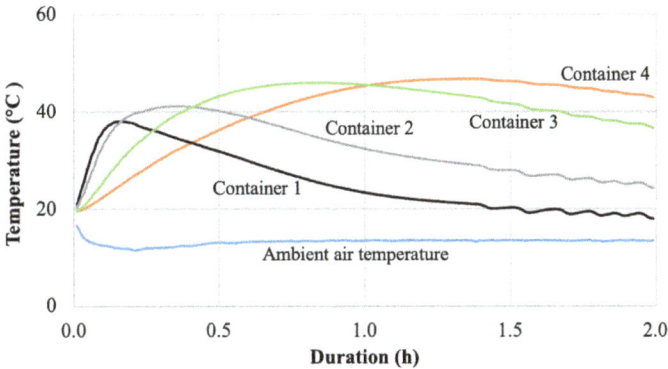

Fig. 5 Outlet and ambient temperature in the discharging process 1 (Case 2)

is the first to reach to its peak temperature, followed by Containers 2, 3, and 4. This indicates that the temperature of a following container is being lifted by the previous one. Specifically, for Containers 2 and 3, the temperature difference between the two is the largest at about 15 °C. Considering the temperature of ambient air and Container 4, the average temperature lift for the reactor is 25 °C.

When completing the discharging test in Case 2, the reactor is fully charged and cooled down to the ambient temperature. Then, a discharging test has been conducted under the condition of Case 3. With the inlet air temperature at 50 °C, the peak zeolite temperature has reached 68 °C. To obtain the water temperature, water pump is switched on at the peak temperature. Figure 6 illustrates the inlet and outlet water temperature. The test has witnessed an unstable temperature profile because water circulation has been turned off before 0.55 h and heat from the zeolite has been transferring to the water. After the transition period, the water outlet temperature profile stables at 30 °C. The value is relatively low than the level for building's

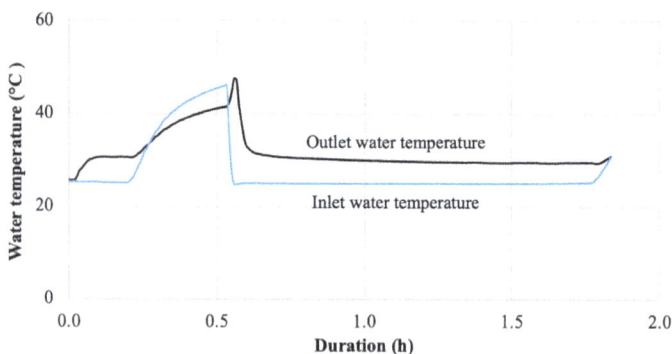

Fig. 6 Inlet and outlet water temperature during the discharging process (Case 3)

application, indicating the potential in optimising the reactor operation strategy and also enhancing the heat transfer between the zeolite and the water pipe.

4 Conclusions

This paper illustrates the experimental investigation of a three-phase thermochemical energy storage reactor. Experimental set-up, facilities, and reactor design have been demonstrated in Sect. 2. In Sect. 3, experimental test results have been presented including charging and discharging processes. According to the tests, the reactor achieves an average air temperature lift at 25 °C with the specific humidity at 12.27 g/kg and airflow rate at 0.024 kg/s. Outlet water temperature at 30 °C has been measured at water flow rate at 0.5 L/min. Overall, this paper provides meaningful insight for achieving an optimal reactor performance.

References

1. H. Garg, S. Mullick, V. Bhargava, in *Solar Thermal Energy Storage* (Springer Science & Business Media, Berlin, 2012)
2. L. Scapino, H.A. Zondag, J. Van Bael, J. Diriken, C.C.M. Rindt, Sorption heat storage for long-term low-temperature applications: A review on the advancements at material and prototype scale. Appl. Energy **190**, 920–948 (2017). https://doi.org/10.1016/j.apenergy.2016.12.148
3. B. Michel, P. Neveu, N. Mazet, Comparison of closed and open thermochemical processes, for long-term thermal energy storage applications. Energy (2014). https://doi.org/10.1016/j.energy.2014.05.097
4. R. van Alebeek, L. Scapino, M.A.J.M. Beving, M. Gaeini, C.C.M. Rindt, H.A. Zondag, Investigation of a household-scale open sorption energy storage system based on the zeolite 13X/water reacting pair. Appl. Therm. Eng. **139**, 325–333 (2018). https://doi.org/10.1016/j.applthermaleng.2018.04.092

5. P. Tatsidjodoung, N. Le Pierrès, J. Heintz, D. Lagre, L. Luo, F. Durier, Experimental and numerical investigations of a zeolite 13X/water reactor for solar heat storage in buildings. Energy Convers. Manag. **108**, 488–500 (2016). https://doi.org/10.1016/j.enconman.2015.11.011

Thermal and Electrical Performance Evaluation and Design Optimization of Hybrid PV/T Systems

Moustafa Al-Damook, Mansour Al Qubeissi, Zinedine Khatir, Darron Dixon-Hardy, and Peter J. Heggs

1 Introduction

Photovoltaic (PV) systems have witnessed exceptional development in the last two decades where it has been shown that a PV panel can absorb more than 75% of the insolation. Only a limited percentage of light waves, however, can be transformed into electricity (10–18%), with the rest wasted as heat into the cells, thus increasing the PV temperature dramatically. As a result, the cell efficiency is predicted to drop by 0.4–0.65% for each 1 °C, when the temperature is above standard conditions (25 °C and $1000\,W\,m^{-2}$) [1, 2]. Therefore, it is important to establish a mechanism to address these issues. A hybrid solar energy system known as photovoltaic/thermal (PV/T) utilizes the features of thermal and solar means to produce thermal and electrical energies simultaneously. The exploitation of the hybrid photovoltaic/thermal (PV/T) technology aims to tackle this problem by keeping the PV cell temperature in its optimal range, which in turn improves the efficiency and produces heat and electricity simultaneously.

In order to improve efficiency of PV/T air collector, several attempts have been made to examine the effects of extended surfaces on improving the overall (hybrid) efficiency of a PV/T air system. Extended surfaces can be classified into three main

M. Al-Damook (✉)
Renewable Energy Research Center, University of Anbar, Ramadi, Iraq
e-mail: mustafa.adil@uoanbar.edu.iq

M. Al-Damook · M. Al Qubeissi
Faculty of Engineering, Environment and Computing, Coventry University, Coventry CV1 2JH, UK

Z. Khatir
School of Engineering and the Built Environment, Birmingham City University, Birmingham B4 7XG, UK

D. Dixon-Hardy · P. J. Heggs
School of Chemical and Process Engineering, University of Leeds, Leeds LS2 9JT, UK

© Springer Nature Singapore Pte Ltd. 2021
C. Wen and Y. Yan (eds.), *Advances in Heat Transfer and Thermal Engineering* ,
https://doi.org/10.1007/978-981-33-4765-6_137

categories: (1) traditional fins [3–5], (2) interposition of a thin metallic sheet [6–9] (TMS), and (3) obstacles or ribs [10–13]. Little attention however has been paid to study the impact of offset strip fins within PV/T air systems using optimization strategy [14–16]. This study is aimed to evaluate the performance of a PV module, without active cooling (i.e. PV module is subject to ambient conditions) and with active cooling a single-duct single-pass hybrid PV/T air collector.

2 Numerical Investigation

Three-dimensional models of single-pass PV/T solar air designs and a PV module have been constructed using the COMSOL Multiphysics V5.3a CFD modelling software. All dimensions of the two systems are listed in Tables 1 and 2, and schematics are provided in Figs. 1 and 2.

The governing equations for the air velocity $\vec{V}(x, y, z) = u, v, w$ and temperature T are based on the conservations of mass, momentum, and energy inferred from [18, 19]. A CFD-based analysis is carried out to improve the PV system efficiency by minimising its temperature as a function of two design variables: the length of the collector (L) and the depth of the flow (δ_D). The design of experiment is constructed with 30 points in the two dimensions corresponding to their respective design variable ranges of 1.2 m $\leq L \leq$ 1.6 m and 0.01 mm $\leq \delta_D \leq$ 0.04 m. The uniform distribution of the DoE points is shown in Fig. 3.

Table 1 Dimension of PV/T air collectors in m

Collector length L (m)	1.332
Collector width w (m)	0.666
Duct depth δ_D (m)	0.025
Bottom absorber plate thickness (m)	0.001
Absorber plate thickness (m)	0.001

Table 2 Physical properties of the PV module layers BP 585 [17].[*]

Layer	t (mm)	k (W m^{-1} K^{-1})	ρ (kg m^{-3})	c (J kg^{-1} C^{-1})	ε
PV glass	3	1.8	3000	500	0.84
EVA	0.5	0.35	960	2090	–
PV cells	0.3	148	2330	677	0.7
Tedlar	0.5	0.2	1200	1250	0.87

[*]t is the thickness of PV module layers, k is the thermal conductivity of PV module layers, ρ is the density of PV module layers, c is the specific heat capacity of PV module layers, and ε is the emmisivity of PV module layers

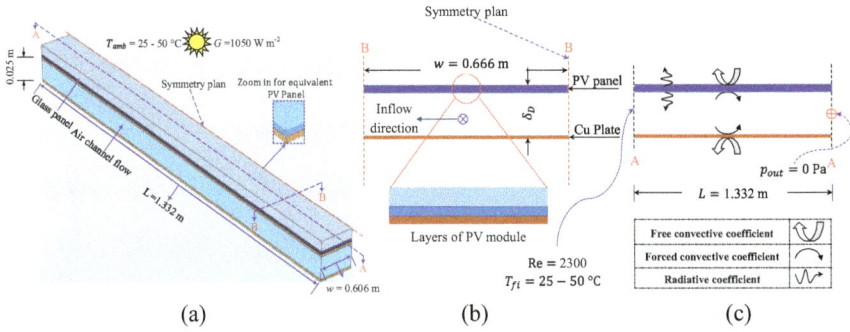

Fig. 1 Schematics of the PV/T air collector: in (**a**) 3D of single-duct single-pass (**b**) 2D cross-sectional front view and (**c**) 2D cross-sectional side view (not to scale)

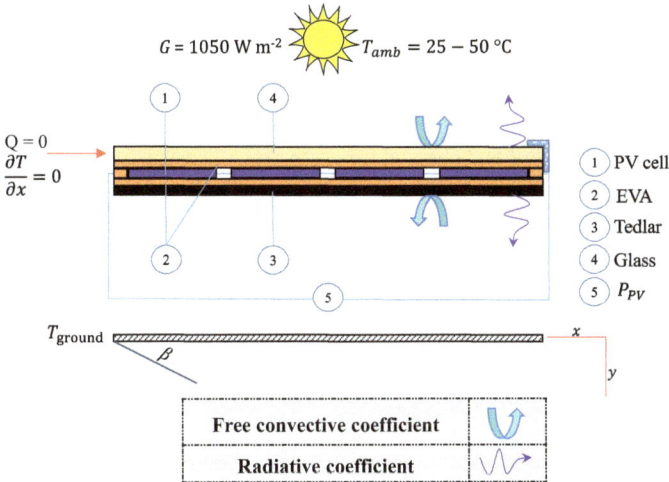

Fig. 2 Description of the main heat transfer modes and components of standard PV module

3 Results

Numerical investigations were performed using COMSOL Multiphysics V5.3a to assess the influence of the PV panel temperature on the monocrystalline PV. To ensure the accuracy of the results, two validations with an experimental set-up were conducted by Amori and Abd-AlRaheem [20] and Amori and Al-Najjar [21], and a good agreement was obtained. The first validation between current PV/T model and Amori and Abd-AlRaheem [20] is given in Table 3. The second validation between PV module (subject to ambient conditions), Amori and Abd-AlRaheem [20], and current CFD model is depicted in Fig. 4.

Figure 5 presents the effect of PV temperature on the PV output power under

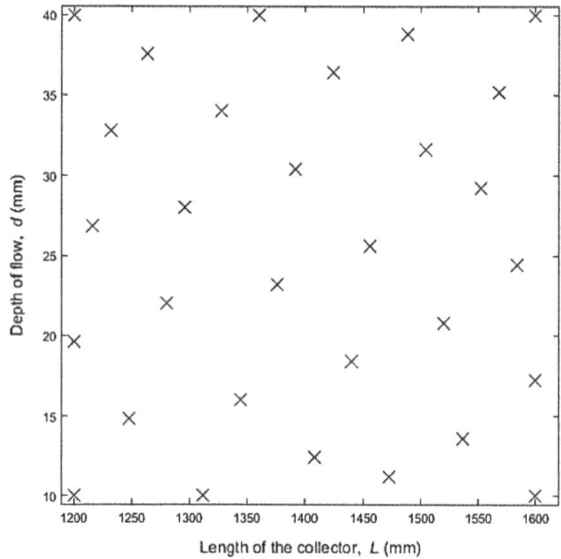

Fig. 3 Design of experiments of single-duct signal-pass PV/T air collector

Table 3 Comparison between CFD model and measurement of [20]

Type	Ref. [20]	CFD
T_g	79.49	80.7
T_{pv}	81.22	80.1
T_{Tedlar}	74.51	79.63
T_{fm}	53.64	48.62
T_{fo}	61.51	52.1

different ambient conditions for standard PV module without active cooling. Predictably, the PV power decreases with increasing PV module temperature. Figure 6 shows the influence of PV temperature on the PV output power under different ambient conditions for PV/T air system. The PV power also decreases with increasing PV module temperature. Figure 7 also reveals the influence of the PV and ambient temperatures (inlet fluid temperatures) on the thermal, electrical, and combined efficiency. It is apparent from this figure that electrical efficiency decreases with increasing PV panel temperature; however, the thermal and combined efficiency increases. Figure 8 presents the temperature gradient across the PV module. It can be seen that the PV cell has the maximum temperature since the electrical power generated in this domain. Figure 9 presents the temperature gradient along the PV/T collector. The temperature in the inlet region is low compared to the rest of the length as this region represents the non-fully developing region, which has a higher heat transfer coefficient. A comparison was made between PV and PV/T air system performances for weather conditions: $G = 1050$ W m^{-2}, T_{amb} 25–50 °C, and Re = 2300, as shown in Fig. 10. The resultant data shows that PV power generation for

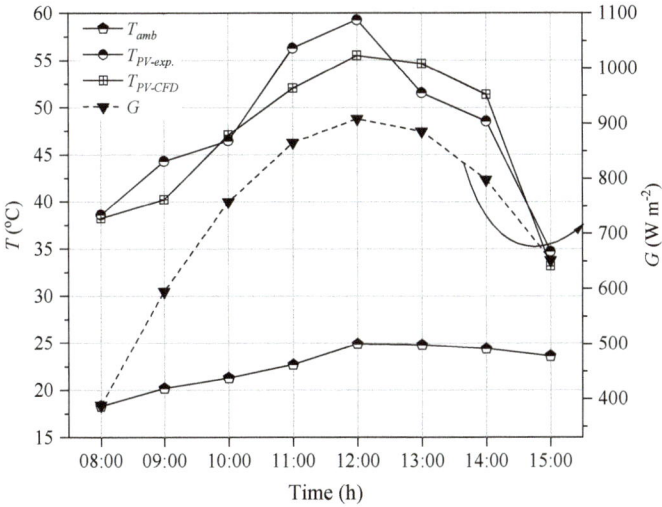

Fig. 4 Validation between experimental and current CFD models for standard PV module subject to ambient conditions

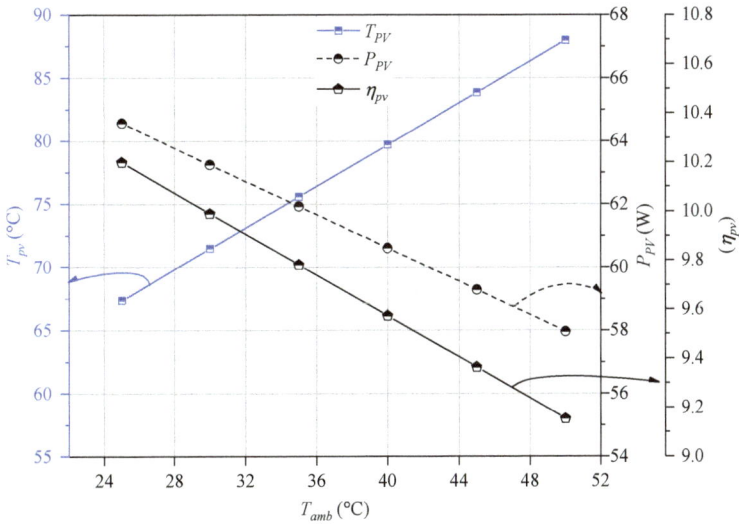

Fig. 5 Effect of different ambient temperatures under same insolation on power of standard PV module

PV/T air system is better than PV without active cooling even at higher temperature (50 °C), since the PV panel temperature is lower for PV/T air collector.

810 M. Al-Damook et al.

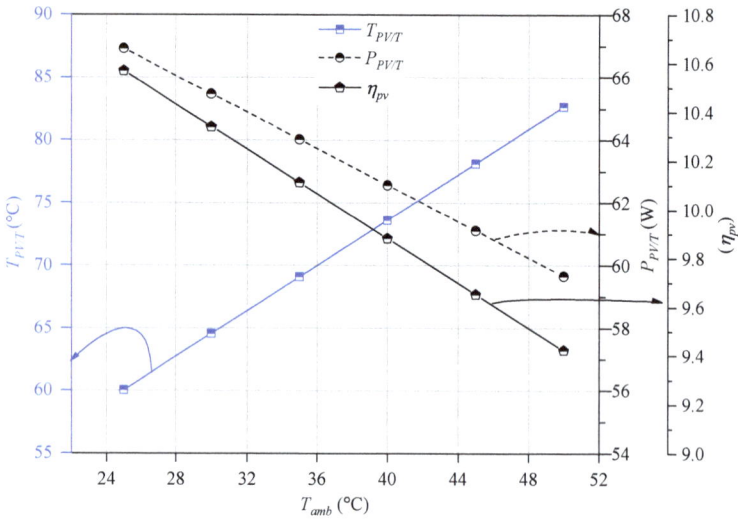

Fig. 6 Effect of different ambient temperatures under same insolation on power of PV/T air system

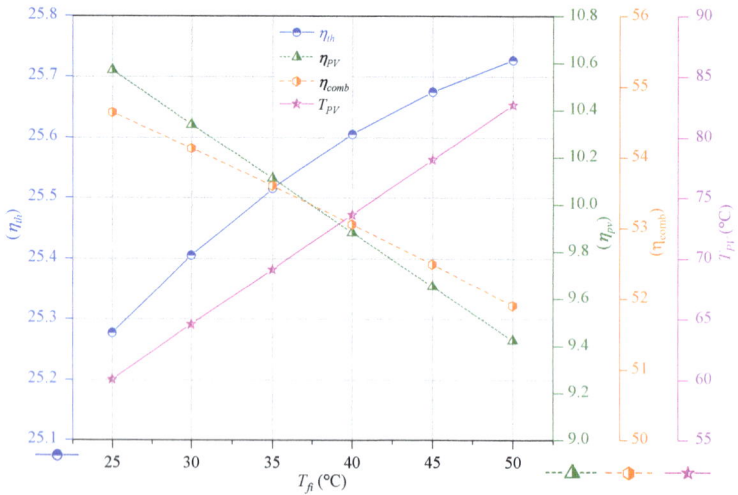

Fig. 7 Influence of PV temperature and inlet fluid temperature on thermal, electrical, and combined efficiency

4 Conclusion

The numerical investigation was conducted to evaluate the performance of two PV thermal designs. The first design was a PV module without active cooling (subjected to the ambient conditions) while the second design was a single duct single pass

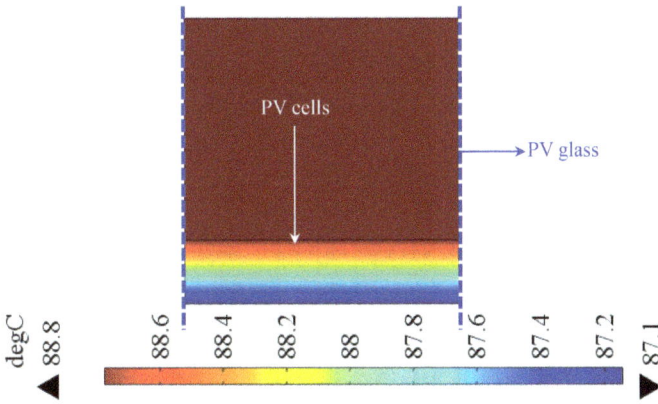

Fig. 8 Temperature surface gradient of standard PV module (section side view) at the ambient conditions of $T_{amb} = 50\,°C$, Re $= 2300$, $G = 1050$ W.m^{-2}

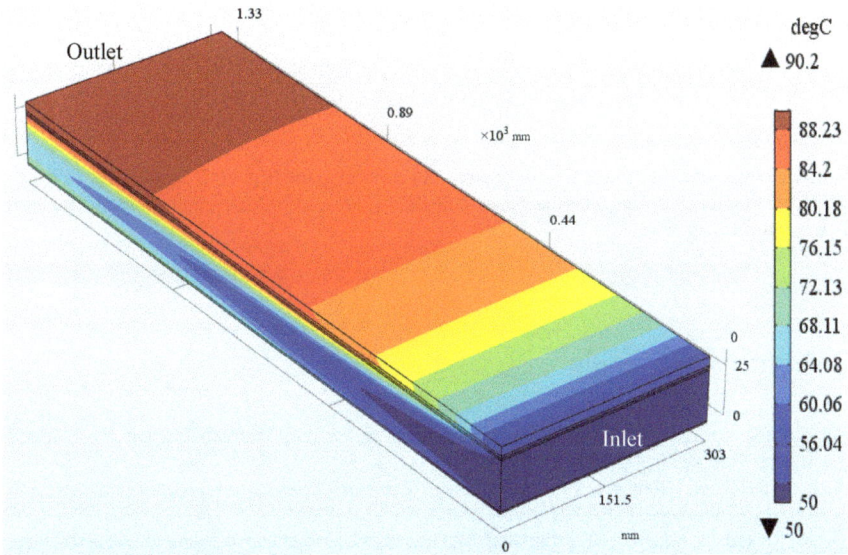

Fig. 9 Temperature surface gradient of isometric 3D view of the PV/T system, for the same ambient conditions as Fig. 8

PV/T design with active cooling. The two systems were evaluated under the same conditions (under the worst-case scenario (1050 W m^{-2} and 50 °C) and lower fan consumption (laminar flow condition Re=2300). Based on the findings, the PV/T design can provide higher performance than the PV system. The air cooling system is effective in reducing the PV temperature (increase PV efficiency). Besides, the overall collector efficiency increases.

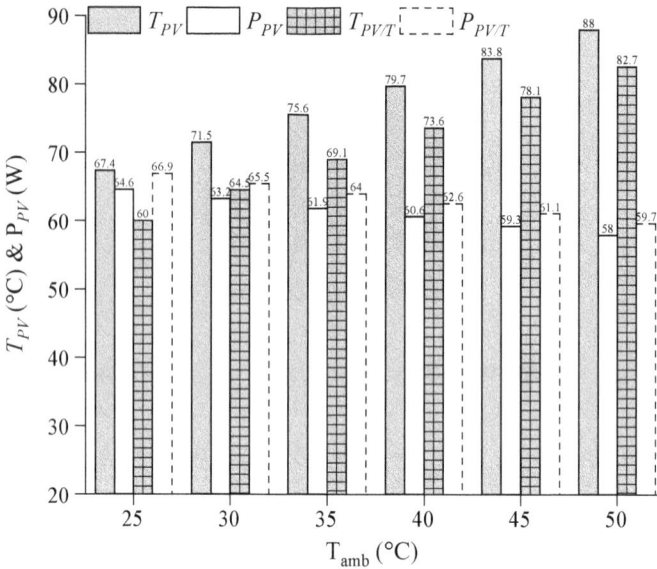

Fig. 10 Comparison between PV and PV/T air system performances for ambient temperature range $T_{amb} = 25$–$50\,°C$, and the same other conditions as those of Figs. 8 and 9

Acknowledgements The first author is extremely grateful to the Higher Committee for Education Development (HCED), Iraq, and the Renewable Energy Research Center, University of Anbar, for their support to the project.

References

1. E. Radziemska, The effect of temperature on the power drop in crystalline silicon solar cells. Renew. Energy **28**, 1–12 (2003)
2. H.R.T Haisler, Latent heat storage on photo-voltaie, in *A. Sixteen European Photovoltaic Solar Energy Conference. C. UK: Glasgow* (2000)
3. M. Masud, R. Ahamed, M. Mourshed, M. Hossan, M. Hossain, Development and performance test of a low-cost hybrid solar air heater. Int. J. Ambient Energy **40**, 40–48 (2019)
4. M. Al-Damook, D. Dixon-Hardy, P. J. Heggs, M. Al Qubeissi, K. Al-Ghaithi, P. E. Mason, et al. CFD analysis of a one-pass photovoltaic/thermal air system with and without offset strip fins, in *MATEC Web of Conferences* (2018)
5. C. Kalkan, M. A. Ezan, J. Duquette, Ş. Yilmaz Balaman, A. Yilanci, Numerical study on photovoltaic/thermal systems with extended surfaces. Int. J. Energy Res. (2019)
6. A. Shahsavar, M. Ameri, Experimental investigation and modeling of a direct-coupled PV/T air collector. Sol. Energy **84**, 1938–1958 (2010)
7. A. Shahsavar, P.T. Sardari, S. Yasseri, R.B. Mahani, Performance evaluation of a naturally ventilated photovoltaic-thermal (PV/T) solar collector: A case study. Int.. J. Energy Environ. **9**, 455–472 (2018)
8. Y. Tripanagnostopoulos, Aspects and improvements of hybrid photovoltaic/thermal solar energy systems. Sol. Energy **81**, 1117–1131 (2007)

9. J. Tonui, Y. Tripanagnostopoulos, Improved PV/T solar collectors with heat extraction by forced or natural air circulation. Renew. Energy **32**, 623–637 (2007)
10. A. Fudholi, M. Zohri, N.S.B. Rukman, N.S. Nazri, M. Mustapha, C.H. Yen et al., Exergy and sustainability index of photovoltaic thermal (PVT) air collector: A theoretical and experimental study. Renew. Sustain. Energy Rev. **100**, 44–51 (2019)
11. M. Al-Damook, Z. H. Obaid, M. Al Qubeissi, D. Dixon-Hardy, J. Cottom, P.J. Heggs, CFD modelling and performance evaluation of multi-pass solar air heaters, in *Numerical Heat Transfer; Part A: Applications,* (In-press) (2019)
12. S. Şevik, M. Abuşka, Thermal performance of flexible air duct using a new absorber construction in a solar air collector. Appl. Therm. Eng. **146**, 123–134 (2019)
13. A. Abdullah, M.A. Al-sood, Z. Omara, M. Bek, A. Kabeel, Performance evaluation of a new counter flow double pass solar air heater with turbulators. Sol. Energy **173**, 398–406 (2018)
14. W. Fan, G. Kokogiannakis, Z. Ma, A multi-objective design optimisation strategy for hybrid photovoltaic thermal collector (PVT)-solar air heater (SAH) systems with fins. Sol. Energy **163**, 315–328 (2018)
15. M. Yang, X. Yang, X. Li, Z. Wang, P. Wang, Design and optimization of a solar air heater with offset strip fin absorber plate. Appl. Energy **113**, 1349–1362 (2014)
16. I.K. Karathanassis, E. Papanicolaou, V. Belessiotis, G.C. Bergeles, Multi-objective design optimization of a micro heat sink for Concentrating Photovoltaic/Thermal (CPVT) systems using a genetic algorithm. Appl. Therm. Eng. **59**, 733–744 (2013)
17. K. Kant, A. Shukla, A. Sharma, P.H. Biwole, Thermal response of poly-crystalline silicon photovoltaic panels: Numerical simulation and experimental study. Sol. Energy **134**, 147–155 (2016)
18. J. A. Jr, Governing equations of fluid dynamics, in *Computational fluid dynamics*, ed Springer, pp. 15–51 (2009)
19. M. Al Hamdani, M. Al Qubeissi, M. Al-Damook, D. Dixon-Hardy, P. J. Heggs, Thermal Optimisation of Fin Clusters for Heat Sink Purposes, in *ICTEA: International Conference on Thermal Engineering* (2018)
20. K.E. Amori, M.A. Abd-AlRaheem, Field study of various air based photovoltaic/thermal hybrid solar collectors. Renew. Energy **63**, 402–414 (2014)
21. K.E. Amori, H.M.T. Al-Najjar, Analysis of thermal and electrical performance of a hybrid (PV/T) air based solar collector for Iraq. Appl. Energy **98**, 384–395 (2012)

Energy conversion and storage

Investigations on the Thermophysical Properties of an Organic Eutectic Phase Change Material Dispersed with GNP–AG Hybrid Nanoparticles

Neeshma Radhakrishnan and C. B. Sobhan

1 Introduction

Phase change materials or PCMs are efficient thermal storage systems due to their capacity to store and release a large amount of heat in its latent form during phase change processes. Many PCMs such as paraffin waxes, fatty acids, metals, metallic salts and salt hydrates are available for various applications. The major disadvantage of most of the low-temperature PCMs is their low thermal conductivity which affects the heat transfer rates in such systems. Many thermal conductivity enhancement techniques have been explored in the past decades, and the use of nano-additives in PCMs was found to be a promising method [1]. The common nano- additives used as thermal conductivity enhancers are metal and metal oxide-based, carbon-based and semiconductor oxide-based. While metallic nanoparticles improve the thermal conductivity of the PCM, undergoing multiple cycles of heating and freezing may cause the highly dense particles to agglomerate. Carbon-based nanoparticles are the most explored category of thermal conductivity enhancers but are very expensive. Even though researches have been conducted to explore the synergistic effects of hybrid nanoparticles on the heat transfer rates of base fluids [2], few studies focused on the effect of hybrid nanoparticles on the thermophysical properties of PCMs.

In the present work, silver nanoparticles decorated on graphene nanoplatelet aggregates are used as the thermal conductivity enhancers, and the prepared nano-enhanced PCMs (NEPCMs) are investigated for their thermophysical properties such as the latent heat capacity, specific heat, degree of supercooling and thermal conductivity. Conventionally, differential calorimetric and thermal analyses techniques are used for the analysis of latent heat characteristics of phase change materials. The thermal conductivities of these materials are measured using transient techniques such as transient hot wire and transient planar source method. The present work uses

N. Radhakrishnan (✉) · C. B. Sobhan
School of Materials Science and Engineering, National Institute of Technology, Calicut, India
e-mail: neeshma_p160050ns@nitc.ac.in

© Springer Nature Singapore Pte Ltd. 2021
C. Wen and Y. Yan (eds.), *Advances in Heat Transfer and Thermal Engineering* ,
https://doi.org/10.1007/978-981-33-4765-6_138

the T-history method [3] to analyze all the thermophysical properties of the prepared nano-enhanced PCMs.

2 Methodology

The base PCM used in the present study is an organic eutectic mixture of technical grade paraffin wax and stearic acid in the ratio 98:2. Two sets of nano-enhanced PCMs are prepared with GNP and GNP–Ag (hybrid nanoparticles or HNP) nanoparticles as additives in three loading levels (0.1, 0.5 and 1.0% by weight) using the melt blend technique. The hybrid nanoparticles are synthesized using a simple chemical synthesis route. Silver nitrate, the precursor for the formation of silver nanoparticles, is added to the GNP dispersion in DI water (50 mg/ 50 ml) and is reduced using sodium borohydride. The nanoparticles obtained after ultracentrifugation of the dispersion and oven drying are characterized using scanning electron microscopy (SEM) and X-ray diffraction (XRD).

T-history method [3] is used for the analysis of thermophysical properties of the PCMs. Test tubes filled with PCM and a reference fluid are heated well above its melting point and are suddenly exposed to ambient temperature. The cooling curves are plotted, and lumped capacitance method is used to apply energy balance in liquid, phase change and solid regions of PCM and corresponding regions of reference fluid cooling curve. The melting temperature and degree of supercooling can be directly obtained from the cooling curve. In the present work, the modifications introduced by Hong et al. [4] are included in the calculations. Specific heat and latent heat of PCM are calculated using the following equations:

$$C_{p,l} = \frac{(m_{t,w}C_{p,t} + m_w C_{p,w})}{m_{pcm}} \frac{I_1}{I_1'} - \frac{(m_{t,pcm}C_{p,t})}{m_{pcm}} \tag{1}$$

$$C_{p,s} = \frac{(m_{t,w}C_{p,t} + m_w C_{p,w})}{m_{pcm}} \frac{I_3}{I_3'} - \frac{(m_{t,pcm}C_{p,t})}{m_{pcm}} \tag{2}$$

$$H_m = \frac{(m_{t,w}C_{p,t} + m_w C_{p,w})(T_m - T_s)}{m_{pcm}} \frac{I_2}{I_2'} - \left[\frac{(m_{t,pcm}C_{p,t})}{m_{pcm}} + C_{p,pcm}\right](T_m - T_s) \tag{3}$$

The thermal conductivity is measured using the one-dimensional analysis of a test tube filled with PCM subjected to initial heating (T_o) and sudden exposure to a stirred water bath with $T_{\infty,w}$ [3]. The solid-state thermal conductivity (k_s) is calculated using the following expression:

$$k_s = \left[1 + \frac{C_p(T_m - T_{\infty,w})}{H_m}\right]/4\left(\frac{t_f(T_m - T_{\infty,w})}{\rho_p R^2 H_m}\right) \tag{4}$$

Few samples are also tested in differential scanning calorimetry (DSC) instrument and transient planar source (TPS) to verify the results obtained from the T-history method.

3 Results

SEM and XRD are used to characterize the prepared nanoparticles. The SEM image of HNP in water dispersion clearly shows the presence of silver nanoparticles (spherical particles with an average size of 50 nm) decorated on GNP aggregates. XRD analysis of both GNP aggregates and HNP shows a sharp peak at 26.6° that represents highly crystalline GNP (0 0 2). XRD pattern of HNP shows additional clear sharp peaks at 2 Theta of 38.1, 44.4, 64.5 and 77.7° which confirm the presence of FCC-structured silver nanoparticles decorated on GNP.

Cooling curve of each NEPCM is compared with that of DI water to analyze the thermophysical properties. From Fig. 1, it is observed that as the loading levels increase, the area under the phase change process in the cooling curves becomes less, which indicates an obvious reduction in the latent heat. There is a reduction of 10% in the latent heat of NEPCMs up to 0.5% loading level, irrespective of the nano-additives used. At 1% loading level of GNP and HNP, the decrease in latent heat is found to be much more significant, while there are no considerable changes observed in other phase change characteristics. Table 1 lists the values of latent heat capacity and other thermophysical properties of the base PCM and NEPCMs at different loading levels obtained via T-history method.

From Table 1, it is observed that nano-additives increase the solid-state specific heat of paraffin wax with increasing concentrations. Similar trends were reported in studies with small amounts of nanoparticles dispersed in ionic liquids and oils and were explained using concepts of nano-layer formation and enhancement in surface energy with nanoparticle addition.

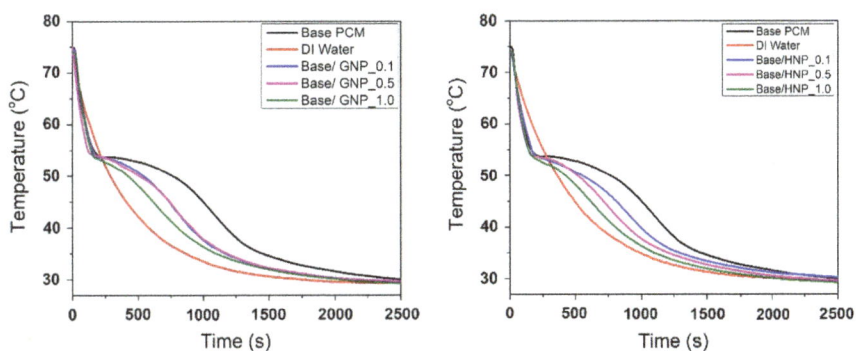

Fig. 1 Cooling curves of base PCM dispersed with 0.1, 0.5 and 1.0 wt.% of GNP and HNP

Table 1 Thermophysical parameters as obtained from T-history method

Samples	Specific Heat		Latent heat	Thermal Conductivity
	$C_{p,l}$	$C_{p,s}$	H_m	k_s
	(J/gK)		(J/g)	(W/mK)
Paraffin wax	3.3	4.2	170	0.268
Base PCM	3.1	4.2	168	0.267
Base PCM/ GNP_0.1%	3.4	5.1	156	0.295
Base PCM/ GNP_0.5%	3.0	5.3	152	0.326
Base PCM/ GNP_1.0%	2.9	5.4	137	0.348
Base PCM/ HNP_0.1%	2.9	5.8	158	0.318
Base PCM/ HNP_0.5%	2.3	5.9	149	0.361
Base PCM/ HNP_1.0%	2.3	5.6	136	0.379

The base PCM dispersed with HNP is found to substantially improve the thermal conductivity of organic PCM, with an increase in additive loading levels. While base PCM with 0.1 wt. % GNP shows an increase of ~10%, at loading levels of 0.1 and 0.5 wt. % HNP, the solid-state thermal conductivity is observed to be augmented by approximately 18% and 35%, respectively. A similar study was conducted by Yarmand et al. [5] to study the effect of hybrid nanoparticles on the effective thermal conductivity of distilled water and its heat transfer efficiency. This is in good agreement with the thermal conductivity enhancement observed in the present study. Even though 1.0 wt.% of HNP shows significant enhancement in thermal conductivity (~41%), the decrease in latent heat capacity of the PCM becomes more pronounced.

Few samples were also analyzed using DSC and TPS techniques to verify the results obtained from the T-history method. While no variation was observed in the latent heat capacity of base PCM with DSC measurement, a deviation of 10% was observed in its thermal conductivity when measured using TPS instrument. This can be attributed to the assumptions and approximations while using the T-history method. Even though minimal deviations were observed in the values of thermophysical properties of NEPCMs, similar trends were obtained for latent heat capacity and thermal conductivity for all samples analyzed via different methods.

4 Conclusions

Organic eutectic PCMs dispersed with GNP–Ag hybrid nanoparticles at low loading levels are analyzed for their enhancement in thermophysical properties such as specific heat, latent heat, other phase change characteristics and solid-state thermal conductivity using T-history method. The thermophysical characteristics of PCM dispersed with HNP are found to be substantially improved compared to base PCM and PCM dispersed with GNP aggregates. Even at low loading levels of 0.5 wt. %, the solid-state thermal conductivity is augmented by 35% with less than 10% reduction in latent heat capacity. GNP aggregates have the ability to form good percolation networks, and silver nanoparticles are very good thermal conductivity enhancers. The significant enhancement in thermal conductivity may be due to the synergistic characteristics of the hybrid nanoparticles. This enhancement is comparable to similar PCMs in literature dispersed with multi-walled carbon nanotubes and carbon nanofibers at 1 wt.% loading level. Thus, the results clearly indicate that the eutectic PCM dispersed with low concentrations of GNP–Ag hybrid nanoparticles can be regarded as an ideal system for cost-effective thermal energy storage applications.

References

1. Y. Lin, Y. Jia, G. Alva, G. Fang, Review on thermal conductivity enhancement, thermal properties and applications of phase change materials in thermal energy storage. Renew. Sustain. Energy Rev. **82**, 2730–2742 (2018)
2. J. A. Ranga Babu, K. K. Kumar, S. Srinivasa Rao, State-of-art review on hybrid nanofluids. Renew. Sustain. Energy Rev. **77**, 551–565 (2017)
3. Z. Yinping, J. Yi, A simple method, the T-history method, of determining the heat of fusion, specific heat and thermal conductivity of phase-change materials. Meas. Sci. Technol. **10**(3), 201–205 (1999)
4. H. Hong, S.K. Kim, Y.S. Kim, Accuracy improvement of T-history method for measuring heat of fusion of various materials. Int. J. Refrig. **27**(4), 360–366 (2004)
5. H. Yarmand et al., Graphene nanoplatelets-silver hybrid nanofluids for enhanced heat transfer. Energy Convers. Manag. **100**, 419–428 (2015)

Thermal Analysis of a 10 Ah Sodium Sulphur Battery (NaS) Cell

Ebikienmo E. Peters, Peter J. Heggs, and Darron W. Dixon-Hardy

1 Introduction

This research is primarily on the sodium sulphur (NaS) battery. The sodium sulphur battery (NaS) is a molten salt battery, and it is one of the main batteries used for energy storage in medium- to large-scale storage [1]. There has been a surge in popularity of NaS batteries also known as a molten metal battery in recent years owing to its high specific energy density capacity and a long-life cycle in energy storage. It is also renowned for its high efficiency and economical use of space because its high-energy density enables it to occupy around a third of the space of other kinds of battery like the lead–acid battery [2]. Baxter in 2007 and later in 2010 Sarasua assert that the NaS battery is one of the most technologically advanced in energy storage applications because of its strong operating qualities [2, 3]. The benefits of using NaS batteries are immense as the materials are inexpensive, relatively non-hazardous and easily obtainable [4].

This research is centred on a feasibility study of an electrical energy storage receiver/solar parabolic trough collector developed at the University of Leeds. The receiver is a heat collecting device with integrated thermal storage of NaS batteries [5]. The aim of the present study is to mitigate the cell limitation of electrolyte membrane fracture during phase change when heating/cooling and charging/discharging and was undertaken in collaboration with Ionotec [6].

E. E. Peters (✉) · P. J. Heggs · D. W. Dixon-Hardy
School of Process and Chemical Engineering, University of Leeds, Leeds LS2 9JT, UK
e-mail: pml1eep@leeds.ac.uk

© Springer Nature Singapore Pte Ltd. 2021   823
C. Wen and Y. Yan (eds.), *Advances in Heat Transfer and Thermal Engineering* ,
https://doi.org/10.1007/978-981-33-4765-6_139

2 Methodology

The investigation of the battery thermal properties was conducted using a 10Ah NaS cell and an Elite Thermal Systems TSV 12/55/250 furnace and data loggers. The cell undergoing test was equipped with two thermocouples spot welded at the middle section of the cell case and the top off the cell case. Furthermore on the case are two electrodes located at the middle section and top of the case, respectively. Figure 1 shows a cross-sectional diagram of the furnace and cell.

To study the thermal properties of the 10Ah NaS cell, owing to high operating temperature and the safety issues of the NaS battery cell, a heating regime as shown in Table 1 was first devised to avoid cell fracture which was a key objective of the study. The heating rate of 1 °C/min was devised and adopted, also at that rate, heat transfer

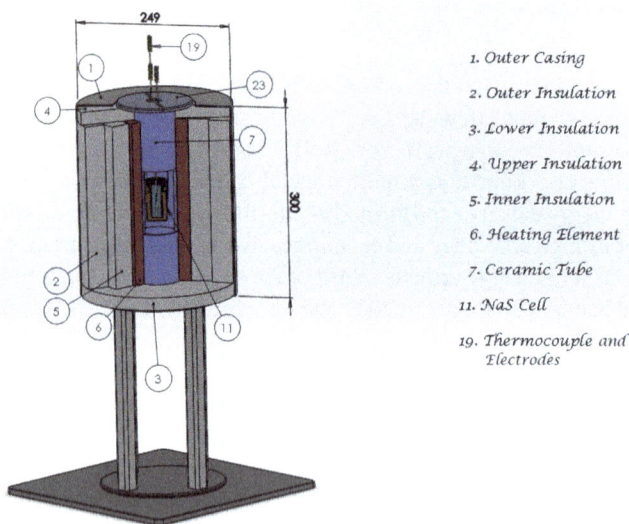

1. Outer Casing
2. Outer Insulation
3. Lower Insulation
4. Upper Insulation
5. Inner Insulation
6. Heating Element
7. Ceramic Tube
11. NaS Cell
19. Thermocouple and Electrodes

Fig. 1 Cross section of NaS and furnace

Table 1 Heating regime

Stages	Temperature parameters
Stage 1	Heat from ambient temperature to 100 °C
Stage 2	Hold at 100 °C for 1 h
Stage 3	Heat from 100 to 160 °C over a period of 1 h
Stage 4	Hold at 160 °C for 1 h
Stage 5	Heat to 330/350 °C over a period of 3 h or until battery temp gets to 350 °C
Stage 6	Hold at operating temperature and measure open-circuit cell voltage
Stage 7	Heating is stopped, and the furnace is left to cool down

by radiation from the furnace to the NaS cell will be absorbed at a slightly lower rate to avoid stress on the electrolyte during melting owing to thermal expansion.

During the warm-up, it is recommended to continuously monitor the open-circuit voltage, which should increase to about 2.076 V o/c at 350 °C if the cell is working properly. The cell should be held at its operating temperature until this voltage is attained. Holding the temperature at various stages in the experiment was to give sodium and sulphur enough time to melt and liquify and reach its operating temperature, owing to their different melting points.

3 Results

The first phase of the thermal cycling trials investigated the thermal properties of the 10Ah NaS cell NaS cell, specifically how it retains heat and loses heat over a complete cycle which consists of the warm-up stage and the cool down stage. Using a vertical tube furnace because of the configuration of the cell and heating at a rate 1 °C/min, the 10Ah NaS battery cell was gradually heated from room temperature and left to cool naturally after attaining its operating temperature of 330/350 °C. Data was collected and logged for a period of 23 h for a complete cycle.

The experimental results shown in Fig. 2 illustrate the phase change of both sodium and sulphur with temperature in the cell with phase change of sodium occurring at 98 °C with a cell voltage of 1.27 V and sulphur at 115 °C with a cell voltage of 1.92 V and attaining open cell voltage at 2.07 V at 330 °C. This result is only preliminary, but it gives a better understanding of the conduct of a NaS battery cell with temperature.

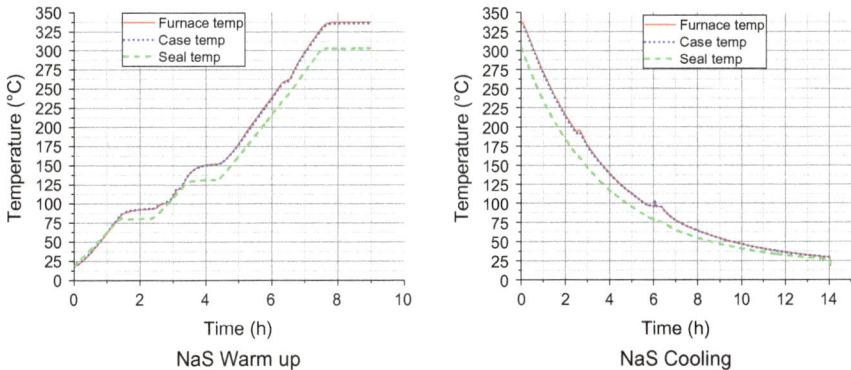

Fig. 2 NaS complete cycle profile

4   Conclusions

A heating regime of 1 °C/min was developed and adopted for the safe heating of the cell to mitigate fracturing. Furthermore, a clear temperature profile of the 10 Ah NaS cell was obtained, the results clearly indicating the phase changes of sodium and sulphur, leading to a better understating of the changes in the battery cell with temperature. The test was repeated several times, and obtaining very similar results it can be inferred that the experiment was successful owing to the fact the cell recorded an open circuit at each run without fracturing the cell membrane.

Acknowledgements Special thanks to Ionotec and my sponsors the Petroleum Development Fund, Nigeria.

Reference

1. C. Menictas, M. Skyllas-Kazacos, T. M. Lim, in *Advances in Batteries for Medium—and Large—Scale Energy Storage.* Elsevier (2015)
2. A. E. Sarasua, M. G. Molina, D. E. Pontoriero, P. E. Mercado, Modelling of NAS energy storage system for power system applications, in *2010 IEEE/PES Transmission and Distribution Conference and Exposition: Latin America (T&D-LA).* pp. 555–560 (2010)
3. R. Baxter, *Energy storage—A Nontechnical Guide*, PennWell (2007)
4. X. Lu et al., Advanced intermediate-temperature Na-S battery. Energy Environ. Sci. **6**(1), 299–306 (2013)
5. D. D. Nation, in *A Conceptual Electrical Energy Storage (EES) Receiver for Solar Parabolic Trough Collector (PTC) Power Plants.* University of Leeds (2013)
6. Ionotec Ltd, Ionotec.com, (2019) [Online]. Available: www.ionotec.com. Accessed 12 May 2019

The Effect of Air Distribution Modes and Load Operations on Boiler Combustion

Yingai Jin, Cong Tian, Yaohong Xing, Mingyu Quan, Jiwei Cheng, Yuying Yan, and Jiatong Guo

1 Introduction

This paper presents part of the UK Royal Academy of Engineering sponsored research which is concerned with reducing air pollution from coal-fired power station in Northeast of China. It is still commonly accepted that the primary structure of energy production and consumption in North Cities of China mainly based on coal combustion will remain no change in the next decade. Therefore, how to improve the thermal efficiency of boilers and reduce the emission of nitrogen oxides has always directed the research of coal production personnel and researchers. To solve above problems, this paper takes a 200 MW four-corner tangential boiler of a power plant as the research object. Based on the study of different methods of air distributing, the adjustment of different tertiary air, variable load and different air distribution of secondary air were discussed. On the basis of the former, the secondary air in DE and EE layers was replaced by tertiary air, and a three-layer separated exhaust air was added to the upper part of the compact exhaust air. Firstly, the research object in the boiler combustion study was modeled, and then cold state simulation analysis and cold state test were done. Secondly, the differences in the concentration of nitrogen oxides before and after the transformation were compared. The analysis shows that the concentration of NO at the outlet of the reformed boiler is lower than that before the reformation. Then the influence of three layers of air on the NO emission of

Y. Jin (✉) · C. Tian · Y. Xing · M. Quan · J. Cheng
State Key Laboratory of Automotive Simulation and Control, Jilin University, Changchun 130022, China
e-mail: jinya@jlu.edu.cn; Yuying.Yan@nottingham.ac.uk

College of Automotive Engineering, Jilin University, Changchun 130022, China

Y. Yan
Faculty of Engineering, University of Nottingham, Nottingham NG7 2RD, UK

J. Guo
Changchun Hecheng Xingye Energy Technology CO.,Ltd, Changchun 130021, China

© Springer Nature Singapore Pte Ltd. 2021
C. Wen and Y. Yan (eds.), *Advances in Heat Transfer and Thermal Engineering* ,
https://doi.org/10.1007/978-981-33-4765-6_140

the reformed boiler was simulated under different loads. It can be concluded that under 165, 140 and 110 MW loads, the molar concentration of NO is the lowest when no coal mill is put into operation, followed by one coal mill and the highest concentration of NO when two coal mills are put into operation.

2 Cold State Test and Cold State Simulation Analysis

After modeling and meshing the reformed boiler, in order to ensure that the pulverized coal in the furnace can fully burn and determine the opening degree of each damper, the reformed damper was tested to meet the needs of the boiler operation, so the cold state aerodynamic field test was carried out [1].

Through this cold state aerodynamic field test, it is known that the deviation of primary air volume on both sides of the dense and lean air is small, which is conducive to ensuring the ignition stability of the burner. From the experimental results, it can be concluded that the horizontal and vertical maximum tangential diameter of the boiler is 6.6 m, which is in line with the design value of the boiler. The diameter ratio of the strong wind ring to the imaginary circle is about 8.98 times, between 6 and 10 times, the strong wind ring within a reasonable range. There is no obvious wall brushing phenomenon in the adherent wind.

In the actual process, the boiler has been tested in the cold state. In this paper, the author used simulation software to simulate the formation process of tangential circle, which provides a theoretical support for the actual cold state test. The partial primary air and secondary air sections of the boiler were analyzed, respectively, in the cold state. From the experimental results, the diameter of the tangent circle is moderate, the velocity is basically symmetric, basically conforms to the actual test data, and there is no obvious adhesion and airflow collision. The phenomenon determines the feasibility of actual boiler operation.

3 Distribution of no in Central Section of Furnace Before and After Transformation

After the transformation, three layers of burnout air were added to the original furnace to reduce the air distribution in the main combustion zone, and a large amount of CO produced by incomplete combustion in the main combustion zone effectively inhibited the formation of nitrogen oxides. With the increase of furnace height, the mass fraction of nitrogen oxides increases, the CO produced by combustion is insufficient, and the amount of the reduction of nitrogen oxides is small, which leads to the increase of the NO mass fraction at the outlet. The input of tertiary air above the furnace height of 17 m brings a large amount of water, which reduces the average temperature of the furnace, reduces the ambient temperature of the formation of

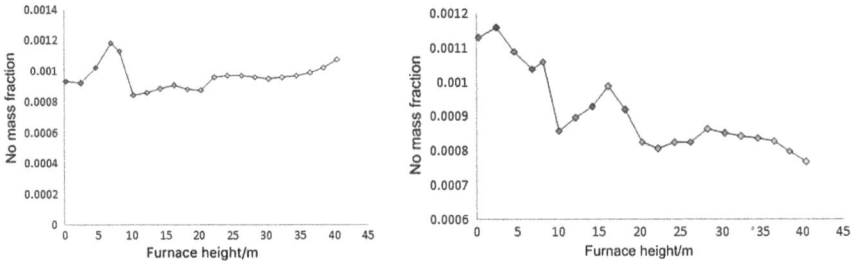

Fig.1 NO mass fraction of central furnace section before and after transformation

nitrogen oxides, and the nitrogen oxides are reduced by reducing carbon produced by the combustion of ultra-fine pulverized coal [2]. By comparing the experimental curves, it is known from Fig. 1 that the mass fraction of NO at the outlet of the reformed boiler is lower than that before the transformation, and the effect of the transformation is obvious.

4 Numerical Simulation and Experimental Analysis of Different Loads of Reformed Boilers

In the actual power plant operation process, the load of the unit is not unchanged, and load changes will have a great impact on the combustion status and emissions of the boiler [3]. Therefore, we have carried out numerical simulation and experimental research on the reformed boiler under 165, 140 and 110 MW load conditions. The influence of tertiary air on NO emission was studied.

The NO_X emission data obtained from the experiment are shown in table. Comparing the numerical simulation results of combustion with the actual measurement of nitrogen oxides, it can be concluded that the molar concentration of NO is the lowest when the no coal mill is put into operation under three loads, followed by one coal mill and the highest concentration of NO when two coal mills were put into operation. Under 100% load, the NO_X concentration at the entrance of SCR decreases, respectively, by 85 mg/m^3 and 5 mg/m^3 when no coal mill operating compared with the single and double coal mill operating, 217 mg/m^3 and 20 mg/m^3, respectively, at 75% load, 195 mg/m^3 and 155 mg/m^3, respectively, at 50% load. Through the analysis of simulation and test data, it can be concluded that the emission trend of nitrogen oxides obtained by theoretical simulation under different working conditions and different loads coincides with the actual measured data. The numerical simulation method has certain theoretical support for the actual operation of the reformed boiler (Table 1).

Table 1 NO$_X$ emission parameters of the experimental program

Parameters	110 MW			140 MW			165 MW		
Working condition	No mill	Single mill	Double mill	No mill	Single mill	Double mill	No mill	Single mill	Double mill
The value of NO$_X$ at SCR entrance/mg/m^3	295.5	373	393	299.5	309.5	408	323	325.5	365.5

5 Conclusions

This paper mainly studies the influence of adding exhaust air and changing part of secondary air to tertiary air on the formation of nitrogen oxides on the basis of the original boiler. The main research results are summarized as follows:

(1) The cold state test and the simulation analysis of the cold state primary air and secondary air of the reformed boiler were carried out. It can be known that the error between the simulated tangential circle and the experimental tangential circle is about 8% by comparing the results of test and numerical simulation analysis, which not only provides theoretical support for the actual test, but also verifies the accuracy of the theoretical model.

(2) The distribution of NO in the furnace before and after transformation was compared. The concentration of NO at the outlet of the reformed boiler is lower than that before transformation, and the effect of transformation is obvious.

(3) Finally, by comparing the results of combustion numerical simulation with the actual measurement of nitrogen oxides, it can be concluded that the concentration of NO molar fraction is the lowest when no coal mill is put into operation, followed by one coal mill and the highest when two coal mills are put into operation.

This paper mainly considers the simulation of warehouse four-corner tangential boilers with different layers of tertiary air under the condition of equal secondary air. In the future, the influence of tertiary air on four-corner tangential boilers and hedge boilers can be considered. The purpose of this paper is to understand and grasp the process and regularity of the reformed four-corner tangential pulverized coal-fired boiler, so as to provide useful reference for improving the design, operation and transformation level of the boiler.

Acknowledgements (1) This work was supported by the Royal Academy of Engineering under the UK-China Industry Academia Partnership Programme scheme (UK-CIAPP\201).

(2) This work was supported by Jilin Province Science and Technology Development Plan Project (No. 20180414021GH).

References

1. L. Tai, T.L. Cui, S.J. Zhang, Numerical simulation of the air distribution of a 1800 t/h wall-type tangential firing boiler. Boiler Technol. (2013)
2. G. Fei, Influence of boiler air distribution on NOX emissions. Chinese Foreign Entrepreneurs, 30, 102 (2015)
3. C.F.M. Coimbra, J.L.T. Azevedo, M.G. Carvalho, 3-D numerical model for predicting NOx, emissions from an industrial pulverized coal combustor. Fuel **73**(7), 1128–1134 (1994)

Transient Performance Improvement for Thermoelectric Generator Used in Automotive Waste Heat Recovery

Kuo Huang, Yuying Yan, Guohua Wang, Bo Li, and Adeel Arshad

1 Introduction

Thermoelectric generator (TEG) system as a burgeoning method for light-duty automotive waste heat recovery has drawn the researcher's attention for the last decade with the advantages of no moving parts and high reliability [1]. However, the low heat to electricity conversion efficiency is the key drawback that prevents the TEG from wildly using in practice. Furthermore, the efficiency of heat to electricity conversion for thermoelectric (TE) material is in the sharp of parabola, which means TE materials have a peak point of conversion efficiency and a best range for working temperatures. As a result, during most of the time in vehicle driving, the TEMs will not perform at their best range for working temperature and affect the efficiency of TEG system subsequently. Although there are plenty of researches focusing on the improvement of efficiency for the (TE) material, there are rare researches regarding the heat transfer enhancement on the system level, especially for the transient performance improvement for TEG.

Phase change materials (PCMs) as one of the prospective techniques for thermal energy storage (TES) have been an attractive research topic in the past two decades. PCM is wiedly used in the area of aerospace, built environment and sustainable energy technologies, because it has the advantages of high-energy storage density, isothermal

K. Huang (✉) · Y. Yan · G. Wang · B. Li · A. Arshad
Fluids and Thermal Engineering Research Group, University of Nottingham, Nottingham NG7 2RD, UK
e-mail: kuo.huang@nottingham.ac.uk

Y. Yan
Research Centre for Fluids and Thermal Engineering, University of Nottingham Ningbo China, Ningbo 315100, China

G. Wang
State Key Laboratory of Automotive Simulation and Control, Jilin University, Changchun 130025, China

© Springer Nature Singapore Pte Ltd. 2021
C. Wen and Y. Yan (eds.), *Advances in Heat Transfer and Thermal Engineering* ,
https://doi.org/10.1007/978-981-33-4765-6_141

storage process and commercially available [2]. Nevertheless, the relatively low heat transfer coefficient is the major weakness of PCM for its applications in the field of heat transfer.

In this paper, Pentaerythritol is introduced as a thermal storage material to boost the transient performance of TEG, according to the properties of bismuth telluride (Bi_2Te_3), which is commercially available in the usage of automotive waste heat recovery. The phase change point of Pentaerythritol is 183.8 °C that is suitable for the best working temperature of Bi_2Te_3 and latent heat is 244.5 J/g [3]. By taking the advantage of PCM, the fluctuating heat from the exhaust gas, under the practical driving condition, can be stored in the PCM and expel to the TEM at a suitable temperature. Furthermore, in order to increase the heat transfer coefficient of PCM, a series of comparison experimental tests are conducted with different integrated ratios of PCM.

2 Methodology

The theoretical maximum efficiency of a TEG, η_{TEG}, can be expressed as a function of ZT and the temperature difference ΔT between its hot (T_h) and cold (Tc) surfaces [4]:

$$\eta_{TEG} = \frac{\Delta T}{T_h} \cdot \frac{\sqrt{1+ZT} - 1}{\sqrt{1+ZT} + T_c/T_h} \tag{1}$$

Therefore, the efficiency of TEG system depends basically on the available temperature difference ΔT and hot surface temperature T_h. Especially when the engine is working under the real-time driving situation of vehicle, the temperature of engine exhaust gas will fluctuate frequently and the total efficient of TEG system will be affected enormously.

The thermal energy storage of PCM is based on the latent heat absorption or release when a storage material undergoes a phase change. The storage capacity of the PCM system can be given by [5]

$$Q = \int_{T_i}^{T_m} mC_{sp}dT + ma_m\Delta h_m + \int_{T_m}^{T_f} mC_p dT \tag{2}$$

According to the equations, a suitable PCM with its phase-changing point in the desired is to improve the waste heat to electricity conversion capacity of TEG system.

In order to investigate the performance of PCM integrated TEG system, a single TEM with PCM test rig was established. The schematic diagram of experiment setup and photo of the main test section are shown in Fig. 1. In the purpose of simulating the fluctuating temperature of exhaust gas, an AC voltage regulator is employed to

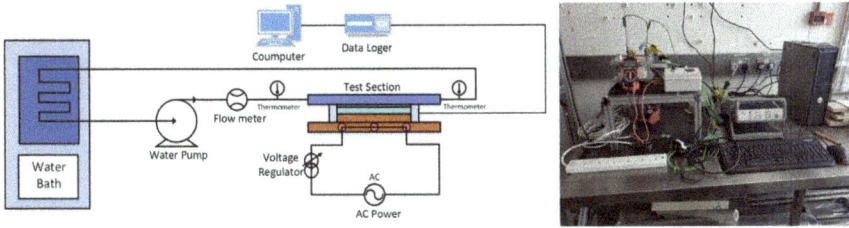

Fig. 1 Schematic diagram and photo of TEG transient performance test rig

adjust the voltage of three cartridge heaters, which will change the power supply to the TEM.

3 Results

In order to investigate heat transfer enhancement for PCM, three different integrated ratios at 65, 80 and 100% aluminum box are employed. Figure 2 shows the experimental results comparison of a single TEM with different PCM integrated ratio. T_1 and T_2 in these figures are the temperatures measured at the surface of heating plate and hot surface of TEM, respectively. The power supplied to the heating plate is in the form of square wave for simulating the fluctuating temperature of exhaust gas. The power of heater is set at 60 W, 70 W, 75 W and 80 w, respectively.

It is shown in the figures that the temperature of TEM hot surface is stabilized at the phase-changing temperature while the TEM is heating and cooling. By inserting PCM can give TEM a relatively steady hot surface temperature within the best working range, thus, it can improve the conversion efficiency of TEM. The average temperature gradients of T_1 and T_2 are reduced by 26.1, 29.1 and 38.5% with the increment of PCM integrated ratio. The less gradient in temperature, the less temperature mismatch

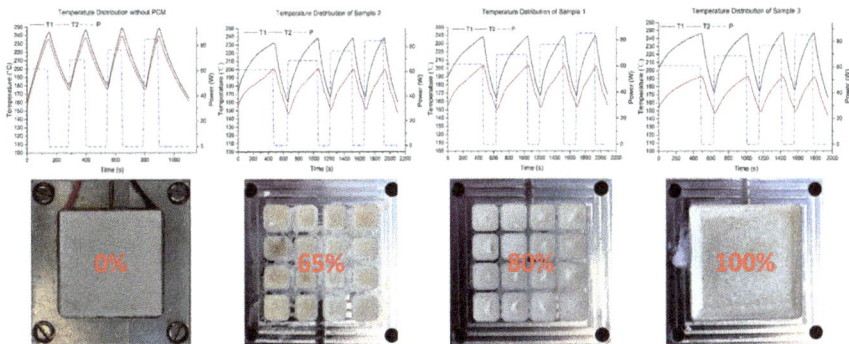

Fig. 2 Result comparison of TEM with different PCM integrated ratios

occurs in TEM, and higher efficiency will be achieved. Furthermore, with the increment of heating power, the stability of PCM becomes larger. Take sample 1 as an example, under the heating power of 60, 70, 75 and 80 W, the average temperature gradients of T_1 and T_2 are reduced by 24.8, 25.3, 26.8 and 27.4%. This is because the PCM absorbed and reserved more latent heat during the heating phase and released while the heating phase ended.

In addition to the stabilizing temperature gradients of PCM, the heat transfer enhancement for PCM is studied as well. As shown in the figures, the heating time of heating plate for reaching the phase-changing temperature of the PCM becomes longer with the increment of PCM integrated ratio. The reason for longer response time is the heat transfer descends with the more PCM added to the sample. The response time of each sample under 60 W heating phase is 140, 430, 460 and 490 s. Moreover, the temperature between heating plate and hot surface of TEM is enlarged, by inserting PCM in the TEG system, because of the low heat transfer coefficient of PCM. The temperature difference is 9.49 °C, 34.45 °C, 34.93 °C and 53.41°C at peak point of each sample, respectively. These results give a method to improve the heat transfer of PCM, which is metal foam combination. By combining metal foam with PCM, the heat transfer can be improved significantly, while there is exclusively minor drawback in temperature stabilization.

4   Conclusions

In this paper, an energy storage method of TEG system for automotive waste heat recovery is presented. Pentaerythritol, as a PCM, is introduced to stabilize the temperature of hot side of TEG system under and improve the efficiency of TEG system. A series of comparison experimental tests are conducted for verifying the concept and seeking the method of heat transfer enhancement for PCM. The experimental results are solid evidence of that Pentaerythritol as PCM can improve the efficiency of TEG system under real-time driving of vehicles. Heat transfer of PCM is enhanced by approximately 30%, after using thin ribbed plate. However, by comparing to none PCM sample, the heat transfer is still not accomplished and needs further improvement.

Acknowledgements This work was supported by H2020-MSCA-RISE-778104-ThermaSMART. The authors appreciate the financial support from Chinese Scholarship Council (CSC) for Kuo Huang.

References

1. K. Huang, B. Li, Y. Yan, Y. Li, S. Twaha, J. Zhu, A comprehensive study on a novel concentric cylindrical thermoelectric power generation system. Appl. Therm. Eng. **117**, 501–510 (2017)

2. B. Zalba, J.M. MaríN, L.F. Cabeza, H. Mehling, Review on thermal energy storage with phase change: materials, heat transfer analysis and applications. Appl. Thermal Eng. **23**, 251–283 (2003)
3. P. Hu, P.-P. Zhao, Y. Jin, Z.-S. Chen, Experimental study on solid–solid phase change properties of pentaerythritol (PE)/nano-AlN composite for thermal storage. Sol. Energy **102**, 91–97 (2014)
4. Y. Tu, W. Zhu, T. Lu, Y. Deng, A novel thermoelectric harvester based on high-performance phase change material for space application. Appl. Energy **206**, 1194–1202 (2017)
5. A. Sharma, V.V. Tyagi, C.R. Chen, D. Buddhi, Review on thermal energy storage with phase change materials and applications. Renew. Sustain. Energy Rev. **13**, 318–345 (2009)

Heat Transfer Analysis of a Liquid Piston Gas Compressor

M. Kaljani, Y. Mahmoudi, A. Murphy, J. Harrison, and D. Surplus

1 Introduction

Renewable energies are clean and unending but most of renewable energies such as solar and wind supply energy periodically which makes them unreliable. Energy storage concept is a solution to make the renewable energy supplies more reliable. This can be achieved by storing the excess energy generated during the low demand of electricity and reused the stored energy during the high demand of electricity [1].

Among different energy storage technologies, compressed air energy storage (CAES) is a technology that has a strong prospect of efficiency improvement. The performance of high-pressure air compressors/expanders is highly important to the economic growth of CAES systems. Any rise in internal energy of air during compression is wasted in the storage tank as the compressed air cools back to ambient temperature [2]. Therefore, the compression/expansion efficiency depends on the rate of heat transfer between the air and surrounding. In other words, a near-isothermal compression is the main aim which can be achieved by increasing the heat transfer from air to the environment during the compression process. In this regard, liquid piston gas compressor (LPGC) was introduced which is a new concept, and the technology is still in development for commercial use. The liquid enters the cylinder at high-pressure driven by a hydraulic pump to compress the air to a desirable pressure in the chamber [3].

M. Kaljani (✉) · Y. Mahmoudi · A. Murphy
School of Mechanical and Aerospace Engineering, Queen's University Belfast, Belfast BT95AH, UK
e-mail: skhaljani01@qub.ac.uk

M. Kaljani · J. Harrison
South West College, Cookstown BT80 8DN, UK

D. Surplus
B9 Energy group, Larne BT40 2SF, UK

© Springer Nature Singapore Pte Ltd. 2021
C. Wen and Y. Yan (eds.), *Advances in Heat Transfer and Thermal Engineering* ,
https://doi.org/10.1007/978-981-33-4765-6_142

As LPGC are more efficient while working isothermally, limited numbers of studies were conducted to study approaches for improving the performance of the LPGC by moving toward near-isothermal compression process. In this regard, different methods such as spraying the water inside the chamber [4] or porous medium inserts [5] were deployed. All these methods are based on increasing the heat transfer surface area in order to increase the rate of heat transfer from air inside the chamber to the surrounding. Control of the water spray in high-pressure operation is the main challenge of this method. Therefore, using porous material inserts is more promising, which a liquid piston that can be coupled with various porous inserts with different material and configuration. A porous medium is a material containing *pores*. The skeletal material is usually *solid*, but structures like *foams* are often analysed using the concept of porous media [6].

The main objective of this work is to investigate the physics of flow and heat transfer in one-stroke compression of air in a LPGC system. For this purpose, first a low-order and one-dimensional (1-D) heat transfer analysis was performed in order to understand the effect of the system's pertinent parameters on the efficiency of the LPGC. Then using a three-dimensional (3-D) modelling platform, a detailed analysis of the fluid flow and heat transfer in a LPGC system is performed. Then five parallel plates with different lengths are used as the porous medium in the LPGC cylinder, in order to examine the role porous materials in achieving a near-isothermal compression process in a LPGC. The results are presented and discussed in the form of compression work, compression efficiency and energy stored.

2   Methodology

In this work, we first conducted a 1D heat transfer and thermodynamic modelling of a LPGC system in EES software [7]. However, in order to gain a deeper insight into the flow and thermal characteristics of the LPGC system with plates (porous medium), a 3D analysis is conducted in the Fluent 19.2 Software. Air is compressed from 8 to 40 bar. Water at constant mass flow rate enters the chamber which acts as a piston to compress the air inside the LPGC chamber. The CFD analysis is conducted for a base case (LPGC with no porous inserts) and two cases where 1-mm-thick aluminium plates are sandwiched in the chamber with two different lengths of 0.35 m and 0.55 m. The compression chamber is made of stainless steel 316 with 1.1 m long and 0.08 m internal diameter. The VOF model is utilized for tracking the interface between the water and air. For turbulence modelling, the k-ε models are deployed. The computation is run on a mesh with 1,900,000 elements for the base case. For cases with plates, mesh with 2,700,000 and 2,900,000 elements is used for the cases with 0.35 m and 0.55 m plates, respectively.

One of the main parameters to be used to evaluate the performance of a LPGC system is the compression efficiency, which is defined as follows [2]:

$$\eta_{Com} = \frac{E_s}{W_C} \tag{1}$$

In this equation, E_s is the maximum energy storage for an isothermal expansion defined as follows:

$$E_s = P_c V_{final} ln|P_c/P_{initial}| \tag{2}$$

W_C in Eq. (1) is the compression work to compress air from initial volume to the final volume for the desired pressure ratio as follows:

$$W_C = (rP_0 - P_0) - \int_{V_{initial}}^{V_{final}} (P - P_0)dv \tag{3}$$

3 Results

The results of the 1D simulation are compared with respect to the experimental data in the literature [2] and shown in Fig. 1. The figure shows at low-power densities there is some 4% difference in efficiency between the current 1D modelling and experimental data. Nonetheless, the results of the 1D modelling are in good agreement with the experimental results.

The 1D computation cannot provide deep understanding of the flow and thermal fields in the system specifically when the aluminium plates are placed in the chamber. Therefore, in the followings, the results of the 3D modelling are presented. Figure 2a shows the contour of temperature for air at the end of the compression phase in the cylinder for the case with and without plates in. It is seen that while air is compressed from 8 to 40 bar, its temperature rises from 298.1 to 473 K for the base case (no plate), from 298.1 to 422 K (for the case with 0.35 m plate) and from 298.1 to 409 K (for the case with 0.55 m plate). This figure clarifies that using plates the final compressed

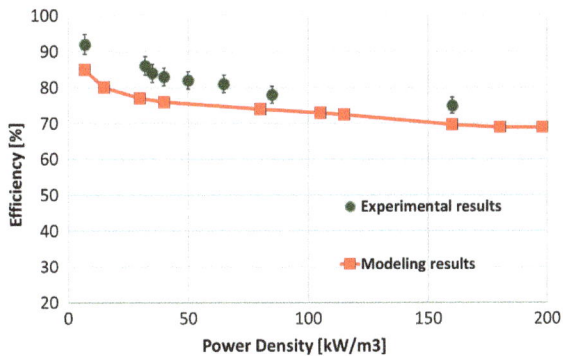

Fig. 1 Compression efficiency as a function of power density in ref [2] and present work

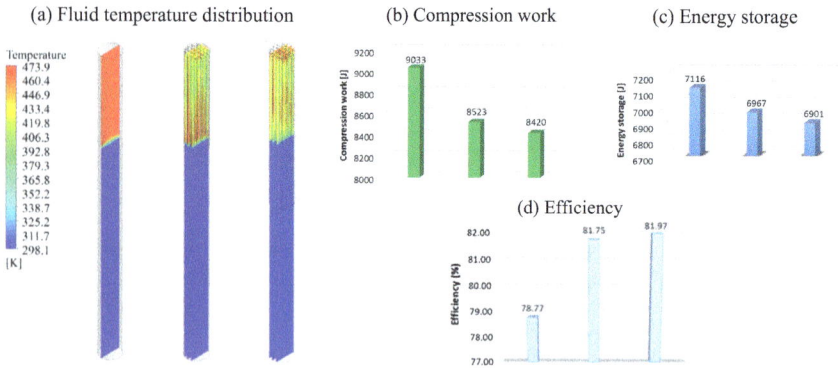

Fig. 2 Transient flow field of an RANS simulation, u = 0.27 m/s, H = 1.1 m, D = 0.08 m

air temperature in the LPGC is significantly lower than that of the LPGC with no plates. Figure 2b compares the compression work in LPGC with and without plates. It is observed that longer plates in the chamber lead to a lower required compression work. Because as Fig. 2a shows for longer plate, the final air temperature is lower, and hence, less power is required to compress the air to the desired pressure. According to Fig. 2c by adding plates inside the chamber, the stored energy decreases. This can be attributed to the lower initial mass of air in the chamber for the plate case compared to the base case. Finally, the compression efficiency is presented in Fig. 2d. It is seen that b insertion the 0.35 m plate the compression efficiency increases by 3%. While an increase in the length of the plates is from 0.35 to 0.55 m, it does not have significant influence on the efficiency, meaning that for fixed systems parameters (e.g. pressure ratio) there is an optimum value for the plate length to yield an optimum compression efficiency.

4   Conclusions

In this work, thermodynamic and heat transfer analyses were conducted to investigate the influence of deploying porous materials in a LPGC systems for the purpose of achieving a near-isothermal compression system and consequently increasing the system efficiency. The porous medium is modelled as aluminium parallel plates inserted in the chamber. The main results of the present modelling are as follows:

- The final temperature of the air at the end of the compression phase with the inclusion of five plates with the length of 0.35 m and 0.55 m reduces by 51 K and 64 K, respectively, compared to the base case.
- By inserting plates into the cylinder, the compression efficiency improves by 3%, but there is optimum value for the plate lengths to achieve an optimum compression efficiency.

Acknowledgements The work is supported by the European Union's INTERREG VA Programmed, managed by the Special EU Program Body (SEUPB), with match funding provided by the Department for the Economy and Department of Jobs, Enterprise and Innovation in Northern Ireland.

References

M. Aneke, M. Wang, Energy storage technologies and real life applications–A state of the art review. Appl. Energy **179**, 350–377 (2016)

B. Yan, J. Wieberdink, F. Shirazi, P.Y. Li, T.W. Simon, D. Van de Ven, Experimental study of heat transfer enhancement in a liquid piston compressor/expander using porous media inserts. Appl. Energy **154**, 40–50 (2015)

B. Yan, Compression/expansion within a cylindrical chamber: Application of a liquid piston and various porous inserts (2013)

M. Saadat, F.A. Shirazi, P.Y. Li, Modeling and trajectory optimization of water spray cooling in a liquid piston air compressor, in *ASME 2013 Heat Transfer Summer Conference collocated with the ASME 2013 7th International Conference on Energy Sustainability and the ASME* (2013)

A.T. Rice, Heat transfer enhancement in a cylindrical compression chamber by way of porous inserts and the optimization of compression and expansion trajectories for varying heat transfer capabilities (2011)

B. Su, C. Sanchez, X. Y. Yang, Hierarchically structured porous materials: from nanoscience to catalysis, separation, optics, energy, and life science, Wiley (2012)

S. Klien, F. Alvarado, *EES Engineering Equation Solver* (F-chart software, Wisconsin, 1997)

Progress with Development of FASTT Technology

W. D. Alexander

1 Introduction

The use of solid materials to transport thermal energy has been used in numerous ways since prehistoric times. Evidence of fire heated stones to boil water in a separate trough has been dated to as early as the Paleolithic period Ref. [1]. There are many references to the storage of ice in Ancient Persia, Ancient Egypt, and in the Roman Empire. Ice from high altitude districts would be brought to urban areas for cooling applications such as food preservation. More up to date, the use of solid belts as a heat transfer medium was the subject of much research by NASA to enable efficient thermal management in outer space Ref. [2]. IBM have also indicated an interest in solid media thermal transport systems as indicated in Ref. [3]. FASTT devices have been developed as part of research into high flux cooling of electronic equipment and employ solid phase material to transfer heat Ref. [4, 5]. They are currently being studied to examine their potential as efficient heat exchangers in a number of application areas.

Global warming and the need to improve the efficiency of energy resource use is attracting increasing attention. Worldwide energy usage (Ref. [6] and Ref. [7]) may be divided into five major sectors: residential (6.8%), commercial (4.7%), industrial (22.7%), transportation (28.0%), and electric power (37.8%). Prime mover class thermodynamic machines such as diesel and petrol engines, for transportation, through to steam and gas turbines, for electricity generation are major energy users. This is also true for the other major class of thermodynamic machines which provide heating and air conditioning plant for the built environment, as well as refrigeration, in domestic, commercial, and industrial environments. Heat exchange components are essential elements in all of these applications, and their efficient design is becoming vital.

W. D. Alexander (✉)
ACPI Ltd., Stafford ST18 0TG, Stafford, United Kingdom
e-mail: william@acpilimited.co.uk

© Springer Nature Singapore Pte Ltd. 2021
C. Wen and Y. Yan (eds.), *Advances in Heat Transfer and Thermal Engineering*,
https://doi.org/10.1007/978-981-33-4765-6_143

Renewable energy resources such as wind, solar, and tidal power can be highly inter-mittent in their output, and energy storage will be required to optimize supply grids. Thermal energy storage is one such means to address this issue. FASTT may provide both the high flux input as well as the thermal storage capabilities desirable for such grid-level energy requirements. However, this paper will focus on the thermal management of small (<1 kW) electronic devices and equipment.

2 Theoretical

Early electronic devices handling only a few watts were cooled using simple air blast, or with extruded finned sections thermally attached to the devices. Simple heatsinks, such as these, were able to cope with modest dissipation up to about 50 W, but the simple finned array was becoming increasingly massive and incapable of handling localized high flux levels. Figure 1 indicates the typical flux limits of simple air blast finned heatsinks. The model on which the figure is based makes the following assumptions: constant air temperature (no allowance for heating of the air as it moves through the fins); constant fin thickness of 40 microns (this was chosen to match current thickness of commercially available pyrolytic graphite); a spacing between fins of 400 microns (chosen as the minimum reasonable spacing to allow adequate air flow while keeping pressure drop to levels which current fan/blowers can provide); 20 °C mean temperature difference between air and fin; simple rectangular fin section. A selection of materials has been analyzed and the maximum flux calculated with respect to fin height. It can be seen that the maximum flux capability, even for a

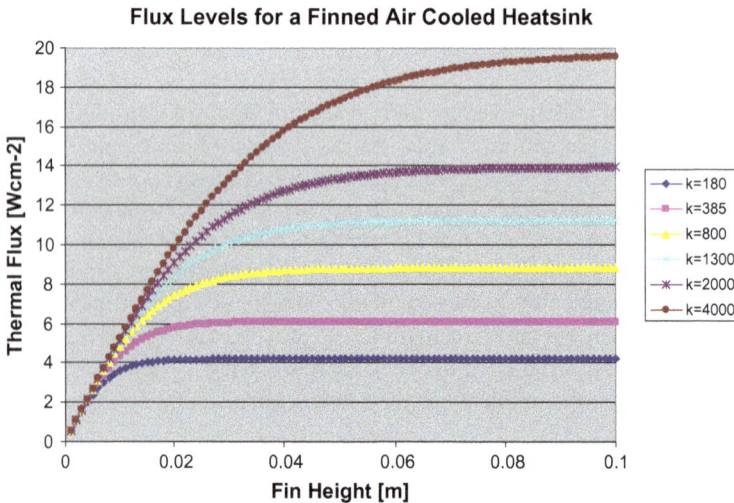

Fig. 1 Input thermal flux capability for a simple air-cooled finned heatsink

material with thermal conductivity of 4000 [Wm^{-1} K^{-1}], will still be capable of less than 20 [Wcm^{-2}] at 20 °C ΔT, implying less than 40 [Wcm^{-2}] at 40 °C ΔT.

The limited thermal input flux capability of simple finned heatsinks is apparent, and it is this which has been the driver to develop more capable heat exchangers for electronic equipment cooling.

High flux cooling of electronic equipment operating around normal room temperatures is currently catered for, by pumped liquid microchannel coolers and by heat pipe and vapor chamber devices. However, all of these devices rely upon relatively large, metal-finned arrays to act as the interface with ambient air, the final sink for many systems. FASTT devices offer an alternative thermal management system which relies on the motion of solids rather than liquids or vapors. Two major features distinguish this new class of heat exchanger from those currently used. The first difference is that in a FASTT device, the solid medium used to convey the thermal energy away from the source can also be used directly as the interface with ambient air in the output section. The foils themselves can be used as the output finning. There is no need for pipework to manipulate the "flow" of the solid foil, and also, there are no additional interfaces such as those which result when heat transfer to and from envelope walls needs to be considered. The second major feature is that the "effective" heat transfer coefficient which the solid foil imposes on the slot walls through which it moves, is velocity independent— heat transfer takes place by conduction across the gap between foil and slot wall; there is no need for high velocity to promote mixing as in a fluid. Furthermore, this heat transfer coefficient "equivalent" is in the range of 625 to 2500 [Wm^{-2} K^{-1}] for typical foil to slot wall gaps of between 10 and 40 microns. Fluid-based microchannel devices are approaching the situation where conduction into the fluid dominates the transfer of heat and this situation is approximately modeled by the "water" foils in Fig. 2.

In pumped two-phase systems, the absorbed thermal energy per unit mass of boiling fluid is many times that absorbed by a single-phase fluid undergoing a modest (say 20 °C) temperature rise. However, the liquid-to-gas phase transition produces

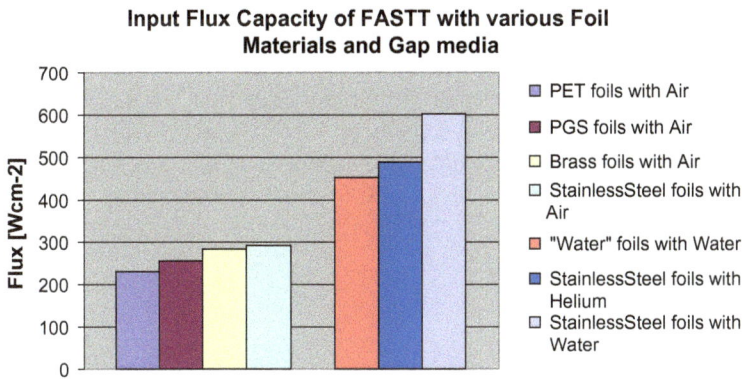

Fig. 2 Input thermal flux capability for FASTT with various materials for foil and gap

a large volume increase which has to be accommodated by the pumping system. This system has to continuously replenish the input with liquid phase, while at the same time ensuring that the gaseous phase is transported to the output stage for condensing—this volume flow requirement dominates over the much smaller liquid phase flow requirement at the input to the system. Heat pipes employing "wicking" structures separate the gas phase and liquid phase flow, and thus, this effect is not apparent for these devices.

The use of nano-particulates in fluids is currently being studied as a means of improving the thermal transport characteristics of fluids. The main property enhancement is in the thermal conductivity, although at high doping levels the density may also be increased. Typical thermal conductivity enhancements of between 10 and 40% over base fluids, such as water, have been achieved (Ref. [8]); however, viscosity is also increased, and the specific heat of the composite fluid will be lower than for water alone. A comparison with circulating solid materials such as in FASTT devices may be drawn, see Fig. 2 and the comments below. The end application as well as the allowable geometry will generally dictate which solution is best suited. FASTT technology offers an alternative heat transfer scheme which may have substantial benefits over other technologies in some situations but pumped fluid systems have more freedom in terms of the geometries which can be served.

In Fig. 2, a number of simplifying assumptions have been made, but they are the same for each of the foil and slot gap combinations. The first three bars on the left of the chart assume the foils are running in air, while those toward the right hand side have either helium or water as media in the gap between foil and slot walls. These latter results based on helium would be for a sealed device only, and the result for the "water" foils with water is included as an interesting comparison. Essentially, a solid material with the properties of water has been included to show the performance of an idealized, water-like solid moving with water as the gap filling medium. This effectively models a sheet of water passing through the slot. Viscous drag and turbulence are assumed absent, with only conduction between the thin sheet/foil and the slot walls. It is worthwhile noting that the performance of the stainless steel/helium combination is superior to the water/water combination. This is largely due to the fact that stainless steel has a (density*specific heat) product close to that of water, but that its thermal conductivity is about twenty times that of water. A mean temperature difference between foils and slot walls of 40 °C has been assumed for all cases, with a foil velocity of only 0.05 [ms^{-1}]. Velocities as low as this would be insufficient to promote adequate levels of h.t.c.s in a liquid; however, they are sufficient in a FASTT device, due to conduction providing the basis of its heat transfer.

3 Experimental

The latest FASTT prototype is of the linear variety where the foils are simple strips moving in a reciprocating fashion through the slotted, thermally conductive core. A high efficiency, low profile reciprocating mechanism has been developed specifically for this application. A positive engagement, special gear and rack system was developed, together with a high efficiency brush motor, operating single direction. Power consumption is about a quarter of that for a similar stepper motor system used on an earlier prototype. Figure 3 shows a displacement vs time trace which indicates the good linearity of the drive. The load for the high flux tests is provided by a single 150 W 50 ohm RF resistor based on a AlN substrate. The slotted core is comprised of a parallel array of precision machined copper sheets, bonded together. Figure 4 shows the slotted core plus the load resistor. The foils are made from polyetheretherketone polymer (PEEK) which is available with high tolerance thickness and which has good mechanical and thermal properties to above 250 °C. The blower used for air flow is a standard "off-the-shelf" unit. Figures 5 and 6 show some views of this prototype. Testing is due to commence shortly, and up to date, results will be presented at conference.

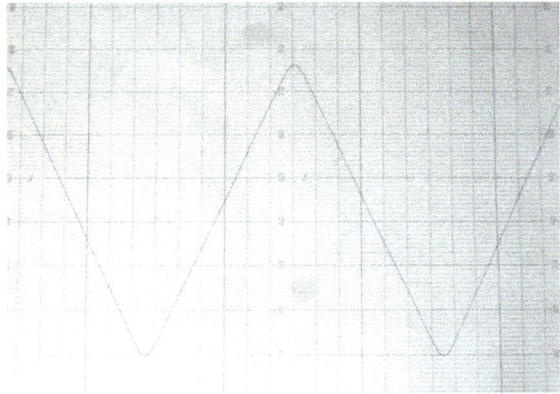

Fig. 3 Linear drive displacement versus time

Fig. 4 Core pack with AlN resistor attached

Fig. 5 FASTT cooler flanked by two standard PC chip coolers

Fig. 6 FASTT linear cooler
showing drive

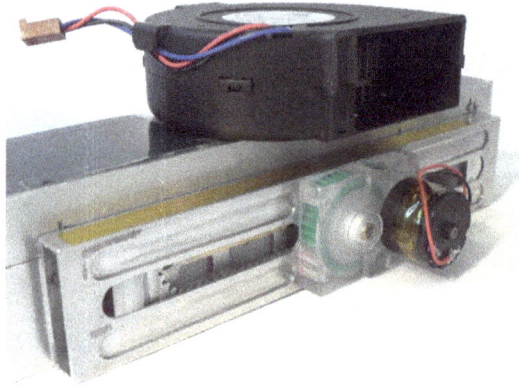

4 Conclusions

The development of FASTT technology continues with increased input thermal flux
capability and the completion of a new linear device. The latest prototype demonstrates the compact form and low mass which this thermal management system can
provide. The prototypes produce to date, have all been low power and the thrust has
been toward the cooling of electronic devices and equipment. However, the application "envelope" which FASTT can achieve, has yet to be determined. Portable power
electronic cooling applications for robotics through to low mass nuclear reactor
systems as well as thermal management of space vehicles are all currently being
considered.

References

1. J. Ó'Néill, in *Burnt Mounds in Northern and Western Europe: A Study of Prehistoric Technology and Society* (VDM Verlag Dr, Müller, 2009)
2. K.A. White, in *Moving Belt Radiator Development Status*, Lewis Research Center, NASA Technical Memorandum 100909, NASA-TM-100909 19880016093, July 1988
3. R.J. Evans, D.L. Gardell, A.M. Palagonia, in *Method and Apparatus for Cooling an Electronic Device*, Armonk, N.Y., US Patent No. 6050326, 2000
4. W. Alexander, Novel Heat Transfer Using Solid Phase Transport Medium, in *14th UKHTC Edinburgh 2015*, Edinburgh University, 7–8 Sept 2015
5. W.D. Alexander, FASTT Technology and some observations relating to other Cooling Technologies, in *15th UKHTC Brunel 2017*, Brunel University London, 4–5 Sept 2017
6. U.S. Energy Information Administration / Monthly Energy Review April 2019. www.eia.gov/totalenergy/data/monthly/pdf/sec2_3.pdf
7. U.S. Energy Consumption 2017, Lawrence Livermore National Laboratory, Data based on DoE/EIA MER (2017)
8. R. Azizian, in *Nanofluids in Electronics Cooling Applications*, Advanced Thermal Solutions, QATS white paper, posted by Perry, J., 4 Oct 2017

A Numerical Model with Experiment Validation for the Melting Process in a Vertical Rectangular Container Subjected to a Uniform Wall Heat Flux

M. Fadl and P. C. Eames

1 Introduction

Thermal energy storage (TES), specifically heat storage, may have a key role to play in supporting the achievement of the UK's future decarbonization targets for heat and electricity. TES systems can be designed to collect and store low-quality heat (such as low-temperature industrial process heat), which can either be used directly or at a later time to produce hot water and space heating for buildings. Thermal energy storage can help mitigate peak loads on the UK electricity grid due to the likely demand patterns of heat pumps and other electricity-based heating technologies. It can also be used to improve utilization of time-constrained heat production, e.g., from solar-based technologies that only generate heat during sunlight hours helping address the mismatch between supply and demand patterns [1].

To effectively design a latent heat storage system for a specific process, the time required for melting and solidification, and thus rates of charge and discharge and heat transfer during the phase change process has to be known [2]. Numerical models used to predict PCM TES system performance should realistically simulate the processes of melting and solidification while being computationally efficient allowing accurate simulation within affordable computational resources.

The objective of the present study was to evaluate the effect different values of A_{mush} that have on the overall simulation of PCM melting within a vertically oriented rectangular cross section storage geometry when subjected to uniform wall heat flux and its effect on the melt fraction development. Model predictions with different values of the A_{mush} parameter and wall heat flux are compared with new experimental data.

M. Fadl (✉) · P. C. Eames
Centre for Renewable Energy Systems Technologies (CREST), Wolfson School of Mechanical, Electrical and Manufacturing Engineering, Loughborough University, Leicestershire L11 3TU, UK
e-mail: m.s.fadl@lboro.ac.uk

© Springer Nature Singapore Pte Ltd. 2021
C. Wen and Y. Yan (eds.), *Advances in Heat Transfer and Thermal Engineering* ,
https://doi.org/10.1007/978-981-33-4765-6_144

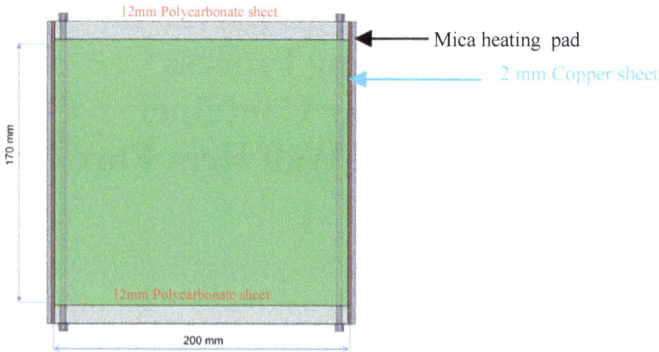

Fig. 1 Schematic diagram of the simulation geometry

2   Methodology

A schematic diagram of the two-dimensional simulation geometry is shown in Fig. 1. The dimensions of the container containing PCM RT44HC used in the simulations are based on those in the experimental study carried out at Loughborough University [3, 4]. In the experimental study, PCM melting was quantified by visually tracking the shape of the solid/liquid interface and how it changed with time. The rectangular test cell was formed from 12 mm thick transparent polycarbonate sheet which allowed observation of the melt fraction development during the experiments. The internal dimensions of the test cell were 176 mm height, 200 mm in width, and 100 mm deep. The depth was considered sufficiently large, so that the boundary effects due to the front and back walls would be negligible at the midplane.

The experiments were conducted at three different input power rates 27, 38.4 and 51.8 W (corresponding to heat fluxes of 675, 960 and 1295 W/m^2), the constant wall heat flux, to the system calculated based on the electric power dissipated over the heat transfer surface area is given in Eq. 1:

$$\tag{1}$$

where A_w is the total surface area of both heaters providing the heat flux to the system.

3   Results

Figure 2 shows the predicted melt fractions for a vertical rectangular enclosure for different values of A_{mush} ranging from 10^5 to 10^8. For $A_{mush} = 10^8$, the melting rate is very slow, and for $A_{mush} = 10^5$, the melting rate increases significantly, being greater than that observed experimentally. An A_{mush} value of 2×10^6 was found to give

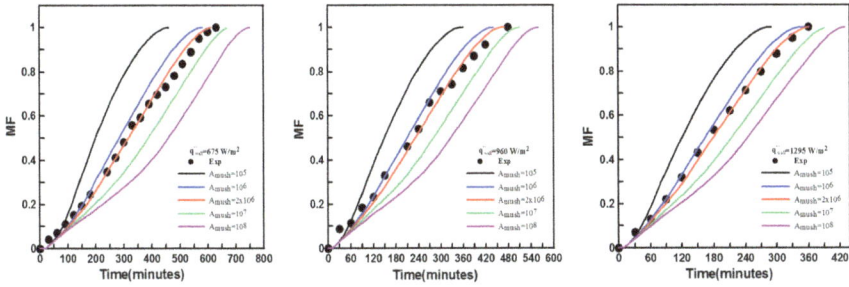

Fig. 2 Melt fraction with time for selected values of the mushy zone parameter, A_{mush} during the melting process

the best agreement with the experimental results for this geometry. During melting, the heat transfer is strongly influenced by convection, and an increase in the value of A_{mush} leads to a decrease in the predicted level of convection with a consequent reduction in the melting rate [5].

4 Conclusions

In the present study, a 2D transient numerical investigation of the melting of RT44HC in a rectangular vertical enclosure has been carried out for different values of the mushy zone parameter A_{mush} ranging from 10^5 to 10^8. Based on this study, it was observed that the selection of the value of A_{mush} is an important parameter if the phase change phenomena is to be accurately predicted. The smallest values of A_{mush} ($<10^5$) used resulted in unrealistically rapid predictions of melting, and higher values of A_{mush} (10^8) resulted in predictions of very slow melting rates. It was clear from the simulations performed that the impact of A_{mush} is less significant in areas at times prior to PCM melting when conductive heat transfer dominates. The effect of A_{mush} was more significant on the PCM further away from the heated surface when more PCM had melted, and heat transfer was dominated by natural convection. This was because with increasing A_{mush}, the predicted convective flow decreased with a concomitant decrease in the heat transfer rate. Proper selection of the mushy zone parameter A_{mush} is essential for the accurate prediction of heat transfer characteristics within a PCM. A further similar study is required for different PCMs with experimental work validation to assess the extent to which the correct selection of A_{mush} can be correlated to material and store geometry.

Acknowledgements The authors are grateful to the Engineering and Physical Sciences Research Council (EPSRC) for funding this work through Grant reference EP/N021304/1.

References

1. Evidence Gathering: Thermal Energy Storage (TES) Technologies
2. A. Erek, Z. Ilken, M.A. Acar, Experimental and numerical investigation of thermal energy storage with a finned tube. Int. J. Energy Res. **29**, 283–301 (2005). https://doi.org/10.1002/er.1057
3. M. Fadl, P.C. Eames, An experimental investigations of the melting of RT44HC inside a horizontal rectangular test cell subject to uniform wall heat flux. Int. J. Heat Mass Transf. **140**, 731–742 (2019). https://doi.org/10.1016/J.IJHEATMASSTRANSFER.2019.06.047
4. M. Fadl, P. Eames, A comparative study of the effect of wall heat flux on melting and heat transfer characteristics in phase change material thermal energy stores arranged vertically and horizontally, in *9th Ed International SOLARIS Conference, Chengdu, China: IOP Conference Series: Materials Science and Engineering (MSE)*(2018). https://doi.org/10.1088/issn.1757-899X
5. S. Arena, E. Casti, J. Gasia, L.F. Cabeza, G. Cau, Numerical simulation of a finned-tube LHTES system: influence of the mushy zone constant on the phase change behaviour. Energy Procedia **126**, 517–524 (2017). https://doi.org/10.1016/j.egypro.2017.08.237

Experimental Study on Heat Transfer Performance of Active Magnetic Regenerator Working at Room Temperature

Georges El Achkar, Bin Liu, Qi Wang, Aiqiang Chen, and Rachid Bennacer

1 Introduction

The growing need for cooling, refrigeration and air conditioning requires an increasing use of two-phase thermal systems, such as heat pumps, capillary pumped loops or loop heat pipes. Nevertheless, these systems have the disadvantage of operating with refrigerants, which often have a negative impact on the ozone layer and/or a significant contribution to the greenhouse effect [1]. The quality of new environmentally friendly refrigerants will probably be to the detriment of the thermal performance and/or the stability of these systems. It is therefore important to search for new cold production solutions ensuring low environmental impact and high thermal efficiency and stability.

In this context, the magnetic refrigeration technology, based on the magnetocaloric effect (MCE), appears as a potentially viable solution [2, 3]. Since the theoretical optimal coefficient of performance (COP) of magnetic refrigeration systems is of the same order as that of two-phase thermal systems and their operation occurs with solid materials coupled with water, the possibilities seem to fit with the current requirements. In addition, this new technology operates at low pressure and frequency, which is convenient for the security, the maintenance cost, the noise and the lifetime.

The MCE, discovered by Weiss and Picard in 1917, is a physical phenomenon occurring in a magnetocaloric material (MCM) under the influence of a varying magnetic field (H) [4, 5]. It is usually expressed as an adiabatic temperature change (ΔT_{ad}) or an isothermal total entropy change (ΔS_{tot}) of the MCM. In the absence of a magnetic field ($H = 0$ T), the magnetic moments in the MCM are disordered

G. El Achkar (✉) · B. Liu · Q. Wang · A. Chen · R. Bennacer
International Centre in Fundamental and Engineering Thermophysics, Tianjin University of Commerce, Tianjin, China
e-mail: georges.elachkar@tjcu.edu.cn

R. Bennacer
LMT/ENS-Cachan/CNRS, Paris-Saclay University, Paris, France

© Springer Nature Singapore Pte Ltd. 2021
C. Wen and Y. Yan (eds.), *Advances in Heat Transfer and Thermal Engineering* ,
https://doi.org/10.1007/978-981-33-4765-6_145

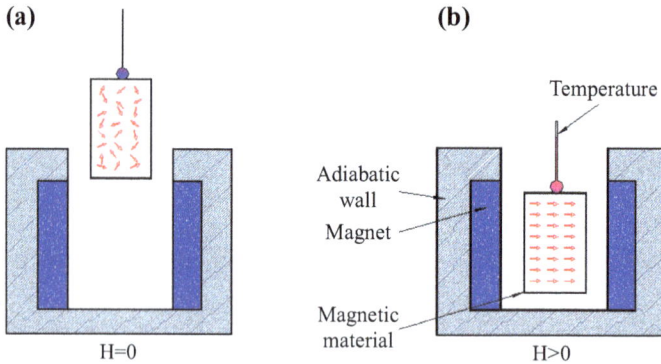

Fig. 1 Schematic diagram of the MCE principle

(Fig. 1a). If a magnetic field is applied ($H > 0$ t), the magnetic moments in the MCM are forced to align in a higher order (Fig. 1b). As a consequence, the magnetic entropy (S_m) decreases. In isentropic (i.e. adiabatic) conditions, the total entropy (S_{tot} = $S_m + S_l$) remains constant. Therefore, the decrease of the S_m manifests itself in an increase of the lattice entropy (S_l). The atoms in the material vibrate more intensively and hence, the temperature of the MCM increases. The opposite occurs when the magnetic field is removed ($H = 0$). The S_m increases and so the S_l decreases, leading to a decrease of the MCM temperature. This MCE is maximum around the Curie temperature (T_C) which marks the change between the ferromagnetic state and the paramagnetic state of the MCM. This transition is the most frequently studied so far.

In this chapter, the heat transfer performance of an active magnetic regenerator (AMR) working at room temperature was experimentally investigated. The effects of different operating conditions on the thermal behaviour of this AMR were studied and presented.

2 Experiments

In order to thermally characterise the AMR, a test rig of a reciprocating magnetic refrigeration system was designed and mounted (Fig. 2). This bench mainly consists of an air-cooled heat exchanger (A), a diaphragm liquid pump (B), four solenoid valves (C), a movable permanent magnet (D), a test section (E) and an acquisition system (Fig. 2a).

The test section consists of two identical cylindrical AMRs, each filled with 900 g of gadolinium (Gd, reference MCM widely used in the literature) particles with a size of 0.3–0.5 mm, connected by a thermally insulated copper tube (Fig. 2b). The permanent magnet of field intensity 1.5 T alternatively moves towards the left and right AMRs, inducing their alternative magnetisation and demagnetisation. When the

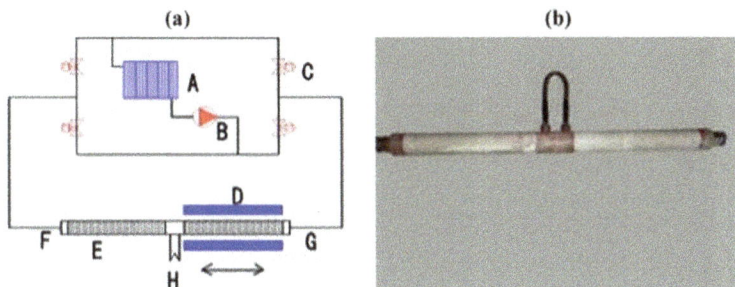

Fig. 2 **a** Schematic diagram of the reciprocating magnetic refrigeration system and **b** photography of the test section

permanent magnet covers the left AMR, its magnetised Gd heats up and the demagnetised Gd of the right AMR cools down. The diaphragm liquid pump (FLOJET, model D1732F5011A) then runs, and the working fluid (water) flowing from the cold end (*H*) to the hot end 1 (*F*) recovers the heat from the left AMR and releases the heat to the right AMR. Similarly, when the right AMR is magnetised, the flow direction is reversed thanks to a set of solenoid valves, and the water flowing from the cold end (*H*) to the hot end 2 (*G*) recovers the heat from the right AMR and releases the heat to the left AMR. Owing to the high flow rate of the pump of 6 l min^{-1} and the small volume of the water circuits, the pumping time is relatively short (of the order of 1 s). The liquid sucked by the diaphragm pump is set at room temperature thanks to the air-cooled heat exchanger placed upstream this pump. Four PT100 probes of accuracy 0.2 °C are used to measure the cold end temperature (H), the hot ends temperatures (F&G) and the room temperature. All the data are collected using an acquisition system consisted of a data logger, an interface card and a computer.

The test section maintained at a room temperature of 18 °C, a campaign of 20 test points combining different water pumping times (500, 600, 700, 800 and 900 ms) and magnet moving speeds (70, 100, 130 and 160 mm s^{-1}) was carried out. The effects of these combinations on the cold end temperature of the AMR were highlighted and analysed.

3 Results and Discussion

Figure 3 shows the evolutions of the cold end temperature of the AMR as a function of time for all water pumping times and magnet moving speeds. The trends of this temperature are qualitatively the same regardless of the water pumping time and the magnet moving speed. This temperature decreases rapidly up to 30 s and then almost stabilises due to the system limitation corresponding to the considered operating conditions. For the same magnet moving speed, the stabilised cold end temperature first decreases and then increases with the water pumping time increase, and its minimum value corresponds to a water pumping time of 700 ms. On the one

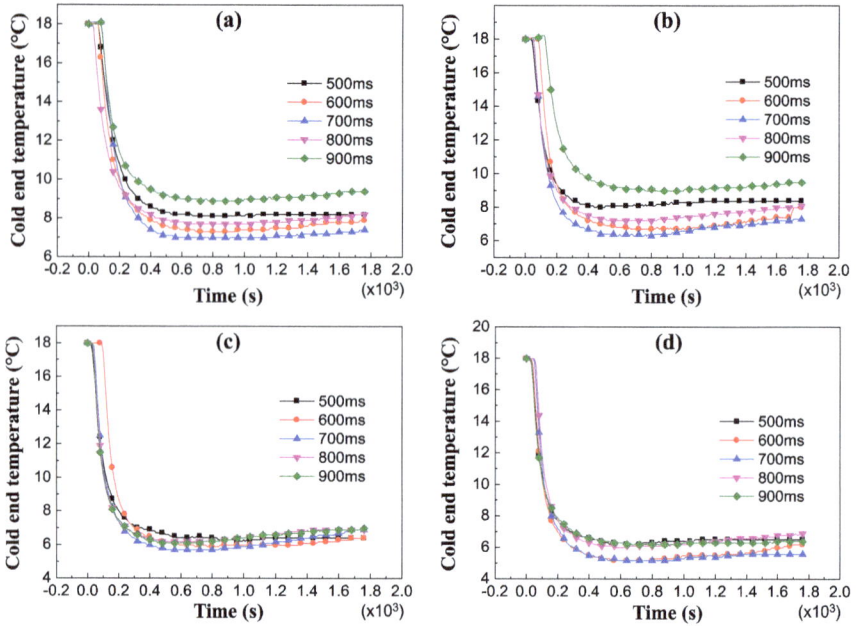

Fig. 3 Evolutions of the cold end temperature of the AMR as a function of time for different water pumping times and magnet moving speeds of **a** 70 mm s^{-1}, **b** 100 mm s^{-1}, **c** 130 mm s^{-1} and **d** 160 mm s^{-1}

hand, during the demagnetisation process, when the water pumping time is smaller than 500 ms, the heat transfer time is so short that the thermal equilibrium between water and gadolinium is not reached, the stabilised cold end temperature remains thus higher than the gadolinium temperature. On the other hand, during the demagnetisation process, when the water pumping time is greater than 900 ms, the heat transfer time is enough to reach the thermal equilibrium between water and gadolinium, but the water heat quantity is greater, the stabilised cold end temperature remains thus higher than the gadolinium temperature. Hence, when the water pumping time is 700 ms, the heat transfer from water to gadolinium is optimal, and the stabilised cold end temperature is minimum. For this water pumping time, the minimum stabilised cold end temperature decreases with the magnet moving speed increase and reaches a lowest value of 5.2 °C for a magnet moving speed of 160 mm s^{-1}. Indeed, the increase of the magnet moving speed (i.e., the magnetic cycles frequency) leads to an increase of the heat quantity absorbed by the gadolinium and hence to a decrease of the minimum stabilised cold end temperature.

4 Conclusions and Perspectives

An experimental study on the heat transfer performance of an AMR working at room temperature, with gadolinium and water respectively used as magnetocaloric material and working fluid, was conducted. A test rig allowing the control of the AMR operating conditions was developed and exploited. The effects of water pumping times (500–900 ms) and magnet moving speeds (70–160 mm s^{-1}) on the cold end temperature of the AMR were determined. A minimum stabilised cold end temperature was found for a water pumping time of 700 ms regardless of the magnet moving speed. For this water pumping time, a decrease of the minimum stabilised cold end temperature with the increase of the magnet moving speed was found, with a lowest value of 5.2 °C reached for a magnet moving speed of 160 mm s^{-1}.

As perspectives to this work, an experimental study on the heat transfer performance of an AMR with a permanent magnet rotary movement is of great interest, because of the limitation of the presented magnetic refrigeration system derived from the mechanical inertia of the permanent magnet translational movement.

Acknowledgements Financial support from the National Natural Science Foundation of China (No. 51706154) is gratefully acknowledged.

References

1. The impact of the refrigeration section on climate change, 35th informatory note on refrigeration technologies, International Institute of Refrigeration (2017)
2. O. Sari, M. Balli, From conventional to magnetic refrigerator technology. Int. J. Refrig. **37**, 8–15 (2014)
3. N.A. Mezaal, K.V. Osintsev, T.B. Zhirgalova, Review of magnetic refrigeration system as alternative to conventional refrigeration system. IOP Conf. Ser. Earth Environ. Sci. **87**, 032024 (2017)
4. E.G. Warburg, Magnetische untersuchungen über einige wirkungen der coerzitivkraft. Annalen Der Physik (Leipzig) **13**, 141–164 (1881)
5. P. Weiss, A. Piccard, Le phénomène magnétocalorique. J. Physique Théorique Et Appliquée **7**(1), 103–109 (1917)

Thermodynamic Analysis of Multi-stage Compression Adiabatic Compressed Air Energy Storage System

Lixiao Liang, Jibiao Hou, Xiangjun Fang, Ying Han, Jie Song, Le Wang, Zhanfeng Deng, Guizhi Xu, and Hongwei Wu

1 Introduction

Over the past few decades, global electricity production increased steadily and has reached 25,592 TWh by the end of 2017. With rapid development of hydro power, solar power, and wind power, the proportion of renewable energy in all energy sources rises remarkable, achieving 24.8% in 2017. However, renewable energy cannot replace all traditional fossil fuels due to the intrinsic intermittence and fluctuation. Electrical energy storage (EES) is becoming a vital aspect to ensure the balance between energy production and demand in order to deal with the intermittent nature of solar and wind energy sources [1, 2]. It is recognized that several EES, i.e., compressed air energy storage (CAES), thermal energy storage (TES), pumped hydroelectric (PHS), chemical or electrochemical batteries, flywheel, capacitor/supercapacitor have been developed. Luo et al. [2] performed a detailed review of the main characteristics of each technology. Each system has certain advantages and limitations based on efficiency, energy density, power range, time of response, and investment cost.

Over the past decades, compressed air energy storage (CAES) has attracted great attention, and many research efforts have been devoted to the CAES. Most recently, Mohamad et al. presented a steady-state thermodynamic model of a simple configuration of the recently proposed concept of trigenerative compressed air energy

L. Liang · J. Hou · J. Song · L. Wang · Z. Deng · G. Xu
State Key Laboratory of Advanced Power Transmission Technology (Global Energy Interconnection Research Institute), Changping District, Beijing 102211, China

X. Fang · Y. Han
School of Energy and Power Engineering, Beihang University, Beijing 100191, China

H. Wu (✉)
School of Physics, Engineering and Computer Science, University of Hertfordshire, Hatfield AL10 9AB, UK
e-mail: h.wu6@herts.ac.uk

© Springer Nature Singapore Pte Ltd. 2021
C. Wen and Y. Yan (eds.), *Advances in Heat Transfer and Thermal Engineering* ,
https://doi.org/10.1007/978-981-33-4765-6_146

storage system (T-CAES) and validated against experimental data [3] during the charge, storage, and discharge stages. Zhang et al. developed a theoretical model under steady-state operating condition, and sensitivity analyses were conducted to investigate the temperature and pressure variations along with exergy destruction in the throttle valve and cavern of a CAES system [4]. In recent years, advanced adiabatic compressed air energy storage (AA-CAES) is now being investigated for its properties of large-scale energy storage and high energy recovery. The biggest issue with the CAES power plant is that the system efficiency is low. The efficiency is about 30-43% for traditional CAES system, while the efficiency is about 50–75% for AA-CAES system. Mozayeni et al. developed a comprehensive thermodynamic model to investigate the thermal performance of an AA-CAES system [5].

It appears from the previous investigation that there were only limited reports on the adiabatic compressed air energy storage system. It can be concluded that the thermodynamic process of the A-CAES is important, and there is still much room to be enhanced in this area. The objective of the current work is to explore the principle of the novel A-CAES system, to reveal the thermodynamic characteristics as well as to explain the reason for better system performance. In addition, the effect of several control parameters, such as the gas storage methods, storage pressures, the inter-stage heat transfer methods, and the numbers of stages of the compressor, and the turbine on the overall performance of the system will be discussed.

2 Methodology

In the current study, a multistage compression–expansion polytropic method is employed, as shown in Fig. 1. Air is cooled after each stage during compression, and heated before each stage during the expansion. In Fig. 1, 1 represents the atmosphere, 2 is cool heat transfer fluid storage tank, 3 is the air storage system, and 4 represents hot heat transfer fluid storage tank. In the current study, the efficiency of CAES system is defined as the ratio of net output energy to net input energy:

$$\eta = \frac{W_{out}}{W_{in}} = \frac{W_t \eta_g - W_{self2}}{W_c \eta_m^{-1} + W_{energy} + W_{self1}} \tag{1}$$

Fig. 1 Polytropic method of the system

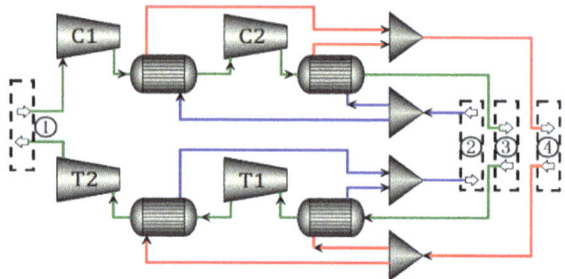

where η is the energy storage efficiency, W_{out} is the net output energy of the system, W_{in} is the input energy of the system, W_t is the total energy output of the turbine, W_c is the total energy consumption of the compressor, η_g is the efficiency of the generator, η_m is the efficiency of the motor, W_{self1} is the electric energy consumption of the system's other parts in the energy storage process, W_{self2} is the electric energy consumption of the system itself in the energy release process, and W_{energy} is energy consumption of other form such as heat energy.

In the following derivations and calculations, W_{self1} and W_{self2} are ignored. For A-CAES system, the polytropic method is applied, $W_{energy} = 0$. Assuming that all efficiencies of m stages compressors are η_c, and the efficiencies of n stages turbines η_t, then η can be expressed as

$$\eta = \frac{W_t \eta_g}{W_c \eta_m^{-1}} = \frac{W_{etn} \eta_t \eta_g}{W_{ecm} \eta_c^{-1} \eta_m^{-1}} \tag{2}$$

where W_{ecm} and W_{etn} are the energy consumption of unit mass air and the energy output of units mass air under polytropic expansion, respectively.

Air is regarded as ideal gas, with its constant specific heat, therefore,

$$W_{ecm} = \frac{k}{k-1} m R T_{cin}^* \left(\pi_c^{\frac{k-1}{mk}} - 1\right) = \frac{k}{k-1} m R \left(T_{cout}^* - T_{cin}^*\right) \eta_c \tag{3}$$

where is T_{cin}^* the inlet total temperature of the compressor, T_{cout}^* is the outlet total temperature of the compression, π is the compression ratio, and k is the adiabatic index of air.

Ignore the loss of the heat exchangers, temperature differences in heat exchangers and loss of gas storage and heat storage, then

$$T_{tin}^* = T_{cout}^* \quad \text{and} \quad \pi_c = \pi_t = \pi \tag{4}$$

where T_{tin}^* is the inlet total temperature of the turbine.
η can be given as

$$\eta = \frac{n \eta_t \eta_m \eta_g \left(\pi^{\frac{k-1}{mk}} + \eta_c - 1\right)\left(1 - \pi^{\frac{1-k}{nk}}\right)}{m\left(\pi^{\frac{k-1}{mk}} - 1\right)} \tag{5}$$

Usually m and n are equal to satisfy the condition of $T_{tin}^* = T_{cout}^*$. Then, Eq. (5) can be simplified into

$$\eta = \frac{\eta_t \eta_m \eta_g \left(\pi^{\frac{k-1}{nk}} + \eta_c - 1\right)\left(1 - \pi^{\frac{1-k}{nk}}\right)}{\left(\pi^{\frac{k-1}{nk}} - 1\right)} \tag{6}$$

Setting $\pi^{\frac{k-1}{nk}} - 1 = x$, $\eta_t\eta_m\eta_g = \eta_{tmg}$, Eq. (6) can be further reduced as

$$\eta = \eta_{tmg} - \frac{\eta_{tmg}(1-\eta_c)}{x+1} \tag{7}$$

Since $\frac{\partial\eta}{\partial x} > 0 (0 < \eta_c, \eta_t < 1)$, $\frac{\partial x}{\partial n} < 0 (\pi > 1, k > 1, n = 1,2,3\ldots)$, there-fore $\frac{\partial\eta}{\partial n} < 0$. Hence, when $m = n = 1$, energy storage efficiency can reach the highest value, which is equivalent to the adiabatic method (actual stage number is not necessarily the 1). When $n \to \infty$, $x \to 0$, then $\eta \to \eta_c\eta_{tmg} = \eta_{ctmg}$, and this is the lowest limit for the efficiency in theory. This conclusion is derived out on the basis of the polytropic method and can be generalized to the three multistage compression–expansion methods, where m and n refer to the number of times of heat exchange during compressions and expansions, respectively.

3 Results

For the purpose of comparison, in the current work, as shown in Fig. 2, the overall efficiencies of the A-CAES system with three different heat transfer modes (adiabatic method, semi-adiabatic method, and polytropic method) are calculated and compared. Results show that the overall efficiency can reach the highest when adopting the adiabatic method, although the method is the most uneconomical for the compression process. This observation consists with the theoretical analysis that discussed previously. Since a large amount of high grade compression heat produced during the adiabatic compression process, the heat could be fully used in the energy release process, thus the output power of the turbine is greatly increased relatively.

4 Conclusions

The current work investigates the theoretical potential of adiabatic compressed air energy storage (A-CAES) system and discusses the importance of process variables. Based on the results of the numerical studies, the following specific conclusions may be made: (1) An in-house code named CAESSC 1.0 is developed, which can successfully evaluate the performance of the proposed A-CAES and power generation system. (2) A multi-mode constant pressure A-CAES system was designed and analyzed. The efficiencies of the real A-CAES system under different configurations were calculated with current code and validated against reference data and commercial software. (3) The effect of several control parameters, such as the gas storage methods, storage pressures, the inter-stage heat transfer methods, and the numbers of stages of the compressor and the turbine on the overall performance of the system is discussed in a systematic manner. (4) Using the constant pressure method of gas

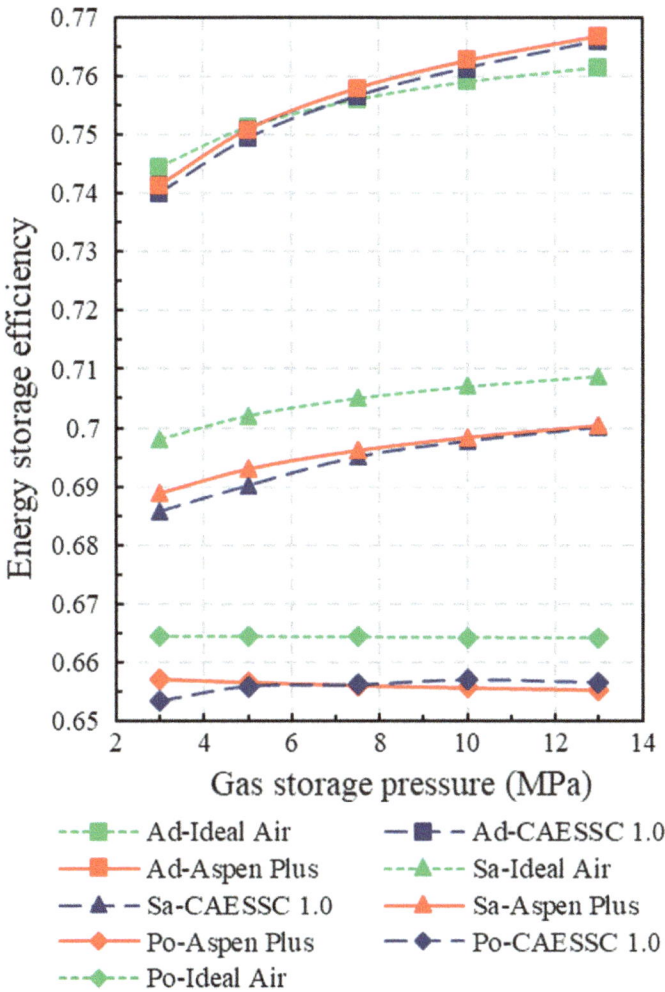

Fig. 2 Overall energy storage efficiencies of adiabatic CAES system with three compression–expansion methods

storage can improve the energy storage efficiency and the energy storage density of the system significantly. The highest efficiency could be obtained with the adiabatic method for the adiabatic CAES system.

Acknowledgements The authors would like to thank the financial support from State Grid Corporation of China Research Program "Preliminary Study of Frequency Modulation Technology for Power Grid Based on Compressed Air Energy Storage" (SGZJ0000KXJS1800283). Project No: GEIRI-DL-71-18-002.

References

1. S. Weitemeyer, D. Kleinhans, T. Vogt, C. Agert, Integration of renewable energy sources in future power systems: the role of storage. Renew. Energy **75**, 14–20 (2015)
2. X. Luo, J. Wang, M. Dooner, J. Clarke, Overview of current development in electrical energy storage technologies and the application potential in power system operation. Appl. Energy **137**, 511–536 (2015)
3. M. Cheayb, M.M. Gallego, M. Tazerout, S. Poncet, Modelling and experimental validation of a small-scale trigenerative compressed air energy storage system. Appl. Energy **239**, 1371–1384 (2019)
4. S. Zhang, H. Wang, R. Li, C. Li, F. Hou, Y. Ben, Thermodynamic analysis of cavern and throttle value in large-scale compressed air energy storage system. Energy Convers. Manage. **183**, 721–731 (2019)
5. H. Mozayeni, M. Negnevitsky, X. Wang, F. Cao, X. Peng, Performance study of an advanced adiabatic compressed air energy storage system. Energy Procedia **110**, 71–76 (2017)

Performance of Triangular Finned Triple Tubes with Phase Change Materials (PCMs) for Thermal Energy Storage

Yan Yang, Xiaowei Zhu, Hongbing Ding, and Chuang Wen

1 Introduction

Energy storage is of great importance for renewable energies, as it provides the potential to solve the mismatches between energy supplies and demands [1]. The thermal energy storage, including sensible, latent and thermochemical, attracts considerable attention. Among these technologies, the latent heat storage with phase change materials (PCMs) is more attractive because of high energy storage densities, almost isothermal natures of the storage process and chemical stabilities. The phage change material has been integrated into the triple tube to increase energy storage performance. However, the low thermal conductivity of most PCMs limits their applications as it may prolong the processes and performance of the melting and solidification processes, as well as decreases energy efficiencies [2]. In this study, we employ triangular fins to enhance the heat transfer during the melting process of the PCM and improve the performance of the triple tubes for latent heat thermal energy storage.

Y. Yang · X. Zhu
Department of Mechanical Engineering, Technical University of Denmark, Nils Koppels Allé,
2800 Kgs. Lyngby, Denmark

H. Ding
School of Electrical and Information Engineering, Tianjin University, Tianjin 300072, China

C. Wen (✉)
Faculty of Engineering, University of Nottingham, Nottingham NG7 2RD, United Kingdom
e-mail: chuang.wen@outlook.com

© Springer Nature Singapore Pte Ltd. 2021
C. Wen and Y. Yan (eds.), *Advances in Heat Transfer and Thermal Engineering* ,
https://doi.org/10.1007/978-981-33-4765-6_147

2 Mathematical Model

For the numerical simulation of PCMs, the following assumption is used [3]: (a) the flow is laminar, (b) the Boussinesq approximation is valid, (c) the viscous dissipation is negligible, (d) thermophysical properties of the heat transfer fluid (HTF) and PCM are independent of the temperature except for the PCM density. The unsteady and incompressible Navier–Stokes equations are employed to govern the flow behaviour during the phase change of the PCM.

$$\frac{\partial \rho}{\partial t} + \frac{\partial (\rho u_i)}{\partial x_i} = 0 \tag{1}$$

$$\frac{\partial}{\partial t}(\rho u_i) + \frac{\partial}{\partial x_j}(\rho u_i u_j) = -\frac{\partial p}{\partial x_i} + \mu \frac{\partial u_i}{\partial x_{jj}} + \rho g_i + S_{ui} \tag{2}$$

$$\frac{\partial}{\partial t}(\rho H) + \frac{\partial}{\partial x_i}(\rho u_i h) = \frac{\partial}{\partial x_i}\left(k \frac{\partial T}{\partial x_i}\right) \tag{3}$$

where ρ, u and p are the density, velocity and pressure, respectively. T is the temperature. μ is the dynamic viscosity, g is the gravity acceleration, and k is the thermal conductivity. H and h are the total enthalpy and sensible enthalpy, which are defined as follows:

$$H = h + \Delta H \tag{4}$$

$$h = h_{\text{ref}} + \int_{T_{\text{ref}}}^{T} Cp\Delta T \tag{5}$$

$$\Delta H = \phi L \tag{6}$$

where h_{ref} is the reference enthalpy at the reference temperature T_{ref}, and Cp is the specific heat. ΔH is the latent heat content changes between zero (for a solid) and L (for a liquid). ϕ is the liquid fraction during the melting process of the PCM, which can be defined:

$$\phi = \begin{cases} 0 & T \leq T_s \\ \frac{(T-T_s)}{(T_l-T_s)} & T_s < T < T_l \\ 1 & T_l \leq T \end{cases} \tag{7}$$

The source term S_{ui} describes the momentum sink due to the reduced porosity in the mushy zone and is defined as follows:

$$S_{ui} = C \frac{(1 - \phi)^2}{(\phi^3 + \varepsilon)} u_i \tag{8}$$

where C is the mushy zone constant and ε is a small number to prevent division by zero.

The numerical simulation is performed based on the commercial platform ANSYS FLUENT 18.0 [4], which uses the enthalpy-porosity technique to model the melting and solidification of the PCM. At the initial time, the temperature of the PCM is 300.15 K to obtain a solid state. The inner and outer walls are assumed with a constant temperature of 363.15 K equalling to the HTF temperature. Therefore, the initial and boundary conditions for the numerical simulation include:

$$\begin{cases} t = 0 \rightarrow T = T_{\text{ini}} = 300.15 \text{ K} \\ r = r_i \rightarrow T = T_{\text{HFT}} = 363.15 \text{ K} \\ t = r_o \rightarrow T = T_{\text{HTF}} = 363.15 \text{ K} \end{cases} \tag{9}$$

3 Results

The RT82 (Rubitherm GmbH-Germany) is employed as the PCM for this numerical study, and the thermodynamic properties can be found in Ref. [5]. Copper pipes and fins are used to achieve high thermal conductivity, and four fins are installed on the surface of inner and outer pipes, respectively, with a stagger angle of 45°, as shown in Fig. 1. The same geometrical model without fines is also performed, and the volume fraction of copper fines is considered during the comparison study. The

Fig. 1 Physical model

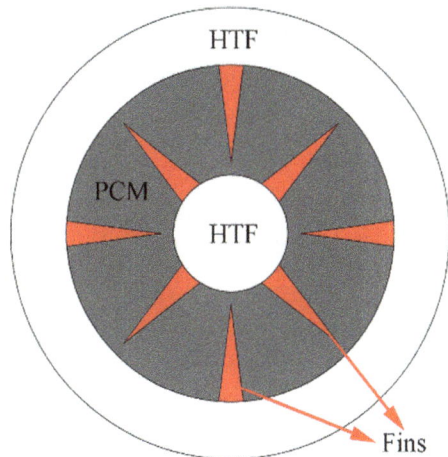

mesh independent study is carried out to test the influence of the grid density on melting processes of RT82, as shown in Fig. 2.

Figure 3 represents the liquid fraction of the PCM RT82 during the melting process with and without triangular fins while the volume fractions of the fins are considered. The numerical result shows that the melting process of RT82 is significantly accelerated by triangular fins. At $t = 30$ min, the triangular fin generates more than 80% of the liquid fraction, which less than 50% of the RT82 is melting without a

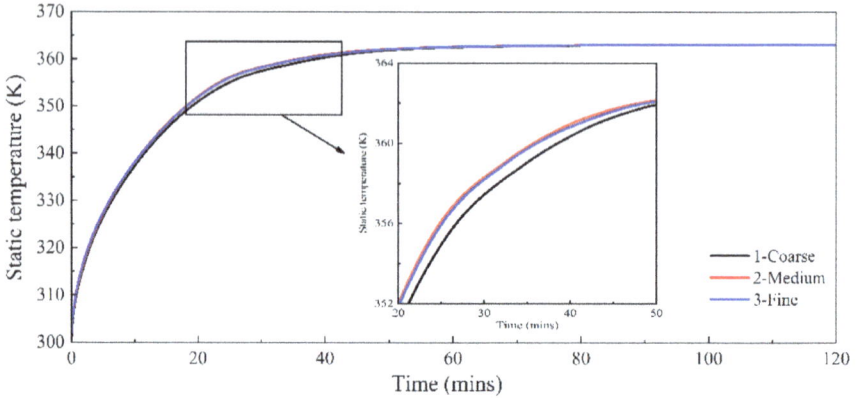

Fig. 2 Effect of grid density on averaged temperature of RT82

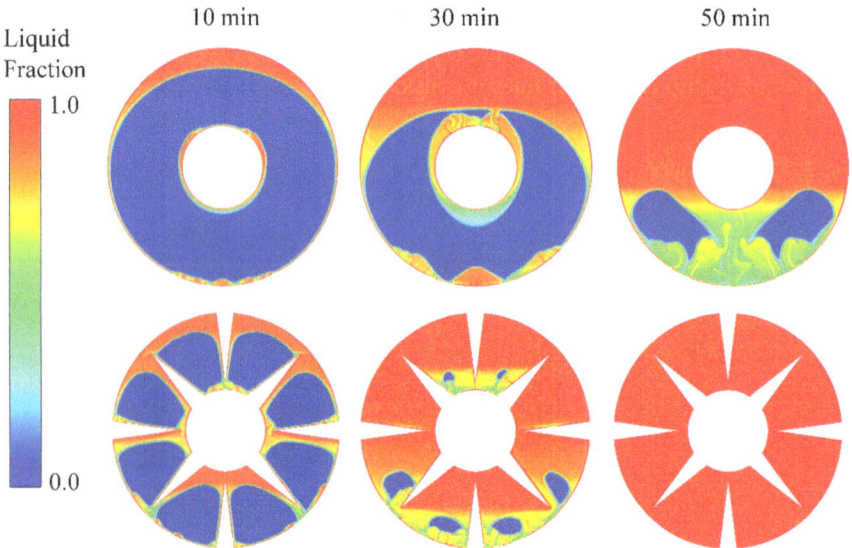

Fig. 3 Liquid fraction during the melting process of RT82

Fig. 4 Averaged liquid fraction during the melting process of RT82 with and without triangular fins

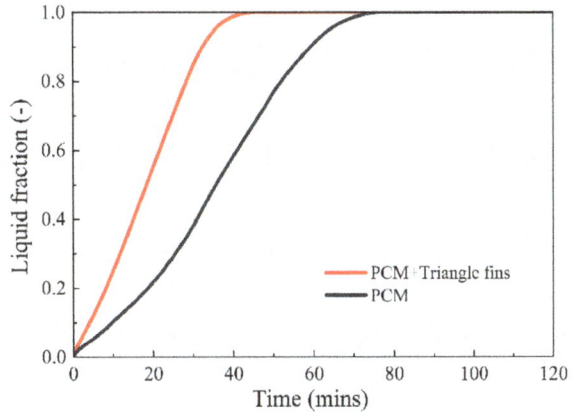

triangular fin. Even after 50 min, the liquid fraction is still less than 80% for pure PCM compared to the completed melting process for the triangular fin case.

Figure 4 shows averaged liquid fractions of RT82 during melting processes with and without triangular fins. With the enhanced heat transfer characteristics of triangular fins, the melting process is completed in 46 min. On the contrary, the pure PCM RT82 takes 77 min to reach a fully melted state, which is 1.67 times compared to the triangular fin enhanced technique. That is, the triangular fin configuration decreases the melting process time to only 60% of the pure PCM system.

4 Conclusions

The melting process of the PCM RT82 within a triple tube is evaluated with and without considering triangular fins as a heat transfer enhancement technology. The liquid fraction reaches 100% of RT82 in 77 min without installing a triangular fin inside the triple tube, while it only needs 46 min to achieve the fully melting state with the enhancement heat transfer of triangular fins. The influence of the configuration, numbers of triangular fins, and the volume fraction of fins will be evaluated in future studies.

References

1. W. Wu, W. Wu, S. Wang, Form-stable and thermally induced flexible composite phase change material for thermal energy storage and thermal management applications. Appl. Energy **236**, 10–21 (2019)
2. A.A. Al-Abidi, S. Mat, K. Sopian, M.Y. Sulaiman, A.T. Mohammad, Numerical study of PCM solidification in a triplex tube heat exchanger with internal and external fins. Int. J. Heat Mass Transf. **61**, 684–695 (2013)

3. J.M. Mahdi, S. Lohrasbi, D.D. Ganji, E.C. Nsofor, Accelerated melting of PCM in energy storage systems via novel configuration of fins in the triplex-tube heat exchanger. Int. J. Heat Mass Transf. **124**, 663–676 (2018)
4. ANSYS Fluent Theory Guide, ANSYS Inc., USA (2017)
5. S. Mat, A.A. Al-Abidi, K. Sopian, M.Y. Sulaiman, A.T. Mohammad, Enhance heat transfer for PCM melting in triplex tube with internal-external fins. Energy Convers. Manage. **74**, 223–236 (2013)

Thermal Analysis of an Earth Energy Bank

Muyiwa A. Oyinlola, Carlos Naranjo-Mendoza, Sakellariou Evangelos, Andrew J. Wright, and Richard M. Greenough

1   Introduction

Buildings account for about 40% of the overall energy demand in the EU, and low grade thermal energy, for space heating and domestic hot water, is a significant percentage of this demand [1]. Solar energy is no doubt a clean and affordable source of low temperature heat; however, its variability, positive correlation with external temperature and lack of dispatchability are problematic to use it for domestic heating. In recent times, however, research and development in thermal energy storage technologies has promoted the viability of solar energy as a domestic heating source.

One of such developments is a novel solar-assisted ground source heat pump system developed by Caplin Homes. The system comprises a geothermal heat exchanger (GHE) which is actually a very shallow borefield (1.5-m depth) of 16 boreholes, called an Earth Energy Bank (EEB), and hybrid photovoltaic/thermal (PVT) collectors. The PVT collectors coproduce electricity which is fed directly into the grid, and heat which is put into the ground via the GHE. The concept of storing thermal energy in the ground using GHEs has been studied by many such as [2–4]; however, the depths of these boreholes are usually in excess of 10 m. The peculiarity of the design in this study is in the very shallow depth, which makes it susceptible to the effects of ambient conditions [5].

This paper reports on experimental and computational studies which were carried out to characterize the thermal performance of a lab scale EEB. This study further

M. A. Oyinlola (✉) · C. Naranjo-Mendoza · S. Evangelos · A. J. Wright · R. M. Greenough
Institute of Energy and Sustainable Development, De Montfort University, Leicester L1 9BH, UK
e-mail: muyiwa.oyinlola@dmu.ac.uk

C. Naranjo-Mendoza
Departamento de Ingeniería Mecánica, Escuela Politécnica Nacional, Ladrón de Guevara
E11-253, Quito, Ecuador

© Springer Nature Singapore Pte Ltd. 2021
C. Wen and Y. Yan (eds.), *Advances in Heat Transfer and Thermal Engineering* ,
https://doi.org/10.1007/978-981-33-4765-6_148

evaluates the influence of various operation parameters, such as flow rate, temperature and operation interval, on radial/axial soil temperature distribution.

2 Methodology

A 3-D conjugate heat transfer model of a single U-tube heat exchanger was conducted using commercial CFD package, ANSYS Fluent. The 25 mm inner diameter U-tube was placed inside a 150 mm diameter cylinder of grout which was placed in a 700 mm by 560 mm by 950 mm container of soil. Simulations were run to study the profile after 8 h of heat transfer to the soil; this was to represent heat transfer on a perfect sunny day. The model was set up to simulate convection from the fluid to the U-tube inner surface, a conjugate heat transfer between the tube outer surface and the grout material as well conduction in the soil.

An experimental test rig was designed, to validate the numerical simulations. The test rig, illustrated in Fig. 1, consisted of a U-tube placed inside a tub of soil, with a grout material as an interface. Water was used as the thermal working fluid; this was supplied at constant temperature and flow rate by a circulating bath. Temperatures of the soil and grout at different points (T_{si}) as well as fluid at inlet (T_{in}) and fluid at outlet (T_{out}) were measured using Type T thermocouples. One thermocouple was placed inside the inlet and outlet ports after elbow fittings (for fluid mixing) to allow accurate readings of bulk fluid temperature. The volumetric flow rate ($\dot{V}$) was measured using an ultrasonic flow meter. All the measured quantities were logged with a 32-bit National Instruments data acquisition system (NI 9174) via LabVIEW.

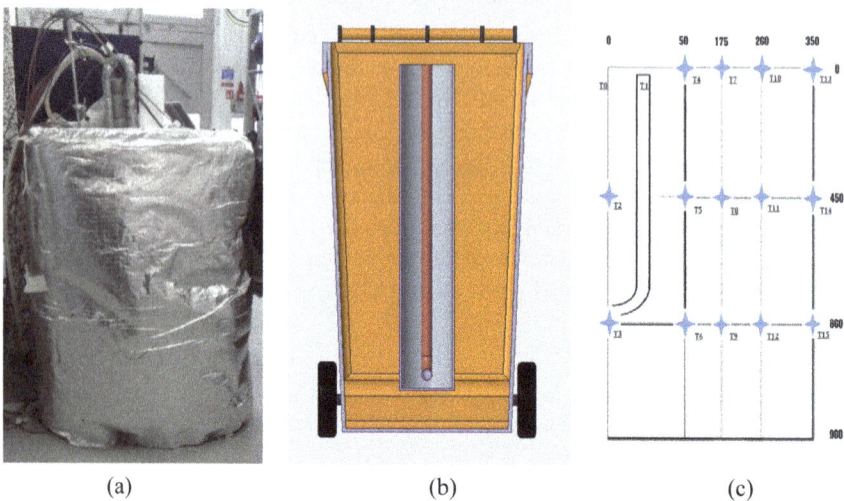

(a) (b) (c)

Fig. 1 a Experimental test rig, b schematic, c thermocouple positions

Thermocouples were connected through a NI 9213 thermocouple interface board. Signals were recorded at 5 min interval. Sensors were calibrated before installation, and signal quality was studied using an oscilloscope to ensure they were clean and free from interference.

3   Results

Temperature and heat flux data from the experiments showed good agreement with results from the computational study. Figure 2 shows an example of vertical and horizontal temperature profile after 8 h. The results showed that in the short term (8 h), the main heat transfer field is within 10 cm, while the thermal interference region is within a radius of 20 cm. The profile between a radius of 20 cm and the boundary, which is at radius 35 cm, showed a relatively constant temperature. This profile suggests significant thermal resistance between grout material and the soil. This indicates that these shallow boreholes should be closely spaced out. A quick thermal response was observed in the grout, with its temperature reaching steady state within 2 h. However, the thermal response of the soil was much slower. An increase in soil temperature improves the overall performance of the system and improves the long-term thermal balance of the soil. A negligible temperature drop in the working fluid, regardless of the flow rate, was observed after about 4 h. This low temperature drop has also been observed in full size installations. Three flow rates, 1, 2 and 4 L per minute were studied, the results showed that the effect of flow rate is only significant in the first hour, as all 3 flow rates had similar results after a few hours. Despite some insulation to the sides, it was observed that ambient conditions had some effect on the temperature profile.

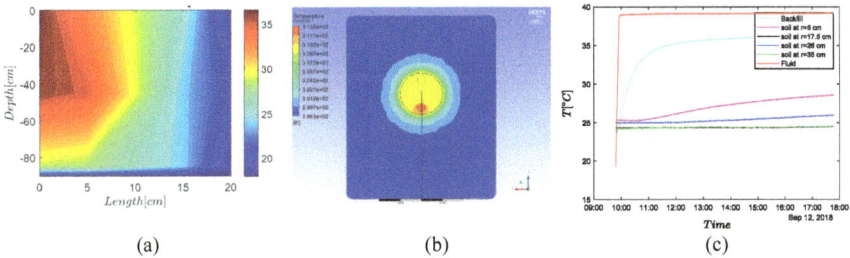

(a)   (b)   (c)

Fig. 2 a Vertical temperature profile from experiment, **b** horizontal temperature profile from simulation, **c** horizontal temperatures profile at 45 cm depth at 4 L per minute flow rate

4 Conclusions

This paper aimed to characterize the performance of very shallow borehole heat exchangers by conducting numerical and experimental analyses. The ANSYS Fluent numerical model was validated with experimental data, which illustrates its suitability for further study, such as investigation of different grout materials. The results showed good heat transfer in the backfill material; however, the temperature profile observed in the soil showed a very narrow region of heat transfer, indicating that boreholes need to be closely spaced in practical installations. The effect of increasing the flow rate was also studied and found to contribute marginal improvements. The results of the study will be significant in the design and operation of future Earth Energy Bank installations.

References

1. A. Bertrand, R. Aggoune, F. Maréchal, In-building waste water heat recovery: an urban-scale method for the characterisation of water streams and the assessment of energy savings and costs. Appl. Energy **192**, 110–125 (2017)
2. L. Zhu, S. Chen, Y. Yang, Y. Sun, Transient heat transfer performance of a vertical double U-tube borehole heat exchanger under different operation conditions. Renew. Energy **131**, 494–505 (2019)
3. L.H. Dai, Y. Shang, X.L. Li, S.F. Li, Analysis on the transient heat transfer process inside and outside the borehole for a vertical U-tube ground heat exchanger under short-term heat storage. Renew. Energy **87**, 1121–1129 (2016)
4. X. Li, C. Tong, L. Duanmu, L. Liu, Study of a U-tube heat exchanger using a shape-stabilized phase change backfill material. Sci. Technol. Built Environ. **23**(3), 430–440 (2017)
5. C. Naranjo-Mendoza, A.J. Wright, M.A. Oyinlola, R.M. Greenough, A comparison of analytical and numerical model predictions of shallow soil temperature variation with experimental measurements. Geothermics**76** (2018)

Numerical Study on Charging Process of Latent Thermal Energy Storage Under Fluctuating Thermal Conditions

Zhi Li, Yiji Lu, Xiaoli Yu, Rui Huang, and Anthony Paul Roskilly

1 Introduction

Waste heat recovery (WHR) is thought as one of the most effective technologies to reduce emissions and improve the energy utilization efficiency of existing energy systems [1, 2]. However, thermal power fluctuations have posed a great challenge on recovering the industrial waste heat due to the fluctuating and intermittent nature of the temperature and flow rate. In recent years, latent thermal energy storage (LTES) has been proposed to overcome and buffer the fluctuation of fluctuating heat sources [3, 4]. Although the heat transfer performance of LTES under constant thermal boundaries has been widely studied, the effects of dynamic thermal boundaries on the heat transfer performance of LTES need to be further revealed in order to design and optimize the LTES [5]. In this paper, the heat transfer performance of a shell-and-tube LTES is analyzed under different time-varying thermal boundaries.

2 Methodology

Figure 1 shows the principle diagram of latent thermal energy storage, which consists of an exhaust channel tube and shell filled with salt PCM.

The heat transfer model for PCM is built based on the enthalpy method. The energy equation, momentum equation and continuity equation are described as follows:

Z. Li · Y. Lu (✉) · X. Yu · R. Huang · A. P. Roskilly
Department of Energy Engineering, Zhejiang University, Hangzhou 310027, China
e-mail: luyiji0620@gmail.com

Z. Li · Y. Lu · X. Yu · A. P. Roskilly
Durham Energy Institute, Durham University, Durham D1 3RU, UK

© Springer Nature Singapore Pte Ltd. 2021
C. Wen and Y. Yan (eds.), *Advances in Heat Transfer and Thermal Engineering* ,
https://doi.org/10.1007/978-981-33-4765-6_149

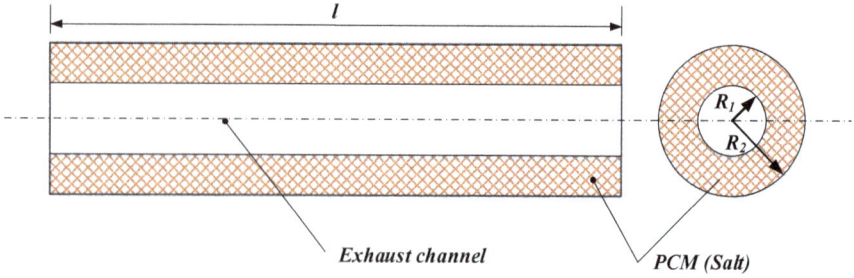

Fig. 1 Principle diagram of latent thermal energy storage

$$\frac{\partial \rho h}{\partial t} + \nabla \cdot (\rho v h) = \nabla \cdot (k \nabla T) - \frac{\partial \rho f L}{\partial t} - \nabla \cdot (\rho v f L) + S \qquad (1)$$

$$\frac{\partial \rho v}{\partial t} + \nabla \cdot (\rho v v) = -\nabla P + \nabla \cdot (\mu \nabla v) + \rho g + \frac{(1-f)^2}{f^3 + \varepsilon} v A_{mush} \qquad (2)$$

$$\frac{\partial \rho}{\partial t} + \nabla \cdot (\rho v) = 0 \qquad (3)$$

The equations are solved based on ANSYS/Fluent. The periodic heat source and a linear-varying heat source are considered, that is, the inlet temperature of exhaust gas is set as periodic function and linear-varying function, which are described by following equations. All of the heat sources are fixed at an average temperature of 300 °C in an hour.

$$T_{f,in} = a \cdot \sin (\pi t / b) + 573 \qquad (4)$$

$$T_{f,in} = a + t / b \qquad (5)$$

3 Results

Figure 2 shows the time-wise liquid volume fraction and total melting time under different heat sources. For a periodic heat source, the effects of the different period are analysed. It can be seen that the liquid volume fraction increases quickly at first and then increases with a lower rate. Smaller periods (1–6 min) almost have no effects on the melting processes while larger periods (20–60 min) can accelerate the melting process to a different extent. With the increase of period, the melting time decreases from 49.8 to 37.3 min, which is reduced by 25% compared to a constant heat source. For a linear-varying heat source, the effects of initial temperature are investigated. The liquid volume fraction increases quickly at first and then increases

Fig. 2 Time-wise liquid
volume fraction and total
melting time

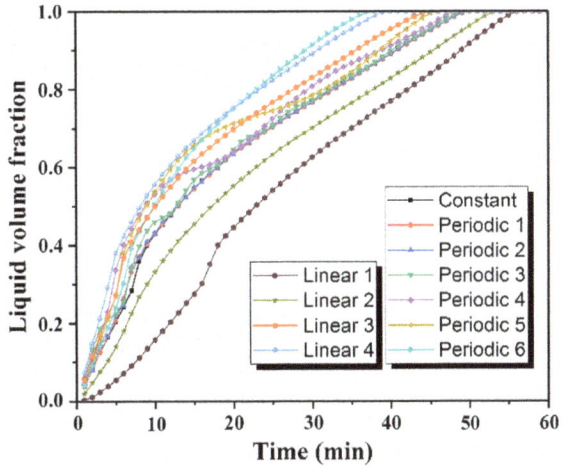

slower, but the melting time decreases with the increase of initial temperature. In detail, the melting time of the constant case ($T = 300$ °C) is 49.7 min. The melting time of case $T = 200$ °C is 55.3 min, while case $T = 400$ °C is 39.5 min, which is 11.3% larger and 20.5% smaller than that of the constant case, respectively.

Figure 3 shows the energy storage capacity of LTES under different heat sources. For a periodic heat source, the energy storage capacity drops with the increase of period. The energy storage capacity of $P = 60$ min (25.66 MJ) is 9.5% less than that of the constant case (28.34 MJ). For a linear-varying heat source, the energy storage

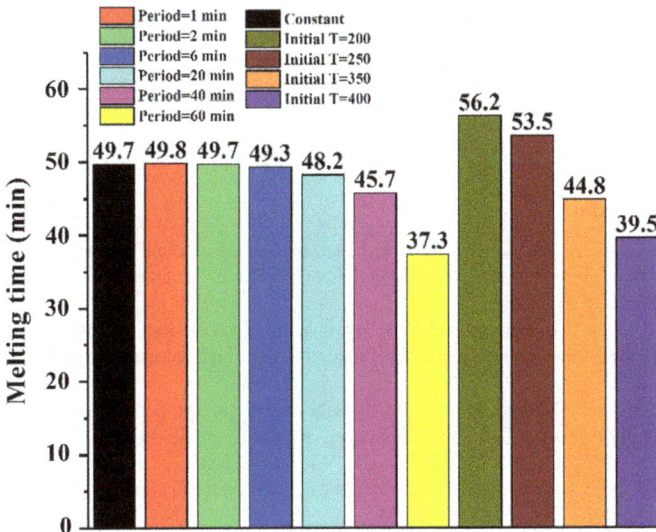

Fig. 3 Energy storage capacity of LTES under different heat sources

capacity decreases with the improvement of initial temperature. The energy storage of case $T = 200\,°C$ (30.89 MJ) and case $T = 400\,°C$ (25.19 MJ) is 9.0% smaller and 11.2% compared to that of the constant case (28.34 MJ), respectively.

4 Conclusions

Based on the above analysis, the conclusions can be drawn

1. For a periodic heat source, the period has a great influence on the heat transfer performance of LTES. The total melting time decreases with the increase of period, while the energy storage capacity shows an opposite trend.
2. For a linear-varying heat source, the initial temperature plays an important role in affecting the melting process of LTES. In detail, the total melting time decreases with the drop of initial temperature, while the energy storage capacity declines with the decrease of initial temperature.
3. In overall, time-varying heat source has significant effects on the heat transfer process of shell-and-tube LTES, which will further influence the design and optimization of LTES for waste heat recovery. More attention should be paid to having a deep insight into this problem.

Acknowledgements Support from the Newton Fund under the UK-China Joint Research and Innovation Partnership fund (Grant Number 201703780098) forms the Chinese Scholarship Council, grants from the National Natural Science Foundation of China (Grant Numbers 51976176 and 51806189), China Science Foundation (Grant Numbers 2019T120514 and 2018M640556) and from Zhejiang Province Science Foundation (Grant Number ZJ20180099) are highly acknowledged. The authors also would like to thank support from the Royal Academy of Engineering (Grant Number TSPC1098).

References

1. X. Li, H. Tian, G. Shu, C. Hu, R. Sun, L. Li, Effects of external perturbations on dynamic performance of carbon dioxide transcritical power cycles for truck engine waste heat recovery. Energy **163**, 920–931 (2018)
2. X. Yu, Z. Li, Y. Lu, R. Huang, A.P. Roskilly, Investigation of organic Rankine cycle integrated with double latent thermal energy storage for engine waste heat recovery. Energy **170**, 1098–1112 (2019)
3. Y.B. Tao, Y.L. He, Numerical study on thermal energy storage performance of phase change material under non-steady-state inlet boundary. Appl. Energy **88**, 4172–4179 (2011)
4. M. Jiménez-Arreola, R. Pili, F. Dal Magro, C. Wieland, S. Rajoo, A. Romagnoli, Thermal power fluctuations in waste heat to power systems: an overview on the challenges and current solutions. Appl. Therm. Eng. **134**, 576–584 (2018)
5. R. Elbahjaoui, H. El Qarnia, Numerical Study of a Shell-and-Tube Latent Thermal Energy Storage Unit Heated by Laminar Pulsed Fluid Flow. Heat Transfer Eng. **38**, 1466–1480 (2017)

Thermal and Electrical Property of Silicon with Metastable Phases Introduced by HPT Process

Masamichi Kohno, Mizuki Kashifuji, Kensuke Matsuda, Harish Sivasankaran, Yoshifumi Ikoma, Makoto Arita, Shenghong Ju, Junichiro Shiom, Zenji Horita, and Yasuyuki Takata

1 Introduction

Thermoelectric materials can convert waste heat directly to electric power, so it is expected as an energy material for a low carbon society. Since the performance of thermoelectric materials is determined by the dimensionless figure of merit $ZT = S^2 \sigma \, T \, k^{-1}$, ($S$: Seebeck coefficient (V/K), σ: electrical conductivity (S/m), T: temperature (K), k: thermal conductivity (W/(m K)), a lot of studies have been conducted to improve ZT. Silicon is an attractive candidate for its highest abundance and the absence of toxicity. However, the problem for bulk Si materials has been its high thermal conductivity 142 W/(m K). Therefore, application of silicon as a thermoelectric material requires reduction in the lattice thermal conductivity while maintaining electrical conductivity. For example, Allon et al. [1] prepared Si nanowire with thermal conductivity of 1.6 W/(m K) and achieved $ZT = 0.6$@RT. Tang et al. [2] fabricated nanoholes into silicon ribbon and then achieved thermal conductivity of 1.5 W/(m K) and $ZT = 0.4$@RT. From an engineering point of view, practical thermoelectric device requires a significant volume of material, and it is difficult to achieve mass production of Si-nanowires or Si-nanoribbon with nanoholes. In

M. Kohno (✉) · M. Kashifuji · K. Matsuda · Y. Takata
Department of Mechanical Engineering, Kyushu University, Fukuoka 819-0395, Japan
e-mail: kohno@mech.kyush

M. Kohno · H. Sivasankaran · Z. Horita · Y. Takata
International Institute for Carbon-Neutral Energy Research (I2CNER),, Kyushu University, Fukuoka 819-0395, Japan

Y. Ikoma · M. Arita · Z. Horita
Department of Materials Science and Engineering, Kyushu University, Fukuoka 819-0395, Japan

S. Ju · J. Shiom
Department of Mechanical Engineering, The University of Tokyo, Tokyo 113-8656, Japan

© Springer Nature Singapore Pte Ltd. 2021
C. Wen and Y. Yan (eds.), *Advances in Heat Transfer and Thermal Engineering* ,
https://doi.org/10.1007/978-981-33-4765-6_150

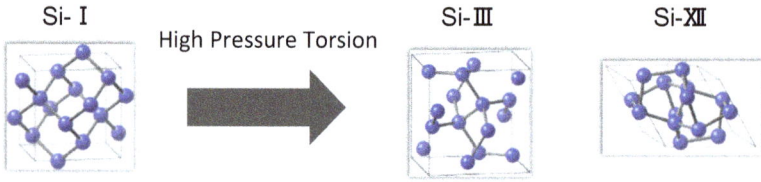

Fig. 1 Fundamental structure (left) and metastable structures (right) of silicon

the present work, we have used high pressure torsion (HPT) [3] to introduce interfaces and create nanograin boundaries to form centered cubic Si-III and rhombohedral Si-XII phases (Fig. 1). HPT is a simple method to introduce interfaces and create nanograin boundaries into bulk materials compared with other methods. These metastable phases effect the thermal transport characteristics and lead to the reduction in thermal conductivity. We have compared the structural distribution induced in metastable phases and material properties of silicon with the aim to understand the crystal structure that is most suitable for reducing the thermal conductivity. In this talk, we introduce our latest results on how the formation of metastable phases in silicon affect the material properties such as thermal conductivity and electrical conductivity.

2 Methodology

Single crystalline Si (100) wafers of size 5×5 mm^2 and thickness 500 µm were subjected to HPT processing. Details of the HPT processing in Si were described elsewhere [4]. Briefly, the HPT facility comprises upper and lower anvil made of tungsten carbide with flat bottomed spherical depressions to mount the samples. During experiments, the test samples were placed in the lower anvil, and the pressure is applied on the upper anvil. The HPT facility was operated at a pressure of 24 GPa (loading time ~7 s and unloading time ~2 s) and at room temperature. Torsional straining is achieved by rotation of the lower anvil with respect to the upper anvil at a rotation speed of 1 rpm. HPT-processed samples with 0, 10 and 20 torsion revolutions were prepared using this process. The samples were further annealed at 873 K for 2 h in nitrogen atmosphere. We performed Raman and X-ray diffraction to examine the structure of processed samples. Thermal conductivity is determined from thermal diffusivity measured by laser flash method, heat capacity (literature value) and density measured by Archimedes method. Electrical conductivity is determined by four probes method.

3 Results

Figure 1 shows structures of Si-1 and centered cubic Si-III and rhombohedral Si-XII phases. As mentioned above, these metastable phases introduced under severe plastic deformation. Figure 2 (left) shows thermal conductivity of HPT process Si from room temperature to 600 °C. Applied pressure during HPT process is 24 GPa, and number of revolutions are 10 and 20, respectively. The thermal conductivity of the sample (before HPT) is around 85 W/(m K) due to doping of impurities, and the thermal conductivity of samples is reduced to around 10 W/(m K) after HPT process. A possible reason for the dramatic reduction in lattice thermal conductivity is due to the decrease in grain size upon increasing plastic deformation. Our previous TEM investigations reported that the grain size of HPT samples reduces to as low as 10 nm during the HPT processing [5]. Phonon scattering at the nanograin boundaries increases as the grain size decreases which leads to the large reduction in the thermal conductivity. In addition, the presence of metastable Si-III/XII phases [5] creates lattice mismatch which further scatters the acoustic phonons. Based on the literature, it is anticipated that the thermal conductivity decreases with decreasing grain size.

Figure 2 (right) shows resistivity changes with respect to HPT processing ($N = 0$ and 10) and successive annealing ($N = 10$) of $n-$ (10 Ω cm) and $n+$ - (0.01 Ω cm) Si. The resistivity of the sample after compressing is increased by formation of coarse grains. The resistivity is decreased with increasing the number of revolutions. It is due to the formation of Si-III having a semi-metallic property. Even though the doping concentration is different, tendency resistivity is similar value. It is due to the grain coarsening is minor factor, and semi-metallic property is major factor under the 10 revolutions condition. By annealing, resistivity is increased. Because the metastable phase transforms to Si-I phase without grain growth.

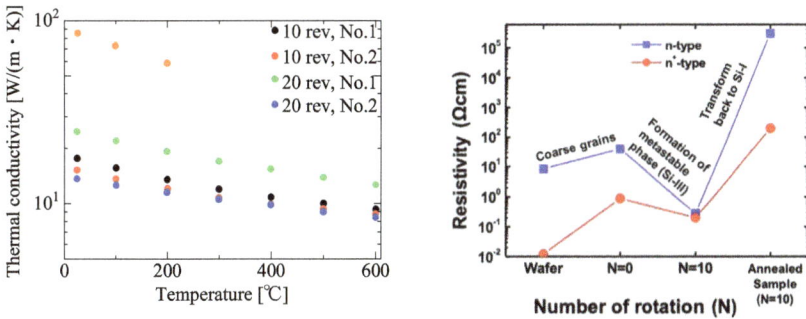

Fig. 2 (Left) Thermal conductivity of HPT process Si from room temperature to 600 °C. Applied pressure during HPT process is 24 GPa, and number of revolutions are 10 and 20, respectively. (Right) Resistivity changes with respect to HPT processing ($N = 0$ and 10) and successive annealing ($N = 10$) of $n-$ (10 Ω cm) and n+ - (0.01 Ω cm) Si

References

1. I. Allon et al., Nature **451**, 163–167 (2008)
2. J. Tang et al., Nano Lett. **10**, 4279–4283 (2010)
3. P.W. Bridgman, *Studies in Large Plastic Flow and Fracture* (McGraw-Hill, New York NY, 1952).
4. Y. Ikoma et al., J. Mater. Sci **49**, 6565 (2014)
5. Y. Fukushima et al., Mater. Charact. **129**, 163–168 (2017)

Micro Gas Turbine Range Extender Performance Analysis Using Varying Intake Temperature

R. M. R. A. Shah, M. A. L. Qubeissi, A. McGordon, M. Amor-Segan, and P. Jennings

1 Introduction

Vehicle electrification has become one of the main requirements for automotive companies due to the stricter global legislation on exhaust tail-pipe emissions and uncertainty in fossil fuel supply. However, one of the issues with the vehicle electrification technology is the range anxiety that can influence the demand for vehicles. There are several solutions to overcome this issue such as higher energy storage density and powertrain hybridization. The powertrain hybridization will be the preferable option for the short and medium-term due to the technology maturity and does not required a full scale of system development that can reduce the overall project cost and the time to market. Several studies have shown that a micro gas-turbine (MGT) can potentially be used as a range extender in a hybrid electric vehicle (HEV) [1, 2]. The MGT has advantages compared to an internal combustion engine in terms of noise, vibration and gaseous emissions reduction as well as the capability to run on multi fuels without hardware changes. Nevertheless, only a limited number of studies have been made on detailed investigations into the technical requirements of the MGT from an automotive perspective such as the effect of intake air temperature and engine bay temperature on the performance and the exhaust emissions [1, 3].

A typical configuration of MGT consists of a single-stage turbine and compressor connected by a shaft. The shaft speed is generally more than 100,000 rpm, which operates at high temperatures within the range of 315 and 370 °C. The performance of MGT however is not in proportion to the full-scale gas turbine. Fluid dynamics of the turbine are controlled by the rotor size, the distance between the compressor and

R. M. R. A. Shah (✉) · M. A. L. Qubeissi
Faculty of Engineering, Environment and Computing, Coventry University, Coventry CV1 2JH, UK
e-mail: ac9217@coventry.ac.uk

A. McGordon · M. Amor-Segan · P. Jennings
Warwick Manufacturing Group, University of Warwick, Coventry CV4 7AL, UK

© Springer Nature Singapore Pte Ltd. 2021
C. Wen and Y. Yan (eds.), *Advances in Heat Transfer and Thermal Engineering* ,
https://doi.org/10.1007/978-981-33-4765-6_151

the turbine and manufacturing tolerances of rotor design properties. Braembussche [4] has indicated that altering these properties can change the characteristics of fluid dynamics across the turbine blade and subsequently decrease the MGT net efficiency. The losses, however, can be compensated by using a recuperator system at the end of the gas turbine cycle.

Recent studies have identified several concerns that influence the behaviour of MGT, particularly the performance output and the NOx gas emission [5]. The performance output of MGT is much influenced by the properties of intake temperature. The MGT produces a constant power output at low ambient temperature before the power output starts to degrade at a *'knee point'* due to the change of air properties. With consideration of the vehicle engine bay and vehicle underfloor architectures, these air properties related concerns can potentially be elevated especially when the vehicle is operated in hot climatic countries. For instance, the small engine bay for passenger vehicles can restrict the air flow rate and increase the engine bay temperature when operating at high ambient temperatures.

In this paper, we study the effects of intake or cell temperature variation on the power output and identify the actual temperature of *'knee point'* and the emission performance, particularly NO_x emission. The other parameters such as exhaust temperature, fuel consumption and intake air flow rate will be monitored to understand the effect of intake air variation.

2 MGT Experimental Set-up

A black box MGT with 25 kW electrical power output has been used to investigate the highlighted concerns in a state-of-the-art vehicle engineering test facility. Based on the specifications, the *'knee point'* temperature of the MGT is at 18 °C. Due to the whole unit configuration, no access to the MGT controller and power electronics has been given except the power demand input. The MGT was instrumented with pressure sensors, resistance temperature detectors and thermocouples to measure all critical boundary conditions. An automotive air intake system and an exhaust system were also mounted to the MGT to mimic vehicle operational conditions [3].To measure the emissions, a pitot tube flow Metre (PTFM) is inserted at an appropriate position along the exhaust pipe to avoid any damage due to the high temperature. The recommended minimum length is 2 m. To reduce the temperature further to the optimum temperature of 150 °C, a straight 0.45 m additional stainless steel pipe is used to connect the PTFM to a gas analyzer cabinet. This length will prevent any damage to the gas analyzer cabinet and inaccurate readings due to gas condensation within the hose. The measurement of the exhaust emissions is based on raw emissions (without exhaust after treatment).

The standard test temperature for Worldwide Harmonized Lightweight Test Protocol (WLTP) is set at 23 °C compared to 20–30 °C of that New European Drive Cycle (NEDC). The value of carbon dioxide (CO_2) is corrected at 14 °C ambient temperature for the WLTP. This temperature setting will be used as the benchmark to measure the characteristics of the MGT.

Diesel fuel is supplied to the MGT through a conditioned unit fuel metre together with 5 bars filtered air to assist the atomization of fuel. DC electrical power output from the power electronics is connected to a battery bank, which consists of 20 unit 12 V 44 Ah lead-acid batteries. A load bank is used to simulate the energy requirement of the vehicle system from the battery based on a duty cycle and to protect damage to the battery bank. A program is downloaded into the load bank controller to absorb any voltage exceeding 276 V. To achieve the objectives of these studies, the intake air temperature was varied between 12 and 24 °C whilst minimizing the test cell pressure depression value to replicate the MGT and the automotive specifications. The proposed temperature range was sufficient to investigate the power degradation and the 'knee point' characteristic. Two fans in the dynamometer cell are initially turned on with appropriate speeds for 30 min until the cell temperature is maintained at 10 °C based on the ambient temperature. A constant power demand at 25 kW is then used to excite the MGT until the cell temperature is stabilized at 12 °C. Subsequently, the cell temperature is increased at 2 °C interval by reducing the fan's speed until it reaches 24 °C. At this temperature setting, the fan speeds are reprogrammed again to reduce the temperature back to 10 °C. This method will allow for the characterization of MGT for sudden temperature change.

3 Results

The MGT baseline performance was evaluated and reported prior to the work undertaken here [3]. It can be observed from Fig. 1a that there is no change in the power output level until it reaches 14 °C. It then starts to drop gradually along with the cell

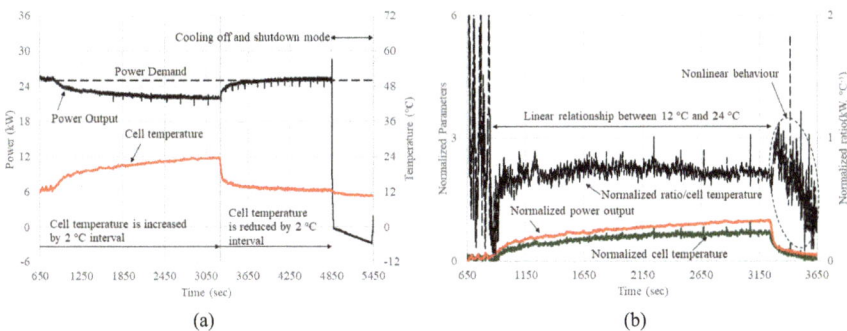

Fig. 1 MGT relationships at cell temperatures between 12 and 24 °C at constant power demand and maximum shaft speed. **a** Power output versus cell temperature, **b** normalized power output versus normalized cell temperature

temperature increase up to 24 °C. Within this cell temperature range, the fuel rate and the air flow rate are also reduced to maintain the correct air fuel ratio.

To understand the relationship between the power output and the cell temperature, both data have been normalized using the Eq. 1 between cell temperature 12 and 24 °C.

$$\text{Data}_{\text{Normalized}} = \frac{\text{abs}(\text{Data}_o - \text{Data}_{n+o})}{\text{max}(\text{Data}_{o \to n+o})} \tag{1}$$

From Fig. 1b, the ratio of both normalized data shows a linear relationship of cell temperature. The nonlinear behaviour when the cell temperature is reduced abruptly shows that the MGT is sensitive in sudden change in intake air temperature. Figure 2a shows the correlation of power output from the MGT specification and the test data. The '*knee point*' temperature is reduced from 18 to 14 °C. As a result, the power output at the standard '*knee point*' temperature is reduced by an approximately 2 kW. There are several potential reasons for this power output behaviour such as

- Compressor or turbine fouls change air aerodynamics.
- Quality of intake air change its properties.
- Recuperator leaks reduce its thermal efficiency.

The test data power output reduction rate is 0.2 kW $°C^{-1}$, the same reduction rate as per specification beyond the '*knee point*' temperature. The consequence of the power output behaviour may give an implication to the HEV energy management strategies. This is because the WLTP automotive test standard operates at 23 °C, where the loss of power output can change the vehicle speed profile and the batteries state-of-charge (SOC). Figure 2b shows the behaviour of NO_x emission. At steady-state power output, NO_x emissions are 50% below the Euro 6c limit based on WLTP. NO_x emission is observed to reduce further after the '*knee point*' temperature due to the reduction of the combustion temperature.

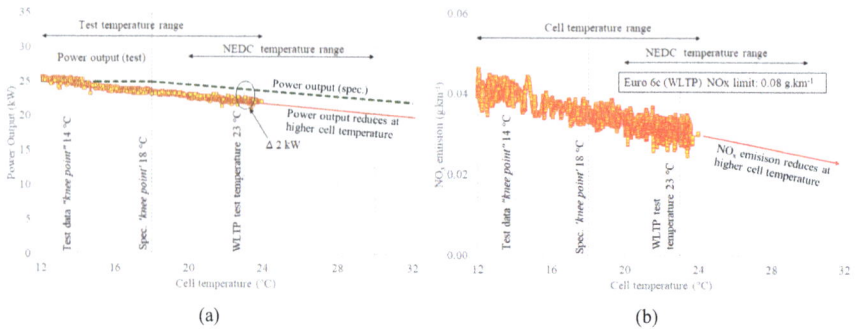

(a)   (b)

Fig. 2 MGT correlation at cell temperature variation between 12 and 24 °C with 25 kW power demand and maximum shaft speed. **a** Power output specification versus test data, **b** NO_x emission and Euro 6c limit based on WLTP.

4 Conclusions

The methods used in these studies have been used successfully to investigate the sensitivity of MGT to the cell temperature. Several findings have been highlighted to understand the implication of temperature variation to the MGT in the automotive application such as

- Linear power output reduction at '*knee point*' 14 °C instead of 18 °C.
- Power output is responsive to the sudden cell temperature change.
- NO_x emission is proportional to the power output and the cell temperature.

These findings will allow detail studies of vehicle packaging for MGT as a range extender in order to prevent any power output surge particularly at extreme hot conditions. For instance, a better air flow circulation within the engine bay using bigger radiator fans or water-cooled air intake system. The other design approach is to oversize the MGT. However, the approach needs to know the vehicle energy requirement as well as the impact of oversizing the MGT to the vehicle system. In summary, the MGT potentially can be adapted as a range extender in HEV if the power output reduction at higher cell temperature can be minimized to cope with the vehicle energy and the vehicle system requirements.

References

1. A. Karvountzis-Kontakiotis et al., Application of micro gas turbine in range-extended electric vehicles. Energy **147**, 351–361 (2018)
2. J. Ribau, et al., Analysis of four-stroke, Wankel, and microturbine based range extenders for electric vehicles. Energy Convers. Manag. 58 (2012)
3. R.M.R.A. Shah, et al., Micro gas turbine range extender—validation techniques for automotive applications, in *IET Hybrid and Electric Vehicles Conference 2013, HEVC 2013* (621 CP ed., vol. 2013), Institution of Engineering and Technology (2014)
4. R.V.D. Braembussche, Micro gas turbines–a short survey of design problems. Educational Notes RTO-EN-AVT-131, (Paper 1), 1–18 (2005)
5. C. FlavioCaresana, et al., Micro gas turbines, in*Gas Turbines*, ed. by G. Injeti (Sciyo, 2010)

Thermoacoustic Electricity Generator for Rural Dwellings in Developing World

Wigdan Kisha and David Hann

1 Introduction

Thermoacoustic seems to be a promising technology for waste heat recovery [1, 2] and are subject to steadily growing interest, particularly in the field of research. The operation of a thermoacoustic device is based on the thermoacoustic effect, which induces a complex interaction between temperature and acoustic fields. A useful thermodynamic process can be undertaken by the working gas to achieve thermal-to-acoustic power conversion when the time-phasing condition between the pressure and velocity of oscillations is successfully met in such interactions. The typical thermoacoustic engine comprises a stack or a regenerator that is placed between two heat exchangers. The three parts constitute the so-called thermoacoustic core, where the thermoacoustic effect takes place. The wave type through the stack or regenerator can be used to classify the thermoacoustic systems into two main types: standing wave systems and travelling wave systems. Extensive research efforts have been devoted to investigating standing wave thermoacoustic systems [3, 4]. However, it has been realized that the standing wave engine works on an intrinsically irreversible thermodynamic process [5]. On the contrary, the travelling wave thermoacoustic engine has a relatively simple configuration and the performance is much better than the standing wave type [6, 7]. The acoustoelectric conversion can be achieved using suitable transduction mechanisms such as linear alternators. However, they can cause reflections in the thermoacoustic field and as a result degrade the performance. Previous research has shown that the changes in the acoustic field due to the existence of the alternator can be partially minimized by installing a phase tuning techniques such as a side branch pipe [8, 9], a combination of a tuning stub and a ball valve [10], a compliance tube or a resistance tube [11], and a side branch volume [12]. A tuning stub technique is likely to look increasingly attractive from a cost point of view. The present

W. Kisha (✉) · D. Hann
Faculty of Engineering, University of Nottingham, Nottingham NG7 2RD, UK
e-mail: wigdan.kisha1@nottingham.ac.uk

© Springer Nature Singapore Pte Ltd. 2021
C. Wen and Y. Yan (eds.), *Advances in Heat Transfer and Thermal Engineering* ,
https://doi.org/10.1007/978-981-33-4765-6_152

research work describes a single-stage looped tube thermoacoustic travelling wave engine with electricity generation for rural communities. In addition, the research explores the feasibility of the stub to tune the acoustic conditions throughout the system.

2 Methodology

Typically, the thermoacoustic engine consists of six main modules: Engine core which has a regenerator in between ambient and hot heat exchangers, a thermal buffer tube, a secondary ambient heat exchanger, a feedback pipe, a loudspeaker functions as a linear alternator, and a tuning stub for phase matching (Fig. 1).

The wave propagation throughout the system captured by a Python-based software called DeltaEC [14]. The feedback pipe filled with atmospheric air and the system worked at 75 Hz frequency. The heat input power was 1.5 kW, which supplied by an electrical heater. At this input power, the optimal load resistance of the speaker is found to be 30 Ω. The stub is a pipe opened from one side and closed from the other. A BRANCH segment was used to simulate the stub, where the acoustic impedance of the stub is estimated using the following equation:

$$|Z|_s = i\omega \left[\frac{L}{3} - \frac{1}{\omega^2 C} \right] \tag{1}$$

Here: C is the acoustic compliance, L is the acoustic inertance, and both are functions of the stub dimensions and the properties of the working gas:

$$C = \frac{Al}{\gamma P_m}, L = \frac{l\rho_m}{A} \tag{2}$$

Fig. 1 General testing outline of the single core thermoacoustic electricity generator showing the main components; **a** the thermoacoustic core, **b** the loudspeaker (L.S), and **c** the water pipes, feedback pipe, and the stub. Dimensions and materials are reported in [13]

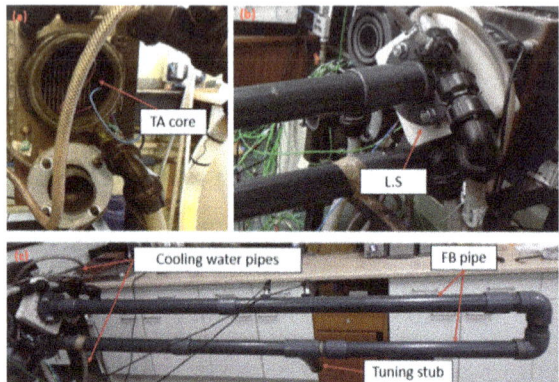

where: l and A are the length and area of the stub, respectively, P_m and ρ_m are the mean pressure and mean density of the gas, respectively.

3 Results

The stub is mainly added to the system to modify the acoustic field changed by the alternator and to reduce the acoustic losses. Therefore, investigations were performed to address the influence of the stub length on the acoustic conditions within the regenerator, the phase angle within the feedback pipe, the pressure drop through the alternator's diaphragm, the electrical power output, and the efficiency. The acoustic conditions within the regenerator and the phase angle within the feedback pipe are plotted as a function of the tuning stub length, as shown in Fig. 2a, b. The acoustic conditions within the regenerator are the impedance $|Z|$ and the time-phasing between the oscillating pressure and velocity $\varnothing$. The acoustic impedance in the regenerator,

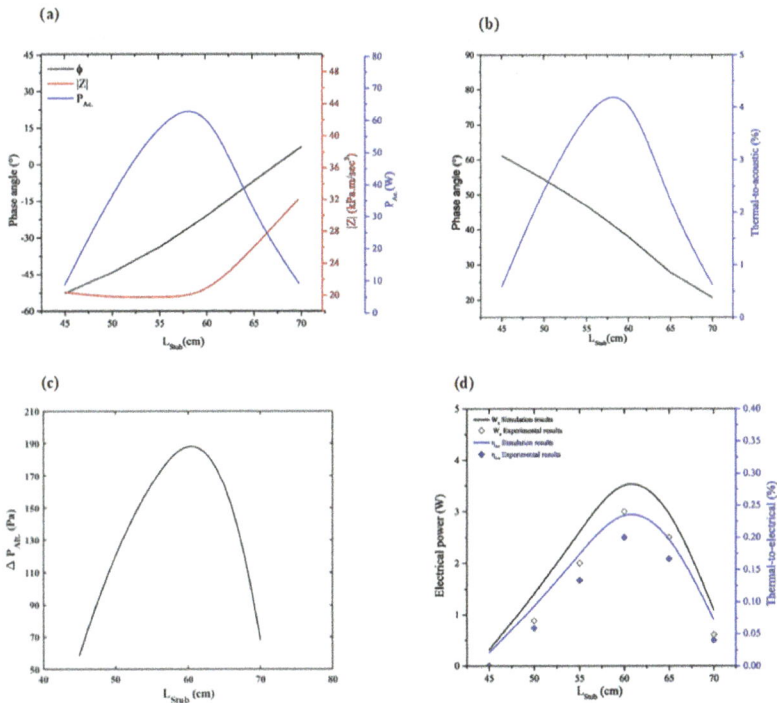

Fig. 2 Performance of the single core thermoacoustic engine with different lengths of the stub branch; **a** at the regenerator, **b** within the feedback pipe, **c** pressure drop across the loudspeaker diaphragm, and **d** electricity output and efficiency

which represents the ratio of the pressure amplitude to the volumetric velocity amplitude, increases as the length of the tuning stub increases from 55 to 70 cm. For stub length of the range 45–55 cm, the local impedance is not improved significantly. The phase angle between the pressure and the volumetric velocity of oscillation in the regenerator and the end of the feedback pipe is affected by increasing the length of the stub.

DeltaEC data showed that, at the optimum electrical power output of 3.5 W, the phase angle in the regenerator is around −21° and about 38° in the feedback pipe. A near travelling wave condition should be achieved within the regenerator and the feedback pipe to improve the acoustic power amplification within the regenerator and tune out the reflections along the feedback loop. As illustrated in Fig. 2c, as the length of the tuning stub increases, the pressure drop across the diaphragm significantly increases until it reaches an optimum value of 188 Pa, when the stub length is 61 cm. Also, as seen in Fig. 2d, an optimum electrical power output of 3.5 W can be achieved with about 0.23% thermal-to-electrical efficiency, when the stub length is 61 cm. The experimental data shows a similar trend to DeltaEC estimation, which indicates that the DeltaEC model has observed the essence of the experimental apparatus. However, the electrical acoustic power achieved in the laboratory is less than the simulation, because there was a high acoustic dissipation as well as the underestimated parasitic losses. The functionality of the stub for phase-changing justifies the dependence of the electrical power output on the length of the tuning stub.

4   Conclusions

Thermoacoustic electricity generator is a type of engine that can convert thermal power into acoustic power, and in turn into electricity. It can potentially be very useful in a wide range of applications, but it can be affected by many design parameters and has a very complex physical behavior. Therefore, designing and developing such a complex energy conversion system is challenging. In this study, a single core thermoacoustic electricity generator in a looped tube configuration, working at a frequency of 75 Hz, has been examined. The experiments were carried out using air at atmospheric pressure as a working fluid. A numerical model of the thermoacoustic engine has been developed with DeltaEC simulation software. The system comprises a single engine stage, a feedback pipe, a loudspeaker, and a matching stub. The regenerator was heated using electricity and cooled using circulated cooling water. It is found that by changing the length of the stub, the acoustic conditions in the regenerator in terms of the local acoustic impedance, and the phase difference between pressure and volumetric velocity of oscillation can be accurately adjusted. It is concluded that 61 cm long stub has enhanced the acoustic conditions and reduced the disturbance in the acoustic field caused by the alternator. The electrical power output is increased from almost zero to about 3.5 W, when the stub length was increased from 45 to 61 cm, which confirms that the stub branch is a good acoustic tuning mechanism.

Acknowledgements The author would like to thank the University of Nottingham and Gordon Memorial College Trust Fund for supporting this research.

References

1. K. De Blok, Low operating temperature integral thermoacoustic devices for solar cooling and waste heat recovery. J. Acoust. Soc. Am. **123**, 3541–3541 (2008)
2. K.O. Abdoulla-Latiwish, X. Mao, A.J. Jaworski, Thermoacoustic micro-electricity generator for rural dwellings in developing countries driven by waste heat from cooking activities. Energy **134**, 1107–1120 (2017)
3. G. Swift, "Analysis and performance of a large thermoacoustic engine," *the Journal of the Acoustical Society of America,* vol. 92, pp. 1551–1563, 1992.
4. J.J. Wollan, G.W. Swift, S.N. Backhaus, D.L. Gardner,*Development of a Thermoacoustic Natural Gas Liquefier* (Los Alamos National Laboratory, 2002)
5. S. Backhaus, G.W. Swift, A thermoacoustic-stirling heat engine: detailed study. J. Acoust. Soc. Am. **107**, 3148–3166 (2000)
6. T. Yazaki, A. Iwata, T. Maekawa, A. Tominaga, Traveling wave thermoacoustic engine in a looped tube. Phys. Rev. Lett. **81**, 3128 (1998)
7. S. Backhaus, G. Swift, A thermoacoustic Stirling heat engine. Nature **399**, 335 (1999)
8. Z. Yu, A.J. Jaworski, S. Backhaus, Travelling-wave thermoacoustic electricity generator using an ultra-compliant alternator for utilization of low-grade thermal energy. Appl. Energy **99**, 135–145 (2012)
9. A. Kruse, A. Ruziewicz, M. Tajmar, Z. Gnutek, A numerical study of a looped-tube thermoacoustic engine with a single-stage for utilization of low-grade heat. Energy Convers. Manage. **149**, 206–218 (2017)
10. H. Kang, P. Cheng, Z. Yu, H. Zheng, A two-stage traveling-wave thermoacoustic electric generator with loudspeakers as alternators. Appl. Energy **137**, 9–17 (2015)
11. T. Jin, R. Yang, Y. Wang, Y. Liu, Y. Feng, Phase adjustment analysis and performance of a looped thermoacoustic prime mover with compliance/resistance tube. Appl. Energy **183**, 290–298 (2016)
12. A. Al-Kayiem, Z. Yu, Using a side-branched volume to tune the acoustic field in a looped-tube travelling-wave thermoacoustic engine with a RC load. Energy Convers. Manage. **150**, 814–821 (2017)
13. P.H. Riley,*Designing a Low-Cost Electricity-Generating Cooking Stove for High-Volume Implementation*(University of Nottingham, 2014)
14. J.P. Clark, W.C. Ward, G.W. Swift, Design environment for low-amplitude thermoacoustic energy conversion (DeltaEC). J. Acoust. Soc. Am. **122**, 3014–3014 (2007)

Heat Transfer Within PCM Heat Sink in the Presence of Copper Profile and Local Element of the Time-Dependent Internal Heat Generation

Nadezhda S. Bondareva and Mikhail A. Sheremet

1 Introduction

Thermal control systems based on phase change materials have recently become more and more common in modern electronic and radio-electronic structures. In a wide range of power and sizes of such structures containing phase change materials, detailed studies are required due to the complexity of the processes occurring in the melt and the variety of shapes of profiles. Studies on heat and mass transfer in the phase change material (PCM) are complicated due to the interaction of convective cells with solid structures and the interphase boundary [1–3]. Numerical problems of melting show that convective heat transfer plays an important role in the heat dissipation process. The intensification of heat transfer owing to the complexity of the design leads to the complication of hydrodynamics and heat transfer inside the system. Thus, Baby and Balaji [4] showed that at the upper temperature limit 42 °C and the source power 8 W the operating time of the device may be increased by more than 4 times using the PCM heat sink in comparison with the heat sink without PCM. In the conditions of the set, maximum temperature of 47 °C and the source power of 7 W using of PCM allows to increase the operating time by eleven times. More additional analysis of PCM behavior in different regions can be found in [5–7].

The present study is devoted to the numerical simulation of the paraffin melting inside a closed radiator heated from a source of time-dependent volumetric heat generation.

N. S. Bondareva · M. A. Sheremet (✉)
Laboratory on Convective Heat and Mass Transfer, Tomsk State University, Tomsk, Russia
e-mail: sheremet@math.tsu.ru

© Springer Nature Singapore Pte Ltd. 2021
C. Wen and Y. Yan (eds.), *Advances in Heat Transfer and Thermal Engineering* ,
https://doi.org/10.1007/978-981-33-4765-6_153

2 Mathematical Formulation and Results

The considered computational domain is presented in Fig. 1. From the beginning of
the source operation, the material was in a solid state, the temperature throughout
the entire system, including the heat source, coincided with the ambient temperature that was below the melting point. Source operation was determined by the
condition $Q = Q_0(1 - \sin(ft))$. At the lateral boundaries and the upper wall, convective heat exchange with the environment was determined by the Biot number, and
the remaining walls were considered to be thermally insulated. The motion of a
viscous Newtonian fluid, which was considered to be a melt, was described by the
system of the Oberbeck–Boussinesq equations using the non-primitive variables such
as stream function and vorticity. The energy equation for paraffin was formulated
taking into account the latent energy of melting and the smoothed transition for the
thermophysical properties. The finite difference method combined with locally one-
dimensional Samarskii scheme was used for numerical analysis. In the radiator and
the heat source, the temperature at each time step was determined employing the
heat conduction equations.

To describe the process of conjugate heat transfer taking into account the natural
convection and phase transformations in dimensionless variables, the following
similarity criteria were chosen: Prandtl number $\mathrm{Pr} = v\rho_l c_l / k_l$, Rayleigh number
$\mathrm{Ra} = g\beta \Delta T H^3 / (v\alpha_l)$, Stefan number $\mathrm{Ste} = L_f / (c_l \Delta T)$, Ostrogradsky number
$\mathrm{Os} = Q H^2 / (k_2 \Delta T)$, Biot number $\mathrm{Bi} = h H / k$, and oscillation parameter f. The
developed computational code was verified using numerical and experimental data
of other authors, and it was shown good accuracy for the melting problems [8, 9].

In this study, the main attention was paid to the influence of the heat generation
mode of the source on the heat transfer performance and hydrodynamic behavior
within the system. The impact of the source power and its frequency was analyzed
and presented in Figs. 2 and 3. The calculations were carried out in a wide range of
the governing parameters. Figure 2 shows the temperature fields for two different
values of the oscillation parameter f. The high thermal conductivity of the radiator

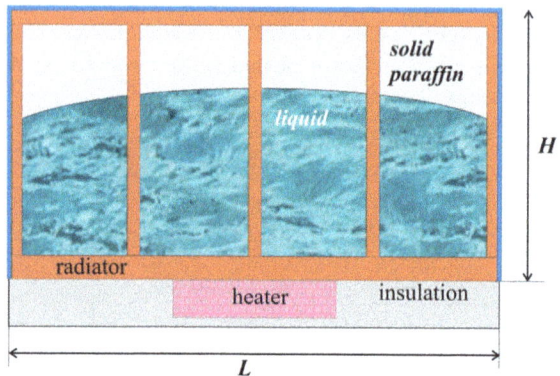

Fig. 1 Considered domain of interest

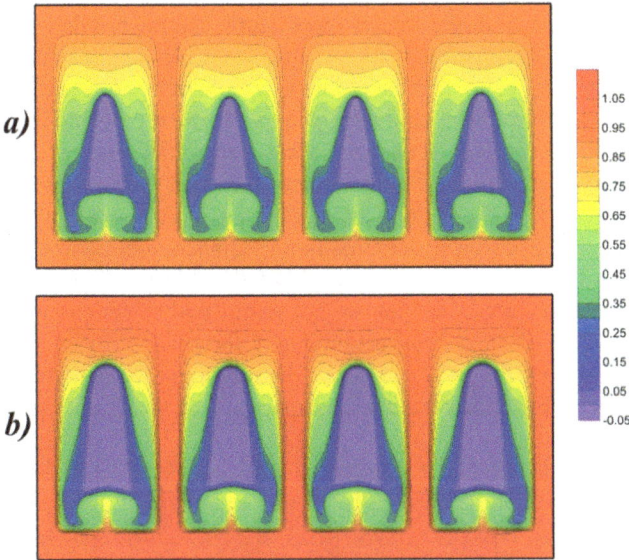

Fig. 2 Temperature field for Ra $= 4.03 \cdot 10^6$, Os $= 0.845$ at $\tau = 509$ a–$f = 0.02\pi$; b–$f = 0.005\pi$

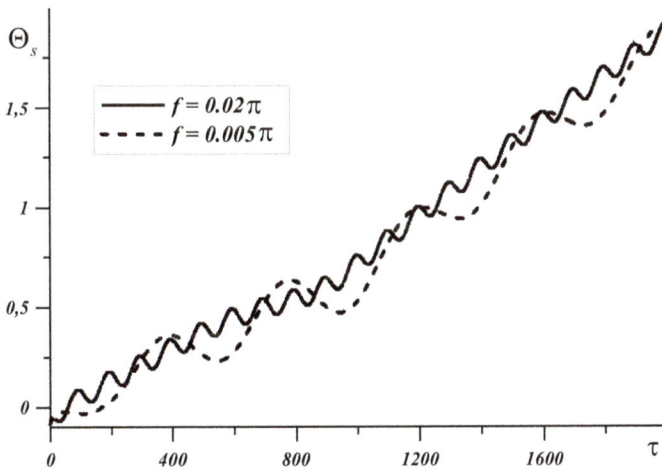

Fig. 3 Average source temperature time profiles for Ra $= 4.03 \cdot 10^6$, Os $= 0.34$ at different parameters f

contributes to the rapid heat dissipation and the temperature within the framework of the profile varies slightly. Longer heat transfer processes occur in compartments filled with phase change material. Natural convection leads to heating of the upper part of the melt region, the heated liquid rises along the side walls, and the cold melt descends to the lower solid surfaces. Periodic change in the source power leads to the oscillatory behavior of the source temperature, and the profile temperature also changes rapidly. In Fig. 2b, it can be seen that the unmelted material takes up more space than in the case of 2a, and the profile temperature in the case of $f = 0.005\pi$ is higher. Along with the change in the temperature of the profile, the intensity of heat exchange with the environment and the melting rate of the material is also changed.

Figure 3 shows the average source temperature as a function of time. It is seen that with an increase in the heating and cooling period the amplitude of the source temperature fluctuations increases; that is, the temperature minima in the case of $f = 0.005\pi$ is lower than in the case of $f = 0.02\pi$. At the same time, it is worth noting that the temperature maxima are not significantly different, which may be a consequence of more intense convection during the heating period at a frequency $f = 0.005\pi$. Therefore, this process should be studied in more detail at different values of the source power and the period of its change.

Acknowledgements This work was supported by the Russian Science Foundation (Project No. 17-79-20141).

References

1. R. Kandasamy, X.Q. Wang, A.S. Mujumdar, Application of phase change materials in thermal management of electronics. Appl. Therm. Eng. **27**(17–18), 2822–2832 (2007)
2. S.C. Fok, W. Shen, F.L. Tan, Cooling of portable hand-held electronic devices using phase change materials in finned heat sinks. Int. J. Therm. Sci. **49**(1), 109–117 (2010)
3. X. Xu, Y. Xiao, Z. Yu, L. Fan, J. Lu, Y. Zeng, Effect of the inclination angle on the transient performance of a phase change material-based heat sink under pulsed heat loads. J. Zhejiang Univ. Sci. A **15**(10), 789–797 (2014)
4. R. Baby, C. Balaji, Thermal performance of a PCM heat sink under different heat loads: an experimental study. Int. J. Therm. Sci. **79**, 240–249 (2014)
5. A. El Ouali, T. El Rhafiki, T. Kousksou, A. Allouhi, M. Mahdaoui, A. Jamil, Y. Zeraouli, Heat transfer within mortar containing micro-encapsulated PCM: numerical approach. Constr. Build. Mater. **210**, 422–433 (2019)
6. R.D. Beltran, J. Martinez-Gomez, Analysis of phase change materials (PCM) for building wallboards based on the effect of environment. J. Build. Eng. **24**, 100726 (2019)
7. M. Bashar, K. Siddiqui, Experimental investigation of transient melting and heat transfer behavior of nanoparticle-enriched PCM in a rectangular enclosure. J. Energy Storage **18**, 485–497 (2018)
8. N.S. Bondareva, M.A. Sheremet, Conjugate heat transfer in the PCM-based heat storage system with finned copper profile: application in electronics cooling. Int. J. Heat Mass Transf. **124**, 1275–1284 (2018)
9. N.S. Bondareva, B. Buonomo, O. Manca, M.A. Sheremet, Heat transfer inside cooling system based on phase change material with alumina nanoparticles. Appl. Therm. Eng. **144**, 972–981 (2018)

Effects of Bionic Models with Simultaneous Thermal Fatigue and Wear Resistance

Dahui Yu, Ti Zhou, Hong Zhou, Haiqiu Lu, Haifeng Bo, and Yuying Yan

1 Introduction

Drum brake is one of important components in braking system. Because of complicated working environment and wrong operation like overload, drum brakes would be damaged earlier than expected [1]. According to interdisciplinary research, bionics have great applications in many fields [3], and bionics has had a profound influence on materials science including grey cast iron. Studies have shown that bionic coupling surface could effectively improve thermal fatigue and wear resistance of grey cast iron.

In this study, enlightened by leaves and dragonfly wings [3], an idea of processing a similar surface, which is soft part like lamina and membrane, hard part like veins, soft and hard structure could make dragonfly wings flap 40 times per second and leaves exposed in wind and rains without being damaged [4]. As such, the bionic coupling surface was processed by pulse laser. The parameters need to be selected for processing good quality bionic coupling unit. A three-level three-factor orthogonal test was designed. However, failure mechanism of drum brake is particular. With regular single model, bionic coupling processing cannot perform its powerful function on drum brake [5].

D. Yu · T. Zhou (✉) · H. Zhou · H. Lu · H. Bo
Key Laboratory of Automobile Materials (Jilin University), Ministry of Education, Changchun 130025, PR China
e-mail: tizhoujlu@163.com

D. Yu · H. Zhou · H. Lu · H. Bo
School of Material Science and Technology, Jilin University, Changchun 130025, PR China

T. Zhou
School of Mechanical Science and Engineering, Jilin University, Changchun 130025, PR China

Y. Yan
Faculty of Engineering, University of Nottingham, Nottingham NG7 2RD, UK

© Springer Nature Singapore Pte Ltd. 2021
C. Wen and Y. Yan (eds.), *Advances in Heat Transfer and Thermal Engineering* ,
https://doi.org/10.1007/978-981-33-4765-6_154

According to the failure mechanism of drum brake in this study, multiple single models were used for various combinations, and the effects of different angles of each double combination model were studied as well. The best double combination model was chosen by considering the best performance of thermal fatigue resistance, wear resistance and processing efficiency. Every single model was also selected by the results of thermal fatigue test and wear test, and the bionic coupling drum brake with the best double combination model was then subjected to bench test.

2 Methdology

Experimental samples of $40 \times 20 \times 6$ mm^3 were cut by electric spark machine; bionic coupling models were processed by Nd-YAG laser of 1.06 μm wavelength and maximum power of 800 W. Six double combination models were made up of five different angels' stripes (0°, 30°, 45°, 60°, 90°) and a 45°mesh with a frame processed along the four edges of samples. Bionic coupling samples were subjected to thermal fatigue cold cycling at temperatures ranging from 25 to 700 °C, where each was heated and cooled for 180 and 5 s, and then bionic coupling samples were tested for wear resistance, using MG-200 wear tester, under the following conditions: load 80 N, speed 700 rpm, and wear time 60 h, with the friction pair was fabricated from a 45# steel with a hardness of 50 HRC. After the above experiments, the best double combination model was processed by pulse laser on the surface of drum brake, and then the bionic coupling drum brake was subjected to bench test, which was speeded up from 0 to 96 km/h and then braking to 0, which took 90 s as one cycle.

3 Results

3.1 Orthogonal Test

Orthogonal test design constitutes an effective scientific method for performing complex multifactor tests. Therefore, a three-level three-factor orthogonal test for drum brakes was designed in this study. The results in Table 1 show current, pulse width, frequency, defocusing amount, and moving speed of mechanical arm.

3.2 Thermal Fatigue Test

The results showed that fewer cracks formed on the bionic coupling surface, but many long cracks appeared on the surface of untreated samples. The stress, material cracking and deformation were led by temperature changes. The surface layer

Table 1 Parameters of orthogonal test

No.	Current (A)	Pulse width (ms)	Frequency (Hz)
1	125	6	10
2	125	7	12
3	125	8	14
4	135	6	12
5	135	7	14
6	135	8	10
7	145	6	14
8	145	7	10
9	145	8	12

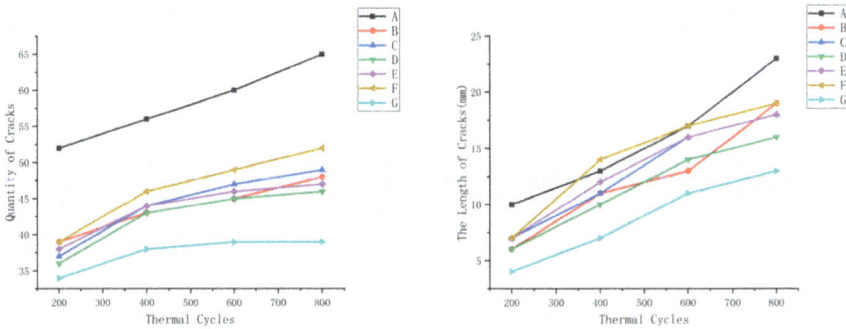

Fig. 1 Quantity and the length of cracks

expanded in dimension at the beginning of cooling, the surface layer of sample shrank firstly while temperature of inter was still high, so the compressive stresses were produced. The best model for thermal fatigue resistance is 0 degree angle which was perpendicular to the direction of cracks' growth with 45 degree mesh double combination as shown in Fig. 1.

3.3 Wear Test

Wear test results showed that less wear tracks on the surface of bionic coupling samples than the normal samples. Due to the higher hardness of the units, hard–soft structure could effectively resist deformation. In addition, the unit itself could eliminate the mechanical stress generated by graphite as the source of cracks. According to the results in Fig. 2, 45° mesh double combination model yielded the largest friction coefficients and minimum weight loss .

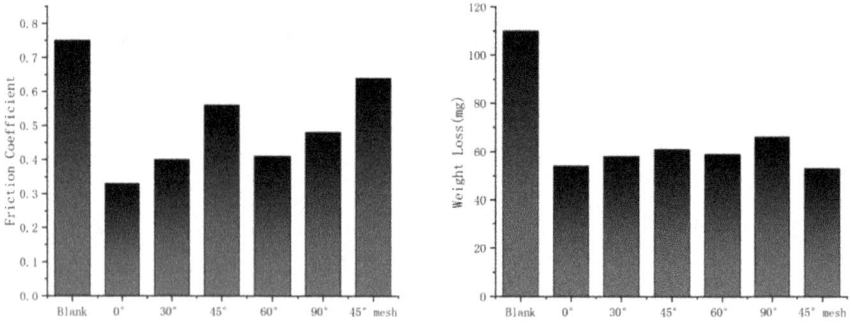

Fig. 2 Friction coefficient and weight loss of each samples

3.4 Bench Test

Two drum brakes were subjected to bench test, combined bionic coupling drum brake and an untreated blank control. Under the same conditions, they both speeded up to 96 km/h and then braking until the speed dropped to 0 km/h every 10 cycles observed the growth of cracks once. By the time the crack propagated through the outer wall of the brake means the brake failed.

According to the results of bench tests, untreated blank control drum brake failed at 370 times, but combined bionic coupling drum brake until 530 failed, and the quantity of cracks on the surface of bionic drum brake was much less than untreated drum brake.

4 Conclusions

The best parameter of processing drum brake is 145 A, 7 ms, 10 Hz.

45° mesh double combination model has the best performance in both thermal fatigue and wear resistance.

Drum brake with bionic double combination model surface has much longer service life than untreated blank control drum brake.

Acknowledgements This work was supported by Project 985-High Performance Materials of Jilin University, Project 985-Bionic Engineering Science and Technology Innovation and double first-class project by Jilin Province and Jilin University (SXGJXX2017-14).

References

1. D.H. Yu, T. Zhou, H. Zhou, H.F. Bo, H.Q. Lu, Non-single bionic coupling model for thermal fatigue and wear resistance of gray cast iron drum brake. Opt. Laser Technol. **111**, 781–788 (2019)
2. C.B. Liu, Y.J. Wang, L.Q. Ren, L. Ren, A Review of Biological Fluid Power Systems and Their Potential Bionic Applications. J. Bionic Eng. **16**(3), 367–399 (2019)
3. L.Q. Ren, B.Q. Li, Z.Y. Song, Q.P. Liu, L. Ren, X.L. Zhou, Bioinspired fiber-regulated composite with tunable permanent shape and shape memory properties via 3d magnetic printing. Compos. Part B-Eng. **164**, 458–466 (2019)
4. J.H. Zhi, L.Z. Zhang, Y.Y. Yang, J. Zhu, Mechanical durability of superhydrophobic surfaces: the role of surface modification technologies. Appl. Surf. Sci. **392**, 286–296 (2017)
5. Y. Liu, X.L. Li, J.F. Jin, J.A. Liu, Y.Y. Yan, Z.W. Han et al., Anti-icing property of bio-inspired micro-structure superhydrophobic surfaces and heat transfer model. Appl. Surf. Sci. **400**, 498–505 (2017)

Experimental Study and Sensitive Simulation of a Heat Pipe Photovoltaic/Thermal System

T. Zhang, Z. W. Yan, and H. D. Fu

1 Introduction

The photovoltaic/thermal (PV/T) collector/system is an integration of PV cells and photothermal technology which can, therefore, translate solar irradiation into electric energy and heat energy simultaneously. PV/T has been extensively explored, and a considerable amount of researches has been carried out during the last decades. Among the proposed studies, a water-based PV/T system is highly complimented because it is able to achieve a higher overall efficiency [1]. However, despite some progress has been made, at least two major obstacles, which are frozen and corrosion problems, must be overcome when a water-based PV/T system is working in winter or high-latitude areas.

Two-phase closed thermosyphons (TPCT) are introduced to overcome the mentioned problems associated with the traditional water-based PV/T collector. At the same time, the high heat transfer efficiency turns the TPCT-PV/T collector into an attractive option for solar application and has drawn significant attention in recent years [2].

In this study, a TPCT–PV/T system was designed and experimentally studied. A dynamic numerical model, which is validated by experimental data, is established to evaluate the photothermal and photovoltaic performance. Based on the validated model, sensitivity analysis of some key design elements is conducted, and the optimal design parameters are finally suggested.

T. Zhang (✉) · Z. W. Yan
College of Energy and Mechanical Engineering, Shanghai University of Electric Power, Shanghai, China
e-mail: zhtyn86@163.com

H. D. Fu
College of Chemistry and Environmental Engineering, Shenzhen University, Shenzhen, China

© Springer Nature Singapore Pte Ltd. 2021
C. Wen and Y. Yan (eds.), *Advances in Heat Transfer and Thermal Engineering* ,
https://doi.org/10.1007/978-981-33-4765-6_155

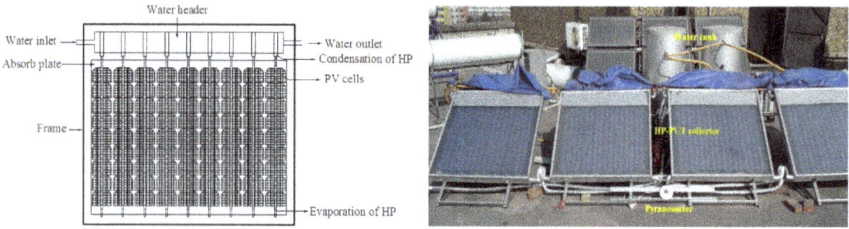

Fig. 1 Schematic diagram of the TPCT-PV/T collector (left) and system (right)

2 Methodology

2.1 Experimental Setup

The schematic diagram of the proposed TPCT-PV/T collector and system is shown in Fig. 1. For the TPCT-PV/T collector, an aluminum plate was employed as the base panel; a transparent tedlar–polyester–tedlar (TPT) was set on the frontage to encapsulate and protect the PV cells; and a black TPT was set on the back side to insulate the PV cells and enhance heat absorption; the transparent TPT, PV cells and base plate were adhesive bonded by ethylene–vinyl acetate (EVA). Evaporation of TPCT was laser welded on the back of aluminum panel, and the condensation of TPCT was inserted into a square water header. For the TPCT-PV/T system, it mainly consists of four TPCT-PV/T collectors, four lead-acid storage batteries, two water tanks, a water circulation pump, a turbine flow meter and a MPPT PV controller; the four TPCT-PV/T collectors were installed in parallel, and a ball valve was set at the inlet of water header of every collector to ensure a constant mass flow.

2.2 Theoretical Model

For the falling liquid film at condensation and evaporation of TPCT, the momentum balance equation is:

$$\mu_l \left(\frac{\partial^2 u_1}{\partial y^2} \right) + (\rho_l - \rho_v) g \sin \theta = 0 \tag{1}$$

The associated boundary conditions are:

$$y = 0 : u_l = 0, \quad T = T_w; \quad y = \delta : \mu_l \left(\frac{\partial u_l}{\partial y} \right) = -\tau_i, \quad T = T_s \tag{2}$$

The shear stress at the liquid–vapor interface can generally be expressed as [3]:

$$\tau_i = \tau_{ia} \pm \tau_{id} = \tau_{ia} \pm (u_{li} + u_v)\frac{d\Gamma}{dx} \tag{3}$$

where the plus and minus signs of the second term on the right-hand side of Eq. (3) account for the effect of condensation and evaporation at the liquid–vapor interface, respectively.

The film heat transfer coefficient of the falling film under laminar and turbulent is, respectively, expressed as:

$$h = {k_l}/{\delta}, \quad h = 0.056 \mathrm{Re}^{-0.2} \, \mathrm{Pr}^{\frac{1}{3}} \, k_l \left({v_l^2}/{g}\right)^{\frac{1}{3}} \tag{4}$$

While the heat transfer coefficient of the pool boiling can be calculated as [4]:

$$\mathrm{Nu}_{nc} = 0.475 \mathrm{Ra}^{0.36} \left({I_m}/{d}\right)^{0.58}$$
$$\mathrm{Nu}_{nb} \left({I_m}/{d}\right)^{0.58} = (1.0 + 4.95\psi)\mathrm{Nu}_{ku}$$
$$\mathrm{Nu}_{cc} = (\mathrm{Nu}_{nc}^4 + \mathrm{Nu}_{nb}^4)^{0.25} \tag{5}$$

The effective heat transfer coefficient of the evaporation of TPCT is expressed as:

$$h_{\mathrm{eva}} = h_{fe}\frac{L_{fe}}{L_{\mathrm{eva}}} + h_{pb}\frac{L_p}{L_{\mathrm{eva}}} \tag{6}$$

After obtaining the heat transfer coefficient of the evaporation and condensation of TPCT, the energy balance equations of the key components should be further established. According to the heat transfer process in order which are the glass cover, PV module, base panel, evaporation of TPCT, condensation of TPCT, water header and water tank, there are summarized as followed:

$$d_g \rho_g c_g \frac{\partial T_g}{\partial t} = h_{a,g}(T_a - T_g) + h_{\mathrm{sky},g}(T_{\mathrm{sky}} - T_g) + h_{g,pv}(T_{pv} - T_g) + G\alpha_g$$

$$\gamma d_{pv} \rho_{pv} c_{pv} \frac{\partial T_{pv}}{\partial t} = h_{g,pv}(T_g - T_{pv}) + (T_b - T_{pv})/R_{b,pv} + G(\tau\alpha)_{pv} - \gamma E_{pv}$$

$$\rho_b c_b \frac{\partial T_b}{\partial t} = k_b \frac{\partial^2 T_b}{\partial x^2} + \frac{1}{d_b}\left[(T_a - T_b)/R_{b,a} + (T_{pv} - T_b)/R_{b,pv}\right]$$
$$+ \frac{1}{d_b}\left[(T_{hp,\mathrm{eva}} - T_b)/(R_{hp,b} \cdot A_{bi})\right]^+$$

$$M_{\mathrm{eva},hp} c_{hp} \frac{\partial T_{\mathrm{eva},hp}}{\partial t} = (T_{\mathrm{con},hp} - T_{\mathrm{eva},hp})/R_{\mathrm{eva,con}} + (T_b - T_{\mathrm{eva},hp})/R_{hp,b}$$
$$+ (T_a - T_{\mathrm{eva},hp})/R_{b,a}$$

$$M_{\mathrm{con},hp} c_{hp} \frac{\partial T_{\mathrm{con},hp}}{\partial t} = (T_{\mathrm{eva},hp} - T_{\mathrm{con},hp})/R_{\mathrm{eva,con}} + (T_w - T_{\mathrm{con},hp})/R_{\mathrm{con},w}$$

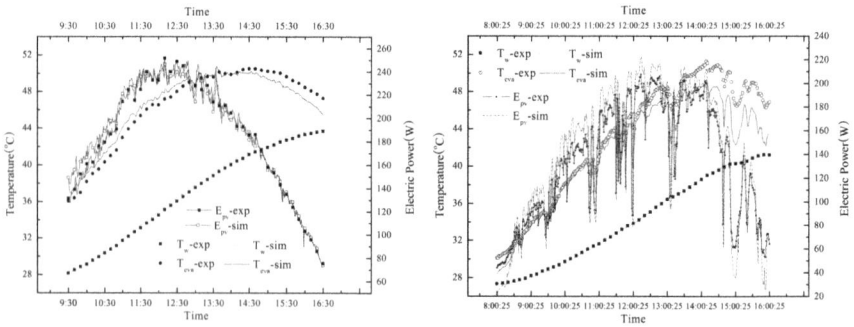

Fig. 2 Simulation and experimental results under a sunny day (left) and a cloudy day (right)

$$M_w c_w \frac{\partial T_{w,j}}{\partial t} + \dot{m} c_w (T_{w,j} - T_{w,j-1}) = \frac{T_a - T_{w,j}}{R_{w,a}} + \frac{T_{con,hp} - T_{w,j}}{R_{con,w}}$$

$$M_w c_w \frac{\partial T_{wt}}{\partial t} = \frac{T_a - T_{wt}}{R_{wt,a}} + n \cdot \dot{m} c_w (T_{w,out} - T_{w,in}) \tag{7}$$

3 Results

3.1 Model Validation

Two days, which was a sunny and a cloudy day, with the different variations of solar irradiation and ambient temperature are invited to validate the numerical model. The initial values in the simulation are set according to the experiments. The comparisons between the simulations and experiments are shown in Fig. 2, in which performances of photothermal and photovoltaic are presented in detail. Temperature of the TPCT's evaporation, which is an important intermediate parameter, is also paid attention to. The mean relative errors between the simulation and experimental results of water temperature, electric generation and temperature under the sunny day are 0.9%, 3.3% and 1.8%, respectively, while that are 0.4%, 4.5% and 3.6%, respectively, under the cloudy day. The simulations agree well with the experiments.

3.2 Sensitive Simulation

Influences of the diameter of the TPCT's condensation and evaporation, the length of condensation and the width of water header to the photothermal and photovoltaic performance of TPCT-PV/T system are plotted in Fig. 3. The results show that system photothermal and photovoltaic performance increase with the increase of the

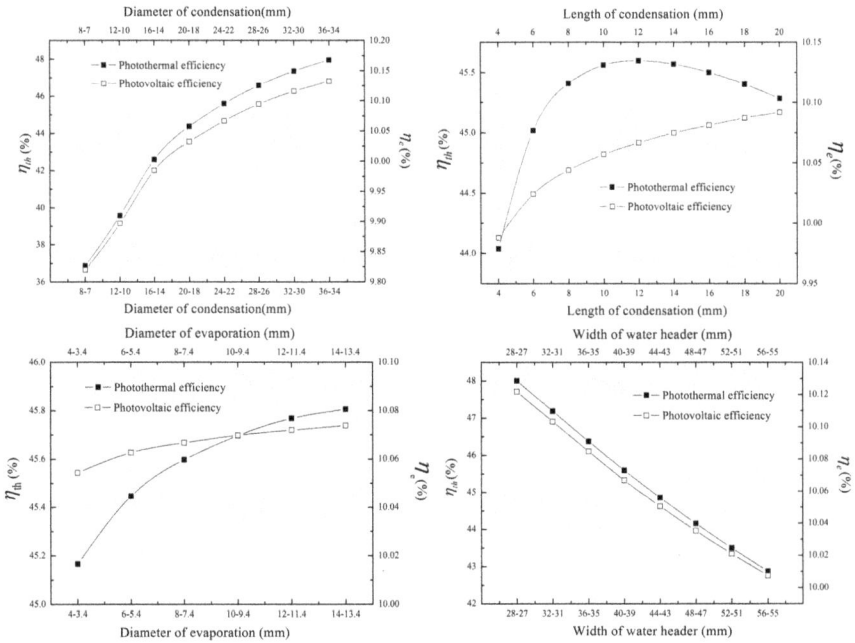

Fig. 3 Variation of efficiency of photothermal and photovoltaic with dimensions of condensation, evaporation and water header

diameter of condensation and decrease with the increase of the width of water header. The length of condensation and diameter of evaporation has a slight influence on the system performance.

4 Conclusions

Introducing two-phase closed thermosyphon (TPCT) to water-based photo-voltaic/thermal (PV/T) system can address the frozen and corrosion problems associated with it. The numerical model of TPCT was firstly established based on the governing equations to acquire the heat transfer coefficient of it. After that, energy balance equations of the key components were further established, and the integral model was validated under a sunny and a cloudy day. Finally, the influence of dimensions of key components to the system performance was sensitive analyzed.

References

1. R. Daghigh, M.H. Ruslan, K. Sopian, Advances in liquid based photovoltaic/thermal (PV/T) collectors. Renew. Sust. Energ. Rev. **15**, 4156–4170 (2011)
2. A. Shafieian, M. Khiadani, A. Nosrati, A review of latest developments, progress, and applications of heat pipe solar collector. Renew. Sust. Energ. Rev. **95**, 273–304 (2018)
3. J.H. Linhan, The interaction of two-dimensional stratified, turbulent air-water and steam-water flows. Ph.D. dissertation. University of Wisconsin (1968)
4. M.S. El-Genk, H.H. Saber, Heat transfer correlations for small, uniformly heated liquid pools. Int. J. Heat Mass. **41**, 261–274 (1998)

Ingram Content Group UK Ltd.
Milton Keynes UK
UKHW022030140323
418549UK00002B/2

9 789813 347670